Particles at Interfaces: Interactions, Deposition, Structure

INTERFACE SCIENCE AND TECHNOLOGY
Series Editor: ARTHUR HUBBARD

In this series:

Vol. 1: Clay Surfaces: Fundamental and Applications
Edited by F. Wypych and K.G. Satyanarayana

Vol. 2: Electrokinetics in Microfluidics
By Dongquig Li

Vol. 3: Radiotracer Studies of Interfaces
Edited by G. Horáyi

Vol. 4: Emulsions: Structure Stability and Interactions
Edited by D.N. Petsev

Vol. 5: Inhaled Particles
By Chiu-sen Wang

Vol. 6: Heavy Metals in the Environment
Edited by H.B. Bradl

Vol. 7: Activated Carbon Surfaces in Environmental Remediation
Edited by T.J. Bandosz

Vol. 8: Tissue Engineering: Fundamental and Applications
By Y. Ikada

Vol. 9: Particles at Interfaces: Interactions, Deposition, Structure
By Z. Adamczyk

INTERFACE SCIENCE AND TECHNOLOGY – VOLUME 9

Particles at Interfaces: Interactions, Deposition, Structure

By

Z. Adamczyk

Institute of Catalysis & Surface Chemistry
Polish Academy of Sciences
Cracow, Poland

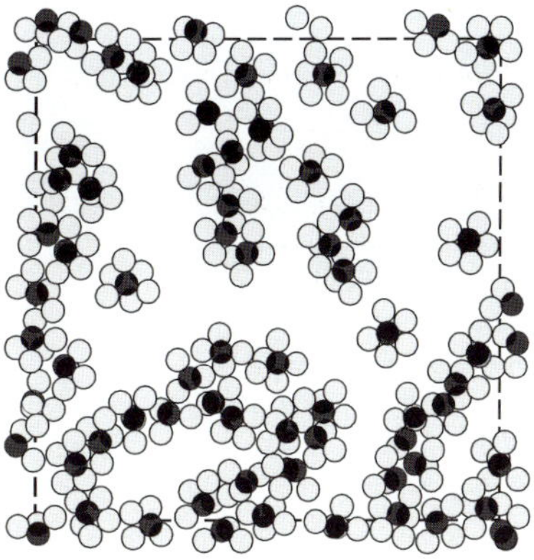

ELSEVIER

Amsterdam • Boston • Heidelberg • London • New York • Oxford
Paris • San Diego • San Francisco • Singapore • Sydney • Tokyo
Academic Press is an imprint of Elsevier

Academic Press is an imprint of Elsevier
84 Theobald's Road, London WC1X 8RR, UK
Radarweg 29, PO Box 211, 1000 AE Amsterdam, The Netherlands
The Boulevard, Langford Lane, Kidlington, Oxford OX5 1GB, UK
30 Corporate Drive, Suite 400, Burlington, MA 01803, USA
525 B Street, Suite 1900, San Diego, CA 92101-4495, USA

First edition 2006

ISBN-13: 978-0-12-370541-9
ISBN-10: 0-12-370541-X
ISSN: 1573-4285

For information on all Academic Press publications visit our website at books.elsevier.com

Printed and bound in The Netherlands

06 07 08 09 10 10 9 8 7 6 5 4 3 2 1

All you need is…charge

To Maria...

Preface

Colloid science, once merely a collection of disconnected facts and observations, matured at the beginning of the 20th century to become a quantitative branch of knowledge.

Undoubtedly, one of the milestones on this road was the mathematical theory of Brownian motion, formulated by Einstein in 1905–1906 and independently at the same time by Smoluchowski, who used kinetic, rather than statistical arguments. This theory, combined with Perrin's precise observations involving latex particles, furnished decisive proof of the existence of molecules, an existence still denied in those days by the phenomenological school of Mach and Ostwald. The humble lesson this theory has to teach us is that no matter how long a Brownian object is observed, one can never predict where its next movement in time will be. The only things that can be known with a defined uncertainty are the time- or ensemble-averaged quantities, for example the distance traveled by the particle after a long time. Philosophically speaking, there are definite limits to our knowledge of the behavior of any system. One can, therefore, construct no more than approximate models of reality, while its true nature remains undisclosed. In this respect, the Brownian motion theory had far reaching consequences, pioneering the contemporary fuzzy-logic way of thinking.

The next major steps in quantifying colloid science were the theory of electrokinetic phenomena, in particular electrophoresis, developed by Smoluchowski in 1903, and the theory of fast coagulation which he elaborated in 1916. Both theories have been successfully used until now to determine the zeta potential of particles and predicting colloid system stability. The range of applicability of the theory of coagulation can be significantly extended by considering the actual profile of particle interaction energy. This profile is derived from another fundamental theory, summarized in

Vervey and Overbeek's book published in 1948, referred to as the DLVO theory (an acronym comprised of the names Derjaguin, Landau, Vervey and Overbeek). The main assumption in formulating the theory was the additivity principle of the electrostatic and van der Waals interactions. Additionally, useful expressions for the double-layer interaction energy between particles were derived using the Poisson–Boltzmann equation, which describes the charge screening effect.

Because of computational limitations, the DLVO theory could not be combined with hydrodynamics and statistical theories concerning particle populations. However, with the advent of the modern computer it became feasible to solve particle transport equations, precisely incorporated long-range hydrodynamic and short-range specific (surface) interactions. A major advancement was achieved in this field in the 1970s and 1980s when particle transport problems were solved for flows of practical interest and interface geometries, such as spherical and cylindrical ones. Further progress was attained in the 1990s when increasing computer efficiency enabled one to perform Monte Carlo or Brownian dynamic simulations, which mimicked real systems more closely. In this way, important clues were gained on the mechanisms of irreversible processes, e.g., the adsorption of particles on interfaces. *Ab initio* type simulations performed for large particle populations enabled one to determine both the kinetics of particle deposition, the structure and the jamming coverage of particle monolayers of various shape.

Yet despite the rapid progress in this field, there are few if any books devoted entirely to the subject of particle transport, deposition and structuring on boundary surfaces. This book attempts to fill this gap by presenting recent developments in this growing field. Combining traditional theories of electrostatics, hydrodynamics and transport with new approaches in a harmonious whole is the major aim of this book. The need for such theoretical reference data obtained for well-defined particle systems and transport conditions is vital in view of the complexity of the problems, which have been studied recently. They involve concentrated systems of polydisperse and non-spherical particles, as well as bio-particles such as DNA fragments, proteins, viruses, bacteria, cells, polymers, etc. These particles are of complex structure and undergo transformations under the action of surface forces. Particle mono- and multilayers are often formed on heterogeneous surfaces, covered by specifically or non-specifically binding sites. There is a possibility that the results can be misinterpreted because the monolayers are usually dried before microscope examination, which disturbs their structure. Even using the in situ AFM methods (with the tapping mode), one may produce some tip-induced artifacts. The situation is especially critical with the subtle problem of protein adsorption, requiring the most refined experimental approaches.

The book, which provides one with readily accessible reference data and equations for estimating basic effects, is mainly addressed to students and young scientists. Consequently, most approaches are of a phenomenological nature, enabling one to derive concrete expressions, which describe the basic physics of the problem under consideration. To facilitate access to the information contained in the book most of the relevant formulae and results are compiled in tables, accompanied with appropriate diagrams. The math is limited to the necessary minimum with emphasis on the physics of the phenomena, defining why they occur and what the kinetics of the processes and the practical implications are. Accordingly, the book, which represents the first part of the saga, is meant as a kind of physical foundation. The next volume will be devoted to experiments and applications. I hope that with such an approach the book, which is also meant to be self-contained, proves user-friendly and will save a good deal of the student's time.

On a personal note, I would like to express my thanks to the various persons who shaped my scientific carrier giving invaluable advice, providing inspiration and motivation in moments of despair.

Looking back to my PhD times I must mention Professor Andrzej Pomianowski, a true scientific father of mine, who taught me that passion is the key when dealing with science, art and related subjects. It was he who first introduced me to the realm of colloids, to the many fascinating phenomena occurring when two interfaces come into intimate contact, foaming and lubrication to name just a couple.

Two friends of mine in those good old days were of invaluable help to me: Tadeusz Dąbroś, with whom I carried out innumerable discussions on the surface of tennis courts, and Jan Czarnecki, who gave me an intense course in the elegant formulation of thoughts. He also convinced me, with a little help of Newtonian liquids, that social activities can be treated as extended science, too.

Then, as a young Post Doc fellow in Montreal, I met another brilliant person, a man of enormous vitality. He opened my eyes to the nuances of scientific reality, taught me that there is industry and practical problems to be solved. Most lessons took place on squash courts where we also practiced elastic two-body collisions, rather painful to Theo van de Ven, just to mention his broken glasses.

Returning to Cracow, I had the chance to collaborate with Piotr Warszyński, first my student, then, quite unexpectedly, my boss. Over the last 20 years we have discussed a myriad of subjects, exchanging thousands of words… one way. Most, of what I have learned of hydrodynamics is to Peter's credit. With Peter my run of good luck still continues. Thanks to him it was proved unequivocally that I am quite an influential partner, as I managed to damage a bone of his over the distance of about 20 m, i.e., the length of a tennis court. Although Peter recovered after a couple of weeks, he became considerably more careful in contact with me.

There are many others to whom I am very indebted, mostly to the members of my and related groups: Basia Jachimska, Kasia Jaszczółt, Marta Kolasińska, Kazimierz Małysa, Aneta Michna, Basia Siwek, Lilianna Szyk-Warszyńska and Maria Zembala. They helped me enormously by providing fascinating experimental results, carrying out numerous discussions, supplying relevant literature, doing artwork and assisting in other technical matters. They suffered a lot for over a year when the book was *in statu nascendi* and I was unable to properly appreciate their hard work.

I am also very thankful to Jakub Barbasz and Paweł Weroński who masterminded the numerical simulations and provided me in emergencies with the required graphs. A significant part of Paweł Weroński's PhD thesis, especially the unpublished results concerning the unoriented adsorption of spheroidal particles, has been exploited in Chapter 5. Jakub Barbasz and Małgorzata Nattich were also kind enough to critically revise the manuscript and in checking the numerical calculations. Another bright person who helped me a lot was Piotr Wandzilak, who performed the linguistic corrections of the manuscript, also giving invaluable advices and suggestions. The list could go on since there were more people who helped me in finishing this project, often suffering from my negligence. My deepest apologies to all of you.

Last but not least there are two very special persons to whom I am indebted eternally: Marta Krasowska and Ela Porębska, true angels who were destined to Earth. Not only have they done all the tedious technical things, patiently dealing with endless corrections and revisions, but also generously offered sustained inspiration and motivation. If you like the look of the book, it is only because they loved what they were doing.

Zbigniew Adamczyk

Table of Contents

CHAPTER 3 DISSIPATIVE INTERACTIONS

CHAPTER 4 TRANSFER OF PARTICLES TO INTERFACES – LINEAR PROBLEMS

Chapter 1

Significance of Particle Deposition

Effective attachment of particles to surfaces, involving transport, adsorption and adhesion steps, is important for many practical processes such as water and waste water filtration, electroflotation, separation of toner and ink particles, coating formation, paper making, xerography, production of magnetic tapes, catalysis, colloid lithography, protein and cell separation (affinity chromatography), food emulsion and foam stabilization, immobilization of enzymes, immunological assays, etc.

Controlled assembly of colloid particles into organized structures has potential applications in the production of nano- and microstructured materials of desired functionality, such as biocompatible coatings. Ordered arrays of colloid particles, for example silver particles, are exploited as narrow-band optical filters; they can be used as optical switches, photonic band gap materials, waveguides and other electrooptical and magnetooptical devices.

An emerging field of application of particle deposition is the "colloid bar coding" technique, which can encode libraries of millions of compounds using a few fluorescent dyes.

In other processes, such as membrane filtration, biofouling of membranes and artificial organs, flotation (slime coating formation) and production of microelectronic or optical devices, particle adsorption is highly undesirable.

Besides these practical applications, studies of colloid particle deposition, carried out using direct experimental methods, can furnish fundamental information on interactions between particles and interfaces, and between adsorbed and moving particles. This is a crucial issue for colloid science, biophysics and medicine, soil chemistry, etc. For example, the symmetry and local intensity of flows can be easily detected by performing deposition experiments and observing the density of particles on the surface with an optical microscope. This is illustrated in Fig. 1.1, showing micrographs of polystyrene latex particles adsorbed on a mica surface exposed to a stagnation point flow of radial symmetry [1] or to an oblique flow [2]. The

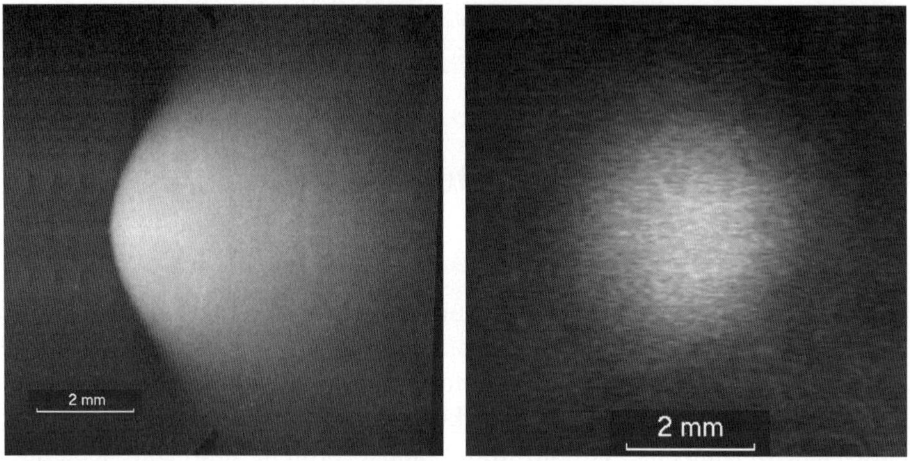

(a) (b)

Fig. 1.1. (a) Flow pattern near a mica/electrolyte interface exposed to an oblique flow, visualized by adsorbed polystyrene latex particles. (b) Flow pattern near a mica/electrolyte interface exposed to a radial stagnation point flow, visualized by adsorbed polystyrene latex particles.

latter flow configuration enables microscopic observations for non-transparent substrates, e.g., metallic surfaces such as gold [3]. However, any quantitative information on the flow pattern can only be extracted from these measurements, if the theoretical predictions of local mass transfer rates are known; they are extensively discussed in Chapters 2 and 3.

Colloid adsorption can also be used to make visible various structures formed near interfaces by aggregating chains of polyelectrolytes, as seen in Fig. 1.2. In a similar way, colloidal particles of silver are used in the staining process to visualize nanosized particles of biological origin.

By measuring particle adsorption for model colloid systems, important information can be gained on the mechanisms and kinetics of molecular adsorption inaccessible for direct experimental studies. In this way the links between irreversible (colloid) and reversible (molecular) systems can be established. For example, by studying the particle distribution on surfaces in the limit of low coverage, when a gas-like phase is formed, one can derive important clues on the dynamic interactions between adsorbed and adsorbing particles, by inverting the Boltzmann distribution. This is illustrated in Fig. 1.3, showing a monolayer of polystyrene latex particles on mica [4] at dimensionless coverage $\Theta = 0.05$.

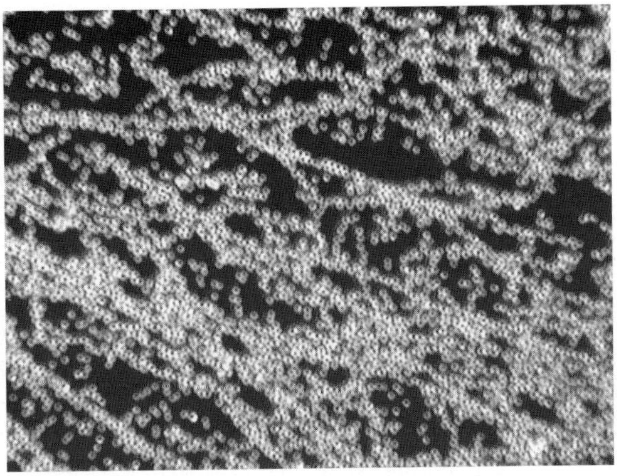

Fig. 1.2. Latex particles attached to polymer fibers formed at a mica/electrolyte interface.

As can be seen, the distribution of particles characterized in terms of the pair correlation function g is well reflected by the Boltzmann distribution. With increasing coverage (see Fig. 1.3.b), the structure of the colloid particle monolayer closely mimics a molecular liquid-phase structure, e.g., liquid argon at 130 K [5]. This fact suggests that colloid particle monolayers can be exploited as a useful reference system for studying fluctuation and structure formation phenomena occurring on a molecular scale, e.g., for proteins.

As a consequence, the possible structure of globular protein monolayers on solid substrates, for example, can be forecasted by using reference data collected for colloids. Experimental data for colloids can be transferred to molecular systems using appropriate theoretical background discussed in Chapters 4 and 5. One of the most efficient ways of obtaining structural data is the Monte-Carlo simulation, described in Chapter 5. Using this method, fascinating structures have been predicted in the case of particle adsorption on substrates bearing isolated adsorption centers of spherical shape. As seen in Fig. 1.4, flower-like structures appear because one adsorption center can coordinate on average 5.5 adsorbing particles [6]. Hence, such theoretical simulations suggest that it would be possible to produce clusters of targeted architecture, e.g., containing a prescribed number of particles. The coordination number can be regulated by changing the particle-to-site size ratio. Indeed, experimental results reported recently [7] seem to confirm this prediction. This is seen in Fig. 1.5, presenting micrographs of polystyrene latex

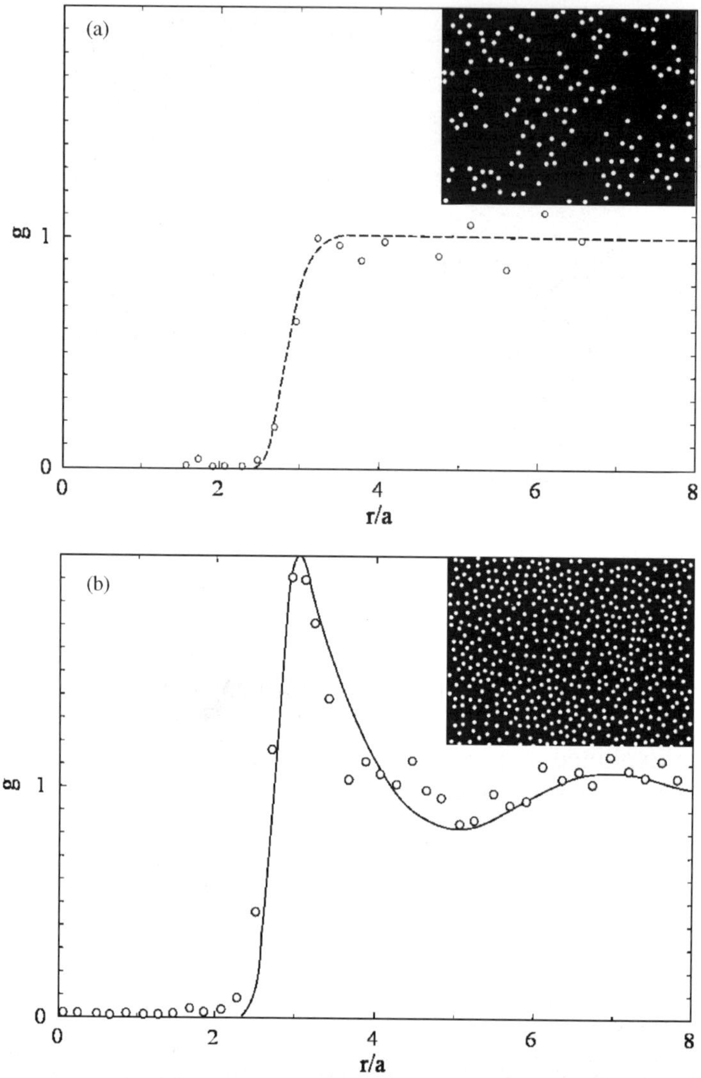

Fig. 1.3. (a) A monolayer of polystyrene latex particles on mica at $\Theta = 0.05$, and the corresponding pair correlation function $g(r)$ (ionic strength of 10^{-5} M). The dashed line shows the Boltzmann distribution. (b) Same as (a) but for $\Theta = 0.24$. The solid lines show the theoretical results obtained from Monte Carlo simulations.

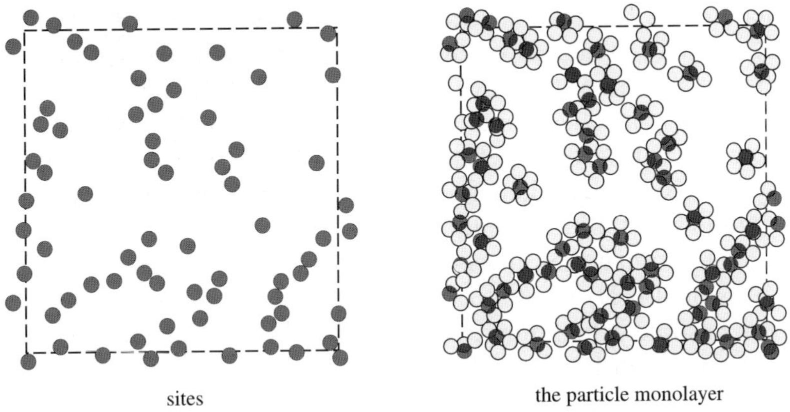

Fig. 1.4. A monolayer of particles adsorbed on spherically shaped sites of the same size, predicted theoretically using the Monte Carlo simulations.

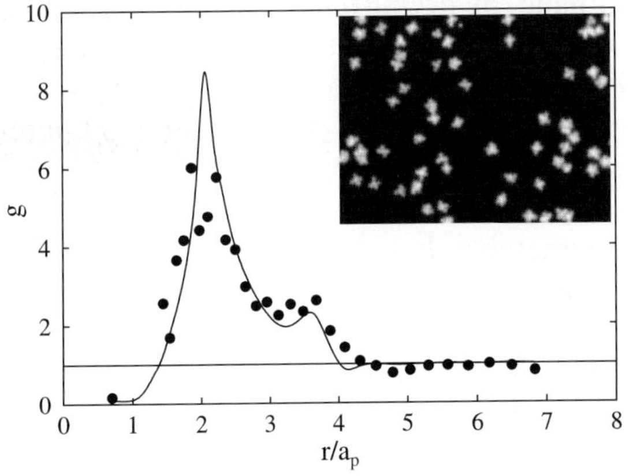

Fig. 1.5. A micrograph of polystyrene latex clusters produced by adsorption on sites (polystyrene particles of opposite charge) and the corresponding pair correlation function g of adsorbed particles (the line shows the results derived from the Monte Carlo simulations).

clusters produced by adsorption on sites (polystyrene particles of opposite charge).

Colloid systems can also be used to calibrate indirect methods of measuring protein surface coverage, such as reflectometry or streaming potential. This is vital in view of the many artifacts that may appear when indirect measurements are carried out, which usually involve the drying step. As illustrated in Fig. 1.6, this may lead to significant alteration of the structure of the monolayer, which is often misinterpreted as a real fact, apparently reflecting the substrate structure. Interestingly, the drying procedure, carried out under well-controlled conditions, for example by evaporation of liquid from the particle film or by letting the liquid/air interface pass through the monolayer, can be exploited to produce ordered structures of particles. These colloid crystalline structures, shown in Fig. 1.7, potentially can be used as photonic material, masks in colloid lithography processes [9]. The two-dimensional (2D) crystallization of particles induced by the meniscus forces can to some extent mimic molecular-scale crystallization phenomena. This is illustrated in Fig. 1.8, showing a 2D polycrystalline sample formed by latex, containing several defects and the solid/liquid interface.

The structures formed upon drying of monolayers can be quite fascinating, often resembling modern art pictures. The colloid particles of micrometer size range shown in Fig. 1.9 may be ordered into caterpillar- or labyrinth-like structures.

On the other hand, by drying a monolayer of nanometer-sized particles (colloid cerium oxide, particle size of about 50 nm) quite interesting structures were created (see Fig. 1.10). What they resemble is up to one's imagination; although these structures do not have any immediate practical applications, they provide one, like art, with esthetic stimuli, indispensable for a continuous research effort.

One should remember, however, that despite the analogies mentioned above, colloid particle adsorption proceeds along much more complicated paths than in the case of molecular adsorption. This is so because particle transfer from the bulk to the interface is affected by a variety of interactions differing widely in magnitude and the characteristic length scale. For

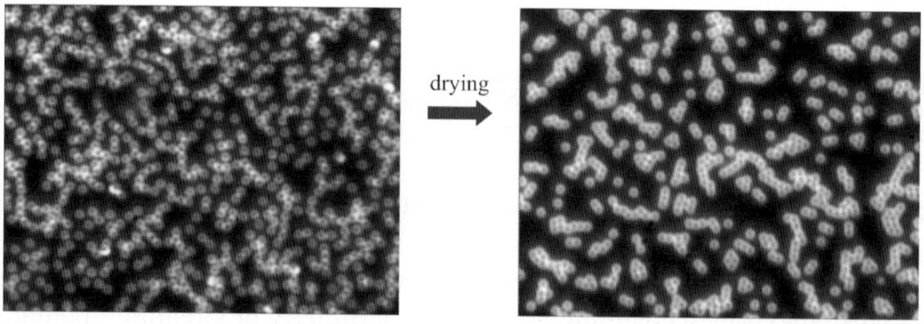

Fig. 1.6. A monolayer of polystyrene latex particles on mica (seen *in situ* under optical microscope) before and after drying, $\Theta = 0.20$.

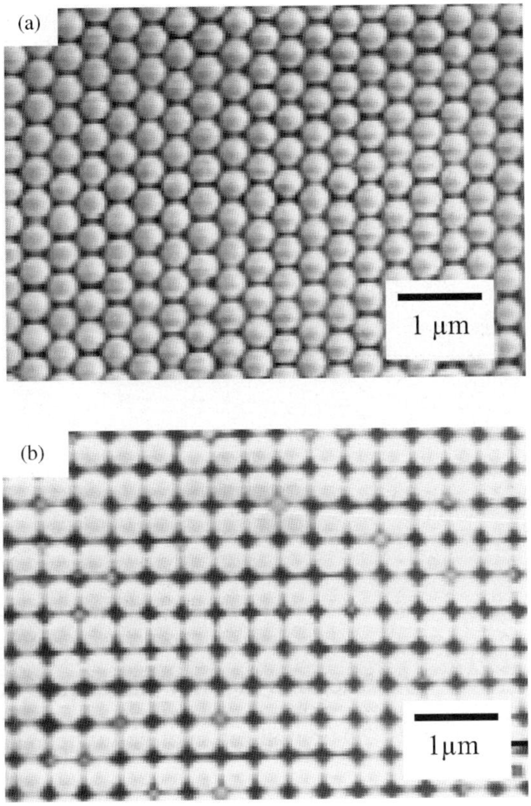

Fig. 1.7. Ordered structures of latex particles formed by drying. (a) The two-dimensional hexagonal crystalline phase. (b) The 2D regular phase. Reprinted with permission from [8]. Copyright 2003 American Chemical Society.

distances exceeding particle dimensions, the most significant transport mechanisms are natural convection, or forced convection applied in a controllable way, using specially designed cells [4] or by stirring. Transfer of particles over macroscopic distances can also be induced by external forces, such as the gravitational force, leading to sedimentation effects, or as electrostatic forces, inducing electrophoresis, which is especially efficient in non-polar media. For distances comparable with particle dimensions, diffusion becomes the most significant transport mechanism, whose rate increases with increased temperature and decreased particle size. The diffusion coefficients needed to evaluate the mass transfer rate can only be derived when we know the hydrodynamic resistance tensor of a particle,

Fig. 1.8. Micrograph showing 2D crystallization of polystyrene latex particles on mica, induced by drying.

which makes it necessary to consider the low Reynolds number hydrodynamics of the system. The diffusion transport mechanism is the most significant one for more concentrated systems in which the transport path remains comparable with particle dimensions.

At distances comparable with their dimensions, adsorbing particles become influenced by specific force fields generated by interfaces. For such small distances, the dominant role is played by the electrostatic interactions resulting from the fixed charges present on particles. Depending on the sign of the charge on particles and the substrate surface, the electrostatic interactions can be either positive (repulsion) or negative (attraction). Additionally, the range and magnitude of the electrostatic interactions can be varied within broad limits by the addition of electrolytes, change of pH and adsorption of surfactants or charged polymers on particles. Another type of interactions occurring universally at separations significantly smaller than particle dimensions are van der Waals interactions, which are

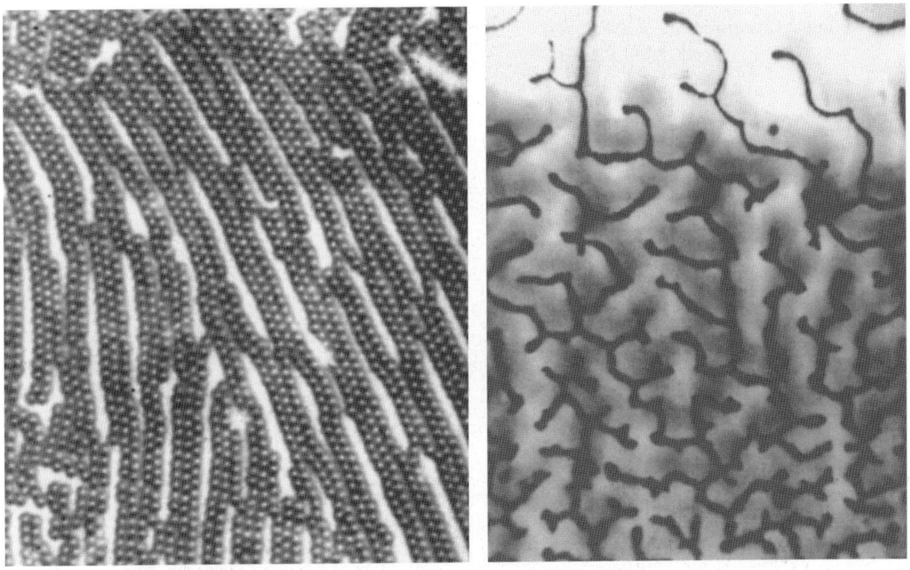

Fig. 1.9. Structures formed upon drying of colloid particle monolayers (latex particles of size 0.9 μm on mica).

mostly attractive. In contrast to electrostatic interactions, their range cannot be regulated in any systematic way.

At nanometer-range distances, when the particles come into physical contact with the interface, many processes occur, often irreversible ones, leading to particle adhesion, deformations, reconformations in the case of proteins and polymers, ion exchange and recrystallization, hydrogen bond formation, chemical reactions leading to sintering and so forth.

Additional complications occur due to the presence of particles accumulated at the interface, which disturb the adsorption pathway of particles coming into their vicinity. The forces between adsorbed and adsorbing particles, especially their hydrodynamic component, depend not only on particle surface properties but also on the topology of particle distribution.

Because transport mechanisms are complicated and often irreversibile, particle adsorption is a highly path-dependent process characterized by a wide spectrum of time scales. Thus, the characteristic adsorption relaxation time may vary from seconds for concentrate colloid suspensions to days for dilute suspensions of larger particles adsorbing under diffusion-controlled transport. In contrast to molecular systems, the kinetic aspects of particle adsorption are, therefore, of primary interest and are extensively discussed

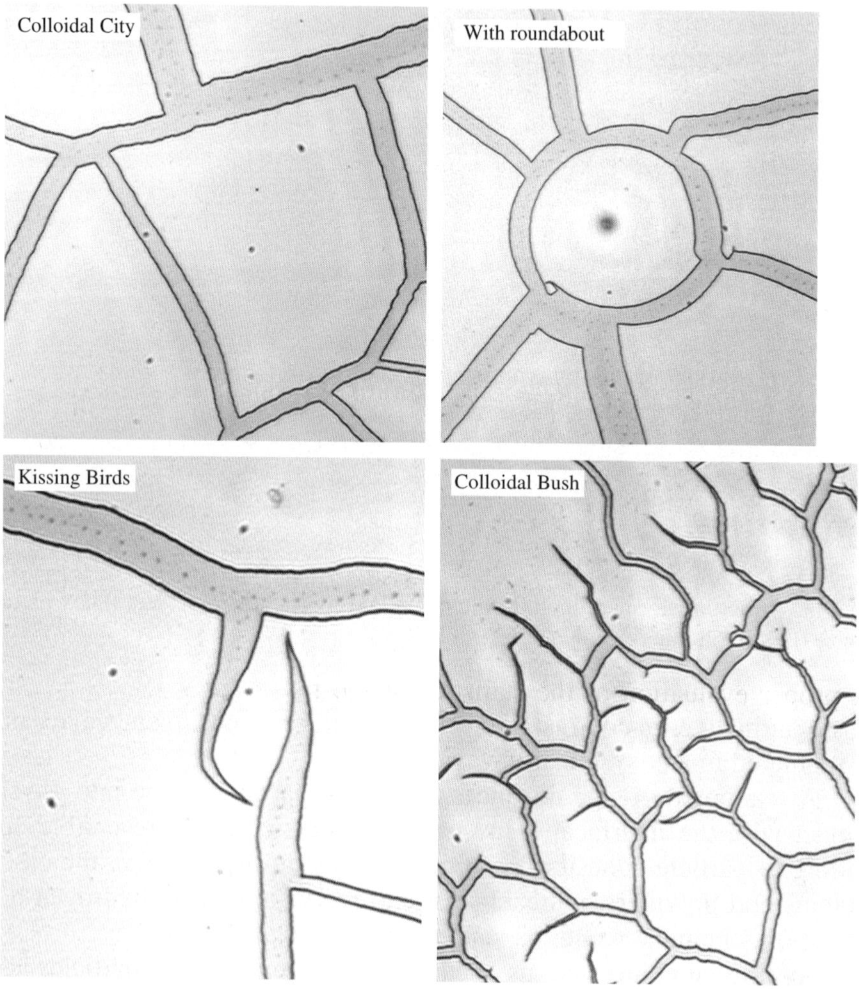

Fig. 1.10. Structures formed upon drying of nanosized particle monolayers (cerium oxide particles of size 50 nm on mica).

in this book. In order to derive meaningful results, it is advisable to consider simultaneously the long-range transport, the short-range particle/surface interactions, and the structure and topology of particles on the surface.

It seems that despite the major significance, there are practically no books that simultaneously analyze the long and short-range interactions leading to particle adsorption, deposition and adhesion and the topological effects influencing the dynamics of these processes.

This book, in which a unified phenomenological approach is adopted, represents an attempt to fill this gap by laying the physicochemical foundation for particle deposition problems. For this aim the book is split into four free-standing chapters, the first devoted to specific interactions between particles, the second to dissipative hydrodynamic interactions, the third to linear transport problems, and the fourth to particle transport under non-linear conditions and to the structure of particle layers. Here the chapters are described in more detail.

1.1. CHAPTER 2

In order to emphasize the common origin of all potential interactions, the basic laws of electrostatics beginning from Coulomb interactions of charges in a vacuum are discussed. Polarization phenomena in dielectric media are described next, and basic equations of electrostatics are formulated, among others the Gauss, Maxwell and Poisson equations. Electric potential and field distributions for simple geometries are analyzed, and expressions describing the electrostatic interactions of particles in dielectric (ion-free) media are given, including the formulae for the dielectrophoretic force in non-uniform fields. The Poisson equation is introduced next and methods for proper evaluation of the statistics of the free charge distribution near charged interfaces are discussed. An extensive part of the analysis is devoted to the electric double layer formed at solid/electrolyte interfaces of various shapes. Methods for evaluating the force and energy of interactions among particles bearing electric double layers are presented. Exact and approximate methods of evaluating these interactions are discussed, including the linear superposition approximation (LSA), the generalized Derjaguin summation method, and the equivalent sphere approach (ESA). The next section is devoted to van der Waals interactions: permanent dipole/permanent dipole (Keesom orientational forces), permanent dipole/induced dipole (inductive Debye forces) and induced dipole/induced dipole interactions resulting from spontaneous fluctuations in local electron density, referred to as the dispersion or London forces. The intermolecular potentials are specified, and the Hamaker theory based on the pairwise additivity is discussed. Explicit expression for dispersion interaction energy of various macrobodies, including deformed and rough objects, are presented. Approximate methods for evaluating dispersion interactions for anisotropic (non-spherical) particles are given, especially the Derjaguin summation method. Then, the macroscopic (continuum) theory of dispersion interactions is presented. Values of Hamaker constants characterizing the magnitude of dispersion interactions in various media are discussed. The section is concluded with analysis of surface deformations due to dispersion interactions and particle–surface adhesion phenomena. In the last section of Chapter 1 the specific energy profiles originating from the superposition of these potential interactions are discussed. The influence of surface roughness and heterogeneity of charge distributions is analyzed. Perturbations of specific energy profiles stemming from external and other types of interactions, such as steric and depletion interactions, are also considered.

1.2. CHAPTER 3

In the first section the equations governing fluid motion arising from the mechanical momentum balance are derived including the Navier–Stokes equation describing flows of Newtonian fluids. The limiting forms of the fluid motion equation are formulated, in particular the quasi-stationary Stokes equation describing creeping motion at low Reynolds number. The boundary conditions at solid and liquid interfaces are specified, and the expressions for hydrodynamic forces and torques acting on interfaces as well as on moving particles. In the second section, macroscopic flows of practical importance are analyzed in detail, especially the confined flows near interfaces, e.g., channel and shearing flows. Stagnation flows are also discussed, including the impinging-jet flows and wall-jet flows met often in experiments and the hydrodynamic boundary layer flows near spherical and cylindrical interfaces. A method of decomposing these macroscopic flows into simple (elementary) flows is presented. The dynamics of particles of various shapes in a quiescent fluid is discussed as well as the wall effects influencing particle motion near interfaces. The hydrodynamic resistance matrix is defined for both the bulk and the interface region. Finally, creeping motion of spherical particles near interfaces is analyzed, as well as and flows past particles attached to interfaces.

1.3. CHAPTER 4

In the first section the force and torque balance equation is formulated and its limiting forms are derived, in particular the inertial and inertia-less trajectory equations. For the latter case, the general mobility relationship, describing particle velocity in terms of the inverted hydrodynamic resistance matrix (called the mobility matrix), and the net force and torque acting on particles are discussed. This universal mobility relationship is used to calculate particle migration velocity under the action of external forces such as gravitational or electrostatic forces. This allows one to determine sedimentation rates of particles of various shapes as well as their migration velocities under electrostatic forces in ion-free media. Further exploiting this method, inertia-less trajectory analysis of particles moving under various flows near interfaces is discussed, with the aim of calculating particle velocities and, consequently, particle deposition rates. In the next section, the force balance equations are extended to incorporate the random force exerted on particles by molecules of the suspending medium. Langevine-type stochastic equations of motion are derived, and the Fokker–Planck and Smoluchowski equations for the case of particle motion in an external force field. Calculated values of diffusion coefficients and the Brownian motion of a single particle in the bulk and near interfaces are analyzed in detail. The general diffusion equation for particle systems is derived using Einstein's method. Limiting forms and particle transport regimes are considered, in particular the convective diffusion regime, as well as the pure diffusion regime. Methods for analytical solution of these equations are discussed, with the emphasis on the Laplace transformation method. One-dimensional transport of particles throughout the surface layer adjacent to interfaces is considered, knowledge of which allows one to formulate the foundations of the surface boundary layer (SFBL) theory and to establish the appropriate boundary conditions for bulk transport equations. The boundary conditions in the form of the perfect sink and the linear adsorption isotherm is derived. In the next section, solved problems concerning linear transport of particles to interfaces are extensively discussed, especially the non-stationary diffusional transport to spherical and planar interfaces, convective diffusion transport for a variety of flows and interfaces, transport to uniformly accessible surfaces, transport in the impinging-jet cell, etc. Analytical expressions for various transport conditions and interface shapes obtained within the framework of convective diffusion theory are tabulated. The influence of electrostatic interactions and external forces on the limiting flux and mass transfer coefficients for various transport conditions is also discussed.

1.4. CHAPTER 5

In the first section, 2D equilibrium systems of particles are considered. Using virial expansion and scaled particle theory (SPT), the chemical potential of particles, the available surface function, adsorption isotherms, density fluctuations and the structure of adsorbed layers are analyzed for convex particles adsorbing side-on: monodisperse and polydisperse spheres, spheroids (ellipses), cylinders (rectangles) spherocylinders, etc. Analytical expressions are derived, useful for testing the accuracy of numerical approaches. In the next section, various random sequential adsorption (RSA) models are considered together with methods of simulating particle configurations. Monolayers of various densities of monodisperse and polydisperse spherical particles are discussed. Jamming coverage results are extensively analyzed, and interpolating functions for calculating them are given. The important case of particle adsorption on a heterogeneous surface, e.g., pre-covered by smaller particles or adsorption centers of various shapes and concentrations, is discussed in detail. Then, side-on and unoriented adsorption of hard (non-interacting) particles of non-spherical shapes is analyzed in terms of the RSA model. Except for the case of jamming coverage, RSA simulations enable one to determine the surface blocking function and the structure of monolayers for any coverage of particles. The important problem of interacting (soft) particle adsorption is discussed. The powerful concept of the effective hard particle is treated. It enables one to formulate limiting analytical expressions for the blocking function and jamming coverage of particles in terms of the effective interaction range, strictly related to the electric double-layer thickness.

REFERENCES

[1] Z. Adamczyk, B. Siwek, P. Warszyński and E. Musiał, J. Colloid Interf. Sci., 242 (2001) 14.
[2] Z. Adamczyk, E. Musiał and B. Siwek, J. Colloid Interf. Sci., 269 (2003) 53.
[3] Z. Adamczyk, E. Musiał, B. Jachimska, L. Szyk-Warszyńska and A. Kowal, J. Colloid Interf. Sci., 254 (2002) 283.
[4] Z. Adamczyk, In "Irreversible Adsorption of particles in Adsorption: Theory, Modeling and Analysis" (J. Toth, ed.), pp 251–374, Marcel-Dekker, New York, 2002.
[5] D. Henderson and S. G. Davison, Equilibrium theory of liquids and liquid mixtures, in "Physical Chemistry, An Advanced Treatise" (H. Eyring ed.), Vol.II, P. 361, Academic Press, New York, 1967.
[6] Z. Adamczyk, K. Jaszczółt, A. Michna, B. Siwek, L. Szyk-Warszyńska and M. Zembala, Adv. Colloid Interf. Sci., 118 (2005) 25.
[7] Z. Adamczyk, K. Jaszczółt, B. Siwek and P. Weroński, Langmuir, 21 (2005) 8952.
[8] H. Cong and W. Cao, Langmuir, 19 (2003) 8177.
[9] J. C. Garno, N. A. Amro, K. Wadu-Mesthrige, G.-Y., Liu, Langmuir, 18 (2002) 8186.

Chapter 2

Potential Interactions Among Particles

2.1. INTRODUCTION

Defining and quantifying interactions are indispensable for predicting the evolution of a system in both its kinetic and structural aspects. In particular, this concerns adsorption, deposition and adhesion phenomena, which are of considerable scientific and practical interest in the various fields listed above. A systematics of interactions (driving forces) governing these processes seems vital in view of persisting controversies and misinterpretations of the true mechanism of particle adsorption.

Therefore, the initial chapter of this book will be devoted to this subject, providing the necessary foundation for the entire problem of predicting the dynamics of particle suspensions near interfaces.

Although it seems quite obvious, let us recall that at present there are only four basic types of interaction known to physics:

(i) gravitational;
(ii) electric and magnetic (electromagnetic);
(iii) weak nuclear; and
(iv) strong nuclear.

All phenomena occurring in nature starting from the origin and evolution of our universe to the appearance and persistence of life are a consequence of the interplay of these interactions.

For any of the problems treated in this book the weak and strong nuclear interactions can be completely ignored, thus we are left with the gravitational and electromagnetic interactions at hand. However, the gravity force between objects met in our environment or in the laboratory can also be ignored. The only effect that matters in particle dynamics is the gravity force on them exerted by the Earth. This leads to migration (sedimentation) effects, whose role obviously increases with the apparent mass of the particles suspended in a medium. Generally, they become significant for dense particles with sizes greater than a micrometer.

Except for sedimentation, other phenomena occurring in particle suspensions are the result of electromagnetic forces, with their origin in the presence and motion of charges. Without exaggerating their role, one can paraphrase the famous song of the Beatles: "*all you need is charge*".

However, describing electromagnetic interactions precisely is not obvious, especially in systems such as particle suspensions near interfaces. Even deducing that many apparently disconnected phenomena have their common origin in subtle variations in the charge distribution is not obvious. For the above example of a particle suspension it is quite easy to understand that their stability is governed by the electrostatic forces due to presence of fixed charges at particle surfaces. However, it is less obvious to deduce that the attractive interactions among particles appearing universally at short separations, called presently van der Waals interactions are the result of spontaneous fluctuations in the charge density of atoms or molecules. In fact, it took hundreds of years and the efforts of such giants of science as Laplace, Gauss, Maxwell, Sutherland, van der Waals, Langevine, Debye and London [1] before the true origin of dispersion interactions was properly evaluated. Analogously, it is not immediately obvious that the Brownian motion of particles and all hydrodynamic interactions, including friction appear because of the repulsion of charge clouds at atoms and molecules, allowing for transfer at momentum from the environment to particles. A proper description of these repulsive interactions appearing at distances comparable with molecular dimensions is still far from being completed.

Going further along the charge interaction landscape in the direction of still smaller distances (below a nanometer) one encounters the hydrogen-bond interactions, occurring in many polar fluids, especially water. This type of interaction is the result of the interference of the electron cloud of a hydrogen atom forming a chemical bond with another atom of high polarity (e.g., oxygen of another molecule). The hydrogen bond exhibits quite variable strength, usually an order of magnitude weaker than a typical chemical bond. However, it plays an essential role in protein chemistry, governing their conformation. One may state once again that all biological life is dependent on this relatively subtle charge transfer effect.

Finally, the chemical bond is the direct effect of increased attraction between nuclei and condensed electron cloud stemming from orbitals of atoms forming the molecule. Hence, a slight shift in the charge cloud density is responsible for the entire chemistry, e.g., the formation of the backbone structure of proteins.

The same electrostatic origin can be attributed to other interactions lingering in the literature under various names. For example, the so-called hydrophobic or hydrophilic interactions are nothing more than the net result of dispersion interactions between hydrocarbon chains and polar head groups with the solvent.

In order to emphasize the common origin of all potential interactions mentioned above, this chapter has been organized as follows: first we discuss the basic laws of electrostatics, starting from Coulomb interactions in a vacuum. Then, the Gauss and Maxwell equations are introduced. Polarization phenomena in dielectric media are discussed next, together with methods of calculating interactions and electrostatic fields. Expressions for calculating the direct electrostatic interactions are given.

The Poisson equation is introduced next, and methods of properly evaluating the statistics of free charge distribution near charged interfaces are discussed. An extensive part of the analysis is devoted to the electric double-layer formed in this way at solid/electrolyte interfaces of various shapes. Methods of evaluating the force and energy of interactions among particles bearing electric double-layers are presented. Exact and approximate methods of evaluating these interactions are discussed, including the generalized Derjaguin summation method.

The next section of this chapter is devoted to dispersion interactions, with the emphasis on the microscopic London theory. The intermolecular potentials are specified, and the Hamaker theory based on the pairwise additivity is discussed. The explicit expression for dispersion interaction energy of various macrobodies including deformed and rough objects are presented. Approximate methods of evaluating dispersion interactions for anisotropic (non-spherical) particles are presented, especially the Derjaguin summation method. Then, the macroscopic (continuum) theory of dispersion interactions is presented. Values of Hamaker constants characterizing the magnitude of dispersion interactions in various media are discussed. The section concludes with analysis of surface deformations due to dispersion interactions and particle–surface adhesion phenomena.

In the last section the specific energy profiles originating from the superposition of these potential interactions are discussed. The influence of surface roughness and heterogeneity of charge distributions are analyzed. Perturbations of specific energy profiles stemming from external and other types of interactions such as steric and depletion forces are also considered.

Although the dissipative hydrodynamic interactions have, in principle, the same physical origin as the potential interactions, for the sake of convenience, they will be discussed in the next chapter.

2.2. ELECTROSTATIC INTERACTIONS

The first observations of the significance of electric phenomena were made already in ancient times, when Tales of Milet noticed that rubbed amber attracts dry grass.

This type of static electricity can be a nuisance often encountered by people combing their hair or cleaning furniture. Consequently, antistatic products preventing the accumulation of charges during washing and cleaning are widespread now a days.

Properly understood in recent times only, lightnings, the most spectacular of all natural phenomena, also originate from static electricity accumulating in clouds.

The vast empirical knowledge gathered over the centuries was led to the formulation of some firm principles governing the nature of charge and its interactions:

(i) There are two types of charge, one of them, by convention, called negative, and the other positive;
(ii) Like charges repel each other; unlike charges attract;
(iii) Charges on material bodies are always discrete (come in portions), there is a smallest charge, defined as the elementary charge; both the negative and positive elementary charges are equal; and
(iv) The amount of charge is conserved; charges cannot be produced or annihilated but merely separated from each other.

There are some comments that naturally come to mind when formulating the list:

(i) The charge division into a negative (acquired by glass when rubbed) and positive (acquired by fur) was due to Franklin. The question of why there are two kinds of charge whereas there is one mass has not yet been satisfactorily answered.
(ii) The elementary charge, determined in the ingenuous experiments of Millikan, equals 1.60218×10^{-19} Coulomb (abbreviated as C, the base unit of the SI system, see Table 2.1). There is an annoying asymmetry: the elementary charge of an electron is associated

Table 2.1
Fundamental physical constants

Quantity	Symbol	Value (SI unit)	Value (C.G.S. unit)
Avogadro constant	N_{A_v}	6.02214×10^{23} (mol^{-1})	6.02214×10^{23}(mol^{-1})
Bohr radius	a_0	0.529177×10^{-10}(m)	5.29177×10^{-9}(cm) 0.5292 (Å)
Boltzmann constant R_g/N_A	k	1.38065×10^{-23} (J K^{-1}) 8.61724×10^{-5} (eV K^{-1})	1.38065×10^{-16} (erg K^{-1})
Electric constant (permittivity of vacuum)	$\varepsilon_0 = 1/\mu_0 c_0^2$ $4\pi\varepsilon_0 = 10^7/c^2$	8.85419×10^{-12} 1.112650×10^{-10} (F m^{-1})=(C V^{-1}m^{-1}) = C^2 N^{-1}m^{-2}	(1)
Elementary charge	$-e$	1.60218×10^{-19} (C)	4.80298×10^{-10} (cm$^{3/2}$ g$^{1/2}$ s^{-1})
Elementary charge to Planck constant ratio	$-e/h_p$	2.41799×10^{14} (A J^{-1})	
Electron mass	m_e	9.10938×10^{-31} (kg)	9.10938×10^{-28} (g)
Electron charge to mass quotient	$-e/m_e$	1.7588×10^{11} (C kg^{-1})	
Faraday constant	F_a	96485.34 (C mol^{-1})	2.8924×10^{14} (cm$^{3/2}$ g$^{1/2}$ s^{-1} mol^{-1})
Magnetic constant	μ_0	$4\pi\times10^{-7}$ = 12.5664×10^{-7} (N A^{-2})	
Molar gas constant	R_g	8.31445 (J mol^{-1}K^{-1})	
Molar volume of ideal gas	v_m	22.4140×10^{-3} (m^3 mol^{-1})	22.4140×10^3 (cm^3 mol^{-1})
Newtonian constant of gravitation	G	6.6731×10^{-11} (m^3 kg^{-1} s^{-2})	6.6731×10^{-8} (cm^3 g^{-1} s^{-2})
Planck constant	h_p $h/2\pi$	6.62607×10^{-34}(J s) 1.05457×10^{-34}(J s) 4.13567×10^{-15} (eV s)	6.62607×10^{-27}(ergs) 1.05457×10^{-27}(ergs)
Proton mass	m_p	1.67262×10^{-27} (kg)	1.67262×10^{-24} (g)
Proton/electron mass quotient	m_p/m_e	1836.15 (1)	1836.15 (1)
Speed of light in vacuum	c_0	2.99792×10^8 (m s^{-1})	2.99792×10^{10} (cm s^{-1})

$kT = 4.0453\times10^{-21}$ J = (4.0453×10^{-14} erg), at T = 293 K.

Source: CRC Handbook of Chemistry and Physics, D.R. Lida (ed.), CRC Press, 2002.

with a much smaller mass m_e than the positive one (proton) m_p, the mass ratio quotient m_p/m_e being exactly 1836.15 (see Table 2.1).

(iii) Charges can be annihilated in the electron positron (positive electron) reaction, however, in the macroscale there is no evidence against the charge conservation principle.

In principle, what has been said above applies to static (motionless) charges, which can easily be detected using an electroscope (Leiden bottle). Also interactions among static charges can easily be measured experimentally, as started by the pioneering works of Ch.A. de Coulomb, who used the torsion balance. In 1785 he first established a mathematical relationship between the force acting on bodies, their charges and mutual distance, known as Coulomb's law. This law, that has been verified with increased accuracy over the centuries, can be used as the starting point for all electrostatics. The Gauss law and Maxwell relationships are just a natural consequences of the Coulomb law and can be easily derived from it.

2.2.1. Basic Electrostatic Relationships

According to the Coulomb law, the force **F** acting on a point charge q_1 located at $\mathbf{r}_1$ due to charge q_2 located at $\mathbf{r}_2$, in a vacuum is described by

$$\mathbf{F} = \frac{q_1 q_2}{4\pi\varepsilon_0 \left| \mathbf{r}_1 - \mathbf{r}_2 \right|^3} (\mathbf{r}_1 - \mathbf{r}_2) \tag{1}$$

where ε_0 is the dielectric permittivity of the vacuum, also called the electric constant (equal 8.854×10^{-12} F m^{-1} = C V^{-1}m^{-1} = C^{2}N^{-1}m^{-2} in the SI system applied in our work, see Table 2.1).

It is interesting to note that the force, which decreases with the square of the distance between charges, is central, i.e., acts on the centers of charges, so there is no torque. It is also reciprocal, i.e., the force acting on one charge is exactly opposite to the force acting on the other charge. This is analogous to the gravitational force given by the formula

$$\mathbf{F} = -G \frac{m_1 m_2}{\left| \mathbf{r}_1 - \mathbf{r}_2 \right|^3} (\mathbf{r}_1 - \mathbf{r}_2) \tag{2}$$

where G is the universal gravitational constant equal to 6.673×10^{-11} m^3 kg^{-1}s^{-2}, and m_1, m_2 are the masses of the interacting bodies.

In contrast to gravitational force, which is always negative (attraction), however, the Coulomb force can be both positive (repulsion for charges of the same sign) or negative (attraction for charges of opposite signs).

The most important property of Coulomb interactions, analogously to gravity interactions, is that they are fully additive, i.e., the net effect is a vectorial sum of contributions stemming from all charges, taken separately. This property is not obvious *a priori*, just considering that the negative and positive charges have different masses and, consequently, volumes. However, it has strong experimental support, at least in the macro scale. By exploiting additivity (also called the superposition rule), Eq. (1) can be generalized for the system of charges to

$$\mathbf{F} = \sum_i \mathbf{F}_i = \frac{q}{4\pi\varepsilon_0} \sum_i \frac{q_i}{|\mathbf{r} - \mathbf{r}_i|^3} (\mathbf{r} - \mathbf{r}_i) \tag{3}$$

where **F** is the net force on the charge q stemming from all other charges q_i located at distances $\mathbf{r}_i$, and $|\mathbf{r}-\mathbf{r}_i|$ is the length of this vector.

Despite the apparent simplicity, the significance of Eqs. (1), (3) reaches far beyond the simple law of forces acting between charged macrobodies in a vacuum.

The Coulomb expressions can be used in quantum physics problems to express, for example, the potential energy of electron/nuclei and electron/electron in the Hamiltonian operator. This law can also be exploited to estimate the magnitude of cohesive nuclear forces keeping the protons together, and to predict the range of stability of atoms with respect to the number of protons.

For example Eq. (1) can be used to estimate the ratio of the electrostatic to gravitational forces for the hydrogen atom composed of an electron and proton system separated by the distance 5.3×10^{-11} m (see Table 2.1). The electrostatic force equals 8.2×10^{-8} N. On the other hand, by using the proton and electron mass (Table 2.1) one can estimate that the gravitational force equals 3.6×10^{-47}. Thus, the electrostatic force is by more than 39 orders of magnitude larger. Hence, the gravitational force can be completely neglected in problems involving atoms and molecules.

In many electrostatic problems, however, the number of charges and their positions are not known exactly, which severely reduces the range of applicability of the Coulomb law. It is, therefore, advantageous to express this law in an alternative form involving the field strength, denoted by **E**. By definition **E,** which is also a vectorial quantity, is defined as the force acting on a positive elementary charge placed at a given point in space. Hence, by using this definition, Eq. (3) can be expressed in the simple form

$$\mathbf{E} = \frac{1}{q}\mathbf{F} \tag{4}$$

where

$$\mathbf{E}(\mathbf{r}_0) = \frac{1}{4\pi\varepsilon_0}\frac{q}{|\mathbf{r}_0 - \mathbf{r}|^3}(\mathbf{r}_0 - \mathbf{r}) \tag{5}$$

$|\mathbf{r}_0 - \mathbf{r}|$ is the length of the vector $(\mathbf{r}_0 - \mathbf{r})$ connecting the point in space where the field is evaluated (described by the position vector $\mathbf{r}_0$).

Note that the net field vector $\mathbf{E}$ is uniquely defined for every point of space if positions of the charge is known. An important property of $\mathbf{E}$, as the direct consequence of the Coulomb law, is that it is an additive quantity, i.e., fields stemming from separate charges are additive, so

$$\mathbf{E} = \mathbf{E}_1 + \mathbf{E}_2 + \mathbf{E}_3 + \ldots = \sum_i \mathbf{E}_i = \frac{1}{4\pi\varepsilon_0}\sum_i \frac{q_i}{|\mathbf{r}_0 - \mathbf{r}_i|^3}(\mathbf{r}_0 - \mathbf{r}_i) \tag{6}$$

If charges are so small and so numerous that they can be described by a continuous distribution the sum in Eq. (6) can be expressed as the integral

$$\mathbf{E}(\mathbf{r}_0) = \int_v \delta\mathbf{E} = \frac{1}{4\pi\varepsilon_0}\int_v \frac{\delta q}{|\mathbf{r}_0 - \mathbf{r}|^3}(\mathbf{r}_0 - \mathbf{r}) \tag{7}$$

where v is the volume where charges occur and δq is the infinitesimal charge located at the point of space.

Eq. (7) becomes especially useful if the charge distribution is governed by a continuous function of space variables. This is often the case when dealing with macroscopic fields when the range of variations in the charge distribution is much larger than atomic dimensions.

By defining the space charge density $\rho_e(\mathbf{r})$ as

$$\rho_e(\mathbf{r}) = \delta q/\delta v \tag{8}$$

where $\delta q = \Sigma_i \, q_i = e\Sigma_i \, z_i \, N_i$ is the net charge within the volume element $\delta \, v$, z_i is the multiplicity of the ith charge (e.g., ion valency) and N_i is the number of charges of a given kind, one can formulate Eq. (7) in the useful form

$$\mathbf{E}(\mathbf{r}_0) = \frac{1}{4\pi\varepsilon_0} \int_v \frac{\rho_e(\mathbf{r})}{|\mathbf{r}_0 - \mathbf{r}|^3} (\mathbf{r}_0 - \mathbf{r}) dv \tag{9}$$

where dv is the volume element.

In problems involving colloid particles one often encounters the situation where the charge is confined to within a thin layer, whose thickness is much smaller than particle dimensions. It is then useful to define, the surface charge σ instead of the space charge, via the constitutive dependence

$$\sigma(\mathbf{r}_s) = \delta q / \delta S = e \sum_i z_i N_i / \delta S \tag{10}$$

where $\mathbf{r}_s$ is the position vector on the surface (measured relative to a space-fixed coordinate system), and δS is the surface element.

Using this definition one can express Eq. (9) in the form

$$\mathbf{E}(\mathbf{r}_s) = \frac{1}{4\pi\varepsilon_0} \int_S \frac{\sigma(\mathbf{r}_s)}{|\mathbf{r}_0 - \mathbf{r}_s|^3} (\mathbf{r}_0 - \mathbf{r}_s) dS \tag{11}$$

where S is the charged surface area and dS is the surface element.

Eqs. (9), (11) can in principle be used for arbitrary distribution of space or surface charge. However, evaluating this surface integral is efficient for symmetric distributions of charge like spherical, cylindrical and planar, as discussed later on. For more complicated geometries this integration procedure becomes quite awkward. Therefore, it is often more advantageous to calculate electric fields from the electric potential, which is the quantity of primary interest for many practical applications.

The concept of the electric potential can be most naturally introduced, in analogy to the gravitational potential, by realizing that for any vector field of radial symmetry, as is the case for the Coulomb force given by Eq. (1) and the electric field, Eq. (5), its rotation is zero, which can be expressed as

$$\nabla \times \mathbf{E} = \nabla \times \mathbf{F} = 0 \tag{12}$$

This means, according to the vectorial analysis, that both **F** and **E** are gradients of scalar quantities, defined as the electric potential ψ and the interaction energy ϕ, respectively

$$\mathbf{E} = -\nabla\psi$$
$$\mathbf{F} = -\nabla\phi \tag{13}$$

The negative sign appears because, by convention, the energy gained by a system is negative. Similarly, the direction of an electric field is determined by the direction of a positive charge motion (opposite to increased electric potential value).

By integrating Eq. (13) along an arbitrary vectorial path $d\mathbf{l}$ between two arbitrarily chosen points a and b, one obtains the expression

$$\int_a^b \mathbf{E}\cdot d\mathbf{l} = -\int_a^b \nabla\psi d\mathbf{l} = \psi_a - \psi_b$$
$$\int_a^b \mathbf{F}\cdot d\mathbf{l} = -\int_a^b \nabla\phi d\mathbf{l} = \phi_a - \phi_b \tag{14}$$

where ψ_a, ψ_b are the electric potentials at points a and b, respectively, and ϕ_a, ϕ_b are the energies at these points.

By using Eq. (4) connecting the **F** and **E** vectors, one can derive the relationship of basic significance

$$\phi = \phi_0 + q\psi \tag{15}$$

where $\phi_0 = \phi_b - q\psi_b$ is the reference energy.

On the other hand, by using the expression for the field given by Eq. (5), one can evaluate the first line integral appearing in Eq. (14) obtaining an explicit expression for the electric potential in the vicinity of the system of charges

$$\psi(\mathbf{r}_0) = -\int_a^b \mathbf{E}\cdot d\mathbf{l} = -\psi_b + \frac{1}{4\pi\varepsilon_0}\sum_j \frac{q_j}{|\mathbf{r}_0 - \mathbf{r}_i|} \tag{16}$$

By conveniently assuming a point at infinity as the reference point, Eqs. (15), (16) can be simplified because $\psi_0 = 0,\ \psi_b = 0$.

Making use of Eqs. (15), (16), one can express the interaction energy for a system of charges in the form

$$\phi = \frac{1}{4\pi\varepsilon_0}\sum_i \sum_{j<i} \frac{q_i q_j}{|\mathbf{r}_i - \mathbf{r}_j|} \tag{17}$$

Eq. (17) can also be expressed in the alternative form, useful for direct calculations of energy for systems consisting of a limiting number of charges

$$\phi = \frac{1}{2}\sum_{i\neq j}\sum \frac{q_i q_j}{4\pi\varepsilon_0 |\mathbf{r}_i - \mathbf{r}_j|} \tag{18}$$

where the factor 1/2 appears because the interaction energy is calculated twice in the sum, and ψ_i is the potential at the point where the ith charge is located.

Eq. (16) indicates that the electric potential is an additive quantity like force or electric field. This property can be exploited for calculating the potential distribution generated by systems of charges, e.g., for estimating simple ion interaction energy in air and in ionic crystals, or for construction of the Hamiltonian for many-electron atoms.

Let us illustrate this for the simplest case of two equal charges q_1 and $-q_1$ separated by the fixed distance d shown in Fig. 2.1. This system charges, called a dipole, plays a fundamental role in both macroscale problems and microscale problems, when calculating the interaction of molecules. As the most common example, one can mention the water molecule, which is permanently polarized, carrying a negative charge on the oxygen atom and positive charges on hydrogen atoms (obviously the positive and negative charges are of equal magnitude). Alcohols and surfactants are further examples of molecules exhibiting permanent dipole properties.

Let us place the charges on the z-axis of the local coordinate system with its origin in the midpoint between them (see Fig. 2.1). Then the electric potential at the arbitrary point $P(x_0, y_0, z_0)$ is given by the exact formula [2]

$$\psi(\mathbf{r}_0) = \frac{q}{4\pi\varepsilon_0}\left(\frac{1}{(r_0^2 - dz_0 + d^2/4)^{1/2}} - \frac{1}{(r_0^2 + dz_0 + d^2/4)^{1/2}}\right) \tag{19}$$

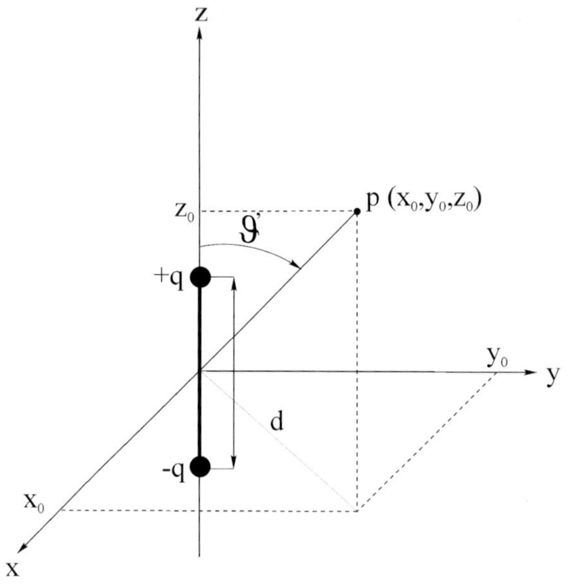

Fig. 2.1. The coordinate system used for calculating the electric potential distribution near a dipole.

where $r_0^2 = x_0^2 + y_0^2 + z_0^2$ is the length of the $\mathbf{r}_0$ vector connecting the origin of the coordinate system with the point P.

At distances $\mathbf{r}_0$ much larger than d, this can be written in the form

$$\psi(\mathbf{r}_0) = \frac{1}{4\pi\varepsilon_0}\frac{p\cos\vartheta'}{r_0^2} = -\frac{\mathbf{p}\cdot\mathbf{r}_0}{4\pi\varepsilon_0 r_0^3} = \mathbf{p}\cdot\nabla\psi_p = -\mathbf{p}\cdot\mathbf{E}_p \tag{20}$$

where $\psi_p(\mathbf{r}_0) = 1/4\pi\varepsilon_0 r_0$ is the potential due to a positive point charge, $\mathbf{E}_p$ the field due to this charge, $p = qd$ the length of the dipole moment vector $\mathbf{p}$ (directed along the z-axis from the negative to positive charge, see Fig. 2.1) and ϑ' the angle between the z-axis and the r_0 direction.

Using Eq. (20) one can deduce that the energy of the dipole interaction with a charge q is given by

$$\phi = q\psi_d = -\mathbf{p}\cdot\mathbf{E} \tag{21}$$

The electric field generated by the dipole is given by the expression

$$\mathbf{E} = q\mathbf{E}_p = \frac{3(\mathbf{p}\cdot\mathbf{r}_0)\mathbf{r}_0 - \mathbf{p}r_0^2}{r_0^5} \tag{22}$$

Eq. (22) can be formulated explicitly in the coordinate system shown in Fig. 2.1.

$$\mathbf{E}_z = p\,\frac{3\cos^2\vartheta' - 1}{4\pi\varepsilon_0 r_0^3}\,\mathbf{i}_z$$

$$\mathbf{E}_\perp = p\,\frac{3\cos\vartheta'\sin\vartheta'}{4\pi\varepsilon_0 r_0^3}\,\mathbf{i}_\perp \tag{23}$$

where $\mathbf{E}_z$ is the field component along the z-axis and $\mathbf{E}_\perp$ is the field component perpendicular to the z-axis and $i_z, i_\perp$ are the unit vectors. The length of the field vector is

$$E = |\mathrm{E}| = (E_z^2 + E_\perp^2)^{1/2} = \frac{\mathrm{p}}{4\pi\varepsilon_0 r_0^3}(3\cos^2\vartheta' + 1)^{1/2} \tag{24}$$

Another interesting example of discrete charge configuration is the linear quadrupole composed of three charges arranged linearly: $-2q_2$ in the middle and $+\,q_2$ separated by the distance d_2 (see Fig. 2.2). This charge configuration

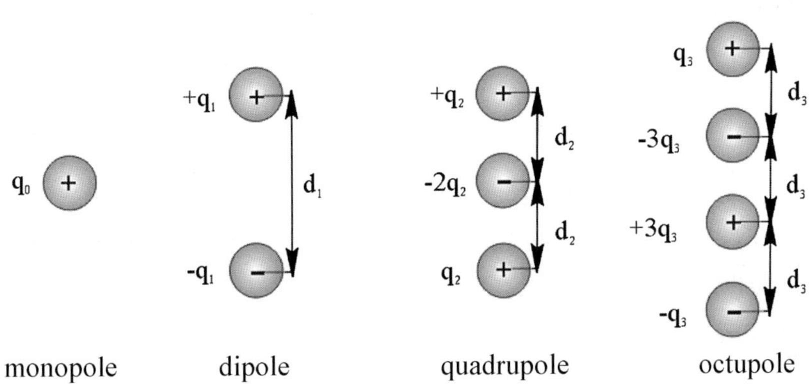

Fig. 2.2. Linear multipoles.

can be treated as two displaced dipoles. In the case of the quadrupole the electric potential at an arbitrary point in space $r_0 \gg d_2$ is given by the formula

$$\psi(r_0) = \frac{qd^2}{4\pi\varepsilon_0}\frac{(3\cos^2\vartheta' - 1)}{r_0^3} = \frac{p_2(3\cos^2\vartheta' - 1)}{4\pi\varepsilon_0 r_0^3} \tag{25}$$

where $p_2 = qd^2$.

Analogously, one can define the linear octupole and higher rank multipoles, which for the sake of brevity, are denoted by $P^{(n)}$, thus

$$P^{(n)} = qd^n \tag{26}$$

where $P^{(0)} = q$ (total charge) is referred to as the monopole, $P^{(1)} = qd$ as the dipole moment, $P^{(2)} = qd^2$ the quadrupole, etc.

The concept of multipoles is useful when dealing with electrostatic interactions of charged particles, or particle interactions with interfaces [4]. In the general case of arbitrary multipoles, spherical harmonic expansion is used [3]. For the axially symmetric case (spherical particles), one can formulate the series expansion in the simple form by exploiting Legendre polynomials P_n [1]

$$\psi(\mathbf{r}_0) = \frac{1}{4\pi\varepsilon}\sum_{n=0}^{\infty}\frac{1}{r_0^{n+1}}\sum_i P_i^{(n)} P_n(\cos\vartheta_i) \tag{27}$$

where $P_n(\cos\vartheta_i)$ are the Legendre polynomials of the order n and ϑ_i is the angle between r_i (position vector of a given charge) and the $\mathbf{r}_0$ vector.

The multipole concept is also useful when dealing with systems of closely spaced charges q_i, (see Fig. 2.3), for example, multielectron molecules. In this case the multipoles are defined in a more general way as [1,3]

$$q^{(0)} = \sum q_i$$

$$\mathbf{p} = \sum q_i \mathbf{r}_i \tag{28}$$

$$\mathbf{Q} = \sum q_i \mathbf{r}_i \mathbf{r}_i$$

where $\mathbf{p}$ is the dipole vector and $\mathbf{Q}$ is the quadrupole tensor, etc.

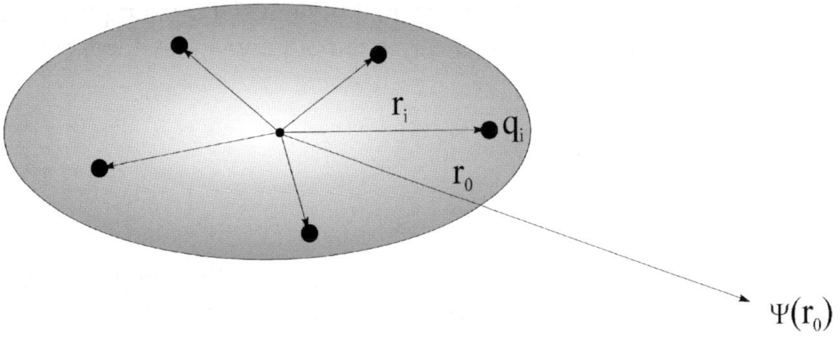

Fig. 2.3. The electric potential near the system of charges.

By using these definitions one can expand Eq. (16) into the Taylor series around r_0 to obtain the result

$$\psi(\mathbf{r}_0)=\frac{q^{(0)}}{4\pi\varepsilon_0 r_0}+\frac{1}{4\pi\varepsilon_0}\mathbf{p}\cdot\nabla\left(\frac{1}{r_0}\right)+\frac{1}{2}\mathbf{Q}:\nabla^2\left(\frac{1}{r_0}\right)+\cdots \tag{29}$$

By substituting the expansion Eq. (29) into Eq. (15), after rearrangement one obtains the following expression for the interaction energy, which is a generalized version of Eq. (21) [3]

$$\phi=q\psi(\mathbf{r}_0)=\frac{qq^{(0)}}{4\pi\varepsilon_0 r_0}-\mathbf{p}\cdot\mathbf{E}-\frac{1}{6}\mathbf{Q}:\nabla\mathbf{E}+\cdots \tag{30}$$

where

$$\mathbf{E}=q\nabla\left(\frac{1}{4\pi\varepsilon_0 r_0}\right) \tag{31}$$

is the electric field at $\mathbf{r}_0$.

However, all the above formulae are valid for discrete charge distributions whose positions are well known and fixed. For macroscopic problems, involving continuous charge distribution, more appropriate is an integral form of Eq. (16), which can be formulated by introducing the space charge concept defined by Eq. (8). In this way one obtains

$$\psi(\mathbf{r}_0) = \frac{1}{4\pi\varepsilon_0} \int_v \frac{\rho_e(\mathbf{r})}{|\mathbf{r}_0 - \mathbf{r}|} \, dv \tag{32}$$

Analogously, using Eq. (11) describing electric field in the case when surface charge is present, one can write for $\psi(\mathbf{r}_s)$

$$\psi(\mathbf{r}_s) = \frac{1}{4\pi\varepsilon_0} \int_s \frac{\sigma(\mathbf{r}_s)}{|\mathbf{r}_0 - \mathbf{r}|} \, dS \tag{33}$$

By assuming again a continuous charge distribution, one can express Eq. (18) in the useful, integrated form

$$\phi = \frac{1}{2} \int_v \rho_e \psi \, dv \tag{34}$$

Analogously, if the surface charge is present, the expression for the interaction energy becomes

$$\phi = \frac{1}{2} \int_s \sigma \psi \, dS \tag{35}$$

In many applications involving continuous electrostatic fields, it is advantageous to express the above relationships in an alternative form. This can be most conveniently done by exploiting the general theorem of Gauss, stating that the streaming of a vector field through a closed surface equals the volume integral of the divergence of this field. This can be expressed in the mathematical form as

$$\int_S \mathbf{E} \cdot d\mathbf{S} = \int_v \nabla \cdot \mathbf{E} \, dv \tag{36}$$

where $d\mathbf{S} = dS\,\hat{\mathbf{n}}$ is the surface element vector and $\hat{\mathbf{n}}$ is the unit vector directed outwards from the surface.

By using the expression for **E** given by Eq. (5), resulting from the Coulomb law, one can demonstrate [2] that the surface integral in Eq. (36) is constant for arbitrary shape of the enclosing surface, i.e.,

$$\int_S \mathbf{E}\cdot d\mathbf{S} = \frac{q}{\varepsilon_0} \tag{37}$$

Eq. (37), usually called the Gauss law of electrostatics, is no more than an alternative formulation of the Coulomb law of forces between charges. However, for many practical applications, especially for calculating fields at charged interfaces of various shapes, Eq. (37) is useful because the surface integral is fully independent of the distribution of the charge within the volume enclosed by the surface S. The Gauss equation can also be used to specify boundary conditions when calculating interactions between particles as discussed later on.

By noticing that the total charge within the volume v is given by the formula

$$q = \sum_i q_i = \int_v \rho_e \, dv \tag{38}$$

one can convert Eq. (36) in to the alternative form

$$\int_v \left(\nabla\cdot\mathbf{E} - \frac{\rho_e}{\varepsilon_0} \right) dv = 0 \tag{39}$$

since this must be valid for arbitrary volume, one can deduce that

$$\nabla\cdot\mathbf{E} = \frac{\rho_e}{\varepsilon_0} \tag{40}$$

Eq. (40), being again an alternative formulation of Gauss's law, has a fundamental significance in electrostatics and is known as the Maxwell equation.

By exploiting the fact that the electric field is a gradient of the electric potential (cf. Eq. (13)) one can express Eq. (40) in the form of major practical significance

$$\nabla\cdot(-\nabla\psi) = -\nabla^2\psi = \frac{\rho_e}{\varepsilon_0} \tag{41}$$

Eq. (41), usually referred to as the Poisson equation, is the starting relationship for any problem dealing with the stability and interactions of particle suspension with surfaces.

It is interesting to note that in the case where the space charge vanishes, Eq. (41) reduces to the familiar form, called the Laplace equation

$$\nabla^2\psi = 0 \tag{42}$$

This form is useful for describing electric potential distribution in the vicinity of charged interfaces (e.g., electrodes) placed in insulating media.

By exploiting the Maxwell equation, one can alternatively express the electrostatic energy formula, Eq. (34) in the following form [2]:

$$\phi = \frac{\varepsilon_0}{2}\int_v \psi\nabla^2\psi\, dv = \frac{\varepsilon_0}{2}\int_v (\nabla\psi)^2\, dv = \frac{\varepsilon_0}{2}\int_v \mathbf{E}\cdot\mathbf{E}\, dv \tag{43}$$

One can deduce from this equation that the existence of an electrostatic field is associated with the energy density ρ_ϕ (energy per unit volume), which equals

$$\rho_\phi = \frac{\varepsilon_0}{2}\mathbf{E}\cdot\mathbf{E} = \frac{\varepsilon_0}{2}E^2 \tag{44}$$

In analyzing colloid particle dynamics one is often interested in calculating electrostatic forces exerted by a field. This can be most conveniently done by transforming Eq. (4) in the following way:

$$\mathbf{F} = q\mathbf{E} = \rho_e \mathbf{E}\, dv \tag{45}$$

where dv is the infinitesimal volume element containing the charge q.

By using Eqs. (40–41) one can transform this in the following way:

$$\rho_e \mathbf{E} = \varepsilon_0 (\nabla \cdot \mathbf{E})\mathbf{E} = \varepsilon_0 \nabla\psi \nabla\psi^2 \tag{46}$$

Eq. (46) can be expressed as the gradient of a tensor, i.e.,

$$\varepsilon_0 (\nabla \cdot \mathbf{E})\mathbf{E} = \nabla \frac{\varepsilon_0}{2}\left[2\mathbf{E}\mathbf{E} - E^2\mathbf{I}\right] \tag{47}$$

where **I** is the unit tensor and

$$\mathbf{T} = \frac{\varepsilon_0}{2}\left(2\mathbf{E}\mathbf{E} - E^2\mathbf{I}\right) \tag{48}$$

is the Maxwell stress tensor.

By combining Eqs. (45) and (48) one can calculate the electric force acting on a charged volume v as the integral

$$\mathbf{F} = \int_v \mathbf{T}\, dv \tag{49}$$

This can be transformed by exploiting the Gauss theorem to the useful form

$$\mathbf{F} = \frac{\varepsilon_0}{2}\int_S \left(2\mathbf{E}\mathbf{E} - E^2\mathbf{I}\right)\, dS \tag{50}$$

Eq.(50) enables one to calculate the electric force on an interface of an arbitrary shape.

For the sake of convenience, the most important electrostatic relationships are compiled in Table 2.2.

All formulae discussed hitherto, especially those in Table 2.2, are strictly valid for interactions in a vacuum. When charged interfaces are placed in material media, e.g., liquids, as is the case for every problem involving particles, their interactions are drastically modified. The effect of matter separating electric charges depends to a significant extent on its conductivity defined as the ability to sustain the free charge motion. In media classified as conductors, the charge carriers, most often electrons, move

Table 2.2
Fundamental relationships of electrostatics

Mathematical form	Traditional name, remarks
Continuous charge distribution:	
$\nabla \cdot \mathbf{E} = \rho_e/\varepsilon_0$	Maxwell's equation
$\int_S \mathbf{E} \cdot d\mathbf{S} = q/\varepsilon_0$	Gauss's law
$\nabla^2 \psi = -\rho_e/\varepsilon_0$	Poisson's equation
$\mathbf{E} = -\nabla\psi = \frac{1}{4\pi\varepsilon_0}\int_v \frac{\rho_e\, dv}{\lvert \mathbf{r}_0 - \mathbf{r} \rvert^3}(\mathbf{r}_0 - \mathbf{r})$	Electric field is conservative
$\nabla \times \mathbf{E} = 0$	Electric field is irrotational
$\mathbf{F} = q\mathbf{E}$	Electric force is proportional to the field
$\psi = -\int_l \mathbf{E} \cdot d\mathbf{l} = \frac{1}{4\pi\varepsilon_0}\int_v \frac{\rho_e\, dv}{\lvert \mathbf{r}_0 - \mathbf{r} \rvert}$	Electric potential definition
$\phi = \frac{1}{2}\int_v \rho_e \psi\, dv$ $\phi = \frac{1}{2}\int_S \sigma\psi\, dS$	Electric interaction energy formulae
Discrete charge distribution	
$\mathbf{F} = \frac{q}{4\pi\varepsilon_0}\sum_i \frac{q_i}{\lvert \mathbf{r}_0 - \mathbf{r}_i \rvert^3}(\mathbf{r}_0 - \mathbf{r}_i)$	Coulomb's law
$\mathbf{E} = \frac{1}{4\pi\varepsilon_0}\sum_i \frac{q_i}{\lvert \mathbf{r}_0 - \mathbf{r}_i \rvert^3}(\mathbf{r}_0 - \mathbf{r}_i)$	Electric field formula
$\psi = \frac{1}{4\pi\varepsilon_0}\sum_i \frac{q_i}{\lvert \mathbf{r}_0 - \mathbf{r}_i \rvert}$	Electric potential formula
$\phi = \frac{1}{8\pi\varepsilon_0}\sum\sum_{i \neq j} \frac{q_i q_j}{\lvert \mathbf{r}_0 - \mathbf{r}_i \rvert}$	Interaction energy formula

freely under applied external electric field over macroscopic distances. The flux of carriers is the net result of the electrostatic force given by Eq. (4) and the electric resistance of the medium, which remains very low for metals and concentrated electrolyte solutions.

On the other hand, for insulating media (dielectrics) the external electric fields induce merely a charge displacement from its equilibrium position over distances comparable with atomic dimensions. The displacement of the electron cloud of atoms and molecules is usually referred to as molecular polarization. This effect appears universally in all media including gases. In materials composed of polar molecules being permanent dipoles, like water, the local electric field induces also orientation of dipoles. The net result of these effects is characterized by the polarization vector $\mathbf{P}_e$ defined as the dipole moment per unit volume [2]. When the polarization is uniform, its divergence gives the polarization charge per unit volume described by the equation

$$-\nabla \cdot \mathbf{P}_e = \rho_p \tag{51}$$

where ρ_p is the polarization charge density.

By introducing the polarization concept one can quantitatively describe the electric field distribution in dielectric media by using the Maxwell equation, Eq. (40) with the charge density interpreted as the sum of real and polarization charges [2], i.e.,

$$\nabla \cdot \mathbf{E} = \frac{\rho_e + \rho_p}{\varepsilon_0} = \frac{1}{\varepsilon_0}(\rho_e - \nabla \cdot \mathbf{P}_e) \tag{52}$$

This can be written in the form

$$\nabla \cdot \left(\mathbf{E} + \frac{1}{\varepsilon_0}\mathbf{P}_e \right) = \frac{\rho_e}{\varepsilon_0} \tag{53}$$

In materials classified as linear media, the polarization vector is proportional to the local electric field which can be expressed by the formula

$$\mathbf{P}_e = \varepsilon_0 \chi \, \mathbf{E} \tag{54}$$

where χ is defined as the electric susceptibility of the medium [2].

It should be mentioned, however, that this linear relationship is not universally valid. On the contrary, for high electric fields the polarization is expected to be less than proportional due to the saturation effect. Similarly, for small distances between charges (or charged interfaces), comparable with atomic dimensions the discreteness of molecule density distribution will certainly lead to nonlinear dependence of polarization on the electric field produced by charges.

Using the dependence expressed by Eq. (54), one can formulate the Maxwell law for dielectrics in the form

$$\nabla\cdot(1+\chi)\,\mathbf{E} = \frac{\rho_e}{\varepsilon_0} \tag{55}$$

It is interesting to mention that the quantity $(1+\chi)$ is often defined as the dielectric constant of a medium ε and the vector $\varepsilon_0\,\mathbf{E}$ is defined as the displacement vector $\mathbf{D}_p$.

By exploiting the concept of the polarization vector and the generalized Maxwell equation, given by Eq. (55), one can convert other dependencies listed in Table 2.2 in to the form suitable for dielectric media by replacing the permittivity of a vacuum ε_0 by the dielectric permittivity $\varepsilon = (1+\chi)\varepsilon_0$.

In particular, the Coulomb interaction law can be expressed as

$$\mathbf{F} = \frac{q_1 q_2}{4\pi\varepsilon\,|\mathbf{r}_1 - \mathbf{r}_2|^3}(\mathbf{r}_1 - \mathbf{r}_2) \tag{56}$$

Similarly, the Maxwell's and Gauss equations can be formulated as

$$\begin{gathered}\nabla\cdot\mathbf{D}_\mathrm{p} = \nabla\cdot(\varepsilon\mathbf{E}) = \rho_e \\ \int_S \varepsilon\,\mathbf{E}\,dS = q\end{gathered} \tag{57}$$

The expression for the energy of the electric field in a dielectric becomes [5]

$$\phi = \frac{1}{2}\int_v \mathbf{E}\cdot(\varepsilon\mathbf{E})\,dv \tag{58}$$

The Maxwell's stress tensor becomes

$$\mathbf{T} = \frac{\varepsilon}{2}(2\mathbf{EE} - E^2\mathbf{I}) \tag{59}$$

In the case of the field dependent permittivity, one has [5]

$$\mathbf{T} = \frac{1}{2}\left[2\varepsilon\mathbf{EE} - \left(\varepsilon E^2 - \int_0^E \left(\frac{\partial\varepsilon}{\partial E'}\right)_T E'^2\, dE'\right)\mathbf{I}\right] \tag{60}$$

It is worthwhile mentioning once more that Eqs. (56–58) involving the permittivity of a medium are approximate only, being material-dependent quantities. They become especially ill-posed for microscopic problems, e.g., calculating from Coulomb's law the interaction energy between discrete charges separated by a distance comparable with molecular dimensions. The permittivity of various media including water at various temperatures is collected in Table 2.3.

Table 2.3
Permittivities of various substances at T = 293 K

Medium	$(1+\chi)$	$4\pi(1+\chi)\varepsilon_0 \times 10^{10}$ F m^{-1}
Vacuum	1	1.11265
Air	1.00054	1.113
Acetone	1.946	2.165
Benzene	2.275	2.531
Ethanol	26.4	29.37
Formamide	111.5	12.79
Glycerol	42.4	47.18
Isopropanol	18.3	20.36
Methanol	32.65	36.32
Octane	1.946	2.165
Pentane	1.843	2.051
Propanol	19.7	21.92
Styrene	2.431	2.705
Water	80.2	89.23
Bakelite	4.8	5.341
Diamond	5.68	6.320
Ice	72.5	80.67
Mica (ruby)	3.5	3.894

Table 2.3 (Continued)

Medium	$(1+\chi)$	$4\pi(1+\chi)\varepsilon_0 \times 10^{10}$ F m^{-1}
Polyethylene	2.3	2.560
Polystyrene	2.6	2.893
PTFE (polytetra fluoroethylene)	2.1	2.337
Porcelain	6.5	7.232
Pyrex glass	4.5	5.007
Quartz (fused)	3.8	4.220
Ruthile (TiO_2)	100	111.3

Permittivity of water at various temperatures

T (K)	$(1+\chi)$	$4\pi(1+\chi)\varepsilon_0 \times 10^{10}$ F m^{-1}
273.15	87.90	97.80
283.15	83.96	93.42
293.15	80.20	89.23
298.15	78.68	87.54
303.15	76.60	85.23
313.15	73.17	81.41
323.15	69.88	77.75
333.15	66.73	74.25
343.15	63.73	70.91
353.15	60.86	67.72
363.15	58.12	64.67
373.15	55.51	61.76

2.2.2. Electric Potential and Field Distributions for Simple Geometries

In this section we present some concrete solutions of the Gauss and Laplace equations for simple geometries that allow one to determine the electric field, potential, interaction force and energy.

Let us first consider a planar interface of infinitesimal thickness separating two dielectric media with the permittivities ε_1 and ε_2, respectively. The surface charge of the interface equals $\sigma = q/S$. The electric field distribution can be found most efficiently by using the Gauss equation, Eq. (57). Because of the planar geometry the field is perpendicular to the interface at all points, so one can evaluate the surface integral immediately, which results in the formula

$$\varepsilon_1 E_1 + \varepsilon_2 E_2 = q/S = \sigma \tag{61}$$

When the adjacent media have the same permittivity ε, one obtains

$$E = \sigma/2\varepsilon \tag{62}$$

By placing the origin of the Cartesian coordinate system at the plane, one can formulate Eq. (61) in a convenient form

$$\varepsilon_1 E_1 - \varepsilon_2 E_2 = \sigma \tag{63}$$

Eq. (63) represents the general boundary condition for the Laplace equation, valid not only for planar interfaces but also for arbitrary interface shapes, provided that the charge and the normal component of the field are treated as local quantities.

Eq. (63) can be expressed in a more convenient form by introducing the gradient of electric potential in place of the field, so $E = -d\psi/dx$

$$\varepsilon_2\left(\frac{d\psi}{dx}\right)_2 - \varepsilon_1\left(\frac{d\psi}{dx}\right)_1 = \sigma \tag{64}$$

For any metallic interface, the inside field $(d\psi/dx)$ vanishes and Eq. (64) reduces to the simpler form

$$\varepsilon_1\left(\frac{d\psi}{dx}\right)_1 = -\sigma \tag{65}$$

On the other hand, for uncharged interfaces Eq. (64) becomes

$$E_1 = \frac{\varepsilon_2}{\varepsilon_1} E_2 \tag{66}$$

With these boundary conditions one can easily solve the Laplace equation, which for planar geometry becomes simply

$$\frac{d^2\psi}{dx^2} = 0 \tag{67}$$

For the most interesting case of two metallic plates (electrodes) separated by the distance δ and bearing surface charges σ^0 and $-\sigma^0$, respectively, the solution of Eq.(67) with the boundary condition, Eq. (65) becomes

$$\psi = \psi^0 - \frac{\sigma^0}{\varepsilon} x \tag{68}$$

Without loss of generality one may assume that the potential of the second plate is zero (the electrode is grounded), so the potential at the first electrode becomes

$$\psi^0 = \frac{\sigma^0 \delta}{\varepsilon} \tag{69}$$

Knowing the charge and the potential of a system, one can calculate the electrical capacity of two conducting electrodes that is defined as

$$C_e = \frac{q^0}{\Delta\psi} = \frac{S\sigma^0}{\Delta\psi} \tag{70}$$

where q^0 is the charge on the electrode and $\Delta\psi = \psi^0$ is the difference in electric potential between the plates.

In the case of a two-plate system, Eq. (70) becomes simply

$$C_e = \frac{\varepsilon S}{\delta} \tag{71}$$

One can see that the capacity, interpreted as the amount of charge that can be accumulated on electrodes at a given potential, increases proportionally to the relative permittivity and the surface area but inversely proportional to the distance between the plates.

Knowing the dependence between the surface charge and the potential, one can easily calculate the energy of the two electrode system using Eq. (35), i.e.,

$$\phi = \frac{1}{2}\int_S \sigma^0 \psi^0 \, dS = \frac{S\sigma^0\psi^0}{2} = \frac{1}{2}C_e(\Delta\psi)^2 = \frac{1}{2}C_e\psi^{0^2} \tag{72}$$

Analogously, from Eq. (43) one obtains

$$\phi = \frac{\varepsilon}{2}\int_v E^2 dv = \frac{\varepsilon}{2} v_p E^{0^2} = \frac{1}{2} C_e \psi^{0^2} \tag{73}$$

where $v_p = S\delta$ is the volume between the plates.

One can perform similar calculations for a two concentric cylinders electrode system (see Table 2.4) with the outer electrode potential kept at zero (grounded). The Laplace equation for this geometry is given by

$$\nabla^2 \psi = \frac{1}{r}\frac{d}{dr}\left(r\frac{d\psi}{dr}\right) = 0 \tag{74}$$

Solving this with the boundary condition, Eq. (65) gives the result

$$\psi = \frac{\sigma^0 R_1}{\varepsilon}\ln\left(\frac{R_2}{r}\right) = \psi_1^0 \frac{\ln(R_2/r)}{\ln(R_2/R_1)} \tag{75}$$

where $\psi_1^0 = (\sigma_1^0 R_1/\varepsilon)\ln(R_2/R_l) = (q^0/2\pi L\varepsilon)\ln(R_2/R_1)$ is the potential of the inner cylinder and R_1, R_2 are the radii of the inner and outer cylinder, respectively.

In the case of one cylinder of radius R, the potential distribution diverges according to the formula

$$\psi = \frac{\sigma^0 R}{\varepsilon}\left[1 + \ln\left(\frac{R}{r}\right)\right] \tag{76}$$

This describes the electric potential distribution near an isolated cylinder.

From Eq. (75) one can calculate that the electric capacity of a two-cylinder system is given by the formula

$$C_e = \frac{q^0}{\Delta\psi} = \frac{q^0}{\psi_1^0} = \frac{2\pi\varepsilon L}{\ln(R_2/R_1)} \tag{77}$$

Table 2.4
Electric field, potential and capacity for simple electrode geometries

Geometry	Field, **E**	Potential, ψ	Capacity, *Ce*
Two planar electrodes			
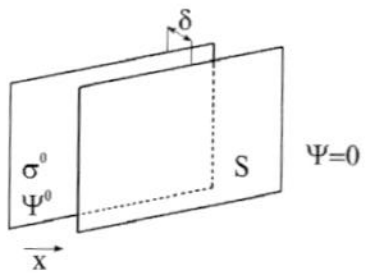	$\frac{\sigma^0}{\varepsilon}\mathbf{i}_\mathbf{x} = \frac{\psi^0}{\delta}\mathbf{i}_\mathbf{x}$	$\frac{\sigma^0\delta}{\varepsilon}(1-x/\delta) = \psi^0(1-x/\delta)$	$\frac{\varepsilon S}{\delta}$
Two concentric spheres			
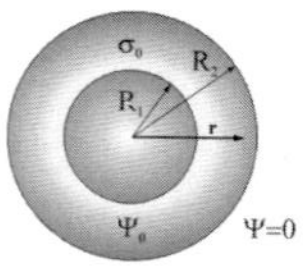	$\frac{\sigma_1^0 R_1^2}{\varepsilon}\frac{1}{r^2}\mathbf{i}_r = \psi_1^0\frac{R_1}{r^2}\mathbf{i}_r$	$\frac{\sigma_1^0 R_1^2}{\varepsilon}\left(\frac{R_2-r}{R_2}\right)\frac{1}{r} = \psi_1^0\left(\frac{R_2-r}{R_2-R_1}\right)\frac{R_1}{r}$	$\frac{4\pi\varepsilon R_1 R_2}{(R_2-R_1)}$
Single sphere			
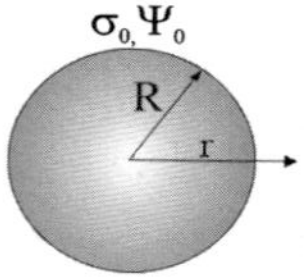	$\frac{\sigma^0 R^2}{\varepsilon}\frac{1}{r^2}\mathbf{i}_r = \psi^0\frac{R}{r^2}\mathbf{i}_r$	$\frac{\sigma^0 R^2}{\varepsilon r} = \psi^0\frac{R}{r}$	$4\pi\varepsilon R$

Two concentric cylinders 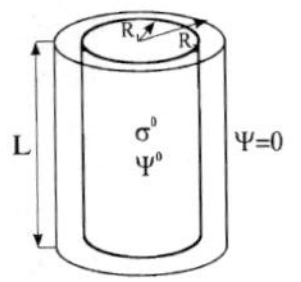	$\frac{\sigma_1^0 R_1}{\varepsilon}\frac{1}{r}\mathbf{i}_r = \frac{\psi_1^0}{\ln(R_2/R_1)}\frac{1}{r}\mathbf{i}_r$	$\frac{\sigma_1^0 R_1}{\varepsilon}\ln\left(\frac{R_2}{r}\right) = \psi_1^0\frac{\ln(R_2/r)}{\ln(R_2/R_1)}$	$\frac{2\pi\varepsilon L}{\ln(R_2/R_1)}$
Isolated cylinder 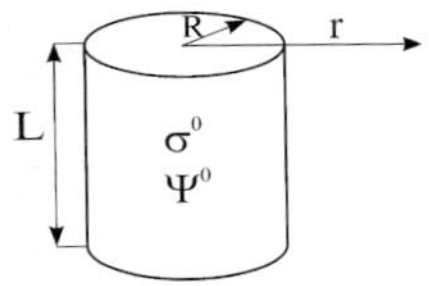	$\frac{\sigma^0 R}{\varepsilon}\frac{1}{r}\mathbf{i}_r = \psi^0\frac{1}{r}\mathbf{i}_r$	$\frac{\sigma^0 R}{\varepsilon}\left[1+\ln\left(\frac{R}{r}\right)\right] = \psi^0\left[1+\ln\left(\frac{R}{r}\right)\right]$	–

The field in the space between the cylinders is given by

$$E = \frac{\psi_1^0}{\ln(R_2/R_1)} \frac{1}{r} \tag{78}$$

Note that the field is nonuniform, characterized by the gradient

$$\nabla E = -\frac{\psi_1^0}{\ln(R_2/R_1)} \frac{1}{r^2} \tag{79}$$

From Eq. (79) one can deduce that the field is strongest at the inner electrode.

Consequently, the quantity $\nabla(E^2) = 2E\nabla E$ used for calculating the dielectrophoretic force is given by

$$\nabla(E^2) = -\frac{2\psi_1^{0^2}}{(\ln R_2/R_1)^2} \frac{1}{r^3} \tag{80}$$

In a similar way, the electric potential distribution and capacity can be calculated for two concentric electrodes of a spherical shape (see Table 2.4).

$$\nabla^2 \psi = \frac{1}{r^2} \frac{d}{dr}\left(r^2 \frac{d\psi}{dr} \right) = 0 \tag{81}$$

Solving this with the boundary condition, Eq. (65) gives the result

$$\psi = \frac{\sigma_1^0 R_1^2}{\varepsilon} \frac{(R_2 - r)}{R_2} \frac{1}{r} = \psi_1^0 \frac{(R_2 - r)}{(R_2 - R_1)} \frac{R_1}{\mathrm{r}} \tag{82}$$

where $\psi_1^0 = \sigma_1^0 R_1 (R_2 - R_1)/\varepsilon R_2 = q_1^0 (R_2 - R_1)/4\pi\varepsilon R_2 R_1$

and R_1, R_2 are the radii of the inner and outer spheres, respectively, and ψ_1^0 is the potential of the inner sphere (the outer remains grounded).

In the case where the outer sphere radius tends to infinity, Eq. (82) reduces to

$$\psi = \frac{\sigma^0 R^2}{\varepsilon} \frac{1}{r} = \frac{q^0}{4\pi\varepsilon r} \tag{83}$$

This describes the electric potential distribution near an isolated sphere.

Note that the electric potential vanishes at separations larger than the sphere radius in contrast to the isolated cylinder where the potential diverges to infinity. It is interesting to observe that the field distribution near a sphere bearing the surface charge q^0 has an identical mathematical shape to the field near a point charge q^0 located at its center.

Using Eq. (82) one can express the electric capacity of a two sphere system by the formula

$$C_e = \frac{4\pi\varepsilon R_1 R_2}{(R_2 - R_1)} \tag{84}$$

For one sphere Eq.(84) simplifies to

$$C_e = 4\pi\varepsilon R \tag{85}$$

Some characteristic values of the electric capacity (expressed in $F = CV^{-1}$ where F is the Farad, C the Coulomb, V the volt) of a single sphere depending on its dimensions are compiled in Table 2.5. One can see that even the capacity of a sphere of the size of the Earth (6.72×10^5 m) remains a fraction of a Farad (precisely 7.5×10^{-4} F). This means that the charge of a Coulomb will load it to the potential of 1300 V. For colloid particles the size of micrometers to nanometers their capacity in water would be $8.7\times10^{-15} \div 8.7\times10^{-18}$ F.

Table 2.5
The electric capacity of spheres of various sizes

Sphere radius (m)	Capacity in air (F)	Capacity in water (F)
10^{-9}	1.1×10^{-19}	8.7×10^{-18}
10^{-6}	1.1×10^{-16}	8.7×10^{-15}
10^{-4}	1.1×10^{-14}	8.7×10^{-13}
6.72×10^{6} (Earth)	7.5×10^{-4}	–

The field in the space between the two concentric spheres is given by

$$E = \psi_1^0 \frac{R_1 R_2}{(R_2 - R_1)} \frac{1}{r^2} \tag{86}$$

From Eq. (86) one can deduce that the field is strongest at the inner electrode. Note that the field is non-uniform, characterized by the gradient

$$\nabla E = -2\psi_1^0 \frac{R_1 R_2}{(R_2 - R_1)} \frac{1}{r^3} \tag{87}$$

Consequently, the quantity $\nabla(E^2) = 2E\nabla E$, which is used for calculating dielectrophoretic force can be expressed as

$$\nabla(E)^2 = -2\psi_1^{02} \left(\frac{R_1 R_2}{R_2 - R_1} \right)^2 \frac{1}{r^5} \tag{88}$$

2.2.3 Particle Interactions in Dielectric Media

Once the above analyzed macroscopic fields are known, one can calculate electrostatic forces acting on particles suspended in dielectric (non-conductive) media. This has a major practical significance in electrofiltration and electrocoalescence (e.g., of water drops in oily phases), xerography and electrophoresis in salt-free media. Usually, the size of particles is much smaller than the characteristic dimensions of the electrode system generating the electric field, therefore, variations of the field in one direction (perpendicular to the electrode) are much larger than in other directions. The field can then be treated as quasi-uniaxial and can be expanded near the center of the particle $z = z_p$ in a Taylor series

$$\mathbf{E} \cong E^0 \mathbf{i}_z + \left(\sum_{n=1}^{\infty} \frac{1}{n!} \frac{\partial^n E}{\partial z^n} \right)_{z=z_p} (z - z_p)^n \mathbf{i}_z \tag{89}$$

where E^0 is the uniform field evaluated at the center of the particle, and z_0 the distance of the particle center from the electrode (see Fig. 2.4).

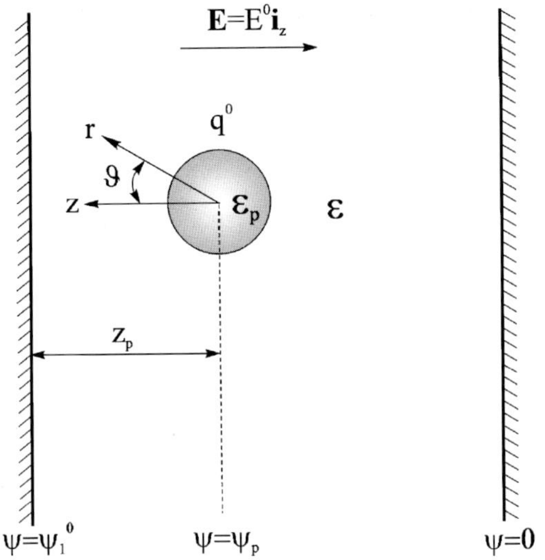

Fig. 2.4. A dielectric sphere of permittivity ε_p in an uniform external field $\mathbf{E} = E^0\, \mathbf{i}_z$.

The electric potential distribution evaluated relative to the particle center is given by the expression

$$\psi = \psi_p + \sum_{n=1}^{\infty} \frac{1}{n!(n+1)!} \left(\frac{\partial^n E}{\partial z^n} \right)_{z=z_p} (z - z_p)^{n+1} \tag{90}$$

where $\psi_p = \psi_1^0 - E^0\, z_p$ is the electric potential at the sphere center.

Obviously, for the planar electrode configuration the field is uniform everywhere (except perhaps in the vicinity of the edges), so all the derivatives in Eqs. (89), (90) vanish. This also concerns the field generated by concentric cylinders and spheres (including the case of a single sphere) that can be treated as locally uniform if the particle size remains much smaller than the characteristic radius of the electrode.

However, the expansion, Eqs. (89), (90), becomes important when the particle approaches the interface and the separation distance becomes comparable with its dimensions. Then, an additional electric field is generated because of the induction effects, which varies fast in space. Let us first analyze the case of an uncharged spherical particle of radius a and the electric permittivity ε_p placed in a non-conductive medium having the permittivity ε

(see Fig. 2.4). The external uniform field $\mathbf{E} = E^0\,\mathbf{i}_z$ polarizes the particle, which creates a counterfield $\mathbf{E}'$ inside it. Because the polarizing field (and the electric potential) is symmetric about the z-axis, the solution of the problem can easily be found using the spherical coordinate system (r, ϑ). The field distribution outside the particle is described by the analytical expression [5]

$$\mathbf{E}_r = -E_0 \cos\vartheta \left(1 + \frac{2C_1}{r^3}\right)\mathbf{i}_r = \mathbf{E}_r^0 + \mathbf{E}_r'$$

$$\mathbf{E}_\vartheta = -E_0 \sin\vartheta \left(\frac{C_1}{r^3} - 1\right)\mathbf{i}_\vartheta = \mathbf{E}_\vartheta^0 + \mathbf{E}_\vartheta' \tag{91}$$

where $C_1 = a^3\,(\varepsilon_p - \varepsilon)/(\varepsilon_p + 2\varepsilon)$, $\mathbf{E}_r$, $\mathbf{E}_\vartheta$ are the r and ϑ components of the net field and $\mathbf{E}$' means the induced field.

From Eq. (91) one can deduce that the induced charge on the particle surface $\sigma(\vartheta) = \varepsilon E_r'$ equals

$$\sigma(\vartheta) = -2\varepsilon E_0 \frac{\varepsilon_p - \varepsilon}{\varepsilon_p + 2\varepsilon} \cos\vartheta \tag{92}$$

Integrating this over the entire particle surface, one can see that the net induced charge is zero, as expected.

The electric potential distribution is given by the formula

$$\psi = \psi_p - E^0 r \cos\vartheta - \psi' \tag{93}$$

where ψ' is the induced potential given by

$$\psi' = -a^3 \left(\frac{\varepsilon_p - \varepsilon}{\varepsilon_p + 2\varepsilon}\right)\frac{E_0 \cos\vartheta}{r^2} = -\frac{3}{4\pi} v_p \left(\frac{\varepsilon_p - \varepsilon}{\varepsilon_p + 2\varepsilon}\right)\frac{E_0 \cos\vartheta}{r^2} \tag{94}$$

where $v_p = (4/3)\pi a^3$ is the particle volume.

It is interesting to note that the ratio of the maximum value of ψ' at the surface (occurring at $\vartheta = 2\pi$) to the polarizing potential drop over the particle, $\psi = 2a\,E^0$, equals

$$\bar{\psi}' = \frac{1}{2}\frac{\varepsilon_p - \varepsilon}{\varepsilon_p + 2\varepsilon} \tag{95}$$

For water drops with $\varepsilon \cong 78\,\varepsilon_0$ (see Table 2.3) dispersed in air ($\varepsilon = \varepsilon_0$), one has $\bar{\psi}' \cong 0.48$. The dependence of the polarized potential at the particle surface on the polar angle ϑ for this situation is plotted in Fig. 2.5.

One can notice from Eq. (94) that ψ' assumes an identical form to the potential distribution around a dipole, given by Eq. (20), with the dipole moment of

$$\mathbf{p} = 4\pi\varepsilon a^3 \frac{\varepsilon_p - \varepsilon}{\varepsilon_p + 2\varepsilon} E^0 \mathbf{i}_z = \alpha\varepsilon\mathbf{E} \tag{96}$$

where $\alpha = 4\pi a^3(\varepsilon_p - \varepsilon)/(\varepsilon_p + 2\varepsilon) = 3v_p(\varepsilon_p - \varepsilon)/(\varepsilon_p + 2\varepsilon)$ is the polarizability of the particle

Eq. (96) has a major significance, indicating that the dipole moment vector $\mathbf{p}$ is parallel to the polarizing uniform field $\mathbf{E}$. Moreover, its magnitude increases as the cube of the particle radius (proportionally to its volume), which means that polarization effects will rapidly increase with the size of the particle.

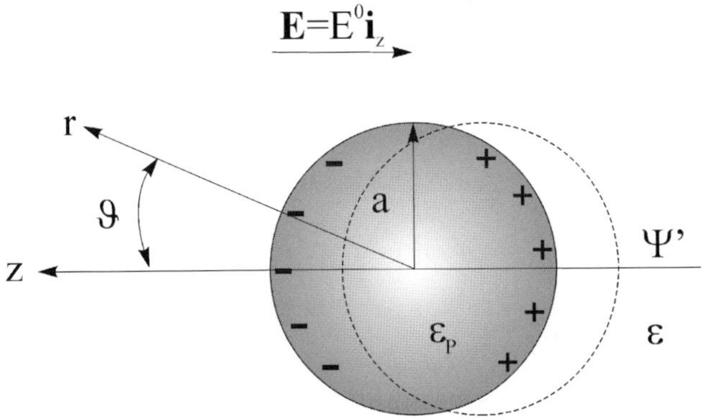

Fig. 2.5. Induced potential distribution ψ' (dashed line) at the surface of a sphere polarized by a uniform external field.

Analogous expressions as that given by Eq. (96) can be derived for particles of more complex shapes like ellipsoids or spheroids [5]. The dipole moment is still given by Eq. (96), with the polarizability of the ellipsoid equal to

$$\mathbf{p} = \varepsilon v_s \frac{\varepsilon_p - \varepsilon}{\varepsilon + (\varepsilon_p - \varepsilon) l_e} \mathbf{E}^0 = \varepsilon \alpha_s \mathbf{E}^0 \tag{97}$$

where $v_s = \frac{4}{3} \pi abc$ is the volume of the ellipsoid and a,b,c are their semiaxes. The coefficient l_e can be evaluated from the elliptic integral

$$l_e = \frac{3}{8} \frac{v_s}{\pi} \int_0^\infty \frac{ds}{(s+a^2)^{3/2}(s+b^2)^{1/2}(s+c^2)^{1/2}} \tag{98}$$

for prolate spheroids ($a > b = c$)

$$l_e = \frac{1-e_x^2}{2e_x^3} \ln\left(\frac{1+e_x}{1-e_x} - 2e_x\right) \tag{99}$$

where $e_x = (1-b^2/a^2)^{1/2}$

For slightly deformed spheres

$$l_e = \frac{1}{3} - \frac{2}{15} e_x^2 \tag{100}$$

For oblate spheroids ($a = b > c$)

$$l_e = \frac{1+e_x^2}{e_x^3} (\mathrm{e}_x - \tan^{-1} e_x) \tag{101}$$

where $e_x = ((a^2/b^2) - 1)^{1/2}$

Knowing the dipole moment and field distribution of the polarized particle, its energy in the uniform external field can be most directly calculated from Eq. (21), i.e.,

$$\phi = -\mathbf{p}\cdot\mathbf{E} + \int \mathbf{E}\cdot d\mathbf{p} = -pE + \varepsilon\alpha \int \mathbf{E}\cdot d\mathbf{E} = -\frac{1}{2}\varepsilon\alpha E^2 \tag{102}$$

The second term in Eq. (102) describes the energy needed to create a dipole by polarization due to the external field.

The force acting on the particle can be calculated as the derivative of the energy with respect to the coordinate z, i.e.,

$$\mathbf{F} = -\nabla\phi = -\frac{1}{2}\varepsilon\alpha\nabla(E^2) = -\varepsilon\alpha(\nabla\mathbf{E}\cdot\mathbf{E}) = -p\cdot\nabla\mathbf{E} \tag{103}$$

In the case of uniform fields, where $\nabla\mathbf{E} = 0$, the particle energy does not depend on its position (z coordinate), so the force exerted on the uncharged dielectric sphere vanishes. However, in the case where the field becomes non-uniform, so that the second and higher terms in the expansion Eq. (89) are non-zero, there appears a force, usually referred to as the dielectrophoretic force, acting in the direction of increased field strength. Analytical expressions for the dielectrophoretic force can be formulated for a simple geometry of the electrodes, e.g., the concentric sphere system considered above [6]. Using Eq. (88) one can derive from Eq. (103) the formula

$$\mathbf{F} = -8\pi\varepsilon\frac{\varepsilon_p - \varepsilon}{\varepsilon_p + 2\varepsilon}a^3\left(\frac{R_1 R_2}{R_2 - R_1}\right)^2 \frac{\psi^{0^2}}{(R_1 + z_p)^5}\mathbf{i}_z \tag{104}$$

In the case when the size of the outer sphere (electrode) tends to infinity, Eq. (104) can be transformed to a simpler form

$$\mathbf{F} = -8\pi\varepsilon\frac{\varepsilon_p - \varepsilon}{\varepsilon_p + 2\varepsilon}\frac{a^3}{R}\frac{E^{0^2}}{(1 + z_p/R)^5}\mathbf{i}_z \tag{105}$$

where $E^0 = \psi^0/R$.

As can be deduced from Eqs. (104), (105), the force is directed to the inner electrode where the field is strongest.

The interaction energy calculated by integration of Eq. (105) is given by the formula

$$\phi = -2\pi\varepsilon \frac{\varepsilon_p - \varepsilon}{\varepsilon_p + 2\varepsilon} \frac{a^3 E^{0^2}}{(1 + z_p/R)^4} \tag{106}$$

Analogous expressions can be derived for the concentric cylinder configuration using Eq. (80).

Consider now the case of a charged particle placed in a uniform electric field. Assume that the surface charge is homogeneous over the sphere surface, characterized by the surface charge density $\sigma^0 = q^0/4\pi\, a^2$ (where q^0 is the net charge on the particle). This charge produces an additional field and the electric potential distribution outside the particle described by Eq. (83). However, because of symmetry, the field inside the particle due to the charge vanishes. By exploiting the additivity of electric fields, the energy of the sphere can be calculated as the sum of those previously calculated (given by Eq. (102)) and the energy due to the presence of the charge. The latter can be determined from Eq. (35) by performing the integration, i.e.,

$$\phi = \frac{1}{2} \int_{S_1} \sigma\psi \, dS_1 + \frac{1}{2} \int_{S_2} \sigma\psi \, dS_2 = \phi_0 + q^0 E^0 z_p \tag{107}$$

where S_1 is the surface of the sphere and S_2 is the surface of the charged electrode (the surface charge of the second electrode is zero, so the integral over its surface vanishes) and ϕ_0 is the reference energy for the zero distance of the particle from the interface.

As can be deduced from Eq. (107), the energy increases linearly with the distance from the electrode.

The force exerted by the field on the particle can be calculated as the derivative of the energy upon the z coordinate, which results in the expression

$$\mathbf{F} = q^0 E^0 \, \mathbf{i}_z \tag{108}$$

Eq. (108), often referred to as the Lorenz equation, states that the force on a particle is constant and identical to the case of a point charge q^0. Moreover, the force does not depend on the distance from the electrode, particle size or its dielectric properties so it remains identical for dielectrics and metals. Eq. (108) is significant because it enables one to calculate the electrophoretic mobility of a particle in dispersing media of a low electric conductivity as discussed in Chapter 4.

By comparing the Lorenz expression for the force with the one previously found for the dielectrophoretic force, given by Eq. (105), one obtains the expression describing their relative significance

$$\bar{F} = -8\pi\varepsilon\left(\frac{\varepsilon_p - \varepsilon}{\varepsilon_p + 2\varepsilon}\right)\frac{a^3 E^0}{q^0 R}\frac{1}{(1+z_p/R)^5} \tag{109}$$

For not too large distances from the electrode, when $z_p/R \ll 1$, Eq. (109) simplifies to the useful form

$$\bar{F} = -C_d\left(\frac{aE^0}{R}\right) \tag{110}$$

where $C_d = 2(\varepsilon_p - \varepsilon)/(\varepsilon_p + 2\varepsilon)/\sigma^0$.
From Eq. (110) one can deduce that the relative significance of the dielectrophoretic force for a fixed charge density increases proportionally to particle size and field strength, so it is expected to play a role for large particles and strong fields, i.e., in non-conductive media able to sustain high voltages. This is best illustrated by considering the example of a water drop of the size 10^{-4}m (100 μm) in air placed in the field of 10^4 V m^{-1}, the radius of the inner electrode is assumed as $R = 10^{-2}$ m. The dielectrophoretic force then equals 2×10^{-12} N, which is much smaller than the gravity force on the particle equal to (4/3) $\pi a^3 \rho_p g \cong 4\times10^{-8}$ N (where g is the acceleration due to gravity).

On the other hand, the Lorenz force for the same field and particle charge of 10^{-9} C (charge density of 10^{-2}C m^{-2}, a typical value met in practice) equals 10^{-5}N. Even for just one elementary charge when $q = 1.6\times10^{-19}$C, the Lorenz force is 1.6×10^{-15} N, which may become comparable with the dielectrophoretic force for a particle size of a micrometer. This simple estimation suggests that the static charge effects described by the Lorenz equation completely dominate over the dielectrophoretic effects, except perhaps of macroscopic particles (size range above 10^{-2}m) and electric fields of a strength higher than 10^5 V m^{-1}.

It is interesting to mention that in the case of a charged particle and an uncharged but conductive electrode, there appears a force stemming from the reflection of the electric field from the interface. Physically this force, usually referred to as the image force, is due to the Coulomb interaction of the charged particle with its image separated by the distance twice as large as the particle distance from the interface z_p. Hence, the force can be calculated from the formula [2,4]

$$\mathbf{F} = -\frac{q^{0^2}}{4\pi\varepsilon(2z_p)^2}\mathbf{i}_z = -4\pi\varepsilon a^2 \frac{\psi^{0^2}}{(2z_p)^2}\mathbf{i}_z \tag{111}$$

By analogy, the interaction energy can be calculated from Eq. (18)

$$\phi = -\frac{q^{0^2}}{8\pi\varepsilon(2z_p)} = -\pi\varepsilon a^2 \frac{\psi^{0^2}}{z_p} \tag{112}$$

These Coulomb image forces play a decisive role in colloid particle adsorption in xerographic processes and in electrofiltration.

When the plane is a dielectric half-space, characterized by the permittivity ε_2, the image energy is given by [5]

$$\phi = -\frac{q^{0^2}}{8\pi(2z_p)} \frac{\varepsilon_2 - \varepsilon}{\varepsilon(\varepsilon + \varepsilon_2)} \tag{113}$$

Note that depending on the sign of $\varepsilon_2 = \varepsilon_1$, the force can be either negative $(\varepsilon_2 - \varepsilon > 0)$ or positive $(\varepsilon_2 - \varepsilon < 0)$. Thus, the sphere will be attracted to the plane of higher dielectric constant than the medium and repealed in the reverse case. This image force is responsible for the attraction of dielectric materials (e.g., paper) to charged surfaces in air.

The above expressions for force and interaction energy remain valid when the sphere is separated from the interface (electrode) by a distance considerably exceeding its size. For smaller distances the situation becomes complicated because the field due to the particle generates image fields in the metallic plate [4], which in turn disturb the initial field and so forth. There are two ways to efficiently deal with this problem (i) the method of images based on multipole expansion given by Eq. (27) and

(ii) the exact method of solution of the governing Laplace equation in the bispherical coordinates.

The image method was applied by Fowlkes and Robinson [4] to calculate the force acting on a dielectric sphere in contact with a metallic interface. Both the cases of an uncharged and charged sphere have been considered. The starting point for the calculations was a generalization of Eq. (96), i.e., an expression for multipoles of arbitrary order, generated in an dielectric sphere by an external electric field. They are given by the expression

$$P^{(n)} = \frac{4\pi\varepsilon}{(n-1)!} \frac{(\varepsilon_p - \varepsilon)a^{2n+1}}{[n\varepsilon_p + (n+1)\varepsilon]} \frac{\partial^{n-1}E_z}{\partial z^{n-1}} \tag{114}$$

where E_z is the non-uniform electric field in the z direction reflected from the interface and given by the equation

$$\mathbf{E}_z = \sum_{n=0}^{\infty} \frac{(n+1)P^{(n)}}{4\varepsilon\pi(2z_p)^2} \mathbf{i}_z \tag{115}$$

As an example let us assume that a dipole (polarized sphere) is oriented perpendicularly to the interface and located at the distance z_p. The axial field reflected from the interface (which can be obtained by differentiation of Eq. (94)) is given by the expression

$$\mathbf{E}_z = -\frac{1}{4\pi\varepsilon} p\left(\frac{3\cos^2\vartheta - 1}{z_p^3}\right)\mathbf{i}_z \tag{116}$$

For angles close to $\vartheta = 0$ this is reduced to the simple form

$$\mathbf{E}_z^0 = -\frac{p}{2\pi\varepsilon} \frac{1}{z_p^3} \mathbf{i}_z \tag{117}$$

The gradient of this reflected field, which travels over the apparent distance $z = 2\,z_p$, is given by

$$\left(\frac{\partial \mathbf{E}^0}{\partial z}\right)_{z=z_p} = \frac{3p}{2\pi\varepsilon} \frac{1}{(2z_p)^4} \mathbf{i}_z \tag{118}$$

This field interacts with the induced dipole, which produces a dielectrophoretic force given by Eq. (103). Substituting Eq. (118) into this expression yields

$$\mathbf{F} = -\frac{3}{2\pi\varepsilon} p^2 \frac{1}{(2z_p)^4} \mathbf{i}_z = -\frac{3\pi\varepsilon}{2} \left(\frac{\varepsilon_p - \varepsilon}{\varepsilon_p + 2\varepsilon} \right)^2 a^2 E^{0^2} \frac{1}{(1 + h_m/a)^4} \mathbf{i}_z \tag{119}$$

where h_m is the surface-to-surface distance between the particle and the electrode.
One can notice that this attractive force, acting perpendicularly to the electrode, plays a significant role for distances comparable to the particle size only. For h_m exceeding the particle dimensions the force becomes negligible in comparison with the Lorenz force or the dielectrophoretic force generated by the macroscopic field gradient.

Integrating Eq. (119) one obtains for the interaction energy the expression

$$\phi = -\frac{\pi}{2} \varepsilon \left(\frac{\varepsilon_p - \varepsilon}{\varepsilon_p + 2\varepsilon} \right)^2 a^3 E^{0^2} \frac{1}{(1 + h_m/a)^3} \tag{120}$$

Analogous formulae can be derived for higher order multipoles as well. In the calculations of Fowlkes and Robinson [4] the highest order multipole was 64.

The same problem of a charged sphere resting on an electrode was considered by Davis [7] who assumed, however, that the sphere charge is located at its center. He solved the governing Laplace equation in the bispherical coordinate system using the method of Legendre harmonic expansion. The cumbersome summation of the series, especially for close separations, limits the precision of these calculations.

Similar calculations have been performed recently using computer-aided summation of the series, which resulted in a higher accuracy [8]. The results are shown in Fig. 2.6. in the form of the dependence of the reduced force $-F/4\pi\varepsilon a^2 E^{0^2}$ on the distance h_m/a (in the logarithmic scale) for various permittivities of an uncharged particle. It can be noticed that the analytical formula Eq. (119) well describes the exact numerical results for the reduced instance $h_m/a > 1$. For smaller distances positive deviations from this formula become significant.

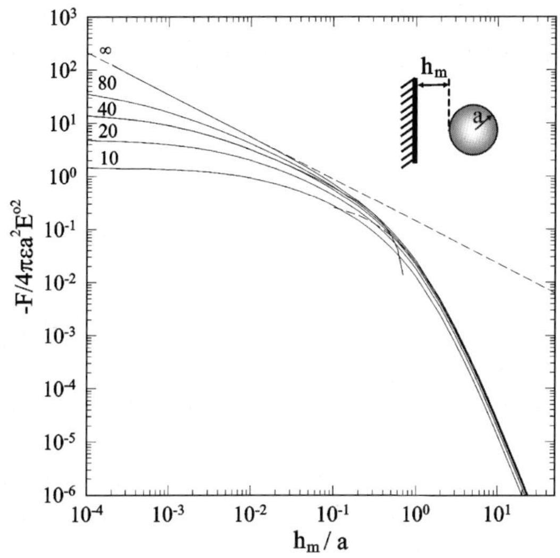

Fig. 2.6. The dependence of the reduced interaction force $-F/4\pi\varepsilon a^2 E^{0^2}$ on the dimensionless distance from the electrode h_m/a calculated numerically for a dielectric particle in the vicinity of metallic electrode. The solid lines denote exact results calculated for $\varepsilon_p/\varepsilon$ equal to 80 (water drop in air), 40, 20, and 10. The dashed lines show the results calculated from Eq. (121) (upper curve) and Eq. (119), lower curve. From Ref [8].

This is caused by the effect of higher order multipole interaction as discussed above. Note that for close separations in the limit of high relative permittivity (e.g., for water drops), the exact results can well be approximated by the formula derived by Davis [7]

$$\mathbf{F} = -1.2\pi\varepsilon a^2 E^{0^2} \frac{1}{(h_m/a)^{0.8}} \mathbf{i}_z \tag{121}$$

Integration of this equation gives the expression for the energy

$$\phi = -6\pi\varepsilon a^2 E^{0^2} \left[1 - (h_m/a)^{0.2}\right] \tag{122}$$

Similar dependencies were derived for a charged sphere in the vicinity of a conducting interface [9]. Numerical calculations showed that the Coulomb formula, Eq. (111), remained quite exact for distances as small as 0.1 of the particle radius. For still smaller distances, positive deviations of the exact data from this formula appeared.

The formulae discussed in this section are useful for describing interactions of isolated particle with bodies of macroscopic dimensions, i.e., electrodes, or with other particles of much larger size in dielectric media when there is no space charge. For the sake of convenience they have been compiled in Table 2.6.

However, due to polarization induced by external fields, considered above, particles will also interact with each other, which leads to their aggregation or coalescence (fusion of liquid drops forming a larger drop). This phenomenon, provoked by electric fields, called electrocoalescence plays an important role in many practical processes, e.g., dewatering of crude oils [10].

Let us now consider interactions between two charged particles of radii a_1, a_2 and permittivities ε_{p1}, ε_{p2} placed in an uniform external field $\mathbf{E} = E^0\mathbf{i}_z$. The center of the second particle is located at the distance $\mathbf{r}_0$ from the first particle and the angle between the $\mathbf{r}_0$ vector direction and the z-axis is ϑ. Due to polarization the particles become dipoles having the moments $\mathbf{p}_1 = \varepsilon\alpha_1\mathbf{E}$ and $\mathbf{p}_2 = \varepsilon\alpha_2\mathbf{E}$, respectively where α_1, α_2 are the polarizabilities of the particles. These dipoles, oriented parallel to each other, generate the electric fields $\mathbf{E_1}$ and $\mathbf{E_2}$, respectively. Using Eq. (102) the interaction energy of a polarized two particle system can be described by the formula

$$\phi = -\frac{1}{2}\mathbf{p}_1 \cdot \mathbf{E}_2'(1) = -\frac{1}{2}\mathbf{p}_2 \cdot \mathbf{E}_1'(1) \tag{123}$$

where

$$\mathbf{E}_2'(1) = \left(\frac{\varepsilon_{p_2} - \varepsilon}{\varepsilon_{p_2} + 2\varepsilon} \right) \frac{a_2^3 E^0}{r_0^3} \; (2\cos\vartheta \mathbf{i}_r + \sin\vartheta \mathbf{i}_r) \tag{124}$$

is the field generated by the second particle at the center of the first particle, $\mathbf{E}_1'$ (2) is the field at the center of the second particle and $\mathbf{i}_r$ is the unit vector of the r-axis.

Table 2.6
Electrostatic interactions of spheres in dielectric media

System	Scheme	Energy, ϕ	Force, F
Two charged spheres	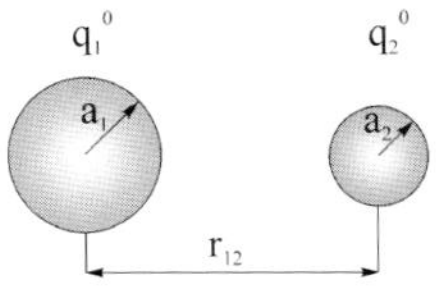	$\frac{q_1^0 q_2^0}{4\pi\varepsilon r_{12}}$	$\frac{1}{4\pi\varepsilon}\frac{q_1^0 q_2^0}{r_{12}^3}\mathbf{r}_{12}$
Charged sphere/ charged plane	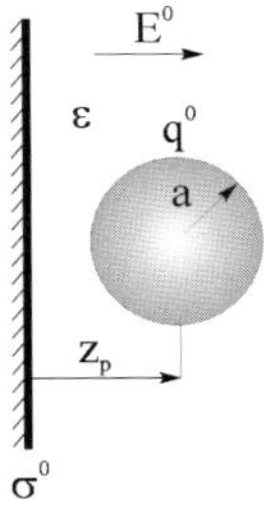	$q^0 E^0 z_p = \frac{q^0 \sigma^0}{\varepsilon} z_p$	$q^0 E^0 = q^0 \frac{\sigma^0}{\varepsilon}\mathbf{i_z}$
Charge sphere/ conductive neutral plane	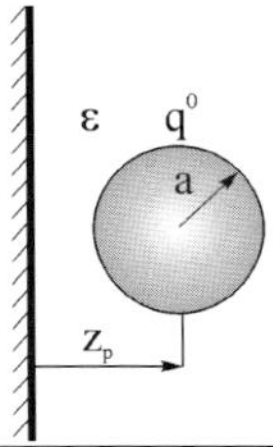	$-\frac{q^{0^2}}{8\pi\varepsilon(2z_p)}$	$-\frac{q^{0^2}}{4\pi\varepsilon(2z_p)}\mathbf{i_z}$

Table 2.6 (continued)

System	Scheme	Energy, ϕ	Force, F
Charged sphere/ dielectric neutral plane	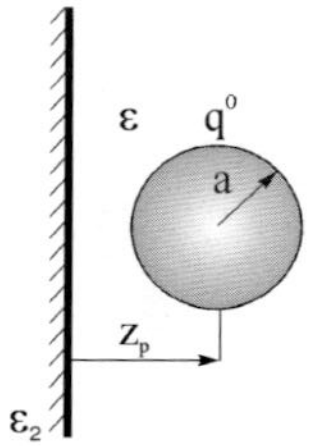	$-\dfrac{(\varepsilon_2-\varepsilon)q^{0^2}}{8\pi\varepsilon(\varepsilon_2+\varepsilon)(2z_p)}$	$-\dfrac{(\varepsilon_2-\varepsilon)q^{0^2}}{4\pi\varepsilon(\varepsilon_2+\varepsilon)(2z_p)}\mathbf{i_z}$
Neutral sphere/ conductive charged plane	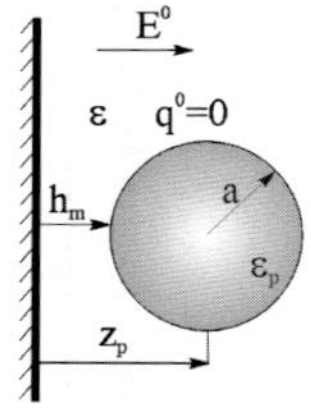	$\dfrac{-\pi\varepsilon}{2}\left(\dfrac{\varepsilon_P-\varepsilon}{\varepsilon_P+2\varepsilon}\right)^2 a^3 E^{0^2}\dfrac{1}{(1+h_m/a)^3}$	$\dfrac{-3\pi\varepsilon}{2}\left(\dfrac{\varepsilon_p-\varepsilon}{\varepsilon_p+2\varepsilon}\right)^2 a^2 E^{0^2}\dfrac{1}{(1+h_m/a)^4}\mathbf{i_z}$
Neutral sphere conductive charged plane (nonuniform field)	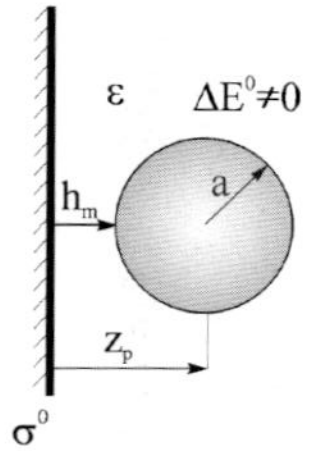	$\dfrac{1}{2}\varepsilon\alpha E^{0^2}$	$\varepsilon\alpha E^0\nabla E^0\mathbf{i_z}$

$\alpha = 4\pi a^3 \dfrac{\varepsilon_p-\varepsilon}{\varepsilon_p+2\varepsilon}.$

By substituting Eq. (123) into Eq. (124) one obtains for ϕ the explicit expression

$$\phi = -4\pi\varepsilon\left(\frac{\varepsilon_{p_1} - \varepsilon}{\varepsilon_{p_1} + 2\varepsilon}\right)\left(\frac{\varepsilon_{p_2} - \varepsilon}{\varepsilon_{p_2} + 2\varepsilon}\right)\frac{a_1^3 a_2^3 E^{0^2}}{r_0^3}P_2(\vartheta) = \phi_0 P_2(\vartheta) \quad (125)$$

where $P_2(\vartheta) = \frac{1}{2}(3\cos^2\vartheta - 1)$ and

$$\phi_0 = -8\pi\varepsilon\left(\frac{\varepsilon_{p_1} - \varepsilon}{\varepsilon_{p_1} + 2\varepsilon}\right)\left(\frac{\varepsilon_{p_2} - \varepsilon}{\varepsilon_{p_2} + 2\varepsilon}\right)\frac{a_1^3 a_2^3 E^{0^2}}{r_0^3} \quad (126)$$

is the interaction energy between dipoles in the in line position ($\vartheta = 0$). As can be seen, ϕ assumes a negative value in this orientation if $\varepsilon_{p_1} > \varepsilon$ and $\varepsilon_{p_1} > \varepsilon$. On the other hand, in the parallel position ($\vartheta = \pi/2$) the interaction energy equals $-\frac{1}{2}\phi_0$, i.e., it becomes positive and two times smaller in the absolute value in comparison with the edge-to-edge orientation. The critical angle where the interaction energy becomes zero is arcos $(1/3)^{1/2} = 54.7^\circ = 0.955$ rad.

For the sake of convenience the angular dependence of the interaction energy for two particle interactions is plotted in Fig. 2.7. It is interesting to note that the angular function $P_2(\vartheta)$ is the same for all particle size ratios. Because of large energy differences, particles will be attracted to each other in the edge-to-edge orientation, so this will be the preferred situation from a thermodynamic point of view.

The angle-averaged energy is given by the formula

$$\phi_\vartheta = \frac{\phi}{\pi}\int_0^\pi P_2(\vartheta)d\vartheta = \phi_0\frac{1}{2\pi}\int_0^\pi (3\cos^2\vartheta - 1)\,d\vartheta = \frac{\pi}{4}\phi_0 \quad (127)$$

As can be seen, the averaged energy also remains negative because ϕ_0 is negative.

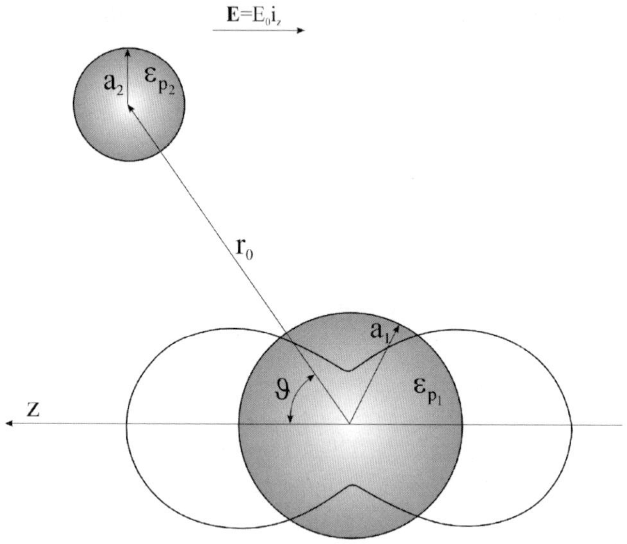

Fig. 2.7. Reduced interaction energy ϕ/ϕ_0 of two polarized particles in the uniform external field E^0. The contour represents the angular part of the interaction energy, i.e., $P_2(\vartheta) = \frac{1}{2}(3\cos^2\vartheta - 1)$ plotted from the sphere surface. The part of the contour outside the circle means positive value (attraction) and the inside part, negative values (repulsion).

The interaction force components obtained from Eq. (125) by differentiation are

$$\mathbf{F}_r = -\frac{d\phi}{dr_0}\mathbf{i}_r = 3\frac{\phi_0}{r_0}P_2(\vartheta)\mathbf{i}_r = \frac{3}{2}\frac{\phi_0}{r_0}\mathbf{i}_r(3\cos^2\vartheta - 1)\mathbf{i}_r$$

$$\mathbf{F}_\vartheta = -\frac{d\phi}{d\vartheta}\mathbf{i}_\vartheta = \frac{3}{2}\frac{\phi_0}{r_0}\sin 2\vartheta\,\mathbf{i}_\vartheta \tag{128}$$

In the edge-to-edge position ($\vartheta = 0$) one has

$$\mathbf{F}_r = 3\frac{\phi_0}{r_0}\mathbf{i}_r \tag{129}$$

$$\mathbf{F}_\vartheta = 0$$

In the parallel position ($\vartheta = \pi/2$)

$$\mathbf{F}_r = -\frac{3}{2}\frac{\phi_0}{r_0}\mathbf{i}_r$$
$$\mathbf{F}_\vartheta = 0 \qquad (130)$$

The large differences in the interaction force between the edge-to-edge and parallel position suggest that there will appear a significant torque tending to align particles in the former position. This will lead to the formation of linear (string like) aggregates, which was indeed observed experimentally [11−12]. It is worthwhile to note that analogous relationships are expected for magnetic particles polarized by external fields.

It is interesting to estimate the magnitude of the interaction energy of polarized particles in different orientations for a typical particle suspension in air (aerosol). Assuming that both particles are water drops of equal size of 10^{-5}m ($10\,\mu$m) one can calculate that the interaction energy at the distance between particles $r_0 = 3$a (so the gap width between two particles equals a) in the field of 10^4 V m^{-1} is 7.4×10^{-19} J. This is more than 180 kT units, equal to 4.04×10^{-21}J at a room temperature of 293 K (where k is the Boltzmann constant and T the absolute temperature). This is appreciable energy, which can induce not only aggregation of droplets but also their coalescence.

If particles are charged, except for the above polarization interactions, there appear Coulomb-type interactions whose energy is described by the expression

$$\phi = 4\pi a_1^2 a_2^2 \frac{\sigma_1^0 \sigma_2^0}{\varepsilon r_{12}} \qquad (131)$$

where r_{12} is the distance between particle centers, carrying surface charges σ_1^0 and σ_2^0

It is to be remembered, however, that Eq. (131) is valid for distances between particles comparable to their dimensions and larger, in the case where the surface charge is uniformly distributed over their surface.

Eq. (131) can alternatively be expressed in the form involving surface potentials of the particle ψ_1^0, ψ_2^0 by using the constitutive dependence

$$\sigma^0 = \frac{\varepsilon\psi^0}{a}$$

$$\phi = 4\pi\varepsilon a_1 a_2 \frac{\psi_1^0 \psi_2^0}{r_{12}} \tag{132}$$

The electrostatic force of interactions between two particles obtained by differentiation of Eq. (131) can be expressed as

$$\mathbf{F} = 4\pi a_1^2 a_2^2 \frac{\sigma_1^0 \sigma_2^0}{\varepsilon |\mathbf{r}_1 - \mathbf{r}_2|^3}(\mathbf{r}_1 - \mathbf{r}_2) = 4\pi\varepsilon a_1 a_2 \frac{\psi_1^0 \psi_2^0}{|\mathbf{r}_1 - \mathbf{r}_2|^3}(\mathbf{r}_1 - \mathbf{r}_2) \tag{133}$$

where $\mathbf{r}_1 - \mathbf{r}_2$ is the vector connecting particle centers.

It can be seen from Eqs. (131), (132) that for two spheres of equal surface charge, as is usually the case in suspensions, the interaction energy is always positive and the interaction force repulsive. Thus, the presence of permanent charges on particle surfaces will lead to appreciable stabilization of the suspension. It is useful to estimate the magnitude of the interaction energy for particle suspensions in air (aerosols), considered above. Assuming again that $a_1 = a_2 = 10^{-5}$ m (10 μm), $\sigma_1^0 = \sigma_2^0 =$ 0.001C m^{-2} and the distance between particles equals $3a$ (so the gap width between them was a) one obtains the value of 4.7×10^{-10} J for the interaction energy. This exceeds by more than 1.16×10^{11} times the thermal energy kT. Even when assuming that there are only 10 elementary charges on each particle, the interaction energy may exceed the kT unit at distances comparable to particle dimensions. These estimations clearly show that the free charge effects play a dominant role, exceeding by orders of magnitude the induced dipole effects considered above. Thus, the presence of free charge can effectively stabilize aerosol systems by preventing their coalescence.

As already mentioned, the above formulae remain valid for an inter-particle separation gap comparable with their dimensions and larger. Then the reflections of the electric fields from other particles leading to image interactions can be neglected and the interactions remain fully additive. This is usually referred to as the non-interactive dipole regime [13,14].

Moreover, for such separations, the effect stemming from the free charge present on particle surfaces can be treated independently from the polarization effects since both are decoupled from each other. However, for smaller separations, the electric fields generated by one particle are reflected by others, which leads to nonlinear coupling of the charge and polarization effects and consequently to deviation from additivity. For two different particles bearing free charge and placed in an external field, the electrostatic problem was solved in an exact way by Davis [9] using the bipolar coordinates. In the general case these solutions, available in tabulated form only, are rather cumbersome. However, it was found, in analogy with the results plotted in Fig. 2.6, that more significant deviations from additivity and from the Coulomb law occur for a separation gap between particles smaller than 0.5, hence the above formulae can be treated as quite accurate for problems of practical importance.

The situation becomes even more complicated for multiparticle systems (suspensions) since no exact solutions can be found due to the lack of appropriate coordinate system. This problem can only be treated in terms of multipolar interactions, which are also important in calculating molecular interaction in condensed phases as discussed later on.

2.2.4. The Electric Double-Layer

The above results were derived for dielectric media in which only surface charges were present, generated by a system of external electrodes. Other mechanisms of surface charge generation included field-induced polarization or electrostriction. The electric field distribution was governed by the Laplace equation whose solution in an appropriate coordinate system with boundary conditions resulting from the Gauss law was sufficient to unequivocally formulate the electrostatic problem.

These charging mechanisms of purely physical character can lead to a considerable surface charge density, typical in the range of 0.1–10^{-3} C m^{-2}. It is instructive to estimate the potential and field produced by surface charge of such magnitude. The corresponding data calculated for particles of size 10^{-6} and 10^{-4} m (1 and 100 μm, respectively) immersed both in air and in water are compiled in Table 2.7. As can be seen, for surface charge $\sigma_1^0 = 0.16$ C m^{-2} (which corresponds to an average distance between charges equal to 1 nm), the surface potential acquired by particles, varies between 1.8×10^6 V (for a 100 μm particle in air) to 2.31×10^2 V (for a 1 μm particle in water). The electric field becomes 1.8×10^{10} V m^{-1} (for a 1 μm particle in air) to 2.31×10^8 V m^{-1} (for a 100 μm particle in water). Even at the distance from

Table 2.7
Surface potential ψ_1^0 and field E_1^0(upper number) for particles in air and deionized water (lower numbers)

Distance between charges (nm)	Surface charge (C m^{-2})	$a = 10^{-6}$ m		$a = 10^{-4}$ m	
		ψ_1^0 (V)	E_1^0 (V m^{-1})	ψ_1^0 (V)	E_1^0 (V m^{-1})
1	0.16	1.8×10^4 2.31×10^2	1.8×10^{10} 2.31×10^8	1.8×10^6 2.31×10^4	1.8×10^{10} 2.31×10^8
10	1.6×10^{-3}	1.8×10^2 2.31	1.8×10^8 2.31×10^6	1.8×10^4 2.31×10^2	1.8×10^8 2.31×10^6
10^2	1.6×10^{-5}	1.8 2.31×10^{-2}	1.8×10^6 2.31×10^4	1.8×10^2 2.31	1.8×10^6 2.31×10^4

Remarks: $T = 293$ K, $\varepsilon = 8.85\times10^{-12}$ F m^{-1} (air) $\varepsilon = 6.91\times10^{-10}$ F m^{-1} (water).

the particle of the order of 10^{-2} m (1 cm), the field will still be about 10^6–10^4 V m^{-1}. This estimation clearly indicates that the extension of electric fields generated by surface charges would reach macroscopic dimensions.

In polar media, especially water, when free ions can exist both at the surface and in the bulk, additional mechanisms of charge formation at interfaces appear, most noticeably

(i) ionization of surface groups attached to the surface; and
(ii) irreversible adsorption (interception) of ionic species from the solution.

The former mechanism plays especially significant role for particles of mineral origin like various oxides (silica) whereas the latter is significant for synthetic colloids like polymeric lattices, or silver iodide.

Such strong fields exert considerable forces on all ions present in the vicinity of the charged particle. This will lead to accumulation of charge of opposite sign near the particle or electrode. Some of these free ions moving as a result of molecular mixing (Brownian motion) will form ion pairs compensating the surface charge to a large extent. This can be easily predicted by noting that the Coulomb interaction energy of two ions separated by the distance of 0.5 nm equals 4.6×10^{-19} J in air, i.e., 114 kT units at a room temperature. A comparable value can be predicted for water because at such small distances the dielectric permittivity will approach that of free space.

However, a part of the surface charge remain uncompensated. Therefore, the rest of free ions will concentrate near the interface forming the space charge of a diffuse character quite analogous to the Earth's atmosphere. The system consisting of localized surface charge at interfaces and the diffuse space charge of opposite sign is referred to as the electric double-layer. It can be best imagined as a condensor with one plate composed of surface charge and the other of space charge.

2.2.4.1. The Poisson Equation-Solutions for an Isolated Double-Layer

The presence of space charge suggests that the best way to quantitatively describe the electric double-layer (hereafter abbreviated as edl) would be to use the Poisson equation written in the form

$$\mathbf{\nabla}\cdot(\varepsilon\mathbf{\nabla}\psi) = -\rho_e \tag{134}$$

It should be mentioned that the basic difficulty in formulating this equation explicitly is in finding a unique functional dependence connecting the local charge density of free ions with the electric potential prevailing at this point of space. A dichotomy appears because the ions react to the local value of the electric potential that fluctuates in space and time due to the motion of other ions in their vicinity. On the other hand, the potential appearing in the Poisson equation Eq. (134) is the mean field potential that has a physically sound interpretation when averaged over macroscopic volumes only [15–19]. The same concerns the dielectric permittivity, which has unclear interpretation in the microscale, especially when large electric field gradients occur. These problems become especially severe at the interface because the surface charge distribution is by definition discrete in the microscale. Discreteness leads to considerable local fluctuations whose relative significance increases with decreasing length scale. Moreover, for all real surfaces there exist surface nonuniformity and heterogeneity both of a geometrical and chemical nature. This makes it rather difficult to unequivocally define the distance from the interface, a prerequisite to formulating the Poisson equation in any concrete coordination system. Because of these complications, no coherent theory of the edl has emerged yet and it is doubtful that it can ever be formulated. One has to rely therefore, on simplified models, which can be quite efficient, especially for predicting double-layer interactions at distances comparable with colloid particle dimensions, i.e., 10–100 nm. The range of validity of these approximate models can now be confirmed not only by experiments but also by numerical simulations performed by, e.g., Monte-Carlo methods [15–19].

Before discussing these approaches, in the next section we present solutions of the Poisson equation in the case of predetermined (fixed) distribution of the space charge. This can provide one with useful reference states necessary for the analysis of the double-layer for diffuse charge distribution.

Let us consider a macroscopic interface (particle) of a spherical shape with the radius a and immersed in an polar medium with the dielectric permittivity ε. The particle bears a charge q_1^0, which is assumed to be uniformly spread over its surface, so the size and discreteness of distribution of the charge are neglected. The surface charge σ_1^0 then equals $q_1^0/4\pi a^2$. Under these assumptions the Poisson equation, Eq.(134), becomes radially symmetric and can be expressed as

$$\frac{1}{r^2}\frac{d}{dr}\left(r^2\varepsilon\frac{d\psi}{dr}\right)=-\rho_e \tag{135}$$

Assuming that there is no space charge within the sphere so the electric field vanishes, one can formulate the general boundary condition at its surface (derived from the Gauss law as previously discussed) as

$$E_1^0=-\left(\frac{d\psi}{dr}\right)_{r=a}=\frac{\sigma_1^0}{\varepsilon_i} \tag{136}$$

where ε_i is the dielectric permittivity at the particle surface.

If the space charge outside the sphere is radially symmetric and independent of the potential (this would physically correspond to a frozen charged layer covering the sphere), i.e., $\rho_e = f(r)$ the first integration of Eq. (135) gives the following expression for the field:

$$E(r)=-\frac{d\psi}{dr}=\frac{a^2\left[\sigma_1^0+\sigma(r)\right]}{\varepsilon(r)r^2} \tag{137}$$

where

$$\sigma(r)=\frac{1}{a^2}\int_0^r \rho(r')r'^2\,dr' \tag{138}$$

A second integration of Eq.(137) gives the electric potential distribution

$$\psi = \Delta\psi^0 - a^2 \int_0^r \frac{\sigma_1^0 + \sigma(r')}{\varepsilon(r')r'^2} dr' \qquad (139)$$

where $\Delta\psi^0$ is the electric potential drop through the layer, given by the expression

$$\Delta\psi^0 = a^2 \int_0^\delta \frac{\sigma_1^0 + \sigma(r')}{\varepsilon(r')r'^2} dr' \qquad (140)$$

where δ is the thickness of the charged layer.

Eq. (140) is the general expression for the potential drop through a spherical shell layer for a distance-dependent permittivity. For most situations of practical interest, the thickness of the charged layer remains much smaller than the particle radius, i.e., $\delta << a$. Then, the curvature effects can be neglected and Eq. (139) simplifies to the form

$$\psi = \Delta\psi^0 - \int_0^x \frac{\sigma_1^0 + \sigma(x')}{\varepsilon(x')} dx' \qquad (141)$$

where $x' = r - a$ is the coordinate perpendicular to the interface, the potential drop is given by

$$\Delta\psi^0 = \int_0^\delta \frac{\sigma_1 + \sigma(x')}{\varepsilon(x')} dx' \qquad (142)$$

and

$$\sigma(x) = \int_0^x \rho_e(x') dx' \qquad (143)$$

It is useful to consider some interesting limiting cases that can be derived from the above general solutions

(i) No charge within the layer of thickness di characterized by the permittivity ei (see Fig. 2.8).

In this case Eq.(142) can be evaluated as

$$\Delta\psi^0 = \Delta\psi_i = \frac{\sigma_1^0 \delta_i}{\varepsilon_i} \tag{144}$$

This could physically reflect the finite size of counterions that cannot approach the interface closer than by the distance δ_i equal to their effective radius a_i. By assuming $\delta_i = a_i = 2\times10^{-10}$ m (a typical value for hydrated ion radius) $\sigma_1^0 = 0.16$ C m^{-2} and $\varepsilon_i = 5\varepsilon_0$ one obtains $\Delta\psi_I = 0.72$ V. For $\varepsilon_I = 78$ (water) one has $\Delta\psi_i = 0.046$ V. This is a very small fraction of the potential acquired by a spherical particle carrying the same charge (see Table 2.7).

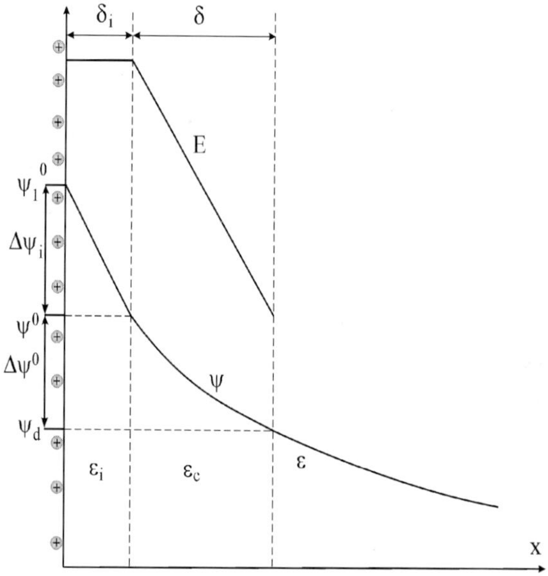

Fig. 2.8. The electric potential ψ and the field E distribution in the layer of fixed space charge adjacent to an interface.

(ii) A uniform space charge distribution within the layer δ (see Fig. 2.8), given by the expression

$$\rho_e(x) = \frac{\sigma^0}{\delta} \tag{145}$$

where σ^0 is the net charge accumulated in the layer per unit surface area of the interface (not necessarily equal to the surface charge σ_1^0). This fixed charge layer can be interpreted as a result of specific ion adsorption at the interface.

Eq. (143) becomes in this case

$$\sigma(x) = \frac{\sigma^0}{\delta} x \tag{146}$$

$$\Delta\psi^0 = \int_0^\delta \frac{\sigma_1^0 + (\sigma^0/\delta)x'}{\varepsilon(x')} dx' \tag{147}$$

It is instructive to consider the effect of a distance-dependent permittivity given for example by the expression

$$\varepsilon(x) = \varepsilon_i \left(1 + c_\varepsilon \frac{x}{\delta}\right) = \varepsilon_i \left(1 + \frac{\varepsilon - \varepsilon_i}{\varepsilon_i} \frac{x}{\delta}\right) \tag{148}$$

where ε is the permittivity of the bulk phase and $c_\varepsilon = (\varepsilon - \varepsilon_i)/\varepsilon$ the dimensionless constant.

By substituting this into Eq. (147), one obtains an important result

$$\Delta\psi^0 = \frac{\sigma_1^0 \delta}{\varepsilon_i c_\varepsilon} \ln(1 + c_\varepsilon) + \frac{\sigma^0 \delta}{\varepsilon_i c_\varepsilon^2} [c_\varepsilon - \ln(1 + c_\varepsilon)] \tag{149}$$

It can be deduced from Eq. (149) that the potential drop in the case of a variable permittivity is mostly determined by its maximum value.

When $c_\varepsilon \to 0$ (constant permittivity), Eq. (149) reduces to

$$\Delta\psi^0 = \frac{\sigma_1^0\delta}{\varepsilon_i} + \frac{1}{2}\frac{\sigma^0\delta}{\varepsilon_i} \tag{150}$$

Furthermore, if $\sigma_1^0 = -\sigma^0$, Eq. (150) becomes

$$\Delta\psi^0 = \frac{1}{2}\frac{\sigma_1^0\delta}{\varepsilon_i} \tag{151}$$

One can notice that there appears a finite potential drop through the charged layer even if the sum of the layer and surface charges is exactly zero. Hence, this situation is quite analogous to a dipole producing a finite electric potential difference for zero net charge.

By considering Eqs (144) and (151), one can deduce that the net potential drop through the layer of fixed space charge is given by

$$\Delta\psi^0 = \frac{\sigma_1^0(\delta_i + \delta)}{\varepsilon_i} + \frac{1}{2}\frac{\sigma^0\delta}{\varepsilon_i} \tag{152}$$

The potential at the edge of this layer, denoted by ψ_d (see Fig. 2.8), is given by

$$\psi(\delta_i + \delta) = \psi_d = \psi_1^0 - \Delta\psi^0 = \psi_1^0 - \frac{\sigma_1^0\delta}{\varepsilon_i} - \frac{(\sigma_1^0 + \frac{1}{2}\sigma^0)\delta}{\varepsilon_i} \tag{153}$$

and the field is

$$E_d = (\sigma_1^0 + \sigma^0)/\varepsilon_i \tag{154}$$

Let us assume that the charged layer of the thickness δ is fuzzy with the space charge density decaying exponentially (this type of behavior could pertain to polymeric colloids) and given by

$$\rho_e(x) = \frac{\sigma^0}{Le}\frac{e^{-x/Le}}{(1 - e^{-\delta/Le})} \tag{155}$$

where *Le* is the characteristic length of space charge density decay. Thus,

$$\sigma(x) = \sigma^0 \frac{1 - e^{-x/Le}}{1 - e^{-\delta/Le}} \tag{156}$$

Consequently, the potential drop through the layer is given by

$$\Delta\psi = \frac{\sigma_1^0 \delta}{\varepsilon_i} + \frac{\sigma^0 \left[\delta - Le\left(1 - e^{-\delta/Le}\right)\right]}{\varepsilon_i \left(1 - e^{-\delta/Le}\right)} \tag{157}$$

If $\delta/Le \to 0$ Eq. (157) reduces to the previously derived, Eq. (150). On the other hand if $\delta/Le \gg 1$, Eq. (156) becomes

$$\Delta\psi = \frac{\sigma_1^0 \delta}{\varepsilon_i} + \frac{\sigma^0 \delta}{\varepsilon_i} \tag{158}$$

It should be noted that in order to calculate the potential drop through the fixed charge layer, one needs to know both the surface charge density σ_1^0 and the charge accumulated in this layer σ^0, i.e. the concentration of ions. Since this requires considering adsorption equilibria of ions that may depend on the local electric potential value, we do not intend to go into detail of such complex analysis. This subject was first treated by Stern [20] and in consequence the adsorbed counterion layer is often called the Stern layer, as discussed in [21]. This model was further refined by Grahame [22] who considered the fact that anion and cation accumulations at the interface may proceed on different planes. The hydrated cations form a layer called the outer Helmholtz plane (OHP) [21], where the diffuse double-layer begins.

Because the effects stemming from the Stern layer are of secondary importance for calculating double-layer interactions of colloid particles, we perform further analysis by merely assuming that the Stern layer does not change appreciably during particle encounter. Accordingly, we will focus our attention on calculating the potential drop ψ_d in the diffuse part of an isolated double-layer. In the next section, systems of two overlapping double layers will be considered, allowing one to calculate particle interactions.

2.2.4.2. *The Poisson–Boltzmann Equation*

Calculations of the electric potential distribution in the double-layer can be carried out by exploiting the Poisson equation specified above. As mentioned, the crucial problem in formulating this equation is finding a unique relationship between the space charge density ρ_e and the electric potential ψ.

Despite the continuous effort aimed at a quantitative description of this relationship, no complete theory has yet emerged. At present there are two main paths of thinking about this problem

(i) the statistical-thermodynamic approach [15–19,23–29]; and

(ii) the phenomenological approach based on the local thermodynamic balance, neglecting all ion–ion correlation, finite size effects, etc. [30–35].

Whereas the latter approach is less strict, it is applicable for a broad range of situations of practical interest. By contrast, the more general statistical-thermodynamic approach produces rather specific results which cannot be generally applied for evaluation of particle interactions. Therefore, for the sake of simplicity, in our further considerations we adopt the phenomenological approach based on the Boltzmann statistics. Other, more refined models, e.g., those based on the statistical mechanics approach and computer simulations, will be treated as corrections to the standard Boltzmann model.

Without going into detail, the foundation of the phenomenological approach is the postulate of a local thermodynamic equilibrium, which can be expressed most conveniently as [33]

$$\tilde{\mu}_i = \mu_i^b + kT \ln f_i n_i + z_i e \psi \tag{159}$$

where $\tilde{\mu}_i$ is the electrochemical potential of an ion, z_i the valency of this ion, μ_i^b the reference potential of this ion in the bulk, n_i the local number concentration of a given ion (number of ions per unit volume) and f_i are the mean activity coefficients of ions, which describes the deviation from ideal behavior when, by definition, f_i become unity. Thus, in the general case, each activity coefficient depends on the local concentration of all other ions, which makes Eq. (159) an implicit, non-linear dependences. Experimental evidences based on electric conductivity and osmotic coefficient measurements suggest that for simple (univalent) ions and electrolyte concentrations below 0.1 M (mol dm^{-3}), the activity coefficients decrease by about 10% [36]. This is the result of self-atmosphere effect, i.e., the accumulation of counterions near a given ion. For higher valency ions, the decrease in ion activity for this range of concentration may reach 50% [36]. For still higher electrolyte concentrations the ion activities increase considerably because of the volume exclusion effect [30–33, 37,38]. This has a simple physical interpretation: the local concentration of ions in the region of high electric potential of opposite sign cannot exceed the limiting value determined by the hydrated radius of the ion. This maximum ion packing concentration is of the order of 10 M [38], which is rather large from the point of view of particle interaction problems.

Because Eq. (159) is in fact a set of non-linear coupled equations it cannot be solved in the general case to obtain explicit dependence of the local ion concentration on the electric potential. This can only be done in the limit of dilute solutions when all activity coefficients tend to unity and become independent of each other. Then, Eq. (159) reduces to the well-known Boltzmann distribution law

$$n_i = n_i^b e^{-z_i e\psi/kT} \tag{160}$$

where n_i^b is the bulk concentration of a given ion.

By substituting this into the Poisson equation, Eq.(41), one obtains the formula

$$\nabla^2 \psi = -\frac{e}{\varepsilon} \sum_i z_i n_i^b e^{-z_i e\psi/kT} \tag{161}$$

Eq. (161), called the Poisson–Boltzmann (PB) equation, is the foundation for the colloid science.

As can be noticed, Eq. (161), in contrast to the Laplace equation describing static field distributions, is non-linear with respect to ψ due to the presence of the exponential terms. This is a unpleasant property which prohibits finding solutions of general validity even for such simple cases as a spherical double layer.

However, important limiting forms of the PB equation can be formulated, susceptible to analytical handling. For a symmetric electrolyte composed of two types of ions, when $z_1 = -z_2 = z$, Eq. (161) can be transformed to the useful form

$$\nabla^2\left(\frac{ze\psi}{kT}\right) = \kappa^2 \sinh\left(\frac{ze\psi}{kT}\right) \tag{162}$$

where

$$\kappa^{-1} = Le = \left(\frac{\varepsilon kT}{2e^2 I}\right)^{1/2} \tag{163}$$

is the Debye screening length, a parameter of primary importance for all particle interaction problems, $I = z^2 n^b$ is the ionic strength of the electrolyte solution and n^b is the bulk concentration of the electrolyte.

Another frequently used form of the PB equation can be derived by applying the linearization procedure, which is justified if the maximum term $|z_1 e\psi/kT| < 1$. By expanding the exponential terms in a power series of ψ and exploiting the electroneutrality condition, one obtains the linear PB equation

$$\nabla^2 \psi = \kappa^2 \psi \tag{164}$$

where κ is defined by Eq. (163) and the ionic strength is given now by

$$I = \frac{1}{2}\sum_i z_i^2 n_i^b \tag{165}$$

The significance of Eq. (164) is increased by the fact that it is also applicable for nonlinear systems (high surface charges of particles) at distances greater than Le, where the potential becomes small because of the electrostatic screening [39]. This observation was the basis for the robust linear superposition approach (LSA) [32] discussed later on.

It is interesting to note that the screening length Le in aqueous media at room temperature varies between approximately 0.4 nm (for 0.1 M Na_3PO_4 solution) and 300 nm (for a 10^{-6} M solution of KCl), see Table 2.8.

Table 2.8
Screening length for various electrolytes $Le = \kappa^{-1}$ (nm)

Electrolyte concentration (M)	1:1 (KCl)	2:1 ($CaCl_2$) 1:2 (Na_2SO_4)	2:2 ($NiSO_4$) 1:3 (Na_3PO_4)	3:1 ($AlCl_3$)
10^{-1}	0.9639	0.5565	0.4820	0.3935
10^{-2}	3.048	1.759	1.524	1.244
10^{-3}	9.639	5.565	4.819	3.935
10^{-4}	30.48	17.59	15.24	12.44
10^{-5}	96.39	55.65	48.19	39.35
10^{-6}	304.8	175.9	152.40	124.4
I/n_b	1	3	4	6

T= 293.15K.

2.2.4.3. A Planar Double Layer

One of the few solutions of the nonlinear PB equation can be derived for a planar double layer under the assumption that the surface charge σ^0 is uniformly spread within a planar sheet of negligible dimensions. This was the original assumption made by Gouy [40] and Chapman [41] to describe the double-layer near a polarized mercury electrode. The PB equation, Eq. (161), reduces then to the one-dimensional form

$$\frac{d^2\psi}{dx^2} = -\frac{e}{\varepsilon}\sum_i z_i n_i^b e^{-z_i e\psi/kT} \tag{166}$$

The boundary condition assumes the usual form, derived from the Gaussian equation

$$E^0 = \sigma^0/\varepsilon = -\left(\frac{d\psi}{dx}\right)_{x=0} \tag{167}$$

The first integration of Eq. (166) gives the formula for the field distribution in the double-layer

$$\left(\frac{d\psi}{dz}\right)^2 = E^2(x) = \frac{2kT}{\varepsilon}\sum_i n_i^b (e^{-z_i e\psi/kT} - 1) \tag{168}$$

Using the boundary condition Eq. (167), one obtains from Eq. (168) the relationship connecting the diffuse surface charge σ^0 with the potential ψ_d

$$\sigma^0 = -\varepsilon\left(\frac{d\psi}{dx}\right)_{x=0} = \pm\left[2\varepsilon kT\sum_i n_i^b(e^{-z_i e\psi_d/kT} - 1)\right]^{1/2} \tag{169}$$

where the upper sign denotes a positive value of the potential ψ.

Eq. (169) remains valid for an arbitrary electrolyte including electrolyte mixtures and for the arbitrary potential ψ. In the case of a symmetric electrolyte, when $z_1 = -z_2 = z$, $n_1^b = n_2^b = n^b$, Eq. (169) reduces to the well-known form

$$\sigma^0 = \pm\{4\varepsilon kTn^b[\cosh(ze\psi_d/kT) - 1]\}^{1/2} = \pm 2(2\varepsilon kT\,n^b)^{1/2}\sinh\left(\frac{ze\psi_d}{2kT}\right) \tag{170}$$

As can be deduced from Eqs. (169–170), the functional dependence connecting σ^0 with the potential drop ψ_d is non-linear, which prevents to defining the electric capacity of the double-layer, as it was done before for charged plates. However, one can introduce instead the differential capacity, which plays a significant role in the electrochemistry of the double-layers on metals (e.g., mercury). It is defined as the electric capacity per unit area for an infinitesimal change in potential, i.e.,

$$C_e' = \frac{d\sigma^0}{d\psi_d} \tag{171}$$

Using Eq. (170) one can easily calculate that for a symmetric electrolyte C_e' is given for by the expression

$$C_e' = \frac{\varepsilon}{Le}\cosh\left(\frac{ze\psi_d}{2kT}\right) \tag{172}$$

Eq. (172) indicates that the double-layer system behaves as a nonlinear condensor whose capacity increases with its potential. This can be well illustrated by the following example. Assuming $Le = 10^{-8}$ m (this would correspond approximately to the 1:1 electrolyte concentration of 10^{-3} M) one obtains in the limit of low potentials the double-layer capacity for water of 7.1×10^{-2} F m^{-2}. For $\psi_d = 0.150$ V the differential capacity will increase 10 times. The same effect can be attained by increasing the electrolyte concentration to 0.1 M.

Knowing the dependence of σ^0 on ψ_d one can also calculate the self energy of the double-layer by using Eq. (72). In this way one obtains

$$\frac{\phi}{S} = \psi_d(2\varepsilon kT n^b)^{1/2} \sinh\left(\frac{ze\psi_d}{2kT}\right) \tag{173}$$

where S is the surface area of the double layer.

In the limit of $ze\,\psi_d/2kT < 1$, Eq. (173) simplifies to the well-known form

$$\frac{\phi}{S} = \frac{1}{2}\frac{\varepsilon}{Le}\psi_d^2 \tag{174}$$

For practical purposes it is more interesting to find the dependence of the diffuse potential drop through the double-layer ψ_d on the charge σ^0. This can be done analytically for a symmetric electrolyte only with the result [41]

$$\psi_d = \pm\frac{2kT}{ze}\ln\frac{|\bar{\sigma}^0| + (\sigma^{0^2} + 4)^{1/2}}{2} \tag{175}$$

where the plus sign denotes the positive surface charge, $\bar{\sigma}^0 = \sigma^0/(2\varepsilon kTn^b)^{1/2}$ is the dimensionless surface charge (see figs. 2.9 and 2.10).
Eq. (175) is known in the literature under the name of the Gouy–Chapman formula.

Eq. (169) can also be evaluated in an approximate manner for mixtures of electrolytes if one of the counterions valency is greater than others. In this case the concentration of the counterion is much larger than other ions, and Eq. (169) can be inverted to the form

$$\psi_d = \pm\frac{2kT}{z_{mx}e}\ln|\bar{\sigma}^0| \tag{176}$$

where z_{mx} is the maximum valency.

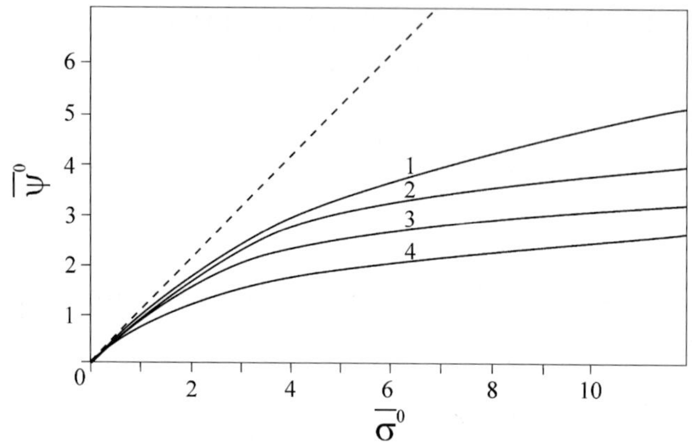

Fig. 2.9. The dependence of the reduced potential drop across the diffuse double layer $\overline{\psi}^0 = \psi_d e/kT$ on the reduced diffuse charge $\overline{\sigma}^0 = \sigma^0/(2\varepsilon\, kT\, n^b)^{1/2}$ calculated for a 1:1 electrolyte (curve 1) and for mixtures of 1:1 electrolyte with 3:1 electrolyte added in the amount of 0.1% (curve 2), 1% (curve 3) and 10% (curve 4), the dashed line shows the values calculated from the linear model, Eq. (177). From Ref [37].

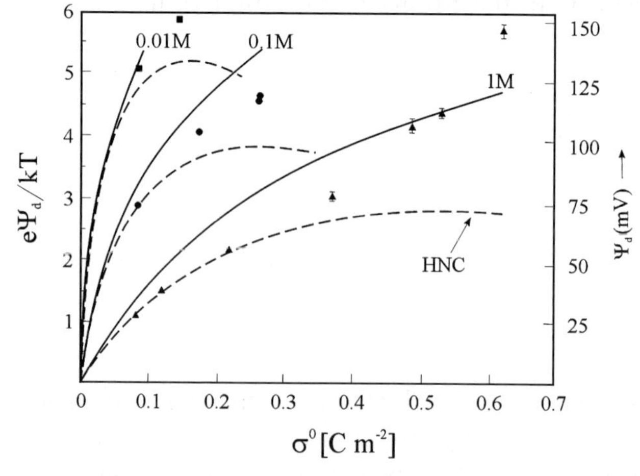

Fig. 2.10. The dependence of the reduced potential drop across the diffuse double layer $\psi^0 = e\psi_d/kT$ on the diffuse charge σ^0 calculated for a 1:1 electrolyte from the GC theory (Eq. (175)), solid lines, from Monte Carlo simulations in the grand canonical ensemble, Torrie and Valleau [15], points, and from the HNC statistical mechanical theory (dashed lines). Reused with permission from G.M. Torrie and J.P. Valleau, J. Chem, Phys. 73, 5807 (1980), Copyright 2002, AIP.

An important solution valid for an arbitrary electrolyte composition can be obtained in the low potential limit, where Eq. (169) reduces to

$$\psi_d = \frac{kT}{e}\overline{\sigma}^0 \tag{177}$$

The dependence of $\psi_d e/kT$ (the reduced ψ potential) on $\overline{\sigma}^0 = \sigma^0/(2\varepsilon kT n^b)^{1/2}$ (the reduced surface charge) for a pure 1:1 electrolyte and 1:1 electrolyte with a small addition of a 3:1 electrolyte is shown in Fig. 2.9. As can be seen, the addition of a few percent of the higher valency electrolyte decreases significantly the diffuse potential drop, especially for dimensionless surface charge $\overline{\sigma}^0 > 1$, in accordance with Eq. (176).

It can be calculated from this equation that for the previously estimated surface charge density 0.16 Cm^{-2} 1:1 electrolyte of the concentration of 10^{-3} M (6.02×10^{23} m^{-3}) $\overline{\sigma}^0 = 8.6$ at a room temperature, so the diffuse potential drop ψ_d calculated from Eq. (175) equals 0.109 V (109 mV) for a 1:1 electrolyte.

It is interesting that the results predicted by the GC theory are in good agreement with the Monte Carlo simulations performed by Thorie and Valleau [15] in the grand canonical ensemble (see Fig. 2.10). On the other hand, the theoretical results derived from the statistical mechanical theory using the hyper netted-chain closure (HNC) deviate significantly from the GC results, especially for higher surface charge values. It has also been noted by Torrie and Valleau that the potential drop for high electrolyte concentrations and surface charges becomes significantly higher than the GC model predicts (Eq. (175)). This increase was interpreted as the result of the volume exclusion effect, which was analyzed in some detail in Ref. [37]. Analytical formulae have been derived by assuming that the ion activity for higher concentration ranges is described by the correcting function

$$f_i = f = 1/\left(1 - \sum_i n_i/n_i^{mx}\right) \tag{178}$$

where n_i^{mx} is the maximum ion packing concentration estimated to be about 10 M, as mentioned previously. With this assumption an analytical expression connecting the reduced surface charge with the double-layer potential was derived in the implicit form

$$\beta\sum_{i=1}\alpha_i e^{-z_i e\psi_d/kT}=\left(1+\beta\sum_i\alpha_i\right)e^{\beta\bar{\sigma}^{0^2}}-1 \qquad (179)$$

where $\beta = n_1^b/n_i^{mx}$ and $\alpha_i = n_i/n_i^b$.
For a symmetric electrolyte, Eq. (179) can be formulated explicitly as

$$\psi_d=\pm\frac{kT}{e|z|}\ln\frac{Q_z+(Q_z^2-4)^{1/2}}{2} \qquad (180)$$

where

$$Q_z=\frac{(1+2\beta)e^{\beta\bar{\sigma}^{0^2}}-1}{\beta} \qquad (181)$$

For an electrolyte mixture containing admission of higher valency ion z_{mx} and $\beta \ll 1$, Eq. (180) becomes

$$\psi_d=\pm\frac{kT}{e|z_{mx}|}\ln\frac{\bar{\sigma}^{0^2}+\sum_i\alpha_i}{\alpha_{mx}} \qquad (182)$$

The effect of the β parameter calculated from Eq. (180) on the double-layer potential is shown in Fig. 2.11.

For many applications involving colloids, especially calculating electrokinetic potentials, it is interesting to know also the entire profile of electric potential in the solution. For the symmetric electrolyte this can be easily done by integrating Eq. (169) with the result

$$\psi(x)=\frac{2kT}{ze}\ln\frac{1+\beta_0 e^{-x/Le}}{1-\beta_0 e^{-x/Le}} \qquad (183)$$

where $\beta_0 = \tanh ze\psi_d/4kT$

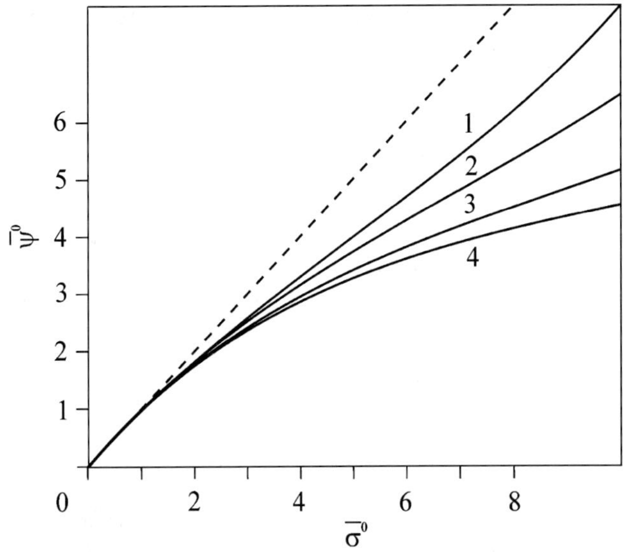

Fig. 2.11. The influence of the volume excluded effect on the reduced potential of the double layer $\bar{\psi}^0 = \psi_d e/kT$. The solid lines represent the results calculated from Eq. (180) for (1) $\beta = 0.05$, (2) $\beta = 0.03$, (3) $\beta = 0.01$, (4) $\beta = 0$ (GC model, no volume exclusion effect), the dashed line denotes the results predicted for the linear model $\bar{\psi}^0 = \bar{\sigma}^0$. From Ref [37].

Eq. (183), first derived by Vervey and Overbeek [42], is the only analytical solution of the nonlinear PB equation.

At larger separations from the interface, when $x/Le \gg 1$, Eq. (183) assumes the asymptotic form

$$\psi(x) = \frac{4kT}{ze}\beta_0 e^{-x/Le} = \bar{Y}^0 e^{-x/Le} \tag{184}$$

where

$$\bar{Y}^0 = \frac{4}{z}\tanh\frac{ze\psi_d}{4kT} \tag{185}$$

can be treated as the effective surface potential of the interface.

Eq. (184), indicating that the electric potential decreases exponentially at larger separations $x/Le \gg 1$ (i.e., 2.718 times when x/Le is increased by

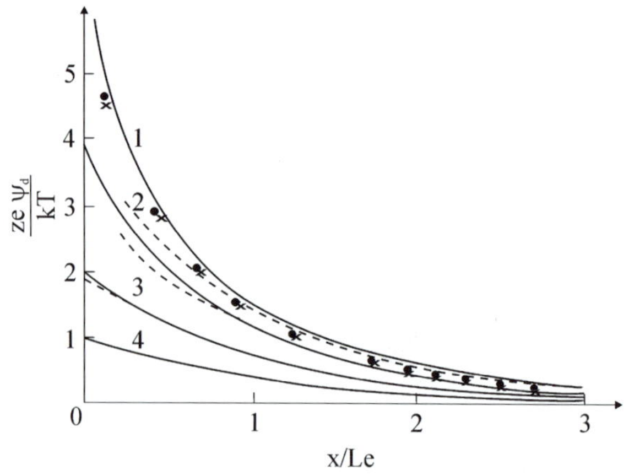

Fig. 2.12. The dependence of the reduced potential $ze\,\psi_d/kT$ on the reduced distance from the interface, calculated from Eq. (183) (solid lines), for a symmetric electrolyte (1) $ze\,\psi_d\,/\,kT = 8$, (2) $ze\,\psi_d/kT = 4$, (3) $ze\,\psi_d/kT = 2$, (4) $ze\,\psi_d/kT = 1$, the points show the results calculated from Monte Carlo simulations of Torrie and Valleau [15], and the dashed lines show the asymptotic results calculated from Eq. (184).

one), has considerable significance for predicting the electrokinetic potential of surfaces (defined as the potential in the slip plane). This is illustrated in Fig. 2.12, showing the dependence of the reduced potential $ze\psi_d/kT$ on the reduced distance *x/Le*, calculated from Eq. (184) for various ψ_d values. One can notice that the asymptotic formula, Eq. (184), describes well the exact data for a broad range of distances with the exception of very high $ze\psi_d/kT > 4$, where close to the interface the potential decreases at a much faster rate. It is also interesting to observe that the data of Torrie and Valleau [15], derived from simulation (depicted by points in Fig. 2.12), are in good agreement with the analytical results derived for the GC theory.

Knowing the electric potential distribution, one can calculate the concentration profiles of ions from the Boltzmann equation and hence the total space charge density ρ_e. The results are presented in Fig. 2.13 for $ze\,\psi_d/kT = 8$. For comparison, the results stemming from the Monte Carlo simulations [15] are shown as well. Again, the agreement between the analytical and simulation data seems quite good. As can be observed, at the charged wall, only the ions of opposite charge (counterions) are present. In our case, the concentration of negative counterions at the interface exceeded its bulk value $e^8 = 2980$ times. Thus, for the bulk electrolyte concentration of 5×10^{-3} M (a typical value for colloid systems), the concentration of counter ions at the

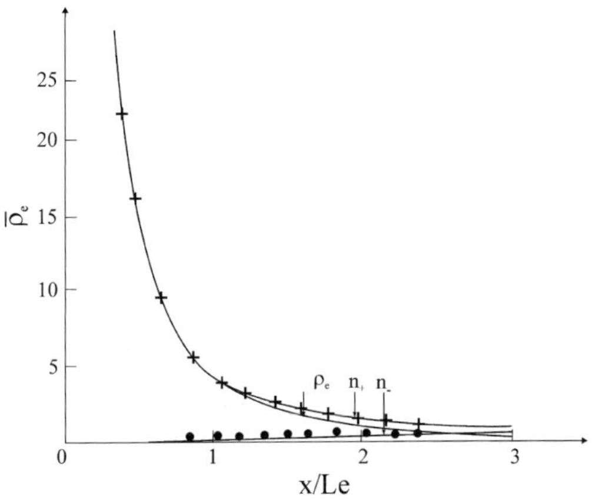

Fig. 2.13. The reduced ion concentration profiles n_+/n_b and n_-/n_b, and the total space charge density $\bar{\rho}_e = \rho_e/n_b$ as a function of the distance from a charged interface *x/Le* calculated from the Boltzmann equation, $ze\,\psi_d/kT = 8$, for a symmetric 1:1 electrolyte. The points show the results calculated from Monte Carlo simulations of Torrie and Valleau [15].

surface should theoretically reach 15 M, larger than the typical value for crystalline salts. This estimation unequivocally shows that for such high potentials the Boltzmann statistics becomes inadequate and assumptions of the Gouy Chapman theory are violated.

2.2.4.4. A Spherical Double Layer

All the solutions derived above are valid for a quasi-planar interface when its radius of curvature *R* remains much larger than the double-layer thickness, so the parameter $Le/R \ll 1$. However, for colloid particles with size range of 100 nm, the double-layer thickness becomes comparable with their dimensions for low electrolyte concentration (see Table 2.8), and the interface curvature effects play a significant role. The spherical double layer arising in this case can, in principle, be analyzed in terms of the PB equation (Eq. (161)). However, because of the non-linearity of this equation, no analytical solutions can be derived. Limiting solutions can only be found in the case of low potentials and a symmetric surface charge distribution when $e\,\psi_d/kT < 1$ and $E_\perp = -\sigma^0/\varepsilon$ remains independent of the position over the sphere. Then the solution of the linear form of the PB equation, Eq.(164), becomes

$$\psi = \psi^0 \frac{a}{r} e^{-(r-a)/Le} \qquad (186)$$

where

$$\psi^0 = \frac{a\,Le}{\varepsilon(a+Le)}\sigma^0 \tag{187}$$

r is the distance from the sphere center and a the sphere radius.

Using this relationship, one can express the differential capacity of a spherical double layer (in the low potential range) in the form

$$C_d' = \frac{\varepsilon(a+Le)}{a\,Le} \tag{188}$$

In the limit $Le \gg a$, $C_d' = \varepsilon/a$ and $C_e = 4\pi a^2 C_d' = 4\pi\varepsilon a$, Eq. (188) transforms into that previously derived for a sphere in an ion-free medium (see Table 2.3).

Eq. (186), valid for an arbitrary electrolyte composition, has major practical significance because it represents the asymptotic form of any solution to the nonlinear PB equation for distances much larger than the screening length, when the potential becomes smaller than kT/e [38]. Hence, at such distances, the electric potential distribution is given by

$$\psi = \frac{kT}{e}\overline{Y}^0 \frac{a}{r} e^{-(r-a)/Le} \tag{189}$$

where the effective potential $\overline{Y}^0$ should be determined by matching this asymptotic solution with exact numerical solutions of the non-linear PB equation for the spherical geometry.

Loeb, Overbeek and Wiersema [43] were the first to obtain such solutions in tabulated form for various electrolyte types, together with empirical formulae connecting the diffuse double-layer charge σ^0 with the potential ψ_d (for a 1:1 electrolyte). Since these tabulated solutions are rather awkward for direct use, other approximate analytical solutions have been derived using the perturbation [44] and variational methods [45]. Ohshima et al. [46] developed useful solutions both for the potential distribution and the effective surface potential $\overline{Y}^0$ by replacing the original PB equation with a simplified version. For a 1:1 electrolyte the following relationship was derived [46]:

$$\bar{Y}^0 = 4\tanh\left(\frac{e\psi_d}{4kT}\right)\frac{2}{1+\left[1-\dfrac{2a/Le+1}{(a/Le+1)^2}\tanh\dfrac{e\psi_d}{4kT}\right]^{1/2}} \tag{190}$$

It can be deduced that in the case of $e\psi_d/kT \gg 1$, Eq. (190) reduces to

$$\bar{Y}^0 = 8\frac{a+Le}{2a+Le} \tag{191}$$

On the other hand, for $a/Le \gg 1$ (planar interface), Eq. (190) reduces to the previously derived relationship given by Eq. (185) (when $z = 1$). It was demonstrated [46] that Eq. (190) gives a few percent accuracy for $a/Le > 0.1$ and $\psi_d\, e/kT < 5$.

The dependence of the effective surface potential $\bar{Y}^0$, calculated from Eq. (190), on the diffuse double-layer potential ψ_d is shown in Fig. 2.14 for various a/Le parameter values. It can be seen that the range of the linear relationship between $\bar{Y}^0$ and ψ increases for smaller a/Le values. The results shown in Fig. 2.14 are especially useful when calculating electrostatic interactions among spherical particles using the approximate linear superposition method, as discussed later on.

For other applications, e.g., for calculating the double-layer potential drop when knowing the double-layer charge, one is interested in finding the relationship between ψ_d and σ^0 which is the generalization of the Gouy–Chapman formula, Eq. (175). This can be done by exploiting the results of Ohshima et al. [46] and Loeb et al. [43] who formulated the following dependence connecting σ^0 with ψ_d for a spherical particle immersed in a 1:1 electrolyte

$$\sigma^0 = 2\sinh\left(\frac{e\psi_d}{2kT}\right) + \frac{4Le}{a}\tanh\left(\frac{e\psi_d}{4kT}\right) \tag{192}$$

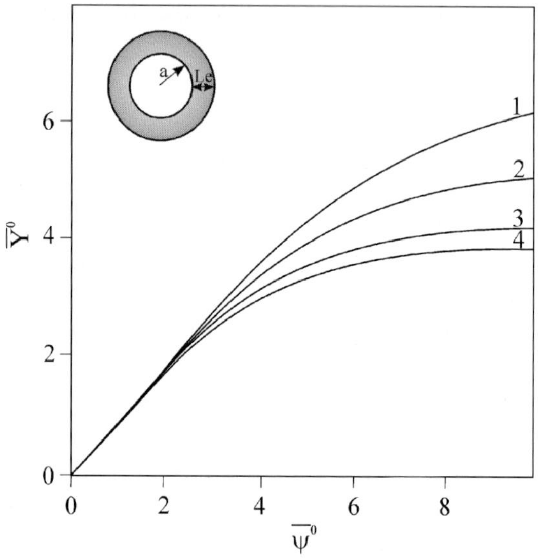

Fig.2.14. The dependence of the effective surface potential $\overline{Y}^0$ of a spherical particle in a 1:1 electrolyte on the reduced double-layer potential $\overline{\psi}^0 = e\psi_d/kT$, (1) $a/Le = 0.5$, (2) $a/Le = 1$, (3) $a/Le = 5$, (4) $a/Le = \infty$ (planar interface).

This can be inverted to the form [38]

$$\psi_d = \frac{kT}{e} 2\ln\left[Q_s + \left(1 + Q_S^2\right)^{1/2}\right] \tag{193}$$

where

$$Q_s = \frac{2\overline{\sigma}^0 B_h^{1/3} - \left(1 + (3)^{1/2} i\right) B_h^{2/3} - \left(1 - (3)^{1/2} i\right) C_h}{6 B_h^{1/3}} \tag{194}$$

$$B_h = -\frac{\overline{\sigma}^0}{2}\left[C_h + 6\frac{Le}{a} - 48\left(\frac{Le}{a}\right)^2\right] - 6(a)^{\frac{1}{2}} \frac{a}{Le} i$$
$$\left\{\frac{16a}{Le}\left(1 + \frac{Le}{a}\right)^3 + \left(\frac{\overline{\sigma}_0}{2}\right)^2 \left[1 + 20\frac{Le}{a} - 8\left(\frac{Le}{a}\right)^2 + \left(\frac{\overline{\sigma}^0}{2}\right)^2\right]^{\frac{1}{2}}\right\}$$

where

$$C_h = \frac{12Le}{a}\left(1+\frac{Le}{a}\right)+\left(\frac{\bar{\sigma}_0}{2}\right)^2$$

and $i = (-1)^{1/2}$ is the imaginary unit.

In Fig. 2.15, the dependence of $e\psi_d/kT$ on σ^0, calculated from Eq. (193) for a/Le ranging from 0.5 to ∞ (planar interface), is plotted. As can be observed, the range of validity of the linear relationship, Eq.(187), increases significantly when a/Le decreases (thick double layers in comparison with the particle radius), thus, for $a/Le = 0.5$ the dependence of ψ_d on σ^0 remains linear for practically the entire range of σ^0.

As can be deduced from relationships discussed above, especially Eq. (189), the electric double layer around a spherical particle exhibits qualitatively the same properties as for the planar interface. In particular, after an

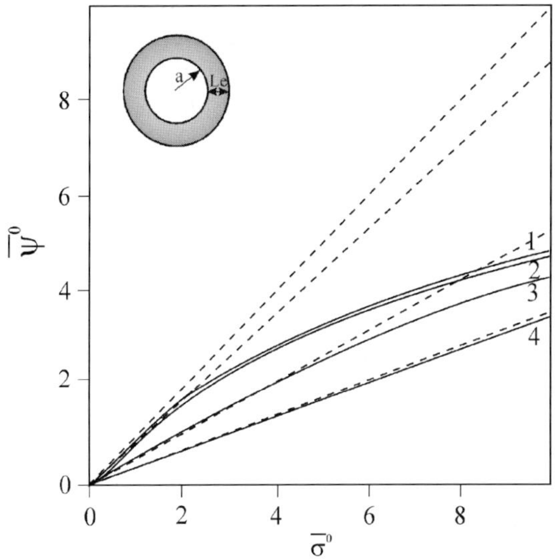

Fig.2.15. The dependence of the reduced potential drop $\bar{\psi}^0 = e\psi_d/kT$ across a spherical double layer on the reduced charge $\bar{\sigma}^0 = \sigma^0/(2\varepsilon\, kT\, n^b)^{1/2}$, calculated from Eq.(193) for a 1:1 electrolyte and various a/Le parameter values, (1) $a/Le = \infty$ (planar interface), (2) $a/Le = 5$, (3) $a/Le = 1$, (4) $a/Le = 0.5$, the dashed lines denote the values calculated from the linear model, Eq. (187).

initial abrupt drop, the electric potential decreases exponentially with a/Le. It is also interesting to note that the potential as well as the field distribution around the spherical particle remains symmetric (for a uniform surface charge distribution), which means that there is no net force acting on the particle. The force only appears when the symmetry is broken, for example, for a two-particle system when Eq. (189) can be exploited for calculating particle interactions approximaly.

2.2.4.5. Two Double-Layer System

For calculating electrostatic interactions for a particle/particle or particle/interface, it is necessary to consider systems of two double layers. However, the nonlinearity of the PB equation prohibits finding any analytical solution even in the simplest case of two planar double layers. Additionally, the problem of specifying the appropriate boundary conditions at the interacting surfaces appears. This is so because upon their approach, the electric potential will change appreciably, which affects ion adsorption equilibrium in the Stern layer as well as ion dissociation equilibria. As a result, a kind of potential-regulating mechanism will appear, changing the surface charge. Obviously, in such a situation, the general electrostatic boundary conditions derived from the Gauss law, Eq. (167), remain valid. However, the surface charge becomes a complicated function of the distance between the interfaces and the velocity of the approach of the interfaces to each other. As a convenient reference situation, one often assumes that the interface potential remains constant, their approach referred to in the relevant literature as the constant potential (c.p.) boundary condition [42].

Another problem in calculating interactions of interfaces immersed in an electrolyte solution arises because ions are subject to a thermodynamic equilibrium. In this case, the knowledge of the electrostatic Maxwell stress tensor, a unique function of the electric field or potential, is not enough to calculate the net stress (pressure tensor). The additional factor is the osmotic pressure tensor whose gradient is given by [33]

$$\Delta \mathbf{P} = -kT \sum_i n_i \nabla \ln f_i\, n_i \tag{195}$$

Because the local concentration of a given ion depends on the activity of all other ions and their local concentrations, the expression for $\Delta \mathbf{P}$ has an implicit form which prohibits its unequivocal evaluation even if the electric potential distribution in space is known. A unique evaluation of **P** can only be achieved, analogously as formulating the PB equation, by assuming the ideal ion behavior (all activity coefficients equal one). Under this assumption, upon integration, Eq. (195) becomes

$$\Delta \mathbf{P} = \mathbf{P} - \mathbf{P}^b = kT \sum_i (n_i - n_i^b)\mathbf{I} \tag{196}$$

where **I** is the unit tensor and $\mathbf{P}^b$ is the uniform bulk value of the osmotic pressure tensor.

In the limit of low potential, Eq. (196) reduces to the linear form

$$\Delta \mathbf{P} = -kT\, I \left(\frac{e\psi}{kT} \right)^2 \mathbf{I} \tag{197}$$

By exploiting the definition of the osmotic pressure tensor, given by Eq. (196), one can formulate the equilibrium condition in the double layer as the postulate that the net electric and osmotic stress vanishes, i.e.,

$$\nabla \cdot (\Delta \mathbf{P} - \mathbf{T}) = 0 \tag{198}$$

where **T** is the Maxwell stress tensor, which for a dielectric medium is given by Eqs. (59–60).

The net force on the double layer is obtained by integrating the net stress over the entire volume of the double layer. By virtue of Eq. (198), one obtains

$$\int_v (\Delta \mathbf{P} - \mathbf{T})\, dv = 0 \tag{199}$$

By using the Gaussian theorem, Eq. (199) can be converted into the form analogous to that previously derived (cnf. Eq. (50))

$$\mathbf{F} = \int_{S_t} \left(\Delta \mathbf{P} + \frac{\varepsilon}{2} E^2 \mathbf{I} - \varepsilon \mathbf{E}\mathbf{E} \right) d\mathbf{S} \tag{200}$$

where S_t is the set of surfaces enclosing the double-layer volume. If one of these surfaces is chosen to be the interface (or particle), Eq. (200) can be converted into the form

$$\mathbf{F} = \int_{S_i} \left(\Delta\mathbf{P} + \frac{\varepsilon}{2} E^2 \mathbf{I} - \varepsilon \mathbf{EE} \right) d\mathbf{S} = \int_{S} \left(\Delta\mathbf{P} + \frac{\varepsilon}{2} E^2 \mathbf{I} - \varepsilon \mathbf{EE} \right) d\mathbf{S} \tag{201}$$

where S_i means the interface and S is an arbitrary closed surface placed in the double-layer region.

Eq. (201) represents the general expression for calculating the interaction force for an arbitrary-shaped interface or particle.

In the limiting case of a flat geometry (two infinite planar interfaces interacting through an electrolyte solution), Eq. (201) reduces to the simple form describing the uniform force per unit area

$$F = \Delta\Pi = kT \sum_i \left[n_i(\psi) - n_i^b \right] - \frac{\varepsilon}{2} \left(\frac{d\psi}{dx} \right)^2 \tag{202}$$

where x is an arbitrary position between the plates, F the force per unit area of the plates, which can be treated as uniform pressure $\Delta\Pi$, often called the disjoining pressure.

As can be noticed, in order to calculate the force, an explicit expression for the potential distribution in the gap between plates is needed. This can be derived from the solution of the PB equation, as discussed below.

The interaction energy ϕ can be most directly obtained by integrating Eq. (201) along a path starting from infinity. This procedure, although generally valid, may become rather cumbersome. A direct use of the general expression for the interaction energy, given by Eq. (35), can be more efficient. Denoting the energy of the double layer system at infinite separation by $\phi = \frac{1}{2} \int_S \sigma^0 \psi_d$, one can express, therefore, the change in the interaction energy upon their approach by the formula

$$\Delta\phi = \frac{1}{2} \int_S \left(\sigma^{0\prime} \psi^{0\prime} - \sigma^0 \psi_d \right) dS \tag{203}$$

where the surface integral extends over the entire interface system, σ^0 is the surface charge of interfaces for a given separation and ψ_d is the diffuse double-layer potential at this distance. Note that in the general case both the charge and potential distributions may depend on the position of the interfaces involved.

For the constant charge boundary conditions, Eq. (203) simplifies to

$$\Delta\phi = \frac{1}{2}\sigma^0 \int_S \left(\psi_d' - \psi_d\right) dS \tag{204}$$

Evaluation of Eqs. (202–204) requires knowledge of the electrostatic potential distribution between the plates. This can be accomplished by solving the PB equation for this geometry.

Let us consider, therefore, the problem of two plates of infinite thickness (half spaces) bearing different charges σ_1^0, σ_2^0 and immersed in an electrolyte solution of arbitrary composition. The general PB equation, Eq. (161), for this geometry assumes the simpler, one-dimensional form

$$\frac{d^2\bar{\psi}}{d\bar{x}^2} = -\frac{1}{2\bar{I}}\sum_i z_i \bar{n}_i e^{-z_i\bar{\psi}} \tag{205}$$

where $\bar{x} = x/Le$ is the dimensionless distance, $\bar{\psi} = e\psi_d/kT$ the dimensionless potential, $\bar{I} = I/n_1^b$ and $\bar{n}_i = \bar{n}_i^b/n_1^b$.
The general electrostatic boundary conditions derived from Eq. (65) are

$$\frac{d\bar{\psi}}{d\bar{x}} = -\bar{\sigma}_1^0 \quad \text{at} \quad \bar{x} = 0$$
$$\frac{d\bar{\psi}}{d\bar{x}} = \bar{\sigma}_2^0 \quad \text{at} \quad \bar{x} = \bar{h} \tag{206}$$

where $\bar{h} = h/Le$ is the dimensionless distance between the plates and the dimensionless charge is defined as follows

$$\bar{\sigma}_1^0 = \sigma_1^0\left(eLe/\varepsilon kT\right)$$

$$\bar{\sigma}_2^0 = \sigma_2^0\left(eLe/\varepsilon kT\right)$$

are the dimensionless surface charges at the plates.

Eq. (206), referred to as constant charge (c.c) boundary conditions, implies that the charge at each plate remains fixed, irrespective of their separation distance. As discussed in [37, 38], this can be thermodynamically unfavorable at close separations due to the considerable increase of the electrostatic potential between the plates. When the system remains in equilibrium at every separation, the plate charges must change upon approach in order to meet the boundary conditions expressed by Eq. (206). In this case, this equation is formulated in a more convenient way as

$$\begin{aligned} &\bar{\psi} = \bar{\psi}_1^0 \quad \text{at} \quad \bar{x} = 0 \quad \text{(surface of the first plate)} \\ &\bar{\psi} = \bar{\psi}_2^0 \quad \text{at} \quad \bar{x} = \bar{h} \quad \text{(surface of the second plate)} \end{aligned} \tag{207}$$

These are the so-called constant potential (c.p.) boundary conditions used commonly in the literature, starting from the work of Vervey and Overbeek [42]. Sometimes a mixed case is considered where one of the plates is postulated to maintain the c.c. conditions, whereas the other fulfils the c.p. conditions [47]. In the case where the surface charge is due to ionizable (amphoteric) groups, the boundary conditions for the PB equation assume the form of nonlinear implicit expressions for the surface potential as a function of ionization constants, pH, etc. [48, 49]. Since these boundary conditions are very specific and system-dependent, they will not be considered in further discussion.

The above boundary conditions should be used for eliminating the constants of integration from the general expression, obtained by a two-fold integration of the PB equation, Eq. (205), i.e.,

$$\int \frac{d\bar{\psi}}{\left[\frac{1}{I}\sum \bar{n}_i\, e^{-z_i\bar{\psi}'} + C_1\right]^{1/2}} = \bar{x} + C_2 \tag{208}$$

Unfortunately, this integral cannot be expressed in any closed form for an arbitrary surface charge.

For a symmetric electrolyte, Eq. (208) simplifies to the form

$$\int \frac{d\bar{\psi}'}{(2\cosh\bar{\psi}' + C_1)^{1/2}} = \bar{x} + C_2 \tag{209}$$

where $\bar{\psi}' = z\bar{\psi}$.

In this case, the integral can be expressed in terms of the elliptic integral of the first kind as done originally by Vervey and Overbeek [42], who also presented graphical solutions of interaction energy for equally charged surfaces and gave approximate solutions for large surface potentials. Jones and Levine [50] formulated approximations valid for small and large distances in the case of an asymmetric electrolyte.

The interactions between identical plates under the c.c. and c.p. boundary conditions were also tabulated by Honig and Mul [51], whereas Devereux and Bruyin [52] extensively tabulated the interactions for dissimilar plates under the c.p. boundary condition. As pointed out by McCormak et al. [49], some results presented in these tables are charged with considerable errors, especially for extreme values of the surface potential. In the latter work, various solutions of Eq. (207) are given in the form of the elliptic integrals and Jacobi elliptic functions for both the c.c., c.p. and mixed boundary conditions. Graphical methods of determining the interaction energy between plates were also presented. During to recent progress in numerical methods, the tabulated and graphical solutions seem less useful than the direct numerical solutions of the nonlinear PB equation, Eq. (205), as done in Ref. [37] by applying the Runge–Kutta method. Examples of such calculations for the two planar double-layer system are shown in Fig. 2.16. As can be observed, upon approach, the plate potential increases in accordance with the analytical formula

$$\bar{\psi} = \pm \frac{1}{|z|} ln \frac{2\bar{I}(\bar{\sigma}_1^0 + \bar{\sigma}_2^0)}{|z| h} \tag{210}$$

where the upper sign denotes $\bar{\sigma}_1^0 + \bar{\sigma}_2^0 > 0$
This formula was derived by observing that, at close plate separation, the electroneutrality condition requires that

$$\bar{\sigma}_1^0 + \bar{\sigma}_2^0 = -\int_0^{\bar{h}} \rho_e(\bar{x})\, d\bar{x} = -\bar{\rho}_e \bar{h} \tag{211}$$

where $\bar{\rho}_e$ is the averaged charge density in the gap between plates, given by the expression

$$\bar{\rho}_e = -\frac{\bar{\sigma}_1^0 + \bar{\sigma}_2^0}{|z| \bar{h}} \tag{212}$$

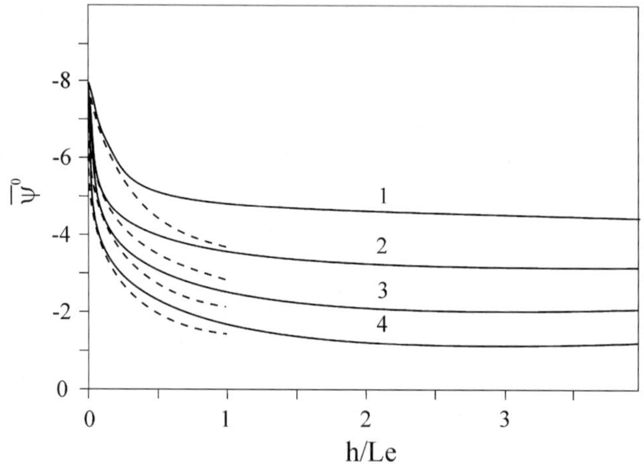

Fig.2.16. The dependence of the potential of the interface $\bar{\psi}^0 = \psi_d e/kT$ on the reduced separation distance between the plates *h/Le*, 1:1 electrolyte, the solid lines denote the exact numerical solutions [37] obtained for: 1. $\bar{\sigma}_1^0 = \bar{\sigma}_2^0 = 8$, 2. $\bar{\sigma}_1^0 = \bar{\sigma}_2^0 = 4$, 3. $\bar{\sigma}_1^0 = \bar{\sigma}_2^0 = 2$, 4. $\bar{\sigma}_1^0 = \bar{\sigma}_2^0 = 1$, the dashed lines denote the approximate analytical results calculated from Eq. (210). From Ref [37].

this is so because the overall charge density is determined by the counterion concentration.

As can be seen in Fig. 2.17, where the exact numerical data obtained for two similar and dissimilar plate systems are presented, Eqs. (210), (212) seem to be quite a good approximation.

By substituting Eq. (210) into Eq. (202), one can express the uniform pressure between plates as

$$\Delta\Pi = kTI \frac{2\,|\bar{\sigma}_1^0 + \bar{\sigma}_2^0|}{|z|\,\bar{h}} \tag{213}$$

The plate interaction energy per unit area obtained by integration of Eq. (213) over *h* is given by

$$\Phi = kT\, I\, Le \frac{2\,|\bar{\sigma}_1^0 + \bar{\sigma}_2^0|}{|z|} \ln \bar{h} \tag{214}$$

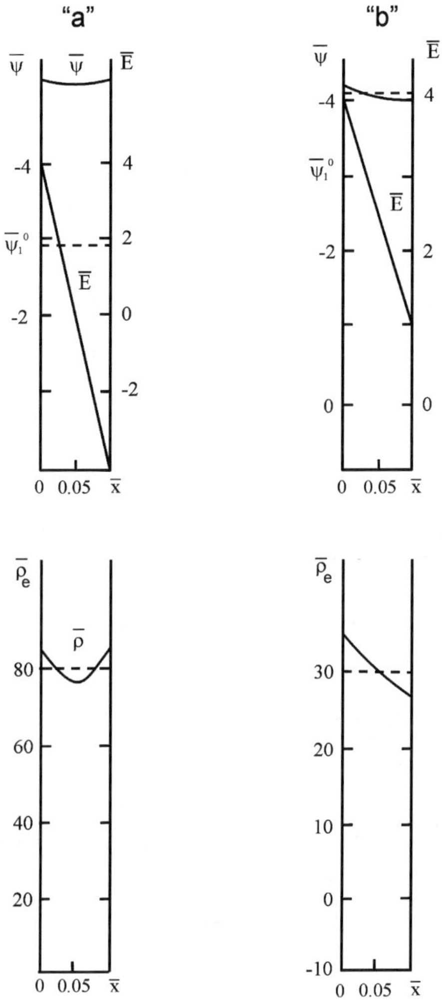

Fig. 2.17. The distribution of the dimensionless potential $\bar{\psi} = e\psi/kT$, field strength $\bar{E} = E(eLe/kT)$ and the charge density ρ_e/n_b in the gap between two dissimilar plates (double layers), 1:1 electrolyte, constant charge model, the solid lines denote exact numerical solutions obtained in Ref. [37] for $\bar{\sigma}_1^0 = \bar{\sigma}_2^0 = -4$, (part "a") and for $\bar{\sigma}_1^0 = -4\ \bar{\sigma}_2^0 = 1$ (part "b"). The dashed lines denote approximate analytical results calculated from Eqs. (210, 212). From Ref [37].

As can be noticed the interaction energy remains positive at short separations and tends to infinity logarithmically.

However, for arbitrary plate separations, the only exact, analytical solutions of the PB equation can be derived for the linear model, when the dimensionless potentials (or surface charges) of both plates remain smaller than unity. Then, Eq. (208) becomes

$$\pm\int \frac{d\bar{\psi}}{(\bar{\psi}^2 + C_1)^{1/2}} = \bar{x} + C_2 \tag{215}$$

By exploiting the boundary conditions, Eq. (206), one can derive the following expression for the electric potential distribution in the gap between the plates:

$$\bar{\psi} = \frac{\bar{\sigma}_2^0 + \bar{\sigma}_1^0 \cosh \bar{h}}{\sinh \bar{h}} \cosh \bar{x} - \bar{\sigma}_1^0 \sinh \bar{x} \tag{216}$$

where in the our case $\bar{\sigma}_1^0 = \bar{\psi}_1^0$, $\bar{\sigma}_2^0 = \bar{\psi}_2^0$.

In the case of the constant potential (c.p.) boundary condition, the potential distribution between the plates assumes the form

$$\bar{\psi} = \bar{\psi}_1^0 \cosh \bar{x} + \frac{\bar{\psi}_2^0 - \bar{\psi}_1^0 \cosh \bar{h}}{\sinh \bar{h}} \sinh \bar{x} \tag{217}$$

Using the above expressions for the potential distribution, one can calculate from Eq. (202) the uniform pressure between plates (force per unit surface), which is given by the equation

$$\Delta\Pi = kT\,I\left[\pm(\bar{\psi}_1^{0^2} + \bar{\psi}_2^{0^2})\frac{1}{\sinh^2 \bar{h}} + 2\bar{\psi}_1^0\bar{\psi}_2^0 \frac{\cosh \bar{h}}{\sinh^2 \bar{h}}\right] \tag{218}$$

where the upper sign denotes the c.c. boundary condition and the lower sign the c.p. boundary conditions.

The interaction energy per unit area is accordingly given by

$$\Phi = Le\int_{\infty}^{\bar{h}} \Delta\Pi \, d\bar{h} = kTLe\,I\left[\mp(1-\coth \bar{h})(\bar{\psi}_1^{0^2} + \bar{\psi}_2^{0^2}) + \frac{2\bar{\psi}_1^0\bar{\psi}_2^0}{\sinh \bar{h}}\right] \tag{219}$$

Eqs. (218), (219) were first derived for the c.p. case by Hogg et al. [53] and will be referred to as the HHF model. Wiese and Healy [54] and Usui [55] considered the c.c. model, whereas Kar et al. [47] derived analogous formula for the interaction energy in the case of the "mixed" case, i.e., c.p. on at plate and c.c. at the other.

It is interesting to note that the limiting forms of Eq. (219) for short separations, i.e., for $\bar{h} \rightarrow 0$ are

$$\Phi = kTLe\,I\left[\frac{(\bar{\psi}_1^0 + \bar{\psi}_2^0)^2}{\bar{h}} + \bar{\psi}_1^{0^2} + \bar{\psi}_2^{0^2}\right] \qquad \text{c.c. model}$$

$$\Phi = -kT\,Le\,I\left[\frac{(\bar{\psi}_1^0 - \bar{\psi}_2^0)^2}{\bar{h}} - \bar{\psi}_1^{0^2} - \bar{\psi}_2^{0^2}\right] \qquad \text{c.p. model} \tag{220}$$

It can be easily deduced that the interaction energy for the c.c. model diverges to plus infinity (repulsion) for short separations, whereas the c.p. model predicts diametrically different behavior, i.e., the interaction energy tends to minus infinity (attraction) for the same combination of surface potentials as for the c.c. case. The divergence between both models appearing at short separations is caused by the violation of the low potential assumption. Indeed, in order to observe the c.c. boundary conditions, the surface potential of the plates should tend to infinity when they closely approach each other, even if at large separations these potentials were very low. As a consequence, $\bar{\psi} \gg 1$ for $\bar{h} \rightarrow 0$ and the linear P.B. equation is not valid. Thus, for such distance range, the force and interaction energy of plates should be approximated in the c.c. model by Eqs. (213), (214).

On the other hand, for larger separations, both models reduce to the same asymptotic form

$$\Phi = 4kT\,Le\,I\,\bar{\psi}_1^0\bar{\psi}_2^0 e^{-h/Le} \tag{221}$$

As one can notice, the interaction energy between plates decreases exponentially at large separations, the rate of decay being proportional to $\kappa = 1/Le$.

It is also worthwhile noting that for equal plate potentials, the expressions for the force and interaction energy Eqs. (218), (219) become,

$$\Delta\Pi = 4kT\, I\, \bar{\psi}^{0^2} \frac{e^{-h/Le}}{\left(1 \mp e^{-h/Le}\right)^2}$$

$$\phi = 4kT\, Le\, I\, \bar{\psi}^{0^2} \frac{e^{-h/Le}}{\left(1 \mp e^{-h/Le}\right)} \qquad (222)$$

where the upper sign denotes the c.c. model. Eq. (222) was derived originally by Derjaguin [56].

All the discussed results are valid for metallic plates or plates of infinite thickness when the inside electric potential remains constant. The influence of the finite plate thickness on their interactions was studied in detail by Ohshima [57], both under linear and nonlinear regimes. It was shown that for situations of practical interest (aqueous solutions) the correction stemming from finite plate thickness remains negligible.

2.2.5. Particle Interactions in Ionic Media

In spite of the major significance of the system of two spherical double layers, no closed-form analytical solution of the nonlinear PB equation has been reported. Only cumbersome numerical solutions are available, which can be derived with excessive computer effort. They can be used, as demonstrated later on, for assessing the range of validity of existing approximate solutions rather than for routine calculations of interaction energy between spherical particles.

The often used simplification, when dealing with the two sphere problem, is again based on the linearization of the governing PB equation to the form given by Eq. (164), which can be solved by applying the perturbation scheme [42–50]. Ohshima and Kondo [58] and Ohshima et al. [59] used this method to calculate interaction energy of ion-penetrable spheres and dissimilar hard spheres with the inclusion of the electric field inside. The resulting analytical expressions in the form of infinite series are, however, rather cumbersome and of limited accuracy only.

McCartney and Levine [60] developed another approximate approach for solving the linearized PB equation, which was extended by Bell et al. [39] and Sader et al. [61] to dissimilar spheres. The method is based on the integral equation governing the distribution of electric dipoles. The main result of this approach was the analytical equation for the interaction energy of spheres expressed in terms of their surface potentials, radii and separation distance. For equal spheres, this formula assumes the form [39]

$$\phi = \phi_0 \left(\frac{a+h}{2a+h} \right) \ln \left(1 + \frac{a}{a+h} e^{-h/Le} \right) \tag{223}$$

where $\phi_0 = \varepsilon a \, (kT/e)^2 \, \bar{\psi}^{0^2}$

The main disadvantage of Eq. (223) is that the geometrical and electrostatic factors stemming from the surface potential are coupled. Moreover, because of linearization, the range of validity of Eq. (223) is restricted to $\bar{\psi}^0 < 1$ and $a/Le > 5$.

Two other methods of calculating interactions of spherical particles more universal, i.e.,

(i) the linear superposition (LSA) method, and
(ii) the Derjaguin summation method.

The advantage of the LSA method is that it can be applicable for arbitrary potential range and the *a/Le* parameter values. The disadvantage consists in the fact that the method gives less accurate results for separation between particles smaller than *Le*. Moreover, it cannot be easily extended to anisotropic (nonspherical) particle interactions.

The disadvantage of the Derjaguin method is that it becomes less accurate for distances larger than *Le* and $a/Le < 5$. However, there are two major advantages of the method that lead to its widespread application, namely the separation of the geometrical and electrostatic contributions and possibility of its application for anisotropic particles. The anisotropic particles are of increasing interest because of the fact that the shape of most bioparticles, e.g., bacteria, viruses, proteins deviates significantly from the spherical shape [8]. Other examples of highly anisotropic particles are the red blood cells, blood platelets, pigments and synthetic inorganic colloids: gold, silver iodide, silver bromide, barium sulfate etc. [62]. The entire variety of nonspherical particles has been synthesized over decades in the well-known school of Matijevic [63].

It seems, therefore, that the LSA and Derjaguin methods are complementary and can be used in combination to describe particle interaction for an arbitrary range of distances.

2.2.5.1. Interaction of Convex Particles – Analytical Solutions

According to the original Derjaguin method [56], the interactions of spheres were calculated as a sum (integral) of corresponding interactions of infinitesimal surface elements (rings) with a planar geometry. The summation was carried out in the region close to the minimum separation distance h_m (see Fig. 2.18) by assuming a fast decay of interactions away from this region. Thus, the Derjaguin method is only valid if the radii of the interacting spheres a_1 and a_2 are both much larger than the double-layer thickness.

By virtue of these assumptions, the interaction energy of two spheres can be calculated as the surface integral of the energy of two planar double-layer discussed above

$$\phi = \int_S \Phi \, dS = \int \Phi(y) 2\pi y \, dy \tag{224}$$

where $y(h)$ is the radius of the ring, depending on the distance h (see Fig. 2.18).

From a simple geometry, one can deduce that

$$h = h_m + a_1 + a_2 - \left(a_1^2 - y^2\right)^{1/2} - \left(a_2^2 - y^2\right)^{1/2} \tag{225}$$

If $y \ll a_2$ Eq. (225) reduces to the quadratic form

$$h = h_m + \frac{1}{2}\frac{y^2}{a_1} + \frac{1}{2}\frac{y^2}{a_2} \tag{226}$$

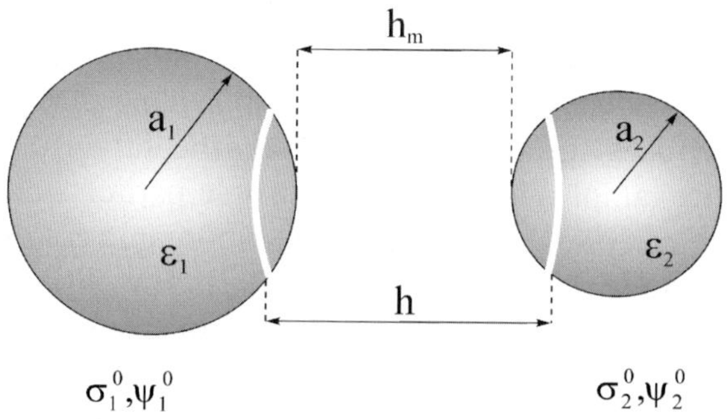

Fig. 2.18. Schematic view of the Derjaguin summation method used for calculating interactions of dissimilar spheres.

Hence, $y\,dy = dh\,a_1a_2/(a_1+a_2) = G_D\,dh$, where $G_D = a_1a_2/(a_1+a_2)$ is the geometrical Derjaguin factor. It is interesting to note that $G_D = 0.5a$ for two equal spheres and a for the plane/sphere configuration.

By considering this, Eq. (224) can be expressed as

$$\phi(h_m) = 2\pi G_D \int_{h_m}^{\infty} \Phi(h)\,dh \tag{227}$$

The interaction force of spheres, the derivative of the interaction energy, can be expressed as

$$\mathbf{F} = 2\pi G_D\,\Phi(h_m)\frac{1}{|\mathbf{r}_1 - \mathbf{r}_2|}(\mathbf{r}_1 - \mathbf{r}_2) \tag{228}$$

Note that the force is acting along the vector connecting the particle centers.

By using Eq. (227) in conjuncture with Eq. (219), the interaction energy of two dissimilar spheres can be expressed as

$$\phi = \pi\varepsilon\left(\frac{kT}{e}\right)^2 G_D\left[\mp\left(\bar{\psi}_1^{0^2} + \bar{\psi}_2^{0^2}\right)\ln\left(1 - e^{-2h_m/Le}\right) + 2\bar{\psi}_1^0\bar{\psi}_2^0\ln\frac{1+e^{-h_m/Le}}{1-e^{-h_m/Le}}\right] \tag{229}$$

where the upper sign denotes the c.c. boundary condition. Note that in contrast to Eq. (219), the interaction energy for spheres does not depend explicitly on the ionic strength *I*.

By differentiation, the expression for the interaction force can be formulated as

$$\boldsymbol{F} = 2\pi G_D\,kT\,Le\,I\left[\mp(1-\coth\bar{h})\left(\bar{\psi}_1^{0^2} + \bar{\psi}_2^{0^2}\right) + \frac{2\bar{\psi}_1^0\bar{\psi}_2^0}{\sinh\bar{h}}\right]\frac{\mathbf{r}_1 - \mathbf{r}_2}{|\mathbf{r}_1 - \mathbf{r}_2|} \tag{230}$$

where $\bar{h} = h_m/Le$.

Eq. (229) was first derived by Hogg et al. [53] for the c.c. model and Wiese and Healy [54] and Usui [55] for the c.p. model.

For equal sphere potential and the c.p. model, Eq. (229) simplifies to the form derived originally by Derjaguin [56]

$$\phi = 2\pi\varepsilon a \left(\frac{kT}{e}\right)^2 \bar{\psi}^{0^2} \ln\left(1+e^{-h_m/Le}\right) \tag{231}$$

Eqs. (229–230) were commonly used in the literature for determining stability criteria of colloid suspension [8,38] and for describing the plane/particle interactions in particle deposition problems.

The Derjaguin method was generalized by White [64] and Adamczyk et al. [37] to convex bodies of arbitrary shape. The first step of these calculations was determining the minimum separation distance h_m between the two particles involved. Then, the four principal radii of curvature R_1', R_1'', R_2' and R_2'', at the minimum separation region, are evaluated. Next, by using the Taylor expansion, one can express the distance between surface elements of the two particles involved (see Fig. 2.19) by the following dependence:

$$h = h_m + \frac{x_1^2}{2R_1'} + \frac{y_1^2}{2R_1''} + \frac{x_2^2}{2R_2'} + \frac{y_2^2}{2R_2''} \tag{232}$$

where x_1, y_1, x_2 and y_2, are the coordinates of the local Cartesian coordinates with the common z-axis. The x_1 and x_2 axes form the angle φ (see

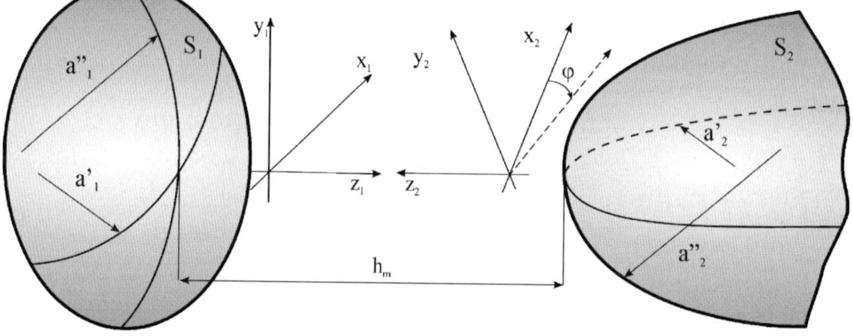

Fig.2.19. Schematic view of the generalized Derjaguin integration method used for calculating interactions of convex particles.

Fig. 2.19). By introducing a new coordinate system x,y, Eq. (232) can be expressed as

$$h = h_m + \frac{\lambda_1}{2} x^2 + \frac{\lambda_2}{2} y^2 \tag{233}$$

where λ_1, λ_2 are functions of the main radii of curvature and the angles and

$$\lambda_1 \lambda_2 = \left(\frac{1}{R_1^{'}} + \frac{1}{R_2^{'}}\right)\left(\frac{1}{R_1^{''}} + \frac{1}{R_2^{''}}\right) + \left(\frac{1}{R_1^{'}} - \frac{1}{R_1^{''}}\right)\left(\frac{1}{R_2^{'}} - \frac{1}{R_2^{''}}\right) \sin^2 \varphi \tag{234}$$

From Eq. (233), one can deduce that the plane h = const produces an ellipse of semiaxes $\left(\frac{2(h-h_m)}{\lambda_1}\right)^{1/2}$ and $\left(\frac{2(h-h_m)}{\lambda_2}\right)^{1/2}$, which has the surface area $2\pi\,(h-h_m)/(\lambda_1\,\lambda_2)^{1/2}$.

Thus, the surface element needed in Eq. (227) can be expressed as

$$dS = 2\pi \frac{dh}{(\lambda_1 \lambda_2)^{1/2}} = 2\pi G_D\, dh \tag{235}$$

The function $1/(\lambda_1\,\lambda_2)^{1/2}$ can be treated as the generalized Derjaguin factor. It is given explicitly by

$$G_D = \left(\frac{R_1^{'} R_1^{''} R_2^{'} R_2^{''}}{(R_1^{'} + R_2^{'})(R_1^{''} + R_2^{''}) + (R_1^{''} - R_1^{'})(R_2^{''} - R_2^{'})\sin^2 \varphi}\right)^{1/2} \tag{236}$$

In the case of particle/plane or two coplanar particle configurations one has and Eq. (236) simplifies to the form derived by Adamczyk et al. [37].

By using Eq. (227), one can formulate a general expression describing interactions of particles of arbitrary shape in the form

$$\phi = 2\pi\, G_D \int_{h_m}^{\infty} \Phi(h)\, dh \tag{237}$$

where $\Phi(h)$ is an arbitrary function describing interactions of two plates.

Eq. (237) has a major significance because it can be used for arbitrary shaped (convex) particles and other types of interactions as well (e.g., for dispersion interaction as discussed later on). The only condition is that the effective range of these interactions should remain much smaller than the particle size involved, so they can effectively be treated as surface interactions.

However, despite the apparent simplicity, it is rather inconvenient to apply Eq. (237) for three-dimensional situations except for the crossed-cylinder problem where G_D can be expressed for orientations close to 90° as

$$G_D = \frac{1}{\sin\beta_c}(R_1 R_2)^{1/2} \tag{238}$$

where β_c is the angle formed by the cylinder axes and R_1 and R_2 the radii of the cylinders.

The main problem when using Eq. (237) is to find the points of the minimum separation of the two bodies involved, as a function of their mutual orientation and consequently to determine h_m. Even for such simple particle shapes as spheroids, one has to solve high-order nonlinear trigonometric equations, which can only be done in an efficient way by iterative methods [65].

However, analytical results can be derived for limiting orientations of prolate and oblate spheroids as shown in Table 2.9. It is interesting to observe that the ratio between the Derjaguin factors (and hence of the interaction energy) for the parallel and perpendicular orientations of prolate spheroids (against a planar boundary) equals $1/A^2$ (where $A = b/a$ is the shorter to longer axis ratio). This means that the electrostatic attraction will be much higher for the parallel orientation (at the same separation distance h_m) so the particles will tend to adsorb parallel.

In the case of electrostatic repulsion (adsorption against an electrostatic barrier), the particles will preferably adsorb under the orientation perpendicular to the surface. The same pertains to the oblate spheroid adsorption.

It is interesting to note that in the case of spheroid/plane interactions, the Derjaguin factor can be evaluated analytically as a function of the orientation angle φ. For prolate spheroids one has [65]

$$G_D = a\frac{A^2}{A^2\cos^2\varphi + \sin^2\varphi} \tag{239}$$

Table 2.9
The Derjaguin $\bar{G}_D$ and the LSA $\bar{G}_e^0$, $\bar{G}_e$ geometrical factors for limiting spheroid orientations

Prolate spheroids

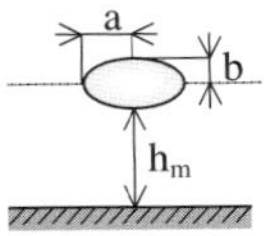	$\bar{G}_D = 1$ $\bar{G}_e^0 = \frac{2A}{A^2+1}$	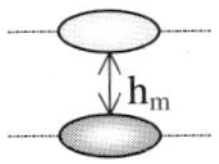	$\bar{G}_D = \frac{1}{2}$ $\bar{G}_e^0 = \frac{A}{A^2+1}$ $\bar{G}_e = \frac{A^2+1}{4A}$	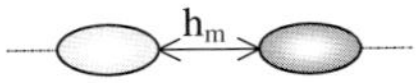	$\bar{G}_D = \bar{G}_e^0 = \frac{1}{2}A^2$ $\bar{G}_e = \frac{1}{2A^2}$
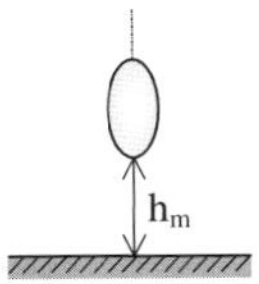	$\bar{G}_D = G_e^0 = A^2$	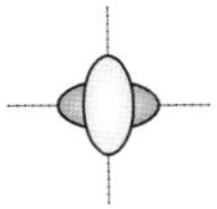	$\bar{G}_D = \bar{G}_e^0 = \frac{A}{A^2+1}$ $\bar{G}_e = \frac{A^2+1}{4A}$	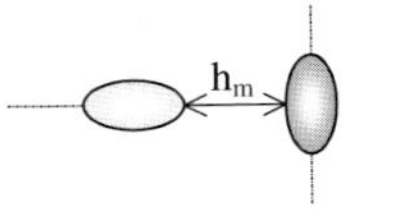	$\bar{G}_D = \frac{A^2}{[(A^3+1)(A+1)]^{1/2}}$ $\bar{G}_e^0 = \frac{2A^2}{A^3+A+2}$ $\bar{G}_e = \frac{A^2+1}{A^4+A^2+2A}$

Table 2.9 (continued)

Oblate spheroids

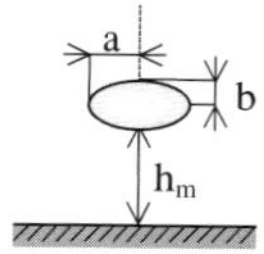	$\bar{G}_D = \bar{G}_e^0 = \frac{1}{A}$	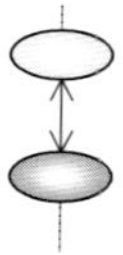	$\bar{G}_D = \bar{G}_e^0 = \frac{1}{2A}$ $\bar{G}_e = \frac{1}{2A}$	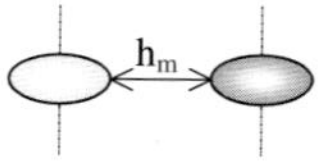	$\bar{G}_D = \frac{1}{2}A$ $\bar{G}_e^0 = \frac{A^2}{A^2+1}$ $\bar{G}_e = \frac{A^2+1}{4}$
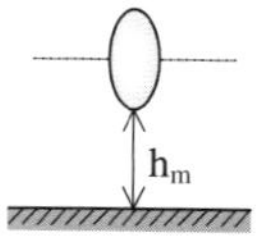	$\bar{G}_D = A$ $\bar{G}_e^0 = \frac{2A^2}{A^2+1}$	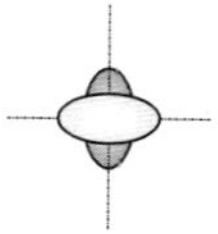	$\bar{G}_D = \bar{G}_e^0 = \frac{A^2}{A^2+1}$ $\bar{G}_e = \frac{A^2+1}{4}$	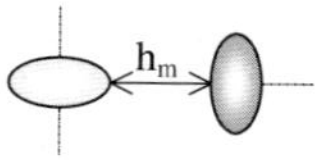	$\bar{G}_D = \frac{A}{[(A^3+1)(A+1)]^{1/2}}$ $\bar{G}_e^0 = \frac{2A^2}{2A^3+A^2+1}$ $\bar{G}_e = \frac{A^2}{2A^3+A+2}$

$\bar{G}_D = G_D/a$, $\bar{G}_e^0 = G_e/a$, $\bar{G}_e = G_e/a$, $A = b/a$.

whereas for the oblate spheroids the solution is

$$G_D = a \frac{A}{A^2 \cos^2 \varphi + \sin^2 \varphi} \tag{240}$$

The dependence of G_D/a on φ determined from these equations is shown in Fig. 2.20 for prolate spheroids and in Fig. 2.21 for oblate spheroids. In accordance with previous discussion, the differences between the perpendicular ($\varphi = 90°$) and parallel ($\varphi = 0°$) orientations increase when the parameter A decreases, e.g., for very elongated or flattened particles. Note also that the most significant changes in G_D (for particle/wall interactions) occur around $\varphi = 0°$, i.e., a slight deviation from the parallel orientation will result in an abrupt change of interactions.

The situation becomes more complicated in the case of particle/particle interactions, occurring for example during slow aggregation, since the

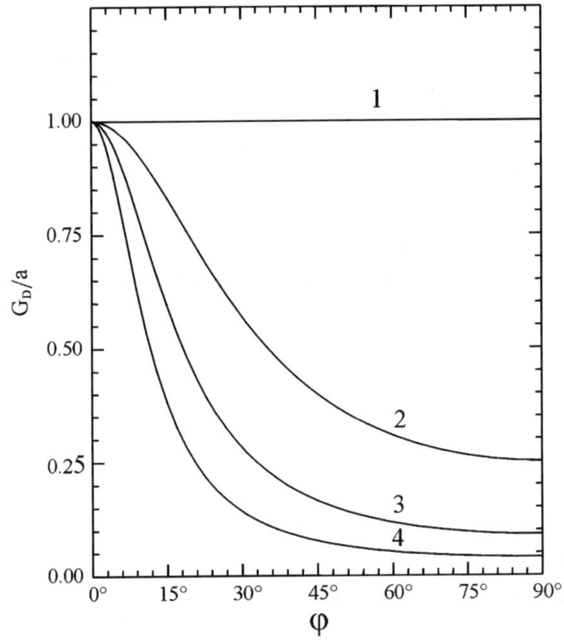

Fig. 2.20. The dependence of the reduced Derjaguin factor G_D/a on the orientation angle φ for a prolate spheroid/plane interactions. 1. $A = 1$ (sphere), 2. $A = 0.5$, 3. $A = 0.3$, 4. $A = 0.2$.

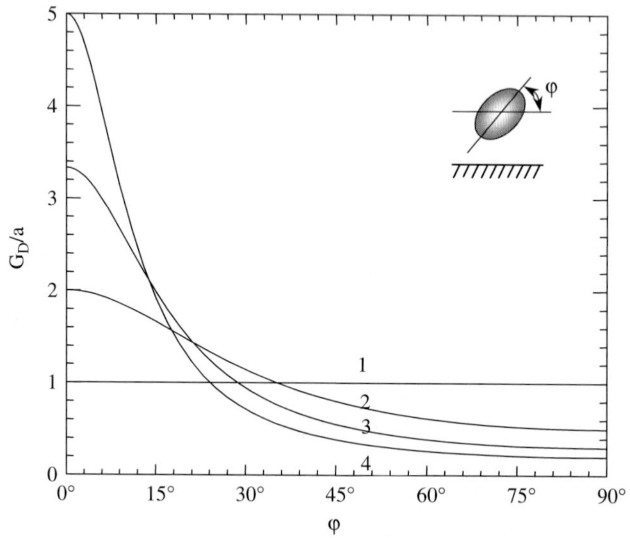

Fig. 2.21. The dependence of the reduced Derjaguin factor G_D/a on the orientation angle φ for an oblate spheroid/plane interactions. 1. $A = 1$ (sphere), 2. $A = 0.5$, 3. $A = 0.3$, 4. $A = 0.2$.

Derjaguin factor will depend (for a fixed h_m) not only on two relative orientation angles but, also on the relative position of the spheroids in space. This makes it difficult to present the results graphically in the general case. However, some limiting cases can be visualized, e.g., for the coplanar orientation of spheroids (when the symmetry axes are parallel to the common adsorption plane) and the crossed-orientation (one particle above the other).

In Fig. 2.22, the interactions between spheroids are visualized in such a way that the length of the line normal to the spheroid surface, connecting the contour represents the Derjaguin factor at this point for a given orientation. As can be easily deduced from Fig. 2.22, the repulsive interactions will be the smallest for the edge-to-edge orientation, which suggests that the slow aggregation of particles will preferably take place in this configuration. Another limiting case of the two spheroid interactions is the crossed particle configuration where the particle centers are above each other and the symmetry axes form the angle φ_c (see Fig. 2.23). For comparison, the results for the cylinder are shown as well. It can be observed that a considerable change in the orientation angle φ_c for crossed particles is predicted to influence the particle interactions in a small way. This quantitatively confirms the

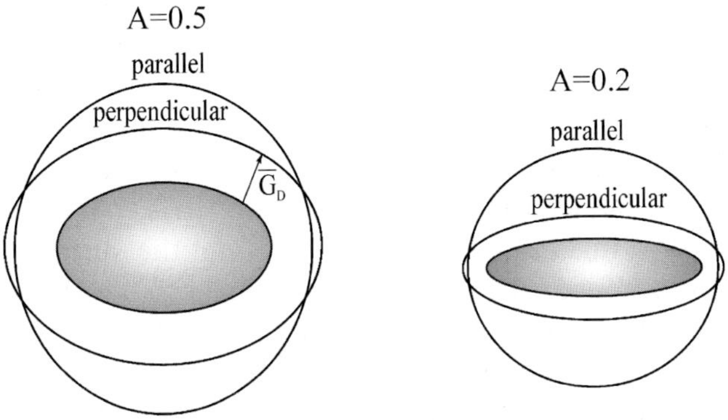

Fig. 2.22. A schematic representation of the spheroid/spheroid interactions (coplanar orientation); the contour shows the geometrical Derjaguin factor G_D/a for $A = 0.5$ (left-hand side) and $A = 0.2$ (right-hand side) orientations.

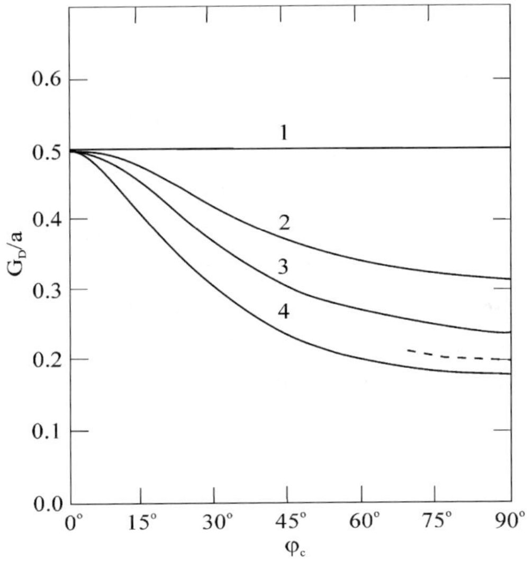

Fig. 2.23. The reduced Derjaguin factor G_D/a for two crossed spheroids forming the angle φ_c, 1. $A = 1$ (sphere), 2. $A = 0.5$, 3. $A = 0.3$, 4. $A = 0.2$, the dashed line denotes the results for crossed cylinders.

usefulness of the crossed cylinder configuration in the direct force measurement technique [22,66].

As mentioned, the Derjaguin method and consequently all the data shown in Table 2.9 are not limited to electrostatic interactions, but also to other interactions whose range remains smaller than the particle dimension (e.g., the dispersion interactions discussed next). However, a limitation of the method is that it becomes less accurate at larger particle/particle or particle/wall separations. This leads to an overestimation of the interactions and to a wrong asymptotic dependence of ϕ on the distance h_m.

Recently, Bhattacharjee and Elimelech [67] have undertaken an attempt to remove this deficiency by developing the surface element integration (SEI) procedure. Their approach is similar to the Derjaguin method, but the integration domain extends over the entire surface of the interacting particle, including the region opposite to the interface, where the sign of interactions is assumed to change. It was argued in [67] that the method predicts results in good agreement with the numerical solution of the linearized PB equation for *a/Le* as low as 0.3. It seems that despite some sporadic success, the usefulness of all integration methods, and the SEI approach, in particular for $\kappa a < 1$, seems rather doubtful because the surface elements cannot be treated as isolated entities in this case.

Particle interactions at larger separations can be better described in terms of the LSA, introduced originally by Bell et al. [39]. The main postulate of this method is that the solution of the PB equation for a two-particle system can be constructed as a linear superposition of the solutions for isolated particles. This is justified because the electrostatic potential at separations larger than *Le* decreases to small values and its distribution can be described by the linear version of the PB equation. As a consequence, the solution of the PB equation in this region for a two-particle configuration can be obtained by postulating the additivity of potentials and fields stemming from isolated particles (see Fig. 2.24)

$$\begin{aligned} \psi &= \psi_1 + \psi_2 \\ \mathbf{E} &= \mathbf{E}_1 + \mathbf{E}_2 \end{aligned} \tag{241}$$

The LSA method can, in principle, be applied to arbitrary particle shape provided a solution of the PB for isolated particles exists. At present, however, such solutions are known for spheres in a simple $1-1$ electrolyte only when the potential distribution is governed by Eq. (189) and the effective potential given by Eq. (190).

Using this solution, with the field calculated as the gradient of the electric potential, one can calculate the interaction force between two spheres using Eq. (201). The surface integration is conveniently carried out over the plane located between particles (see Fig. 2.24). In this way, Bell et al. [39]

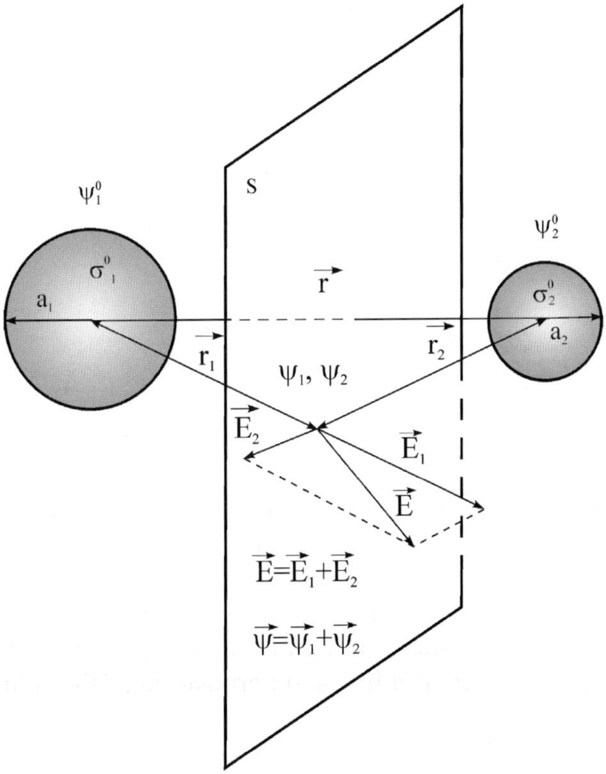

Fig. 2.24. A schematic representation of the linear superposition approach (LSA) for two unequal spheres.

derived the following analytical expression for the interaction force between two dissimilar particles:

$$\mathbf{F} = \phi_0 a_1 \frac{1 + r/Le}{r^2} e^{-h_m/Le} \frac{1}{|\mathbf{r}_1 - \mathbf{r}_2|}(\mathbf{r}_1 - \mathbf{r}_2) \tag{242}$$

where $\phi_0 = 4\pi\varepsilon\, a_2\,(kT/e)^2\, \overline{Y}_1^0 \overline{Y}_2^0$, $r = a_1 + a_2 + h_m$ is the distance between the particle centers and $\mathbf{r}_1 - \mathbf{r}_2$ is the vector connecting particle centers.

Eq. (242) indicates that the force vector is acting along the direction of the relative position vector $\mathbf{r}_1$, i.e., parallel to the line connecting the sphere centers.

The interaction energy can be easily calculated by integration of the force, which results in the expression [39].

$$\phi = \phi_0 \frac{a_1}{r} e^{-h_m/Le} \tag{243}$$

In the case of sphere/plane interactions, where $a_1 \rightarrow \infty$ Eqs. (242), (243) reduce to the simpler form

$$F = \frac{\phi_0}{Le} e^{-h_m/Le} \tag{244}$$

$$\phi = \phi_0 \, \mathrm{e}^{-h_m/Le}$$

It is interesting to mention that the LSA expression for two plate interactions has an identical form to Eq. (244), with $\phi_0 = \Phi_0 = kT\,Le I$.

As can be noted, Eqs. (242), (244) assume a simple two-parametric form, analogous to the Yukawa potential used widely in statistical mechanics. Unlike the previously derived expression stemming from Coulomb law, Eqs. (243), (244) predict an exponential decay of the interaction force and energy with the distance, thus for $h_m/Le \gg 1$ they become negligible. An additional advantage of Eq. (243) is that it does not diverge to infinity in the limit $h_m \rightarrow 0$, but approaches the constant value $\phi_{mx} = a_1\phi_0/(a_1 + a_2)$ which can be treated as the energy at contact. It can be calculated that for two equal spheres with the radius of 10^{-7} m (100 nm) and potential $\psi_d = 100$ mV, the energy at contact (at room temperature) equals 40 kT units. As can be noticed from Eq. (243), this value increases proportionally to the particle size. It is also worthwhile noting that in contrast to the unscreened interactions described by Coulomb's law the interaction energy ϕ_0, attains the limiting value of $4\pi\varepsilon a_2\,(kT/e)^2$ at high potentials.

Because of their simple mathematical shape, Eqs. (242), (244) are extensively used in numerical simulations of colloid particle adsorption problems.

However, the disadvantage of the LSA method is that it can be used in the original form exclusively for spherical particles. Because of the increasing importance of anisotropic particle interactions, an approximate method to solve this problem has been proposed in [65]. The essence of this approach, in principle a mutation of the LSA, consists in replacing the

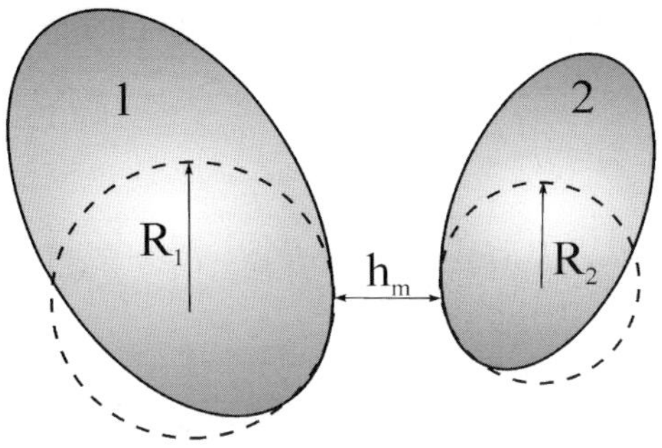

Fig. 2.25. A schematic view of the equivalent sphere approach (ESA) for two convex particles.

interactions of convex bodies by analogous interactions of spheres with appropriately defined radii of curvature R_1, R_2, see Fig. 2.25. As postulated [65], these radii should be calculated as the geometrical means of the principal radii of curvature evaluated at the point of minimum separation between the bodies, i.e.,

$$R_1 = \frac{2R_1^{'}R_1^{''}}{R_1^{'} + R_1^{''}}$$
$$R_2 = \frac{2R_2^{'}R_2^{''}}{R_2^{'} + R_2^{''}} \tag{245}$$

The advantage of this method, referred to as the equivalent sphere approach (ESA), consists in that the known numerical and analytical results concerning sphere interactions can be directly transferred to anisotropic particles. Thus, the LSA results, Eq. (243), can be expressed for spheroidal particles in the form

$$\phi = \phi_0 \frac{R_1 R_2}{a(R_1 + R_2 + h_m)} e^{-h_m/Le} = \phi_0 \frac{\overline{G}_e^0}{1 + \overline{G}_e h_m / a} e^{-h_m/Le} \tag{246}$$

where $\phi = 4\pi\varepsilon_0\varepsilon a\,(kT/e)^2\,\bar{Y}_1^0\bar{Y}_2^0$ and a the longer semiaxis of the spheroid and

$$\bar{G}_e^0 = \frac{R_1R_2}{a(R_1+R_2)} = \frac{2R_1'R_1''R_2'R_2''}{a\left[R_1'R_1''\left(R_2'+R_2''\right)+R_2'R_2''\left(R_1'+R_1''\right)\right]}$$

$$\bar{G}_e = \frac{a}{(R_1+R_2)} = \frac{a\left(R_1'+R_1''\right)\left(R_2'+R_2''\right)}{2\left[R_1'R_1''\left(R_2'+R_2''\right)+R_2'R_2''\left(R_1'+R_1''\right)\right]}$$

are the two geometrical factors.

Although Eq. (246) has the simple Yukawa-type form, its application in the general case of spheroid interaction in space is not straightforward due to the necessity of a numerical evaluation of the geometrical functions $\bar{G}_e^0$ and $\bar{G}_e$ [65]. However, analogously to the Derjaguin model, these functions can be evaluated analytically for some limiting orientations compiled in Table 2.9.

It is interesting to note that for the spheroid/plane interactions, due to the fact that $\bar{G}_e = 0$, the energy is described by the equation analogous to the LSA, Eq. (243)

$$\phi = \phi_0 = \bar{G}_e^0\, e^{-h_m/Le} \tag{247}$$

where the geometrical factor $\bar{G}_e^0$ can be evaluated analytically for prolate spheroids in terms of the inclination angle φ as [65].

$$\bar{G}_e^0 = 2A\frac{(G_D/a)^{3/2}}{G_D/a + A^2} \tag{248}$$

and G_D is the Derjaguin factor given by Eq. (236).
The dependence of G_D/a on the angle φ is plotted in Fig. 2.26.

As mentioned, in the case of arbitrary orientation of spheroids, one has to use numerical method for evaluating the minimum separation distance and calculating the radii of curvature. The use of efficient iterative schemes makes this task quite simple, so tedious simulations for spheroids become feasible [65]. Even with this complication, the use of the ESA is

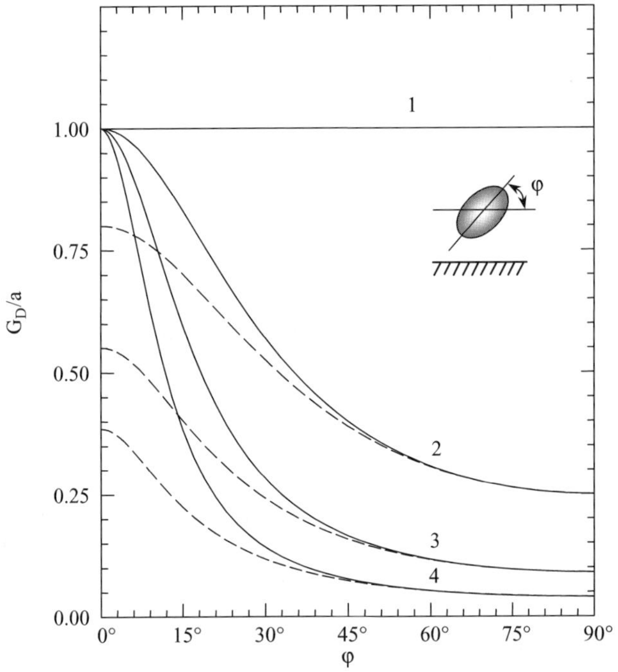

Fig. 2.26. Comparison of the Derjaguin (solid lines) and ESA (dashed lines) geometrical factors for prolate spheroid/plane interactions. 1. $A = 1$ (sphere), 2. $A = 0.5$, 3. $A = 0.3$, 4. $A = 0.2$.

more efficient than any attempt at solving the PB equation for the anisotropic particles.

For the sake of convenience, the most useful analytical expressions for interaction energy are compiled in Table 2.10.

2.2.5.2. *Comparison of Exact and Approximate Results*

It is not a trivial task to estimate the range of validity of the approximate approaches discussed above, especially the Derjaguin method, because of the lack of analytical solutions of the nonlinear PB equation for the sphere/sphere or sphere/plane geometry.

As already mentioned, the approximation often used for describing interactions of spheres consists in the linearization of the PB equation, which is then solved for two equal spheres by the perturbation techniques [42,44]. The resulting analytical expressions are too cumbersome for direct use, however. McCartney and Levine [60] developed the approximate surface dipole

Table 2.10
Double-layer interaction energy expressions for various configurations

System	Expression	Remarks, References
Plate/plate	$\Phi_0 \bar{Y}_1^0 \bar{Y}_2^0 e^{-\bar{h}}$	LSA [37] valid for arbitrary potential, $\bar{h} > 1$
Plate/plate	$\Phi_0 \left[\mp(\bar{\psi}_1^{0^2} + \bar{\psi}_2^{0^2})(1-\coth\bar{h}) + \dfrac{2\bar{\psi}_1^0 \bar{\psi}_2^0}{\sinh\bar{h}} \right]$	Linear model [53] $\lvert\bar{\psi}_1^0\rvert < 1$, $\lvert\bar{\psi}_2^0\rvert < 1$ upper sign c.c. lower sign c.p.
Plate/plate	$\Phi_0 \dfrac{2\lvert\sigma_1^0 + \sigma_2^0\rvert}{z_{mx}} \ln\bar{h}$ $-\Phi_0 \left[\dfrac{(\bar{\psi}_1^0 - \bar{\psi}_2^0)^2}{\bar{h}} - \bar{\psi}_1^{0^2} - \bar{\psi}_2^{0^2} \right]$	Nonlinear c.c. [37] valid for arbitrary $\lvert\bar{\sigma}_1^0\rvert$, $\lvert\bar{\sigma}_2^0\rvert$, $\bar{h} > 1$ nonlinear c.p. [65] $\bar{h} < 1$
Sphere/plate	$\phi_0' \bar{Y}_1^0 \bar{Y}_2^0 e^{-\bar{h}_m}$	LSA [39] valid for arbitrary potential, $\bar{h}_m > 1$
Sphere/sphere	$\phi_0' \bar{Y}_1^0 \bar{Y}_2^0 \dfrac{a_1}{a_1 + a_2 + h_m} e^{-\bar{h}_m}$	LSA [39] valid for arbitrary potential, $\bar{h}_m > 1$
Sphere/plate	$\dfrac{1}{4}\phi_0' \Big[\mp\left(\bar{\psi}_1^{0^2} + \bar{\psi}_2^{0^2}\right) \ln\left(1 - e^{-2\bar{h}_m}\right) +$ $2\bar{\psi}_1^0 \bar{\psi}_2^0 \ln\dfrac{1+e^{-\bar{h}_m}}{1-e^{-\bar{h}_m}} \Big] = \dfrac{1}{4}\phi_0' f(\bar{h}_m)$	Linear [53] $\lvert\bar{\psi}_1^0\rvert < 1$, $\lvert\bar{\psi}_2^0\rvert < 1$ upper sign c.c. lower sign c.p.
Sphere/ sphere	$\dfrac{1}{4}\phi_0' \dfrac{a_1}{a_1 + a_2} f(\bar{h}_m)$	Linear c.c., c.p. [53] valid for $\lvert\bar{\psi}_1^0\rvert < 1$, $\lvert\bar{\psi}_2^0\rvert < 1$
Identical spheres	$\dfrac{1}{2}\phi_0' \bar{\psi}_0^2 \ln\left(1 + e^{-\bar{h}_m}\right)$	Derjaguin formula

Table 2.10 (continued)

System	Expression	Remarks, References
Convex particles/ plate	$2\pi G_D \int \Phi(h)\,dh$	Generalized Derjaguin model valid for arbitrary interaction law $\Phi(h)$[65] $a/Le > 1$
Sphere/plate	$-\phi_0' \dfrac{(\bar{\psi}_1^0 - \bar{\psi}_2^0)^2}{\bar{h}_m}$	Nonlinear c.p. model valid for $\bar{h}_m \ll 1$
Prolate spheroid/ plate	$\dfrac{1}{4}\phi_0' \dfrac{(b/a)^2}{(b/a)^2\cos^2\varphi + \sin^2\varphi} f(\bar{h}_m)$	Linear model [65] valid for $\bar{\psi}_1^0 < 1, \bar{\psi}_2^0 < 1$

$$\bar{\psi}_1^0 = \psi_1^0 \frac{e}{kT},\ \bar{\psi}_2^0 = \psi_2^0 \frac{e}{kT},\ \bar{\sigma}_1^0 = \bar{\sigma}_1^0/(2\varepsilon kT n^b)^{1/2},\ \bar{\sigma}_2^0 = \sigma_2^0/(2\varepsilon kT n^b)^{1/2}$$

$$\bar{h} = h/Le, \bar{h}_m = h_m/Le, \qquad Le = \left(\frac{\varepsilon kT}{2e^2 I}\right)^{1/2}$$

$$\Phi_0 = kT\ Le\ I = \left[\frac{\varepsilon (kT)^3 I}{2e^2}\right]^{1/2}, \phi_0' = 4\pi\varepsilon \left(\frac{kT}{e}\right)^2 a_2$$

$$\bar{Y}_1^0 = 4\tanh\left(\frac{e\psi_1^0}{4kT}\right), \bar{Y}_2^0 = 4\tanh\left(\frac{e\psi_2^0}{4kT}\right)\text{(plates)},$$

for spheres see Eq. (190)

$$f(\bar{h}_m) = \left[\mp(\bar{\psi}_1^{0^2} + \bar{\psi}_2^{0^2})\ln(1 - e^{-2\bar{h}_m}) + 2\bar{\psi}_1^0 \bar{\psi}_2^0 \frac{1 + e^{-\bar{h}_m}}{1 - e^{-\bar{h}_m}}\right]$$

$$f(\bar{h}_m) = 4\bar{\psi}_1^0 \bar{\psi}_2^{0^2} e^{-\bar{h}_m} \quad \text{for } \bar{h}_m \gg 1$$

integration method which was extended by Bell [39] and Sader et al. [61] to a system of dissimilar sphere. The disadvantage of the analytical solutions, valid for low surface potentials, is that the geometrical and electrostatic factors stemming from surface potentials are coupled in a nonlinear way.

Hoskin [68] and Hoskin and Levine [69] were the first to solve the nonlinear PB equation for two equal-sized spheres in an exact way, using the finite difference method. They applied the bispherical coordinate system, which allowed one to accurately formulate the boundary conditions at particle surfaces. This coordinate system (with more mesh points) was subsequently used by Carnie et al. [70], who performed calculations of the interaction force for two spherical particles in a 1-1 electrolyte. Taking into account, in a rigorous manner, the electrostatic field distribution within the particles, the authors proved that this exerted a negligible effect on the interaction force of particles characterized by $\varepsilon < 5\varepsilon_0$, e.g., polystyrene latex particles. Performing numerical calculations, the authors also determined the range of validity of the Derjaguin approach as a function of κa and particle surface potentials. It has been concluded that the Derjaguin method is a good approximation with only a few per cent error for $a/Le > 5$ and particle/particle separation $h_m / Le < 0.2$, especially for the constant potential (c.p.) double-layer model.

More extensive calculations of this type have been performed by Stankovich and Carnie [71], who treated the case of two dissimilar spheres including sphere/plane interactions. They concluded that the sphere/plane attractive interactions in the c.p. case are very well approximated by the linear Derjaguin model (HHF) even for a/Le as small as unity. For the constant charge (c.c.) model and the sphere/sphere interactions, much better agreement with the exact calculations was attained by using the nonlinear Derjaguin model. This model was also used by Adamczyk et al. [37] to evaluate interactions of convex bodies, in particular the effect of the excluded volume that was found rather insignificant. It is to remember, however, that the implementation of the nonlinear Derjaguin model requires a numerical solution of the nonlinear PB equation for the planar geometry. This is a rather complex matter because of problems with the stability of numerical schemes encountered at large separations between the plates. It seems, therefore, that the use of the linear HHF model based on the Derjaguin approximation, offering concrete analytical expressions, is far more efficient. However, the linear Derjaguin model fails at large separations and high potentials as it tends to overestimate the energy. In these cases the use of the LSA model is more efficient.

This was confirmed by extensive calculations [72]. Electrostatic potential distribution and the energy of interaction for a system of dissimilar spheres, including the important subcase of sphere interaction with a plane, was calculated by using the bispherical coordinate system. Calculations were performed for a 1-1 electrolyte, κa changed from 0.25 to 10, and the dimensionless surface potential reaching 4.

A quantitative comparison of the numerical results obtained in [72] with various approximate expressions is shown in Figs. 2.27–2.30 where the normalized interaction energy is plotted as a function of $\bar{h} = h_m/Le$ for both the two particle system $\bar{\phi} = \phi/\left(\frac{1}{2}\varepsilon a \psi_2^{0^2}\right)$ (upper part of Figs. 2.27–2.30) and the particle-interface system $\bar{\phi} = 2\phi/\varepsilon a\ |\psi_1^0 \psi_2^0|$ (lower parts of these figures). The exact numerical data obtained in [72] were compared with the LSA model (given by Eq. (243)) and the linear HHF model (Eq. (229)), for both the c.c., c.p. and the mixed case.

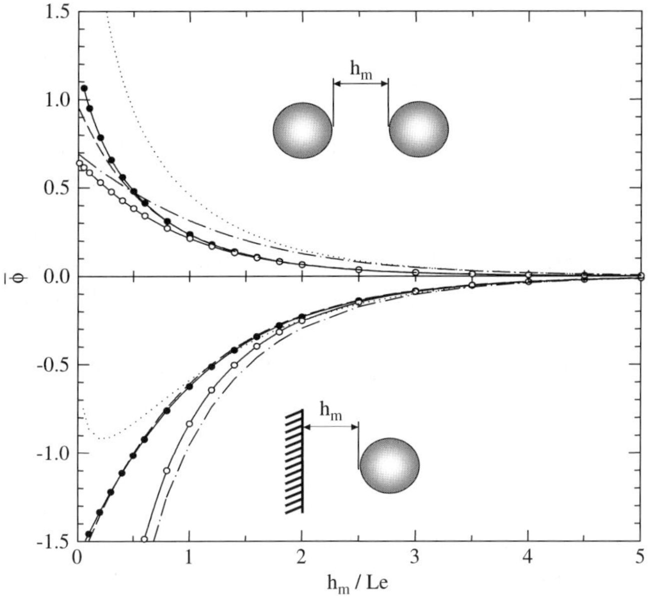

Fig. 2.27. The reduced interaction energy between two identical particles (upper part) $\bar{\phi} = 2\phi/\varepsilon\ a\psi_2^{0^2}$ and the particle/interface $\bar{\phi} = 2\phi/\varepsilon\ a\ |\psi_1^0 \psi_2^0|$ (lower part) calculated from various models for $\varepsilon = 78$, $\varepsilon_2 = 2.5$, $\psi_1^0 e/kT = 3$ (interface), $\psi_2^0 e/kT = -1.5$ (particle), $a/Le = 5$. (-•-•-•-) Exact numerical solution for the c.c. model; (-o-o-o-) Exact numerical solution for the c.p. model; (- - -) LSA model (analytical) Eqs. (243) and (244); (···) HHF model c.c. Eq.(229); (-·-·-) HHF model c.p. (calculated from Eq. (229)) [47].

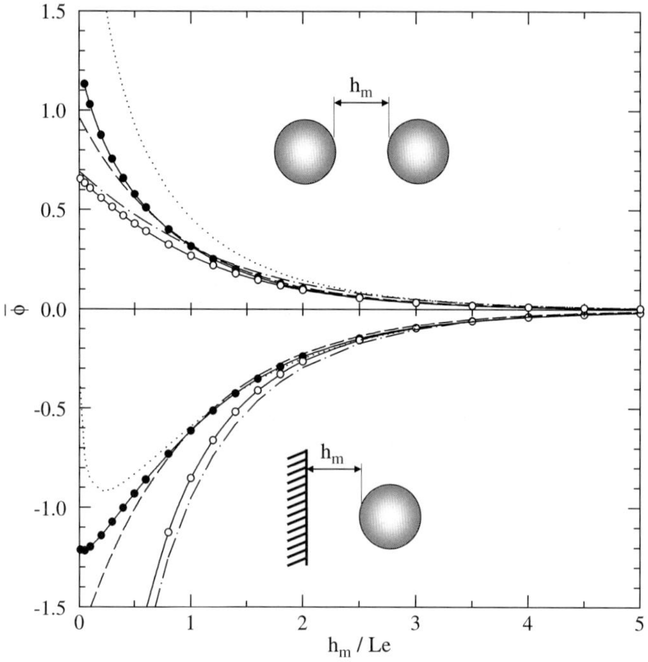

Fig. 2.28. The reduced interaction energy between two identical particles (upper part) $\overline{\phi} = 2\phi/\varepsilon\, a\psi_2^{0\,2}$ and the particle/interface $\overline{\phi} = 2\phi/\varepsilon\, a\, |\psi_1^0\, \psi_2^0|$ (lower part) calculated from various models for $\varepsilon = 78$, $\varepsilon_2 = 2.5$, $\psi_1^0\, e/kT = 3$ (interface), $\psi_2^0\, e/kT = -1.5$ (particle), $a/Le = 1$. (-•-•-•-) Exact numerical solution for the c.c. model; (-o-o-o-) Exact numerical solution for the c.p. model; (- - -) LSA model (analytical) Eqs.(243) and (244); (....) HHF model c.c. Eq.(229); (-·-·-) HHF model c.p. (calculated from Eq.(229)) [47].

As can be seen in Fig. 2.27, the particle/particle energy interaction profile, determined for $a/Le = 5$, is well reflected by the LSA model, whereas the HHF model shows a definite tendency to overestimate the interactions for the c.c. boundary conditions. Also, the particle/interface energy profiles are fairly well reflected by the LSA model with slightly lower accuracy for the c.p. boundary conditions.

Similar conclusions can be drawn from a comparison of the data shown in Fig. 2.28, obtained for $a/Le = 1$. Much larger deviations of the HHF model from the exact data are observed for a/Le as low as 0.25 with the LSA model performing very well again.

An interesting feature of the exact numerical results shown in Figs. 2.27–2.30 is that the c.c. and c.p. model give very similar interaction energy values for the particle/particle case (identical surface potentials), except for

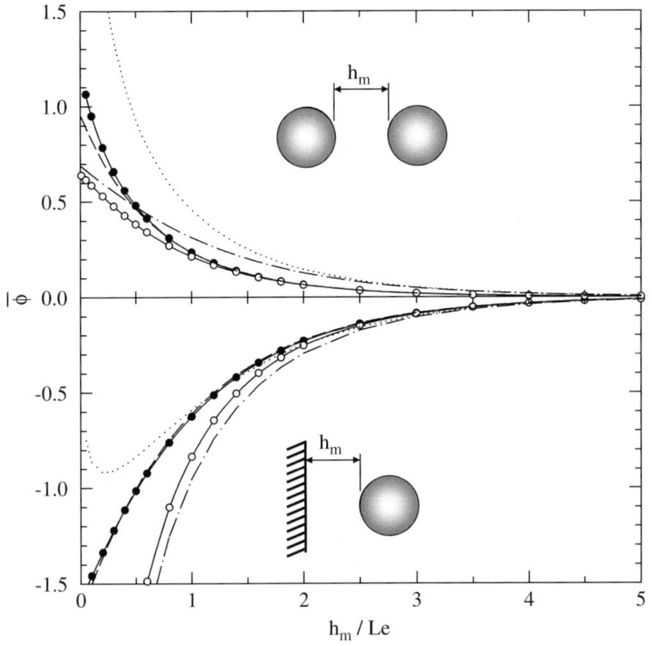

Fig. 2.29. The reduced interaction energy between two identical particles (upper part) $\bar{\phi} = 2\phi/\varepsilon\, a\psi_2^{0^2}$ and the particle/interface $\bar{\phi} = 2\phi/\varepsilon\, a\, |\psi_1^0\, \psi_2^0|$ (lower part) calculated from various models for $\varepsilon = 78$, $\varepsilon_2 = 2.5$, $\psi_1^0\, e/kT = 3$ (interface), $\psi_2^0\, e/kT = -1.5 =$ (particle), $a/Le = 0.25$. (-•-•-•-) Exact numerical solution for the c.c. model; (-o-o-o-) Exact numerical solution for the c.p. model; (- - -) LSA model (analytical) Eqs. (243) and (244); (···) HHF model c.c. Eq.(229); (-•-•) HHF model c.p. (calculated from Eq. (229)) [47].

very short distances $h_m/Le < 0.25$. Moreover, in the c.c. model, the exact energy value remains finite in the limit $h_m/Le \to 0$, which contrasts with the linear HHF model predicting a logarithmically diverging interaction energy in the limit $h_m \to 0$ (cf. Eq. (220)). In view of these results, the controversy in accepting the c.c. or c.p. models seems rather immaterial.

It should be mentioned, however, that the accuracy of the LSA approximation is strongly influenced by the surface potential asymmetry in the case of particle/wall interactions.

This is illustrated in Fig. 2.30 where one can see that the LSA reflects the exact results well if the absolute values of the surface potentials of the particle and the interface do not differ too much. For the case of asymmetry of potential exceeding 2:1, the LSA overestimates the attraction energy for distances $a/Le < 1$.

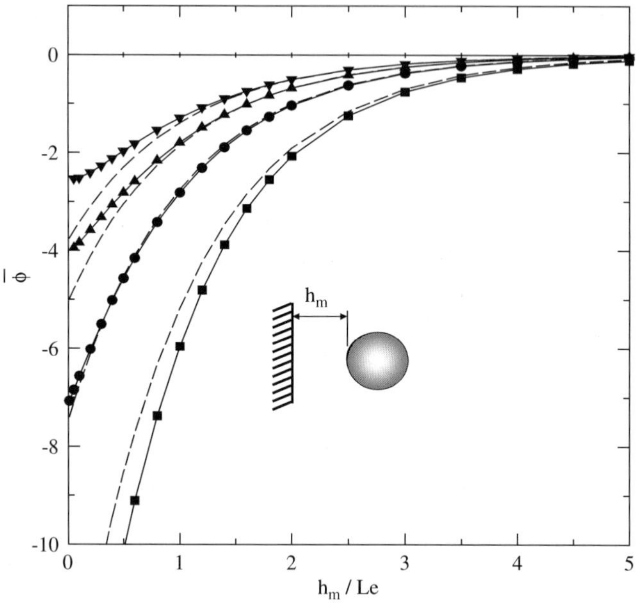

Fig. 2.30. The reduced interaction energy particle/interface $\overline{\phi} = 2\phi/\varepsilon a \,\psi_2^{0^2}|$ calculated from various models for $\varepsilon = 78$, $\varepsilon_2 = 2.5$, $a/Le = 1$, $\psi_1^0 e/kT = 3$ (interface),. The points denote the exact numerical solutions of the PB equation for the c.c. model and for $\phi_2^0 = -3$ (squares), $\phi_2^0 = -1.5$ (circles), $\phi_2^0 = -0.75$ (triangles), $\phi_2^0 = -0.6$ (reversed triangles). The dashed lines the analytical LSA model [47].

It may be concluded from the data presented in Figs. 2.27–2.30 and from similar results discussed in [70,71] that more significant differences between the LSA and the exact results occur at distances smaller than *a/Le* only where the interaction energy assumes large values either positive (similar surface potentials) or negative (opposite surface potentials). In both cases a relatively large uncertainty in ϕ can be tolerated. Moreover, the surface deformation, roughness and charge heterogeneity effects are expected to play a decisive role at such small separations, as discussed in the next section.

2.2.6. Concluding Remarks, Limitations of the Classical Double-Layer Model

It is worth while mentioning that the above results obtained by exploiting the classical models of the double layer are well suited for describing attractive interactions of macroscopic (colloid) particles at distances exceeding nanometers. This is so because of the number of simplifying assumptions made by formulating the PB equation, in particular the postulate of a continuous charge distribution in the bulk (mean field approximation) and on interacting surfaces. Other simplifications of the classical edl model are:

(i) lack of ion/ion correlations (neglecting the self-atmosphere effect);
(ii) lack of specific interactions of ions with surfaces;
(iii) neglecting all ion size effects, in particular the excluded volume effect;
(iv) neglecting the dielectric saturation effect; and
(v) ideally smooth and rigid surfaces.

It seems that the most restrictive is the assumption of continuous charge distribution and neglecting charge fluctuations, especially on interacting surfaces. As a result, the classical edl interaction theory cannot furnish any quantitative information about the behavior of a system in a microscale, at separations below a nanometer, especially if they are of a repulsive character. The discreteness in the charge distribution and local heterogeneities of chemical and geometrical nature will then play a decisive role rather than the macroscopic volume averaged interactions.

Moreover, when dealing with real systems, e.g., polymeric particles and deformable surfaces, many additional effects appear that require modifications and extensions of the classical edl theory such as

(i) Heterogeneity of charge distribution at interacting surfaces, which can be of a microscopic scale (of chemical origin) or macroscopic, patchwise scale; considerable differences within particle populations are also expected to appear in respect to, e.g., average charge (zeta potential).
(ii) Surface roughness, either due to isolated, well-defined asperities of different shape or of a statistical nature where the regular particle profile is perturbed.
(iii) Surface deformations upon approach, which seem particularly important for polymeric colloids (latexes) characterized by a low value of Young modulus.
(iv) Dynamic relaxation phenomena of the double layer upon contact (aging effects) due to ion migration along the surface or from one surface to another and eventually due to ion transfer into the bulk; these processes are again expected to appear for polymeric colloids with a random coil structure.

Despite the practical significance of these dynamic phenomena, little effort has been devoted to quantifying them by developing theoretical approaches. The effect of microscopic non-uniformity of charge distribution (discrete charge effect) was studied by Levine [73]. A similar problem was considered in [74,75] where the interaction energy between parallel plates with a discrete charge distribution forming two-dimensional lattices was studied using the linearized PB equation. It was found that the discrete charges generated a larger interaction potential in comparison with uniformly charged surfaces.

The effect of heterogeneity of charge distribution on particle/surface interactions studied by Song et al. [76] by considering two simple models of a macroscopic, patchwise heterogeneity, and the microscopic model where the charge distribution was described by the Gaussian probability distribution. The interactions were simply calculated from the usual expressions stemming from the edl theory by introducing the local values of the surface potential. A very similar approach was used in [77] by postulating that the surface charge (characterized by the zeta potential) of particles is described by the Gaussian distribution.

The effect of geometrical surface roughness has been studied in more detail. Krupp [78] was probably the first to qualitatively consider the effect of a hemispherical asperity on the adhesion force of a smooth colloid particle. It was concluded that the attractive electrostatic interactions will decrease less than the Van der Waals forces, so the net adhesion force should be determined by the electrostatic component.

A similar model was used in [77] to graphically determine the maximum size of a rough particle which can adhere to an interface under hydrodynamic shearing forces.

Elimelech and O'Melia [79] used an analogous model for simulating interaction of a smooth particle with a flat plate containing a single hemispherical asperity. The electrostatic interaction energy was calculated by using the Derjaguin method as a sum of particle/smooth wall and particle/asperity contributions.

The energy additivity rule was also exploited by Herman and Papadopolous [80,81] who determined the effect of conical and hemispherical asperities on the electrostatic (and van der Waals) interactions between flat plates, using the LSA approach combined with the Derjaguin summation method. It was demonstrated that due to asperities, the repulsive interaction energy was increased over the smooth plate case, especially for large κa values. This approach was generalized to the case of rough colloid particle/smooth surface [82-83]. The following formula for the electrostatic interaction energy was derived, based on the LSA/ Derjaguin approach:

$$\phi = \phi_0(1-\Theta)e^{-h_m/Le} + \phi'_0\, e^{-(h_m - a_s)/Le} \tag{249}$$

where $\phi_0 = 4\pi\varepsilon a\,(kT/e)^2\,\bar{Y}_1^0\bar{Y}_2^0$ as previously defined, $\phi'_0 = 4\pi\varepsilon a\,(kT/e)^2\bar{Y}_1^0\,\bar{Y}_s^0$, $\Theta = \pi a_s^2 N$ is the surface coverage of the asperities, a_s the asperity radius and N their surface concentration, $\bar{Y}_1^0$ $\bar{Y}_2^0$ and $\bar{Y}_s^0$ effective surface potentials of the interface, particle and asperity, respectively.

This model is applicable for high electrolyte concentration where $a_sL_e \gg 1$ and for low coverages θ when the number of asperities within the contact area remains low, so the uncovered surface areas can be treated as isolated patches.

It can be deduced from Eq. (249) that due to asperities the repulsive interaction energy is much higher (at the same distance between the smooth surface and the plate) than in the case of smooth objects. On the other hand, when defining h_m as the distance between the asperity surface and the smooth boundary (which has a more natural physical interpretation in the case of attractive interactions leading to adsorption and adhesion), one can deduce from Eq. (249) that the absolute value of ϕ should become much smaller than for bare particles at the same separation.

In an attempt to develop improved models, Kostaglou and Karabelas [84] considered the problem of electrostatic interactions of two infinite surfaces exhibiting periodic (sinusoidal) surface roughness. The linear PB equation was applied which was then solved by the perturbation, boundary collocation and boundary integral methods. It has been found that the interaction energy of rough surfaces is at all separations larger than that for smooth surfaces. However, it was claimed that the theory predicts a decrease in the electrostatic interaction energy upon contact. Obviously, further studies using more realistic surface roughness distributions (e.g., of stochastic nature) are needed to resolve this discrepancy.

2.3. MOLECULAR–VAN DER WAALS INTERACTIONS

In order to properly understand interactions of macroscopic objects, e.g., colloid particles, it is necessary to first learn about interactions of atoms and molecules forming condensed phases. Then, by using various approximations, most often the pairwise additivity principle, formulae can be derived describing the interaction of macrobodies of various shapes.

The existence of intermolecular forces should be obvious for any careful observer who sees small drops violating the law of gravity, i.e., assuming a spherical shape rather than spreading over surfaces. One needs, however, a great deal of imagination to deduce that this behavior is caused by some kind of attractive forces analogous to the gravitational forces making stars and planets (that were once in a liquid state), spherical. Indeed, the first attempt to introduce the concept of intermolecular forces, although the notion of a molecule was not clearly established in those days, was made by Clairault in 1743 who investigated the shape of small liquid masses under the action of gravity [1].

Other striking examples of macroscopic effect caused by these interactions are the capillary phenomena, especially the capillary rise (or depression) in thin tubes, wetting or dewetting, depending on the character of the interfaces, adhesion, cohesion, lubrication, formation of emulsions, coagulation of colloids, aggregation of suspensions, coalescence, filtration, flotation, etc.

In the microscale, these interactions manifest themselves in the Joule–Thompson effect, i.e., cooling of expanding gases and other deviations of real gases from ideality (non-hyperbolic pressure vs. volume dependence) accounted for by van der Waals famous v^{-2}, i.e., r^{-6} correction, where v is the volume of the gas.

The intermolecular forces also drive adsorption phenomena, e.g., gases on solids, surfactants on liquid/solid or liquid/liquid interfaces, adsorption of polymers, proteins and colloids, which is the prerequisite of efficient chromatographic separation. Other examples are the nucleation and condensation phenomena, epitaxial growth of crystals, formation of van der Waals solids such as graphite, etc.

At present, the true origin of intermolecular forces is fairly well understood and interpreted as the net result of electrostatic interactions between ions, permanent or induced dipoles and multipoles (e.g., for symmetric molecules like hydrogen). Except for the Coulomb interactions between ions analyzed before, one can divide the intermolecular forces into the following main categories [85,86]:

(i) permanent dipole or multipole/permanent dipole or multipole interactions called the Keesom orientational forces;
(ii) permanent dipole or multipole/induced dipole interactions called the inductive Debye forces; and
(iii) induced dipole/induced dipole interactions resulting from spontaneous fluctuation in local electron density, called London dispersive forces.

The first two types of interactions of the list can be well described in terms of the classical electrostatics exposed above, by analyzing the multipole interactions given by Eq. (30). The only problem arises when calculating proper averages over molecule orientation. If all are available with equal probability, the net interaction energy should be zero. Non-vanishing contributions arise if some orientations are preferred because they are energetically favorable (Boltzmann statistics). By considering this, one can predict that both the Keesom and Debye energy decays as r_0^{-6}. An important property of these interactions is that they are temperature dependent because the degree of orientational order falls abruptly with the temperature.

The third category of interactions, i.e., the dispersive ones cannot be analyzed in terms of classical electrostatics, which does not predict spontaneous fluctuations in the local charge density leading to temporary dipoles. They can only be properly described by using concepts of quantum physics, in particular the perturbation theories.

2.3.1. Dipolar Interactions – Keesom Forces

In this section we discuss interactions between permanent and induced dipoles and multipoles that are relevant for calculating molecular interactions. It should be remembered, however, that the concept of a dipole in relation to molecules is rather approximate due to their complex shapes and charge distribution, even for the water molecule. The validity of the dipolar approximation becomes, therefore, satisfactory only for distances larger than molecular dimensions.

The electric potential distribution in the vicinity of a permanent dipole is given by Eq. (20), derived previously. Therefore, the interaction energy of the dipole with a charge (monopole) $q = ze$ is given by

$$\phi dl = -\mathbf{p}\cdot\mathbf{E} = pE\cos\vartheta' = \frac{pq\cos\vartheta'}{4\pi\varepsilon r_0^2} \tag{250}$$

where p is the length of the dipole moment vector $\mathbf{p}$, ϑ' is the angle between the $\mathbf{p}$ and $\boldsymbol{E}$ vectors, and r_0 the distance between the dipole center and the charge center.

As can be noticed in Eq. (250) and in further equations derived in this section, the permittivity of the medium where the dipoles interact was introduced rather than the permittivity of the free space. Although the concept of dielectric permittivity may break down for distances comparable with molecular dimensions, it is useful to remember that the dipole/ion and other interactions involving dipoles are strongly modified by the medium, in analogy to macroscopic dipole interactions analyzed earlier (cf. Eq. (123–125).

Note also that the absolute value of the interaction energy between a dipole and a charge decreases as the square of their mutual distance. As can be seen from Eq. (250), the strongest repulsion (if q is positive) occurs for $\vartheta = 0$ (in-line position of the dipole and charge) and the strongest repulsion for $\vartheta = \pi$.

It is interesting to estimate the energy of the water dipole interaction with simple ions. By considering that p for water equals 1.85 Da (a dipole moment unit equal to 3.336×10^{-30} C m) and the equivalent radius is 1.4×10^{-10} m, one can calculate from Eq. (250) that for the lithium Li^+ ion with the bare radius of 0.68×10^{-10} m [85], the maximum attractive energy equals -2.05×10^{-15} J, i.e., -51 kT. For the Na^+ ion with the radius of 0.95×10^{-10} m, the energy equals -40 kT. For small divalent cations (like Mg^{2+}, radius 0.68×10^{-10}, the energy reaches -102 kT units. It should be remembered, however, that these values are strictly valid for a vacuum. For aqueous solutions this ion/dipole energy will be much smaller (in absolute values) because of increased permittivity.

The ion/permanent dipole interactions are responsible for solvation phenomena occurring in polar solvents, especially water (hydration). A few shells of solvent are usually coordinated to the ions because the hydrated ion radii exceed the bare ion radii many times. For Na^+ ion, for example, the hydrated radius equals 3.6×10^{-10} m, almost six times larger than the bare radius [85]. It should be remembered, however, that the net interaction energy of dipoles with the central ion in the shell is reduced because of the dipole/dipole interactions that depend on their mutual orientations, as demonstrated later on.

It is useful to compare the dipole/charge interactions with the charge/charge interactions calculated from the Coulomb law. The quotient of these two interactions for $\vartheta = 0$ is given by the expression

$$\phi_{dc}/\phi_{cc} = \frac{p}{q r_0} \tag{251}$$

For two natrium ions and water dipoles, taking the data used before, one can estimate that the dipole/ion interaction energy is only 0.26 of the ion/ion interaction energy (at the distance of 2.35×10^{-10} m), i.e., almost four times smaller. Because this quotient decreases with the distance, the charge/charge interactions will dominate over charge/dipole interactions.

Let us now consider two permanent dipole interactions. The first is oriented for the sake of convenience along the z-axis (see Table 2.11) forming the angle ϑ_1 with the $\mathbf{r}_0$ vector connecting the centers of the dipoles. The second forms the angle ϑ_2 with $\mathbf{r}_0$ and the angle ϑ_2 with the $\mathbf{r}_0$, $\mathbf{p}$ plane. Considering that the field generated by the first dipole at the center of the second dipole is given by

$$\mathbf{E}_1^0 = \frac{3(\mathbf{p}_1 \cdot \mathbf{r}_0)\mathbf{r}_0 - \mathbf{p}_1 r_0^2}{r_0^5} \tag{252}$$

for their interaction energy one obtains the expression

$$\phi_{dd} = -\frac{3(\mathbf{p}_1 \cdot \mathbf{r}_0)(\mathbf{p}_2 \cdot \mathbf{r}_0) - \mathbf{p}_1 \cdot \mathbf{p}_2 r_0^2}{4\pi\varepsilon r_0^5} = -\frac{p_1 p_2}{4\pi\varepsilon r_0^3} \tag{253}$$

$$[2\cos\vartheta_1 \cos\vartheta_2 - \sin\vartheta_1 \sin\vartheta_2 \cos\vartheta]$$

Note that the energy of the interaction of two dipoles decreases as the cube of their mutual distance r_0.

If $\vartheta_1 = \vartheta_2 = 0$, $\varphi = 0$ (both dipoles in line), the attraction between the dipoles is maximal, given by the expression

$$\phi = -\frac{2 p_1 p_2}{4\pi\varepsilon r_0^3} = 2\phi_{dd}^0 \tag{254}$$

where $\phi_{dd}^0 = -p_1 p_2 / 4\pi\varepsilon r_0^3$

On the other hand, for the parallel orientation of the dipoles $\vartheta_1 = \vartheta_2 = \pi/2$, $\varphi = 0$ there appears repulsion described by

$$\phi = \frac{p_1 p_2}{4\pi\varepsilon r_0^3} = -\phi_{dd}^0 \tag{255}$$

For water dipole interactions in a vacuum, it can be calculated from Eqs. (254–255) that the maximum attractive energy (at the distance 2.8×10^{-10} m) equals -3.12×10^{-20} J, i.e., $-7.8\ kT$ and the maximum repulsive energy 1.56×10^{-20} J, i.e., $3.9\ kT$. At a twice large a distance, equal to

Table 2.11
The dipolar interactions

System	Scheme	Energy, ϕ	Boltzmann averaged energy
Charge/dipole	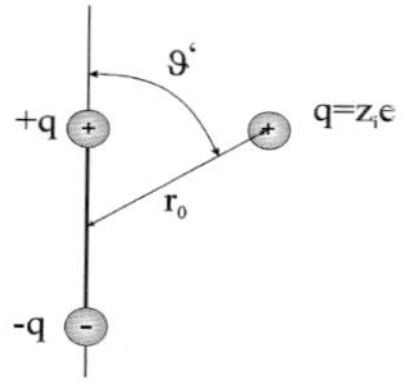	$\frac{p(z_i e)\cos\vartheta'}{4\pi\varepsilon r_0^2}$	$-\frac{p^2(z_i e)^2}{3kT(4\pi\varepsilon)^2}\frac{1}{r_0^4}$
Charge/induced dipole	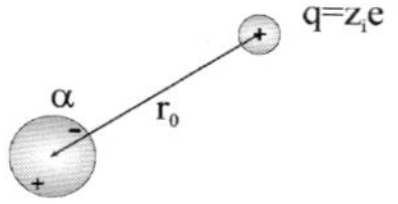	$-\frac{(z_i e)^2\,\varepsilon\alpha}{2(4\pi\varepsilon)^2\,r_0^4}$	$-\frac{(z_i e)^2\,\varepsilon\alpha}{2(4\pi\varepsilon)^2\,r_0^4}$
Dipole/dipole (permanent)	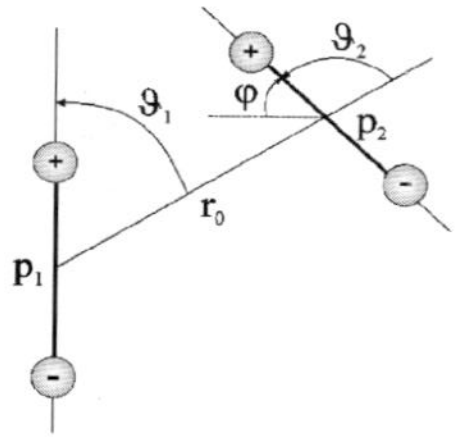	$\frac{p_1 p_2[2\cos\vartheta_1\cos\vartheta_2 - \sin\vartheta_1\sin\vartheta_2\cos\varphi]}{4\pi\varepsilon r_0^3}$	$-\frac{p_1^2 p_2^2}{3kT(4\pi\varepsilon)^2}\frac{1}{r_0^6}$

Dipole/induced dipole	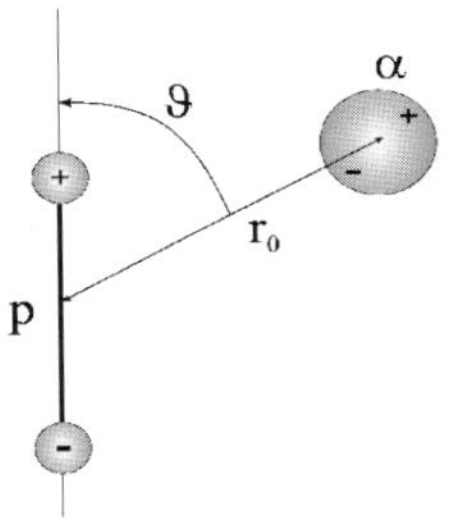	$\frac{\varepsilon p^2 \alpha (3\cos^2 \vartheta_1 + 1)}{2(4\pi\varepsilon)^2 r_0^6}$	$-\frac{p^2 \varepsilon a}{(4\pi\varepsilon)^2}\frac{1}{r_0^6}$
Induced dipole/induced dipole	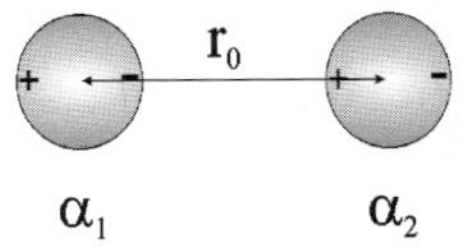	$\phi = -\frac{3}{2}\frac{\alpha_{0_1}\alpha_{0_2}}{(4\pi\varepsilon)^2}\frac{h_p \tilde{\nu}_{e_1}\tilde{\nu}_{e_2}}{(\tilde{\nu}_{e_1} + \tilde{\nu}_{e_2})}\frac{1}{r_0^6}$	$\phi = -\frac{3}{2}\frac{\alpha_{0_1}\alpha_{0_2} h_p \tilde{\nu}_{e_1}\tilde{\nu}_{e_2}}{(4\pi\varepsilon)^2(\tilde{\nu}_{e_1} + \tilde{\nu}_{e_2})}\frac{\lambda}{r_0^6}$

$(Z_i\, e) = q$

5.6×10^{-10} m, these interactions will fall well below 1 kT. This comparison indicates that the dipole/dipole interactions are much weaker than the dipole/ion or ion/ion interactions. However, in the absence of charges they play a role for highly polar media over distances smaller than 5×10^{-10} m. An additional factor that reduces their significance is orientational averaging promoted by the thermal motion (rotation) of molecules.

Let us analyze this effect in some detail. In the general case, the orientational averaging of any relevant function $f(r_0,\Omega)$ (dipole moment or energy, for example) is described by the formula

$$\langle f(r_0)\rangle = \int_{\Omega} f(r_0,\Omega)\bar{\rho}_p(\Omega)\,d\Omega \tag{256}$$

where $\langle f(r_0)\rangle$ is the function averaged over the orientational space Ω determined by the angles ϑ_1, ϑ_2, φ etc., and $\bar{\rho}_p$ (Ω) is the probability density of a given state in the orientaional space. Obviously, $\bar{\rho}_p$ (Ω) is subject to the normalization condition

$$\int_{\Omega} \bar{\rho}_p(\Omega)\,d\Omega = 1 \tag{257}$$

If the probability density $\bar{\rho}_p$ is assumed independent of the orientations (uniform), the integration of Eq.(257) gives Ω, consequently Eq.(256) becomes

$$\langle f(r_0)\rangle = \frac{1}{\Omega}\int_{\Omega} f(r_0,\Omega)\,d\Omega \tag{258}$$

In the case of the dipole/ion interactions (or dipole/field interactions), by substituting Eq.(250) into Eq.(258) one obtains

$$\langle \phi\rangle_{\Omega} = \int_{\Omega}\mathbf{p}\cdot\mathbf{E}\,d\Omega = \frac{pE}{4\pi}\int_0^{2\pi} d\varphi\int_0^{\pi}\cos\vartheta\sin\vartheta\,d\vartheta = 0 \tag{259}$$

where $d\Omega = \sin\vartheta\, d\vartheta\, d\varphi$ and $\Omega = 4\pi$.

As indicated by Eq. (259), the averaged energy (or the dipole moment) vanishes because

$\int_0^{\pi} \cos \vartheta \sin \vartheta \, d\vartheta = 0.$

An analogous result is obtained for the dipole/dipole interactions because:

$$\langle \Phi \rangle_{\Omega} = \frac{p_1 p_2}{4\pi\varepsilon r_0^3} \frac{1}{8\pi} = \int_0^{2\pi} d\varphi \int_0^{\pi}\int_0^{\pi} 2\cos\vartheta_1 \cos\vartheta_2 - \sin\vartheta_1 \sin\vartheta_2 \cos\varphi)$$

$$\sin\vartheta_1 \sin\vartheta_2 d\vartheta_1 d\vartheta_2 = 0 \tag{260}$$

Thus, for uniform probability density the permanent dipole interactions would give no contribution to the interaction energy.

However, in molecular systems, the probability of finding a given state is governed by the Boltzmann statistics and $\bar{\rho}_p$ is given by the expression

$$\bar{\rho}_p(\Omega) = \frac{e^{-\phi(r_0,\Omega)/kT}}{\int_{\Omega} e^{-\phi(r_0,\Omega)/kT} \, d\Omega} = \frac{e^{-\phi(r_0,\Omega)/kT}}{\left\langle e^{-\phi/kT} \right\rangle_{\Omega}} \tag{261}$$

By substituting Eq.(261) into Eq.(256), one obtains

$$\left\langle f(r_0) \right\rangle_{\Omega} = \frac{1}{\left\langle e^{-\phi/kT} \right\rangle_{\Omega}} \int_{\Omega} f(r_0,\Omega) e^{-\phi(r_0,\Omega)/kT} \, d\Omega \tag{262}$$

Unfortunately, in the general case, this integral cannot be evaluated analytically. However, for weak interactions when $\phi/kT < 1$ (the case for larger separations between dipoles as estimated above), one can express Eq. (262) in the following simpler form by expanding ϕ in a power series and considering the fact that $\langle \phi \rangle_{\Omega} = 0$

$$\left\langle f(r_0) \right\rangle_{\Omega} \cong \frac{1}{\Omega} \int_{\Omega} f(r_0,\Omega) \left[1 - \frac{\phi}{kT} + \frac{\phi^2}{2(kT)^2} + \cdots \right] d\Omega \tag{263}$$

Eq. (263) is useful for evaluating various averages. For example, the averaged dipole moment is given by

$$\langle p\rangle_{\Omega} = \frac{1}{4\pi}\int_{0}^{2\pi} d\varphi \int_{0}^{\pi} p\cos\vartheta\left(1 + \frac{pE}{kT}\cos\vartheta\right)\sin\vartheta\, d\vartheta = \frac{p^2 E}{3kT} \tag{264}$$

Analogously, the averaged energy of the dipole in the external field (or the dipole/charge interactions) is given by

$$\langle \phi\rangle_{\Omega} = \frac{1}{4\pi}\int_{0}^{2\pi} d\varphi \int_{0}^{\pi} - pE\cos\vartheta\left(1 + \frac{pE}{kT}\cos\vartheta\right)\sin\vartheta d\vartheta$$

$$= -\frac{p^2E^2}{3kT} = -\frac{p^2q^2}{3(4\pi\varepsilon)^2kT}\frac{1}{r_0^4} \tag{265}$$

Eq. (265) describes the Keesom orientation interactions.

In a similar way, the averaged energy for the dipole/dipole interactions in the thermodynamic limit can be calculated by substituting Eq.(253) into Eq. (263)

$$\langle \phi\rangle_{\Omega} = \frac{\phi_{dd}^{02}}{8\pi kT}\int_{0}^{2\pi} d\varphi \int_{0}^{\pi}\int_{0}^{\pi} (2\cos\vartheta_1\cos\vartheta_2$$

$$- \sin\vartheta_1 \sin\vartheta_2 \cos\varphi)^2 \sin\vartheta_1 \sin\vartheta_2 d\vartheta_1 d\vartheta_2$$

$$= -\frac{2\phi_{dd}^{02}}{3kT} = -2p_1^2p_2^2/3kT(4\pi\varepsilon)^2r_0^6 \tag{266}$$

As can be deduced, the averaged energy is temperature dependent and decreases as T^{-1}.

It should be mentioned that Eq. (265), (266) describe the internal energy of dipoles in the thermodynamic limit. The free energy of dipoles can be obtained by considering the entropic contribution needed for orienting the dipoles. It can be shown [85] that because of the inverse temperature dependence of the energy, the free energy ϕ_f dipolar interactions becomes simply

$$\phi_f = \frac{1}{2[1+(T/\varepsilon)\partial\varepsilon/\partial T]}\langle\phi\rangle = \langle\phi\rangle f' \tag{267}$$

and

$$f' = \frac{1}{2[1+(T/\varepsilon)\partial\varepsilon/\partial T]}$$

where $\partial\varepsilon/\partial T$ is the temperature gradient of the medium permittivity. In the case of the interactions in a vacuum $f' = \frac{1}{2}$.

Eq. (266) predicting that the interaction energy vanishes with the distance in the six power, describes the Keesom orientational forces. For polar substances such as water, it may contribute up to 80% of total interactions [85,86].

In the same way, interactions of higher order multipoles, e.g., quadrupoles, can be considered. They play a significant role in interactions of symmetric molecules without permanent dipole moment like hydrogen. Although the interaction energy vanishes at a much faster rate than for dipoles, i.e., r_0^{-8}, this can have a significant contribution to the virial coefficients of non-polar gases [87]

2.3.2. Induced Dipole Interactions – Debye Forces

The second important contribution to the intermolecular forces stems from polarization effects discussed before by analyzing the behavior of dielectric media. The main postulate of this dielectric media theory was the linear correspondence between the polarization vector $\mathbf{P}_e$ (defined as the induced dipolar moment per unit volume) and the electric field in the medium (cf. Eq. (54)).

An analogous result was obtained by solving the problem of a dielectric sphere or spheroid placed in an uniform field. It was shown that in both cases the dipole moment generated by the uniform electric field was given by the expression

$$\mathbf{p} = \varepsilon\alpha\mathbf{E} \tag{268}$$

where the polarizability α, with the dimension of m^3, was found proportional to the volume of the particle placed in the field.

It was demonstrated by Debye [88] that Eq. (268) is a valid approximation when dealing with atoms and molecules as well. For non-polar molecules, the polarizability arises because of displacement of electron clouds relative to nuclei under the action of the external field. Hence, it is referred to as the electronic polarizability. It often remains proportional to the volume of the molecule [85]. For water, $\alpha = 1.48 \times 4\pi \times 10^{-30}$ m^3 [85], and for CH_3OH $3.2 \times 4\pi \times 10^{-30}$ m^3 [85].

Apart from the electronic polarizability due to the induction effect, in the case of polar molecules, an orientational component also appears. It arises because of the Boltzmann statistics, which prefers the dipole orientations along the electric field. Using Eq. (264), it can be calculated that the orientational polarizability is given by the expression

$$\alpha_{0r} = \frac{p^2}{3\varepsilon kT} \tag{269}$$

Hence, the total polarizability becomes

$$\alpha = \alpha_0 + \frac{p^2}{3\varepsilon kT} \tag{270}$$

Eq. (270) is known in the literature under the name of the Debye– Langevin equation.

Taking as an example water, with the permanent dipole moment of $1.85 \times 3.336 \times 10^{-30}$ C, one can calculate that for the temperature of 293 K, the orientational component of the polarizability will be 3.6×10^{-28} m^3, almost 20 times larger than the electronic polarizability component.

Knowing the polarizability, one can calculate the interaction energy for various fields from the formula derived earlier (cf. Eq. (102))

$$\phi_{di} = -\frac{1}{2}\alpha E^2 \tag{271}$$

For example, in the case of ion uncharged molecule interaction, by using the expression for the field generated by the ion, one obtains the formula

$$\phi = -\frac{1}{2}\left(\frac{q}{4\pi\varepsilon}\right)^2\left(\alpha_0 + \frac{p^2}{3kT\varepsilon}\right)\frac{1}{r_0^4} \tag{272}$$

As can be noticed, the energy does not depend on the orientation of the dipole.

For permanent/induced dipole interactions, using Eq.(24) for the square of the field one obtains the formula

$$\phi_{\text{ind}} = -\frac{1}{2}\ \varepsilon\alpha_2 \mathbf{E}_1 \cdot \mathbf{E}_1 = -\frac{\varepsilon\alpha_{0_2} p_1^2 (3\cos^2\vartheta + 1)}{2(4\pi\varepsilon)^2 r_0^6} \tag{273}$$

One can notice that the interaction energy of this system remains negative (attraction) for all orientations of the permanent dipole and decreases as the sixth power of the distance r_0. As expected, the maximum value of the interaction energy appears for the in-line orientation of dipoles.

It can be estimated that for water molecules being in contact when $r_0 = 2.8 \times 10^{-10}$ cm, the maximum interaction energy equals 2×10^{-21} J, i.e., 0.5 kT at a room temperature. This is much smaller than the previous values estimated for dipole/ion and dipole/dipole interactions. This means that the orientational distribution of molecules will be quasi-continuous, so the averaging of ϕ over all angles gives

$$\langle\phi_{\text{ind}}\rangle = -\frac{\varepsilon\alpha_{0_2} p_1^2}{(4\pi\varepsilon)^2 r_0^6}\frac{1}{2\pi}\int_0^{\pi}(3\cos^2\vartheta + 1)\sin\vartheta d\vartheta = -\frac{\varepsilon\alpha_{0_2} p_1^2}{(4\pi\varepsilon)^2 r_0^6} \tag{274}$$

As can be noticed, in contrast to the situation with two permanent dipoles, the averaged energy does not vanish. In consequence, this can be responsible for a significant part of polar molecule interactions.

In the case of two different molecules having permanent dipole moments p_1 and p_2, mutual induction occurs, so the angle averaged interaction energy is given by

$$\langle\phi_{ind}\rangle = -\frac{\varepsilon\left(p_1^2\alpha_{0_2} + p_2^2\alpha_{0_1}\right)}{(4\pi\varepsilon)^2}\frac{1}{r_0^6} \tag{275}$$

Eq. (275) describes the Debye interactions, called also the induction interaction.

It is interesting to mention that except for dipolar interactions, a contribution to the molecular potential may arise from quadrupole and higher multipole interactions. The molecular quadrupoles Q_1, Q_2 give the following contribution to the interaction energy [1]:

$$\langle \phi_Q \rangle = -\frac{3}{2}\frac{\varepsilon\left(Q_1^2\alpha_{0_2} + Q_2^2\alpha_{0_1}\right)}{(4\pi\varepsilon)^2 r_0^8} \tag{276}$$

As can be seen, these interactions vanish with the distance at a much faster rate than the dipolar interactions: proportionally to r_0^{-8}.

We can summarize the results obtained in this section by formulating the general equation describing intermolecular interactions in the form

$$\phi_{dip} = -\frac{C_{dip_1}}{(4\pi\varepsilon)^2 r_0^6} - \frac{C_{dip_2}}{(4\pi\varepsilon)^2 r_0^8} \tag{277}$$

where the constants describing the Keesom orientational and Debye induction interactions are given by the expressions

$$\begin{aligned} C_{dip_1} &= \frac{2}{3} f' \frac{p_1^2 p_2^2}{kT} + \varepsilon\left(p_1^2 \alpha_{0_1} + p_2^2 \alpha_{0_2}\right) \\ C_{dip_2} &= \frac{3}{2}\varepsilon\left(Q_1^2\alpha_{0_1} + Q_2^2\alpha_{0_2}\right) \end{aligned} \tag{278}$$

For the sake of convenience, various energy expressions derived for dipolar interactions are compiled in Table 2.11.

With the dipole-induced dipole or multipole interactions, one approaches the limit of classical electrostatics, which does not predict any interactions in a system when there are no charges or permanent dipoles present. Hence, the molecular dispersion interactions observed for symmetric molecules devoid of any dipole moment, like rare gases, methane, etc., can only be explained within the framework of quantum mechanics as discussed next.

2.3.3. The Dispersion Interactions – London Forces

Because the subject of dispersion interactions has been extensively treated in many excellent monographs [1,85,89] and review articles [78,86,90], it will be covered here in a rather condensed way.

The dispersion interactions play the most important role because they occur universally for all material particles, even those devoid of any permanent dipolar or multipolar moment. They originate from spontaneous appearance of dipoles in atoms or molecules, governed by the Heisenberg's uncertainty principle. It states that it is impossible to know the product of the momentum and the position of any particle with an accuracy better that $h/4\pi$ (where h is the Planck constant equal to 6.626×10^{-34} J s). For an electron this can be written in the simple form as

$$\Delta p_m = m\,\Delta V > h_p/4\pi\,\Delta x \tag{279}$$

where Δp_m is the uncertainty of the electron momentum, ΔV the uncertainty of its velocity and Δx the uncertainty of its position.

Therefore, if we confine the electron, for example to the space within the atomic distance a_0 (where $a_0 = e^2/2h\tilde{v}_e 4\pi\varepsilon_0$ is the Bohr radius of hydrogen (see Table 2.1) and $\tilde{v}_e$ is the characteristic frequency) its momentum must increase accordingly, setting the electron in permanent motion. The system made up of an electron in an instantaneous position and the nucleus (proton) forms a fluctuating dipole with the dipolar moment of the order of $\underline{e}a_0$ equal to 1.6×10^{-19} C $\times$ 0.53×10^{-10} m $= 8.5\times10^{-30}$ C m, i.e., 2.5 D, which is larger than the water molecule permanent dipole moment. Such a dipole will polarize other atoms in its vicinity leading to mutual attraction described by Eq. (274), which can be expressed in this case as

$$\phi = -\frac{(a_0 e)^2\,\alpha_0\varepsilon}{(4\pi\varepsilon)^2}\frac{1}{r_0^6} \tag{280}$$

By substituting $\alpha_0 = 4\pi a_0^3$, one can express Eq. (280) as

$$\phi = -\frac{h_p\tilde{v}_e\alpha_0^2}{(4\pi\varepsilon)^2 r_0^6} \tag{281}$$

As can be seen, the interaction energy is negative (attraction) and falls as r_0^6, similarly to the orientational and inductive contributions.

This estimate seems quite crude, though. However, quantum chemical calculations of London [91, 92] based on the perturbation method showed that Eq.(281) can be quite an accurate approximation. In particular, it was demonstrated that the numerical coefficient in Eq.(281) is 3/4 instead of 2. It was also shown that interactions of two different atoms with characteristic field adsorption frequencies ν_{e1}, ν_{e2} (in the visible region) can be approximated by the formula

$$\phi = -\frac{3}{2}\frac{\alpha_{0_1}\alpha_{0_2}}{(4\pi\varepsilon)^2}\frac{h_p\tilde{\nu}_{e_1}\tilde{\nu}_{e_2}}{(\tilde{\nu}_{e_1}+\tilde{\nu}_{e_2})}\frac{1}{r_0^6} = -\frac{3}{2}\frac{\alpha_{0_1}\alpha_{0_2}}{(4\pi\varepsilon)^2}\frac{I_{0_1}I_{0_2}}{(I_{0_1}+I_{0_2})r_0^6} = -\frac{C_{disp_1}}{r_0^6} \tag{282}$$

where $I_{0_1} = h_p\tilde{\nu}_{0_1}$, $I_{0_2} = h_p\tilde{\nu}_{0_2}$ are the first ionization potentials of atoms or molecules and C_{disp_1} is the dispersion interaction constant.

Because I_0 and the polarizabilities of many atoms and compounds are fairly well known [1,85], Eq. (282) can be used for estimating their interactions in a vacuum.

It should be mentioned, however, that Eq. (282) underestimates dispersion interactions because contributions stemming from multipolar interactions were neglected in its derivation. More accurate calculations [1,85] showed that intermolecular potentials can be better approximated by series expansion such as

$$\phi_{disp} = -\frac{C_{disp_1}}{r_0^6} - \frac{C_{disp_2}}{r_0^8} - \frac{C_{disp_3}}{r_0^{10}} \tag{283}$$

Values of the constants C_{disp_1}, C_{disp_2} calculated for some common atoms and molecules are compiled in Table 2.12. As can be noticed, the contribution stemming from higher order terms (describing the effect of multipoles) remains comparable with the leading term for distances below 10 a_0. For example in the case of atomic hydrogens at the distance $r_0 = 5a_0 = 2.6\times10^{-10}$ m, the first term gives the interaction energy -1.8×10^{-21} J $= -0.44\,kT$. The second term in Eq. (283) gives $\phi_{disp} = -5.1\times10^{-21}$ J $= -1.3\,kT$. As can be seen the interaction energy assumes quite appreciable values. However, it falls abruptly with the distance, becoming just $-1.2\times10^{-2}\,kT$ at $r_0 = 10\,a_0$ (5.3×10^{-10} m). For methane, one can calculate using the data

Table 2.12
Dispersion energy constants for various molecule pairs

Molecular systems	C_{disp_1} J m^6×10^{79}	C_{disp_2} J m^8×10^{99}
H–H	6.2	124
H_2–H_2	11.4	31.1
He–He	1.47	14.2
N_2–N_2	57.5	119
O_2–O_2	40.3	96
CH_4–CH_4	113	310
H_2O–H_2O	33	—
Li–Li	500	6430
Na–Na	817	11 000
K–K	1630	26 700

From Ref. [1], Chapters 2–4.

from Table 2.12 that the interaction energy equals $-0.68\ kT$ at $r_0 = 4\times10^{-10}$ m, i.e., when the molecules are in contact) when the first term is considered. The second term gives the correction of 17%. At a distance twice as large the energy becomes practically negligible (0.01 kT). Even for sodium atoms, with exceptionally large values of C_{disp_1}, C_{disp_2}, (see Table 2.12), the energy of interactions at $r_0 = 10^{-9}$ m becomes negligible.

This comparison indicates quite unequivocally that the dispersion interactions between atoms and single molecules are exceptionally short-ranged: practically all measurable effects occur at distances smaller than 5×10^{-10} m.

Even with the correction for multipole interactions, the London theory has other shortcomings such as:

(i) the assumption of only one radiation absorption frequency whose effect is described by the static value of the atomic polarizability,
(ii) the interaction in a continuous medium (water) cannot be properly handled.

It seems that the macroscopic theory of dispersion interactions formulated by Lifshitz et al. [93–94], treating the interacting objects as continuous phases, is more suitable for evaluating interactions of particles dispersed in solvents [95].

Another problem that was not considered in the London theory of dispersion forces is the retardation effect stemming from the finite speed of electromagnetic waves in a medium. Hence, this is, in principle, a relativistic effect. The electric field produced by the fluctuating dipole of an atom reaches it again upon reflection from other atoms after some finite time, so a phase shift occurs between them. This results in a decrease in the induced dipole strength and consequently in the energy of interaction. The retardation effect appears for larger distances only, comparable with the characteristic wavelength λ_r being of the order 100 nm. Thus, it is expected to play a role in larger molecules or colloid particle interactions. It was shown in [96] that the retardation effect can be accounted for by the expression valid for $r/\lambda_r > 0.5$

$$\phi = -\frac{C_{disp1}}{r_0^6}\left[\frac{2.45\lambda_r}{2\pi r_0} - \frac{2.17\lambda_r^2}{4\pi^2 r_0^2} + \frac{0.59\lambda_r^3}{8\pi^3 r_0^3}\right] \tag{284}$$

At larger distances, the leading term obviously dominates and one arrives at the formula derived originally by Casimir and Polder [97].

On the other hand, for very short distances, comparable with atom or molecule dimensions, repulsive interactions stemming from overlapping of electron clouds of most external orbitals occur. Despite extensive efforts, no sound quantum chemical theory for describing these interactions occurred. It is usually assumed that they increase upon decreasing the distance between molecules as r_0^{-12} [1]. In combination with the London expression, Eq. (282) the so-called Lennard–Jones (6–12) molecular potential is obtained, given by the expression

$$\phi = -\frac{C_{disp_1}}{r_0^6} + \frac{C_L}{r_0^{12}} \tag{285}$$

Eq. (285) with the C_L coefficient determined in a semiempirical way has been widely used in statistical mechanics to calculate virial coefficients, structure factors, phase transitions etc. [98–100].

It seems, however, that in view of the ambiguity in defining the intermolecular distance, especially in the case of non-spherical particles, it is conceptually more useful to postulate that the repulsive branch of the intermolecular potential is simply the hard particle potential. This means that upon approaching a critical distance, usually called the van der Waals distance, the interaction energy tends to plus infinity and particles become totally impenetrable to each other.

The dispersion (London) interactions are the dominating contribution to the overall molecular interactions for the non-polar or slightly polar substances like hydrocarbons. On the other hand, for polar molecules, especially water, alcohols, etc., the orientation and induction effects described by Eq. (277) play an essential role. By assuming that the dipolar and dispersion interactions are additive one can express the net (van der Waals) interaction potential in the form

$$\phi_a = \phi_{dip} + \phi_{disp} = -\sum_n \frac{C_n}{r_0^n} \tag{286}$$

where $C_n = C_{dip_n} + C_{disp_n}$.

This form of intermolecular potential is particularly suitable for evaluating interactions of macrobodies (particles) by using the pairwise summation method discussed next.

2.3.4. van der Waals Interactions of Macrobodies – Hamaker Theory

In the classical microscopic approach developed by Hamaker [101], one exploits the energy additivity principle to calculate interaction energy of macroscopic particles from solutions obtained for atoms and molecules. A major advantage of the microscopic approach presented in this section is that quite handy analytical expressions can be derived for complicated geometries of the interacting particles, including the case of rough surfaces. On the other hand, a shortcoming of this method is that for condensed phases the additivity principle is an approximation only, because the permanent and induced dipole position are not fixed. On the contrary, they tend to stay in the most thermodynamically favorable orientations. This exerts a feedback effect because two interacting particles affect other particles in their vicinity, which in turn affects interactions of the former and so on. This situation is not unlike the image reflection method discussed earlier for electrostatic interactions of a charged sphere with metallic interfaces.

According to the additivity principle, the interactions of two bodies 1 and 2 (see Fig. 2.31) can be calculated by summing them up in a pair for all atoms (molecules) similarly as it was done for electrostatic interactions (cf. Eq. (18)).

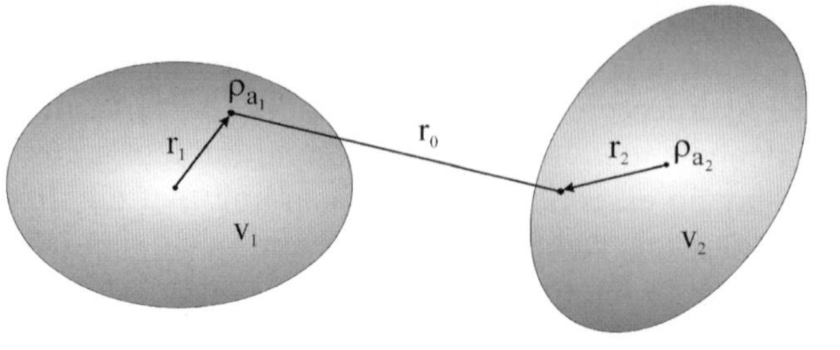

Fig. 2.31. Scheme of calculating van der Waals interactions for two macrobodies via the pairwise summation procedure.

For macrobodies containing many atoms, the summation procedure can be replaced with integration introducing the density of atoms (similarly as done previously by introducing the charge density). In the general case, these densities are position-dependent quantities (functions of the space variables in both bodies). Accordingly, the general expression for interaction of two particles of arbitrary shape can be formulated as the volume integral

$$\phi = \int_{v_1}\int_{v_2} \phi_a \rho_{a_1} \rho_{a_2} dv_1 dv_2 = -\sum_n C_n \int_{v_1}\int_{v_2} \frac{\rho_{a_1}(r_1)\rho_{a_2}(r_2)}{r_0^n} dv_1 dv_2 \tag{287}$$

where ρ_{a_1}, ρ_{a_2} are the number densities of atoms (molecules) in the two bodies involved, and v_1, v_2 are the volumes of the bodies.

Various cases of great practical interest can be derived from Eq. (287) by evaluating the volume integrals. Let us start from the atom interactions with interfaces of various shapes with uniform densities ρ_{a_1}. If the interface has the form of a circular disk of radius R and the infinitesimal thickness dh (see Table 2.13), the volume element becomes $2\pi r dr dh$ (where r is the radial distance from the disk center) and the interaction energy is given by

$$\phi = -2\pi\rho_{a_1} dh \sum_n \frac{C_n}{n-2}\left[\frac{1}{h^{n-2}} - \frac{1}{(h^2+R^2)^{\frac{n}{2}-1}}\right] \tag{288}$$

where h is the distance of the atom from the disk surface.

It is seen that the second term in Eq. (288) can be neglected if $R/h > 1$. For example, taking $n = 6$ and $R/h = 2$, the second term in Eq. (288) is

Table 2.13
General van der Waals energy expressions for various configurations

Configuration	Energy expression	Leading term	Force expression (leading term)
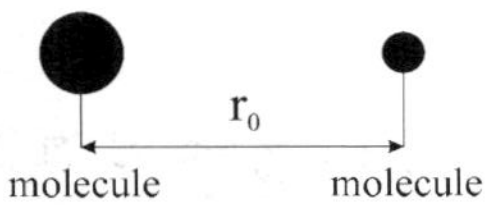	$-\sum_n \frac{C_n}{r_0^n}$	$-\frac{C_1}{r_0^6}$	$-\frac{6C_1}{r_0^7}$
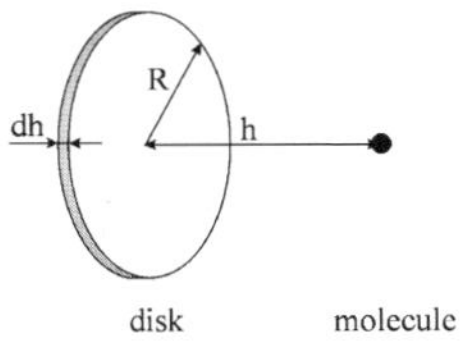	$-2\pi\rho_{a_1}\, dh \sum_n \frac{C_n}{(n-2)} \left[\frac{1}{h^{n-2}} - \frac{1}{(h^2+R^2)^{(n/2)-1}}\right]$	$-\frac{\pi}{2}\rho_{a_1}\, dhC_1 \left[\frac{1}{h^4} - \frac{1}{(h^2+R^2)^2}\right]$	$-2\pi\rho_{a_1}\, dhC_1 \left[\frac{1}{h^5} - \frac{h}{(h^2+R^2)^3}\right]$
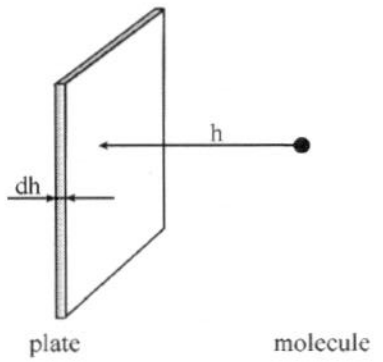	$-2\pi\rho_{a_1}\, dh \sum_n \frac{C_n}{(n-2)} \frac{1}{h^{n-2}}$	$-\frac{\pi}{2}\rho_{a_1}\, dh \frac{C_1}{h^4}$	$-2\pi\rho_{a_1}\, dh \frac{C_1}{h^5}$
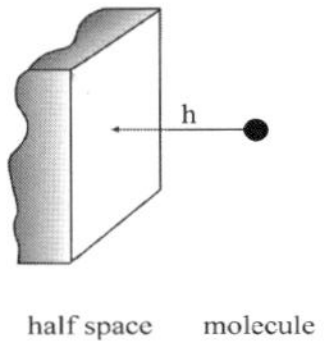	$-2\pi\rho_{a_1} \sum_n \frac{C_n}{(n-2)(n-3)} \frac{1}{h^{n-3}}$	$-\frac{\pi\rho_{a_1}}{6}\frac{C_1}{h^3}$	$\frac{\pi\rho_{a_1}}{2}\frac{C_1}{h^4}$

Table 2.13 (continued)

Configuration	Energy expression	Leading term	Force expression (leading term)
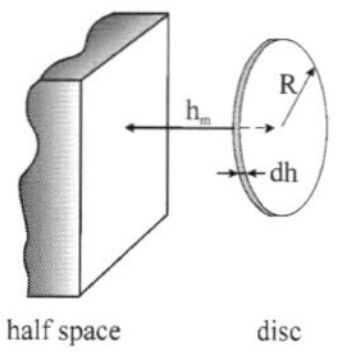	$-2\pi^2 \rho_{a_1}\rho_{a_2}\, dh\, R^2 \sum \frac{C_n}{(n-2)(n-3)} \frac{1}{h_m^{n-3}}$	$-\frac{1}{6}\pi^2 \rho_{a_1}\rho_{a_2}\, dh\, R^2 \frac{1}{h^3}$	$-\frac{1}{2}\pi^2 \rho_{a_1}\rho_{a_2}\, dh\, R^2 \frac{1}{h^4}$
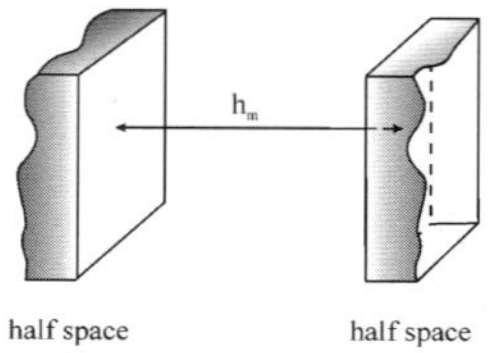	$-2\pi \rho_{a_1}\rho_{a_2} \sum \frac{C_n}{(n-2)(n-3)(n-4)} \frac{1}{h_m^{n-4}}$	$-\frac{1}{12}\pi \rho_{a_1}\rho_{a_2} \frac{C_1}{h_m^2}$	$-\frac{1}{6}\pi \rho_{a_1}\rho_{a_2} \frac{C_1}{h_m^3}$

only 11% of the leading term. This means that the disk can be treated as an interface of an infinite extension and the interaction energy is given by

$$\phi = -2\pi\rho_{a_1} dh \sum_h \frac{C_n}{(n-2)} \frac{1}{h^{n-2}} \tag{289}$$

By integrating Eq. (289) one can derive the expression for interaction of an atom with an infinite interface (foil) of thickness δ_{f_0}

$$\begin{aligned}\phi &= -2\pi\rho_{a_1} \sum \frac{C_n}{n-2} \int_{h_m}^{h_m+\delta_{f0}} \frac{dh}{h^{n-2}} \\ &= -2\pi\rho_{a_1} \sum_n \frac{C_n}{(n-2)(n-3)} \left[\frac{1}{h_m^{\,n-3}} - \frac{1}{(h_m+\delta_{f_0})^{n-3}} \right]\end{aligned} \tag{290}$$

where h_m is the distance between the atom and the disk surface.

If $\delta_{f_0} \to \infty$, Eq. (290) simplifies to

$$\phi = -2\pi\rho_{a_1} \sum_n \frac{C_n}{(n-2)(n-3)} \frac{1}{h_m^{n-3}} \tag{291}$$

This describes interaction of an atom with a half-space.

If one is interested in interactions of particles of arbitrary shape with the half-space, it is necessary to evaluate the second volume integral within body 2, characterized by the atom density ρ_{a2}. If the particle is a disk of arbitrary cross section area *S(h)* and the thickness *dh*, the interaction energy is given by

$$\phi = -2\pi\rho_{a_1}\rho_{a_2} \sum_n \frac{C_n}{(n-2)(n-3)} \frac{S(h)dh}{h^{n-3}} \tag{292}$$

For a circular disk with the radius *R(h)*, Eq. (292) becomes

$$\phi = -2\pi^2\rho_{a_1}\rho_{a_2} \sum_n \frac{C_n R^2(h)}{(n-2)(n-3)} \frac{dh}{h^{n-3}} \tag{293}$$

If the disk has a finite thickness L, one can integrate Eq. (293) obtaining the expression for a cylinder oriented perpendicularly to an interface

$$\phi = -2\pi\rho_{a_1}\rho_{a_2}\sum_n \frac{C_n}{(n-2)(n-3)(n-4)}\left[\frac{1}{h_m^{n-4}} - \frac{1}{(h_m+L)^{n-4}}\right] \tag{294}$$

Finally, when the length L and radius R of the cylinder tend to infinity, Eq.(294) can be transformed to the form describing the energy of interaction of two half-spaces per unit area

$$\Phi = \phi/\Delta S = -2\pi\rho_{a_1}\rho_{a_2}\sum_n \frac{C_n}{(n-2)(n-3)(n-4)}\frac{1}{h^{n-4}} \tag{295}$$

where ΔS is the surface area of the half space (plate).

It is interesting to mention that the coefficient $B_n = (n-2)(n-3)(n-4)$ equals 24 for $n = 6$, 60 for $n = 7$, 120 for $n = 8$, and 720 for $n = 12$ (for the sake of convenience, expressions for interaction energy for various configurations are compiled in Table 2.13).

Eq. (295) has major significance because by knowing interactions of half-spaces, one can calculate interactions of arbitrary shaped convex bodies by using the Derjaguin method described previously (see Eq. (227)). Thus, the van der Waals interaction energy of two convex bodies is given by the general expression

$$\begin{aligned}\phi &= -2\pi G_D \int_{h_m}^{\infty} \Phi(h)dh \\ &= -4G_D\pi^2\rho_{a_1}\rho_{a_2}\sum \frac{C_n}{(n-2)(n-3)(n-4)(n-5)}\frac{1}{h_m^{n-5}}\end{aligned} \tag{296}$$

where G_D is the geometrical factor shown for spheroids in Figs. 2.20 and 2.21 and in Table 2.9.

The interaction force (derivative of the energy upon the separation h) is given by the expression

$$F = -4G_D\pi^2\rho_{a_1}\rho_{a_2}\sum \frac{C_n}{(n-2)(n-3)(n-4)}\frac{1}{h_m^{n-4}} \tag{297}$$

the coefficients $B_n = (n-2)(n-3)(n-4)(n-5)$ occurring in Eqs. (296) are 24 for $n = 6$, 120 for $n = 7$ and 360 for $n = 8$.

Let us now consider some concrete cases that can be derived from the above formulae valid for an arbitrary intermolecular potential. For $n = 6$ (leading term in the power-series expansion, Eq. (286)), the half-space interaction energy per unit area is given by the expression

$$\Phi = -\frac{\pi \rho_{a_1} \rho_{a_2} C_1}{12 h_m^2} = -\frac{A_{12}}{12 \pi h_m^2} \tag{298}$$

where $A_{12} = \pi^2 C_1 \rho_{a_1} \rho_{a_2}$ is the Hamaker constant for two bodies 1 and 2 interacting across a vacuum.

Accordingly, the Derjaguin expression Eq. (296) assumes the simple form

$$\phi = -G_D \frac{A_{12}}{6} \frac{1}{h_m} \tag{299}$$

The force is given by the expression

$$F = -G_D \frac{A_{12}}{6} \frac{1}{h_m^2} \tag{300}$$

On the other hand, the interactions of a disk of infinitesimal thickness *dh* and the surface area *S(h)* with a half space in the case of $n = 6$ are described by the formula

$$\phi = -\pi \rho_{a_1} \rho_{a_2} \frac{C_1}{6} \frac{S(h)dh}{h^3} = -\frac{A_{12}}{6\pi} \frac{S(h)dh}{h^3} \tag{301}$$

Eq. (301) can be used for evaluating interactions of arbitrary-shaped bodies with planar interfaces if the dependence of its cross section area on the distance h is known. For example, for a spherical cup with thickness δ_c, $S(h)$ is given by the expression

$$S(h) = a^2 \left[1 - \left(\frac{h - h_m + a}{a} \right)^2 \right] \tag{302}$$

Using this the interaction energy of the cup with the half space upon integration of Eq. (301) can be evaluated as

$$\phi = -\frac{A_{12}}{6}\left\{ \ln\frac{h_m}{h_m+\delta_c} - (h_m+2a)\left[\frac{1}{h_m+\delta_c} - \frac{1}{h_m}\right] + \frac{1}{2}h_m(h_m+2a)\left[\frac{1}{(h_m+\delta_c)^2} - \frac{1}{h_m^2}\right]\right\} \tag{303}$$

When δ_c = 2a, one arrives at the case of the sphere/half-space interactions (see Table 2.14)

$$\phi = -\frac{A_{12}}{6}\left[\frac{2a(a+h_m)}{h_m(2a+h_m)} + \ln\frac{h_m}{(h_m+2a)}\right] = \phi_s(h_m, a) \tag{304}$$

If $h_m \ll a$, Eq. (304) reduces to the simple form

$$\phi = -\frac{A_{12}}{6}\frac{a}{h_m} \tag{305}$$

The interaction force is given by

$$F = -\frac{A_{12}}{6}\frac{a}{h_m^2} \tag{306}$$

Eqs. (305), (306), which can also be derived from Eqs. (299), (300), using the Derjaguin method have major significance and are widely used for calculating particle interactions with interfaces as well as for calculating adhesion forces.

Because of the frequently occurring flattening of adhering particles [78], it may be interesting to calculate interactions of a flattened sphere (see Table 2.14). In this case the integration of Eq. (301) gives

$$\begin{aligned}\phi = &-\frac{A_{12}}{6}\left\{ \ln\frac{h_m}{H_m+a} + \frac{H_m(2a-\delta_f)}{h_m(H_m+a)} + (2a-\delta_f)(a^2-H_m^2)\frac{(H_m+h_m+a)}{2h_m^2(H_m+a)^2}\right\} \\ &= \phi_{sf}(h_m, a, \delta_f)\end{aligned} \tag{307}$$

Table 2.14
van der Waals energy expressions for various macrobodies $A_{12} = \pi^2 \rho_{a_1} \rho_{a_2} C_1$ (Hamaker constant for interactions in vacuum)

Configuration	Energy (leading term)	Energy expression short separations	Energy expression Derjaguin
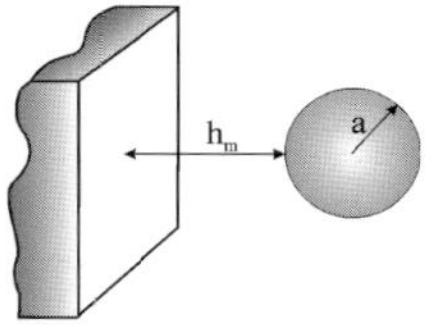	$-A_{12}\left[\frac{2a(h_m+a)}{h_m(h_m+2a)}+\ln\frac{h_m}{h_m+2a}\right] = \phi_s(h_m,a)$	$-\frac{A_{12}a}{6h_m}$	$-\frac{A_{12}a}{6h_m}$
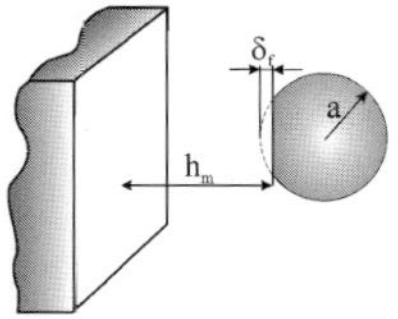 half space flattened sphere	$-\frac{A_{12}}{6}\left\{\ln\frac{h_m}{H_m+a}+\frac{H_m(2a-\delta_f)}{h_m(H_m+a)} +(2a-\delta_f)(a^2-H_m^2)\frac{(H_m+h_m+a)}{2h_m^2(H_m+a)^2}\right\} = \phi_{sf}(h_m,a,\delta_f)$, $H_m = h_m + a - \delta_f$	$-\frac{A_{12}a}{6h_m}\left(1+\frac{\delta_f}{h_m}\right)$	
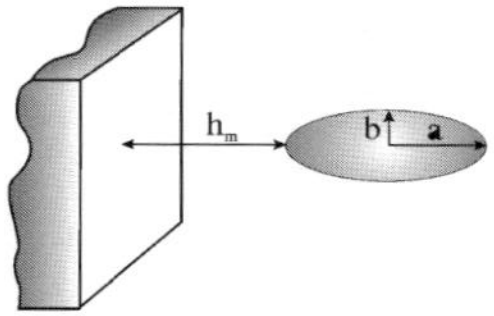 half space prolate spheroid	$\frac{b^2}{a^2}\phi_s(h_m,a)$	$-\frac{A_{12}b^2}{6h_m a}$	$-\frac{A_{12}b^2}{6h_m a}$

Table 2.14 (continued)

Configuration	Energy (leading term)	Energy expression short separations	Energy expression Derjaguin
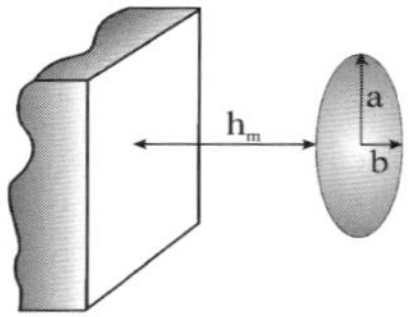 half space, prolate spheroid	$\frac{a}{b}\phi_s(h_m,b)$	$-\frac{A_{12}a}{6h_m}$	$-\frac{A_{12}a}{12h_m^2}$
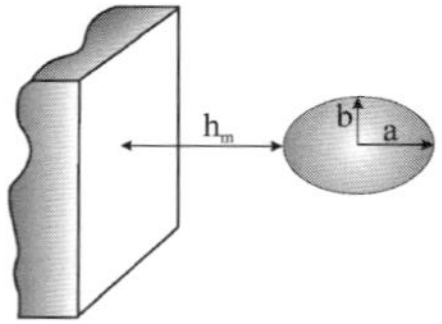 half space, oblate spheroid	$\frac{b}{a}\phi_s(h_m,a)$	$-\frac{A_{12}b}{6h_m}$	$-\frac{A_{12}b}{6h_m}$
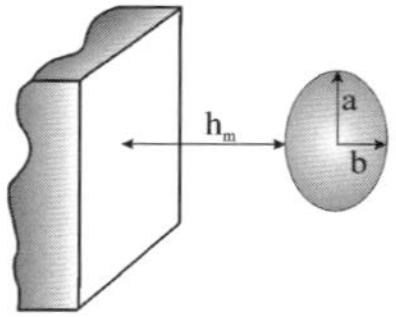 half space, oblate spheroid	$\frac{a^2}{b^2}\phi_s(h_m,b)$	$-\frac{A_{12}a^2}{6h_m b}$	$-\frac{A_{12}a^2}{6h_m b}$

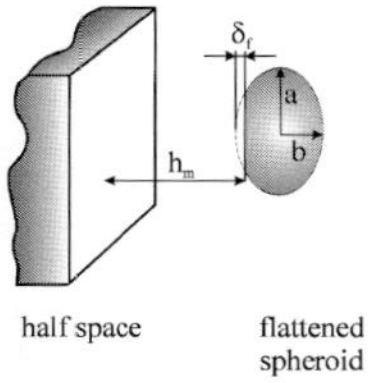

$$\frac{a^2}{b^2}\phi_{sf}(h_m,b,\delta_f)$$

$$-\frac{A_{12}a^2}{6h_m b}\left(1+\frac{\delta}{h_m}\right)$$

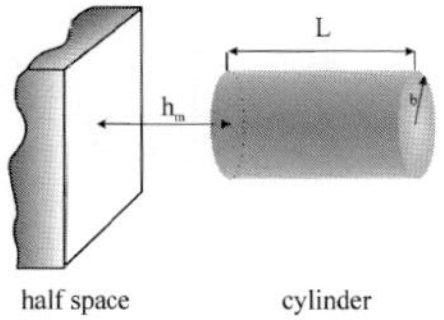

$$-\frac{A_{12}b^2}{12h_m^2}\left[1-\left(\frac{h_m}{L}\right)^2\right]$$

$$-\frac{A_{12}b^2}{12h_m^2}$$

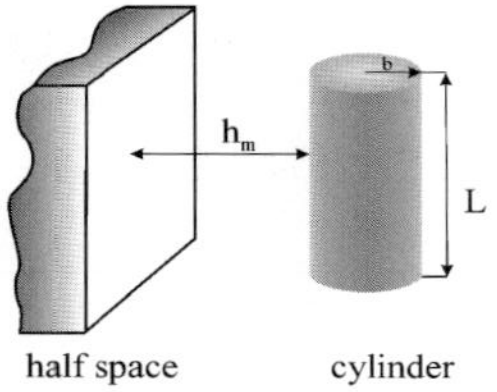

$$\frac{-A_{12}b^2L}{3\pi\left[h_m(h_m+2b)\right]^{3/2}}\left[\tan^{-1}\frac{h_m+2b}{\left[h_m^{\frac{1}{2}}(h_m+2b)^{\frac{1}{2}}\right]}+\tan^{-1}\frac{h_m}{\left[h_m^{\frac{1}{2}}(h_m+2b)^{\frac{1}{2}}\right]}\right]$$

$$-\frac{A_{12}b}{6h_m}\frac{L}{\left(h_m^{1/2}b^{1/2}2^{3/2}\right)}$$

Table 2.14 (continued)

Configuration	Energy (leading term)	Energy expression short separations	Energy expression Derjaguin
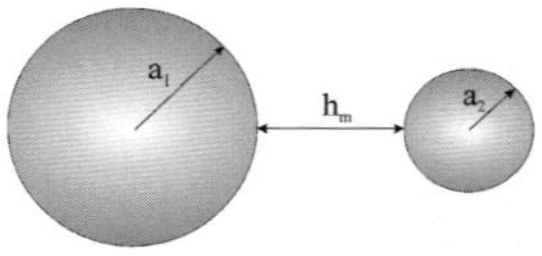 sphere sphere	$-\frac{A_{12}}{6}\left[\frac{2a_1a_2}{h_m(h_m+2a_1+2a_2)} + \frac{2a_1a_2}{(h_m+2a_1)(h_m+2a_2)} + ln\frac{h_m(h_m+2a_1+2a_2)}{(h_m+2a_1)(h_m+2a_2)}\right] = \phi_{ss}(h_m,a_1,a_2)$	$-\frac{A_{12}}{6h_m}\left(\frac{a_1a_2}{a_1+a_2}\right)$	$-\frac{A_{12}}{6h_m}\left(\frac{a_1a_2}{a_1+a_2}\right)$
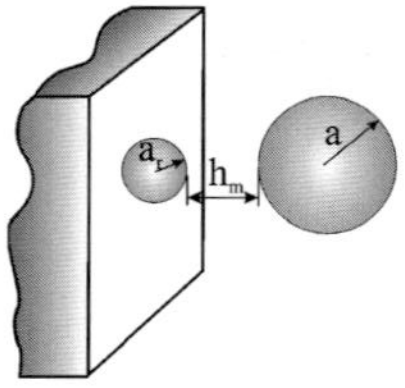 rough surface sphere	$\phi_s(h_m+2a_r,a)$ $\phi_{ss}(h_m,2a_r,a)$	$-\frac{A_{12}a}{6(h_m+2a_r)}$ $-\frac{A_{r_2}}{6h_m}\left(\frac{a_ra}{a_r+a}\right)$	

where $H_m = h_m + a - \delta_f$, and δ_f is the thickness of the flattened area (assumed to have the shape of a spherical cup).

If the distance of the sphere from the interface becomes much smaller than its radius and $\delta_f < a$, Eq. (307) reduces to the simple form

$$\phi = -\frac{A_{12}}{6}\frac{a}{h_m}\left(1+\frac{\delta_f}{h_m}\right) \tag{308}$$

The interaction force is given by

$$F = -\frac{A_{12}}{6}\frac{a}{h_m^2}\left(1+\frac{2\delta_f}{h_m}\right) \tag{309}$$

These formulae indicate that the adhesion energy or force may increase many times even for a small flattening. For typical values of h_m = 1 nm (10^{-8} m) and a = 100 nm, a 1% flattening increases the energy twice, a 5% flattening 6 times, etc.

Using Eq. (301), one can also evaluate interactions of spheroids, both prolate and oblate by simple integration. For a prolate spheroid oriented with its longer axis perpendicularly to an interface (see Table 2.14) with the cross section area $S(h) = b^2\,[1-(h-h_0)^2/a^2]$ (where $h_0 = h_m + a$ is the position of the spheroid center, a the longer axis and b the shorter axis. Integration of Eq. (301) with this expression for $S(h)$ gives

$$\phi = \frac{b^2}{a^2}\phi_s(h_m, a) \tag{310}$$

Eq. (310) representing an exact result, indicates that the interaction of a prolate spheroid with a flat interface is described by the same mathematical formula as that of a sphere, except for the geometrical prefactor b^2/a^2. Obviously, the same concerns the interaction force, which has the same scaling prefactor b^2/a^2.

For short separations, Eq. (310) reduces to the form

$$\phi = -\frac{A_{12}}{6h_m}\frac{b^2}{a} \tag{311}$$

This formula is identical to the Derjaguin expression, which can be derived from Eq. (299) by using the Derjaguin geometrical factor given in Table 2.9.

Analogously, the interaction energy of a flattened spheroid with an interface can be expressed, using Eq. (310), in the form

$$\phi = \frac{b^2}{a^2}\phi_{sf}(h_m, a, \delta_f) \tag{312}$$

In the case where a prolate spheroid is oriented parallel to the interface, the expression for the interaction energy becomes

$$\phi = \frac{a}{b}\phi_s(h_m, a) \tag{313}$$

For oblate spheroids with the shorter axis b perpendicular to the interface the energy is given by

$$\phi = \frac{a^2}{b^2}\phi_s(h_m, b) \tag{314}$$

For short separations, Eq. (314) becomes

$$\phi = -\frac{A_{12}}{6h_m}\frac{a^2}{b} \tag{315}$$

This is again identical to the Derjaguin formula.

It can be demonstrated by using Eq. (314) that interactions of oblate spheroids characterized by slight deviation from the spherical shape when $1 - b/a \ll 1$ (this would correspond to interactions of slightly deformed spheres), are given by the expression

$$\phi = \phi_s(h_m, a) + 0\left[1 - \left(\frac{b}{a}\right)\right]^2 \tag{316}$$

Eq. (316) suggests that small deformations of spheres (provided they are uniform not the flattening considered above) exert a second-order effect on their interactions with surfaces.

The interaction energy of an oblate spheroid with its shorter axis parallel to the interface is given by the formula

$$\phi = \frac{b}{a}\phi_s(h_m, a) \tag{317}$$

It is interesting to note that in all the above formulae, the asymptotic form of the interaction energy diverged to infinity proportionally to h_m^{-1} in the limit of $h_m \to 0$. Accordingly, the interaction force diverged as h_m^{-2}. Moreover, both quantities were proportional to the interacting particle dimension. Quite a different situation arises for cylindrically shaped particles. In the case of a cylinder oriented perpendicularly to the interface, the interaction energy is given by

$$\phi = -\frac{A_{12}b^2}{12h_m^2}\left[1 - \left(\frac{h_m}{L}\right)^2\right] \tag{318}$$

Thus, the energy diverges to minus infinity as h_m^{-2}, which means that attractive interactions of the cylinder with the same radius becomes much stronger at short separations. This again demonstrates the significance of the flattening effect.

When the cylinder is oriented parallel to the interface, the expression for the interaction energy becomes

$$\phi = -\frac{A_{12}b^2 L}{3\pi\left[h_m(h_m+2b)\right]^{3/2}} \left[\tan^{-1}\frac{h_m+2b}{\left[h_m(h_m+2b)\right]^{1/2}} + \tan^{-1}\frac{h_m}{\left[h_m(h_m+2b)\right]^{1/2}}\right] \tag{319}$$

Note that the Derjaguin method fails in the case of cylinder/interface interactions.

The same integration procedure can be applied for deriving exact formulae describing retarded interactions of particles with interfaces. For the sphere/half-space interactions, take Eq. (284) for the retarded potential which gives the exact result

$$\phi = -\frac{A_{12}}{\pi h_m^2}\left\{\frac{2.45\lambda_r}{60}\left[(a-h_m)+\frac{(h_m+3a)}{(h_m+2a)^2}\right]\right.$$
$$-\frac{2.17\lambda_r^2}{720\pi}\left[\frac{2a-h_m}{h_m}+\frac{(h_m+4a)h_m^2}{(h_m+2a)^3}\right] \tag{320}$$
$$\left.+\frac{0.59\lambda_r^3}{5040\pi^2}\left[\frac{3a-h_m}{h_m^2}+\frac{(h_m+5a)h_m^2}{4(h_m+2a)^4}\right]\right\}$$

Eq. (320), valid for distances $h_m > 100$ nm, can also be used for evaluating interactions of various spheroids by multiplying geometrical prefactors as demonstrated above.

Since Eq. (320) is rather cumbersome for direct use, e.g., in numerical simulations requiring repeated evaluations of interaction potential, Suzuki [102] derived an approximate equation having the simpler form

$$\phi = -\frac{A_{12}}{6}\frac{a}{h_m(1+11.11h_m/\lambda_r)} \tag{321}$$

The integration procedure can also be applied for deriving exact formulae describing interactions of two particles. The starting point could be, e.g., Eq. (301) describing interactions of an atom with a disk, then two disks, etc. For two dissimilar spheres, the integration method (performed in the spherical coordinate system) was first applied by Hamaker [101] who derived the formula

$$\phi = -\frac{A_{12}}{6}\left[\frac{2a_1a_2}{h_m(h_m+2a_1+2a_2)}+\frac{2a_1a_2}{(h_m+2a_1)(h_m+2a_2)}\right.$$
$$\left.+\ln\frac{h_m(h_m+2a_1+2a_2)}{(h_m+2a_1)(h_m+2a_2)}\right] \tag{322}$$

At short separations, when $h/a \ll 1$, Eq. (322) reduces to

$$\phi = -\left(\frac{a_1a_2}{a_1+a_2}\right)\frac{A_{12}}{6h_m} \tag{323}$$

This is again identical to the Derjaguin result because for two dissimilar spheres, G_D equals $a_1 a_2/(a_1 + a_2)$.

For more complicated geometries of particles like spheroids under arbitrary orientations, evaluation of exact results by integration becomes too complicated. In this case the Derjaguin method becomes especially useful because the geometrical factors are known for various orientations (see Table 2.9). The Derjaguin method seems adequate for calculating retarded interactions as well. By using the intermolecular potential given by Eq. (284), the interactions of arbitrary-shaped particles can be calculated from the formula

$$\phi = -G_D \frac{A_{12}}{\pi h_m^2}\left[\frac{2.45\lambda_r}{60} - \frac{2.17\lambda_r^2}{360\pi h_m} + \frac{0.59\lambda_r^3}{1680\pi^2 h_m^2}\right] \tag{324}$$

Another area of application of the Derjaguin method is the calculation of interactions of rough particles or interfaces. This problem has vital practical significance because all colloid particles, whether of natural origin or synthesized, exhibit an appreciable degree of surface heterogeneity of geometrical nature. In order to illustrate the significance of this effect, let us consider a planar interface with a small particle of the size δ_f attached to it (this can be of arbitrary convex shape) (see Table 2.14). The interaction of a larger particle of arbitrary shape with the interface bearing this model roughness element can be expressed according to the Derjaguin formula as

$$\phi = -G_D \frac{A_{12}}{6(h_m + \delta_r)} - G_D' \frac{A_{12}}{6h_m} \tag{325}$$

h_m is measured between the particle and the roughness of the size δ_r where G_D' is the Derjaguin factor for the interactions of a large particle with roughness, h_m the minimum separation distance between the particle and the roughness.

Using Eq. (325), one can calculate that the quotient of the interaction energy of the particle with rough to smooth surfaces at the same minimum separation distance from the surface is

$$\frac{\phi}{\phi_s} = \frac{h_m}{h_m + \delta_r} + \frac{G_D'}{G_D} \cong \frac{h_m}{h_m + \delta_r} \tag{326}$$

because $G_D' \ll G_D$

One can deduce from this simple formula that for $h_m = 1$ nm, a microroughness of the size of 1 nm will reduce the interaction energy of a larger sphere two times. For a 10 nm roughness, this reduction is 11 times. This means that the van der Waals interactions of rough particles are considerably reduced over smooth particles, which can prevent their efficient adhesion [77].

This problem has been studied in much detail in the literature [78,80–84,103–106]. Model rough surfaces were generated either by depositing spherical particles of polydisperse size distribution or by covering the smooth core with a rough shell of well-defined density [104,105]. Although exact results for arbitrary statistical distribution of microroughness were found rather complicated, Czarnecki [104] and Czarnecki and Dąbroś [105] have derived a simple interpolating function for ϕ_r valid for both sphere/sphere and sphere/plane interactions

$$\phi_r = \phi_s \left(\frac{h_m}{H_r} \right)^l \tag{327}$$

where ϕ_s is the energy for the smooth particle interface and the distance h_m is measured between the two outermost points at the particle surfaces, $H_r = h_m + (b_1 + b_2)/2$, b_1 and b_2 are the thicknesses of the rough layer at particles 1 and 2, respectively, and l is the exponent, close to 1 for the unretarded case and 1.5 for the retarded case [104,105]. Thus, in the limit $h_m \to 0$, Eq. (327) reduces to the simple form (for unretarded interactions)

$$\phi_r = \phi_s H_r \tag{328}$$

where $H_r = 2h_m/(b_1 + b_2)$ is the scaled distance between particle surfaces.

Hence, Eq. (328) has an analogous form as Eq. (326) derived by indicating that dispersion interactions between rough bodies are substantially reduced at all separations in comparison with smooth particles.

The appropriate expressions for the force of interaction can easily be derived from the above formulae by a simple differentiation with respect to the distance h_m.

2.3.5. Interactions in Dispersing Media, Hamaker Constant Calculations

The expressions presented above are strictly valid for interactions of particles across a vacuum (air) only. In this case, the Hamaker constant can be estimated because the data on atomic densities and the London constant C_1 are known for many substances [85]. Taking for example water, with $C_1 = 140\times10^{-79}$ J m^3 and $\rho_a = 10^3\times6.023\times10^{23}/0.018 = 3.3\times10^{28}$ m^3, the Hamker constant $A_{11} = \pi^2\rho^2 C_1$ becomes 1.5×10^{-19} J. For a hydrocarbon, e.g., hexane $A_{11} = 0.6\times10^{-19}$ J. Larger values are obtained for metal, e.g. potassium, having $C_{displ} = 1630$ and $\rho_a = 8.6\times10^2\times6.023\times10^{23}/0.039$, thus $A_{11} = 2.8\times10^{-19}$ J.

Using these data one can calculate the energy of interaction of colloid particles (drops) made of these substances. For two equal water drops having the radius 100 nm, one can calculate from Eq. (323) that the interaction energy at $h_m = 1$ nm equals -1.25×10^{-18} J $= -310\,kT$. This is a considerable value that could provoke droplet coalescence. Even at the distance $h_m =$ 200 (drop diameter), the energy is still considerable, exceeding $-1.55\,kT$.

However, for most of the experimentally relevant situations, interacting particles are immersed in a dispersing medium (most frequently water), which significantly modifies their van der Waals interactions. As shown already by Hamaker [101], the role of the intervening medium can be evaluated by again exploiting the energy additivity principle. Accordingly, by introducing the following notation:

ϕ_{13} the energy of interaction of the first particle with the medium (solvent)

ϕ_{23} the energy of interaction of the second particle with the medium (solvent)

ϕ_{12} the energy of interaction of the first particle with the second across a vacuum

ϕ_{33} the energy of interaction of the particles made of the solvent across a vacuum

one can formulate the energy balance for two particles 1,2 interacting in a medium 3 in the following form [101]:

$$\phi_{132} = \Delta\phi = \phi_{12} - \phi_{13} - (\phi_{23} - \phi_{33}) = \phi_{12} + \phi_{33} - \phi_{13} - \phi_{23} \qquad (329)$$

Eq. (329) is generally valid for arbitrary interactions, provided that they remain additive, i.e., when the interaction of the first particle with the medium remains unaffected by the interaction of the second particle.

In the case of the van der Waals interactions, the energies ϕ_{13} through ϕ_{33} can be expressed as

$$\phi_{ij} = -A_{ij}\, f(h_m, a) \qquad (330)$$

where A_{ij} are the Hamaker constants characterizing various interactions and $f(h_m, a)$ is the universal function depending solely on the geometry of the system (interparticle distance, particle shape and size). By considering this, one can express Eq. (329) in the form

$$A_{132} = A_{12} + A_{33} - A_{13} - A_{23} \tag{331}$$

where A_{132} is the composite Hamaker constant describing interactions of particle 1 with particle 2 across the medium denoted by the index 3.

In the case of two bodies of the same material, Eq. (331) simplifies to the form

$$A_{131} = A_{11} + A_{33} - 2A_{13} \tag{332}$$

It can be shown, as first done by Hamaker [101], that in the case when the overall van der Waals interactions are dominated by the London dispersion interactions described by Eq. (283) the following inequality is fulfilled:

$$A_{131} \geq 0 \tag{333}$$

In this case, the interaction of two particles of the same material always remains attractive.

However, in the case of different particles, this statement does not need to be true and the A_{132} constant can become theoretically negative, which means that repulsive dispersion interactions between particles may occur.

It should be mentioned, however, that the additivity assumption pertinent to the Hamaker theory in general, and Eqs. (331), (333) in particular has certain limitations, especially when dealing with condensed phases. It is expected that the van der Waals interactions will be underestimated. This is so because the fluctuating electric field generated within the interacting pair of molecules (atoms) is reflected from other molecules that enhances the pair interactions. In principle this effect can be accounted for by considering the multipole interactions governed by additional terms in Eq. (283). However, these constants are known for a limited number of pairs of different molecules.

These many body effects have been considered in a more universal way in the so-called macroscopic theory of van der Waals interactions developed by Lifshitz [93] and Dzialoshinsky and Lifshitz [94]. Since their approach is based on quantum electrodynamics rather than the electrostatic approach exposed above, it will not be analyzed in much detail here. Interested readers are advised to consult the excellent reviews [78,90,107] and books [85,89,108] devoted to this topic.

In the original macroscopic theory [93,94], the case of two planar interfaces (half spaces) separated by a third medium have been considered. It was assumed that interactions between these half spaces are the result of spontaneous electromagnetic fluctuations within the phases involved. These phases have been treated as continuous media characterized by isotropic (position independent) dielectric permittivity. However, the dielectric permittivity was treated as a complex function (having real and imaginary components) of the radiation frequency. The fluctuating fields propagate outside the phases in the form of traveling waves that decay exponentially away from the interfaces (analogously to the so-called evanescent waves). The discrete set of characteristic frequencies of these waves (eigen frequencies) has been calculated by solving the boundary value problem consisting of Maxwell's equations with the boundary conditions postulating the continuity of the tangential component of electric fields and discontinuity of perpendicular component governed by Eq. (66) (with the static permittivity replaced by the complex dielectric permittivity). After determining the set of characteristic frequencies, the interaction energy was calculated from the dispersion relation in terms of the imaginary frequency $i\tilde{\nu}$ (where i is the imaginary unit). In the general case, the energy formula derived was too cumbersome for a direct use [86,108]. However, a useful limiting form can be derived in the case of close separation between phases (but obviously larger than molecular dimensions in order to comply with the continuum phase assumption). The interaction energy between two half spaces is given in this limit by the expression analogous to that previously derived using the Hamaker theory, cf. Eq. (298)

$$\Phi_{132} = -\frac{A_{132}}{12\pi}\frac{1}{h_m^2} \tag{334}$$

where the composite Hamaker constant A_{132} can be explicitly evaluated from the formula [85]

$$A_{132} = \frac{3}{2}kT\sum_n \left(\frac{\varepsilon_1(i\tilde{\nu}_n) - \varepsilon_3(i\tilde{\nu}_n)}{\varepsilon_1(i\tilde{\nu}_n) + \varepsilon_3(i\tilde{\nu}_n}\right)\left(\frac{\varepsilon_2(i\tilde{\nu}_n) - \varepsilon_3(i\tilde{\nu}_n)}{\varepsilon_2(i\tilde{\nu}_n) + \varepsilon_3(i\tilde{\nu}_n}\right)$$

$$= \frac{3}{2}kT\sum_n \bar{\varepsilon}_{13}(i\tilde{\nu}_n)\bar{\varepsilon}_{23}(i\tilde{\nu}_n) \tag{335}$$

where $\varepsilon_1(i\tilde{\nu}_n)$, $\varepsilon_2(i\tilde{\nu}_n)$, $\varepsilon_3(i\tilde{\nu}_n)$ are the real components of dielectric permittivities of the three phases involved, which are functions of the imaginary frequency $i\tilde{\nu}_n$.

The advantage of Eq. (335) is that dielectric permittivities of many substances of practical interest are quite known from experiments [85]. Usually, these data can be fitted by approximate analytical expressions having the following form [85]:

$$\varepsilon\,(i\tilde{\nu}) = 1 + \frac{\epsilon - n^2}{(1+\tilde{\nu}/\tilde{\nu}_{rot})} + \frac{(n_r^2-1)}{(1+\tilde{\nu}^2/\tilde{\nu}_e^2)} \tag{336}$$

where ε is the static dielectric permittivity of a phase (in the limit of $\tilde{\nu} \to 0$), n_r the refractive index of the medium in the visible region, $\tilde{\nu}_{rot}$ the rotational relaxation frequency, typically in the microwave region, and $\tilde{\nu}_e$ is the main electronic adsorption frequency in the UV region. The second term in Eq. (336) describes the orientational (permanent dipole) relaxation for polar media, and the third term describes the electronic oscillation relaxation.

One can easily note that because $\tilde{\nu}_{rot}$ is usually of the order of $10^{12}\,\mathrm{s}^{-1}$ and $\tilde{\nu}_e$ of the order of $3\times10^{15}\,\mathrm{s}^{-1}$, so $\tilde{\nu}_e/\tilde{\nu}_{rot} \gg 1$, and the second term in Eq. (336) can be neglected. Moreover, the first electronic adsorption frequency, typically $4\times10^{13}\,\mathrm{s}^{-1}$ (in the infrared region), is much smaller than $\tilde{\nu}_e$ so one can replace the summing procedure in Eq. (335) by integration [85]. By assuming further that the characteristic frequency $\tilde{\nu}_e$ for the three media does not differ too much, one can derive the following expression for the Hamaker constant [80]:

$$A_{132} = \frac{3}{4}kT\left(\frac{\varepsilon_1-\varepsilon_3}{\varepsilon_1+\varepsilon_3}\right)\left(\frac{\varepsilon_2-\varepsilon_3}{\varepsilon_2+\varepsilon_3}\right) + \frac{3}{8}\frac{h\bar{\nu}_e}{(2)^{1/2}} \frac{\left(n_1^2-n_3^2\right)\left(n_2^2-n_3^2\right)}{\left[\left(n_1^2+n_3^2\right)\left(n_2^2+n_3^2\right)\right]^{1/2}\left[\left(n_1^2+n_3^2\right)^{1/2}+\left(n_2^2+n_3^2\right)^{1/2}\right]} \tag{337}$$

where n_1, n_2, n_3 are the refractive indices of the media.
As can be deduced from Eq. (337), in the case of three various phases (e.g., two particles in a solvent) A_{132} can in principle become negative (meaning repulsive van der Waals interactions), e.g., in the case when the static permittivities are: $\varepsilon_1 = 5$, $\varepsilon_2 = 4$, $\varepsilon_3 = 2$ and when the second term in Eq. (337) becomes much smaller than the first term.

For identical substances, Eq. (337) reduces to the form

$$A_{131} = \frac{3}{4} kT \left(\frac{\varepsilon_1 - \varepsilon_3}{\varepsilon_1 + \varepsilon_3} \right)^2 + \frac{3h_p \tilde{\nu}_e}{16(2)^{1/2}} \frac{(n_1^2 - n_3^2)^2}{(n_1^2 + n_3^2)^{3/2}} \tag{338}$$

As can be noticed from Eq. (338), for identical substances interacting through a solvent $A_{131} > 0$, which agrees with the previously derived result, Eq. (333). As discussed in [85], this means that a thin liquid film immersed in arbitrary fluid (e.g., air) will always exhibit a tendency to thinning under the action of van der Waals forces alone.

In the case of metals, when $\varepsilon\,(i\tilde{\nu}) = 1 + \tilde{\nu}_e^2/\tilde{\nu}^2$ (where $\tilde{\nu}_e$ is the characteristic frequency of the free electron gas)

$$A_{132} = \frac{3}{16(2)^{1/2}} h_p \tilde{\nu}_e \tag{339}$$

The A_{131} and A_{132} Hamaker constants for some common substances of practical interest are compiled in Table 2.15. Other values not available in the table can be calculated by exploiting the so-called combining relation [85] being, in principle, semi-empirical interpolating functions. Its form is suggested by Eq. (335), which can be formulated as

$$A_{132} = \frac{3}{2} kT \sum_n \overline{\varepsilon}_{13}(i\tilde{\nu}_n) \overline{\varepsilon}_{23}(i\tilde{\nu}_n) \cong \frac{3}{2} kT \sum_n \overline{\varepsilon}_{13}(i\tilde{\nu}_n) \overline{\varepsilon}_{23}(i\tilde{\nu}_n) \tag{340}$$

Considering that $\frac{3}{2} kT \sum_n \overline{\varepsilon}_{13}(i\tilde{\nu}_n) = (A_{131})^{1/2}$ and $\frac{3}{2} kT \sum_n \overline{\varepsilon}_{23}(i\tilde{\nu}_n) = (A_{232})^{1/2}$, Eq. (340) becomes

$$A_{132} = \pm \left(A_{131} A_{232} \right)^{1/2} \tag{341}$$

where the minus sign concerns the case where either of A_{131} or A_{232} becomes negative.

From Eq. (341) one also can deduce that

$$A_{13} = \left(A_{11} A_{33} \right)^{1/2} \tag{342}$$

Table 2.15
Calculated Hamaker constants for various systems

System	$A_{123}(10^{-20}$ J)
Water/air/water	3.7
n-pentane/air/*n*-pentane	3.8
Cyclohexane/air/cyclohexane	5.2
Ethanol/air/ethanol	4.2
Polystyrene/air/polystyrene	6.5
Polystyrene/water/polystyrene	1.4
PTFE/air/PTFE	3.8
PTFE/water/PTFE	0.33
Fused quartz/air/fused quartz	6.3
Mica/air/mica	10.0
Au/air/Au	25–40
Water/hydrocarbon/water	0.3–0.5
Quartz/water/quartz	0.83
Mica/water/mica	2.0
Au/air/Au	40
Au/water/Au	30
Ag/air/Ag	50
Ag/water/Ag	40

PTFE, – poly(tetra-fluoroethylene).
Data taken from Ref. [85, Chapter 11].

This relationship can be exploited to convert Eq. (332) into the useful form

$$\begin{aligned} A_{131} &= A_{11} + A_{33} - 2A_{13} \\ &\cong A_{11} + A_{33} - 2\left(A_{11}A_{33}\right)^{1/2} \\ &= \left[\left(A_{11}\right)^{1/2} - \left(A_{33}\right)^{1/2}\right]^2 \end{aligned} \tag{343}$$

Eq. (343) can be used for calculating interactions of particles composed of the same material across an arbitrary solvent when their interactions across a vacuum (air) are known.

In an analogous way one can express the composite Hamaker constant for interactions of two different particles across an arbitrary solvent when the corresponding interactions across vacuum (air) are known

$$A_{132} = \left[\left(A_{11}\right)^{1/2} - \left(A_{33}\right)^{1/2}\right]\left[\left(A_{22}\right)^{1/2} - \left(A_{33}\right)^{1/2}\right] \tag{344}$$

It should be mentioned, however, that the accuracy of Eq. (343) or (344) can be rather limited in the case of highly polar liquids like water [85]. In this case the use of Eq. (337) is more recommended. This is precisely the major advantage of the macroscopic approach that allows one to evaluate in a well-defined way the Hamaker constant for systems consisting of three different phases. However, the deficiency of the macroscopic approach is inherent inability to properly evaluate interactions for more complicated geometries than the two half-space case. For example, the sphere/plane or two sphere cases (the most important for colloid applications) have not been solved yet. Moreover, the assumption of homogeneous dielectric permittivity within each phase seems doubtful for separations between interacting bodies of the order of a nanometer. It seems, therefore, that for practical purposes the most efficient seems the hybrid approach when interaction formulae for complex geometries are derived from the microscopic theory whereas the Hamaker constant, characterizing their strength, is taken from the macroscopic theory.

2.4. SUPERPOSITION OF INTERACTIONS – ENERGY PROFILES

Once the electrostatic and van der Waals interactions for various particle/particle and particle/interface configurations are known, one can attempt to construct the overall interaction energy profile. This is a prerequisite for estimating colloid particle stability, adsorption, deposition and adhesion phenomena, which are of vital practical interest. An approach of appealing simplicity would be to treat these interactions as independent from each other and to construct the energy profile as a sum of electrostatic and van der Waals contributions. This was precisely the idea behind the DLVO theory (abbreviation from the Derjaguin and Landau [109] and Vervey and Overbeek [42] who are credited with its foundation) extensively used over the decades in the field of colloid sciences [100,108,110]. The sum of the electrostatic and van der Waals interactions is therefore, often called the DLVO potential, used as a reference potential. In consequence, all interactions except these two are referred to as non-DLVO interactions, examples being the Born repulsion, steric interactions due to adsorbed polymer layers, depletion interactions, hydrogen and chemical bonding, external forces like gravity or magnetic forces, hydrodynamic forces, etc.

Let us now analyze in a qualitative manner the DLVO energy profiles arising from the simple addition of electrostatic and van der Waals contributions. As seen from Table 2.15, for most known cases, the Hamaker constant is positive and confined within the range 10^{-21}–10^{-20}J (with the exception of metals when it becomes slightly larger). One can therefore deduce from equations derived in the previous section (see Table 2.14) that the van der Waals contribution to the interaction energy becomes significantly larger than a kT unit for particle/particle or particle/interface separations much smaller than particle dimensions, approximately 10–100 nm. For larger distances, the retardation effect makes these interactions negligible.

In contrast, the electrostatic interactions can be either positive or negative depending on surface potentials of particles, boundary conditions at their surfaces and on the separation distance. Additionally, their range can be varied between broad limits (1–1000 nm) by simply changing the ionic strength of electrolyte solutions (cf. Table 2.8). As a result, at separations comparable with particle dimensions, the electrostatic interactions dominate over the van der Waals contribution except for high ionic strength solutions. Because of the diversified range and magnitude of electrostatic interactions, the resulting energy profiles may become quite complicated, as discussed in Refs. [8,110]. However, for the sake of convenience, these profiles can be classified into some basic categories.

The Type I profiles (see Fig. 2.32) reflect interactions of similarly charged particles in not too concentrated electrolyte solutions when the overall interaction energy is dominated by the electrostatic interactions. In this case, the energy increases monotonically when the particles approach each other attaining values much higher than the kT unit for $h_m \to 0$. Appearance of this profile, typical for particle size above 100 nm, excludes

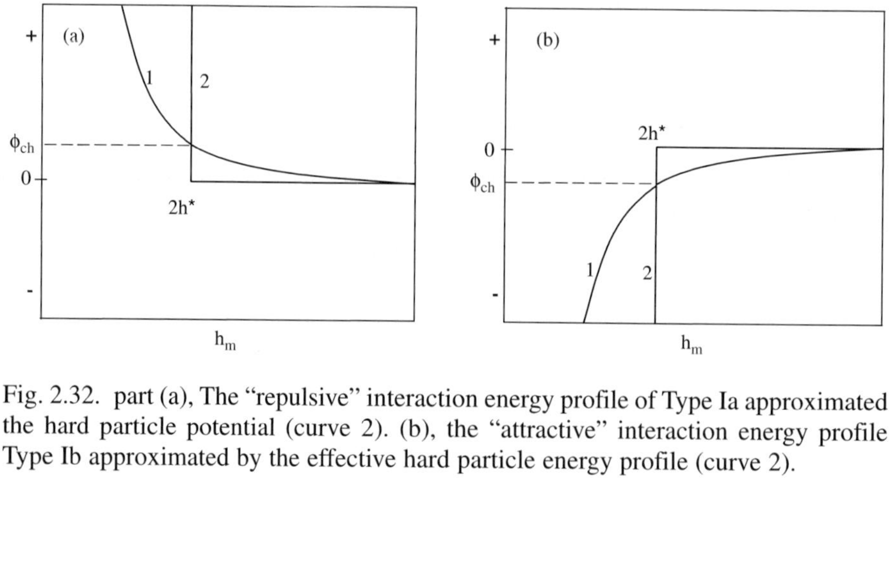

Fig. 2.32. part (a), The "repulsive" interaction energy profile of Type Ia approximated by the hard particle potential (curve 2). (b), the "attractive" interaction energy profile of Type Ib approximated by the effective hard particle energy profile (curve 2).

the possibility of particle aggregation, so the colloid suspension remains indefinitely stable in time. This is advantageous for performing particle adsorption experiments since the bulk particle concentration remains constant and no aggregates appear in the colloid suspension. Obviously, the profile of this type may appear for the particle/interface interactions when both are equally charged.

In this case, particle adsorption remains negligible unless charge heterogeneity (either natural or artificially introduced) appears at the surface.

It is often convenient when analyzing particle adsorption under non-linear conditions (surface blocking effects) to replace the energy profile of Type I (soft interaction profile) by the idealized profile called the hard-particle interaction (see Fig. 2.32). According to this model, introduced originally by Barker and Henderson [111], the interaction energy remains zero except for the critical distance $2h^*$ where it tends to infinity. Physically this means that the interacting particles can be treated as hard ones having the equivalent dimensions increased over the true geometrical dimensions by the small value h^* (skin), which can be treated as the effective interaction range. Therefore, this concept is often referred to as the effective hard particle (EHP) model. It was demonstrated [9,38,65,110] that h^* is strictly related to the double-layer thickness Le. In practice, for $a/Le \ll 1$, h^* remains proportional to Le with the proportionality constant being about 2 for colloid suspensions [65]. The numerical calculations discussed later on demonstrate that the EHP concept is a powerful method of analyzing adsorption of interacting particles of spherical and non-spherical shape.

On the other hand, particle adsorption phenomena are reflected by the energy profile of type Ib, shown in Fig. 2.32(b), when the interaction energy decreases monotonically from zero attaining large negative values (attraction) when $h_m \rightarrow 0$. This profile appears in systems when particle and interface bear opposite surface charges. In order to simplify the mathematical analysis of particle transport phenomena, this energy profile is often idealized by introducing the perfect sink (PS) model, as done originally by Smoluchowski [112] in his fast coagulation theory. According to this approach, the interaction energy remains zero up to a characteristic distance δ_m where it becomes minus infinity. Smoluchowski originally assumed $\delta_m = 0$. This model can be improved upon by identifying δ_m with $2h^*$, analogously to that of the particle/particle interactions. One can interpret in this way the enhanced particle transport due to attractive double-layer interactions as discussed in Refs. [8, 110].

A better, although more tedious, alternative is to consider the true interaction energy distribution up to the point $h = \delta_m$ where it is assumed to become minus infinity. This model has often been used in numerical calculations of colloid particle deposition at various macroscopic surfaces [8,113,114].

Obviously, both the energy profile of Type I and the PS model should be treated as an idealization of any real situation because, at very small separations, the interaction energy must become positive due to the Born repulsion preventing particle/wall penetration. In the DLVO theory, these repulsive interactions were not considered. Even at present, no quantitative theory of these interactions for macroscopic objects has been developed. Assuming that the repulsive part of the potential is described for atoms and molecules by the 12 power law, cf. Eq. (285), one may expect that for the particle/wall interactions $\phi \sim r^{-7}$. These interactions seem, therefore, very short ranged, probably not exceeding 0.5–1nm.

In any case, the appearance of the repulsive interactions fixes the minimum value of the interaction energy, which remains finite in accordance with intuition. This minimum energy value is often referred to as the primary minimum denoted by ϕ_m and the distance where it appears is called the primary minimum distance δ_m (see Fig. 2.33). One may expect that δ_m is of the order of the range of the Born repulsive forces, i.e., 0.5–1 nm. However, the extension of the region where the interaction energy assumes a negative value can be much larger, comparable with the Debye screening length, i.e., about 96 nm for a 10^{-5} M electrolyte solution.

The appearance of the primary energy minimum reflects physically reversible adsorption of the colloid particle, at least under static (no-flow) conditions. However, for a flowing colloid system, as is the case for most practical applications, the situation becomes conceptually more complicated because neither the classical DLVO, nor the theory with inclusion of Born repulsion would explain particle immobilization under vigorous shearing forces. Thus, the particles accumulated at the interface could easily be removed by the tangential fluid flow analyzed in detail in the next section. One has to accept somehow *ad hoc* the appearance of strong tangential interactions most probably due to short ranged geometrical and charge heterogeneities [115,116], as well to hydrogen or chemical bonding. It is difficult to estimate the magnitude of the local energy sinks, responsible for irreversible adsorption of particles, although it can be predicted that their depth will be a fraction of ϕ_m. They are expected, however, to influence the adhesion force and consequently the maximum size of a particle

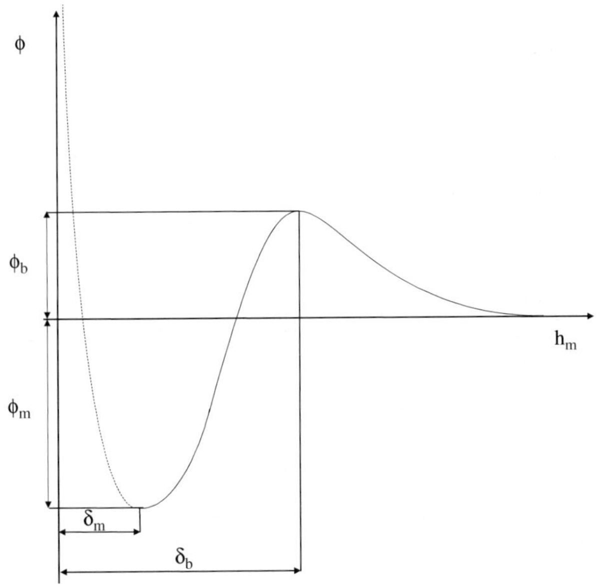

Fig. 2.33. The energy profile of Type II, characterized by the energy minimum of depth ϕ_m and the energy barrier of height ϕ_b.

attached to the surface under a given flow shear rate. The adhesion phenomena will be discussed next.

In the case where the interface and particle bear a surface charge of a similar sign, except for the above monotonic energy profiles, another, more complicated energy profile of Type II is likely to appear for higher ionic strengths. The characteristic feature of the profile, shown schematically in Fig. 2.33, is the presence of a maximum energy (barrier) of height ϕ_b at the distance $\delta_{b.}$ This energy profile corresponds to the activated transport conditions in chemical kinetics. Obviously, the height of the barrier is very sensitive to the electrolyte concentration and composition (presence of polyvalent ions), the Hamaker constant, particle size, shape and orientation. The presence of the energy barrier considerably reduces the particle adsorption rate, which may become inaccessible for accurate measurements. Therefore, in this case, the experimental measurements are often difficult to interpret in terms of the classical DLVO theory. It should also be mentioned that due to a large transport resistance provoked by the barrier, the bulk transport conditions are less important in this case.

For low electrolyte concentration and large particles (micrometer size range), a situation may arise when the so-called secondary minimum appears at the distance δ_{sm} significantly exceeding δ_m. This minimum is usually shallower than the primary minimum due to smaller dispersion energy contribution at this distance. This type of energy profile will be referred to as IIa. Fundamentally, there is not much difference in the II and IIa energy profiles. In the latter case, additional accumulation of particles around δ_{sm} is expected. The concentration peak within this region does not become significant, however, due to flow [115,116].

As this qualitative analysis suggests, it is generally more efficient to analyze the influence of the energy profile on particle adsorption phenomena than to analyze the influence of numerous physicochemical parameters influencing ϕ_m, δ_m, and δ_{sm}. This approach will be generally adopted in further parts of our work dealing with quantitative analysis of particle adsorption kinetics.

It ought to be remembered, however, that the results discussed above are valid for an idealized system only, characterized by perfectly smooth surfaces, uniform charge distribution and the lack of deformation upon approach. When dealing with real systems, e.g., colloid suspensions, these assumptions are unlikely to be met because of many complicating effects:

(i) Heterogeneity of charge distribution at interacting surfaces, which can be of a microscopic scale (of chemical origin) or macroscopic, patchwise scale; also significant differences within particle populations are expected to appear in respect to, e.g., the average charge.
(ii) Surface roughness, either of a well-defined geometrical shape or of a statistical nature.
(iii) Surface deformation upon approach, which can be important for soft polymeric colloids characterized by a low Young's modulus value.
(iv) Dynamic relaxation phenomena of the double layer upon contact (aging effects) due to ion migration along the surface or from one surface to another; these processes are again expected to appear for polymeric colloids. These effects will be discussed briefly in the next section.

2.5. PARTICLE ADHESION PHENOMENA AND OTHER NON-DLVO INTERACTIONS

The particle/particle or particle/interface interactions and energy profiles analyzed above are strictly valid for perfectly rigid bodies. However, in reality, the appearance of forces is inevitably connected with deformations of material bodies, which can be of elastic type (reversible) or plastic type (irreversible). Particle deformations can in turn significantly modify both electrostatic and dispersion interactions as suggested by the previously derived formulae. Hence, a unique solution of this problem seems rather difficult, especially in the case of soft colloid particles whose deformations may even become time-dependent [117]. However, for most particles of practical interest, e.g., polystyrene latex spheres, the extent of deformations remains much smaller than their dimensions and can be described in terms of well-established theoretical approaches. This can be estimated by using the Young law stating that the elastic deformation of a body (a rod) is described by the linear relationship

$$\delta/L = k_e P_0 \tag{345}$$

where δ is the decrease in the length L of the rod under the applied pressure P_o and k_e the elasticity constant given by

$$k_e = \frac{1 - v_p^2}{\pi E y} \tag{346}$$

where Ey is the Young modulus and v_p the Poisson ratio of the material.

The Young modulus for glass equals 7×10^{10} Pa (N m^{-2}) and for polystyrene 3×10^9 Pa [117]. Only for such a soft polymer as polyurethane, the Young modulus becomes much smaller, about 5×10^6 Pa. Using these data one can estimate from Eq. (345) that a 1% deformation of a glass is produced under the pressure of 7×10^8 Pa, i.e., 7×10^3 atm. Analogous value for polystyrene equals 3×10^7 Pa $\cong 3 \times 10^2$ atm and for polyurethane 5×10^4 Pa $\cong 0.5$ atm.

It is interesting to compare this limiting pressure with values stemming from the van der Waals interactions, given by the formula

$$P_v = -\frac{A_{132}}{6\pi h_m^3} \tag{347}$$

By taking $A_{132} = 1.4 \times 10^{-20}$ J (polystyrene in water) and the usually assumed value of $h_m = 0.4$ nm, one can calculate that $P_v = 1.2 \times 10^7$ Pa = 120 atm. However, the van der Waals pressure decreases rapidly with the separation distance between plates. Accordingly, for the distance $h_m = 4$ nm (a typical value of surface roughness for smooth surfaces) one obtains $|P_v| = 1.2\times10^4$ Pa $= 0.12$ atm. This comparison unequivocally shows that the deformations of particles induced by the van der Waals forces are very small indeed. For example, in the case of polystyrene, they are expected to be about 0.3% of the particle size for very smooth surfaces. For rougher surfaces ($h_m = 4$ nm) the surface deformation amounts to 0.0003%, becoming completely negligible. However, even such small deformations, leading to particle flattening may exert an influence on the interaction energy as suggested by Eq. (308). This is so because, the factor δ/h_m can be comparable with unity even if the relative deformations remain much smaller than the particle size.

The above analysis is valid for rods, however, when the stress distribution remains quasi homogeneous. In the case of spherical particles, the analysis becomes quite complicated. As a result, no complete theory describing particle deformations for the entire range of loads and Young's modulus has yet been developed.

Historically, the first attempt at solving the problem of deformation of two spheres pressed together by the uniform force F_0 was made by Hertz [118]. The basic assumptions of his theory were that deformations remain purely elastic, and small in comparison with the sphere dimensions. The main result of his theoretical work proved that the flat contact area between spheres after applying the mechanical load F_0 has a circular shape with the radius described by the formula

$$R_f^3 = \frac{3}{4}\pi(k_{e_1} + k_{e_2})\bar{a}\, F_0 \tag{348}$$

where $\bar{a} = a_1\, a_2/(a_1 + a_2)$ is the mean radius of the spheres and $k_{e_1} = (1 - v_{p_1}^2)/\pi\, Ey_1$, $k_{e_2} = (1-v_{p_2}^2)/\pi\, Ey_2$, are the elasticity coefficients of spheres 1 and 2, respectively.

By noting that the surface area of the contact zone equals πR_f^2, one can express Eq. (348) in a simpler form

$$R_f/\bar{a} = \frac{3}{4}\pi^2(k_{e_1} + k_{e_2})P_0 \tag{349}$$

where $P_0 = F_0/\pi\, R_f^2$ can be treated as the uniform pressure exerted by the external load.

According to the Hertz theory, the depth of penetration of spheres δ_f (elastic displacement) is given by the formula

$$\delta_f = \frac{R_f^2}{\bar{a}} \tag{350}$$

By exploiting Eq. (348) this can be expressed as

$$\delta_f^3 = \frac{1}{\bar{a}}\left[\frac{3}{4}\pi(k_{e_1} + k_{e_2})\right]^2 F_0^2 \tag{351}$$

For soft sphere/hard interface interactions (when $k_{e1} > k_{e2}$) pertinent to adhesion problems Eq. (351) assumes a simpler form

$$\delta_f^3 = \frac{1}{a}\left(\frac{3}{4}\pi k_e\right)^2 F_0^2 \tag{352}$$

where $k_e = (1 - v_p^2)/\pi\ Ey$ and a is the sphere radius.

Let us assume that the load is due to the specific forces alone consisting of the van der Waals and electrostatic contributions, so that

$$F_0 = -2\pi a \Phi_p = -2\pi a w_A \tag{353}$$

where Φ_p is the energy per unit area of plates often denoted in the relevant literature dealing with adhesion as w_A (work of adhesion) and $\gamma = w_A/2$ is the surface energy of one plate.

Using Eq. (298) for the van der Waals contribution and Eq. (220) for the electrostatic contribution valid in the limit of short separations (for the c.p. model more appropriate in this case) one can express Eq. (353) as

$$\begin{aligned} F_0 &= -2\pi a(\Phi_v + \Phi_e) \\ &= 2\pi a\left[\frac{A_{132}}{12\pi h_m^2} + \frac{\varepsilon}{2} Le \frac{(\bar{\psi}_1^0 - \bar{\psi}_2^0)^2}{h_m^2}\right] \end{aligned} \tag{354}$$

It is useful to estimate the relative significance of the van der Waals and electrostatic contributions. Taking again $A_{132} = 1.4 \times 10^{-20}$ J (polystyrene particles in water), one obtains $\Phi_v = 2.3\times10^{-3}$ J m^{-2} (for $h_m = 0.4$ nm). On the other hand, the electrostatic interactions are calculated by assuming $I = 10^{-3}$ M (6.023×10^{23} m^{-3}) and $\psi_1^0 - \psi_2^0 = 0.05$ V, a typical value for polystyrene suspension interacting with glass. Then the electrostatic energy $\Phi_e = 15.2\times10^{-2}$ J (for $h_m = 0.4$ nm) that is longer than the van der Waals contribution. However, the electrostatic contribution will dominate for dilute electrolyte solutions and higher potentials. This becomes obvious when considering the quotient of the van der Waals to electrostatic contributions that is given by the expression

$$\frac{\Phi_v}{\Phi_e} = \frac{A_{132}}{6\pi\varepsilon Le(\psi_1^0 - \psi_2^0)^2} \tag{355}$$

However, assuming that the load is solely due to the van der Waals forces as is usually done in the literature, one can transform Eq. (352) to the explicit form

$$\delta_f = a^{1/3}\left(\frac{\pi k_e A_{132}}{8h_m^2}\right)^{2/3} \tag{356}$$

By taking the previously used data for polystyrene adhering to glass, i.e., $A_{132} = 1.4\times10^{-20}$ J and $k_e = 1 - v_p^2/\pi Ey = 10^{-10}$ Pa one obtains $\delta_f = 0.1$ nm, from Eq. (356), for a particle of the radius 100 nm, i.e., 0.1% of its radius. For $a = 1000$ nm, $\delta_f = 0.23$ nm, i.e., 0.022 % of particle size. The respective values of R_f calculated from Eq. (350) are 3.3 and 14.8 nm, respectively. One can, therefore, conclude that particle deformations induced by the van der Waals interactions are practically negligible in comparison with particle dimensions. It is to remember, however, that these estimates are valid for solid polystyrene particles. In the case of lattices their Young modulus is expected to become significantly lower, probably close to the soft polymeric particles. It this case these deformations may be more significant. Indeed, for polyurethane, having $k_e = 5.7\times10^{-7}$ Pa^{-1} one can calculate from Eq. (356) that $\delta_f = 34$ nm for a particle of the radius 100 nm, i.e., 34% of its radius and $\delta_f = 73$ nm, for $a = 1000$ nm, i.e., 7.3% of particle radius. These deformations are significantly larger than the minimum distance between the particle and the interface taken to be 0.4 nm. This means, according to the formula derived for the flattened sphere, Eq. (308), that the van der Waals energy of interaction will increase manifold over the undeformed sphere case. However, since this energy is stored in the particle because of elastic deformations it can be regained when removing the particle from the surface in the pull-out (detachment) process. This suggests that the pull-out force will not be affected by sphere deformations, i.e., by the Young modulus of the particle.

This assumption was exploited in the theory of Derjaguin, Muller and Toporov [119] referred to in the literature as the DMT theory. It was elaborated by assuming that particle deformations remain small in comparison with their dimensions and that the stress distribution profile under the van der Waals load remains the same as for an external load, i.e., the Hertzian. It also was postulated in the DMT theory that the van der Waals forces are acting outside the contact area as well. The theory was further refined in the later works [120,121] by considering the repulsive interactions in the form of the Lennard-Jones potential. The main result of this complex theory was deriving Eq. (356) and proving that the pull-out force (defined as the force needed to remove the spherical particle from the surface) $F_p = 2\pi\Phi_v a = 2\pi a w_A = F_v$. Additionally, the detachment of a particle was found to occur at zero contact radius. The same result was deduced much earlier by Bradley [122]. Moreover, as pointed out by Maugis [123] it is not physically consistent to have only compressive (Hertz) stress in the contact zone area and adhesion forces outside this zone.

For larger deformations, involving complaint materials, a more appropriate seems the Johnsonn, Kendall and Roberts theory [124] referred widely to as the JKR theory. It was assumed that the surface interactions are due to van der Waals forces alone. The change in the surface energy induced by the contact of the particle with the interface was assumed to be

$$\phi = \pi R_f^2 w_A = -\pi R_f^2 \frac{A_{132}}{12\pi h_m^2} = -a\delta_f \frac{A_{132}}{12 h_m^2} \tag{357}$$

It is obvious that Eq. (357) should be treated as an approximation only because it predicts no interaction energy between the particle and an interface in the case of no deformations.

The surface force was then calculated from Eq. (353) as the derivative of the interaction energy on the deformation δ_f, i.e.,

$$F = -\frac{d\phi}{d\delta_f} = a\frac{A_{132}}{12h_m^2} = \frac{1}{2}F_v \tag{358}$$

As can be seen, the force is two times smaller than the van der Waals force for a undeformed sphere given by Eq. (353).

The main difference in comparison with the DMT theory was the postulate that the van der Waals interactions not only increase the contact area but also modify the stress distribution and stored elastic energy. Thus, except for the van der Waals component, the total energy included the elastic component and the mechanical potential energy. From the equilibrium postulate (stating that the derivative of the total energy upon the force equals zero) it was then deduced that the apparent Herzian load is given by the expression

$$F_a = F_0 + 3\pi a w_A + \left[6\pi a w_A F_0 + (3\pi a w_A)^2\right]^{1/2} = F_0 + \frac{3}{2}|F_v| + \left[3|F_v|F_0 + \left(\frac{3}{2}F_v\right)^2\right]^{1/2} \tag{359}$$

The radius of the contact area and the deformation thickness δ_f are given by the formulae

$$R_f^3 = \frac{3}{4}\pi k_e a \left\{F_0 + \frac{3}{2}|F_v| + \left[3|F_v|F_0 + \left(\frac{3}{2}F_v\right)^2\right]^{1/2}\right\} \tag{360}$$

$$\delta_f = \frac{R_f^2}{a} - \left(\frac{4}{3}\pi k_e |F_v| \frac{R_f}{a}\right)^{1/2} \tag{361}$$

For zero external load one has

$$R_f = \left(\frac{9}{4}\pi k_e |F_v| a\right)^{1/3} = a^{2/3}\left(\frac{3}{8}\frac{\pi k_e A_{132}}{h_m^2}\right)^{1/3} \tag{362}$$

$$\delta_f = \frac{\left(\frac{9}{4}k_e \pi |F_v|\right)^{2/3}}{a^{2/3}} - \left(\frac{4}{3}\pi k_e |F_v| \frac{R_f}{a}\right)^{1/2} \tag{363}$$

By taking the previously used data for polystyrene adhering to glass, i.e., $A_{132} = 1.4\times10^{-20}$ J and $k_e = 10^{-10}$ Pa one obtains the JKR theory, Eqs. (362,363), that $\delta_f = 0.050$ nm for a particle of the radius 100 nm, and $\delta_f = 0.11$ nm for $a = 1000$ nm. The respective values of R_f are 14.7 and 21.8 nm. As can be seen, these data are significantly smaller than previously estimated from the simple DMT theory.

The main result of the JKR theory was showing that separation of the sphere occurs if the pull-out force F_p becomes larger than

$$F_p > \frac{3}{2}\pi a w_A \cong \frac{4}{3}F_v = -\frac{2A_{132}a}{9h_m^2} \tag{364}$$

The range of applicability of various theories was estimated by Johnson and Greenwood [125] by introducing two parameters characterizing the load (expressed in the reduced form by using the van der Waals force as scaling variable) and the extent of deformation governed by the Young modulus representing the ratio of the elastic displacement to the minimum distance of the adhering surfaces.

Summing up the results derived from the above theories of adhesion phenomena under the elastic regime one can derive two interesting conclusions:

(i) The deformations of typical colloid particles under the van der Waals forces are negligible in comparison with their dimensions, remaining of the order of a percent.
(ii) The pull-out force needed to remove adhering particles is independent of surface deformations and elasticity of the particles.

One also can conclude that any appreciable increase in the adhesion force (sticking strength of particles to surfaces), especially for rough surfaces, can only be achieved via irreversible processes connected with plastic deformations of particles [78,117] or accompanied by molecular transport or chemical bond formation. These processes, often referred to as aging, are usually dependent on the time of the contact of particles with interfaces. Because of their specificity and system-dependent properties they will not be further analyzed here.

Additional effects leading to non-DLVO interactions appear in systems containing polymer solutions [108]. Let us assume for simplicity that the polymers are non-ionic and dissolved in a good solvent, so their shape in solution can be described by the random coil model [108]. Then, the mean square end-to-end distance of the polymer coil, by assuming no interactions between segments is given by the expression

$$\bar{r}_p = \langle r_p^2 \rangle^{1/2} = l_{po} (N_p)^{1/2} \tag{365}$$

where N_p is the number of segments in the polymer chain (blobs) and l_{po} the length of one segment.

The end-to-end distance is an important scaling length variable enabling one to describe quantitatively interactions of particles in polymeric solutions, the most important of them being [108]:

(i) the depletion interactions for non-adsorbing polymers;
(ii) steric interactions for adsorbing polymers.

The former interactions, first described by Asakura et al. [126] are due physically to the reduced osmotic pressure of the polymer in the gap between two interacting colloid particles or the particle/interface system. This is so because for a critical distance between particles $h < r - 2a + L_p$ (where h is the distance between particle surfaces and L_p is the mean size of the polymer chain strictly related to $\bar{r}_p$) the polymer chains cannot enter the space between particles. Thus, their concentration in this space is reduced that results in decreased osmotic pressure there. In the general case, the osmotic pressure of polymers depends in a nonlinear way on their concentration that can be described in terms of the virial expansion [108]. However, for dilute solutions the osmotic pressure can be described in terms of the linear expression, analogous to previously used for electrolyte solutions, i.e., Eq. (196). For polymer solutions it can be written as

$$\Delta P = kT(n_e - n_p) \tag{366}$$

where n_p is the bulk concentration of the polymer and n_e the concentration in the gap h between particles. It is usually assumed that the concentration of polymers is negligible in the exclusion zone having the volume [127]

$$v_e = \frac{4}{3}\pi (a + L_p)^3 \left[1 - \frac{3(2a+h)}{4(a+L_p)} + \frac{(2a+h)^3}{16(a+L_p)^3} \right] \tag{367}$$

Knowing the osmotic pressure and the volume of the exclusion zone, one can calculate the isothermal work as $\Delta P\, v_e$, so the energy of the system can be expressed as [128]

$$\phi = -kTv_e n_p \tag{368}$$

As can be deduced from Eqs. (367,368), the interaction energy is negative at all separations h that means attraction between particles.

One can also notice that the maximum attractive energy (appearing at $h = 0$) is given by

$$\phi_{mx} = -kT\frac{4}{3}\pi L_p^3\left(1+\frac{3a}{2L_p}\right)n_p \tag{369}$$

In the case when $a/L_p \gg 1$ (a typical situation met under experimental conditions) Eq. (369) simplifies to

$$\phi_{mx} = -2kT\,\pi L_p^2 a n_p \tag{370}$$

Analogous expressions for the case of two particles at the interface have been derived in Ref. [129].

Taking typical data, i.e., $a = 5\times10^{-7}$ m (500 nm) a polymer of molecular weight 10^2 kg mol^{-1}, $L_p = 10^{-8}$ m (10 nm) and polymer aqueous solution of the weight concentration 0.1 kg m^{-3} (100 ppm) one can calculate from Eq. (370) that $\phi_{mx} = 0.19\,kT$ unit. For polymer of concentration of 10 kg m^{-3} (1000 ppm = 1%) one obtains $\phi_{mx} = 19\,kT$. From this estimation one can, therefore, conclude that except for very high polymer concentration, the depletion interaction energy contribution remains less important than the electrostatic and van der Waals energy contribution. However, the depletion interactions, producing a shallow secondary minimum [128,129] may play a more significant role in phase transitions occurring in colloidal systems in a long time scale [130].

Although this hypothesis has not been checked yet, it seems that a complete theoretical treatment of depletion interactions should also involve repulsion. This is so because when particles approach each other there is a finite probability of finding polymer chains in the gap between them. This will increase the interaction energy of particles that will partially compensate the effect of the attraction due to diminished polymer concentration. As a result, the net maximum of the depletion energy is expected to be much smaller than that predicted in Eq. (369).

The repulsive interaction in a more pronounced form also appears in the case of adsorbed polymer layers [108]. The structure of these layers strongly depend on the polymer solvent interactions and the adsorbed polymer concentration. It is quantitatively characterized in terms of the segment density profile and the overall layer thickness. When the distance between two surfaces bearing adsorbed polymer layers becomes smaller than twice the layer thickness, the polymer chains start to interact. In good solvents these interactions are of purely steric type, originating from hard body interactions (repulsive part of the Lennard-Jones potential). Then the increase in the chemical potential of chain segments is solely due to the increase in their concentration. For ideal polymer layers characterized by uniform segment density profile and the thickness given by Eq. (365) the interaction energy per unit area is given by [108]

$$\Phi_{pol} = kT\,\theta_{pol}\left[\frac{4\pi}{3h^2} + \frac{4}{\pi L_p^2}\ln\left(\frac{3h^2}{8\pi L_p^2}\right)\right] \tag{371}$$

where $\theta_{pol} = \pi L_p^2/4$ is the coverage (dimensionless surface concentration of the polymer). Eq. (371), valid for separations between plates smaller than 1.5 h/L_p, indicates that there appears to be a strong repulsion when the surfaces approach each other. When $h \ll L_p$ Eq. (371) reduces to

$$\Phi_{pol} = \frac{4\pi kT\theta_{pol}}{3h^2} = \frac{A_{pol}}{12\pi h^2} \tag{372}$$

where $A_{pol} = 16\pi^2\, kT\theta_{pol}$

Using the Derjaguin method one can specify the expression describing interaction energy of polymer-coated particles in the form

$$\phi_{pol} = G_D\,\frac{A_{pol}}{6h} \tag{373}$$

where G_D is the Derjaguin geometrical factor discussed previously. The use of the Derjaguin method implies that Eq. (373) remains valid if L_p remains smaller than particle radii. Obviously, Eq. (373) also describes particle/wall interactions.

As can be noticed, Eqs. (372), (373), with A_{pol} treated as the apparent Hamaker constant possessing a negative value, have the same mathematical shape as Eqs. (298), (299) describing the van der Waals interactions. Thus, the quotient of the steric interactions to van der Waals interactions is given by

$$\left|\frac{\phi_{pol}}{\phi_v}\right| = \frac{A_{pol}}{A_{132}} = \frac{16\pi^2 kT\theta_{pol}}{A_{132}} \tag{374}$$

By taking the Hamaker constant $A_{132} = 1.4\times10^{-20}$ J as previously and $\Theta_{pol} = 0.025$ (2.5% coverage of the surface by polymer) one can estimate from Eq. (374) that the quotient equals 1.1. This means that for a polymer coverage larger than 0.1, one can totally compensate van der Waals interactions, obtaining the overall energy profile similar to Type Ia (see Fig. 2.32). This effect, called steric stabilization, is often used in practice to prevent coagulation of colloid suspensions [108,131].

For higher coverage of adsorbed polymer and for non-uniform segment distribution profiles Eq. (371) should be replaced by more accurate expressions [131–133]. However, the essential physics of this problem remains the same, i.e., the presence of adsorbed neutral polymer layers generates energy barriers that effectively eliminate particle aggregation or deposition on surfaces.

Other interactions neglected entirely in the DLVO theory are the inertia forces (produced, for example, by centrifugation of particle suspensions) and gravity forces due to the difference in density between particles and the suspending media, causing sedimentation effects. Also the hydrodynamic forces exerted by fluid streams on particles are not considered. These interactions may severely disturb the DLVO energy profiles, especially for particles of the size of 100 nm and larger. Additionally, the hydrodynamic interactions may induce particle detachment from surfaces, which is of vital importance for surface cleaning processes. Because of their significance the hydrodynamic effects will be treated extensively in the next chapter.

LIST OF SYMBOLS

Symbol	Definition	Character	Unit	Numerical value
A	Shorter to longer spheroid axis ratio	S	[1]	
A_{ij}	Hamaker constants for various media interactions	S	J	
A_{pol}	Apparent Hamaker constant for depletion interactions	S	J	
A_{12}, A_{13}, A_{23} A_{22}, A_{23}, A_{33}	Hamaker constants for interactions in vacuum	S	J	
A_{131}	Hamaker constants for interactions of two identical particles in medium	S	J	
A_{132}	Hamaker constants for interactions of two different particles in medium	S	J	
a	Particle radius, spheroid semi-axis	S	m	
$\bar{a}$	Mean radius of spheres	S	m	
a_0	Bohr radius	S	m	0.529177×10^{-10}
a_r	Surface roughness radius	S	m	
a_1, a_2	Particle radii	S	m	
B_h	Dimensionless parameter	S	[1]	
B_n	Coefficients in energy expansion	S	[1]	
b	Spheroid semi-axis	S	m	
C_d	Constant	S	N^{-1}	
C'_d	Differential capacity of spherical double layer	S	$C\ V^{-1}\ m^{-2}$	
C_{dip_1}	Constant describing orientational and induction interactions	S	$J\ m^6$	
C_{dip_2}	Constant describing orientational and induction interactions	S	$J\ m^8$	
C_{disp_1}	Dispersion interaction constant	S	$J\ m^6$	
C_{disp_2}	Dispersion interaction constant	S	$J\ m^8$	
C_{disp_3}	Dispersion interaction constant	S	$J\ m^{10}$	

List of Symbols (continued)

Symbol	Definition	Character	Unit	Numerical value
C_{disp_n}	Dispersion interaction constant	S	J m^6	
C_e'	Electric capacity	S	F = C V^{-1}	
	Differential electric capacity	S	C V^{-1} m^{-2}	
C_h	Dimensionless parameter	S	[1]	
C_L	Lennard-Jones interaction constant	S	J m^{12}	
C_n	Constant of van der Waals interactions	S	J m^{6+2n}	
C_1	Leading constant of van der Waals interactions	S	J m^6	
c	Ellipsoid semi-axis	S	m	
c_0	Speed of light in vacuum	S	m s^{-1}	2.99792×10^8
c_ε	Dimensionless constant	S	[1]	
$\mathbf{D}_p$	Electric displacement	**V**	C m^{-2}	
d, d_1, d_2, d_3	Distances between charges	S	m	
$\mathbf{E}$, $\mathbf{E}_1$, $\mathbf{E}_i$	Electric field strength	**V**	V m^{-1}	
$\mathbf{E}_z$, $\mathbf{E}_\perp$	Components of electric field strength vector	**V**	V m^{-1}	
E, E^0	Electric field (scalar)	S	V m^{-1}	
$\overline{E}$	Dimensionless electric field	S	[1]	
Ey, Ey_1,Ey_2	Young's moduli	S	N m^{-2}	
E_z, $E_\perp$	Length of electric field components	S	V m^{-1}	
e	Elementary charge	S	C	1.60218×10^{-19}
e_x	Excentricity parameter of spheroid	S	[1]	
$\mathbf{F}$	Force vector	**V**	N = kg·m·s^{-2}	
F	Force of interactions (scalar)	S	N	
Fa	Faraday constant	S	C mol^{-1}	96485.34
F_0	Uniform force load	S	N	
F_p	Pull-off force	S	N	
F_v	van der Waals force loud of interactions	S	N	
$f(r_0\ \Omega)$ $\langle f(\Omega r_0)\rangle$	Dimensionless functions	S	[1]	
$f(h_m)$	Dimensionless functions	S	[1]	
f_i	Activity coefficient of ion	S	[1]	
G	Newtonian constant of gravitation	S	m^3·kg^{-1}·s^{-2}	6.6731×10^{-11}
G_D	Derjaguin factor	S	m	

List of Symbols (continued)

Symbol	Definition	Character	Unit	Numerical value
G'_D	Derjaguin factor for particle/roughness interactions	S	m	
$\overline{G}_e, \overline{G}_e^0$	Geometrical factor describing electrostatic interactions	S	[1]	
$g, g(r)$	Pair correlation function	S	[1]	
H_m	Distance parameter	S	m	
H_r	Distance between rough surfaces	S	m	
$\overline{H}_r$	Scaled distance between rough surfaces	S	[1]	
dh	Thickness of disk	S	m	
h_p	Planck constant	S	J s	6.62607×10^{-34}
$\overline{h}$	Dimensionless distance between particles	S	[1]	
h^*	Effective interaction range	S	m	
h_m	Distance between particle and surface	S	m	
h_0	Distance parameter	S	m	
$\mathbf{I}$	Unit tensor	**T**	[1]	
I	Ionic strength	S	m^{-3}	
$\overline{I}$	Dimensionless ionic strength	S	[1]	
I_{01}, I_{02}	First ionization potentials of atoms	S	J	
$\mathbf{i}, \mathbf{i}_r, \mathbf{i}_z, \mathbf{i}_\perp$	Unit vectors	**V**	[1]	
i	Imaginary unit	S	[1]	
k	Boltzmann constant	S	$J\,K^{-1}$	1.38065×10^{-23}
k_e, k_{e1}, k_{e2}	Elasticity constants	S	$N^{-1}m^2$	
L	Length of cylinder, particle	S	m	
Le	Diffuse double-layer thickness	S	m	
L_p	Mean size of polymer chain	S	m	
$d\mathbf{l}$	Vectorial path	**V**	m	
l_e	Coefficients of spheroids	S	[1]	
l_{Po}	Length of polymer segment	S	m	
l_r	Dimensionless exponent	S	[1]	
m_e	Electron mass	S	kg	9.10938×10^{-31} kg

List of Symbols (continued)

Symbol	Definition	Character	Unit	Numerical value
m_p	Proton mass	*S*	kg	1.67262×10^{-27} kg
m, m_1, m_2	Mass of bodies	*S*	kg	
N_{Av}	Avogadro constant	*S*	[1]	6.02214×10^{23}
N_p	Number of polymer segments	*S*	[1]	
$\hat{\mathbf{n}}$	Unit surface vector	**V**	[1]	
n_e	Polymer concentration in gap between particles	*S*	m^{-3}	
n_i^b	Bulk concentration of ion	*S*	m^{-3}	
n_p	Polymer concentration in bulk	*S*	m^{-3}	
n_r, n_1, n_2, n_3	Refractive indices of various media in visible region	*S*	[1]	
n_1^{mx}	Maximum ion packing	*S*	m^{-3}	
$\mathbf{P}$	Osmotic pressure	**T**	$N\,m^{-2}$	
$\mathbf{P}^b$	Bulk osmotic pressure	**T**	$N\,m^2$	
$\mathbf{P}_e$	Electric polarization vector	**V**	$C\,m^{-2}$	
$P^{(0)}$	Monopole (electric charge)	*S*	C	
$P^{(1)}$	Electric dipole	*S*	C m	
$P^{(2)}$	Electric quadrupole	*S*	$C\,m^2$	
P_n	Legendre polynomials			
P_0	Uniform pressure	*S*	$N\,m^{-2}$	
P_v	van der Waals pressure	*S*	$N\,m^{-2}$	
$\mathbf{p}$	Electric dipole vector	**V**	C m	
p, p_1, p_2	Length of electric dipole vectors	*S*	C m	
Δp_m	Uncertainty of electron momentum	*S*		
$\mathbf{Q}$	Electric quadrupole tensor	**T**	$C\,m^2$	
Q_s, Q_z	Dimensionless parameter	*S*	[1]	
Q_1, Q_2	Molecular quadrupoles	*S*	$C\,m^{-2}$	
$q, q_1 q_2, q^0$	Electric charge	*S*	C	
δq	Infinitesimal charge	*S*	C	
$R(h)$	Radius of disk	*S*	m	
R_f	Radius of contact area	*S*	m	
R_g	Molar gas constant	*S*	$J\,K^{-1}\,mol^{-1}$	8.31445
R, R_1, R_2	Radii of bodies	*S*	m	
R_1', R_1'', R_2', R_2''	Radii of curvature of bodies	*S*	m	
$\mathbf{r}, \mathbf{r}_1, \mathbf{r}_2, \mathbf{r}_i, \mathbf{r}_0$	Position vectors	**V**	m	
$\mathbf{r}_s$	Surface position vector	**V**	m	

List of Symbols (continued)

Symbol	Definition	Character	Unit	Numerical value
r, r_0	Length of position vectors	*S*	m	
$\bar{r}_p$	Mean square end-to-and distance of polymer	*S*	m	
$d\mathbf{S}$	Surface element vector	**V**	m^2	
S	Surface area	*S*	m^2	
$\delta S, dS$	Surface element	*S*	m^2	
S_1, S_2	Electrode areas	*S*	m^2	
$\mathbf{T}$	Maxwell stress tensor	**T**	$N\ m^{-3}$	
T	Absolute temperature	*S*	K	
V	Velocity	*S*	$m\ s^{-1}$	
ΔV	Uncertainty of velocity	*S*	$m\ s^{-1}$	
v	Volume	*S*	m^3	
δv	Volume element	*S*	m^3	
v_e	Volume of exclusion zone	*S*	m^3	
v_n	Molar volume of ideal gas	*S*	$m^3\ mol^{-1}$	22.4140×10^{-3}
v_p	Particle volume	*S*	m^3	
v_s	Volume of spheroid	*S*	m^3	
w_A	Work of adhesion	*S*	J	
x	Distance	*S*	m	
$\bar{x}$	Dimensionless distance	*S*	[1]	
Δx	Uncertainty of electron position	*S*	m	
$\bar{Y}, \bar{Y}_1^0, \bar{Y}_2^0$	Effective surface potentials	*S*	[1]	
z_i	Multiplicity of charge, ion valency	*S*	[1]	
z_p	Distance from electrode	*S*	m	
Greek				
α	Polarizability of particle	*S*	m^3	
α_i	Ion concentration ratio	*S*	[1]	
α_s	Polarizability of spheroid	*S*	m^3	
$\alpha_0, \alpha_{0_1}, \alpha_{0_2}$	Molecular polarizabilities	*S*	m^3	
β	Dimensionless ion concentration ratio	*S*	[1]	
β_c	Angle between cylinders	*S*	[1]	
β_0	Dimensionless parameter	*S*	[1]	
δ	Distance between plates, thickness of fixed charge layer	*S*	m	

List of Symbols (continued)

Symbol	Definition	Character	Unit	Numerical value
δ_b	Energy barrier distance	S	m	
δ_c	Spherical cup thickness	S	m	
δ_f	Thickness of flattened area, depth of penetration	S	m	
δ_i	Thickness of layer devoted of charge	S	m	
δ_m	Primary minimum distance	S	m	
δ_γ	Deformation of rod	S	m	
ε_i	Permittivities of various layers	S	$\mathrm{F\,m^{-1}}$	
ε_0	Permittivity of vacuum	S	$\mathrm{F\,m^{-1}}$	8.85419×10^{-12}
$\varepsilon_p, \varepsilon_{p_1}, \varepsilon_{p_2}$	Permittivities of particles	S	$\mathrm{F\,m^{-1}}$	
$\varepsilon_1, \varepsilon_2, \varepsilon_3$	Permittivities of media	S	$\mathrm{F\,m^{-1}}$	
Θ	Surface coverage of particles	S	[1]	
Θ_{pol}	Polymer coverage	S	[1]	
$\kappa = Le^{-1}$	Reciprocal double-layer thickness	S	$\mathrm{m^{-1}}$	
λ_r	Characteristic wavelength	S	$\mathrm{m^{-1}}$	
λ_1, λ_2	Functions of radii of curvature	S	$\mathrm{m^{-1}}$	
μ_i^b	Chemical potential of ion in bulk	S	J	
$\tilde{\mu}_i^b$	Electrochemical potential of ion	S	J	
μ_0	Magnetic constant	S	$\mathrm{N\,A^{-2}}$	12.5664×10^{-7}
$\tilde{\nu}_e, \tilde{\nu}_{e1}, \tilde{\nu}_{e2}$	Characteristic atomic frequencies	S	$\mathrm{s^{-1}}$	
ν_n	Imaginary frequency	S	$\mathrm{s^{-1}}$	
ν_p	Poisson's ratio of material	S	[1]	
ν_{rot}	Rotational relaxation frequency	S	$\mathrm{s^{-1}}$	
$\Delta\Pi$	Uniform pressure between plates	S	$\mathrm{N\,m^{-2}}$	
$\rho_a, \rho_{a_1}, \rho_{a_2}$	Number densities of atoms	S	$\mathrm{m^{-3}}$	
ρ_e	Electric charge density	S	$\mathrm{C\,m^{-3}}$	
ρ_{e_p}	Polarization charge density	S	$\mathrm{C\,m^{-3}}$	
$\overline{\rho}_p$	Probability density	S	[1]	
$\sigma, \sigma(r_s)$	Surface charge	S	$\mathrm{C\,m^{-2}}$	
$\overline{\sigma}^0, \overline{\sigma}_1^0, \overline{\sigma}_2^0$	Dimensionless surface charge	S	[1]	

List of Symbols (continued)

Symbol	Definition	Character	Unit	Numerical value
Φ, Φ_p	Interaction energy of plates per unit area	S	J m^{-2}	
Φ_v	van der Waals interaction energy of plates	S	J m^{-2}	
Φ_{132}	Interaction energy of two different plates in a medium	S	J m^{-2}	
ϕ	Interaction energy	S	J	
ϕ_b	Energy barrier height	S	J	
ϕ_{dd}^0	Characteristic dipolar interaction energy	S	J	
ϕ_{dip}	Interaction energy of dipoles	S	J	
ϕ_e	Double-layer energy of plates	S	J	
ϕ_m	Depth of the energy minimum	S	J	
ϕ_{mx}	Maximum attractive energy	S	J	
ϕ_0, ϕ_a, ϕ_b	Reference energies	S	J	
ϕ_s	Energy of sphere/plane van der Waals interaction	S	J	
ϕ_{12}, ϕ_{23}, ϕ_{13}, ϕ_{33}, ϕ_{132}	Energies of van der Waals interaction in media	S	J	
φ	Angle between spheroid and plane	S	[1]	
χ	Electric susceptibility	S	[1]	
ψ	Electric potential	S	V	
$\overline{\psi}$	Dimensionless potential	S	[1]	
ψ_a, ψ_b	Reference potentials	S	V	
ψ_d	Diffuse double-layer potential	S	V	
ψ_p	Induced potential	S	V	
ψ_1, ψ_1^0, ψ_2^0, ψ^0	Surface potentials	S	V	
$\overline{\psi}_0$, $\overline{\psi}_1^0$, $\overline{\psi}_2^0$	Dimensionless double-layer potential	S	[1]	
	Orientational space	S	[1]	

Note: S = scalar; **V** = vector; **T** = tensor (matrix); [1] = dimensionless.

REFERENCES

[1] H. Morgenau and N.R. Kestner, Theory of Intermolecular Forces, Pergamon Press, Oxford, 1969.
[2] R.P. Feynman, R.B. Leighton and M. Sands, The Feynman Lectures on Physics, Vol. 2, Addison-Wesley, Reading, MA, 1964.
[3] J.D. Jackson, Classical Electrodynamics, Wiley, New York, 1975.
[4] Wm.Y. Fowlkes and K.S. Robinson, The electrostatic force on a dielectric sphere resting on a conducting substrate, K.L. Mittal, (in ed.). Particles on Surfaces, Detection, Adhesion and Removal", p. 143, Marcel Dekker, New York, 1988,
[5] L.D. Landau and E.M. Lifshitz, Electrodynamics of Continuous Media, Pergamon Press, Oxford, 1960.
[6] H.A. Pohl, J. Appl. Phys., 29 (1958) 1182.
[7] M.H. Davis, Am. J. Phys., 37 (1969) 26.
[8] Z. Adamczyk, Irreversible Adsorption of Particles, p. 251 in "Adsorption: Theory, Modeling and Analysis" (J. Toth, ed.), Marcel Dekker, New York, 2002.
[9] M.H. Davis, Quart. J. Mech. Applied Math., 17 (1964) 500.
[10] P. Atten, J. Electrostat. 30 (1993) 259.
[11] P.A. Arp and S.G. Mason, Colloids Polym. Sci., 255 (1977) 980.
[12] M. Fermiger and A.P. Gast, J. Colloid Interface Sci., 154 (1992) 522.
[13] R.D. Stoy, J. Electrostat. 33 (1994) 385.
[14] R.D. Stoy, J. Appl. Phys., 69 (1991) 2800.
[15] G.M. Torrie and J.P. Valleau, J. Chem. Phys., 73 (1980) 5807.
[16] G.M. Torrie and J.P. Valleau, J. Phys. Chem., 86 (1982) 3251.
[17] J.P. Valleau, R. Ivkov and G.M. Torrie, J. Chem. Phys., 95 (1991) 520.
[18] J.A. Greathouse and D.A. McQuarrie, J. Phys. Chem., 100 (1996) 1847.
[19] J.A. Greathouse and G. Sposito, Electrical double layer at particles, in "Encyclopedia of Surface and Colloid Science" (A.T. Hubbard, ed.), p. 1642, Marcel Dekker, New York, 2002.
[20] O. Stern, Z. Elektrochem., 30 (1924) 508.
[21] R.J. Hunter, Foundations of Colloid Science, 2nd ed., Oxford University Press, Oxford, 2001, p. 326.
[22] D.C. Grahame, Chem. Rev., 41 (1947) 441.
[23] F.H. Stillinger and J.G. Kirkwood, J. Chem. Phys., 33 (1960) 1282.
[24] F.P. Buff, F.H. Stillinger, J. Chem. Phys., 39 (1963) 1911.
[25] F.P. Buff and N.S. Goal, J. Chem. Phys., 51 (1969) 4983.
[26] C.W. Outhwaite, Chem. Phys. Lett., 7 (1970) 636.
[27] L. Blum, J. Phys. Chem., 81 (1977) 136.
[28] C.W. Outhwhaite and L.B. Bhuiyan, J. Chem. Soc. Faraday Trans. II, 79 (1983) 707.
[29] M. Lozada-Cassou, R.Saavedra-Barrera and D. Henderson, J. Chem. Phys., 77 (1982) 5150.
[30] H. Brodowsk and H. Strehlow, Z. Electrochem., 63 (1950) 262.
[31] E. Wicke and M. Eigen, Z. Electrochem., 56 (1952) 551.
[32] M. Eigen, E. Wicke, J. Phys. Chem., 58 (1954) 702.

[33] G.M. Bell and S. Levine, Trans. Faraday Soc., 53 (1957) 143.
[34] G.M. Bell and S. Levine, Trans. Faraday Soc., 54 (1958) 785.
[35] G.M. Bell and S. Levine, Trans Faraday Soc., 54 (1958) 975.
[36] J. Koryta, J. Dvorak and V. Bohackova, Lehrbuch der Electrochemie, Springer, Berlin, 1975.
[37] Z. Adamczyk, P. Belouschek and D. Lorenz, Ber. Bunse. Phys. Chem., 94 (1990) 1483.
[38] Z. Adamczyk and P. Warszyński, Adv. Colloid Interface Sci., 63 (1996) 41.
[39] G.M. Bell, S. Levine and L.N. Mc Cartney, J. Colloid Interface Sci., 33 (1970) 335.
[40] G. Gouy, J. Phys. Theor. Appl., 9 (1910) 457.
[41] D.L. Chapman, London, Edinburgh Dublin, Phil. Mag. J. Sci., 25 (1913) 475.
[42] E.J.W. Vervey and J. Th.G. Overbeek, Theory of the Stability of Lyophobic Colloids, Elsevier, Amsterdam, 1948.
[43] A.L. Loeb, J.Th.G. Overbeek and P.H.Wiersema, The Electrical Double Layer around a Spherical Colloid Particle, MIT Press, MA, USA, 1961.
[44] B.A. Schrauner, J. Colloid Interface Sci., 44 (1973) 79.
[45] S.L. Brenner, R.E. Roberts, J. Chem. Phys., 77 (1973) 2367.
[46] H. Ohshima, T.W. Healy, and L.R. White. J. Colloid Interface Sci., 90 (1982) 17.
[47] G. Kar, S. Chander and T.S. Mika, J. Colloid Interface Sci., 44 (1973) 347.
[48] D.Y.C. Chan and D.J. Mitchel, J. Colloid Interface Sci., 95 (1983) 193.
[49] D. Mc Cormack, S.L. Carnie and D.Y.C. Chan, J. Colloid Interface Sci., 169 (1995) 177.
[50] J.E. Jones, S. Levine, J. Colloid Interface Sci., 30 (1969) 241.
[51] E.P. Honig and P.M. Mul, J. Colloid Interface Sci., 36 (1971) 258.
[52] O.F. Devereux and P.L. de Bruyn, Interactions of Plane-Parallel Layers, MIT Press, MA, USA, 1963.
[53] R. Hogg, T.W. Healy and D.W. Fuerstenau, Trans. Faraday Soc., 62 (1973) 1638.
[54] G.R. Wiese and T.W. Healy, Trans. Faraday Soc., 66 (1970) 490.
[55] S. Usui, J. Colloid Interface Sci., 44 (1973) 107.
[56] B.V. Derjaguin, Kolloid Z., 69 (1934) 155; Trans. Faraday Soc., 36 (1940) 203.
[57] H. Ohshima, Colloid Polymer Sci., 252 (1974) 158; *ibid.* 253 (1975) 150; *ibid.* 254 (1976) 484; *ibid.* 257 (1979) 630.
[58] H. Ohshima and T. Condo, J. Colloid Interface Sci., 155 (1993) 499.
[59] H. Ohshima, J. Colloid Interface Sci., 170 (1995) 432.
[60] L.N. McCartney and S. Levine, J. Colloid Interface Sci., 30 (1969) 345.
[61] J.E. Sader, S.L. Carnie and D.Y.C. Chan, J. Colloid Interface Sci., 171 (1995) 46.
[62] T. Sugimoto, Adv. Colloid Interface Sci., 28 (1987) 65.
[63] E. Matijevic, R.S. Sapieszko, in "Fine Particles, Synthesis, Characterization, and Mechanism of Growth, Science Series" (T. Sugimoto, ed.), p.2, vol. 92, Marcel Dekker, New York, 2000.
[64] L.R. White, J. Colloid Interface Sci., 95 (1983) 286.
[65] Z. Adamczyk and P. Weroński, Adv. Colloid Interface Sci., 83 (1999) 137.
[66] V.E. Shubin and P. Kekicheff, J. Colloid Interface Sci., 155 (1993) 108.
[67] S. Bhattacharjee and M. Elimelech, J. Colloid Interface Sci., 193 (1997) 273.
[68] N.E. Hoskin, Philos. Trans. Roy. Soc. London, Ser.A. 248 (1956) 433.
[69] N.E. Hoskin and S. Levine, Philos. Trans. Roy. Soc. London, Ser.A., 248 (1956) 449.

[70] S.L. Carnie, D.Y.C. Chan and J. Stankovich, J. Colloid Interface Sci., 165 (1994) 116.
[71] J. Stankovich and S.L. Carnie, Langmuir, 12 (1996) 1453.
[72] P. Warszyński and Z. Adamczyk, J. Colloid Interface Sci., 187 (1997) 283.
[73] P.L. Levine, J. Colloid Interface Sci., 51 (1975) 72.
[74] P. Richmond, JCS Faraday Trans. II, 70 (1974) 1066.
[75] P. Richmond, JCS Faraday Trans. II, 71 (1975) 1154.
[76] L. Song and P.R. Johnson and M. Elimelech, Environ. Sci. Technol., 28 (1994) 1164.
[77] Z. Adamczyk, Colloids Surf., 39 (1989) 1.
[78] H. Krupp, Adv. Colloid Interface Sci., 1 (1967) 111.
[79] M. Elimelech and C.R.O. Melia, Langmuir, 6 (1990) 1153.
[80] M.C. Herman and J.Y. Papadopoulos, J. Colloid Interface Sci., 136 (1990) 385.
[81] M.C. Herman and J.Y. Papadopoulos, J. Colloid Interface Sci., 142 (1991) 331.
[82] L. Suresh and K.Y. Walz, J. Colloid Interface Sci., 183 (1996) 199.
[83] L. Suresh and K.Y. Walz, J. Colloid Interface Sci., 196 (1997) 177.
[84] M. Kostaglou and A.J. Karabelas, J. Colloid Interface Sci., 151 (1992) 34.
[85] J.N. Israelachvili, Intermolecular and Surfaces Forces, 2nd ed, Academic Press, New York, 1992.
[86] S.-J. Park, Van der Waals interactions at surfaces, in "Encyclopedia of Surface and Colloid Science" (A.T. Hubbard, ed.), p. 5570, Marcel Dekker, New York, 2002.
[87] W.H. Keesom, Phys. Zeit., 22 (1921) 129.
[88] P. Debye, Phys. Z., 21 (1920) 178.
[89] J. Mahanty and B.V. Ninham, Dispersion Forces, Academic Press, New York, 1976.
[90] J.N. Israelachvili and D. Tabor, Prog. Surf. Membr. Sci., 7 (1973) 1.
[91] F. London, Phys. Z., 63 (1930) 245.
[92] F. London, Trans. Faraday Soc., 33 (1937) 8.
[93] E.M. Lifshitz, Soviet Phys. JETP, 2 (1956) 73.
[94] I.E. Dzyaloshinski, E.M. Lifshitz and L.P. Pitaevski, Adv. Phys., 10 (1961) 165.
[95] B.W. Ninham, V.A.Parsegian and G. Weiss, J. Statist. Phys., 2 (1970) 323.
[96] J.H. Schenkel and J.A. Kitchener, Trans. Faraday Soc., 56 (1960) 161.
[97] H.B.G. Casimir and D. Polder, Phys.Rev., 73 (1948) 86.
[98] D.A. Mc Quarrie, Statistical Mechanics, Harper & Row, New York, 1977.
[99] L. Verlet, Phys. Rev., 165 (1968) 201.
[100] D. Henderson and S.G. Davison, Equilibrium Theory of Liquids and Liquid Mixtures, In: H. Eyring D. Henderson and W. Jost (eds), Physical Chemistry, an Advanced Treatise Vol. II, Academic Press, NY (1967) p.339.
[101] H.C. Hamaker, Physica, 4 (1937) 1058.
[102] A. Suzuki, N.F.H. Ho and W. I. Higuchi, J. Colloid Interface Sci., 29 (1969) 552.
[103] J.L. M.J. Van Bree, J.A. Poulis and B.J. Verhaar, Physica, 78 (1974) 187.
[104] J. Czarnecki, J. Colloid Interface Sci., 72 (1979) 361.
[105] J. Czarnecki and T. Dabroś, J. Colloid Interface Sci., 78 (1980) 25.
[106] J.Y. Waltz, Adv. Colloid Interface Sci., 74 (1978) 119.
[107] J. Gregory, Adv. Colloid Interface Sci., 2 (1969) 396.
[108] W.B. Russel, D.A. Saville and W.R. Schowalter, Colloidal Suspensions, Cambridge University Press, Cambridge, MA, 1993.
[109] B.V. Derjaguin, L.D. Landau, Acta Physicochem., URSS 14 (1941) 733.

[110] Z. Adamczyk, Adv. Colloid Interface Sci., 100–102 (2003) 267.
[111] J.A. Barker and D. Henderson, J. Chem. Phys., 47 (1967) 4714.
[112] M. Smoluchowski, Phys. Zeit., 17 (1916) 557.
[113] Z. Adamczyk, T. Dąbroś, J. Czarnecki and T.G.M. van de Ven, Adv. Colloid Interface Sci., 19 (1983) 183.
[114] Z. Adamczyk, B. Siwek, M. Zembala and P. Belouschek. Adv.Colloid Interface Sci., 48 (1994) 151.
[115] Z. Adamczyk and T.G.M. van de Ven, J. Colloid Interface Sci., 97 (1984) 68.
[116] Z. Adamczyk, T. Dąbroś, J. Czarnecki and T.G.M. van de Ven, J. Colloid Interface Sci., 97 (1984) 91.
[117] D.S. Rimai, L.P.DeMejo, R.Bowen and J.D. Morris, Particles at Surfaces: Adhesion Induced Deformations, in "Particles on Surfaces, Detection, Adhesion and Removal" (K.L. Mittal, ed.), Marcel Dekker, Inc., New York, 1995.
[118] H. Hertz and J. Reine, Angew. Math., 92 (1882) 156.
[119] B.W. Derjaguin, W.M. Muller and J.P. Toporow, J. Colloid Interface Sci., 53 (1975) 314.
[120] W.M. Muller, W.S. Juszczenko and B.W. Derjaguin, J. Colloid Interface Sci., 77 (1980) 91.
[121] W.M. Muller, W.S. Juszczenko and Derjaguin, J. Colloid Interface Sci., 92 (1983) 92.
[122] R.S. Bradley, Philos. Mag., 13 (1932) 853.
[123] D. Maugis, J. Colloid Interface Sci., 150 (1992) 243.
[124] K.L. Johnson, K. Kendall and Robersts, Proc. R. Soc. Lond.A, 24 (1971) 301.
[125] K.L. Johnson and J.A. Greenwood, J. Colloid Interface Sci., 192 (1997) 326.
[126] S. Asakura and F. Oosawa, J. Chem. Phys. 22 (1954) 1255.
[127] A. Vrij, Pure Appl., Chem. 48 (1976) 471.
[128] K.Y. Waltz and A. Sharma, J. Colloid Interface Sci., 168 (1994) 485.
[129] M. Piech, P. Weronski, X. Wu and J.Y. Walz, J. Colloid Interface Sci., 247 (2002) 327.
[130] H. De Hek and A. Vrij, J. Colloid Interface Sci., 84 (1981) 409.
[131] D.H. Napper, J. Colloid Interface Sci., 58 (1977) 390.
[132] J.M.A.M. Scheutjens and G.J. Fleer, Adv.Colloid Interface Sci., 16 (1982) 361.
[133] G.J. Fleer, J.M.A.M. Scheutjens, M.A. Cohen-Stewart, T. Cosgrove and B. Vincent, Polymers at Interfaces, Chapmann & Hall, London, 1993.

Chapter 3

Dissipative Interactions

3.1. INTRODUCTION

As demonstrated in the previous chapter, the electrostatic and van der Waals interactions are short-ranged, hence they can influence particle motion over microscopic distances only. Therefore, for an efficient transfer of particles over macroscopic distances – required in adsorption, deposition, and filtration – other modes of transportation have to be applied. The most efficient in this respect seems convection, defined as macroscopic motion of the dispersing medium usually provoked by external forces. Fluid streams exert hydrodynamic forces and torques on particles dragging them in a well-defined way toward interfaces.

It is useful to define the forces that drive the flows which often have a complex character. Take for example waves over an ocean, clouds, winds, including hurricanes and tornados, rivers and streams, waterfalls, etc. Yet, despite enormous diversity of these flows, they all are driven by gravity alone (density differences due to heating of air masses for example). Only in the microscale does the action of van der Waals interactions become visible, making droplets round and decorating waves with foamy layers. Another important effect due to gravity is sedimentation of particles, producing flows of a very complicated nature, especially in concentrated systems.

On the other hand, various flows exploited in the laboratory and in the industry to enhance mass transfer efficiency, most often ordinary stirring, are driven by electromagnetic forces (electric motors) or chemical forces (combustion motors) originating from electron transfer as mentioned before. Macroscopic flows in channels and ducts, pertinent to microfluidics, can also be generated by the electric Lorenz force (discussed in the previous section) applied via the system of electrodes. This electroosmotic flow appears because of the space charge in the diffuse part of the double layer near interfaces.

Similarly, the force exerted by fluid streams on stationary or slower moving objects is transmitted via the short-ranged Lennard-Jones interactions, which occur because of electron cloud repulsion.

Once the forces are defined, one can quantitatively analyze fluid motion by formulating some basic relationships which are, in essence, the momentum and mass conservation postulates. Because of nonlinearity of these equations and high dimensionality of hydrodynamic problems there exist few exact solutions. Most of the problems of vital practical significance, e.g., particle motion in suspensions, can only be solved in an approximate way with the assumption of slow motion.

Accordingly, the organization of this chapter is as follows: in the first section the governing fluid motion, as well as their limiting forms, will be formulated. Also, the boundary conditions at solid and liquid interfaces will be specified together with forces and torques acting on interfaces as well as on moving particles. In the second section the macroscopic flows of practical importance will be analyzed in detail, especially the confined flows near interfaces (e.g., channel flows), the shearing and stagnation flows including the impinging-jet flows often used in experiments. A method of decomposing these macroscopic flows into simple (elementary) flows is presented. The dynamics of particles of various shapes in a quiescent fluid is discussed as well as the wall effects influencing particle motion near interfaces. The hydrodynamic resistance matrix is defined for both the bulk and the interface region. Finally, creeping flows past particles attached to interfaces are discussed.

3.2. BASIC HYDRODYNAMIC EQUATIONS

3.2.1. General Equations of Fluid Motion

The governing equations of fluid motion can be derived from the general conservation law of a physical quantity **f** (either scalar or vectorial) stating that

$$\frac{\partial \mathbf{f}}{\partial t} + \int_S \mathbf{j}_f \cdot \hat{\mathbf{n}} dS = \mathbf{Q}_f \tag{1}$$

where the first term describes the net temporal change of the quantity **f** in the control volume δv, t is the time, the second describes the rate of transport through the surface S enclosing this volume, $\mathbf{j}_f$ the flux of this quantity, $\hat{\mathbf{n}}$ the unit normal vector, dS the surface element and $\mathbf{Q}_f$ the source term describing the overall rate of production of **f** within the volume.

By applying the Gauss divergence theorem, Eq. (1) can be expressed in the alternative form as

$$\frac{\partial \mathbf{f}}{\partial t} + \nabla \cdot \mathbf{j}_f = \mathbf{Q}_f \tag{2}$$

By assuming that **f** is the mass of the fluid element one obtains

$$\mathbf{f} = m = \int_{\delta v} \rho \, dv$$
$$\mathbf{j}_f = m\mathbf{V} = \mathbf{p}_m = \left(\int_{\delta v} \rho \, dv \right) \mathbf{V} \tag{3}$$

where ρ is the fluid density and $\mathbf{p}_m$ is the momentum vector. In this case, $\mathbf{Q}_f = 0$ since mass cannot be produced or annihilated.

Accordingly, Eq. (2) becomes

$$\int_{\partial v} \left[\frac{\partial \rho}{\partial t} + \nabla \cdot (\rho \mathbf{V}) \right] dv = 0 \tag{4}$$

Since this must be valid for an arbitrary space element, one can deduce that

$$\frac{\partial \rho}{\partial t} + \nabla \cdot (\rho \mathbf{V}) = 0 \tag{5}$$

In an analogous way the momentum balance can be performed by assuming that

$$\mathbf{f} = \mathbf{p}_m = m\mathbf{V}$$
$$\mathbf{j}_f = (m\mathbf{V}\,)\,\mathbf{V} \tag{6}$$
$$\mathbf{Q} = \mathbf{F}$$

In this case, the momentum source is the net force **F** acting on the volume that is the postulate of thc sccond law of mechanics (Newton's law).

By substituting the definitions, Eq. (6) into Eq. (1), one can express the momentum balance equation as

$$\frac{\partial(\rho \mathbf{V})}{\partial t} + \nabla\cdot(\rho \mathbf{V}\mathbf{V}) = \bar{\mathbf{F}} \tag{7}$$

where $\bar{\mathbf{F}} = \mathbf{F}/\delta V$ is the force per unit volume.

Let us consider the forces acting on the fluid element. Generally, they can be divided into body forces acting on the entire volume and the hydrodynamic force transmitted through the boundary of the volume element as a result of molecular interactions. Examples of the former are the gravitational force $\rho\mathbf{g}$ (where $\mathbf{g}$ is the gravity acceleration vector) and the Lorenz electric force $\rho_e\mathbf{E}$. On the other hand, the hydrodynamic force $\bar{\mathbf{F}}$ (per unit volume) can be expressed as

$$\bar{\mathbf{F}} = \nabla\mathbf{\Pi} \tag{8}$$

where $\mathbf{\Pi}$ is the hydrodynamic stress tensor [1,2]

By considering the hydrodynamic forces definition given by Eq. (8) one can formulate the momentum balance, Eq. (7), as

$$\frac{\partial(\rho \mathbf{V})}{\partial t} + \nabla\cdot(\rho \mathbf{V}\mathbf{V}) = \nabla\cdot\mathbf{\Pi} + \rho\mathbf{g} + \rho_e\mathbf{E} \tag{9}$$

The main problem in applying Eq. (9) to real fluid flows is evaluating the stress tensor $\mathbf{\Pi}$ that cannot be done from first principles alone. One can only deduce from thermodynamics that for a fluid at equilibrium the stress tensor equals the isotropic pressure p governed solely by state variables (temperature, density, concentration). This can be written as

$$\Pi = -p\mathbf{I} \tag{10}$$

where $\mathbf{I}$ is the unit tensor.

However, under flow conditions one can expect that the stress tensor may depend not only on the instantaneous fluid velocity but also upon the history of the fluid motion. Thus, without accepting a simplifying hypothesis based on an empirical knowledge, one cannot formulate explicitly $\mathbf{\Pi}$ in the general case. One such hypothesis postulates that $\mathbf{\Pi}$ depends solely on the local deformation gradient. Fluids that fulfill such a condition are called simple fluids [1]. By exploiting the empirical knowledge it is postulated that this dependence becomes linear for a broad class of fluids. One can then write the constitutive equation for $\mathbf{\Pi}$ in the form [1,2]

$$\mathbf{\Pi} = -p_h \mathbf{I} + 2\eta\,\Delta \tag{11}$$

where p_h is the mean pressure and Δ is the rate of deformation (strain) tensor given by [1–3].

$$\Delta = \frac{1}{2}(\nabla\mathbf{V} + \nabla\mathbf{V}^{\dagger}) - \frac{1}{3}\mathbf{I}(\nabla\mathbf{V}) \tag{12}$$

where † means the transpose operation.

The proportionality constant η in Eq. (11), which is a specific property of a fluid, is called the dynamic viscosity or shear viscosity. Water, air and other fluids composed of simple molecules exhibit Newtonian behavior with η independent of shear rate (fluid velocity). On the other hand, fluids formed by elongated molecules (molten or dissolved polymeric species) as well as fluids with microstructure (particle suspensions) exhibit non-Newtonian properties with η strongly dependent on shear rate (usually the apparent viscosity decreases with the shear rate that is called the shear thinning). The unit of dynamic viscosity in the SI system is kg m^{-1} s^{-1} called the Stokes. However, traditionally one often uses the CGS unit g cm^{-1} s^{-1} called the Poise (abbreviation P). The viscosity of water is 1.002×10^{-3} St $= 1.002\times10^{-2}$ P of the temperature 293K. On the other hand, the data for other temperatures are given in Table 3.1, as well as the values of the dynamic viscosity for other common fluids.

In practical applications it is useful to introduce the kinematic viscosity $\nu = \eta/\rho$ having the unit m^2 s^{-1} (or cm^2 s^{-1} in the CGS unit system). Values of the kinematic viscosity for water and other fluids are also given in Table 3.1.

Table 3.1
The dynamic η [kg $m^{-1}s^{-1}$] and the kinematic ν [m^2 s^{-1}] viscosity of common fluids at T = 293 K

Fluid	η [kg $m^{-1}s^{-1}$]	ρ[kg m^{-3}]	ν[m^2 s^{-1}]
Air (293 K)	181×10^{-5}	1.204	1.50×10^{-5}
Air (313 K)	1.91×10^{-5}	1.127	1.69×10^{-5}
Acetone	3.25×10^{-4}	784.5	4.14×10^{-7}
Alcohol (ethyl)	1.20×10^{-3}	789.4	1.52×10^{-6}
Alcohol (methyl)	5.97×10^{-4}	791.5	7.54×10^{-7}
Alcohol (isopropyl)	2.39×10^{-3}	803.5	2.97×10^{-6}
Benzene	5.59×10^{-4}	878.9	6.36×10^{-7}
Cyclohexane	8.26×10^{-4}	778.5	1.06×10^{-6}
Glycol (ethylene)	1.99×10^{-2}	1113	1.79×10^{-6}
Glycerol	1.49	1261	1.18×10^{-3}
n-Pentane	2.4×10^{-4}	619.6	3.87×10^{-7}
n-Hexane	3.07×10^{-4}	659.5	4.66×10^{-7}
n-Decane	9.07×10^{-4}	729.8	1.24×10^{-6}
Pyridine	9.74×10^{-4}	978.2 (298.15K)	9.96×10^{-7}
Toluene	5.17×10^{-4}	867.0	5.96×10^{-7}
Water (293.15 K)	1.002×10^{-3}	998.2	1.004×10^{-6}

The dynamic η [kg $m^{-1}s^{-1}$] and the kinematic ν [m^2 s^{-1}] viscosity of water at various temperatures

T [K]	η [kg $m^{-1}s^{-1}$]	ρ[kg m^{-3}]	ν[m^2 s^{-1}]
273.15	1.793×10^{-3}	999.8	1.793×10^{-6}
283.15	1.307×10^{-3}	999.7	1.307×10^{-6}
293.15	1.002×10^{-3}	998.2	1.004×10^{-6}
298.15	8.879×10^{-4}	997.05	8.91×10^{-7}
303.15	7.977×10^{-4}	995.6	8.012×10^{-7}
313.15	6.532×10^{-4}	992.2	6.583×10^{-7}
323.15	5.470×10^{-4}	988.0	5.536×10^{-7}
333.16	4.665×10^{-4}	983.2	4.745×10^{-7}
343.15	4.040×10^{-4}	977.8	4.132×10^{-7}
353.15	3.544×10^{-4}	971.8	3.647×10^{-7}
363.15	3.145×10^{-4}	965.4	3.258×10^{-7}
373.15	2.818×10^{-4}	958.4	2.940×10^{-7}

The hydrodynamic pressure occurring in Eq. (11) is given in the general case by [1]

$$p_h = p - \kappa_h \mathbf{V} \cdot \mathbf{V} \tag{13}$$

where the constant of proportionality κ_h (having the dimension kg s^{-1} m^{-1}) is called the dilatational viscosity or bulk viscosity or alternatively second viscosity [1–3].

By using the expression for the stress tensor, Eq. (11), one can express the momentum balance equation, Eq. (9) in the form

$$\begin{aligned}\frac{\partial(\rho\mathbf{V})}{\partial t}+\nabla\cdot(\rho\mathbf{V}\mathbf{V}) &= -\nabla p+\eta\nabla^2\mathbf{V}+\frac{1}{3}\eta\nabla(\nabla\cdot\mathbf{V})+2(\nabla\eta)\cdot\nabla\mathbf{V}\\ &\quad+\nabla\eta x(\nabla\times\mathbf{V})-\frac{2}{3}(\nabla\eta)(\nabla\cdot\mathbf{V})\\ &\quad+\kappa_h\nabla(\nabla\cdot\mathbf{V})+\nabla\kappa_h(\nabla\cdot\mathbf{V})+\rho\mathbf{g}+\rho_e\mathbf{E}\end{aligned}\tag{14}$$

Eq. (14) is the general expression for a compressible fluid flow. However, for completing the boundary value problem one additionally needs the equation of state $p = p(\rho, T)$, the dependence of the shear viscosity on the density and temperature $\eta = \eta(\rho, T)$, the energy conservation equation and the appropriate boundary and initial conditions [1]. Since in further analysis the focus is on isothermal isobaric systems, we will not analyze this problem in more detail. Instead, some limiting forms of Eq. (14) will be derived, useful for practical applications. For incompressible fluids, whose density is not dependent on pressure $\nabla\rho = 0$, the mass conservation postulate, Eq. (5) simplifies to

$$\nabla\cdot\mathbf{V}=0\tag{15}$$

Consequently, Eq. (14) becomes

$$\rho\left(\frac{\partial\mathbf{V}}{\partial t}+\mathbf{V}\cdot\nabla\mathbf{V}\right)=-\nabla p+\eta\nabla^2\mathbf{V}+\rho\mathbf{g}+\rho_e\mathbf{E}\tag{16}$$

When there is no external electric field, Eq. (16) reduces to

$$\rho\left(\frac{\partial\mathbf{V}}{\partial t}+\mathbf{V}\cdot\nabla\mathbf{V}\right)=-\nabla p+\eta\nabla^2\mathbf{V}+\rho\mathbf{g}\tag{17}$$

Eq. (16) is traditionally called the Navier-Stokes equation, which was derived in 1827 [4].

By using the vectorial identity $\nabla^2 \mathbf{V} = -\nabla\times\boldsymbol{\omega}$ (where $\boldsymbol{\omega} = \nabla\times\mathbf{V}$ is the fluid vorticity), one can formulate Eq. (17) in the alternative form [2]

$$\rho\left(\frac{\partial \mathbf{V}}{\partial t}+\mathbf{V}\cdot\nabla\mathbf{V}\right)=-\nabla p+\eta\nabla\times\boldsymbol{\omega}+\rho\mathbf{g} \tag{18}$$

Eq. (18) indicates that the viscous effects are only important in regions of the fluid where the vorticity becomes significant. This means that irrotational flows behave like inviscid (ideal fluid devoid of viscosity) flows governed by the equation (characterized by zero vorticity)

$$\rho\left(\frac{\partial \mathbf{V}}{\partial t}+\mathbf{V}\cdot\nabla\mathbf{V}\right)=-\nabla p+\rho\mathbf{g} \tag{19}$$

Eq. (19) is called Euler's equation. It is of the first order in respect to velocity in contrast to the Navier–Stokes equation being of the second order. This facilitates significantly its mathematical handling.

The Navier–Stokes equation can also be expressed in the alternative form [2]

$$\frac{\partial \mathbf{V}}{\partial t}+\nabla\left(\frac{1}{2}\mathbf{V}\cdot\mathbf{V}+\frac{p}{\rho}-\mathbf{g}\cdot\mathbf{r}\right)=\mathbf{V}\times\boldsymbol{\omega}-\nu\nabla\times\boldsymbol{\omega} \tag{20}$$

where $\mathbf{r}$ is the position vector measured relatively to a fixed point, and

$$\frac{1}{2}\mathbf{V}\cdot\mathbf{V}+\frac{p}{\rho}-\mathbf{g}\cdot\mathbf{r}=\Phi_B \tag{21}$$

is the Bernoulli function. It represents the energy of the fluid stream per unit mass consisting of kinetic energy, internal energy due to pressure and potential energy due to gravity. For many flows of practical significance (pipe or channel flows), the function Φ_B remains almost constant over the entire domain of flow that can be exploited for estimating variations in the mean fluid velocity. For example, one can deduce from Eq. (21) that the local value of the hydrodynamic pressure in the absence of gravity $p/\rho = \Phi_B - \frac{1}{2}\mathbf{V}\cdot\mathbf{V}$ decreases in places where the flow rate increases. This is the case when the pipeline cross-section is reduced (the Venturi effect). Obviously, in regions where the flow rate becomes smaller, the hydrodynamic pressure increases. This effect is exploited, for example, to create the lift force by a special configuration of airplane wings.

3.2.2. Stokes Equation – Creeping Flows

As can be deduced, the Navier–Stokes equation, Eq. (17), is nonlinear in respect to fluid velocity, that makes its analytical solutions in the general case cumbersome. In most cases it is necessary to apply sophisticated numerical methods. However, for many situations of practical interest fluid motion becomes slow or the spatial extension of the flow is very limited. If this is the case, one can carry out the mathematical analysis of flows by using a simplifying version of the Navier–Stokes equation. These limiting forms can be derived by applying the dimensional analysis using scaling variables such as the characteristic length L_{ch}, characteristic fluid velocity V_{ch}, and characteristic time T_{ch}. Accordingly, one can define the dimensionless variables in the Navier–Stokes equation

$$\bar{\mathbf{V}} = \frac{\mathbf{V}}{V_{ch}}, \quad \bar{\mathbf{r}} = \frac{\mathbf{r}}{L_{ch}}, \quad \bar{t} = \frac{t}{T_{ch}}, \quad \bar{p} = p\left(\frac{L_{ch}}{\eta V_{ch}}\right)$$
$$\bar{\nabla} = L_{ch}\nabla, \quad \bar{\nabla}^2 = L_{ch}\nabla^2, \quad \bar{\mathbf{g}} = \frac{\mathbf{g}}{|\mathbf{g}|}, \quad \bar{\rho}_e = \frac{e\,Le^2}{\varepsilon kT}\rho_e, \quad \bar{\mathbf{E}} = \frac{e\,Le}{kT}\mathbf{E} \tag{22}$$

where $\bar{\rho}_e$ is the reduced charge density, $\bar{\mathbf{E}}$ is the reduced field strength. By exploiting these definitions one can express the Navier–Stokes equation in the dimensionless form

$$\frac{1}{St}\frac{\partial \bar{\mathbf{V}}}{\partial \bar{t}} + Re\,\bar{\mathbf{V}}\cdot\bar{\nabla}\bar{\mathbf{V}} = -\bar{\nabla}\bar{p} + \bar{\nabla}^2\bar{\mathbf{V}} + \frac{Re}{Fr^2}\frac{\mathbf{g}}{|\mathbf{g}|} + El\bar{\rho}_e\bar{\mathbf{E}} \tag{23}$$

The dimensionless criterion numbers are

$$St = \nu\frac{T_{ch}}{L_{ch}^2}, \quad Re = \frac{V_{ch}L_{ch}}{\nu}$$
$$Fr = \frac{V_{ch}}{(gL_{ch})^{1/2}}, \quad El = \varepsilon\left(\frac{kT}{e}\right)^2\frac{L_{ch}^2}{V_{ch}\,\Delta\eta Le^3} \tag{24}$$

where St is the Stokes number characterizing the quotient of the convection time L_{ch}/V_{ch} to the diffusion time L_{ch}^2/ν, Re the Reynolds number reflecting the quotient of the nonlinear inertia term $\rho V_{ch}^2/L_{ch}$ to the viscous term $\eta V_{ch}/L_{ch}^2$, Fr the Froude number characterizing the ratio of the inertia forces to the gravitational force [2] and El is the electrostatic number characterizing the quotient of the electrostatic body force $\varepsilon\,(kT/e)^2/Le^3$ to the viscous force.

The most important one seems to be the Reynolds number which enables one to define various flow regimes. For Re comparable to, and smaller than unity, the flow becomes practically inertia-less and laminar. This means that the fluid velocity field is a deterministic function of the space and the time variables. For $Re > 1$, the inertia effects start to play a more significant role and the flow often becomes unsteady.

On the other hand, for $Re \gg 1$ the viscous effects play a role in a thin layer adjacent to solid interfaces only, called the hydrodynamic boundarylayer [5]. Its thickness decreases if $Re \ll 1$ with the Re number. Outside the boundary layer, the viscous effects are neglected and the flow is approximated by the Euler' ideal fluid equation, Eq. (19). If Re becomes very large, exceeding a critical value that depends on the geometry of the flow, the turbulent flow regime appears. In that case the velocity field, at least in some regions of space, usually close to interfaces (obstacles) placed in the flow, becomes stochastic [1–5].

Values of the Re number calculated for various flows and objects, varying from molecules and colloid particles the size of nanometers to macroscopic objects, are collected in Table 3.2. The maximum velocity of molecules and colloids was calculated by averaging the thermal motion over the distance of their diameter, and the minimum value was derived by considering a typical migration velocity of particles occurring in experiments.

As can be deduced from Table 3.2, the Re number for objects smaller than 100 μm remains smaller than unity in water and in air.

If the Re number of a flow becomes smaller than unity, the inertia term can be neglected in Eq. (23), which becomes (after reverting to the dimensional variables)

$$\rho\frac{\partial \mathbf{V}}{\partial t} = -\nabla p + \eta\nabla^2\mathbf{V} + \rho\mathbf{g} + \rho_e\mathbf{E}$$

$$\nabla\cdot\mathbf{V} = 0 \qquad (25)$$

Table 3.2
Reynolds number (Re) for various objects

Flow configuration (object) L_{ch} [m], V_{ch} [m s^{-1}]	Re, water $\nu = 10^{-6}$ m^2 s^{-1}	Re, air $\nu = 1.5 \times 10^{-5}$ m^2 s^{-1}
Molecule, ion $L_{ch} = 5 \times 10^{-10}$ (0.5 nm) $V_{ch} = 10^{-6} \div 2$	$5 \text{ x } 10^{-10} \div 10^{-3*}$	$3.3 \times 10^{-11} \div 6.7 \times 10^{-5}$
Macromolecule, protein $L_{ch} = 10^{-8}$ (10 nm) $V_{ch} = 10^{-6} \div 10^{-1}$	$10^{-8} \div 10^{-3*}$	$6.7 \times 10^{-10} \div 6.7 \times 10^{-5}$
Colloid (virus) $L_{ch} = 10^{-7}$ (100 nm) $V_{ch} = 10^{-6} \div 10^{-2}$	$10^{-7} \div 10^{-3*}$	$6.7 \times 10^{-9} \div 6.7 \times 10^{-5}$
Bacterium (cell) $L_{ch} = 5 \times 10^{-6}$ (5 μm) $V_{ch} = 5 \times 10^{-6} \div 10^{-3}$	$2.5 \times 10^{-5} \div 5 \times 10^{-3*}$	$1.7 \times 10^{-6} \div 3.3 \times 10^{-4}$
Dust particle $L_{ch} = 10^{-4}$ (100 μm) $V_{ch} = 2 \times 10^{-3}$	0.2	1.3×10^{-2}
Gas bubble $L_{ch} = 10^{-3}$ (1 mm) $V_{ch} = 0.2$	200	×
Fly $L_{ch} = 2 \times 10^{-3}$ (2 mm) $V_{ch} = 1 \div 10$	×	$1.3 \times 10^{2} \div 1.3 \times 10^{3}$
Bullet $L_{ch} = 10^{-2}$ (1 cm) $V_{ch} = 2 \times 10^{2} \div 5 \times 10^{2}$	×	$1.3 \times 10^{5} \div 3.3 \times 10^{5}$
Tennis ball $L_{ch} = 6.5 \times 10^{-2}$ (6.5 cm) $V_{ch} = 20 \div 60$	×	$8.7 \times 10^{4} \div 2.6 \times 10^{5}$
Bird (falkon) $L_{ch} = 0.2$ (20 cm) $V_{ch} = 10 \div 50$	×	$1.3 \times 10^{5} \div 6.7 \times 10^{5}$

Table 3.2 (continued)

Flow configuration (object) L_{ch} [m], V_{ch} [m s^{-1}]	*Re*, water $\nu = 10^{-6}$ m^2 s^{-1}	*Re*, air $\nu = 1.5 \times 10^{-5}$ m^2 s^{-1}
Fish (shark) $L_{ch} = 1$ $V_{ch} = 1 \div 20$	$10^6 \div 2 \times 10^7$	×
Ship $L_{ch} - 10^2$ $V_{ch} = 10–20$	$10^9 \div 2 \times 10^9$	×
Airplane $L_{ch} = 20$ $V_{ch} = 10^2 \div 10^3$	×	$1.3 \times 10^8 \div 1.3 \times 10^9$
The Earth $L_{ch} = 1.3 \times 10^7$ $V_{ch} = 4.6 \times 10^2$ (rotary motion)	×	4×10^{14}

$T = 293$K, $p = 1$atm.

*The lower velocity limit was derived from the typical electrophoretic migration velocity, the upper limit from the diffusion time over the distance of particle diameter.

In the absence of external body forces (gravitational or electrostatic), Eq. (25) is further reduced to the form

$$\rho \frac{\partial \mathbf{V}}{\partial t} = -\nabla p + \eta \nabla^2 \mathbf{V}$$
$$\nabla \cdot \mathbf{V} = 0 \tag{26}$$

Eq. (26) is often called the unsteady Stokes equation, useful for describing a certain type of flows converted with sudden acceleration or deceleration, as is the case, e.g., of bubble rise or particle impaction against interfaces.

On the other hand, if the characteristic time of flow variation T_{ch} becomes large, so $St \gg 1$, and Eq. (25) simplifies to

$$\eta \nabla^2 \mathbf{V} = \nabla p - \rho \mathbf{g} - \rho_e \mathbf{E}$$
$$\nabla \cdot \mathbf{V} = 0 \tag{27}$$

In the case where there are no external body forces, Eq. (27) assumes the common form

$$\eta \nabla^2 \mathbf{V} = \nabla p \qquad \nabla \cdot \mathbf{V} = 0 \tag{28}$$

Eq. (28) is often called the creeping flow or the Stokes equation [1,2]. Because of the absence of the explicit time derivative and the nonlinear inertia term, its analytical solutions are feasible for many situations of practical interest, e.g., for the motion of colloid particles of various shapes in a quiescent fluid, flow in capillaries (micro-channels), natural convection flows, etc. The significance of the Stokes equation is enhanced by the fact that in the vicinity of solid surfaces, all flows, even very vigorous ones at large separations, reduce to creeping flows, because the fluid velocity vector vanishes. This is the result of the no-penetration boundary condition discussed later on. Thus, the inertia terms become insignificant and the Stokes equation can be exploited to derive asymptotic expressions for the fluid velocity field.

It is interesting to point out that the lack of the explicit time derivative in Eq. (28) does not mean that the velocity field is constant in time. On the contrary, the Stokes flows are of a quasi steady character, meaning that the velocity field at a given point of space adjusts fast enough to the instantaneous position of a moving particle.

Another important property of Stokes flows is that they are symmetric with respect to the inversion of the sign of velocity, pressure and the electrostatic field. This can be verified by substituting into Eq. (27), $\mathbf{V}' = -\mathbf{V}$, $p' = -p$ and $\mathbf{E}' = -\mathbf{E}$ (obviously the direction of gravity cannot be inverted). One obtains the same equation for the primed variables that is a direct consequence of the lack of the explicit time derivative in Eq. (28). From this property one can deduce that the creeping flows, governed by the Stokes equation, are reversible. In particular, the flows involving particles having the forafter symmetry (either stationary or moving in a quiescent fluid) are symmetric. This means that fluid element trajectories (streamlines) are symmetric as well.

It is interesting to note that both Eq. (26) and the steady Stokes equation, Eq. (28), are linear, which allows for the analytical description of complex flows, e.g., the motion of spherical and anisotropic particles near interfaces. One can also use the powerful superposition method (described previously for electrostatic fields) to obtain solutions for more complicated flows, e.g., translation and rotation, by adding solutions for simple flows, e.g., either rotation or translation.

The analytical solution of the Stokes equation, Eq. (28), is facilitated by applying to it the divergence operator, which results in

$$\nabla \cdot (\eta \nabla^2 \mathbf{V} - \nabla p + \rho \mathbf{g}) = \nabla^2 p = 0 \tag{29}$$

Eq. (29), which is, in essence, the Laplace equation, can be solved for many flows of practical significance by expressing the pressure in terms of solid spherical harmonics [1,6]. Then by inserting this solution into the original Stokes equation, the velocity field can be evaluated as originally done by Lamb [7].

On the other hand, by treating the Stokes equation by the Laplacian and using Eq. (29), one can show that the velocity field fulfills the biharmonic equation

$$\nabla^4 \mathbf{V} = 0 \tag{30}$$

Eq. (29) can be further simplified in the case of axis-symmetric Stokes flows by introducing the stream function defined in such a way that the fluid velocity components are calculated as its derivatives [1–2].

3.2.3. Boundary Conditions and Hydrodynamic Forces

Any explicit formulation of the Navier–Stokes or Stokes equations for three-dimensional (3D) flows leads to four partial differential equations of the second order (three for the velocity components and one for the pressure). In order to complete this boundary problem, one has to specify appropriate boundary and initial conditions (in the case of unsteady flows). The simplest situation arises for boundaries impermeable to fluid molecules. These can be either rigid boundaries, e.g., solid/fluid interfaces or immiscible fluid/fluid interfaces. By assuming this, one can postulate that the normal component of the fluid velocity vector must be equal to the normal component of the boundary velocity. This can be expressed as

$$\mathbf{V} \cdot \hat{\mathbf{n}} = (\mathbf{U} + \boldsymbol{\Omega} \times \bar{\mathbf{r}}) \cdot \hat{\mathbf{n}} \tag{31}$$

where $\mathbf{U}$ is the boundary translation velocity vector, $\boldsymbol{\Omega}$ (is the angular velocity vector evaluated about a point within the body and $\bar{\mathbf{r}}$ is the position vector measured relative to this point.

Another boundary condition can be formulated by exploiting the vast empirical knowledge indicating that there is no slip of molecules on liquid/solid or liquid/liquid interfaces. The slip of the fluid velocity may only occur in very diluted gaseous media, which are outside our interest in this book. The no-slip conditions require that the tangential velocity component on a stationary solid boundary vanishes. Combining this with the previous requirement, one can deduce that

$$\mathbf{V} = 0 \tag{32}$$

for any stationary solid interface.

For a moving interface, one has accordingly

$$\mathbf{V} = \mathbf{U} + \mathbf{\Omega} \times \mathbf{r} \tag{33}$$

In the case of liquid interfaces, one cannot formulate such a simple expression because the interface velocity may depend on the position, so there can be local tangential motion of the interface even if it is not moving as a whole. This is the case, for example, with a liquid drop or a bubble moving in an immiscible liquid.

Instead, for fluid/fluid interfaces one can formulate the boundary condition by postulating that the tangential component of the hydrodynamic force vanishes, rather than the fluid velocity. This is so because such interfaces, often referred to as free interfaces, are unable to sustain any tangential stress. This postulate can be written in the general form [2]

$$(\mathbf{\Pi}\cdot\hat{\mathbf{n}}) \times \hat{\mathbf{n}} = \eta(\Delta\cdot\hat{\mathbf{n}})\times\hat{\mathbf{n}} = 0 \tag{34}$$

where $\mathbf{\Pi} = \mathbf{\Pi}_1 + \mathbf{\Pi}_2$ is the overall stress at the free interface, and $\mathbf{\Pi}_1$ and $\mathbf{\Pi}_2$ are the stresses in the contacting phases.

For planar interfaces, Eq. (34) can be written in the useful form

$$\eta_1\left(\frac{\partial V_1}{\partial z}\right)_1 - \eta_2\left(\frac{\partial V_2}{\partial z}\right)_2 = 0 \tag{35}$$

where η_1, η_2 are viscosities of adjacent phases, V_1, V_2 are the tangential components of fluid velocity vectors and z is the coordinate perpendicular to the interface.

It should be mentioned that in the case of curved free interfaces, there appears a pressure jump because of the appearance of van der Waals interactions discussed in the previous section. This pressure difference between phases is governed by the Laplace equation derived by realizing that the isothermal work done by the pressure $(p_1^0 - p_2^0)v$ equals the work of expanding the interface, i.e., $2\bar{R}\gamma\ S$, where γ is the specific surface energy per unit area (interfacial tension), $\bar{R} = R_1R_2/(R_1+R_2)$ and R_1, R_2 are the principal radii of curvature of the interface. Therefore, this pressure difference can be expressed as

$$p_1^0 - p_2^0 = \Delta p^0 = 2\gamma\left(\frac{1}{R_1} + \frac{1}{R_2}\right) \tag{36}$$

Using Eq. (36), one can express the overall hydrodynamic pressure jump at curved interfaces as

$$p_1 - p_2 = 2\gamma\left(\frac{1}{R_1} + \frac{1}{R_2}\right) + 2\left[(\eta_1\nabla\mathbf{V}_1 - \eta_2\nabla\mathbf{V}_2)\cdot\hat{\mathbf{n}}\right]\hat{\mathbf{n}} \tag{37}$$

It should be mentioned that Eq. (37) is valid for free interfaces under thermodynamic equilibrium conditions. This may not be the case for problems connected with bubble or droplet motion, when transport of surface-active substances to interfaces occurs. This produces interfacial tension gradients, which results in additional stress at the interface. The stress due to interfacial tension gradients, often leading to macroscopic motion of interfaces, is called the Marangoni effect. Since these problems are very specific and system-dependent (they are a function of diffusion coefficients of the surfactant involved) they will be not analyzed in more detail here. An interested reader is asked to consult the specialized monographs [8,9].

In order to complete the boundary value problem represented by the Navier–Stokes or Stokes equations, one also has to specify the conditions in the bulk (far from the interfaces) as well. Obviously in quiescent media, one can immediately write $\mathbf{V} = 0$ far away from moving interfaces. In the case of flows near stationary interfaces (particles), one can express the boundary condition far from the interface in the more general form

$$\frac{|\mathbf{U}|}{|\mathbf{V}_\infty|} \to 0 \tag{38}$$

where $\mathbf{V}_\infty$ is the magnitude of the macroscopic flow at infinity and $\mathbf{U}$ is the asymptotic velocity of the disturbing flow stemming from the interface.

Eq. (38) does not require that the disturbing flow vanishes at infinity but merely that it becomes negligible in comparison with the macroscopic (driving) flow. The critical point by formulating Eq. (38) is to define how far the infinity is in terms of the characteristic length scale of a given flow. Often in numerical calculations one has to truncate the physical domain where the flow is considered. Then a careful asymptotic analysis of the far field velocity field is required to eliminate the possible source of artifacts.

Once the boundary conditions at interfaces confining the flow are specified, the boundary value problem represented by the fluid motion equations becomes complete. It can be then solved, either by analytical or numerical methods, producing components of the fluid velocity and their gradients in space, as well as the hydrodynamic pressure distribution. The knowledge of these parameters enables one to evaluate forces and torques on fluid element and interfaces, which are quantities of a primary practical interest, for calculating resistance coefficients of various particles. The formulae for the force can be most directly derived by integrating the hydrodynamic stress tensor and the body force over the volume element

$$\mathbf{F} = \int_{\delta_v} (\nabla\cdot\mathbf{\Pi} + \rho\mathbf{g} + \rho_e\mathbf{E})dv = m_f\mathbf{g} + q\mathbf{E} + \mathbf{F}_h \tag{39}$$

where m_f is the mass of the fluid element, q the charge of the fluid, and $\mathbf{F}_h = \int_{\delta v} \nabla\Pi dv$ the hydrodynamic force.

Eq. (39) remains valid if the electric charge distribution is not affected by the flow. Otherwise, when the flow perturbs the equilibrium charge distribution in the diffuse part of the double layer, the ion transport has to be considered together with the Navier–Stokes equation [6].

By exploiting the Gauss theorem, one can express the hydrodynamic force in the useful form involving a surface integral

$$\mathbf{F}_h = \int_S \mathbf{\Pi}\cdot\hat{\mathbf{n}}dS = -\int_S p\mathbf{I}\cdot\hat{\mathbf{n}}dS + 2\eta\int_S \mathbf{\Delta}\cdot\hat{\mathbf{n}}dS \tag{40}$$

where S is the entire surface enclosing the fluid.

The first term on the right-hand side of Eq. (40) describes the form drag, the second the viscous drag (skin friction).

The hydrodynamic torque is given by

$$\mathbf{To}_h = \int_S (\mathbf{F}\times\bar{\mathbf{r}})\cdot\hat{\mathbf{n}}dS \tag{41}$$

where **r** is the radial position vector measured against a reference point.

3.3. MACROSCOPIC FLOWS NEAR INTERFACES

In this section we discuss in some detail macroscopic flows for simple geometries of the interfaces when analytical solutions of the Navier–Stokes equation can be derived. Emphasis will be placed on fluid velocity fields near solid/liquid interfaces. Evaluating velocity components is necessary for calculating mass transfer rates to solid interfaces, in particular colloid particle deposition kinetics as discussed in the next section. These flows also have major significance in nature, various technological processes and in experimental studies concerned with protein, polymer and colloid adsorption.

3.3.1. Laminar Flows in Channels

We consider first the unidirectional flows in channels, distinguished by the fact that in a certain coordinate system all components of the fluid velocity vanish except for one that is constant in the direction of flow. By denoting this velocity component, directed along the x axis as V_x (see Fig. 3.1) one can express the Navier–Stokes equation (with the electrostatic term independent of the longitudinal coordinate x) in the form

$$\rho\frac{\partial V_x}{\partial t} = -\nabla p(t)+\eta\nabla^2 V_x+\rho_e E_x \tag{42}$$

where ∇p is the pressure gradient along the x axis, which may depend on time in the general case, E_x is the external electric field in the x direction and ∇^2 is the two-dimensional Laplace operator given for a planar (rectangular) geometry by the expression

$$\nabla^2 = \frac{\partial^2}{\partial y^2}+\frac{\partial^2}{\partial z^2} \tag{43}$$

For many situations occurring in practice, the width of channel $2c$ is much greater than its height $2b$, so the Laplace operator becomes simply $\frac{\partial^2}{\partial y^2}$). In this case the fluid velocity depends on the y coordinate alone.

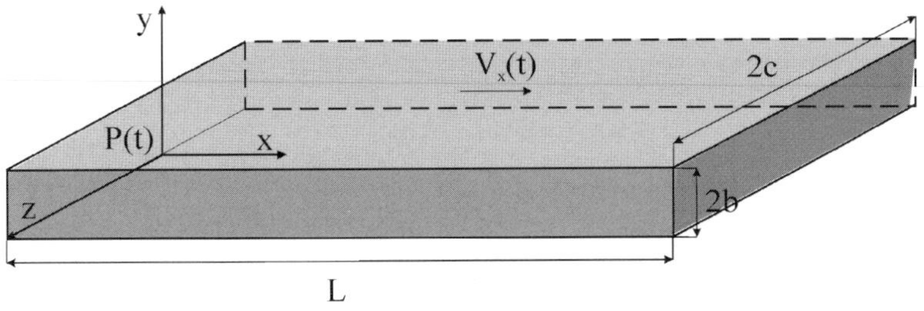

Fig. 3.1. A schematic view of the unidirectional flow in a parallel-plate channel.

An analogous situation occurs for circular channel flows (see Fig. 3.2), where the Laplace operator becomes $(1/r)(\partial/\partial r)r(\partial/\partial r)$ and the velocity depends solely on the radial coordinate r.

Note that in both cases the continuity equation $\nabla \cdot \mathrm{V} = (\partial V_x/\partial x)$ is fulfilled automatically because V_x is independent on x.

Eq. (42) is valid outside the entrance region whose length for thin channels remains a small fraction of the entire channel length. It can be estimated by comparing the momentum diffusion time b^2/ν with the convection time L_t/V_{ch} that the transition length $L_t \sim bRe$ (where $Re = bV_{ch}/\nu$. A more precise value of L_t can be calculated by using the boundary layer theory [5] as discussed later on.

The flow in the channel can be driven either by gravity (hydrostatic pressure gradient), the external pressure gradient (pumping), the motion of the side walls of the channel (uniform or oscillatory) or by the external electrostatic field. In the latter case fluid motion, called electroosmotic flow [10], appears because of the existence of the diffuse double layer adjacent to the walls of the channel.

Let us first consider stationary flows in channels that are established after a short transition time (estimated later on) in the case where the pressure $P(t)$ is constant over time. For a rectangular channel, the stationary flow distribution can be most directly found using the method of separation of variables. In this way one obtains the solution in the form of an infinite series [11,12]

$$\frac{V_x}{V_{mx}} = 1 - \left(\frac{y'}{b}\right)^2 - \frac{32}{\pi^3}\sum_{n=0}^{\infty}\frac{(-1)^n}{(2n+1)^3}\cos\frac{(2n+1)\pi y'}{2b}\,\frac{\cosh((2n+1)\pi z/2b)}{\cosh((2n+1)\pi c/2b)} \tag{44}$$

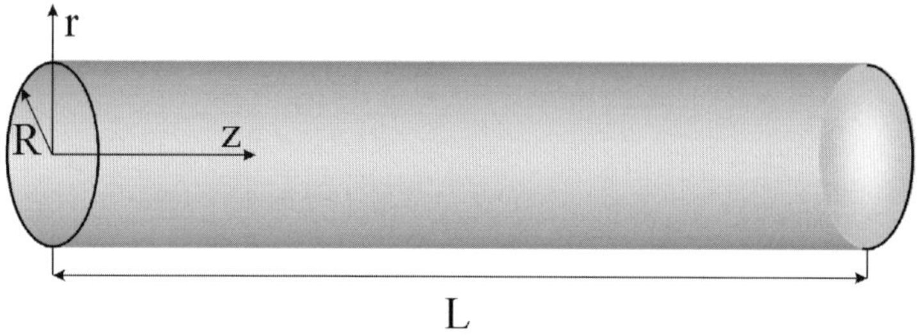

Fig. 3.2. A schematic view of the unidirectional flow in a cylindrical channel (a capillary).

where $V_{mx} = Pb^2/2\eta L$ is the maximum velocity in the middle of the channel, $y' = y - b$, P/L is the uniform pressure gradient along the channel and L is the channel length (see Fig. 3.1).

The gradient of the velocity at the (x,z) plane (horizontal walls, $y' = \pm b$) is given by [12]

$$\left(\frac{\partial V_x}{\partial y}\right)_{\pm b} = G_y^0(z) = \mp\frac{2V_{mx}}{b}\left[1 - \frac{8}{\pi^2}\sum_{n=0}^{\infty}\frac{1}{(2n+1)^2}\frac{\cosh((2n+1)\pi z/2b)}{\cosh((2n+1)\pi c/2b)}\right] \quad (45)$$

Analogously, for the x,y plane (side walls, $z = \pm c$), one has

$$\left(\frac{\partial V_x}{\partial z}\right)_{\pm c} = G_z^0(y') = \frac{V_{mx}}{b}\frac{16}{\pi^2}\sum_{n=0}^{\infty}\frac{(-1)^n}{(2n+1)^2}\tanh\frac{(2n+1)\pi c}{2b}\cos\frac{(2n+1)\pi\, y'}{2b} \quad (46)$$

The velocity gradient averaged over the base and side walls is

$$\begin{aligned}\langle G_y^0\rangle &= \frac{1}{2c}\int_{-c}^{c} G_y^0(z)\,dz = \frac{2V_{mx}}{b}\left[1 - \frac{16b}{\pi^3 c}\sum_{n=0}^{\infty}\frac{1}{(2n+1)^3}\tanh\frac{(2n+1)\pi c}{2b}\right] \\ \langle G_z^0\rangle &= \frac{1}{2b}\int_{-b}^{b} G_z^0(y')\,dy' = \frac{V_{mx}}{b}\frac{32}{\pi^3}\sum_{n=0}^{\infty}\frac{1}{(2n+1)^3}\tanh\frac{(2n+1)\pi c}{2b}\end{aligned} \quad (47)$$

Hence, the hydrodynamic force on the walls of the channel is

$$F_h = 4\eta L\left[c\langle G_y^0\rangle + b\langle G_z^0\rangle\right] = 8\eta V_{mx} L\frac{c}{b} \tag{48}$$

On the other hand, the averaged linear velocity in the channel is [12]

$$\begin{aligned}\langle V_x\rangle &= \frac{1}{4bc}\int_{-b}^{b}\int_{-c}^{c} V(y',z)\,dy'dz' \\ &= V_{mx}\left[\frac{2}{3} - \frac{128b}{c\pi^5}\sum_{n=0}^{\infty}\frac{1}{(2n+1)^5}\tanh\frac{(2n+1)\pi c}{b}\right]\end{aligned} \tag{49}$$

Consequently, the volumetric flow rate Q is given by

$$Q = 4bc\langle V\rangle = \frac{4}{3}\frac{Pb^3c}{\eta L}\left[1 - \frac{192b}{c\pi^5}\sum_{n=0}^{\infty}\frac{1}{(2n+1)^5}\tanh\frac{(2n+1)\pi c}{b}\right] \tag{50}$$

Eqs. (44–50) are valid for an arbitrary b/c ratio, including the case $b/c=1$ (a channel of square cross-section). However, under usual experimental conditions $b/c \ll 1$, and the expressions for the wall shear rate and the averaged fluid velocity simplify to

$$\langle G_y^0\rangle = \frac{2V_{mx}}{b}\left(1 - \frac{16b}{\pi^3 c}\sum_0^{\infty}\frac{1}{(2n+1)^3}\right) = \frac{2V_{mx}}{b}\left(1 - 0.540\frac{b}{c}\right)$$

$$\langle G_z^0\rangle = \frac{V_{mx}}{b}\frac{32}{\pi^3}\sum_0^{\infty}\frac{1}{(2n+1)^3} = 1.08\frac{V_{mx}}{b} \tag{51}$$

$$\langle V\rangle = V_{mx}\left(\frac{2}{3} - \frac{128b}{\pi^5 c}\right) = \frac{2}{3}\left(1 - \frac{128}{\pi^5}\frac{b}{c}\sum_0^{\infty}\frac{1}{(2n+1)^3}\right) = \frac{2}{3}V_{mx}\left(1 - 0.438\frac{b}{c}\right)$$

Eq. (51) allows one to estimate corrections due to the finite width of the channel.

In the important limiting case of $b/c \to 0$, the fluid velocity distribution in the channel depends parabolically on the y' coordinate

$$V_x = \frac{Pb^2}{2\eta L}\left[1-\left(\frac{y'}{b}\right)^2\right] = V_{mx}\left[1-\left(\frac{y'}{b}\right)^2\right] \tag{52}$$

It is interesting to note that the flow in the region close to the channel center (where $y'/b \ll 1$) can be treated as a uniform flow with velocity independent of position and equal to V_{mx}.

On the other hand, close to the channel walls, where $y' \rightarrow b$, the fluid velocity increases linearly with the distance, according to the expression

$$V_x = \frac{2V_{mx}}{b} h \tag{53}$$

where $h = y' - b$ is the distance from the wall.

The flow field described by Eq. (53), which has practical significance, is called the simple shear flow. The y' derivative of this flow is defined as the shear rate given explicitly by the expression

$$G_{y'}^0 = \left(\frac{\partial V}{\partial y'}\right)_0 = \frac{2V_{mx}}{b} = \frac{Pb}{\eta L} \tag{54}$$

A very similar flow arises in the channel of a circular cross-section (capillary) shown schematically in Fig. 3.2, driven by the external pressure gradient P/L. For the critical longitudinal distance larger than $R^2 V_\infty / \nu$ (where V_∞ is the fluid velocity of the plug flow at the entrance), the flow becomes fully developed, with the velocity having an axial component only, [2]

$$V_z = \frac{PR^2}{4\eta L}\left[1-\left(\frac{r}{R}\right)^2\right] = V_{mx}\left[1-\left(\frac{r}{R}\right)^2\right] \tag{55}$$

where $V_{mx} = Pb^2 / 4\eta L$ is the maximum velocity in the middle of the channel, and r is the distance from its axis. The flow expressed by Eq. (55), called the Poiseuille flow, occurs in capillary viscometers used to determine the viscosity of simple fluids, polymer solutions and suspensions [13]. The viscosity can be calculated by realizing that the averaged velocity in the capillary is given by the formula

$$\langle V_z \rangle = \frac{V_{mx}}{\pi R^2} \int_0^R \left[1 - \left(\frac{r}{R} \right) \right]^2 2\pi r \, dr = \frac{1}{2} V_{mx} \tag{56}$$

Thus, the volume of fluid flowing through the channel during the time t is given by

$$v = \pi R^2 \langle V_z \rangle t = \frac{\pi R^4 P}{8\eta L} t \tag{57}$$

By measuring the time experimentally one can determine the fluid viscosity from the formula

$$\eta = \frac{\pi R^4 P}{8 v L} t \tag{58}$$

Often the pressure gradient P/L is due to the hydrostatic pressure of the fluid itself, so $P \sim \rho$.

One can deduce from Eq. (55) that in the region close to the channel wall, where $r \to R$, the flow reduces to the simple shear flow

$$V_z = \frac{PR}{2\eta L} h \tag{59}$$

where $h = r - R$ is the distance from the wall.

The wall shear rate is constant, and given by the formula

$$\left(\frac{\partial V_z}{\partial r} \right)_0 = G^0 = \frac{PR}{2\eta L} \tag{60}$$

Other Poiseuille flows in channels of elliptical, triangular and more complicated geometry are discussed in Ref. [2].

The flow in the parallel-plate and cylindrical channels can be generated even if there is no pressure gradient but the walls are moving. By assuming that the upper wall is moving with a constant velocity V_0 in the x

direction, and that $b/c << 1$, one obtains the stationary solution of the flow in the simple form

$$V_z = \frac{V_0}{2b} y \tag{61}$$

The flow described by Eq. (61) is called the Couette flow. The shear rate $V_0 / 2b$ is constant throughout the entire channel. This means that by moving the upper wall, one can produce simple shear flows over macroscopic volumes. Interestingly enough, the shear rate is independent of the fluid velocity, in contrast to pressure-driven flows, where the velocity is inversely proportional to the viscosity.

The simple shear flow generated by plate motion over solid/liquid interface was exploited in the colloid collider set-up used for measuring particle trajectories in the vicinity of spheres attached to the surface [14].

Another type of stationary flow of practical significance can be generated by an external electric field. In the absence of the pressure gradient, a flow of this type, known as electroosmotic flow, is governed by the equation

$$\eta \nabla^2 V + \rho_e E_x = 0 \tag{62}$$

By utilizing the Poisson equation discussed in the previous chapter, one can express Eq. (62) in the simple form by assuming that ε and E_x do not depend on the spatial coordinates

$$\nabla^2 (\eta V + \varepsilon \psi E_x) = 0 \tag{63}$$

By integrating Eq. (63) twice with the boundary conditions $V=0$, $\psi=\zeta$ at the wall (where ζ is the zeta potential of the wall) and $\psi=0$, $d\psi/dy = 0$ in the bulk, one can show that the fluid velocity is given by the formula [10]

$$V = \frac{\varepsilon E_x}{\eta} (\zeta - \psi) \tag{64}$$

This equation, first derived by Smoluchowski [15], is valid for channels of an arbitrary cross-section. It can be easily deduced from Eq. (64) that in the region outside the double layer the flow becomes uniform (often called plug flow) with the velocity

$$V = \frac{\varepsilon E_x}{\eta}\zeta \tag{65}$$

On the other hand, the flow rate vanishes to zero in the region of the double layer. By using the linearized potential distribution in the double layer, given by Eq. (184) in chapter 2, one can express Eq. (65) explicitly as

$$V = \frac{\varepsilon E_x}{\eta}(1 - e^{-y/Le}) \tag{66}$$

Eq. (66) suggests that the uniform flow velocity is attained exponentially. It is interesting to note that from a practical viewpoint, the fluid in the electroosmotic flow appears to slip past the channel surface with a constant velocity proportional to the zeta potential of the interface and the applied field. It can be estimated from Eq. (65) that for a field of 100 V m^{-1}, zeta potential of 0.05 V, the electroosmotic flow rate in water is 3.5×10^{-6} m s^{-1} ($\eta = 1.002 \times 10^{-3}$ kg$m^{-1}s^{-1}$). This value can be neglected for channels of macroscopic dimensions because the usual flow rates are of the order of 10^{-2} to 1m s^{-1}.

The solutions described by Eqs. (64)–(66) are valid if the channel dimensions are much smaller than the electrical double-layer thickness *Le*, varying between 1 and 100 nm as previously estimated. For smaller channel dimensions, more complex flows appear, being the domain of microfluidics treated extensively in a recent monograph [16].

Having established the steady-state flow distributions in the channels, one can analyze various non-stationary flows of practical significance, e.g., oscillatory flows. This can be done by solving the non-steady Navier–Stokes, which upon neglecting the electrostatic term assumes the form

$$\frac{\partial V}{\partial t} = -\frac{\nabla p(t)}{\rho} + \nu \nabla^2 V \tag{67}$$

The initial and boundary conditions are

$$V = 0 \quad \text{for} \quad t = 0 \quad \text{everywhere in the channel}$$

$$\begin{aligned} V &= 0 \quad \text{for} \quad y = 0 \\ V &= V_0(t) \quad \text{for} \quad y = 2b \end{aligned} \tag{68}$$

A general solution of Eqs. (67),(68) can be derived in terms of the Laplace transformation [17]. Using this transformation, Eq. (67) becomes

$$s\tilde{V}(s) = -\frac{\nabla\tilde{p}(s)}{\rho} + \nu\nabla^2\tilde{V}(s) \tag{69}$$

where s is the Laplace transformation parameter, and the transformed variables are defined as [17]

$$\tilde{V}(s) = \int_0^\infty V(t)e^{-st}\,dt$$

$$\frac{\nabla\tilde{p}(s)}{\rho} = \int_0^\infty \frac{\nabla p(t)}{\rho}\,e^{-st}dt \tag{70}$$

Because Eq. (69) is a linear differential equation of a second order, it can be solved analytically for both planar and cylindrical geometry. Usually, the solution can be expressed as the sum of the particular solution of the inhomogeneous equation, Eq. (69), and the general solution of the homogeneous equation

$$\nabla^2\tilde{V}(s) = \frac{s}{\nu}\tilde{V}(s) \tag{71}$$

Eq. (71), which has the form of a linearized version of the Poisson–Boltzmann equation, can be solved by the method of separation of variables for the rectangular geometry, or can be expressed in terms of modified Bessel functions of the first order in the case of a circular channel.

By knowing $\tilde{V}(s)$, one can obtain the general solution for $V(t)$ using the inversion theorem for the Laplace transformation [17]

$$V(t)=\frac{1}{2\pi i}\int_{\gamma_i-i_\infty}^{\gamma_i+i_\infty} e^{st}\,\tilde{V}(s)ds \tag{72}$$

However, in the general case the inversion procedure can be quite awkward, especially for two-dimensional problems (flow in the rectangular channel). Let us, therefore, consider some limiting cases of practical significance. For the Couette flow of planar, one-dimensional geometry, driven by the upper plate motion $V_0(t)$, the transformed velocity is given by

$$\tilde{V}(s)=V_0(s)\;\frac{\sinh\left(\frac{s}{v}\right)^{1/2}y}{\sinh\left(\frac{s}{v}\right)^{1/2}2b} \tag{73}$$

When the distance between plates b increases to infinity, Eq. (73) becomes

$$\tilde{V}(s)=V_0(s)e^{-(s/v)^{1/2}y} \tag{74}$$

where y is now measured from the upper plate.
The inversion of Eq. (74) gives [17]

$$V(y,t)=\frac{y}{2(\pi v)^{1/2}}\int_0^t V_0(t)\frac{e^{-y^2/4v(t-\tau)}}{(t-\tau)^{3/2}}d\tau \tag{75}$$

In the case when the upper plate is suddenly set in motion with constant velocity V_0, Eq. (75) simplifies to

$$V(y,t)=V_0\,erfc\frac{y}{2(vt)^{1/2}} \tag{76}$$

where

$$\operatorname{erfc} x = 1 - \operatorname{erf} x = 1 - \frac{2}{(\pi)^{1/2}} \int_0^x e^{-\xi^2}\, d\xi \tag{77}$$

and erf is the error function.

The shear rate at the upper wall is

$$\left(\frac{\partial V}{\partial y}\right)_{2b} = \frac{V_0}{(\pi \nu t)^{1/2}} \tag{78}$$

It can be seen that the shear rate decreases with $t^{1/2}$ because of diffusion of momentum into the fluid.

In the case of a finite distance between the walls, the solution for the velocity (when the upper plate is suddenly set in motion) is given by the infinite series [17]

$$V = V_0 \sum_{n=0}^{\infty} \left[\operatorname{erfc} \frac{(2n+1)2b - y}{2(\nu t)^{1/2}} - \operatorname{erfc} \frac{(2n+1)2b + y}{2(\nu t)^{1/2}} \right] \tag{79}$$

where the distance y is now measured from the lower wall.

For longer times it is more efficient to use the alternative solution [17]

$$V(y,t) = \frac{V_0 y}{2b} + \frac{V_0}{\pi} \sum_{n=1}^{\infty} \frac{(-1)^n}{n} e^{-n^2 \pi^2 \nu t/(2b)^2} \sin \frac{n\pi y}{2b} \tag{80}$$

The shear rate at the lower wall is

$$\left(\frac{\partial V}{\partial y}\right)_0 = \frac{V_0}{2b} + \frac{V_0}{b} \sum_{n=1}^{\infty} \frac{(-1)^n}{n} e^{-n^2 \pi^2 \nu t/(2b)^2} \tag{81}$$

As can be noticed from Eq. (78) and Eq. (80), in the case of non-stationary flows the shear rate is dependent on the fluid viscosity. However, the non-stationary Couette flow prevails for a short transition time only, which can be easily estimated from Eq. (80). By considering the leading term in the series expansion, which is the largest for all times, one can deduce that the steady-state velocity profile is established when

$$t > \frac{(2b)^2}{\nu\pi^2} \sim \frac{b^2}{\nu} \tag{82}$$

Values of this relaxation time for water and air are collected in Table 3.3. As can be noticed, for a distance of the order of 100 nm (characteristic of colloid dimensions), the hydrodynamic relaxation time is extremely small, of the order of 10^{-8} s in water. This means that the momentum diffusion over such small distances is so fast, that the fluid velocity field will adjust almost immediately to the local position of a moving particle. The relaxation time only becomes appreciable of the order of a second and greater for distances of the order of 10^{-3} m (1 mm).

Solutions for non-stationary flows of other types can be derived form Eq. (74). For example, in the case of oscillatory motion of the upper plate, where $V(t) = V_0 \sin(\omega t)$ (where ω is the angular velocity), $\tilde{V}(s) = V_0\omega$ $(s^2 + \omega^2)$ sin (ωt) and the inversion of Eq. (74) gives the solution [17]

$$V = V_{st} + \frac{2\nu V_0 \omega}{\pi} \int_0^{\infty} \frac{e^{-\nu\xi^2 t}}{\omega^2 + \nu^2\xi^4} \sin(\xi y)\xi d\xi \tag{83}$$

After a transition time of the order of b^2/ν, the second term on the r.h.s of Eq. (83) vanishes, and the flow is described by the quasi-stationary solution

$$V_{st} = V_0 e^{-\left(\frac{\omega}{2\nu}\right)^{1/2} y} \sin\left(\omega t - \left(\frac{\omega}{2\nu}\right)^{1/2} y\right) \tag{84}$$

Table 3.3
The transition times for the Couette (simple shear) flows

Distance between plates [m]	Relaxation time [s] water ($\nu = 10^{-6}\,m^2\,s^{-1}$)	Relaxation time [s] air ($\nu = 1.5\times10^{-5}\,m^2\,s^{-1}$)
10^{-9} (1 nm)	10^{-12}	6.7×10^{-14}
10^{-7}	10^{-8}	6.7×10^{-10}
10^{-6} (1 μm)	10^{-6}	6.7×10^{-8}
10^{-3} (1 mm)	1	6.7×10^{-2}
10^{-2} (1 cm)	10^{2}	6.7

This velocity profile assumes, therefore, the form of a damped wave with the wavelength equal to $2\pi(2\nu/\omega)^{1/2}$ propagating in the y direction with the phase velocity of $(2\nu\omega)^{1/2}$. The amplitude of the fluid velocity decays exponentially with the distance and becomes practically negligible outside the layer of thickness $(2\nu/\omega)^{1/2}$, called the Stokes boundary layer [2].

By differentiating Eq. (84), one finds that the shear rate at the upper wall $y = 0$ is given by

$$\left(\frac{\partial V}{\partial y}\right) = G_y^0 = V_0\left(\frac{\omega}{\nu}\right)^{1/2} \sin\left(\omega t + \frac{\pi}{4}\right) \tag{85}$$

As can be noticed, the shear rate is shifted in phase by $(\pi/4)$ against the velocity oscillation of the plate.

For a finite distance between plates, equal to $(2b)$, assuming that the upper is oscillating in the same way as previously, the solution for the non-stationary flow is [17, p. 105]

$$V = V_{st} - 8\pi\nu\omega b^2 \sum_{n=1}^{\infty} n \frac{(-1)^n \sin\dfrac{n\pi y}{2b}}{\nu^2 n^4 \pi^4 + 4\omega^2 b^2} e^{-n^2\pi^2\nu t/4b^2} \tag{86}$$

where the quasi-stationary flow $V_{st}(y,t)$ is given by

$$V_{st} = V_0 A(y) \sin\left[\omega t + \varphi(y)\right] \tag{87}$$

the amplitude and the phase shift of this flow are

$$\begin{aligned} A(y) &= \left|\frac{\sinh \bar{\omega}(1+i)y}{\sinh \bar{\omega}(1+i)2b}\right| \\ \varphi &= \arg\left|\frac{\sinh \bar{\omega}(1+i)y}{\sinh \bar{\omega}(1+i)2b}\right| \end{aligned} \tag{88}$$

where $\bar{\omega} = (\omega/2\nu)^{1/2}$ and i is the imaginary unit.

Similarly, non-stationary flows in the rectangular or cylindrical channel can be driven by pressure variations instead of wall velocity. However, solutions derived by the inversion of the Laplace transformation are rather cumbersome. Simpler solutions can be derived in the one-dimensional case for parallel-plate or cylindrical channels. For example the transient flow, occurring in a parallel-plate channel after a sudden application of the pressure gradient *P/L*, is given by

$$V = \frac{Pb^2}{2\eta L}\left(1-\frac{y'^2}{b^2}\right)-\frac{16Pb^2}{\eta L\pi^3}\sum_{n=1,3,..}^{\infty}\frac{1}{n^3}\sin\frac{n\pi(y'+b)}{2b}e^{-(n^2\pi^2 \nu t/4b^2)} \tag{89}$$

As can be noticed, after a transition time of the order of b^2/ν, the flow becomes stationary, with the velocity distribution given by Eq. (52) derived previously.

On the other hand, in the case where the pressure in a parallel-plate channel is oscillating harmonically according to the function $\frac{P}{L}\sin\omega t = -i\frac{P}{L}e^{-i\omega t}$, the quasi-stationary solution for fluid velocity becomes [2]

$$\begin{aligned} V &= -\frac{P}{\rho L\omega}\mathrm{e}^{-i\omega t}\left[1-\frac{\cosh\varpi}{\cosh\varpi}\left(\frac{(1-i)y'}{(1-i)b}\right)\right] \\ &= -\frac{P}{\rho L\omega}A(y)\sin[\omega t+\varphi(y')] \end{aligned} \tag{90}$$

where

$$A(y') = A'(y')\cos\varphi'(y')$$

$$\varphi'(y') = \tan^{-1}[B'(y')/A'(y')]$$

$$A'(y') = 1-\frac{\cosh\varpi y'\cos\varpi y'\cosh\varpi b\cos\varpi b+\sinh\varpi y'\sin\varpi y'\sinh\varpi b\sin\varpi b}{(\cosh\varpi b\cos\varpi b)^2+(\sinh\varpi b\sin\varpi b)^2}$$

$$B'(y') = \frac{\cosh\varpi y'\cos\varpi y'\sinh\varpi b\sin\varpi b-\sinh\varpi y'\sin\varpi y'\cosh\varpi b\cos\varpi b}{(\cosh\varpi b\cos\varpi b)^2+(\sinh\varpi b\sin\varpi b)^2} \tag{91}$$

In the limit of high frequencies, where $\varpi b \gg 1$, Eq. (90) simplifies to

$$V = -\frac{P}{L\rho\omega}e^{-i\omega t}\left[1-e^{-\varpi(y'+b)(1-i)}-e^{-\varpi(b-y')(1-i)}\right] \tag{92}$$

Eq. (92) indicates that the flow is composed of the core flow, performing a rigid-body motion (analogous to the electroosmotic plug flow), and two Stokes boundary layers attached to each wall.

In the limit of higher frequencies, the velocity gradient at the surface is given by the expression

$$\left(\frac{\partial V}{\partial y'}\right)_{y'=b} = G_0(\omega t) = -\frac{P}{\rho L(2\nu\omega)^{1/2}}\sin\left(\omega t + \frac{\pi}{4}\right) \tag{93}$$

As can be seen, the gradient (wall shear rate) is oscillating in time with the same frequency as the pressure but with the phase shift $\pi/4$. The amplitude of these oscillations vanishes with frequency as $1/\omega^{1/2}$.

By comparing Eq. (93) with the stationary wall velocity gradient, given by Eq. (54) one obtains the expression

$$\frac{G^0(\omega t)}{G^0} = \left(\frac{\nu}{2\omega b^2}\right)^{1/2} \sin\left(\omega t + \frac{\pi}{4}\right) \tag{94}$$

By taking $\nu = 10^{-6}$ m^2 s^{-1}, $b = 10^{-4}$ m, $\omega = 100$ rad s^{-1} (16 Hz), one obtains $G^0(\omega t) / G^0 = 1/(2)^{1/2} = 0.71$. This means that amplitude of the oscillatory wall shear rate is comparable with the stationary wall shear rate. For $\omega = 10^4$ rad s^{-1} (1600 Hz) the amplitude is reduced to 7% of the stationary shear rate.

In the case of a cylindrical channel, the transient flow appearing after sudden application of a pressure gradient is given by [2]

$$V = \frac{PR^2}{4\eta L}\left(1 - \frac{r^2}{R^2}\right) - \frac{2PR^2}{\eta L}\sum_{n=1}^{\infty}\frac{J_0(\alpha_n r/R)}{\alpha_n^3 J_1(\alpha_n)} e^{-(\alpha_n^2 \nu t/R^2)} \tag{95}$$

where J_0 and J_1 are the Bessel functions of the zeroth and first orders and α_n are the real positive roots of J_0, the first three being 2.405, 5.52 and 8.65 [18].

Again, it can be noticed that the stationary Poiseuille's velocity profile in the channel is established after the transition time of R^2/ν.

Analogously, the pulsating flow in the cylindrical channel due to an oscillatory pressure gradient $P/L = \sin \omega t = -i\,(Pe^{-i\omega t}/L)$ is described by the real part of the expression

$$V = -\frac{P}{\rho L\omega} e^{-i\omega t}\left(1 - \frac{J_0[\varpi(1-i)r]}{J_0[\varpi(1-i)R]}\right) \tag{96}$$

The Bessel function of a complex argument can be evaluated using the subsidiary Kelvin functions ber and bei [2]. For higher frequencies, when the condition $\left(\frac{\omega}{2\nu}\right) R^2 \gg 1$ is fulfilled, the $J_0(z_i)$ function becomes $(2/\pi\, z_i)^{1/2} \cos\left(z_i - \frac{\pi}{4}\right)$ and Eq. (96) simplifies to the form

$$V = -\frac{P}{\rho L\omega}\left\{\cos\omega t - \left(\frac{R}{r}\right)^{1/2} e^{-\varpi(R-r)} \cos[\varpi(R-r) + \omega t]\right\} \tag{97}$$

The first term describes the rigid body type oscillations of the core region, and the second, vanishing at distances from the wall $(R - r)$ greater than $(2\nu/\omega)^{1/2}$, describes the oscillating boundary layer at the surface of the channel.

The velocity gradient at the surface is given by

$$\left(\frac{\partial V}{\partial r}\right)_{r=R} = G^0(\omega t) = -\frac{P}{\rho L(\omega \nu)^{1/2}} \sin\left(\omega t + \frac{\pi}{\alpha}\right) \tag{98}$$

Analogously to the parallel-plate channel, the gradient (wall shear rate) oscillates with the same frequency as the pressure, with the phase shift $\pi/4$. The amplitude of these oscillations vanishes with frequency as $1/(\omega)^{1/2}$.

3.3.2. Stagnation-point and impinging-jet flows

Another class of flows having practical significance, which can be calculated analytically, is that of stagnation flows, either of planar or radial symmetry. For sake of convenience, both will be referred to as stagnation-point flows, the former as a two-dimensional flow, the latter as an axisymmetric flow. Such flow configurations often occur in the vicinity of solid surfaces exposed to external fluid streams of simple geometry, a common example being a stationary cylinder or sphere placed in a uniform flow, or a plate exposed to impinging-jet flows, often used in particle deposition experiments. However, the size of the area where stagnation point flows occurs is a small fraction of the entire surface area exposed to the flow.

The solution of the Navier–Stokes equation for the stagnation point flows can be derived by matching the inner flow at the interface with the outer flow, usually described in terms of the potential (Euler's) flow. The mathematical shape of functions describing the inner flow can be guessed from the outer flow by exploiting the no-penetration boundary conditions at solid surfaces. In this way a set of nonlinear, ordinary differential equations is obtained, which are solved numerically to obtain the velocity components and the pressure distribution.

Let us first consider the oblique two-dimensional flow (see Fig. 3.3) in which the x and y velocity components of the outer flow, of a potential character, are given by

$$V_x^\infty = G_f(x\sin\varphi + y\cos\varphi)$$
$$V_y^\infty = -G_f\, y\sin\varphi \tag{99}$$

where G_f is the shear rate of the stagnation flow and φ is the parameter connected with the angle defining the outer flow direction β (see Fig. 3.3) by the expression

$$\varphi = \tan^{-1}\left(\frac{1}{2}\tan\beta\right) \tag{100}$$

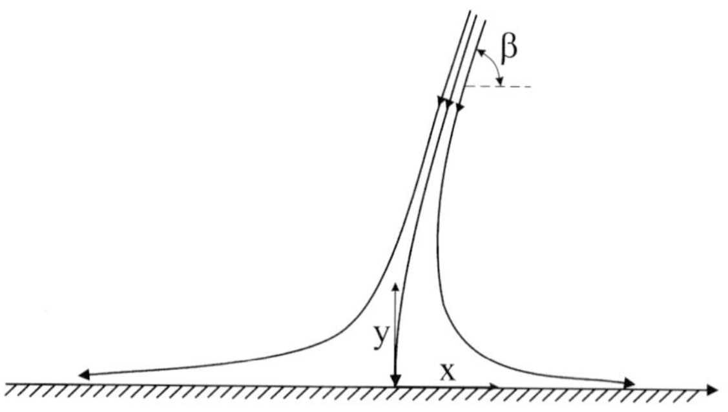

Fig. 3.3. Schematic view of the oblique two-dimensional stagnation-point flow.

For $\varphi = \pi$ the outer flow becomes the unidirectional simple shear flow in the positive x direction and $\varphi = \pi/2$ corresponds to the irrotational, orthogonal stagnation-point flow in the upper part of the half-space with the shear rate equal to G_f.

As can be deduced from Eq. (99), the outer flow does not vanish at the solid interface. In order to fulfill the no-penetration boundary condition, it is assumed that the viscous flow at the surface (inner flow) is described by a modified version of Eq. (99) [2]

$$V_x = x f'(y) + g'(y)$$
$$V_y = -f(y) \tag{101}$$

where $f(y)$ and $g(y)$ are functions to be determined by solving the Navier–Stokes equation.

The boundary conditions for Eq. (101) are

$$f = 0 \quad \text{for} \quad y = 0$$
$$x f' + g' = 0 \quad \text{for} \quad y = 0 \tag{102}$$

Moreover, it is required that for $y \to \infty$, $f \to G_{fy} \sin \varphi$

It can be noticed from Eq. (101) that the continuity equation $(\partial V_x/\partial x) + (\partial V_y/\partial y) = 0$ is fulfilled.

By using the similarity transformation

$$y = \left(\frac{\nu}{G_f \sin\varphi} \right)^{1/2} \xi \tag{103}$$
$$f(y) = (G_f \nu \sin\varphi)^{1/2} F(\xi)$$

One can convert the Navier–Stokes equation to the nonlinear differential equation for F [2,5]

$$F''' + FF'' - F'^2 + 1 = 0 \tag{104}$$

With the boundary conditions:

$$F = 0 \quad F' = 0 \quad \text{for} \quad \xi = 0$$
$$F' = 1 \quad F = \xi \quad \text{for} \quad \xi \to \infty \tag{105}$$

The boundary value problem represented by Eqs. (104,105) was first solved by Hiemenz [19]. Numerical calculations revealed that

$$F''(0) = 1.2326 = f(0)$$
$$F(\xi) \cong \frac{1}{2} f(0)\xi^2 - \frac{1}{6}\xi^3 \quad \text{for} \quad \xi < 1 \tag{106}$$

On the other hand, for greater distances from the solid surface, the inner flow quickly approaches the outer flow. Thus, for $\xi = 2.4$, the normal component of the inner flow assumes 99% of the outer flow. For $\xi = 3.8$, the difference between the inner and outer flows is only 0.01% [5]. If the hydrodynamic boundary layer thickness δ_h is defined as the distance where the characteristic velocity attains 99% of the outer flow velocity, one can deduce that for the stagnation-point flow

$$\delta_h = 2.4\left(\frac{\nu}{G_f \sin\varphi}\right)^{1/2} \tag{107}$$

It can be estimated by taking $\nu = 10^{-6}\,\text{m}^2\,\text{s}^{-1}$, $G_f = 100\ \text{s}^{-1}$, $\varphi = \pi/2$ that $\delta_h = 2.4\times10^{-4}$ m (240 μm). For $\varphi = \pi/6$, $\delta_h = 3.4\times10^{-4}$ m (340 μm). As can be noticed, δ_h, although considerably smaller than the typical size of macroscopic boundaries, is three orders of magnitude greater than typical colloid dimensions (100 nm). One can, therefore, conclude that colloid particles are deeply immersed in the hydrodynamic boundary layer, where the fluid velocity components, by exploiting Eqs (101, 103, 106), can be approximated by

$$V_x \cong G_f\, \nu \sin\varphi\, \xi\left(f(0) - \frac{1}{2}\xi^2\right)x + g'(y)$$
$$= \frac{(G_f \sin\varphi)^{3/2}}{\nu^{1/2}}\, yx\left[f(0) - \frac{1}{2}\left(\frac{G_f \sin\varphi}{\nu}\right)^{1/2} y\right] + g'(y)$$

$$V_y \cong -\frac{1}{2}(G_f\, \nu \sin\varphi)^{1/2}\xi^2\left(f(0)-\frac{1}{3}\xi\right)$$
$$= -\frac{1}{2}\frac{(G_f \sin\varphi)^{3/2}}{\nu^{1/2}}y^2\left[f(0)-\frac{1}{3}\left(\frac{G_f \sin\varphi}{\nu}\right)^{1/2} y\right] \quad (108)$$

As can be noticed, Eq. (108) contains the unknown function $g'(y)$, which vanishes in the case of the orthogonal flow. By integrating the second equation for the vorticity, it was demonstrated in [2] that the $g(y)$ function is given by

$$g(y) = \nu \cot\varphi\, G(\xi) \quad (109)$$

where G(ξ) is the universal function of the reduced distance, known in a graphical form only [2]. For small distances from the interface, when $\xi \to 0$, $G'' \to 1.4$. By exploiting this property, one can formulate the expression for the $g'(y)$ function in the form

$$g'(y) = 1.4 G_f \cos\varphi\, y \quad (110)$$

Using Eq. (110), one can formulate the following expression for V_x at the boundary by retaining the lowest order terms

$$V_x = f(0)\frac{(G_f \sin\varphi)^{3/2}}{\nu^{1/2}}xy + 1.4 G_f \cos\varphi y \quad (111)$$

It can be noticed from Eq. (111) that the tangential velocity component vanishes at the distance

$$x_s = -\frac{1.4\nu^{1/2}\cos\varphi}{f(0)G_f^{1/2}(\sin\varphi)^{3/2}} \quad (112)$$

Thus, x_s can be treated as the stagnation point. Expressing the fluid flow relative to this point, one obtains

$$V_x = f(0)\frac{(G_f \sin\varphi)^{3/2}}{\nu^{1/2}}x'y$$
$$V_y = -\frac{1}{2}f(0)\frac{(G_f \sin\varphi)^{3/2}}{\nu^{1/2}}y^2 \quad (113)$$

where x' is the tangential distance measured from x_s.

Using Eq. (113) the shape of the fluid streamlines at the boundary can be expressed by the formula

$$y = y_0 \left(\frac{x_0}{x'} \right)^{1/2} \tag{114}$$

where x_0,y_0 are the coordinates of an arbitrary point far from the boundary, crossed by the streamline.

Obviously, Eq. (114) is olso valid for two-dimensional, orthogonal flows.

The distribution of streamlines calculated numerically for the orthogonal and oblique stagnation-point flows [1] are shown in Figs. 3.4 and 3.5.

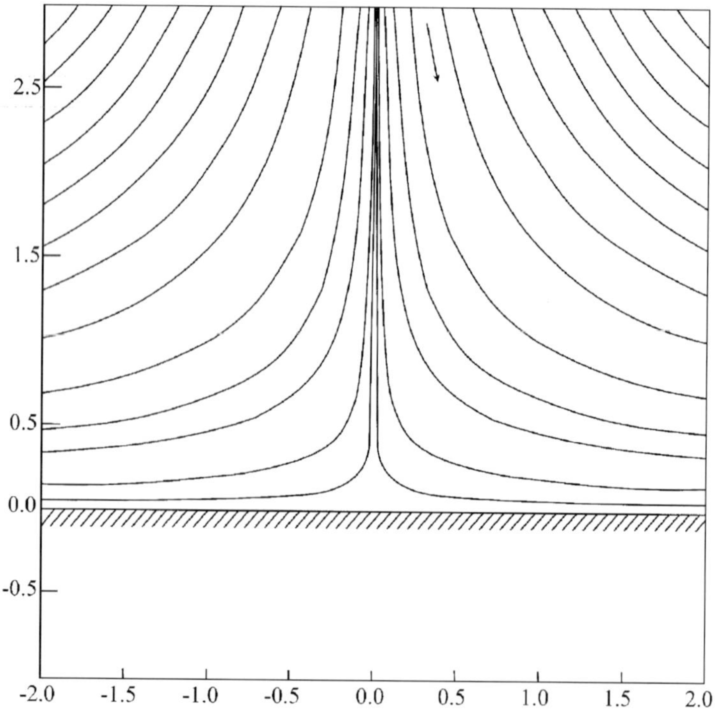

Fig. 3.4. Fluid streamline distribution calculated for the two-dimensional, orthogonal stagnation-point flow [2].

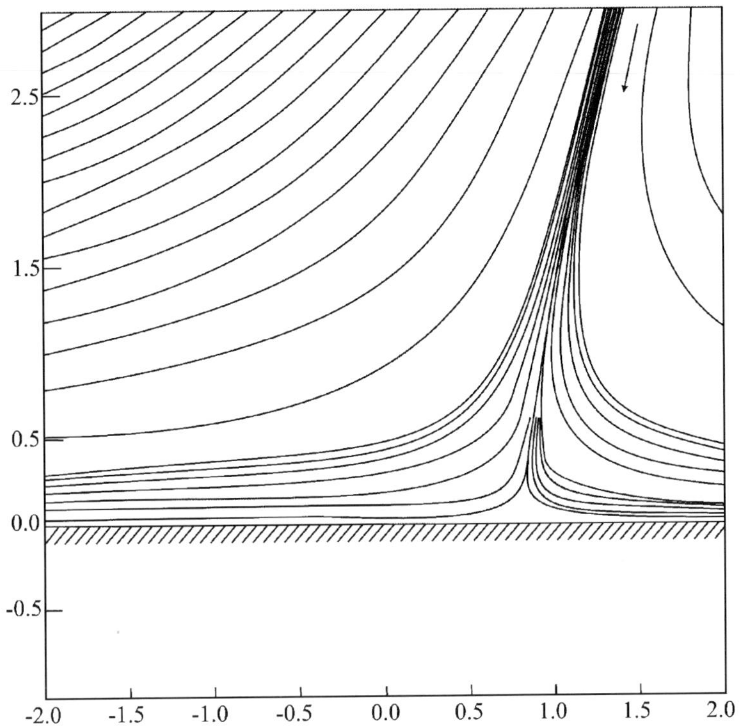

Fig. 3.5. Fluid streamline distribution calculated for the two-dimensional, oblique stagnation-point flow [2].

In an analogous way the axisymmetric stagnation-point flow can be analyzed. In this case, the outer flow at greater distances from the interface is given by [5]

$$V_r = G_f\, r \qquad V_z = -2G_f z \tag{115}$$

where V_r, V_z are the radial and axial components of the outer flow.

Accordingly, it is assumed that the viscous flow at the surface (inner flow) is described by

$$V_r = G_f r\, f'(z) = G_f r\, F(\xi) \qquad V_z = -2G_f f(z) = -2(G_f \nu)^{1/2}\, F(\xi) \tag{116}$$

where $\xi = (G_f/\nu)^{1/2} z$ and $F(\xi)$ is the universal function to be determine from the Navier–Stokes equation assuming the form

$$F''' + 2FF'' - F'^2 + 1 = 0 \tag{117}$$

The boundary conditions for Eq. (117) are

$$\begin{aligned} &F = 0 \quad \text{for} \quad \xi = 0 \\ &F' = 1 \quad \text{for} \quad \xi = \infty \end{aligned} \tag{118}$$

The boundary value problem represented by Eqs. (117–118) was first solved by Homann [20]. Numerical calculations revealed that

$$\begin{aligned} &F''(0) = 1.312 = f(0), \quad \text{for} \quad \xi = 0 \\ &F'(\xi) = \frac{1}{2} f(0)\xi^2 - \frac{1}{6}\xi^3 \quad \text{for} \quad \xi < 1 \end{aligned} \tag{119}$$

Consequently, the fluid velocity components at small distances from the boundary, $\xi < 1$, can be expressed as

$$\begin{aligned} &V_r = \frac{G_f^{3/2}}{\nu^{1/2}} rz \left[f(0) - \frac{1}{2}\left(\frac{G_f}{\nu}\right)^{1/2} z \right] \\ &V_z = -\frac{G_f^{3/2}}{\nu^{1/2}} z^2 \left[f(0) - \frac{1}{6}\left(\frac{G_f}{\nu}\right)^{1/2} z \right] \end{aligned} \tag{120}$$

The boundary layer thickness δ_h for the axisymmetric flow is

$$\delta_h = 2\left(\frac{\nu}{G_f}\right)^{1/2} \tag{121}$$

By exploiting Eq. (116), the fluid streamline shape can be expressed as

$$f(z) = \frac{C_0'}{r^2} \tag{122}$$

where C_0' is the constant of integration. For the region near the boundary, where $f(z) = z^2$, Eq. (122) can be expressed explicitly as

$$z = \frac{z_0 r_0}{r} \tag{123}$$

where r_0 and z_0 are the coordinates of an arbitrary point far from the boundary, crossed by the streamline. In contrast to the two-dimensional case, the fluid streamlines have now the hyperbolic shape, approaching the boundary faster for growing tangential distance. This can be noticed by comparing Fig. 3.4 and Fig. 3.6, where streamlines calculated numerically for the axisymmetric stagnation-point flows are shown.

The above analysis of the stagnation-point flow was based on the postulate that there exists an outer flow of a potential character, without going into detail how this can be done in practice. One of the few cases where the axisymmetric stagnation-point flow can be created over macroscopic volumes is the flow due to a rotating disk shown schematically in Fig. 3.7. A useful property of this arrangement is that the intensity of the flow and the hydrodynamic boundary layer thickness can be regulated within broad limits by changing the angular velocity of the disk, Ω. The flow is truly three-dimensional, having the radial V_r axial V_z and angular V_φ components. The governing Navier–Stokes equations in the case of the disk flow can be reduced to a set of ordinary differential equations by expressing these velocity components in the form [5]

$$V_r = \Omega r F(\xi)$$

$$V_\varphi = \Omega r G(\xi) \tag{124}$$

$$V_z = (\Omega \nu)^{1/2} H(\xi)$$

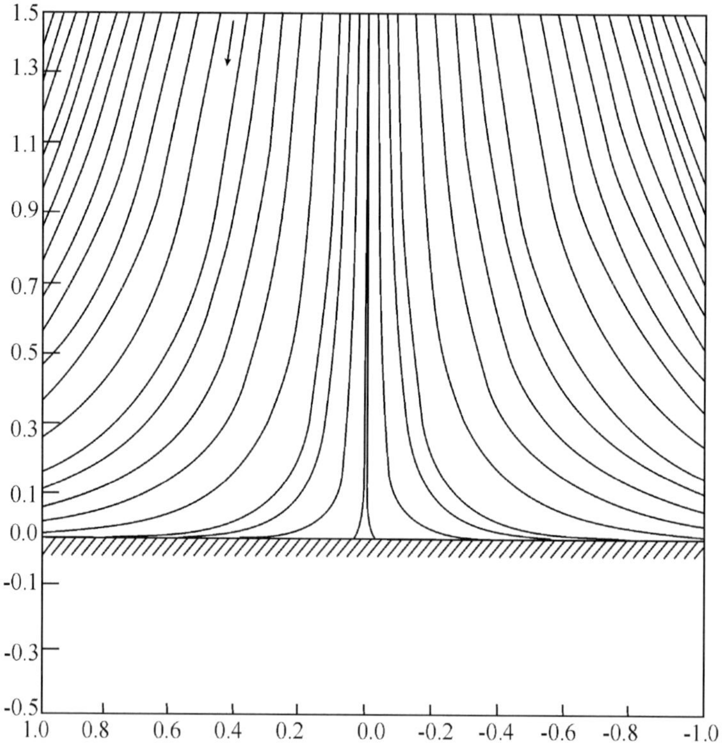

Fig. 3.6. Fluid streamline distribution calculated for the axisymmetric, orthogonal stagnation-point flow [2].

where F, G and H are unknown functions of the reduced distance from the disk only, given by

$$\xi = \left(\frac{\Omega}{\nu} \right)^{1/2} z \tag{125}$$

Thus, the Navier–Stokes equation becomes [5]

$$\begin{aligned} &2F + H' = 0 \\ &F^2 + F'H - G^2 - F'' = 0 \\ &2FG + HG' - G'' = 0 \end{aligned} \tag{126}$$

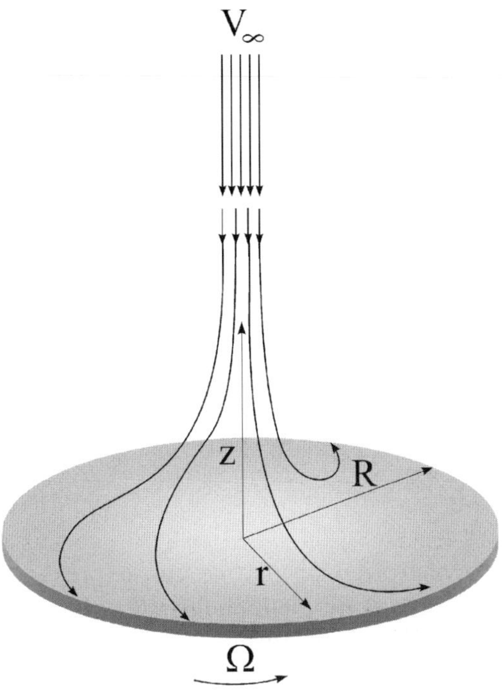

Fig. 3.7. A schematic view of the flow near the rotating disk.

The no-slip and far-field boundary conditions at the interface are

$$F=0,\quad G=1\quad H=0\quad \text{for}\quad \xi=\infty$$
$$F=0,\quad G=0\quad H'=0\quad \text{for}\quad \xi\to\infty \tag{127}$$

The boundary value problem represented by Eqs. (126), (127) was first solved by von Karman [21] and later on in a more exact way by Cochran [22]. Numerical calculations revealed that

$$H(0)=-2f(0)=-1.0204,\quad H(\infty)=-0.886$$
$$G(0)=g(0)=0.6159 \tag{128}$$

Fluid velocity components at small distances from the boundary $\xi<1$ can be expressed as

$$V_r = \frac{\Omega^{3/2}}{\nu^{1/2}} rz \left[f(0) - \frac{1}{3}\left(\frac{\Omega}{\nu}\right)^{1/2} z - \frac{1}{3} g(0)\left(\frac{\Omega}{\nu}\right) z^2 + \dots \right.$$

$$V_\varphi = \Omega r \left[1 - g(0)\left(\frac{\Omega}{\nu}\right)^{1/2} z + \frac{1}{3} f(0)\left(\frac{\Omega}{\nu}\right) z^3 + \dots \right. \tag{129}$$

$$V_z = -\frac{\Omega^{3/2}}{\nu^{1/2}} z^2 \left[f(0) - \frac{1}{3}\left(\frac{\Omega}{\nu}\right)^{1/2} z + \dots \right.$$

By retaining the lowest order terms one can express the radial and axial components of the fluid velocity, which are the only ones influencing mass transfer to the disk, in the simple form

$$V_r = 0.5102 \frac{\Omega^{3/2}}{\nu^{1/2}} rz$$
$$\tag{130}$$
$$V_z = -0.5102 \frac{\Omega^{3/2}}{\nu^{1/2}} z^2$$

As can be seen, the V_r and V_z components assume a form analogous to the axisymmetric stagnation-point flow, given by Eq. (120). Moreover, the axial velocity V_z does not depend on the radial coordinate, which is unique for any macroscopic flow. The boundary layer thickness δ_h for the rotating disk is given by

$$\delta_h \cong 5\left(\frac{\nu}{\Omega}\right)^{1/2} \tag{131}$$

By taking $\nu=10^{-6}$ m^2 s^{-1} and $\Omega = 25$ rad s^{-1} one can estimate that $\delta_h=10^{-3}$ m (1000 μm). For $\Omega=1000$ rad s^{-1}, $\delta_h=1.58\text{x}10^{-4}$ m (120 μm). At distances from the disk greater than this limiting value, the axial velocity component assumes a constant value independent of the position over the disk

$$V_\infty = -0.886(\Omega\nu)^{1/2} \tag{132}$$

By taking $\nu = 10^{-6}\,\mathrm{m^2\,s^{-1}}$ and $\Omega = 25$ rad $\mathrm{s^{-1}}$, one has $V_\infty = -4.43\times10^{-3}$ $\mathrm{ms^{-1}}$. For $\Omega = 1000$ rad $\mathrm{s^{-1}}$, $V_\infty = 2.8\mathrm{x}10^{-2}$ m $\mathrm{s^{-1}}$.

The hydrodynamic torque acting on the disk is given by the expression [5]

$$To_h = \frac{\pi}{2} g(0)\rho R^4 (\Omega^3 \nu)^{1/2} \tag{133}$$

All the above formulae are valid if the flow near the disk is laminar. As demonstrated experimentally by measuring the torque, the flow due to a rotating disk, at least in the region near the interface, remains laminar for Reynolds number $Re = R^2\,\Omega/\nu$ (where R is the disk radius) below 3×10^5 [5]. By taking the above data $\nu = 10^{-6}\,\mathrm{m^2\,s^{-1}}$, $\Omega = 25$ $\mathrm{rads^{-1}}$ and $R = 10^{-2}$ m a typical value used in the laboratory) one has $Re = 2.5\mathrm{x}10^3$. For $\Omega = 1000$ rad $\mathrm{s^{-1}}$, $Re = 10^5$. This means that for this angular velocity range, the flow is laminar.

The rotating disk is often used for enhancing mass transfer rates in industrial processes (extraction, mixing) in electrochemistry as a convenient working electrode [9] or substrate for particle deposition studies [23].

The disadvantage of flow due to the rotating disk is the swirling character of the motion, which eliminates the possibility of direct observations of the disk surface when used as a substrate for particle deposition studies. Such observations become feasible, however, by exploiting the impinging-jet flows matching the above stagnation-point flows quite well. In this case the interface remains stationary, see Fig. 3.8, and the flow is driven by the hydrostatic pressure difference between the upper level of the fluid and the outlet level. Obviously, the flow can also be driven by pumping (e.g., using a peristaltic pump) but this is less advantageous from the experimental point of view because of flow pulsations. The first experimental cell of this type, often referred to as the radial impinging jet (RIJ) cell, was constructed by Dąbroś and van de Ven [24]. Unfortunately, the two-dimensional flow distribution in the RIJ can only be calculated numerically. Dąbroś and

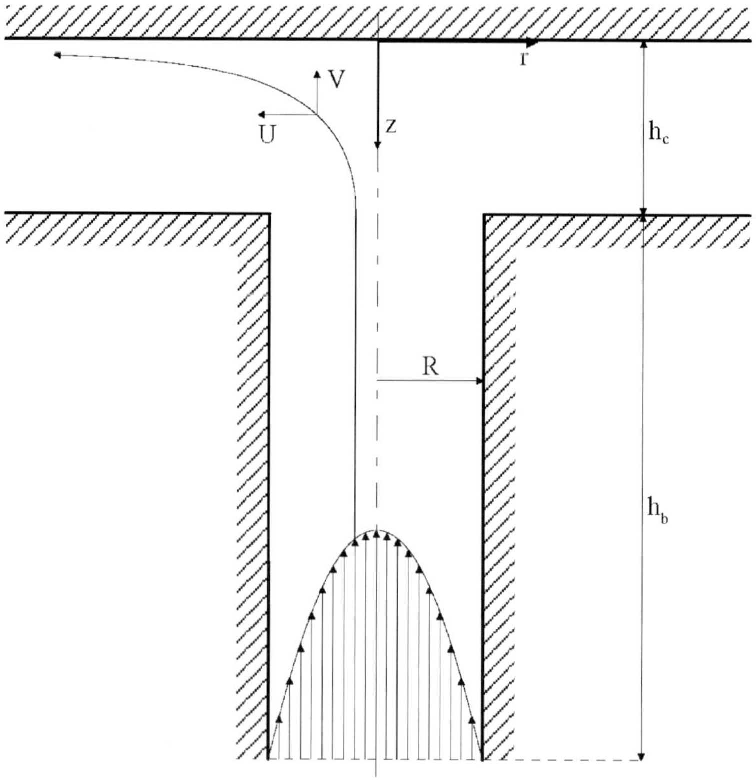

Fig. 3.8. A schematic view of the laminar flow in the radial impinging-jet (RIJ) cell. From Ref. [25].

van de Ven [24] used the finite-difference method expressing the vorticity over a mesh of 41 nodal points. The iterative over-relaxation method was used to solve the resulting set of nonlinear equations. The velocity profile in the inlet tube was assumed to be parabolic, in accordance with the fully developed Poiseuille flow.

There are two dimensionless parameters governing the flow distribution in the cell: the Reynolds number $Re = V_\infty R/\nu = Q/\pi R\nu$ (where $V_\infty = Q/\pi R^2$ is the mean fluid velocity in the tube and Q is the volumetric flow rate) and the geometrical parameter h_c/R. Dąbroś and van de Ven [24] carried out calculations for $0 < Re < 50$ and $h_c/R = 1$ and 1.6.

More extensive calculations of flow distribution for the RIJ cell have been performed in Refs. [25,26]. The finite-difference split method of

Harlow and Welch [2,27] was applied to the 2D Navier–Stokes equation expressed in terms of fluid velocity components and pressure rather than vorticity.

Although the flow distribution for the RIJ cell is rather complex, it was shown that for small distances from the interface $h/R<0.2$, the fluid velocity components can be approximated by the expression [25] (Fig. 3.9)

$$V_r = \alpha_r(Re, h_c/R)\frac{V_\infty}{R} z S(r/R)$$

$$V_z = -\alpha_r(Re, h_c/R)\frac{V_\infty}{R^2} z^2 C(r/R) \tag{134}$$

where α_r is the flow parameter depending on the cell geometry and the Reynolds number, $C_r(r/R)$ and $S(r/R)$ are the dimensionless correction functions.

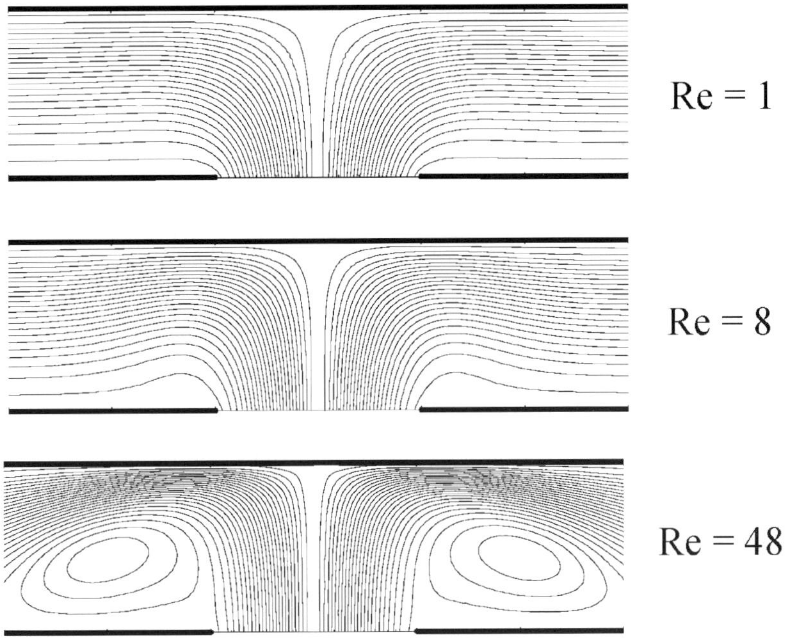

Fig. 3.9. Fluid streamlines in the RIJ cell for Re = 1, 8 and 48 (h/R = 1.6) [25].

When $r \to 0$, the $C_r(r/R)$ function tends to unity and $S_r(r/R)$ approaches r. Thus, in the region close to the center of the cell, the flow distribution is given by the expression

$$
\begin{aligned}
V_r &= \alpha_r \frac{V_\infty}{R} z r \\
V_h &= \alpha_r \frac{V_\infty}{R^2} z^2
\end{aligned}
\tag{135}
$$

As can be noticed, this flow distribution in the RIJ cell is analogous to the radial stagnation-point flow or the flow in the vicinity of the rotating disk. The flow parameter α_r can be calculated from the fitting function [25]

$$
\alpha_r = c_0 + c_1 Re + c_2 Re^2 \tag{136}
$$

where c_0, c_1, c_2 are dimensionless constants.

The range of validity of Eq. (135) was estimated from the numerical results shown in Figs. 3.10 a,b, where these functions are plotted for various Re number values (1 through 48). It is interesting to note that the S function characterizes the shear rate at the interface because

$$
G_0 = \left(\frac{\partial V_r}{\partial z} \right)_0 = \alpha_r \frac{V_\infty}{R} S(r/R) \tag{137}
$$

As can be seen in Fig. 3.10, the $C(r/R)$ and $S(r/R)$ functions can be well approximated for $r/h < 0.5$ by the simple relationship

$$
\begin{aligned}
C(r/R) &= \cos\left(\frac{\pi r}{2R} \right) \\
S(r/R) &= \frac{2}{\pi} \sin\left(\frac{\pi r}{2R} \right)
\end{aligned}
\tag{138}
$$

On the other hand, the dependence of the flow intensity parameter α_r on the Re number is plotted in Figs. 3.11 and 3.12. The exact numerical results (points) can be fitted by Eq. (136) valid for $Re < 20$, with

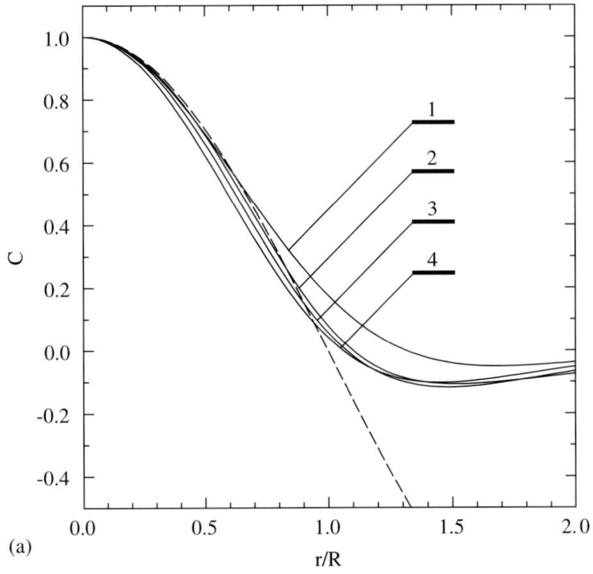

(a)

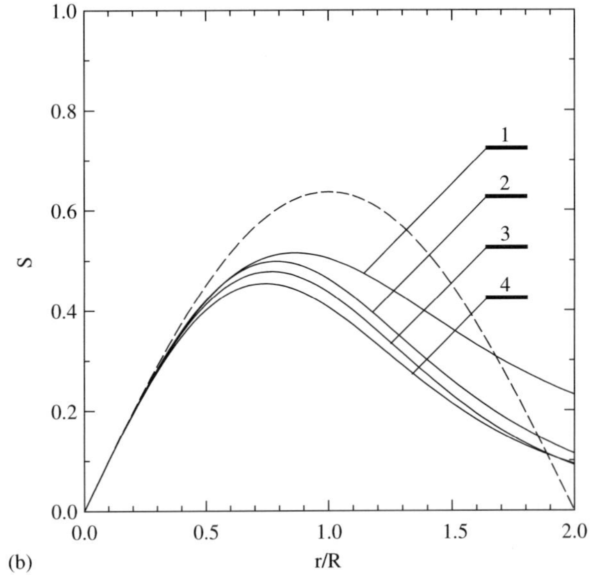

(b)

Fig. 3.10. (a) The dependence of the C function (reduced axial velocity component in the RIJ cell) on the distance from the center r/R determined numerically; (b) the dependence of the S function (reduced shear rate at the interface) on r/R: 1. $Re = 4$, 2. $Re = 48$, 3. $Re = 30$, 4. $Re = 16$, the dashed line denotes the analytical results calculated from $C = \cos(\pi r/2R)$, $S = (2/\pi)\sin(\pi r/2R)$ [26].

$c_0 = 1.78$, $c_1 = 0.186$, $c_2 = 0.034$. For $Re > 20$, the fitting function was found to be

$$\alpha_r = c_3 Re^{0.5} + c_4 \tag{139}$$

where $c_3 = 4.96$ and $c_4 = -8.41$.

Numerical data obtained for other values of the h_c/R ranging from 1 to 3 are shown in Fig. 3.12. As calculated in Ref. [28], they can be fitted well for $Re < 20$ by Eq. (136) with

$c_0 = 4.030$, $c_1 = 0.628$, $c_2 = -1.895\times10^{-3}$ for $h_c/R = 1$
$c_0 = 0.438$, $c_1 = 0.337$, $c_2 = 1.254\times10^{-2}$ for $h_c/R = 2$
$c_0 = 0.081$, $c_1 = 0.081$, $c_2 = 1.736\times10^{-2}$ for $h_c/R = 3$

For $Re > 20$, constants appearing in Eq. (139) are

$c_3 = 4.60$, $c_4 = -4.84$ for $h_c/R = 1$
$c_3 = 5.27$, $c_4 = -11.43$ for $h_c/R = 2$
$c_3 = 5.78$, $c_4 = -17.37$ for $h_c/R = 3$

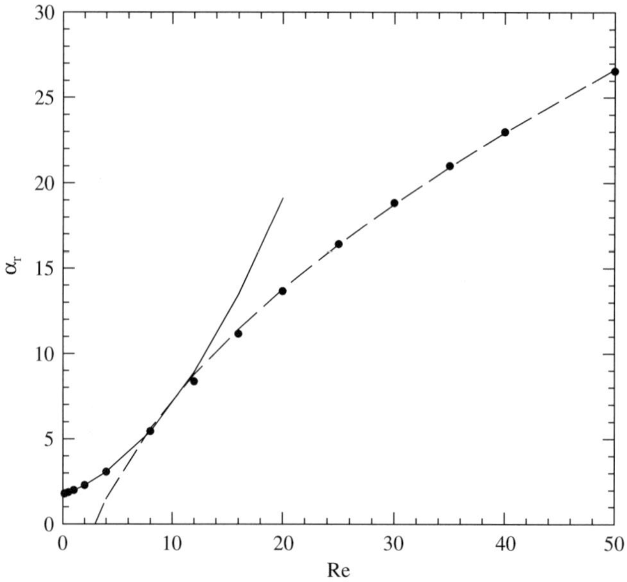

Fig. 3.11. The dependence of the parameter on the Reynolds number (Re) calculated numerically for the RIJ cell ($h_c/R = 1.6$), the solid line represents the low Re fit, Eq. (136), the dashed line represents the high Re fit, Eq. (139) [28].

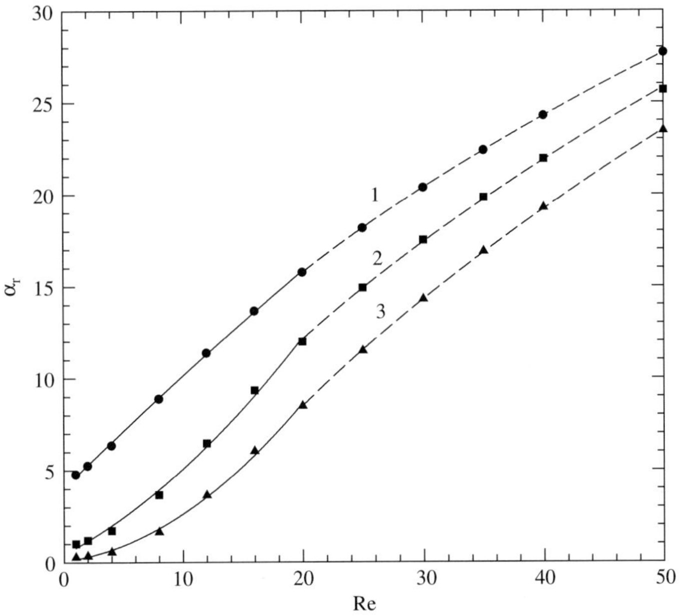

Fig. 3.12. The dependence of the α_r parameter on the Reynolds number *Re* calculated numerically for the RIJ cell 1. $h_c/R = 1$, 2. $h_c/R = 2$, 3. $h_c/R = 3$, the solid line represents the low *Re* fit, Eq. (136) and the dashed line represents the high *Re* fit, Eq. (139) [28]

As can be noticed from the results shown in Fig. 3.12, the increase in the α_r parameter with h_c/R is most abrupt for $Re < 5$. One can therefore expect that the RIJ cells characterized by a small h_c / R value are most efficient for studying the particle deposition because of increased mass transfer efficiency.

The surface area available for observations can also be increased by using the two-dimensional impinging-jet cell of a plane-parallel geometry [29]. The flow field in the cell, called the slot impinging-jet cell (SIJ), was evaluated numerically for various values of the *Re* number and the h_c/d parameter (where $2d$ is the slot width). The streamline pattern was found to be analogous to the radial jet, with an eddy formed in the region near the tip of the slot for $Re > 8$.

The dependence of fluid velocity components on the reduced distance from the symmetry plane is shown in Figs. 3.13 for $Re = 2$ and 16. From these graphs, one can notice that the tangential and normal velocity components are proportional to y and y^2, respectively, if the reduced distance from

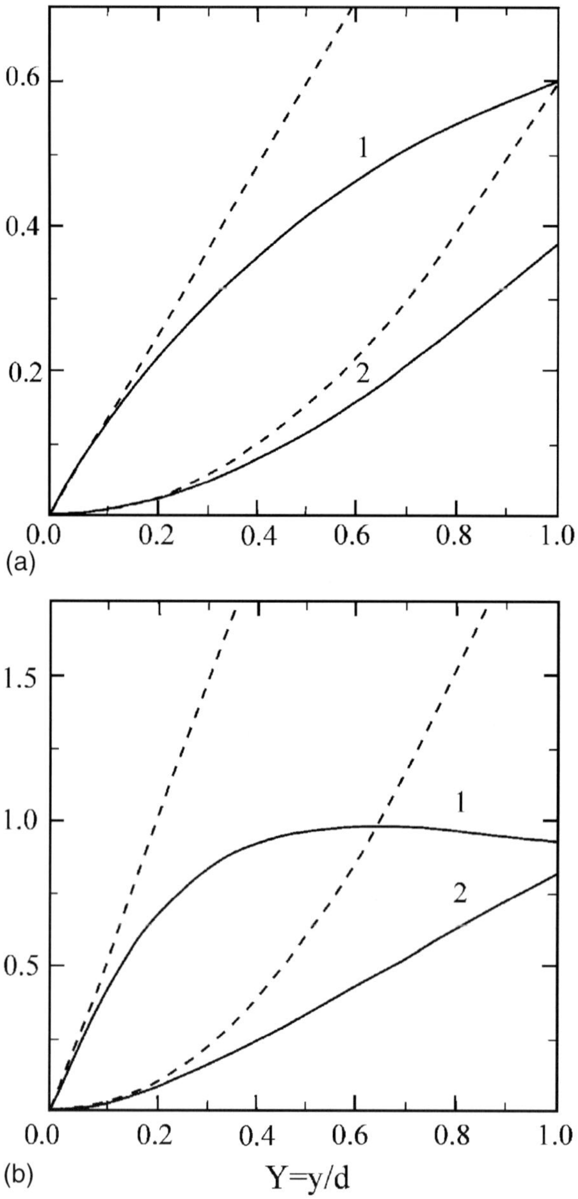

Fig. 3.13. The dependence of $V_x\, d/V_\infty y$ (curve 1) and $-V_y\, /V_\infty$ (curve 2) on the reduced distance $Y=y/d$. Solid lines show numerical solutions obtained for $Re = 2$ (part "a") and 16 (part "b"). The dashed lines represent the limiting analytical solutions obtained from Eq. (140) with $S = C = 1$ [29].

the interface $y/d < 0.1$. In this region the flow distribution can be described by an equation analogous to Eq. (134)

$$V_x = 2\alpha_s(Re, h_c/d)\frac{V_\infty}{d}\, y\, S(x/d)$$

$$V_y = -\alpha_s(Re, h_c/d)\frac{V_\infty}{d^2}\, y^2\, C(x/d) \tag{140}$$

where $V_\infty = Q/2dl$, d is the depth of the parallel-plate channel in which the impinging-jet is produced, l is the width of the channel, x is the distance from the symmetry plane and y is the distance from the interface.

Similar to the RIJ cell, the C and S functions can be approximated by

$$C(x/R) = \cos\left(\frac{\pi x}{2d}\right)$$

$$S(x/R) = \frac{2}{\pi}\sin\left(\frac{\pi x}{2d}\right) \tag{141}$$

The range of validity of Eq. (141) can be estimated from the results shown in Fig. 3.14. The comparison of the exact numerical results obtained in Ref. [29] and the analytical results (dashed lines) indicates that Eq. (141) remains a valid approximation for $x/d < 0.7$.

The dependence of the flow intensity parameter on the Reynolds number is plotted in Fig. 3.15 for various h_c/d values. As in the previous case, the numerical results were fitted by Eq. (136) for $Re < 10$ and Eq. (139) for $Re > 10$.

The constants $c_0 - c_4$ are

$c_0 = 3.18$	$c_1 = 0.156$,	$c_2 = -1.86\times10^{-3}$,	$c_3 = 0.91$	$c_4 = 1.48$ for $h_c/d = 1$
$c_0 = 1.64$	$c_1 = 0.139$,	$c_2 = -1.61\times10^{-3}$,	$c_3 = 0.82$	$c_4 = 0.09$ for $h_c/d = 1.5$
$c_0 = 0.86$	$c_1 = 0.113$,	$c_2 = -4.52\times10^{-4}$,	$c_3 = 0.88$	$c_4 = -1.04$ for $h_c/d = 2$
$c_0 = 0.28$	$c_1 = 0.069$,	$c_2 = -1.30\times10^{-3}$,	$c_3 = 0.95$	$c_4 = -2.10$ for $h_c/d = 3$

The advantage of the SIJ cell is that it has a much larger observation area in comparison with the RIJ cell, which facilitates a statistical analysis of the distribution of adsorbed particle.

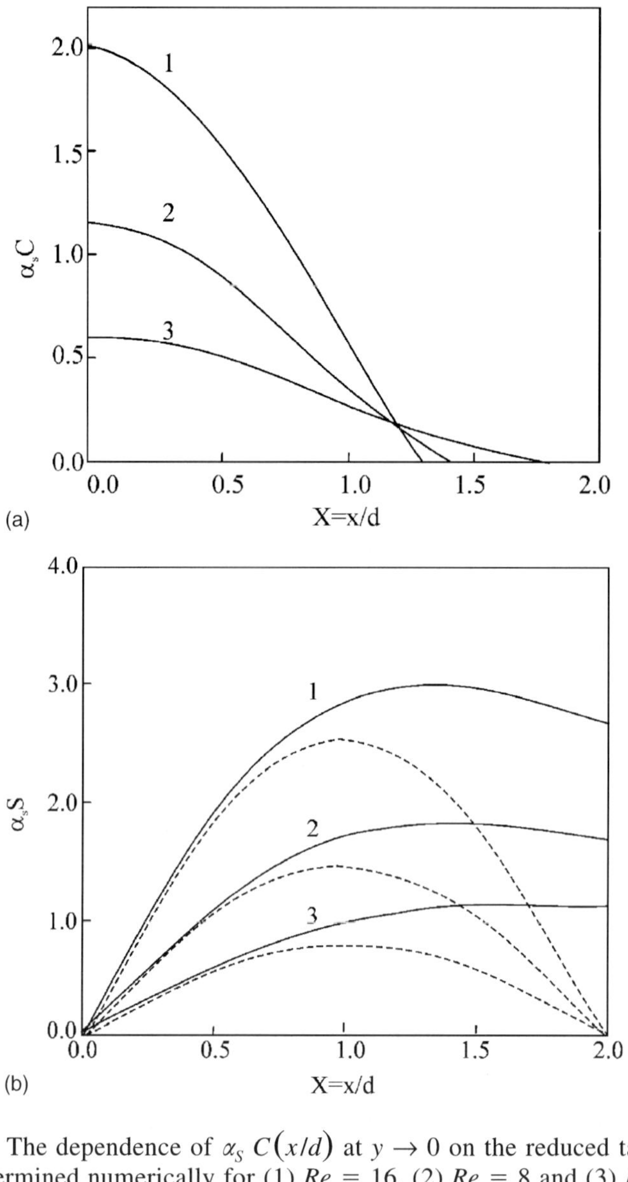

Fig. 3.14. (a) The dependence of $\alpha_S\ C(x/d)$ at $y \rightarrow 0$ on the reduced tangential distance x/d determined numerically for (1) $Re = 16$, (2) $Re = 8$ and (3) $Re = 2$ (solid lines); (b) The dependence of $\alpha_S\ S(x\ /\ d)$ on the reduced distance x/d determined numerically for (1) $Re = 16$, (2) $Re = 8$ and (3) $Re = 2$ (solid lines). The dashed lines represent the limiting analytical solutions obtained from the equation: $\alpha_S S = (2/\pi)\ \alpha_S \sin\ (\pi x/2d)$ [26].

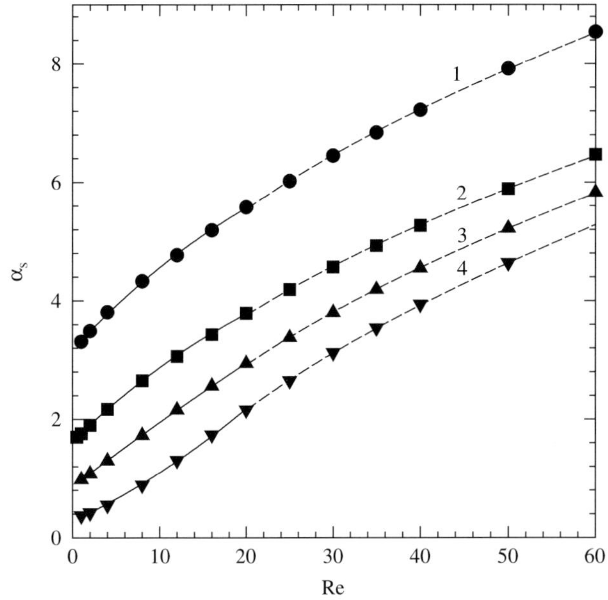

Fig. 3.15. The dependence of α_S on *Re* for the SIJ cell: 1. h_c/d = 1, 2. h_c/d = 1.52, 3. h_c/d = 2, 4. h_c/d = 3, the solid line represents the low *Re* fit, Eq. (136) and the dashed line represents the high *Re* fit, Eq. (139) [28].

3.3.3. Flows Past Stationary Interfaces – Boundary Layer Flows

Due to the nonlinearity of the Navier–Stokes equation, other flows in the vicinity of solid boundaries, associated with particle deposition problems, can be analyzed in an approximate manner only. This is so even for apparently the simplest case of uniform flow past bodies of various shapes, such as plates, cylinders, spheres, etc. One of the most efficient approximate approaches to dealing with such problems is the hydrodynamic boundary layer theory, whose foundations have been formulated by Prandtl, Blasius and later on by von Karman and Pohlhausen [2,5]. Because of its major practical significance, this subject was extensively treated in monographs by Schlichting [5], Rosenhead [30] and Evans [31].

The basic assumption which was made in developing this theory is that the overall flow can be divided into two components: (i) the outer flow of an inviscid (irrotational) character, governed by the Euler equation, and (ii) the inner flow adjacent to boundaries where the viscosity effects dominate. The thickness of the boundary layer δ_h is much smaller than the characteristic dimension L of the boundary, see Fig. 3.16.

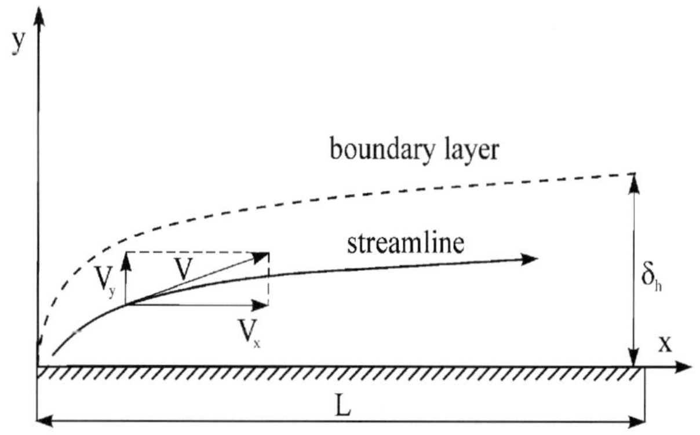

Fig. 3.16. A schematic view of the flow in the hydrodynamic boundary layer.

Because of this, most of the boundary layer flows can be effectively described in terms of two spatial coordinates, one tangential to the surface and the other perpendicular. In the case of planar interfaces, the flow can be described in terms of the two Cartesian coordinates (*x*, *y*) (see Fig. 3.16). The fluid continuity equation then assumes the form

$$\frac{\partial V_x}{\partial x} + \frac{\partial V_y}{\partial y} = 0 \tag{142}$$

By realizing that the tangential velocity gradient equals approximately V_∞ / L (where L is the length of the interface and V_∞ is the characteristic velocity of the outer flow), and the normal velocity gradient equals approximately V_∞ / δ_h, one can deduce from Eq. (142) that

$$V_y = V_\infty \frac{\delta_h}{L} \tag{143}$$

For a two-dimensional flow, the balance equation of the tangential momentum component under a steady state becomes

$$V_x \frac{\partial V_x}{\partial x} + V_y \frac{\partial V_y}{\partial y} = -\frac{1}{\rho}\frac{\partial p}{\partial x} + \nu \frac{\partial^2 V_x}{\partial x^2} + \nu \frac{\partial^2 V_y}{\partial y^2} \tag{144}$$

By using Eq. (143), one can deduce that the viscous term $\nu\,(\partial^2 V_x/\partial x^2)$ is of the order of $\nu V_\infty/L^2$, which is negligible in comparison with the second viscous term $\nu\,(\partial^2 V_x/\partial y^2)$ and can be neglected. Similarly, by comparing the inertia term in Eq. (144) V_∞^2/L with the viscous term $\nu\; V_\infty/\delta_h{}^2$, one obtains

$$\delta_h = \left(\frac{\nu L}{V_\infty}\right)^{1/2} = L\,Re^{-1/2} \tag{145}$$

where $Re = V_\infty L/\nu$ is the Reynolds number of the outer flow.

One can notice from Eq. (145) that the thickness of the boundary layer vanishes as $Re^{-1/2}$. This means that the boundary layer theory becomes an accurate approximation for $Re >> 1$.

The tangential pressure gradient can be estimated by considering the outer potential flow. From the Euler equation, one has

$$-\frac{1}{\rho}\frac{\partial p}{\partial x} = V_\infty \frac{\partial V_\infty}{\partial x} \tag{146}$$

By considering Eq. (146) and dropping the first viscosity term, in Eq. (144), one can formulate the hydrodynamic boundary-layer equation in the form

$$V_x \frac{\partial V_x}{\partial x} + V_y \frac{\partial V_x}{\partial y} = V_\infty \frac{\partial V_\infty}{\partial x} + \nu \frac{\partial^2 V_x}{\partial y^2} \tag{147}$$

It can be noticed that Eq. (147) assumes the form of a parabolic partial differential equation, in contrast to the original Navier–Stokes equation, which is of the elliptic character. This property significantly facilitates mathematical handling of Eq. (147), which can often be reduced to ordinary differential equation by using similarity transformation.

In an analogous way, the scaling of the equation describing the balance of the normal momentum component can be carried out. It can be easily shown that all terms are δ_h times smaller than for the tangential direction, in particular

$$\frac{\partial p}{\partial y} \cong 0 \tag{148}$$

This means that the pressure variation across the boundary layer can be neglected, and the pressure within the boundary layer is a function of the tangential coordinate only.

Let us now consider some applications of the boundary-layer theory, which have practical significance. The first one is the uniform flow along a stationary flat plate (of negligible thickness), see Fig. 3.17. Since the outer flow is uniform $\partial p / \partial x = 0$ and Eq. (147) becomes

$$V_x \frac{\partial V_x}{\partial x} + V_y \frac{\partial V_x}{\partial y} = \nu \frac{\partial^2 V_x}{\partial y^2} \tag{149}$$

The continuity equation is given by Eq. (142). The boundary conditions for Eq. (149) are $V_x = V_y$ at $y = 0$, $V_x \to V_\infty$ for $y \to \infty$.

The boundary value problem represented by Eq. (149) with these boundary conditions was first solved by Blasius [32] and later on by Howarth [33] using the similarity transformation

$$\xi = y\left(\frac{V_\infty}{\nu x}\right)^{1/2} = y/\delta_h'(x) \tag{150}$$

where

$$\delta_h'(x) = \left(\frac{\nu x}{V_\infty}\right)^{1/2} \tag{151}$$

can be treated as the local thickness of thc boundary layer.

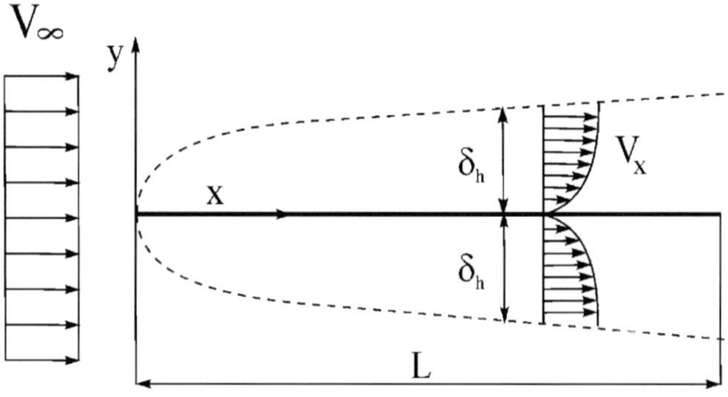

Fig. 3.17. The flow in the vicinity of a stationary rigid plate in uniform incident stream.

The continuity equation is automatically satisfied by introducing the stream function

$$\psi_s(\xi) = (\nu x V_\infty)^{1/2} f(\xi) \tag{152}$$

Consequently, the fluid velocity components become

$$\begin{aligned} V_x &= \frac{\partial \psi_s}{dy} = \frac{\partial \psi}{\partial \xi}\frac{d\xi}{dy} = V_\infty f'(\xi) \\ V_y &= -\frac{\partial \psi_s}{\partial x} = \frac{1}{2}\left(\frac{\nu V_\infty}{x}\right)^{1/2}(\xi f' - f) \end{aligned} \tag{153}$$

Substituting these expressions into Eq. (149), one obtains the third-order nonlinear differential equation for the unknown function f

$$2f''' + f f'' = 0 \tag{154}$$

with the boundary conditions $f' = 0$, $f = 0$ at $\xi = 0$ and $f = 1$ for $\xi \to \infty$. Unfortunately, Eq. (154) can only be solved numerically as originally done by Blasius [32], who showed that $f'' = f(0) = 0.3321$ in the limit of $\xi \to 0$.

This allows one to calculate the local stress on the plate surface, which is given by the expression

$$\Pi_0(x) = \eta\left(\frac{\partial V_x}{\partial y}\right)_{y=0} = \eta f(0) V_\infty \left(\frac{V_\infty}{\nu x}\right)^{1/2} = f(0)\left(\eta \rho V_\infty^3\right)^{1/2}\frac{1}{x^{1/2}} \tag{155}$$

The drag force on the plate that is given by the formula

$$F_h = 2d_p \int_0^L \Pi_0(x)dx = 4f(0)d_p L^{1/2}\left(\eta \rho V_\infty^3\right)^{1/2} \tag{156}$$

where d_p is the plate width and L is the plate length.

For larger distances from the plate, where $\xi > 1$, Schlichting [5] derived the following series expansion for the f function

$$f = \sum_{n=0}^{\infty}\left(-\frac{1}{2}\right)^n \frac{f(0)^{n+1} C_n}{(3n+2)!}\xi^{3n+2} \tag{157}$$

where C_n are the coefficients of this expansion equal to $C_0 = C_1 = 1$, $C_2 = 11$, $C_3 = 375$ [5].

Using Eq. (157) and Eq. (153), the fluid velocity components can be expressed as

$$V_x = V_\infty \sum_{n=0}^{\infty}\left(-\frac{1}{2}\right)^n \frac{f(0)^{n+1} C_n}{(3\mathrm{n}+1)!}\xi^{3n+1} \tag{158}$$

$$V_y = \frac{1}{2}\left(\frac{\nu V_\infty}{x}\right)^{1/2} \sum_{n=0}^{\infty}\left(-\frac{1}{2}\right)^n \frac{f(0)^{n+1} C_n}{3n!(3n+2)}\xi^{3n+2} \tag{159}$$

By retaining the leading term only, one obtains the simple expression

$$V_x = f(0)\frac{V_\infty^{3/2}}{\nu^{1/2}}\frac{y}{x^{1/2}}$$
$$V_y = f(0)\frac{V_\infty^{3/2}}{4\nu^{1/2}}\frac{y^2}{x^{3/2}} \tag{160}$$

Eq. (157) gives a 5% accuracy of velocity estimation for $y < \left(\frac{\nu x}{V_\infty}\right)^{1/2}$.

It is to mention, however, that Eq. (157) and Eq. (159), are not appropriate for tangential distances comparable to and smaller than ν/V_∞ because of the singularity which appears for $x \to 0$.

From Eq. (160) one can deduce that fluid streamlines are described by the formula

$$y = y_0\left(\frac{x}{x_0}\right)^{1/4} \tag{161}$$

where x_0, y_0 are the coordinates of an arbitrary point crossed by a streamline.

Blasius [32] also showed that the thickness of the hydrodynamic boundary layer (defined as the distance from the plate where the tangential velocity attains 99% of the uniform velocity of the outer flow) is given by

$$\delta_h = 5\left(\frac{\nu x}{V_\infty}\right)^{1/2} \tag{162}$$

The above solution for a plate can be exploited to analyze the flow in the entrance region of the parallel-plate channel, and in particular to estimate the length of the transition region where a stationary velocity profile develops. This can be done by expanding the stream function in terms of the entrance length parameter defined as

$$\bar{l} = \left(\frac{\nu x}{b^2 V_\infty}\right)^{1/2} \tag{163}$$

where V_∞ is the uniform velocity at the entrance to the channel and $2b$ is the thickness of the channel. The similarity variable ξ is the same as that defined by Eq. (150).

By postulating that $\bar{l}<1$, the stream function for this flow can be expanded in terms of the series [34,35]

$$\psi_s(\xi) = V_\infty b\left[\bar{l}\, f(\xi) + \bar{l}^2 f_2(\xi) + \bar{l}^3 f_3(\xi) + \cdots\right. \tag{164}$$

where f is the Blasius function describing the flow near the plate discussed above, and f_2 and f_3 are the perturbing functions.

Using Eq. (164) one can express the velocity components in the form [35]

$$\begin{aligned} V_x &= V_\infty\left[f' + \bar{l}\, f_2' + \bar{l}^2 f_3' + \cdots\right. \\ V_y &= -\frac{1}{2}\left(\frac{V_\infty \nu}{x}\right)^{1/2}\left[f - \xi f' + \bar{l}\left(f_2 - \xi f_2'\right) + \bar{l}^2\left(f_3 - \xi f_3'\right) + \cdots\right. \end{aligned} \tag{165}$$

The function f_2 can be calculated from the linear differential equation of third order

$$2f_2''' + ff_2'' - f'f_2' + 2f''f_2 = -1.73 \tag{166}$$

with the boundary conditions $f_2=0, f_2'=0$ at $\xi=0$, $f_2'=-1.73$ for $\xi \to \infty$. Analogous equations can be formulated for f_3. It has been calculated that the characteristic length of the entrance region is given by the expression [5,35]

$$L_c = 0.16\frac{V_\infty b^2}{\nu} = 0.08(2b)Re \tag{167}$$

where $Re = Vb/\nu$ is the channel's Reynolds number.

Taking typical data occurring in experiments, $2b=10^{-3}$ m and $V_\infty = 0.1\text{m s}^{-1}$, one obtains $L_c = 4\times10^{-3}$ m in water. For $2b=10^{-4}$ m and the same velocity $L_c=4\times10^{-5}$ m.

For distances from the entrance to the channel that are smaller than this critical value, the fluid velocity components can be calculated from Eq. (158), derived previously for the plate in uniform flow.

An analogous analysis can be performed for the boundary layer flow generated by a continuous surface (sheet) moving in an otherwise quiescent fluid (Fig. 3.18). The constant translation velocity of the sheet equals V_s, and its length equals L_s. Thus, the Reynolds number of the flow is $Re = L_s V_s/\nu$. It is assumed that the Reynolds number is much greater than unity in order to justify the boundary-layer approximation. The analysis of the flow due to the moving sheet was done by Sakiadis [36,37], who showed that the flow remains laminar for $Re < 5\times10^5$. Because the flow is two-dimensional and there is no pressure gradient, one can use Eq. (149) together with the similarity transformation Eq. (150) and the velocity components defined by the stream function given by Eq. (153). The boundary conditions at the surface are $V_x=V_s$, $V_y=0$ and in the bulk $V_x=V_y=0$. Sakiadis [36] showed that the stream function for this flow can be approximated for the entire range of ξ by the formula

$$\psi_s = 1.616\left[1-(1+0.3812\xi+0.0185\xi^2-0.00543\xi^3)e^{-\xi}\right] \tag{168}$$

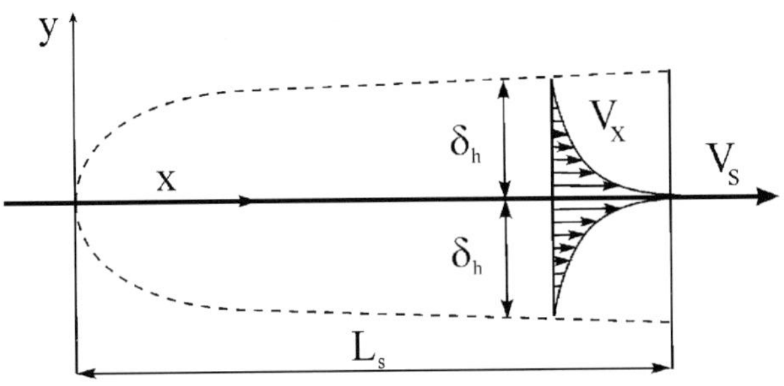

Fig. 3.18. A schematic view of the flow generated by a sheet moving with a constant velocity through a quiescent fluid.

For $\xi < 1$, ψ_s can be approximated by

$$\psi_s = -\frac{1}{2} f(0) \frac{V_s^{3/2}}{\nu^{1/2}} \frac{y^2}{x^{1/2}} + V_s\, y \tag{169}$$

where $f(0) = 0.444$.
Consequently, the fluid velocity components are given by [38]

$$\begin{aligned} V_x &= V_s - f(0) \frac{V_s^{3/2}}{\nu^{1/2}} \frac{y}{x^{1/2}} \\ V_y &= -\frac{f(0)}{4} \frac{V_s^{3/2}}{\nu^{1/2}} \frac{y^2}{x^{3/2}} \end{aligned} \tag{170}$$

Eq. (170) is accurate for the distance from the plate $y < (\nu\, x/V_s)^{1/2}$.

The boundary layer theory can also be exploited to analyze the impinging-jet flows at greater distances from the stagnation point when the so-called wall-jet is formed. This kind of flow, especially for 2D geometry (parallel plane jet), is often used in various drying processes of technological significance [39]. The wall-jet flow configuration is also used to create well-defined conditions of colloid particle deposition on surfaces [40]. In electrochemistry, the electrode based on the wall-jet flow model is used as a detector in liquid chromatography [41]. The appearance of the wall-jet region in the impinging jet flow could already be observed in Figs. 3.10 and 3.14, by noting that the normal fluid velocity component changed sign, and was directed outwards from the surface, at distances from the stagnation point $r/R > 1.5$.

Fluid velocity distribution in the wall-jet, shown schematically in Fig. 3.19, was first analyzed by Glauert [42] both in the radial and planar (2D) cases. The flow is governed by the boundary-layer equation, Eq. (149), for radial and planar jets. However, the fluid continuity equation for the radial jet has the alternative form

$$\frac{\partial(rV_r)}{\partial r} + \frac{\partial(rV_z)}{\partial z} = 0 \qquad (171)$$

This equation is automatically satisfied if the fluid velocity components are calculated as

$$V_r = \frac{1}{r}\frac{\partial \psi_s}{\partial z}$$
$$V_z = -\frac{1}{r}\frac{\partial \psi_s}{\partial r} \qquad (172)$$

where the stream function is expressed in the form [42]

$$\psi_s = \left(\frac{\nu^5 r^3}{V_\infty}\right)^{1/4} f(\xi) \qquad (173)$$

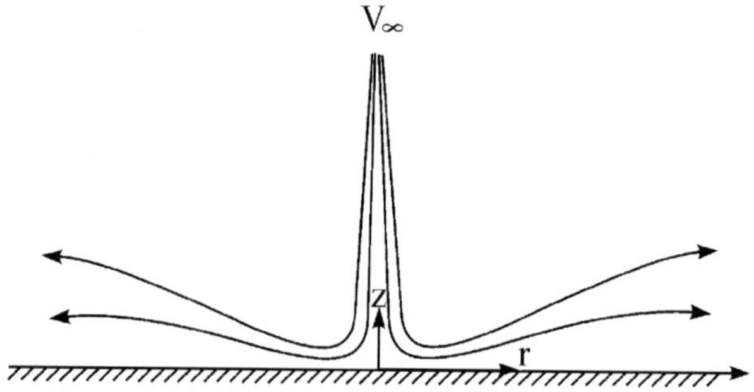

Fig. 3.19. A schematic view of the wall-jet flow generated by a liquid jet striking a solid surface.

and V_∞ is the characteristic fluid velocity.
The similarity variable ξ is defined as

$$\xi = \frac{3}{4}\left(\frac{\nu}{V_\infty r^5}\right)^{1/4} z \tag{174}$$

Using Eqs. (172,173), the fluid velocity components can be expressed in the form

$$\begin{aligned} V_r &= \frac{3}{4}\left(\frac{\nu^6}{V_\infty^2}\right)^{1/4} \frac{1}{r^{3/2}} f'(\xi) \\ V_z &= \frac{3}{4}\nu\left(\frac{\nu}{V_\infty}\right)^{1/4} \frac{1}{r^{5/4}}\left[f - \frac{15}{16}\left(\frac{\nu}{V_\infty}\right)^{1/4} \frac{z}{r^{5/4}} f'\right] \end{aligned} \tag{175}$$

The function f can be calculated from the nonlinear differential equation

$$f''' + f f'' + 2f'^2 = 0 \tag{176}$$

The boundary conditions at the surface are $f = f' = 0$ at $\xi = 0$. On the other hand, far from the surface, where $\xi \to \infty$, both velocity components must vanish because there is no pressure difference and no outer flow, so $f' = 0$ at $\xi \to \infty$ (in the bulk). Because V_r becomes zero at the surface and in the bulk it must attain a maximum value in between.

By solving Eq. (176) Glauert showed that the function f can be evaluated from the implicit dependence

$$\xi = \ln\frac{\left[1+(f)^{1/2}+f\right]^{1/2}}{1-(f)^{1/2}} + (3)^{1/2}\tan^{-1}\frac{(3f)^{1/2}}{2+(f)^{1/2}} \tag{177}$$

It can be calculated from Eq. (177) that $f(0) = 2/9$ and the maximum value of $f = 2^{-5/3} = 0.315$. Moreover, for $\xi < 1$, one can approximate the function f by the series expansion

$$f = \frac{1}{9}\xi^2 - \bar{C}_3\xi^3 \tag{178}$$

where $C_3 = 0.0017$

By retaining the first term of this expansion, fluid velocity components can be expressed as

$$V_r = \frac{2}{9}\left(\frac{15F_w}{2\nu r^3}\right)^{1/2}\xi = \frac{2}{9}\left(\frac{15F_w}{2\nu}\right)^{1/2}\left(\frac{135F_w}{32\nu^3}\right)^{1/4}\frac{z}{r^{11/4}} = \frac{C_w z}{r^{11/4}}$$

$$V_z = \frac{7}{8}\frac{C_w z^2}{r^{15/4}} \tag{179}$$

where

$$C_w = \frac{5}{3}\left(\frac{3}{40}\frac{F_w^3}{\nu^5}\right)^{1/4} = 0.872\left(\frac{F_w}{\nu^5}\right)^{1/4}$$

$$F_w = \int_0^\infty rV_r\left[\int_r^\infty r'V_{r'}^2\,dr'\right]dr = \frac{3\nu^4}{40V_\infty} \tag{180}$$

Glauert [42] postulated that the function F_w can be approximated by

$$F_w = \frac{1}{8\pi^2}V_i Q^2 \tag{181}$$

where V_i is the characteristic velocity of the impinging jet and Q is the volume flow rate in the jet. Using Eq. (181) one can express the constants C_w as

$$C_w = \frac{5}{3(8\pi^2)^{3/4}}\left(\frac{3}{40}\frac{V_i^3Q^6}{\nu^5}\right)^{1/4} = 0.0329\left(\frac{V_i^3Q^6}{\nu^5}\right)^{1/4} \tag{182}$$

As can be noticed from Eq. (179), at small distances from the surface, the radial component of the flow and the wall shear rate $G^0 = \partial V_r/\partial z$ vanish with $r^{-11/4}$. This means that at larger distances from the center, the wall shear rate becomes negligible in comparison with the impinging-jet region. The normal component (directed outwards from the surface) vanishes with $r^{-15/4}$.

Using Eq. (179), one also can deduce that the streamline pattern in the region close to the interface is described by

$$z = z_0 \left(\frac{r}{r_0} \right)^{7/8} \tag{183}$$

where r_0, z_0 are the coordinates of a point crossed by the streamline.

Note, however, that the above expressions for the velocity components are strictly valid only at distances exceeding significantly the impinging-jet diameter, i.e., for $r/R >> 1$.

It was further shown by Glauert [42] that the function f is the same for a two-dimensional flow, so the stream function and velocity components in this case become

$$\psi_s = (40 F_w \nu)^{1/4} f(\xi)$$

$$V_x = \left(\frac{5F_w}{2\nu x} \right)^{1/2} f'(\xi) \tag{184}$$

$$V_y = \frac{1}{4}(40 F_w \nu)^{1/4} \frac{1}{x^{3/4}} \left[f(\xi) - 3\left(\frac{5F_w}{32\nu^3} \right)^{1/4} \frac{yf'}{x^{3/4}} \right]$$

where $\xi = \left(\dfrac{5F_w}{32\nu^3 x^3} \right)^{1/4} y$

$$F_w = \frac{\nu^2 V_\infty}{40}$$

By retaining the first term of this expansion, the fluid velocity components can be expressed as

$$V_x = \frac{2}{9} \left(\frac{5F_w}{2\nu} \right)^{1/2} \left(\frac{5F_w}{32\nu^2} \right)^{1/4} \frac{y}{x^{5/4}} = \frac{C_w y}{x^{5/4}}$$

$$V_y = \frac{5}{8} \frac{C_w y^2}{x^{9/4}} \tag{185}$$

where $C_w = \dfrac{1}{9} \left(\dfrac{125 F_w^3}{8\nu^5} \right)^{1/4} = 0.221 \left(\dfrac{F_w^3}{\nu^5} \right)^{1/4}$

The function F_w can be approximated for a two-dimensional flow by

$$F_w = \frac{1}{2} V_i Q^2 \tag{186}$$

where Q is the volume flow rate in the slot impinging jet.

As can be noticed from Eq. (185), at small distances from the surface, the tangential velocity component of the flow vanishes with $x^{-5/4}$ and the normal component (directed outwards from the surfacc) vanishes with $r^{-9/4}$. Using this equation one can also deduce that the streamline pattern in the region close to the interface is described by

$$y = y_0 \left(\frac{x}{x_0} \right)^{5/8} \tag{187}$$

where (x_0, y_0) are the coordinates of a point crossed by the streamline.

It is to mention, however, that the above expressions for velocity components are strictly valid at distances exceeding significantly the impinging jet width, i.e., for $x/d >> 1$. In order to increase the accuracy of Eqs. (184),(185) Schwartz and Caswell [39] introduced the virtual origin of the flow l, which was approximated by the expression

$$l = 0.05 d \, Re \tag{188}$$

where $Re = Vd/\nu$ is the Reynolds number and d is the width of the slot. When $Re = 20$, $l = d$.

The above formulae for the wall-jet velocity distribution are strictly valid for laminar jets, i.e., for small-scale flows usually found in experimental measurements of particle deposition kinetics. For large-scale flows, occurring in industrial applications, the flow becomes turbulent and can be analyzed in terms of the approximate theory developed by Glauert in the same paper [42].

The boundary-layer approximation can also be used to describe two-dimensional flows past curved bodies when the outer flow velocity becomes dependent on the longitudinal coordinate. In this case, one can use the Blasius' series expansion method with the outer flow given by [5]

$$V(x) = \sum_{n=1}^{\infty} \bar{V}_n x^n \tag{189}$$

where the expansion coefficients $\bar{V}_n$ depend solely on the shape of the body.

The dimensionless coordinate perpendicular to the direction of the flow is defined as

$$\xi = \left(\frac{\bar{V}_1}{\nu}\right)^{1/2} y \tag{190}$$

where $\bar{V}_1$ is the leading term of the expansion, Eq. (189). Consequently, the stream function can be expressed as

$$\psi_s = \left(\frac{\nu}{\bar{V}_1}\right)^{1/2} [\bar{V}_1 x F_1(\xi) + 4\bar{V}_3 x^3 F_3(\xi) + 6\bar{V}_5 x^5 F_5(\xi) + \cdots \tag{191}$$

where $F_1(\xi) \div F_5(\xi)$ are functions to be determined from the configuration outer flow.

By knowing the stream function, the fluid velocity components can be expressed as

$$V_x = \frac{\partial \psi_s}{\partial y} = \bar{V}_1 x F_1'(\xi) + 4\bar{V}_3 x^3 F_3'(\xi) + 6\bar{V}_5 x^5 F_5'(\xi) + \cdots$$

$$V_y = -\frac{\partial \psi_s}{\partial x} = -\left(\frac{\bar{V}_1}{\nu}\right)^{1/2} [\bar{V}_1 F_1(\xi) + 12\bar{V}_3 x^2 F_3(\xi) + 30\bar{V}_5 x^4 F_5(\xi) + \cdots \tag{192}$$

The F_1 function is identical to that previously determined for the two-dimensional stagnation-point flow, given by Eq. (106). Other functions are given in tabulated form in [5]. It is worth pointing out that for $\xi < 0.5$, these functions can be approximated by

$$F_n = \frac{1}{2} f_n(0)\xi^2 \tag{193}$$

where $f_1(0) = f(0) = 1.2326, f_3(0) = 0.7244, f_5(0) = 1.0321, f_7(0) = 2.059$.

The above method can be applied to the interesting case of a cylinder immersed in uniform flow V_∞. At large distances from the cylinder, the flow can be approximated by the potential flow governed by the Euler equation. The tangential and normal velocity components of this flow are given by [2]

$$V_{\vartheta} = V_{\infty}\left(\frac{1+R^2}{r^2}\right)\sin\vartheta$$

$$V_r = -V_{\infty}\left(\frac{1+R^2}{r^2}\right)\cos\vartheta \tag{194}$$

where R is the cylinder radius and $\vartheta = x/R$ is the angle measured against the direction of the flow.

From Eq. (194), one can deduce that the outer flow near the cylinder, needed to formulate Eq. (189), can be approximated by

$$V_{\vartheta} = V_{\infty}'\sin\vartheta \tag{195}$$

where $V_{\infty} = 1 + R^2/r_0^2$ is the characteristic velocity to be matched by the inner flow and r_0 is the characteristic distance from the cylinder.

By expanding the $\sin\vartheta$ into the Taylor series, one can express Eq. (195) in the form

$$V(x) = V'_{\infty}\left[\frac{x}{R} - \frac{1}{3!}\left(\frac{x}{R}\right)^3 + \frac{1}{5!}\left(\frac{x}{R}\right)^5 - \dots\right. \tag{196}$$

The two terms of this expansion assure good accuracy for $\vartheta < \pi/2$, and three terms, for $\vartheta < 3\pi/4$. Using the expansion, Eq. (196), one can deduce that

$$\bar{V}_1 = \frac{V'_{\infty}}{R} \quad V_3 = -\frac{1}{3!}\frac{V'_{\infty}}{R^3} \quad V_5 = \frac{1}{5!}\frac{V'_{\infty}}{R^5}$$

$$\xi = \left(\frac{V_{\infty}' R}{\nu}\right)^{1/2}\frac{y}{R} \tag{197}$$

Hence, the stream function and the velocity components near the cylinder are given by

$$\psi_s = \left(\nu R V'_\infty\right)^{1/2}\left[\frac{V'_\infty}{R} x F_1 - \frac{4}{3!}\frac{V'_\infty}{R^3} x^3 F_3 + \frac{6}{5!}\frac{V'_\infty}{R^5} x^5 F_5 + \cdots\right.$$

$$V_x = V'_\infty\left[\frac{x}{R}F'_1 - \frac{4}{3!}\left(\frac{x}{R}\right)^3 F'_3 + \frac{6}{5!}V'_\infty\left(\frac{x}{R}\right)^5 F'_5 + \cdots\right. \qquad (198)$$

$$V_y = -\left(\frac{\nu V'_\infty}{R}\right)\left[F_1 - 2\left(\frac{x}{R}\right)^2 F_3 + \frac{1}{4}\left(\frac{x}{R}\right)^4 F_5 + \cdots\right.$$

For small distances from the cylinder surface, where $y = (r - R)/R \ll 1$, the fluid velocity components calculated from Eq. (198) are

$$V_x = 2A_f V'_\infty \frac{xy}{R^2}\left[1 - \frac{4}{3!}\frac{f_3(0)}{f(0)}\frac{x^2}{R^2} + \frac{6}{5!}\frac{f_5(0)}{f(0)}\frac{x^4}{R^4} + \cdots\right.$$

$$V_y = -A_f V'_\infty \frac{y^2}{R^2}\left[1 - 2\frac{f_3(0)}{f(0)}\frac{x^2}{R^2} + \frac{1}{4}\frac{f_5(0)}{f(0)}\frac{x^4}{R^4} + \cdots\right. \qquad (199)$$

where $A_f = (f(0)/2(2)^{1/2})\ Re^{1/2} = 0.435\ Re^{1/2}$ is the dimensionless flow parameter and $Re = 2V_\infty R/\nu$ is the Reynolds number based on the cylinder diameter.

The tangential stress at the cylinder surface is

$$\Pi_0 = \eta G^0(x) = \eta\left(\frac{\partial V_x}{\partial y}\right)_0$$

$$= \frac{2A_f V'_\infty}{R}\left[\frac{x}{R} - \frac{4}{3!}\frac{f_3(0)}{f(0)}\frac{x^3}{R^3} + \frac{6}{5!}\frac{f_5(0)}{f(0)}\frac{x^5}{R^5} + \cdots\right. \qquad (200)$$

Using five terms of the expansion, Schlichting demonstrated [5] that for an angle $\vartheta > 1.22\pi$ (110°), the tangential stress at the wall vanishes. This leads to flow separation and formation of vortices at larger distances from the cylinder, quite analogously to the impinging-jet flow. This phenomenon is well illustrated in Fig. 3.20, which shows the wake formation behind the cylinder placed in a uniform flow [43].

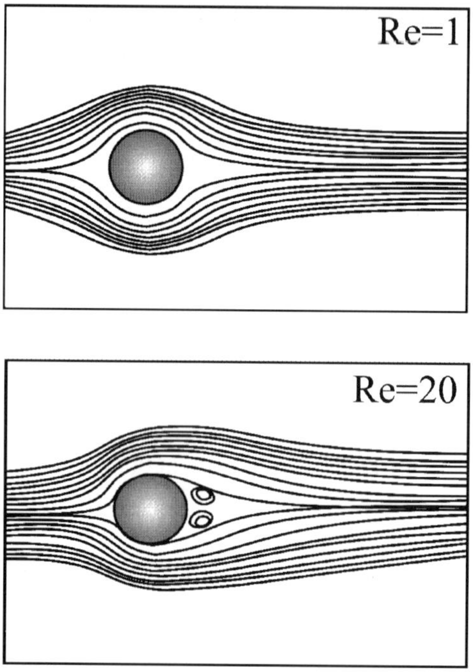

Fig. 3.20. A schematic view of a flow past a cylinder: $Re = 1$ and 20 From Ref. [43].

Eq. (199) has a practical significance in view of the widespread use of filters packed with cylindrically shaped fibers (filtration mats). Moreover, it can be used for testing the applicability of approximate models for which the flow near a cylinder is described. For example, Weber and Paddock [44] assumed that the fluid velocity components are for close separations given by the formula

$$V_x = 2A_f V_\infty \frac{y}{R}\sin\left(\frac{x}{R}\right)$$
$$V_y = -A_f V_\infty \frac{y^2}{R^2}\cos\left(\frac{x}{R}\right) \tag{201}$$

with the semiempirical expression for the flow parameter

$$A_f = 0.44\,Re^{0.52} \tag{202}$$

valid for $Re < 200$.

By comparing Eq. (199) with Eq. (201) it can be deduced that a good agreement between both (at least in the region before the critical separation distance) is attained if $V'_\infty = V_\infty$.

An analogous boundary-layer analysis can be performed for curved bodies exhibiting rotational symmetry (see Fig. 3.21), e.g., spheres, which is of practical interest in various filtration procedures. Let us assume that the radius r of a cross-section of a body is the following function of the x coordinate, tangential to the body's surface

$$r(x) = \bar{r}_1 x + \bar{r}_3 x^3 + \bar{r}_5 x^5 + \cdots \tag{203}$$

where $\bar{r}_1$, $\bar{r}_3$, $\bar{r}_5$, are the expansion coefficients.
Similarly, the velocity of the outer (potential) flow is expanded as

$$V(x) = \bar{V}_1 x + \bar{V}_3 x^3 + \bar{V}_5 x^5 + \cdots \tag{204}$$

Using these expansions, Eqs. (203),(204), one can express the stream function as [5]

$$\psi_s = \left(\frac{\nu}{2\bar{V}_1}\right)^{1/2} [\bar{V}_1 x F_1(\xi) + 2\bar{V}_3 x^3 F_3(\xi) + 3\bar{V}_5 x^5 F_5(\xi) + \cdots \tag{205}$$

where $F_1(\xi) \div F_5(\xi)$ are functions of the reduced distance $\xi = (2\bar{V}_1/\nu)^{1/2}\, y$. $F_1(\xi)$ is the function determined previously for the stagnation-point flow. The remaining functions F_3 through F_n depend not only on ξ but also on the parameters $\bar{V}_1$ and $\bar{V}_n$ and $\bar{r}_1$ through $\bar{r}_n$.

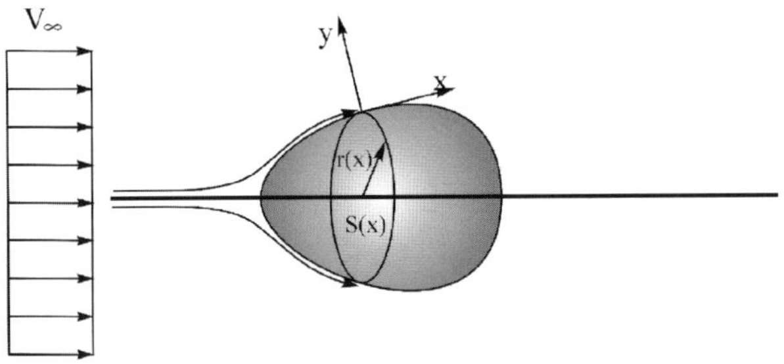

Fig. 3.21. A schematic view of the boundary-layer flow near bodies of revolution (axisymmetrical flows).

Having defined ψ_s, the fluid velocity components can be expressed as

$$V_x = \frac{\partial \psi_s}{\partial y}$$
$$V_y = -\frac{1}{r}\frac{\partial (r\psi_s)}{\partial x} = -\frac{\partial \psi_s}{\partial x} - \frac{1}{r}\frac{dr}{dx}\psi_s \tag{206}$$

The above general method can be applied to the interesting case of a flow past a sphere in a uniform flow. The flow at large distances from the sphere can be approximated by the potential flow governed by Euler's equation, with the tangential velocity component given by [2]

$$V(x) = \frac{1}{2}V_\infty\left(2 + \frac{R^3}{r^3}\right)\sin\left(\frac{x}{R}\right) \tag{207}$$

where R is the radius of the sphere.

In the region close to the interface one has

$$V(x) \cong \frac{3}{2}V_\infty \sin\left(\frac{x}{R}\right) \tag{208}$$

By expanding $V(x)$ and $\bar{r} = (1/R)\sin(x/R)$ into Taylor series, one can deduce that for the sphere

$$\bar{V}_1 = \frac{3}{2}\frac{V_\infty}{R} \quad \bar{r}_1 = 1$$
$$\bar{V}_3 = -\frac{1}{4}\frac{V_\infty}{R^3} \quad \bar{r}_3 = -\frac{1}{3!R^2}$$
$$\bar{V}_5 = \frac{1}{80}\frac{V_\infty}{R^5} \quad \bar{r}_5 = \frac{1}{5!R^4} \tag{209}$$
$$\xi = \left(\frac{3V_\infty R}{\nu}\right)^{1/2}\frac{y}{R}$$

Hence, the stream function and the fluid velocity components near the sphere are given by

$$\psi_s = \left(\frac{\nu R}{3V_\infty}\right)^{1/2} V_\infty \left[\frac{3}{2}\frac{x}{R}F_1(\xi) - \frac{1}{2}\left(\frac{x}{R}\right)^3 F_3(\xi) + \frac{3}{80}\left(\frac{x}{R}\right)^5 F_5(\xi) + \cdots\right.$$

$$V_x = V_\infty \left[\frac{3}{2}\frac{x}{R}F'_1(\xi) - \frac{1}{2}\left(\frac{x}{R}\right)^3 F'_3(\xi) + \frac{3}{80}\left(\frac{x}{R}\right)^5 F'_5(\xi) + \cdots\right. \quad (210)$$

$$V_y = -\left(\frac{3\nu V_\infty}{R}\right)^{1/2}\left[F_1(\xi) - \tilde{F}_3(\xi)\left(\frac{x}{R}\right)^2 + \tilde{F}_5(\xi)\left(\frac{x}{R}\right)^4 + \cdots\right.$$

where $\tilde{F}_3(\xi) = (1/6)\,F_1(\xi) + (2/3)\,F_3(\xi)$ and $\tilde{F}_5(\xi) = (1/90)\,F_1(\xi) + (1/18)\,F_3(\xi) + (3/40)\,F_5(\xi)$.

Calculations involving higher terms revealed that for an angle $\vartheta > 1.22\pi$ (110°), the flow at larger distances from the sphere tends to separate [5] similarly as in the case of the cylinder.

For small distances from the surface, when $y = (r-R)/R \ll 1$, the fluid velocity components, calculated from Eq. (210) become

$$V_x = \frac{3}{2}A_f V_\infty \frac{xy}{R^2}\left[1 - \frac{1}{3}\frac{f_3(0)}{f(0)}\left(\frac{x}{R}\right)^2 + \frac{1}{40}\frac{f_5(0)}{f(0)}\left(\frac{x}{R}\right)^4 + \cdots\right.$$

$$V_y = -\frac{3}{2}A_f V_\infty \frac{y^2}{R^2}\left[1 - \frac{\tilde{f}_3(0)}{f(0)}\left(\frac{x}{R}\right)^2 - \frac{\tilde{f}_5(0)}{f(0)}\left(\frac{x}{R}\right)^4 + \cdots\right. \quad (211)$$

$$A_f = \left(\frac{3}{2}\right)^{1/2} f(0)\,Re^{1/2}, \quad Re = 2V_\infty R/\nu$$

where A_f is the dimensionless flow parameter, $Re = 2V_\infty R/\nu$ is the Reynolds number based on the sphere diameter and $f(0) = 0.9277$, $f_3(0) = 1.0923$, $f_5(0) = 1.561$, $\tilde{f}_3 = 0.8828$, $\tilde{f}_5(0) = 0.1675$.

The velocity components given by Eq. (211) agree well, in the region $\vartheta < \pi$ with the semiempirical velocity profiles given by Weber and Paddock [44]

$$V_r = V_h = -\frac{3}{2} A_f V_\infty \left(\frac{y}{R}\right)^2 \cos\vartheta \tag{212}$$

$$V_\vartheta = \frac{3}{2} A_f V_\infty \frac{y}{R} \sin\vartheta$$

where the flow parameter A_f is given by

$$A_f = \frac{3}{2}\left(1 + \frac{(3/16)Re}{1 + 0.249\,Re^{0.56}}\right) \tag{213}$$

Eq. (213) is valid for $0 < Re < 300$.

All expressions discussed in this section derived within the framework of the hydrodynamic boundary-layer theory are valid for *Re* numbers much larger than unity. For *Re* approaching unity or smaller, fluid velocity distribution near macroscopic interfaces can be obtained by solving the linear Stokes equation discussed in the next section.

3.3.4. Decomposition of Macroscopic Flows into Simple Flows

For the sake of convenience, various macroscopic flow configurations analyzed above are collected in Table 3.4 together with fluid velocity distribution near surfaces. This is the most important region from the viewpoint of particle transport because colloid particles have typical dimensions of 10–100 nm, whereas the typical macroscopic flow length scale L_{ch} is of the order of 10^{-3} to 10^{-1} m. Hence the ratio of particle size to L_{ch} is of the order of 10^{-4} to 10^{-7}. The large disparity in the particle and flow length scales means that macroscopic flow variations over distances comparable to particle dimension are very small. Hence, by introducing local coordinate systems each macroscopic flow can be decomposed into the sum of simpler flows. This is important because these "elementary" flows can be analytically evaluated using the Stokes equation, as demonstrated later on. It then becomes possible to calculate the hydrodynamic forces and torques on a particle, which is the necessary condition for formulating the particle transport equations.

Table 3.4
Stationary macroscopic flows

Flow configuration	Velocity components at the wall	Shear rates at the wall G_{Sh}	G_{St}
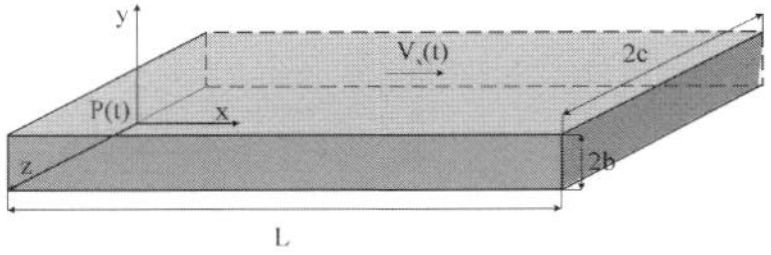 Parallel-plate channel	$V_x = \frac{Pb}{\eta L} y$ $V_y = 0$	$\frac{Pb}{\eta L}$	0
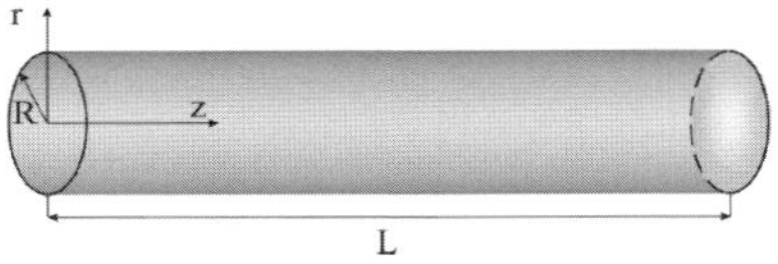 Cylindrical channel (capillary)	$V_z = \frac{PR}{2\eta L}(R-r)$ $V_r = 0$	$\frac{PR}{2\eta L}$	0
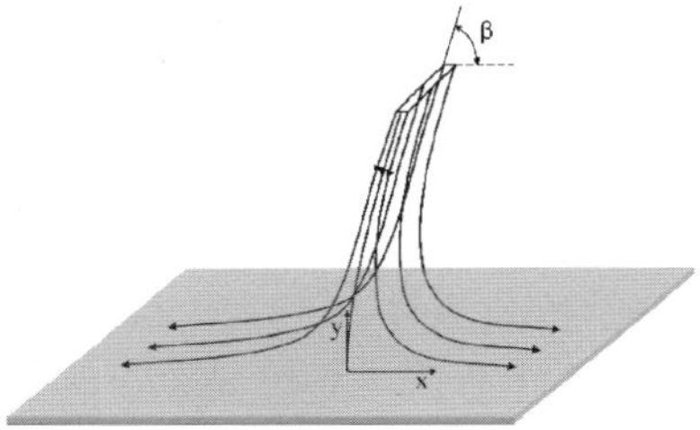 Oblique stagnation–point flow (2D)	$V_x = f(0)\frac{(G_f \sin\varphi)^{3/2}}{v^{1/2}} x'y$ $V_y = -\frac{1}{2} f(0)\frac{(G_f \sin\varphi)^{3/2}}{v^{1/2}} y^2$ $\varphi = -\frac{1}{2}\tan^{-1}\left(\frac{\tan\beta}{2}\right)$ $f(0) = 1.2326$	$f(0)\frac{(G_f \sin\varphi)^{3/2}}{v^{1/2}} x_p'$	$\frac{1}{2} f(0)\frac{(G_f \sin\varphi)^{3/2}}{v^{1/2}}$

Table 3.4 (continued)

Flow configuration	Velocity components at the wall	Shear rates at the wall G_{Sh}	G_{St}
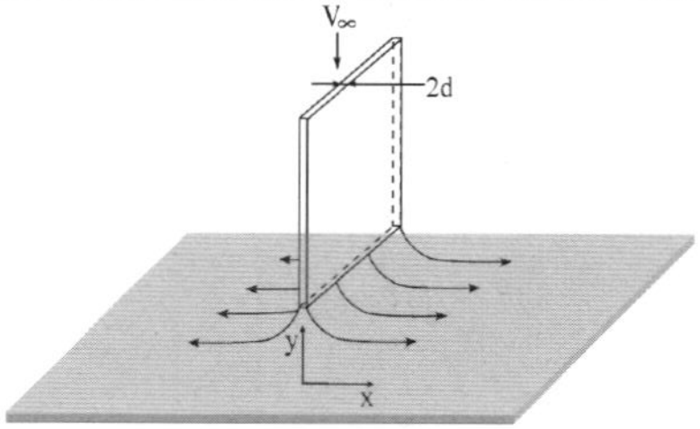 Slot impinging-jet (SIJ)	$V_x = 2\alpha_s \frac{V_\infty}{d} y S(x/d)$ $V_y = -\alpha_s \frac{V_\infty}{d^2} y^2 C(x/d)$ $V_\infty = Q/2dl$	$2\alpha_s \frac{V_\infty}{d} S(x_p/d)$	$\alpha_s \frac{V_\infty}{d^2} C(x_p/d)$
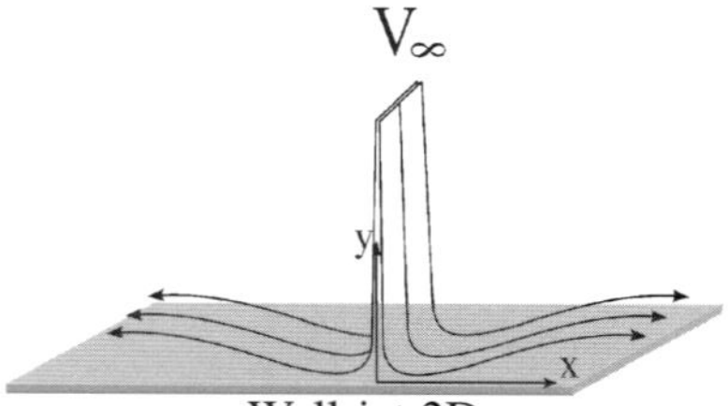 Wall-jet 2D	$V_x = \frac{C_w}{x^{5/4}} y$ $V_y = \frac{5}{8} \frac{C_w}{x^{9/4}} y^2$ C_w is given by Eq. (185)	$\frac{C_w}{x_p^{5/4}}$	$\frac{5}{8} \frac{C_w}{x_p^{9/4}}$
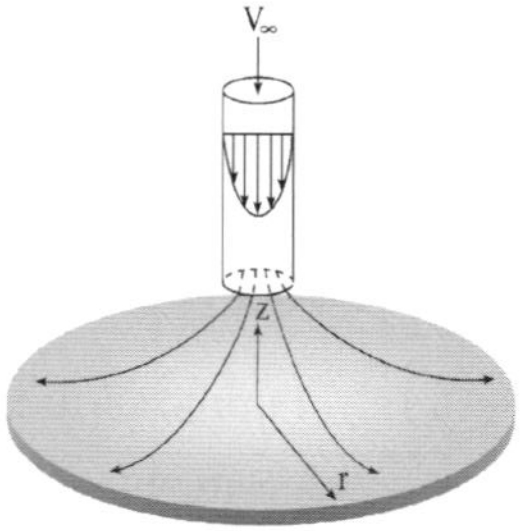 Radial impinging-jet (RIJ)	$V_r = \alpha_r \frac{V_\infty}{R} z S(r/R)$ $V_z = -\alpha_r \frac{V_\infty}{R^2} z^2 C(r/R)$ $V_\infty = Q/\pi R^2$	$\alpha_r \frac{V_\infty}{R} S(r_p/R)$	$\alpha_r \frac{V_\infty}{R^2} C(r_p/R)$

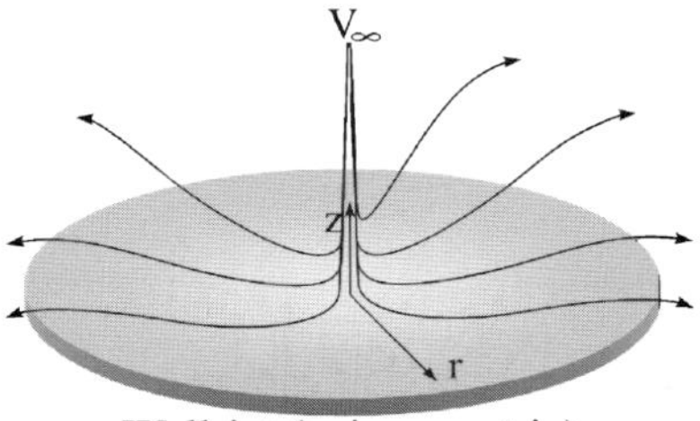 Wall-jet (axisymmetric)	$V_r = \frac{C_w z}{r^{11/4}}$ $V_z = \frac{7}{8}\frac{C_w z^2}{r^{15/4}}$ C_w is given by Eq. (182)	$\frac{C_w}{r_P^{11/4}}$	$\frac{7}{8}\frac{C_w}{r_P^{15/4}}$
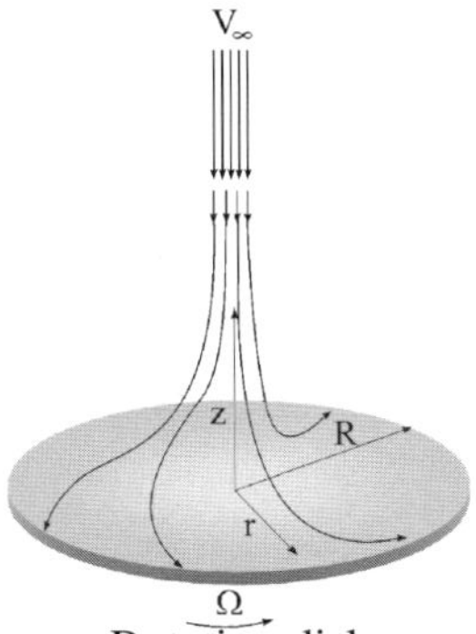 Rotating disk	$V_r = f(0)\frac{\Omega^{3/2}}{\nu^{1/2}} r z$ $V_\varphi = \Omega r\left[1 - g(0)\left(\frac{\Omega}{\nu}\right)^{1/2} z\right]$ $V_z = -f(0)\frac{\Omega^{3/2}}{\nu^{1/2}} z^2$ $f(0) = 0.5102$ $g(0) = 0.616$	$f(0)\frac{\Omega^{3/2}}{\nu^{1/2}} r_p$ $g(0)\frac{\Omega^{3/2}}{\nu^{1/2}} r_p$	$f(0)\frac{\Omega^{3/2}}{\nu^{1/2}}$
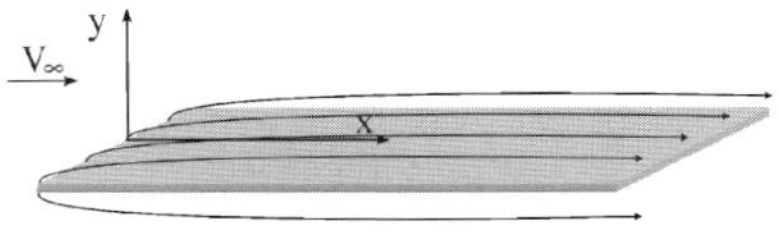 Plate in uniform flow	$V_x = f(0)\frac{V_\infty^{3/2}}{\nu^{1/2} x^{1/2}} y$ $V_y = f(0)\frac{V_\infty^{3/2}}{4\nu^{1/2} x^{3/2}} y^2$ $f(0) = 0.3321$	$f(0)\frac{V_\infty^{3/2}}{\nu^{1/2} x_p^{1/2}}$	$\frac{f(0)}{4}\frac{V_\infty^{3/2}}{\nu^{1/2} x_p^{3/2}}$

Table 3.4 (continued)

Flow configuration	Velocity components at the wall	Shear rates at the wall G_{Sh}	G_{St}
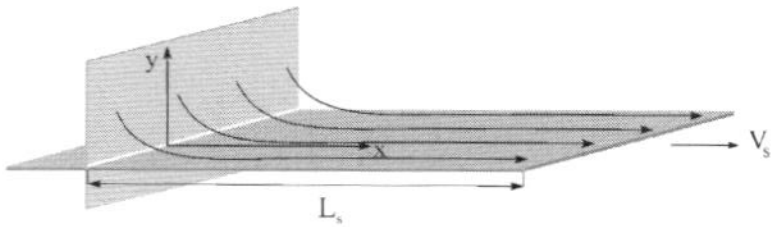 Continous moving plate	$V_x = V_s - f(0)\frac{V_s^{3/2}}{\nu^{1/2}x_p^{1/2}}y$ $V_y = -\frac{f(0)}{4}\frac{V_s^{3/2}}{\nu^{1/2}x_p^{3/2}}y^2$ $f(0) = 0.444$	$f(0)\frac{V_s^{3/2}}{\nu^{1/2}x_p^{1/2}}$	$f(0)\frac{V_s^{3/2}}{\nu^{1/2}x_p^{3/2}}$
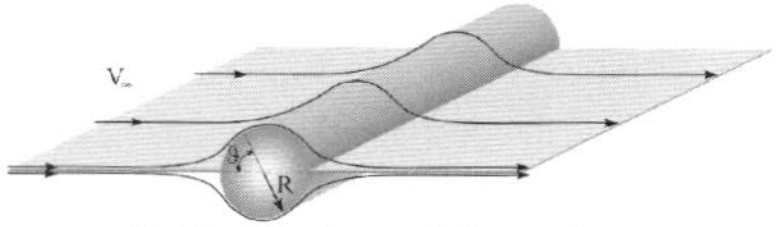 Cylinder in uniform flow	$V_x = 2A_f V_\infty \frac{y}{R^2} S(x/R)$ $V_y = -A_f V_\infty \frac{y^2}{R^2} C(x/R)$ approximately $S=\sin\vartheta$; $C=\cos\vartheta$; $\vartheta=x/R$ $A_f = 0.44\ Re^{0.52}$	$2A_f \frac{V_\infty}{R} S(x_p/R)$	$A_f \frac{V_\infty}{R^2} C\left(\frac{x_p}{R}\right)$
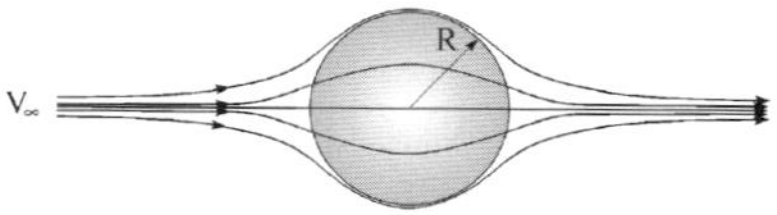 Sphere in uniform flow	$V_x = \frac{3}{2} A_f V_\infty \frac{y}{R} S\left(\frac{x}{R}\right)$ $V_y = -A_f V_\infty \frac{y^2}{R^2} C\left(\frac{x}{R}\right)$ approximately $S=\sin\vartheta$; $C=\cos\vartheta$; $\vartheta=x_p/R$ $A_f = \frac{3}{2}\left(1+\frac{(3/16)Re}{1+0.249Re^{0.56}}\right)$	$\frac{3}{2} A_f \frac{V_\infty}{R} S\left(\frac{x_p}{R}\right)$	$\frac{3}{2} A_f V_\infty \frac{V_\infty}{R^2} C\left(\frac{x_p}{R^2}\right)$

As an example of the flow decomposition procedure let us first consider the two-dimensional macroscopic flows that can be expressed by the stream function in the factorized form

$$\psi_s = f(y)g(x)$$
$$V_x = \frac{\partial \psi_s}{\partial y} = f'g \tag{214}$$
$$V_y = -\frac{\partial \psi_s}{\partial x} = -f\,g'$$

By introducing the local Cartesian coordinate system (x', y') at the point $(x_p, 0)$ (where x_p is the temporary position of the particle), see Fig. 3.22 the fluid velocity components become

$$V_x = f'(y)g(x'+x_p) \cong f'(y)g(x_p) + f'(y)g'(x_p)x'$$
$$V_y = -f(y)g(x'+x_p) \cong -f(y)g'(x_p) - f(y)g''(x_p)x' \tag{215}$$
$$\cong -f(y)g'(x_p)$$

Thus, for distances, $x'/x_p << 1$, the macroscopic velocity field $\mathbf{V}$ can be decomposed into two parts

$$\mathbf{V} = V_x\mathbf{i}_x + V_y\mathbf{i}_y = f'(y)g(x_p)\mathbf{i}_x + g'(x_p)\left[f'(x') - f(y)\right]\mathbf{i}_y \tag{216}$$

where $\mathbf{i}_x$, $\mathbf{i}_y$ are unit vectors of the x and y axes, respectively.

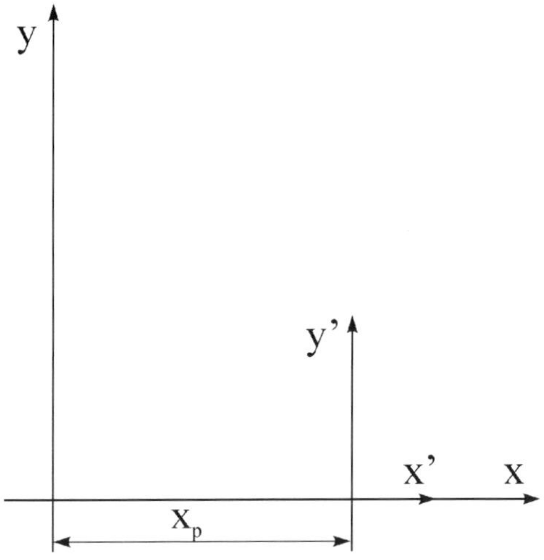

Fig. 3.22. The coordinate system used for decomposition of macroscopic flows.

Except for channel flows, $f(y) = C_f y^2$ for all other flows in the region close to the interface. Therefore, Eq. (216) can be written as

$$\mathbf{V} = 2C_f g(x_p) y \mathbf{i}_x + C_f g'(x_p)\left[2yx'\mathbf{i}_x - y^2\mathbf{i}_y\right] = \mathbf{V}_{Sh} + \mathbf{V}_{St} \tag{217}$$

The component $\mathbf{V}_{Sh} = 2C_f g(x_p) y\, \mathbf{i}_x$ can be identified as a simple shear flow in the x direction with the shear rate $G^0 = G_{sh} = 2C_f g\,(x_p)$ and the component $\mathbf{V}_{St} = C_f g'(x_p)\,[2yx'\mathbf{i}_x - y^2\mathbf{i}_y]$, a two-dimensional stagnation-point flow with the stagnation shear rate $G_{st} = C_f g'(x_p)$. In the case of the slot impinging-jet, one has explicitly

$$\begin{aligned} G_{Sh} &= 2\alpha_S \frac{V_\infty}{d} S(x_p/d) \\ G_{S_t} &= \alpha_S \frac{V_\infty}{d^2} C(x_p/d) \end{aligned} \tag{218}$$

For the plate in a uniform flow, one has

$$\begin{aligned} G_{Sh} &= 0.3321\left(\frac{V_\infty^3}{x_p \nu}\right)^{1/2} \\ G_{S_t} &= 0.3321\frac{1}{4}\left(\frac{V_\infty^3}{x_p^3 \nu}\right)^{1/2} \end{aligned} \tag{219}$$

In the case of the channel flow $G_{Sh} = Pb/\eta L$ and $G_{St} = 0$.
Other examples for two-dimensional flows are collected in Table 3.4.

Quite analogous analysis can be performed for other flows of radial symmetry and three-dimensional flows. For example in the case of the rotating disk, by introducing the local coordinate system (x',y',z') with its origin at the point $(r_p, \varphi_p, 0)$ (Fig. 3.23) one can represent the flow field at the surface as a sum of four elementary flows

$$\mathbf{V} = \mathbf{V}_1 + \mathbf{V}_2 + \mathbf{V}_3 + \mathbf{V}_4 \tag{220}$$

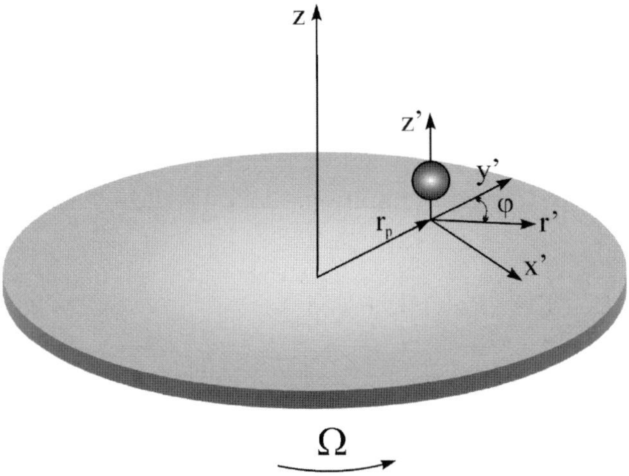

Fig. 3.23. The coordinate system used for decomposition of the macroscopic flow near the rotating disk.

where

$$\mathbf{V}_1 = G_{St}(x'z'\mathbf{i}_{x'} + y'z'\mathbf{i}_{y'} - z'^2\mathbf{i}_{z'}) = G_{St}(r'z'\mathbf{i}_{r'} - z'^2\mathbf{i}_z)$$

$$\mathbf{V}_2 = -G_{Sh}z'\mathbf{i}_{x'}$$

$$\mathbf{V}_3 = G_{Sh}z'\mathbf{i}_{y'}$$

$$\mathbf{V}_4 = \frac{G_{Sh}^{'}}{r_p}(x'\mathbf{i}_{y'} - y'\mathbf{i}_{x'})\,z' = \frac{G_{Sh}^{'}}{r_p}z\mathbf{i}_y$$

$$G_{St} = 0.5102\,\frac{\omega^{3/2}}{\nu^{1/2}} \tag{221}$$

$$G_{sh}^{'} = 0.5102\,\frac{\omega^{3/2}}{\nu^{1/2}}r_p$$

$$G_{_{Sh}}^{'} = 0.616\,\frac{\omega^{3/2}}{\nu^{1/2}}r_p$$

(r',φ,z') are the coordinates of the local cylindrical coordinate system (see Fig. 3.23) and $\mathbf{i}_r$, $\mathbf{i}_y$, $\mathbf{i}_z$ are the unit vectors of this coordinate system.

$\mathbf{V}_1$ in Eq. (221) is the axisymmetric stagnation-point flow, $\mathbf{V}_2$, $\mathbf{V}_3$ are simple shear flows in the x and y directions and $\mathbf{V}_4$ is the shear flow in the φ direction.

In the case of the radial impinging-jet only the $\mathbf{V}_1$ and $\mathbf{V}_3$ flows arise, given by

$$\mathbf{V}_1 = \mathbf{V}_{St} = \alpha_r \frac{V_\infty}{R^2} C(r_p/R)(r'z'\mathbf{i}_{r'} - z'^2\mathbf{i}_z)$$
$$\mathbf{V}_3 = \mathbf{V}_{Sh} = \alpha_r \frac{\mathbf{V}_\infty}{R} S(r_p/R)z'\mathbf{i}_{y'} \tag{222}$$

Values of the parameters G_{St}, G_{Sh} for other flows are collected in Table 3.4. In the next section hydrodynamic forces and torques exerted on particles by these flows will be analyzed.

3.4. FLOWS INVOLVING A SINGLE PARTICLE

As demonstrated above, there are few exact solutions of three-dimensional flows governed by the nonlinear Navier–Stokes equation containing inertia terms. The radial stagnation-point flow and the flow due to the rotating disk are few exceptions. In contrast, there is no exact solution for the important case of flows generated by a spherical particle moving in an unbounded fluid or near interfaces. In this case, solutions can only be derived in two limiting cases (i) the flows at high Reynolds numbers governed by the Euler equation (complicated however, by the appearance of instabilities and then turbulence) and (ii) the flows at low Reynolds numbers governed by the Stokes equation, derived previously. The major advantage of the Stokes equation is its linearity which enables one to derive solutions of general validity by the superposition of solutions for simpler cases. For example, one can deal with the translation and rotation of a particle separately and then simply add the solutions. Similarly, particle motion in various directions near boundaries (parallel or perpendicular) can be dealt with separately. The linearity of the Stokes equation also enables one to formulate integral representation of complex flows in terms of simple tensors and their derivatives. On the other hand, a disadvantage of the Stokes approximation is the difficulty in properly treating the transient motion of particles and velocity fields at greater distances from moving particles. This leads to quite unexpected results and paradoxes; for example, there is no physically sound solution of the flow field due to a translating cylinder at large distances.

Even though the Stokes equation is linear, there still remains an elliptic partial differential equation that requires boundary conditions at all the interfaces involved. This eliminates practically the possibility of deriving exact solutions for interfaces of complex shapes because of the lack of appropriate coordinate systems, and for multi-particle systems (except two particle ones) in an arbitrary configuration.

There are three main ways of analyzing fluid velocity field distribution governed by the Stokes equation

(i) the spherical harmonic expansion of the pressure and velocity components proposed by Lamb

(ii) the stream function approach, valid for axisymmetric quasi two-dimensional flows

(iii) the integral representation approach exploiting singularity solutions for a point force and its moments (multipoles).

The latter method is quite general and can be effectively used to evaluate the hydrodynamic forces and torques without even evaluating explicitly the velocity field. It can also be extended to analyze particle motion near solid and free interfaces. On the other hand, the advantage of the stream function approach is the direct access to the physical structure of the flow via the streamline pattern.

3.4.1. Stokes' Flows Near a Spherical Particle

Let us first consider the integral method that is based on the flow superposition principle expressed as

$$\mathbf{V} = \mathbf{V}_\infty(\mathbf{x}) + \sum \mathbf{V}_i(\mathbf{x}) \tag{223}$$

where $\mathbf{V}_\infty(\mathbf{x})$ is the ambient (external) velocity field of the fluid in which the particle is immersed, fulfilling the Stokes equation, and $\mathbf{V}_i$ are velocity fields derived from the solutions of fundamental problems, called singularity solutions.

Note that the simple flows at interfaces, defined previously, such as the shearing or stagnation-point flows, obey the Stokes equation as well as any linear or uniform flow.

On the other hand, one of the singularity fields $\mathbf{V}_i$ can be derived from the solutions of the Euler equation describing potential flow produced by a source of strength q_h located at the point $\mathbf{x} = \mathbf{x}_0$, where $\mathbf{x}$ is the position vector in the Cartesian coordinate system. In this case the velocity field assumes the simple form [2]

$$\mathbf{V}(x) = \frac{q_h}{4\pi}\frac{\overline{\mathbf{x}}}{r^3} \tag{224}$$

where $r = |\mathbf{x} - \mathbf{x}_0|$and $\overline{\mathbf{x}} = \mathbf{x}–\mathbf{x}_0$ are the relative position vectors. Note that Eq. (224) is analogous to the equation describing the electric field distribution near a charge in free space (cf. Eq. (2.5)).

Other singularities can be derived from the Stokes equation, written as

$$\eta \nabla^2 \mathbf{V} = \nabla p - \mathbf{f}_p\, \delta_D(\overline{\mathbf{x}}) \tag{225}$$

where $\mathbf{f}_p$ is the body force applied at the point $\mathbf{x} = \mathbf{x}_0$, and $\delta_D\,(\overline{\mathbf{x}})$ is the Dirac delta function.

By taking the Fourier transformation of Eq. (225), it can be shown that the velocity field produced by this force in an unbounded fluid is given by the expression [2]

$$\mathbf{V}(\mathbf{x}) = \mathbf{G}\cdot\mathbf{f}_p \tag{226}$$

where

$$\mathbf{G} = \frac{1}{8\pi\eta}\left(\frac{\mathbf{I}}{r} + \frac{\overline{\mathbf{x}}\overline{\mathbf{x}}}{r^3}\right) \tag{227}$$

is the Oseen tensor for an unbounded fluid (sometimes the factor $1/8\pi\eta$ is omitted when referring to the Oseen tensor), and $\mathbf{I}$ is the unit tensor.

The pressure and hydrodynamic stress tensors are

$$\begin{aligned} p(\overline{\mathbf{x}}) &= \frac{1}{4\pi r^3}\overline{\mathbf{x}}\cdot\mathbf{f}_p \\ \boldsymbol{\Pi} &= -\frac{3}{4\pi r^5}\overline{\mathbf{x}}\overline{\mathbf{x}}\overline{\mathbf{x}}\cdot\mathbf{f}_p \end{aligned} \tag{228}$$

The components of **G**, p and **Π** are explicitly

$$G_{ij} = \frac{1}{8\pi\eta}\left(\frac{\delta_{ij}}{r} + \frac{\bar{x}_i\bar{x}_j}{r^3}\right)$$

$$p_i = \frac{x_i f_i}{4\pi r^3} \tag{229}$$

$$\Pi_{\mathrm{ij}} = -\frac{3}{4\pi r^5} x_i x_j x_k f_j$$

where k is the dummy index.
The flow due to a point force in an unbounded fluid described by Eqs. (226)–(228) is called the Stokeslet [2].

For problems involving particle motion near interfaces, it is useful to define the modified Oseen tensor $\mathbf{G_w}$ vanishing at the wall located at $x = x_w$. It was shown by Blake [45] that in this case $\mathbf{G_w}$ can be constructed from the Stokeslet and a few image singularities (reflection tensors), i.e.,

$$\mathbf{G_w} = \mathbf{G}(\bar{\mathbf{x}}) - \mathbf{G}(\bar{\mathbf{x}}_{im}) + 2h_0^2\,\mathbf{D}(\bar{\mathbf{x}}_{im}) - 2h_0\mathbf{G_d}(\bar{\mathbf{x}}_{im}) \tag{230}$$

where $\bar{\mathbf{x}}_{im} = \mathbf{x}-\mathbf{x}^0_{im}$, $\mathbf{x}^0_{im}\,(2x_w - x_0, y_0, z_0)$ is the image of x_0 with respect to the wall, $h_0 = x_0 - x_w$ and **D**, $\mathbf{G_d}$ are the reflection tensors derived from the potential dipole and the Stokeslet doublet, defined later on.

Other singularities, which by analogy to electrostatics are called dipoles, quadrupoles, etc., can be derived from the potential source defined above and from the Stokeslet singularity by differentiating with respect to x_0. In this way one can define the potential dipole tensor $\mathbf{D_p}$ by differentiating Eq. (224)

$$\mathbf{D_p} = \nabla\left(\frac{\bar{x}}{r^3}\right) = -\frac{\mathbf{I}}{r^3} + 3\frac{\bar{\mathbf{x}}\bar{\mathbf{x}}}{r^5} \tag{231}$$

whose components are

$$D_{p_{ij}} = -\frac{\delta_{ij}}{r^3} + 3\frac{\bar{x}_i\bar{x}_j}{r^5} \tag{232}$$

The flow generated by the potential dipole is given by the expression

$$\mathbf{V}(x) = \mathbf{D_p} \cdot \mathbf{a} \tag{233}$$

where **a** is an arbitrary constant vector to be determined from the boundary conditions.

On the other hand, by differentiating the Stokeslet (Oseen tensor) with respect to $\mathbf{x}_0$, one can generate a series of derivative singularities, the first being the point-force dipole [2] whose components are

$$G_{d_{ijk}} = \frac{1}{8\pi\eta}\left(\frac{\delta_{ij}x_k - \delta_{ik}\overline{x}_j - \delta_{jk}\overline{x}_i}{r^3} + 3\,\frac{\overline{x}_i\overline{x}_j\overline{x}_k}{r^5}\right) \tag{234}$$

The flow generated by the point force dipole is given by the expression

$$\mathbf{V}(x) = \mathbf{G_d} \cdot \mathbf{b} \tag{235}$$

where **b** is an arbitrary constant vector to be determined from the boundary conditions.

By splitting the point force dipole into a symmetric and an antisymmetric component, one can construct another singularity tensor **C,** called a couplet or rotlet [2], whose components are

$$C_{ij} = \frac{e_{ijk}}{8\pi\eta}\frac{\overline{x}_k}{r^3} \tag{236}$$

where e_{ijk} is the alternating matrix equal to zero if two indices are equal and $e_{ijk}=1$ if the indices are in cyclic order; $e_{ijk}=-1$ otherwise.

The flow field of the couplet is given by

$$\mathbf{V}(x) = \mathbf{C} \cdot \mathbf{To} \tag{237}$$

where **To** is an arbitrary constant vector to be determined from the boundary conditions.

Another second-order singularity tensor, called a stresslet, is defined as [2]

$$G_{s_{ijk}} = \frac{3}{8\pi\eta}\frac{\overline{x}_i\overline{x}_j\overline{x}_k}{r^5} \tag{238}$$

For analyzing flows inside fluid particles (inside flows), it is useful to define another tensor called the Stokeson, G_{so}, whose components are [2]

$$G_{so_{ijk}} = \frac{1}{8\pi\eta}(2r^2\delta_{ij} - \overline{x}_i\overline{x}_j) \tag{239}$$

The flow field of the Stokeson is given by

$$\mathbf{V}(x) = \mathbf{G}_{so}\cdot\mathbf{c} \tag{240}$$

where **c** is the arbitrary constant vector to be determined from the boundary conditions.

In an analogous way, using the Laplace operator, one can define third-order tensors (quadrupoles). Consequently, the point-force quadrupole is given by

$$G_{q_{ij}} = \nabla_0^2 G_{ij} = -\frac{1}{4\pi\eta} D_{p_{ij}} \tag{241}$$

Using the above set of singularities one can derive useful expressions for various velocity fields produced by a moving sphere. For example, in the case of a solid particle translating with a constant velocity in a quiescent (unbounded) fluid, the net flow field can be expressed as the sum of the point force and potential dipole flows. The velocity components are [2]

$$\mathbf{V}(\mathbf{x}) = \mathbf{D}\cdot\mathbf{a} + \mathbf{G}\cdot\mathbf{b} = \frac{1}{4\pi}\left(-\frac{\mathbf{I}}{r^3} + 3\frac{\overline{\mathbf{xx}}}{r^5}\right)\cdot\mathbf{a} + \frac{1}{8\pi\eta}\left(\frac{\mathbf{I}}{r} + \frac{\overline{\mathbf{xx}}}{r^3}\right)\cdot\mathbf{b} \tag{242}$$

The unknown constants appearing in Eq. (242) can be found from the boundary condition **V** = **U** (where **U** is the constant translation velocity of the sphere, a is the sphere radius). In this way one obtains

$$\mathbf{a} = -\pi a^3\mathbf{U} \tag{243}$$

$$\mathbf{b} = 6\pi\eta a\mathbf{U}$$

By substituting Eq. (243) into Eq. (242), one obtains the velocity field near the translating sphere

$$\mathbf{V}(\overline{x}) = \frac{a}{4}\left[3\left(\frac{\mathbf{I}}{r} + \frac{\overline{\mathbf{xx}}}{r^3}\right) + a^2\left(\frac{\mathbf{I}}{r^3} - 3\frac{\overline{\mathbf{xx}}}{r^5}\right)\right]\cdot\mathbf{U} \tag{244}$$

The hydrodynamic pressure near the sphere is given by

$$p = p_0 + \frac{3}{2} a\eta \frac{\bar{\mathbf{x}} \cdot \mathbf{U}}{r^3} \tag{245}$$

If the sphere is moving with a constant velocity U in the positive direction of the x_1 axis, the fluid velocity components and the pressure are given by the expression

$$\begin{aligned}
V_1 &= \frac{aU}{4r}\left[\left(3+\frac{a^2}{r^2}\right)+3\left(1-\frac{a^2}{r^2}\right)\frac{x_1^2}{r^2}\right] \\
V_2 &= \frac{aU}{4r}\left[\left(3+\frac{a^2}{r^2}\right)+3\left(1-\frac{a^2}{r^2}\right)\frac{x_1 x_2}{r^2}\right] \\
V_3 &= \frac{aU}{4r}\left[\left(3+\frac{a^2}{r^2}\right)+3\left(1-\frac{a^2}{r^2}\right)\frac{x_1 x_3}{r^2}\right] \\
p &= p_0 + \frac{3}{2} a\eta U \frac{x_1}{r^3}
\end{aligned} \tag{246}$$

where x_1, x_2, x_3 are measured relatively to the sphere center.

The force acting on the sphere can be evaluated using Eq. (40) as

$$\mathrm{F}_h' = \int_{S_s} \mathbf{\Pi} \cdot \hat{\mathbf{n}} dS = \int_{S_s} \mathbf{\Pi}_d \cdot \hat{\mathbf{n}} dS + \int_{S_s} \mathbf{\Pi}_s \cdot \hat{\mathbf{n}} dS \tag{247}$$

where $\mathbf{\Pi}_d$, $\mathbf{\Pi}_s$, are stress tensors produced by the potential dipole and the Stokeslet and S_s is the sphere surface.

It can be shown [2] that

$$\mathbf{F}_h' = -\mathbf{b} = -6\pi\eta a \mathbf{U} \tag{248}$$

Eq. (248) expresses the well-known Stokes' law. It indicates that the force, referred to as the hydrodynamic resistance force, is directed oppositely to the direction of the particle velocity vector. Also, the force is proportional to the particle size rather than its cross-section area, which is a direct consequence of the viscous forces domination under the Stokes flow regime.

It is interesting to note that because of the symmetry of the flow, the hydrodynamic torque on the translating sphere vanishes.

In an analogous way, other types of flows can be analyzed via the singularity representation, for example the rotary motion of a sphere in a linear ambient flow or in quiescent fluid. In the former case, the solution can be constructed by superposition of the Stokeslet doublet and a potential quadrupole placed at the sphere center. For a simple shear flow in the $(x_2 - x_3)$ plane when $\boldsymbol{V}_\infty = Gx_2\,\mathbf{i}_3$ where G is the shear rate and $\mathbf{i}_3$ is the unit vector of the x_2 axis), the solution derived by assuming no net torque on the sphere is [46,47]

$$V_i = Gx_2\delta_{i3} + \frac{1}{2}Ga\left[\frac{5a^2x_1x_2x_3(a^2-r^2)}{r^7} - \frac{a^4(\delta_{i2}x_3+\delta_{i3}x_2)}{r^5}\right] \tag{249}$$

where δ_{ij} is the kronecker's delta function ($\delta_{ij} = 1$ for $i=j$ and zero otherwise and the (x_1, x_2, x_3) coordinate system has the origin at the sphere center. As can be deduced [46], the sphere rotates with the constant angular velocity $(1/2)G$.

On the other hand, the rotation of a sphere in a quiescent fluid can be simply derived by considering the couplet singularity given by Eq. (236). In this way one obtains the solution for the fluid velocity components in the form [2]

$$V_i = a^3\varepsilon_{ijk}\Omega_j\frac{x_k}{r^3} \tag{250}$$

where Ω_j is the component of the angular velocity vector of the sphere. When the sphere is rotating about the x_1 axis with the constant angular velocity Ω, the fluid velocity components become explicitly

$$V_1 = 0$$

$$V_2 = -\frac{a^3\Omega}{r^3}x_3 \tag{251}$$

$$V_3 = -\frac{a^3\Omega}{r^3}x_2$$

The hydrodynamic torque on the sphere calculated relative to its center is

$$\mathbf{To}'_h = \int_{S_s} (\mathbf{\Pi} \times \mathbf{r}_0) \cdot \hat{\mathbf{n}} dS = -8\pi a^3 \Omega \tag{252}$$

As can be seen, the torque is proportional to the angular velocity of the sphere and its volume.

Other cases, for example liquid drop motion can also be analyzed by the singularity method [2,6]. However, the disadvantage of solutions derived by the superposition of singularities is that there is a lack of a general criterion for their selection and proving of the uniqueness of the solutions obtained.

More related to physical situations seems the method based on the stream function approach, especially efficient for axisymmetric flows [1]. In the spherical coordinate system (r, ϑ, φ), the velocity components can be derived from the stream function ψ_s using the constitutive relationships [1,2,6]

$$\begin{aligned} V_r &= -\frac{1}{r^2 \sin\vartheta} \frac{\partial \psi_s}{\partial \vartheta} \\ V_\vartheta &= \frac{1}{r \sin\vartheta} \frac{\partial \psi_s}{\partial r} \\ V_\varphi &= 0 \end{aligned} \tag{253}$$

From Eq. (253) one can deduce that for a uniform ambient flow V_∞, directed against the symmetry axis, the stream function becomes

$$\psi_s = \frac{1}{2} V_\infty r^2 \sin^2\vartheta \tag{254}$$

An interesting property of the stream function is that the lines drawn for a constant ψ_s are the streamlines. For example using Eq. (254), one can show that the streamlines for the uniform ambient flow are described in spherical by the equation

$$r = \frac{C}{\sin\vartheta} \tag{255}$$

where C is a constant.

By exploiting the stream function concept one can express the Stokes equation in the compact form

$$E_s^4 \psi_s = 0 \tag{256}$$

where the stream-function operator E_s in spherical coordinates has the form [1]

$$E_s^2 = \frac{\partial^2}{\partial r^2} + \frac{\sin \vartheta}{r^2} \frac{\partial}{\partial \vartheta} \frac{1}{\sin \vartheta} \frac{\partial}{\partial \vartheta} \tag{257}$$

The hydrodynamic force (drag) on the body, directed along the symmetry axis, can be expressed in terms of the stream function as [1]

$$F_h = \pi \eta \int_0^{\pi} (r \sin \vartheta)^3 \frac{\partial}{\partial r} \left(\frac{E_s^2 \psi_s}{r^2 \sin^2 \vartheta} \right) r \, d\vartheta \tag{258}$$

The mathematical form of the stream function for uniform flow, given by Eq. (254), suggests that in the general case it can be expressed in the factorized form $f(r) \sin^2 \vartheta$. This implies that all flows are symmetric with respect to equatorial plane ($\vartheta = \pi/2$) as should be expected.

By applying the E_s operator, one obtains the following differential equation for $f(r)$

$$\begin{aligned} f'' - \frac{2f}{r^2} &= g \\ g'' - \frac{2g}{r^2} &= 0 \end{aligned} \tag{259}$$

An integration of Eq. (259) leads to the general expression for $g(r), f(r)$

$$\begin{aligned} g &= \frac{E_s^2(\psi_s)}{\sin^2 \vartheta} = C_1 r^2 + \frac{C_2}{r} \\ f &= \frac{C_1}{10} r^4 - \frac{C_2}{2} r + C_3 r^2 + \frac{C_4}{r} \end{aligned} \tag{260}$$

where C_1 through C_4 are constants to be determined from the boundary conditions.

Using Eq. (260) and Eq. (258), one can easily show that the hydrodynamic force is given by [1]

$$F_h' = \pi\eta \int_0^{\pi} (r\sin\vartheta)^3 \frac{\partial}{\partial r}\left[\frac{(C_1 r^2 + C_2 r)}{r^2}\right] r d\vartheta = 4\pi\eta\, C_2 \tag{261}$$

Eq. (261) can be used for anarbitrary axisymmetric flow. For example, in the case of sphere translating with a constant velocity through a quiescent fluid one has $C_1 = C_3 = 0$ (because the flow should vanish at large separations) and $C_2 = (3/2)\, Ua$, $C_4 = (1/4)\, Ua^3$ (this can be derived from the postulate of a constant velocity U at the sphere surface). Hence, the stream function becomes

$$\psi_s = \frac{1}{4} Ua^2 \sin^2\vartheta \left(\frac{a}{r} - 3\frac{r}{a}\right) \tag{262}$$

The fluid velocity components are

$$V_r = \frac{1}{2} U \left[\left(\frac{a}{r}\right)^3 - 3\frac{a}{r}\right] \cos\vartheta \tag{263}$$

$$V_\vartheta = -\frac{1}{4} U \left[\left(\frac{a}{r}\right)^3 + 3\frac{a}{r}\right] \sin\vartheta$$

For small distances from the interface, when $(r/a) - 1 \ll 1$, Eq. (263) simplifies to

$$V_r = \frac{3}{2} U \left(\frac{h}{a}\right)^2 \cos\vartheta - U\cos\vartheta \tag{264}$$

$$V_\vartheta = \frac{3}{2} U \frac{h}{a} \sin\vartheta - U\sin\vartheta$$

By considering that $C_2= (3/2)Ua$, one can calculate from Eq. (261) that the hydrodynamic force on particle is given by

$$F_h' = -6\pi\eta a U \tag{265}$$

This is the same as previously obtained, cf. Eq. (248). Because the force on the moving particle has a negative sign (it is acting against the direction of the flow) an external force is needed to maintain particle motion. Usually gravity is the driving force, and appears as if the sphere density differs from the fluid density.

In a similar way the flow past a stationary sphere of radius R immersed in uniform ambient flow V_∞ directed opposite to the z axis can be analyzed. In this case $\psi_s=(1/2)\, V_\infty r^2 \sin^2\vartheta$ when $r \to \infty$, so $C_1=0$, $C_3=(3/4)\, V_\infty R^3$. The second boundary condition at the surface is $\psi_s=0$, which requires that $C_2=(3/2)V_\infty R$, $C_4=(1/2)\, V_\infty R$. Hence, the stream function for the flow past the stationary sphere is given by

$$\psi_s = \frac{1}{4} V_\infty R^2 \sin^2\vartheta \left[\frac{R}{r} - 3\frac{r}{R} + 2\left(\frac{r}{R}\right)^2 \right] \tag{266}$$

The streamline pattern for the uniform flow near a stationary sphere is shown in Fig. 3.24.

The components of the fluid velocity vector are

$$V_r = -\frac{1}{2} V_\infty \left[2 - \frac{3R}{r} + \frac{R^3}{r^3} \right] \cos\vartheta \tag{267}$$

$$V_\vartheta = \frac{1}{4} V_\infty \left[4 - \frac{3R}{r} - \frac{R^3}{r^3} \right] \sin\vartheta$$

For small distances from the interface, when $(r/R) - 1 \ll 1$, Eq. (267) simplifies to

$$V_r = V_h = -\frac{3}{2} V_\infty \left(\frac{h}{R}\right)^2 \cos\vartheta \tag{268}$$

$$V_\vartheta = \frac{3}{2} V_\infty \frac{h}{R} \sin\vartheta$$

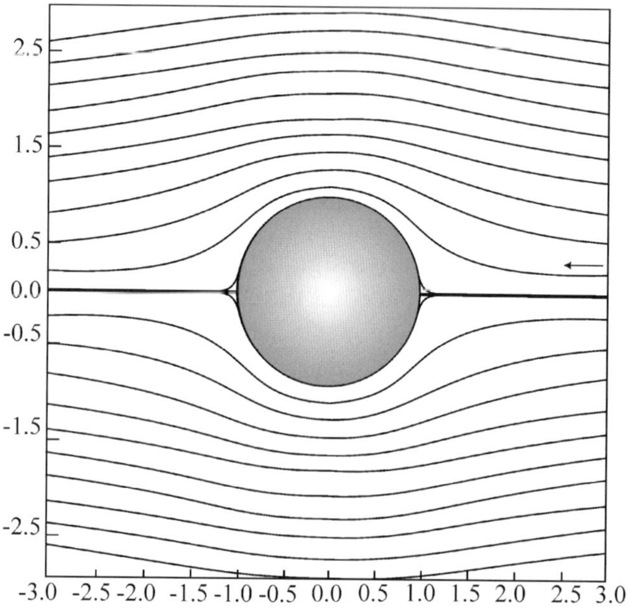

Fig. 3.24. Streamline pattern for the uniform flow past a stationary solid sphere From Ref [43].

As can be noticed, the velocity field relative to the surface has the same mathematical form as for the moving sphere.

Because $C_2=(3/2)\ V_\infty R$, one can deduce from Eq. (261) that the hydrodynamic force on a particle (drag force) is

$$F_h = -6\pi\eta R V_\infty \tag{269}$$

If the sphere immersed in the uniform ambient flow is not solid (a liquid drop or an air bubble) the flow pattern becomes more complicated. In addition to the outside flow, a circulating flow inside the sphere appears. This problem was first treated independently by Rybczyński [48] and Hadamard [49]. The flow pattern can be efficiently analyzed in terms of the above stream function approach by assuming that the sphere shape deformations due to the flow are negligible. This sets an upper limit on the sphere size, which should remain smaller than about 10^{-2}m [50].

By assuming a lack of deformations, of the sphere one can express the two stream functions for the outside and inside flows as

$$\psi_s = f(r)\sin^2\vartheta = \left(-\frac{1}{2}C_2 r + \frac{1}{2}V_\infty r^2 + \frac{C_4}{r}\right)\sin^2\vartheta \tag{270}$$

$$\psi_{sf} = f_f(r)\sin^2\vartheta = \left(\frac{1}{10}C_{1_f} r^4 + C_{3_f} r^2\right)\sin^2\vartheta$$

The four constants can be evaluated by considering the no-penetration boundary condition at the sphere surface and the postulate that the tangential velocity V_ϑ is continuous at the fluid–fluid interface. This can be written as

$$\begin{aligned} &f(R)=0 \\ &f_f(R)=0 \\ &f'(R)=f_f'(R) \end{aligned} \tag{271}$$

The fourth boundary condition at the fluid interface can be derived from the postulate of the continuity of the tangential stress, which can be expressed in terms of the stream function as

$$\eta_f \frac{\partial}{\partial r}\left(\frac{1}{r^2}f_f'\right)_{r=R} = \eta\frac{\partial}{\partial r}\left(\frac{1}{r^2}f'\right)_{r=R} \tag{272}$$

where η is the dynamic viscosity of the outer fluid and η_f is the dynamic viscosity of the fluid constituting the sphere.

This set of boundary conditions allows one to evaluate the constants C_2 through C_4, C_{1f} through C_{3f}, are given by

$$C_2 = \frac{3}{2}V_\infty R\frac{\eta_f + (2/3)\eta}{\eta_f + \eta} \qquad C_4 = \frac{1}{4}V_\infty R^3\frac{\eta_f}{\eta + \eta_f} \tag{273}$$

$$C_{1_f} = \frac{5}{2}\frac{V_\infty}{R^2}\frac{\eta}{\eta_f + \eta} \qquad C_{3_f} = -\frac{1}{4}V_\infty\frac{\eta}{\eta_f + \eta}$$

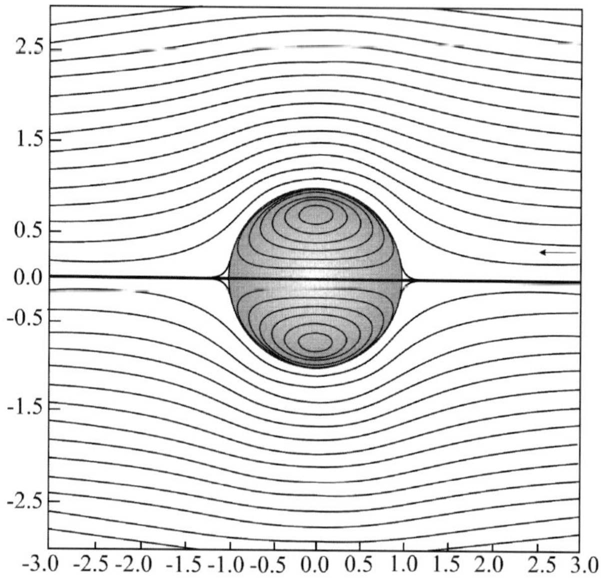

Fig. 3.25. Streamline pattern for the uniform flow past a stationary, fluid particle, $\eta_f / \eta = 0.1$ From Ref [43].

Consequently, the stream functions for the inside and outside flows can be calculated from Eq. (270). The streamline pattern for the uniform flow near a stationary sphere when $\eta_f/\eta = 0.1$ is shown in Fig. 3.25.

The components of the fluid velocity inside and outside the sphere are

$$\left.\begin{aligned} V_r &= -\frac{2f(r)}{r^2}\cos\vartheta = -\frac{1}{2}V_\infty\left[2-3\frac{R}{r}\overline{\eta}_1+\left(\frac{R}{r}\right)^3\overline{\eta}_2\right]\cos\vartheta \\ V_\vartheta &= -\frac{f'(r)}{r}\sin\vartheta = \frac{1}{4}V_\infty\left[4-3\frac{R}{r}\overline{\eta}_1-\left(\frac{R}{r}\right)^3\overline{\eta}_2\right]\sin\vartheta \end{aligned}\right\}\text{outside}$$

$$\left.\begin{aligned} V_r &= -\frac{2f_f(r)}{r^2}\cos\vartheta = -\frac{1}{2}V_\infty\left[\left(\frac{r}{R}\right)^2-1\right]\overline{\eta}_3\cos\vartheta \\ V_\vartheta &= -\frac{f_f'(r)}{r}\sin\vartheta = \frac{1}{2}V_\infty\left[2\left(\frac{r}{R}\right)^2-1\right]\overline{\eta}_3\sin\vartheta \end{aligned}\right\}\text{inside} \qquad (274)$$

where $\bar{\eta}_1 = \dfrac{\eta_f + \frac{2}{3}\eta}{\eta_f + \eta}, \bar{\eta}_2 = \dfrac{\eta_f}{\eta_f + \eta}, \bar{\eta}_3 = \dfrac{\eta}{\eta_f + \eta}$

For small distances from the interface, when $h/r = r/R - 1 \ll 1$, the velocity components of the outside and inside flows are

$$\left.\begin{aligned} V_r &= V_h - \frac{3}{2} V_\infty \left[(\bar{\eta}_1 - \bar{\eta}_2) \frac{h}{R} + (2\bar{\eta}_2 - \bar{\eta}_1) \left(\frac{h}{R}\right)^2 \right] \cos\vartheta \\ V_\vartheta &= \frac{1}{4} V_\infty \left[4 - 3\bar{\eta}_1 - \bar{\eta}_2 + 3(\bar{\eta}_1 + \bar{\eta}_2) \frac{h}{R} \right] \sin\vartheta \end{aligned}\right\} \text{outside} \tag{275}$$

$$\left.\begin{aligned} V_r &= V_\infty \bar{\eta}_3 \frac{h}{R} \cos\vartheta \\ V_\vartheta &= \frac{1}{2} V_\infty \left[1 - 4\frac{y}{R} \right] \bar{\eta}_3 \sin\vartheta \end{aligned}\right\} \text{inside}$$

Knowing that $C_2 = (3/2)\, V_\infty R \bar{\eta}_1$, one can deduce from Eq. (261) that the hydrodynamic force on a particle (drag force) is

$$F_h = -6\pi\eta V_\infty R(1 + 2\eta/3\eta_f)/(1 + \eta/\eta_f) \tag{276}$$

In the case of an emulsion droplet when $\eta / \eta_f \cong 1$, one can deduce from Eq. (276) that the resistance coefficient is 5/6 = 0.833... times smaller than for a solid particle. For $\eta / \eta_f \gg 1$ (an air bubble in a liquid), this correction equals 2/3 = 0.666...

Similarly, for a fluid sphere moving in a quiescent fluid with a constant velocity U the steam function for the outside and inside flows are

$$\psi_s = \left(-\frac{1}{2} C_2' r + \frac{C_4'}{r} \right) \sin^2\vartheta \tag{277}$$

$$\psi_{sf} = \left(\frac{1}{10} C_1' r^4 + C_{3f}' r^4 \right) \sin^2\vartheta$$

the components of the fluid velocity inside and outside the sphere are [6]

$$\left.\begin{aligned} V_r &= V_h - \frac{1}{2}U\left[\left(\frac{a}{r}\right)^3\bar{\eta}_2 - (3\bar{\eta}_1)\frac{a}{r}\right]\cos\vartheta \\ V_\vartheta &= -\frac{1}{4}U\left[\left(\frac{a}{r}\right)^3\bar{\eta}_2 + (3\bar{\eta}_1)\frac{a}{r}\right]\sin\vartheta \end{aligned}\right\}\text{outside}$$

$$\left.\begin{aligned} V_r &= -\frac{1}{2}U\left[\left(\frac{r}{a}\right)^2\bar{\eta}_3 - \bar{\eta}_3 + 2\right]\cos\vartheta \\ V_\vartheta &= \frac{1}{2}U\left[2\left(\frac{r}{a}\right)^3\bar{\eta}_3 - \bar{\eta}_3 + 2\right]\sin\vartheta \end{aligned}\right\}\text{inside} \quad (278)$$

The flow field relative to the surface in the region attached to the interface is the same as that described by Eq. (275), with U in place of V_∞ and a in place of R. The resistance force is given by

$$F_h' = -6\pi\eta a U \frac{(1+2\eta/3\eta_f)}{(1+\eta/\eta_f)} \quad (279)$$

It should be noted, however, that the motion of fluid particles (drops or bubbles), and consequently the fluid velocity fields inside and outside them, are inherently non-stationary because of adsorption phenomena occurring at the moving interface. The relaxation time needed to saturate the surface with surface active substances varies from minutes to seconds for surfactant concentrations of 10^{-5} and 10^{-4}M, respectively [50], as discussed later on. Surfactant adsorption leads to a considerable increase in the surface elasticity of the interface, which prohibits its motion. Consequently, the interface behaves as a solid–liquid interface and the internal circulation inside the particle vanishes [51].

The above solutions, strictly valid for *Re* numbers much smaller than unity, can be treated as the leading-order approximation to real flows at finite Reynolds numbers. Indeed, the boundary-layer solutions analyzed before suggest that the drag force on a stationary spherical particle increases with *Re* as $Re^{1/2}$ for $Re \gg 1$. For Re comparable to unity the correction can be calculated from the Oseen approach, by using the stream function concept. Oseen replaced the Stokes equation with the linearized Navier–Stokes equation [52,53]

$$\rho \mathbf{V}_{\infty} \cdot \nabla \mathbf{V} = -\nabla p + \eta \nabla^2 \mathbf{V} \tag{280}$$

with the boundary condition postulating that $\mathbf{V}$ vanishes at the sphere surface. Analogously, for a sphere translating in a quiescent fluid, the Oseen equation becomes

$$\rho \mathbf{U} \cdot \nabla \mathbf{V} = -\nabla p + \eta \nabla^2 \mathbf{V} \tag{281}$$

with the boundary condition $\mathbf{V} = \mathbf{U}$ at the sphere surface and $\mathbf{V} \to 0$ for $r \to \infty$. The stream function for the flow past a stationary sphere is

$$\begin{aligned} \psi_s = V_{\infty} R^2 \Bigg[\left(\frac{1}{2} \frac{r^2}{R^2} \right) + \frac{1}{4} \frac{R}{r} \Bigg] \sin^2 \vartheta \\ - \frac{3}{Re} (1 + \cos\vartheta) \left(1 - \mathrm{e}^{-\frac{1}{4} Re \frac{r}{R} (1 - \cos\vartheta)} \right) \Bigg] \end{aligned} \tag{282}$$

Using this expression, the correction to the drag force that results from the Oseen approach can be expressed as [2]

$$F_h = -6\pi \eta a V_{\infty} \left(1 + \frac{3}{16} Re \right) \tag{283}$$

Eq. (283) can be used for $Re < 10$.

3.4.2. Transient Motion of a Sphere

All results discussed above concerned the quasi-stationary motion of spherical particles governed by the Stokes equation, Eq. (28). In this case the hydrodynamic resistance force was dependent solely on the instantaneous particle velocity rather than acceleration (derivative of velocity) or any previous history of particle motion. However, there is a certain class of particle motions having practical significance such as oscillatory motion or sudden acceleration due to external forces when the non-stationary term $\rho(\partial \mathbf{V}/\partial t)$ remains important but the inertia term $\rho \mathbf{V} \cdot \nabla \mathbf{V}$ is negligible. According to the dimensional analysis performed above, expressed by

Eq. (23) this can be the case if the product $St\ Re = T_{ch}V_{ch} / L_{ch} << 1$. For example, if a particle is executing oscillatory motion with the angular velocity Ω, one can assume $T_h \sim 1/\Omega$, so for higher frequencies and small amplitude of oscillations the product $StRe$ indeed becomes much smaller than unity. In this case the fluid velocity field is governed by the non-stationary Stokes equation, Eq. (26) having the form

$$\rho \frac{\partial \mathbf{V}}{\partial t} = -\nabla p + \eta \nabla^2 \mathbf{V} \tag{284}$$

For a harmonic oscillatory motion of the sphere when

$$\mathbf{U} = \mathbf{U}_0 e^{-i\Omega t} \tag{285}$$

Eq. (284) becomes

$$-i\Omega\rho\mathbf{V} = -\nabla p + \eta \nabla^2 \mathbf{V} \tag{286}$$

The boundary conditions for Eq. (286) are

$$\begin{aligned} &\mathbf{V} = \mathbf{U}_0 \quad \text{on the sphere surface} \\ &\mathbf{V} = 0 \quad \text{far from the surface} \end{aligned} \tag{287}$$

The solution of Eq. (286,287) can be derived in terms of the singularity method as a sum of the unsteady Stokeslet and a symmetric Stokeslet quadrupole [2,6]. The fluid velocity field is given by

$$\begin{aligned} \mathbf{V} = \frac{3}{4}\Bigg\{ &\frac{\mathbf{xx}}{r^2}\left[\frac{6C_1}{\lambda_i^2} - 2\left(\frac{C_1}{\lambda_i^2} + C_2\right)e^{-\lambda_i}\left(3 + 3\lambda_i + \lambda_i^2\right)\right] \\ &+\mathbf{I}\left[-\frac{2C_1}{\lambda_i^2} + 2\left(\frac{C_1}{\lambda_i^2} + C_2\right)e^{-\lambda_i}\left(1 + \lambda_i + \lambda_i^2\right)\right]\Bigg\}\cdot\mathbf{U} \end{aligned} \tag{288}$$

where

$$\lambda_i = a\left(\frac{-i\Omega}{\nu}\right)^{1/2} = \left(\frac{a^2\Omega}{2\nu}\right)^{1/2}(1-i) \tag{289}$$

and the constants C_1, C_2 are

$$C_1 = 1+\lambda_i+\frac{1}{3}\lambda_i^2$$
$$C_2 = \lambda_i^{-2}(e^{\lambda_i}-C_1) \tag{290}$$

The hydrodynamic resistance force on the oscillating sphere is given by [6]

$$\mathbf{F}_h' = -6\pi\eta a\left(1+\lambda_i+\frac{1}{9}\lambda_i^2\right)\mathbf{U} \tag{291}$$

The first term in Eq. (291) describes the stationary resistance force governed by the Stokes' formula (being exactly in phase with the particle velocity), the second proportional to $(a^2\Omega / \nu)^{1/2}$ is shifted in phase by $\pi/4$, and the third is the acceleration term shifted in phase by $\pi/2$. For small angular velocity of the sphere, when the condition $a^2\Omega/\nu \ll 1$ is met, the second and third terms can be ignored and one recovers from Eq. (291) the Stokes' formula for the resistance force on a spherical particle. It can be estimated that for colloid particles, when $a=10^{-7}$ m (100 nm) and $\nu = 10^{-6}$ m^2 s^{-1}, the second and third term in Eq. (291) are negligible for oscillation frequencies smaller than 10^8 s^{-1}, which is excessively large from a practical viewpoint.

The non-stationary resistance formula, Eq. (291), can be generalized to arbitrary motion of particles, including non-periodic motions by using the Laplace transformation defined before. In this way one obtains [2]

$$\tilde{\mathbf{F}}_h(s) = -6\pi\eta a\left[1+\left(\frac{a^2 s}{\nu}\right)^{1/2}+\frac{a^2 s}{9\nu}\right]\tilde{\mathbf{U}}(s) \tag{292}$$

where $\tilde{\mathbf{F}}_h$ is the transformed force given by

$$\tilde{\mathbf{F}}_h(s) = \int_0^{\infty} \mathbf{F}_h(t) e^{-st} dt \tag{293}$$

It can be easily deduced from Eq. (293) that the non-stationary terms play a role for short-times such that

$$t \ll \frac{a^2}{\nu} \tag{294}$$

By taking $a = 10^{-6}$ m (1 μm) and $\nu = 10^{-6}$ m^2 sec^{-1} the critical time equals 10^{-6} s, whereas for $a = 10^{-7}$ m (0.1 μm) and the same kinematic viscosity ν the critical time equal 10^{-8} s, which is a negligible value from an experimental viewpoint.

Eq. (293) can be inverted by exploiting the Falting theorem [17] that results in the formula first derived by Basset [2]

$$F_h'(t) = -6\pi\eta a\,\mathbf{U}(t) - 6\eta a^2 \left(\frac{\pi}{\nu}\right)^{1/2} \int_{-\infty}^{t} \left(\frac{d\mathbf{U}}{dt}\right)_{t=\tau} \frac{d\tau}{(t-\tau)^{1/2}} - \frac{2}{3}\pi\rho a^3 \left(\frac{d\mathbf{U}}{dt}\right) \tag{295}$$

The first term in Eq. (295) can be identified as the stationary Stokes' drag, the second term described by the convolution integral that involves the history of the particle motion is known as the Basset force (memory integral) and the third term is the added mass term because it is proportional to m_f (where m_f is the mass of the fluid displaced by the particle).

Using the Laplace transformation method, exact analytical solutions describing the particle velocity and resistance force can be derived for various cases, e.g., for a sphere relised from a rest or for a sudden acceleration of a sphere under constant external force applied for $t > 0$. Using these solutions it has been demonstrated that the Basset term for colloid particles gives a 10% correction only, for $t < a^2/\nu$ [6].

3.4.3. Flows Involving Non-spherical (Anisotropic) Particles

The stream function approach can also be used to determine axisymmetric Stokes flows past non-spherical bodies, e.g., oblate and prolate spheroids (see Fig. 3.26). The uniform flow is directed opposite to the z-axis direction. The appropriate curvilinear coordinate system for this problem is the oblate and prolate spheroidal coordinates [1] (ξ_s, η_s), which are connected with the r and z coordinates of the cylindrical system by the relationships [1,54]

$$\begin{aligned} r &= c_f \cosh\xi_s \sin\eta_s = c_f\left(\lambda_s^2+1\right)^{1/2}\left(1-\zeta_s^2\right)^{1/2} \\ z &= c_f \sinh\xi_s \cos\eta_s = c_f\lambda_s\zeta_s \end{aligned} \tag{296}$$

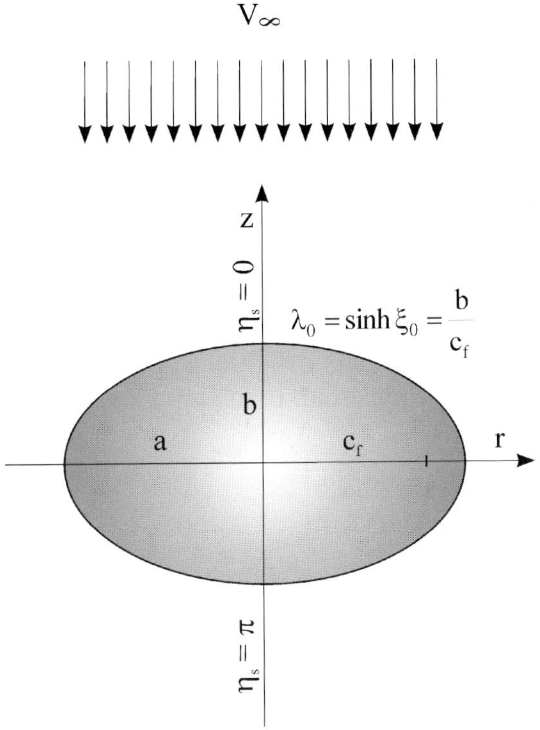

Fig. 3.26. Uniform flow past an oblate spheroid along the symmetry axis.

where $\lambda_s = sinh\ \xi_s$, $\zeta_s = \cos\eta_s$, $c_f=(a^2 - b^2)^{1/2}$ are the positions of the two focal points of the spheroid. The spheroid surface is defined by a constant value of $\lambda_0 = \lambda_s = b/c_f = \sinh\xi_0$.

The fluid velocity components can be evaluated through combined use of the stream function and the constitutive relationships for curvilinear coordinate systems [1]

$$V_{\xi_s} = -\frac{h_2}{r}\frac{\partial\psi_s}{\partial\eta_s} = -\frac{1}{c_f r\left(\lambda_s^2+\zeta_s^2\right)^{1/2}}\frac{\partial\psi_s}{\partial\eta_s} = \frac{\sin\eta_s}{c_f r\left(\lambda_s^2+\zeta_s^2\right)^{1/2}}\frac{\partial\psi_s}{\partial\zeta_s}$$

$$V_{\eta_s} = \frac{h_1}{r}\frac{\partial\psi_s}{\partial\xi_s} = \frac{1}{c_f r\left(\lambda_s^2+\zeta_s^2\right)^{1/2}}\frac{\partial\psi_s}{\partial\xi_s} = \frac{\cos\xi_s}{c_f r\left(\lambda_s^2+\zeta_s^2\right)^{1/2}}\frac{\partial\psi_s}{\partial\lambda_s} \tag{297}$$

where

$$h_1 = h_2 = \frac{1}{c_f(\lambda_s^2 + \zeta_s^2)^{1/2}}$$

Analogously to other axisymmetric flows, the stream function for an oblate spheroid can be expressed in the factorized form

$$\psi_S = \sin^2\eta_s\, g(\lambda_s) = \left(1-\zeta_s^2\right)g(\lambda_s) \tag{298}$$

The unknown function $g(\lambda_s)$ can be determined from the biharmonical equation, Eq. (256), with the boundary conditions $g = 0$ for $\lambda_s = \lambda_0$ and $g \to (1/2)$ $V_\infty c_f^2\,(\lambda_s^2+1)$ far from the spheroid. In this way, one can show that [1]

$$g(\lambda_s) = \frac{1}{2}V_\infty c_f^2\left(\lambda_s^2+1\right)\left[1-\frac{f(\lambda_s)}{f(\lambda_0)}\right] \tag{299}$$

where $f(\lambda_s) = \dfrac{\lambda_s}{\lambda_s^2 + 1} - \dfrac{\lambda_0^2 - 1}{\lambda_0^2 + 1}\cot^{-1}\lambda_s$, $\cot^{-1}\lambda_s = \dfrac{\pi}{2}\tan^{-1}\lambda_s$

Thus, the stream function is given by [1]

$$\psi_S = \frac{1}{2}V_\infty c_f^2\left(1-\zeta_s^2\right)\left(\lambda_s^2+1\right)\left[1-\frac{f(\lambda_s)}{f(\lambda_0)}\right] = \frac{1}{2}V_\infty r^2\left[1-\frac{f(\lambda_s)}{f(\lambda_0)}\right] \tag{300}$$

It is interesting to note that in the case of an oblate spheroid translating with a uniform velocity U in the z direction, the stream function is obtained by subtracting Ur^2 from Eq. (300) and replacing V_∞ with U.

When the axes ratio approaches unity, $c_f \to 0$ and $\lambda_0 \to \infty$. In this case Eq. (300) reduces to that previously derived for a sphere, Eq. (270).

On the other hand, in the limit of small distance from the spheroid surface when $(\lambda_s + \lambda_0)/\lambda_0 = (y/c_f) \ll 1$, the stream function expressed by Eq. (300) becomes approximately

$$\psi_s = \frac{1}{4}V_\infty r^2 \lambda_0^2 \frac{f''(\lambda_0)}{f(\lambda_0)}\left(\frac{y}{b}\right)^2 = V_\infty c_f^2 \frac{\lambda_0(1-\zeta_s^2)}{(\lambda_0^2+l)^3 f(\lambda_0)}\left(\frac{y}{c_f}\right)^2 \tag{301}$$

where y is the distance from the spheroid surface measured along the line perpendicular to its surface.

By considering Eq. (301), one obtains from Eq. (297) the expression for the fluid velocity components

$$\begin{aligned} V_{\xi_s} &= -V_\infty \frac{c_f \zeta_s(\lambda_s^2+1)\sin\eta_s}{r(\lambda_s^2+\zeta_s^2)^{1/2}}\left(1-\frac{f(\lambda_s)}{f(\lambda_0)}\right) \\ &= -V_\infty \frac{(\lambda_s^2+1)^{1/2}}{(\lambda_s^2+\zeta_s^2)^{1/2}}\left(1-\frac{f(\lambda_s)}{f(\lambda_0)}\right)\cos\eta_s \end{aligned} \tag{302}$$

$$V_{\eta_s} = \frac{1}{2}V_\infty \frac{1}{(\lambda_s^2+\zeta_s^2)^{1/2}}\left[2\lambda_s\left(1-\frac{f(\lambda_s)}{f(\lambda_0)}\right)-(\lambda_s^2+1)\frac{f'(\lambda_s)}{f(\lambda_0)}\right]\sin\eta_s$$

where $f'(\lambda_s) = \dfrac{2}{(\lambda_s^2+1)}\left(\dfrac{\lambda_0^2}{\lambda_0^2+1} - \dfrac{\lambda_s^2}{\lambda_s^2+1}\right)$

Using Eq. (301), the velocity components can be expressed as in the region close to the interface

$$\begin{aligned} V_{\xi_s} &= -2V_\infty \frac{\lambda_0^3}{(\lambda_0^2+1)^{5/2} f(\lambda_0)}\left(\frac{y}{b}\right)^2 \frac{\cos\eta_s}{(\lambda_0^2+\cos^2\eta_s)^{1/2}} \\ V_{\eta_s} &= 2V_\infty \frac{\lambda_0^2}{(\lambda_0^2+1)^2 f(\lambda_0)}\left(\frac{y}{b}\right)\frac{\sin\eta_s}{(\lambda_0^2+\cos^2\eta_s)^{1/2}} \end{aligned} \tag{303}$$

In the region close to the forward stagnation point, where $\cos\eta_s \to 1$, the velocity components can be expressed in the form analogous to that for a sphere

$$
\begin{aligned}
V_{\xi_s} &= -2V_\infty \frac{\lambda_0}{(\lambda_0^2+1)^3 f(\lambda_0)} \left(\frac{y}{b}\right)^2 \cos\eta_s \\
&= -\frac{3}{2} V_\infty A_f \left(\frac{y}{b}\right)^2 \cos\eta_s \\
V_{\eta_s} &= 2V_\infty \frac{\lambda_0^2}{(\lambda_0^2+1)^{5/2} f(\lambda_0)} \left(\frac{y}{b}\right)^2 \sin\eta_s \\
&= \frac{3}{2} V_\infty A_f \frac{(\lambda_0^2+1)^{1/2}}{\lambda_0} \left(\frac{y}{b}\right) \sin\eta_s
\end{aligned}
\tag{304}
$$

where the flow parameter is given explicitly by

$$
A_f(\lambda_0) = \frac{4}{3} \frac{\lambda_0^3}{(\lambda_0^2+1)^3 f(\lambda_0)} = \frac{\lambda_0^3}{(\lambda_0^2+1)^{3/2}} K(\lambda_0) \tag{305}
$$

and

$$
K(\lambda_0) = \frac{4}{3} \frac{1}{(\lambda_0^2+1)^{1/2}\left[\lambda_0 - (\lambda_0^2-1)\cot^{-1}\lambda_0\right]} \tag{306}
$$

It can be deduced from Eq. (304) that at the rear of the spheroid, the stagnation-point flow is directed outwards from the surface, which can be exploited in experiments designed to study particle detachment from surfaces.

Identical expression can be derived for the fluid velocity components near an oblate spheroid translating with a uniform velocity U.

The hydrodynamic force on a stationary spheroid in uniform stream is given by [1]

$$
F_h = -\frac{8\pi\eta c_f V_\infty}{\lambda_0 - (\lambda_0^2-1)\cot^{-1}\lambda_0} = -6\pi\eta a V_\infty K(\lambda_0) \tag{307}
$$

As can be noticed, the function $K(\lambda_0)$ accounts for the deviation of the resistance force from Stokes' formula for a sphere having the radius equal to the major semiaxis of the spheroid.

By analogy to the sphere problem, one can deduce that the resistance force of a spheroid translating in a quiescent fluid with a constant velocity U in the z direction is [2]

$$F_h' = -6\pi\eta aU\,K(\lambda_0) \tag{308}$$

For a slightly deformed sphere, when $(a-b)/a - \bar{\delta} \ll 1$, $K = 1-(1/5)\,\bar{\delta}$. For larger deformations, e.g., when $\bar{\delta}=0.5$ ($b/a = 0.5$), the correction factor $K = 0.9053$, for $\bar{\delta} = 0.8$ ($b/a = 0.2$), $K = 0.8614$ and for $\bar{\delta} = 0.9$ ($b/a = 0.1$), $K = 0.8525$. Even in the case of $b/a = 0$, which corresponds to a circular disk of radius a the correction function K equals $8/3\pi = 0.8488$. Thus, the resistance force for a disk translating broadside is

$$F_h' = -16\eta aU \tag{309}$$

As can be noticed, the corrections for oblate spheroids are rather insignificant.

It is interesting to calculate K for a deformed sphere by assuming that its volume remains constant and that its shape can be approximated by an oblate spheroid. The radius of the sphere prior to deformation equals a_s. After deformation, the axes of the spheroid are connected by the formula $a^2b = a_s^3$. Assuming that b/a has a fixed value one can express the correction factor $K(\lambda_0)$ for a deformed sphere as

$$K_s = \frac{6\pi\eta a\,K(\lambda_0)}{6\pi\eta a_s} = \frac{K(\lambda_0)}{(b/a)^{1/3}} \tag{310}$$

with $\lambda_0 = \dfrac{b/a}{(1 - b/a)^{1/2}}$

In the limit for small deformations when $1-(b/a) = \bar{\delta} \ll 1$, $K_s \to 1 + 2\bar{\delta}/15$.

On the other hand, in the limit when the sphere is deformed to a flat disk of the same volume, the resistance force becomes

$$F'_h = -16\eta aU\left(\frac{a}{b}\right)^{1/3} \tag{311}$$

The dependence of the function K on the $\bar{\delta} = 1-(b/a)$ parameter is shown in Fig. 3.27.

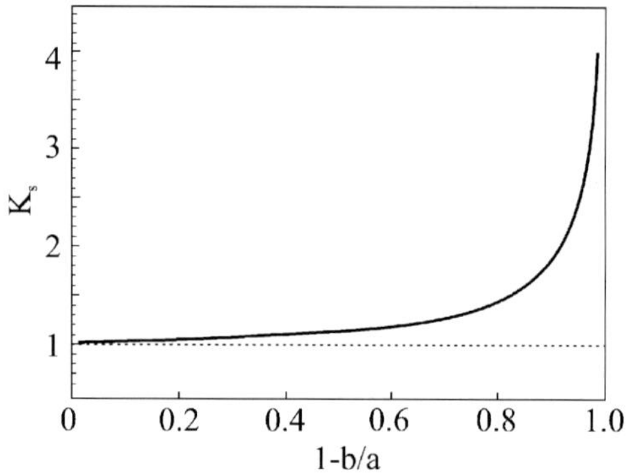

Fig. 3.27. The dependence of the K_s function on $\overline{\delta} = 1-(b/a)$ calculated from Eq. (310).

An analogous analysis can be performed for a uniform flow past the longer axis of a prolate spheroid (see Fig. 3.28). The cylindrical coordinates (r,z) are connected with prolate spheroid coordinates through the relationships

$$r = c_f \sinh\xi_s \sin\eta_s = c_f(\lambda_s^2 - 1)^{1/2}(1-\zeta_s^2)^{1/2}$$

$$z = c_f \cosh\xi_s \cos\eta_s = c_f\lambda_s\zeta_s$$

$$\lambda_s = \cosh\xi_s, \quad \zeta_s = \cos\eta_s, \quad c_f = (a^2 - b^2)^{1/2}, \tag{312}$$

$$\lambda_0 = \cosh\xi_0 = \frac{a}{c_f} = \frac{1}{\left[1-(b/a)^2\right]^{1/2}}$$

The fluid velocity components can be calculated from

$$V_{\xi_s} = -\frac{1}{c_f r(\lambda_s^2 - \zeta_s^2)^{1/2}}\frac{\partial\psi_s}{\partial\eta_s} = \frac{\sin\eta_s}{c_f\, r(\lambda_s^2 - \zeta_s^2)^{1/2}}\frac{\partial\psi_s}{\partial\zeta_s}$$

$$V_{\eta_s} = \frac{1}{c_f r(\lambda_s^2 - \zeta_s^2)^{1/2}}\frac{\partial\psi_s}{\partial\xi_s} = \frac{\sin\xi_s}{c_f\, r(\lambda_s^2 - \zeta_s^2)^{1/2}}\frac{\partial\psi_s}{\partial\lambda_s} \tag{313}$$

where $h_1 = h_2 = \dfrac{1}{c_f(\lambda_s^2 - \zeta_s^2)^{1/2}}$

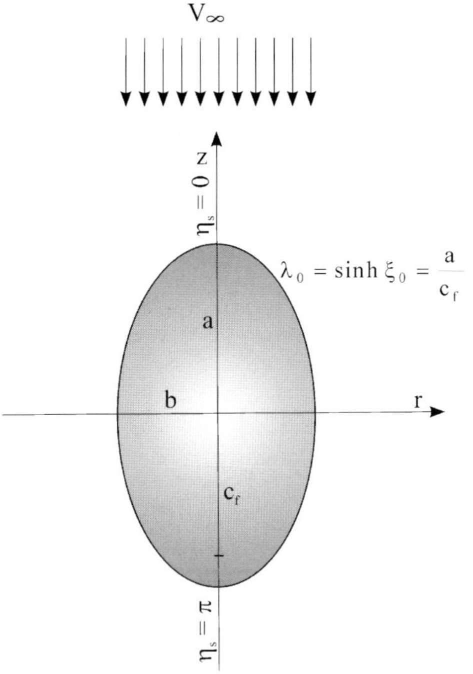

Fig. 3.28. Uniform flow past a prolate spheroid along the symmetry axis.

The stream function is given by [1]

$$\psi_s = \frac{1}{2} V_\infty r^2 \left[1 - \frac{f(\lambda_s)}{f(\lambda_0)}\right] = \frac{1}{2} V_\infty c_f (1 - \zeta_s^2)(\lambda_s^2 - 1)\left[1 - \frac{f(\lambda_s)}{f(\lambda_0)}\right] \tag{314}$$

where $f(\lambda_s) = \dfrac{\lambda_0^2 + 1}{\lambda_0^2 - 1} \coth^{-1} \lambda_s - \dfrac{\lambda_s}{\lambda_s^2 - 1}$; $\coth^{-1} \lambda_s = \dfrac{1}{2} \ln \dfrac{\lambda_s + 1}{\lambda_s - 1}$

It is interesting to note that in the case of a prolate spheroid translating with a uniform velocity U in the z direction, the stream function is obtained by just subtracting Ur^2 from Eq. (314) and replacing V_∞ with U.

In the limit of small distance from the spheroid surface when $(\lambda_s - \lambda_0)$ $\lambda_0 \ll 1$ the stream function expressed by Eq. (314) becomes

$$\psi_s = V_\infty r^2 \frac{\lambda_0}{(\lambda_s^2 - 1)^3 f_0} \left(\frac{y}{c_f}\right)^2 \tag{315}$$

where $f_0 = f(\lambda_0) = \frac{1}{2}\frac{\lambda_0^2+1}{\lambda_0^2-1}\ln\frac{\lambda_0+1}{\lambda_0-1} - \frac{\lambda_0}{\lambda_0^2-1}$

and y is the distance from the spheroid surface measured along the line perpendicular to its surface.

By considering Eq. (314), one obtains from Eq. (313) the expression for fluid velocity

$$V_{\xi_s} = -V_\infty \frac{(\lambda_s^2-1)^{1/2}}{(\lambda_s^2-\zeta_s^2)^{1/2}}\left[1-\frac{f(\lambda_s)}{f(\lambda_0)}\right]\cos\eta_s$$

$$V_{\eta_s} = \frac{1}{2}V_\infty \frac{1}{(\lambda_s^2-\zeta_s^2)^{1/2}} \left[2\lambda_s\left(1-\frac{f(\lambda_s)}{f_0}\right)-(\lambda_s^2-1)\frac{f'(\lambda_s)}{f_0}\right]\sin\eta_s \tag{316}$$

The velocity components in the region close to the interface can be expressed, using Eq. (316), as

$$\begin{aligned} V_{\xi_s} &= -2V_\infty \frac{\lambda_0^3}{(\lambda_0^2-1)^{5/2} f(\lambda_0)}\left(\frac{y}{a}\right)^2 \frac{\cos\eta_s}{(\lambda_0^2-\cos^2\eta_s)^{1/2}} \\ &= -\frac{3}{2}V_\infty A_f\left(\frac{y}{a}\right)^2 \frac{\cos\eta_s}{(\lambda_0^2-\cos^2\eta_s)^{1/2}} \\ V_{\eta_s} &= 2V_\infty \frac{\lambda_0^3}{(\lambda_0^2-1)^2 f(\lambda_0)}\left(\frac{y}{a}\right)\frac{\sin\eta_s}{(\lambda_0^2-\cos^2\eta_s)^{1/2}} \\ &= \frac{3}{2}V_\infty A_f\left(\frac{y}{a}\right)\frac{(\lambda_0^2-1)^{1/2}}{f(\lambda_0)}\frac{\sin\eta_s}{(\lambda_0^2-\cos^2\eta_s)^{1/2}} \end{aligned} \tag{317}$$

where

$$A_f = \frac{4}{3}\frac{\lambda_0^3}{(\lambda_0^2-1)f(\lambda_0)} = \frac{\lambda_0^3}{(\lambda_0^2-1)^{3/2}}K(\lambda_0) \tag{318}$$

$$K = \frac{4}{3(\lambda_0^2 - 1)^{1/2}} \left[\frac{1}{2}(\lambda_0^2 + 1) \ln \frac{(\lambda_0 + 1)}{(\lambda_0 - 1)} - \lambda_0 \right]^{-1}$$

An identical expression can be derived for the fluid velocity components near an oblate spheroid translating with a uniform velocity U.

The hydrodynamic force on a stationary spheroid in uniform stream is given by [1]

$$F_h = -6\pi\eta b V_\infty K = -6\pi\eta a V_\infty K'$$

$$K' = \frac{b}{a} K = \frac{4}{3\lambda_0[(1/2)(\lambda_0^2 + 1)\ln(\lambda_0 + 1)/(\lambda_0 - 1) - \lambda_0]} \tag{319}$$

By analogy to a sphere, one can deduce that the resistance force of a spheroid translating in a quiescent fluid with the constant velocity U is [2]

$$F_h' = -6\pi\eta b U K = -6\pi\eta a U K' \tag{320}$$

where K is the function accounting for the deviation from a sphere of radius equal to b (minor semiaxis of the spheroid) and K' is the function accounting for the deviation from a sphere having the radius a (equal to the longer semiaxis of the spheroid). For b/a=0.5, K=1.204 (K'=0.602), for b/a=0.2, K=1.785 (K'=0.357) and for b/a=0.1, K'=2.647 (K=0.2647). The dependence of the K and K' functions on b/a is shown in Fig. 3.29.

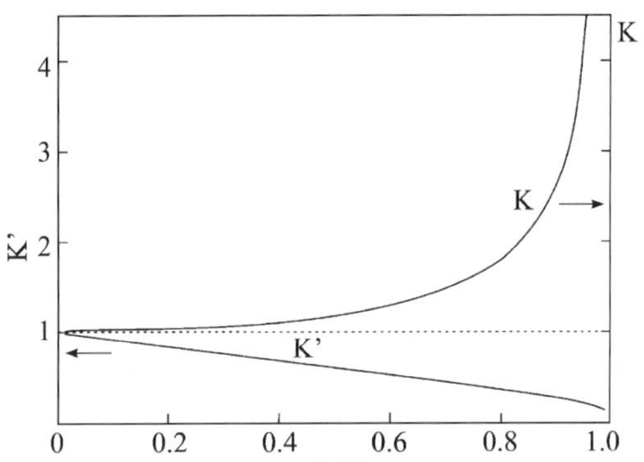

Fig. 3.29. The dependence of the K and K′ functions on the b/a parameter for prolate spheroids.

In contrast to oblate spheroids, for very elongated prolate spheroids, where $b/a \to 0$, the resistance force diverges logarithmically according to the formula

$$F_h' = -\frac{4\pi\eta a U}{\ln(2a/b) - 0.5} \tag{321}$$

Analogously to a sphere, the correction to the drag force resulting from the Oseen approximation of the inertia force is given by [55]

$$K_{\text{Oseen}} = K\left(1 + \frac{3}{16} Re\, K\right) \tag{322}$$

where $Re = 2aU/\nu$ is the Reynolds number based on the longer axis of the spheroid. Masliyah and Epstein [56] performed a numerical study of the flow past spheroids for Reynolds numbers up to 100. They showed that Eq. (322) can be used with a satisfactory accuracy for $Re < 1$.

A limitation of the stream function approach is that it cannot be used for asymmetric flows such as translational motion of an oblate spheroid parallel to its longer axis, or a prolate spheroid perpendicular to its longer axis. These flows, which represent limiting cases of motion of an ellipsoid (having three axes of various length) can be efficiently treated by the method of separation of variables. This was originally done by Oberbeek [57] for translational motion of ellipsoids, and by Jefferey [58] and Stimson and Jefferey [59] for rotational motion. The outline of the method proposed by Lamb [7] is given in [1].

The essence of the method is that the flow field past an ellipsoid or near a translating ellipsoid in the direction parallel to one of its principal axes is expressed in the form

$$\begin{aligned} \nabla^2 V_x &= 2C_s \frac{\partial^2 \chi_s}{\partial x^2} \\ \nabla^2 V_y &= 2C_s \frac{\partial^2 \chi_s}{\partial x \partial y} \\ \nabla^2 V_z &= 2C_s \frac{\partial^2 \chi_s}{\partial x \partial z} \end{aligned} \tag{323}$$

where (*x*, *y*, *z*) are the coordinates of the Cartesian system, with the origin at the spheroid center, C_s is the integration constant to be determined from the boundary condition, and χ_s is the function of the Cartesian coordinates (*x*, *y*, *z*) defined as

$$\chi_s(x,y,z) = abc \int_{\lambda_s}^{\infty} \frac{ds}{\left[(a^2+s)(b^2+s)(c^2+s)\right]^{1/2}} \tag{324}$$

In Eq. (312), (*a*, *b*, *c*) are the semiaxes of the ellipsoid, and $\lambda_s(x, y, z)$ is the ellipsoidal coordinate representing the positive root of

$$\frac{x^2}{a^2+\lambda_s} + \frac{y^2}{b^2+\lambda_s} + \frac{z^2}{c^2+\lambda_s} = 1 \tag{325}$$

The integration constant for the fluid motion along the *x* axis C_s is given by [1]

$$C_s = \frac{V_\infty}{\chi_0 + \alpha_1 a^2} \tag{326}$$

where χ_0 and α_1 are given by the elliptic integrals

$$\chi_0 = abc \int_0^{\infty} \frac{ds}{\left[(a^2+s)(b^2+s)(c^2+s)\right]^{1/2}}$$

$$\alpha_1 = abc \int_0^{\infty} \frac{ds}{(a^2+s)^{3/2}(b^2+s)^{1/2}(c^2+s)^{1/2}} \tag{327}$$

The drag force on the ellipsoid is given by [1]

$$F_1 = -16\pi\eta abc C_s = -6\pi\eta a V_\infty K_1 \tag{328}$$

where

$$K_1 = \frac{8}{3} \frac{bc}{\chi_0 + \alpha_1 a^2} \tag{329}$$

can be treated as correction to the Stokes law for motion of the ellipsoid parallel to the *x* axis.

In the case of an ellipsoid translating parallel to the x axis with a constant velocity U_l in a quiescent fluid, the resistance force is given by

$$F_1' = -6\pi a U_1 K_1 \tag{330}$$

Analogously, the resistance force and Stokes' correction factor for motion along the y and z axes can be expressed as

$$F_2' = -6\pi\eta b U_2 K_2$$

$$F_3' = -6\pi\eta c U_3 K_3$$

$$K_2 = \frac{8}{3}\frac{ac}{\chi_0 + \alpha_2 b^2}$$

$$K_3 = \frac{8}{3}\frac{ab}{\chi_0 + \alpha_3 c^2} \tag{331}$$

$$\alpha_2 = abc\int_0^\infty \frac{ds}{(b^2+s)^{3/2}(a^2+s)^{1/2}(c^2+s)^{1/2}}$$

$$\alpha_3 = abc\int_0^\infty \frac{ds}{(c^2+s)^{3/2}(a^2+s)^{1/2}(b^2+s)^{1/2}}$$

where U_2 and U_3 are the velocities of spheroid motion in the y and z directions, respectively.

Note that all the resistance force and Stokes' correction factors $K_1 - K_3$ for the ellipsoid are different for various directions of motion.

Using the method of separation of variables one can also calculate the torque and resistance coefficients for rotational motion of ellipsoids [58–60]. The torque components for rotation of the spheroid with constant angular velocities Ω_1, Ω_2, Ω_3 about the three principal axes are given by

$$To_1' = -\frac{16\pi\eta abc(b^2+c^2)}{3\left(\alpha_2^2 b^2 + \alpha_3^2 c^2\right)}\Omega_1$$

$$To_2' = -\frac{16\pi\eta abc(a^2+c^2)}{3(\alpha_1^2 a^2+\alpha_3^2 c^2)}\Omega_2$$

$$To_3' = -\frac{16\pi\eta abc(a^2+b^2)}{3(\alpha_1^2 a^2+\alpha_2^2 b^2)}\Omega_3 \qquad (332)$$

The knowledge of the hydrodynamic torque exerted on rotating spheroids is necessary for calculating the rotary diffusion coefficients and viscosity of ellipsoidal particle suspensions [60].

Note that in the case of an ellipsoid, the resistance coefficients are given in the form of elliptic integrals that cannot be calculated puolute explicitly. However, useful analytical formulae can be derived in the case of prolate spheroids whose two axes are equal, e.g., $b = c$. In this case the χ_o and $\alpha_1 \div \alpha_3$ functions defined by Eq. (327) and Eq. (331) are

$$\chi_0 = ab^2\int_0^\infty \frac{ds}{(a^2+s)^{1/2}(b^2+s)} = 2a^2 f_0$$

$$\alpha_1 = ab^2\int_0^\infty \frac{ds}{(a^2+s)^{3/2}(b^2+s)} = \frac{2}{((a^2+b^2)-1)}\left(\frac{a^2}{b^2}f_0-1\right) \qquad (333)$$

$$\alpha_2 = \alpha_3 = ab^2\int_0^\infty \frac{ds}{(b^2+s)^{3/2}(a^2+s)^{1/2}} = \frac{a^2}{a^2+b^2}(1-f_0)$$

where $f_0 = \dfrac{\cosh^{-1}\bar{L}}{\bar{L}(\bar{L}^2-1)^{1/2}} = \dfrac{1}{\bar{L}(\bar{L}^2-1)^{1/2}}\coth\dfrac{\bar{L}}{(\bar{L}^2-1)^{1/2}}$ (334)

$= \dfrac{ln\left[\bar{L}+(\bar{L}^2-1)^{1/2}\right]}{\bar{L}(\bar{L}^2-1)^{1/2}}$, $\bar{L} = a/b$ is the length- to-width parameter

(aspect ratio of the spheroid).

Accordingly, the resistance force and torque components for prolate spheroids are

$$F_1' = -6\pi\eta a U_1 K_1$$

$$F_2' = F_3' = -6\pi\eta b U_2 K$$

$$To_1' = -\frac{16\pi\eta ab^2}{3\alpha_2^2}\Omega_1 \tag{335}$$

$$To_2' = -\frac{16\pi\eta ab^2(a^2+b^2)}{3(\alpha^2 a^2 + \alpha_2^2 b^2)}\Omega_2$$

where the correction functions K_1 and K_2 are

$$K_1 = \frac{8}{3\bar{L}}$$

$$\frac{1}{\left[(2\bar{L}^2-1)/(\bar{L}^2-1)^{3/2}\ln(\bar{L}+(\bar{L}^2-1)^{1/2})/(\bar{L}-(\bar{L}^2-1)^{1/2})-2\bar{L}/(\bar{L}^2-1)\right]}$$

$$= \frac{4}{3\bar{L}\left[(2\bar{L}^2-1)/(\bar{L}^2-1)^{3/2}\cosh^{-1}\bar{L}-\bar{L}/(\bar{L}^2-1)\right]}$$

$$K_2 = K_3 = \frac{8}{3\left[(2\bar{L}^2-3)/(\bar{L}^2-1)^{3/2}\cosh^{\ 1}\bar{L}+\bar{L}/(\bar{L}^2-1)\right]} \tag{336}$$

It is interesting to note that $K_1 = K'$, i.e., equal to the function given by Eq. (319) derived via the stream function approach, which describes the resistance of the spheroid for the motion parallel to its longer axis. On the other hand, the K_2 function shown in Fig. 3.30, describes the resistance coefficient for the motion perpendicular to the longer axis of the spheroid. In the case where $\bar{L}>>1$, for slender (needle-like) bodies, the K_1 and K_2 functions become approximately

$$K_1 = \frac{2}{3}\frac{1}{\left[\ln(2a/b)-0.5\right]} \tag{337}$$

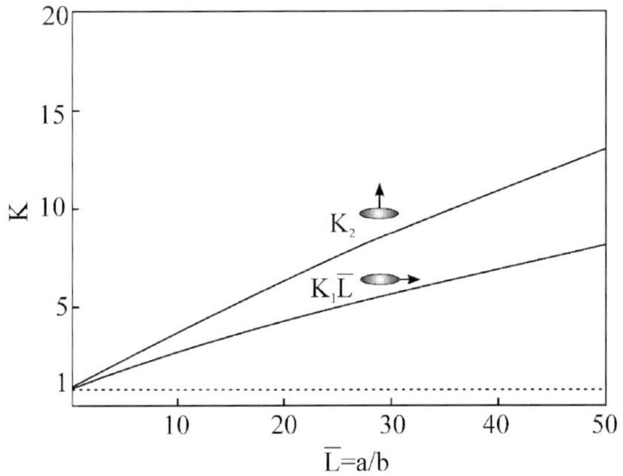

Fig. 3.30. The K_1 and K_2 functions of a prolate spheroid for the motion parallel and perpendicular to the symmetry axis, calculated from Eq. (324).

$$K_2 = \frac{4}{3}\frac{a}{b[\ln(2a/b)+0.5]}$$

Hence, the resistance force for parallel and perpendicular motion equals

$$F_1' = -\frac{4\pi\eta a U_1}{\ln(2a/b)-0.5}$$

$$F_2' = -\frac{8\pi\eta a U_2}{\ln(2a/b)+0.5} \tag{338}$$

Note that the ratio of the resistance coefficients for the parallel to perpendicular motions tends to 2 when $2a/b \to \infty$. It is interesting to point out that Eq. (337) can be used for calculating the resistance coefficients of a long cylinder of length $L = 2a$ and the radius $R = b$ [1]. In the case of the motion parallel to the symmetry axis, the K_1 function is approximated by a more accurate expression [1]

$$K_1 = \frac{2}{3[\ln(2a/b)-0.72]} \tag{339}$$

Analogously, in the case of oblate spheroids when $a = c > b$, the resistance force is given by the expressions

$$F_1' = -6\pi\eta a U_1 K_1$$
$$F_2' = -6\pi\eta b U_2 K_2 \tag{340}$$

where the K_1 and K_2 functions are

$$K_1 = \frac{4}{3}\frac{1}{\left[(1-2\bar{L}^2)/(1-\bar{L}^2)^{\frac{3}{2}}\cos^{-1}\bar{L} + \bar{L}/(1-\bar{L}^2)\right]}$$
$$K_2 = \frac{8}{3\bar{L}}\frac{1}{\left[(3-2\bar{L}^2)/(1-\bar{L}^2)^{\frac{3}{2}}\cos^{-1}\bar{L} - \bar{L}/(1-\bar{L}^2)\right]} \tag{341}$$

where $\bar{L} = a/b = 1$

The dependence of K_1 and K_2 on $\bar{L}$ is plotted in Fig. 3.31.

It is interesting to note that in the case of an infinitesimally thin disk, when $\bar{L} \rightarrow 0$, the K_1 and K_2 coefficients become

$$K_1 = \frac{8}{3\pi} \cong 0.8488$$
$$K_2\bar{L} = \frac{16}{9\pi} \cong 0.5659 \tag{342}$$

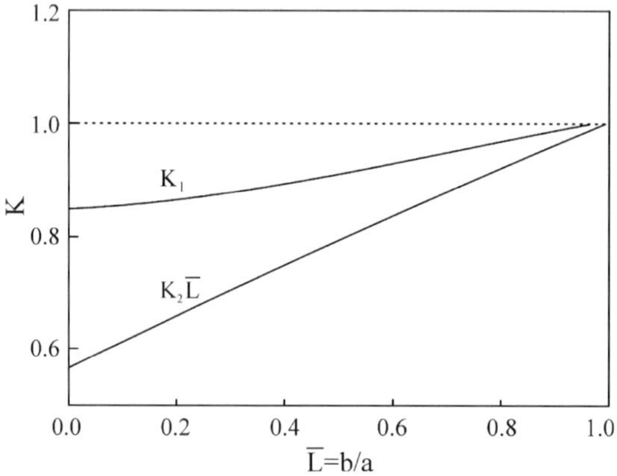

Fig. 3.31. The K_1 and $K_2\bar{L}$ functions for an oblate spheroid for the motion parallel and perpendicular to the symmetry axis calculated from Eq. (341).

Thus, the resistance force for broadside and edgewise motion are

$$
\begin{aligned}
F_1' &= -16\eta bU \\
F_2' &= -\frac{32}{3}\eta bU
\end{aligned}
\tag{343}
$$

Tchen [61] derived expressions for long spheroids that are bent to form of a half circle. This case is of considerable interest for calculating diffusion coefficients of macromolecules (polyelectrolytes), as discussed later on. It was shown that in the limiting case of a half circle, the resistance forces for the motion along the three principal axes are given by

$$
\begin{aligned}
F_1' = F_2' &= -\frac{3\pi\eta l}{\ln(l/b) - 0.68} U_1 \\
F_3' &= -\frac{4\pi\eta l}{\ln(l/b) + 0.56} U_2
\end{aligned}
\tag{344}
$$

where l is the length of the bent spheroid and b is the maximum radius.

It is interesting to note that although the half circle has no symmetry axis (as an ellipsoid) its resistance coefficient has only two different values, analogously to a spheroid.

In the case where the spheroid is bent into a circle tonus, the resistance force components for the broadside and edgewise motion are

$$
\begin{aligned}
F_1' &= -\frac{3\pi\eta l}{\ln(l/b) - 2.09} U_1 \\
F_2' = F_3' &= -\frac{4\pi\eta l}{\ln(l/b) + 0.75} U_2
\end{aligned}
\tag{345}
$$

Note that the resistance force for bent spheroids is larger than that for straight spheroids, for both the parallel and perpendicular motion.

For sake of convenience the resistance force for various objects and modes of motion are collected in Table 3.5. Also the space averaged resistance force expressions are given, which are of primary interest for calculating diffusion coefficients of particles. The averaged resistance coefficient can be calculated from the expression [1]

$$\frac{1}{\langle K\rangle} = \frac{1}{3}\left(\frac{1}{K_1} + \frac{1}{K_2} + \frac{1}{K_3}\right) \tag{346}$$

In the case when $K_2 = K_3$ one has explicitly

$$\langle K\rangle = \frac{3K_1K_2}{2K_1 + K_2}$$
$$\langle F\rangle = \frac{3F_1F_2}{2F_1 + F_2} \tag{347}$$

No analytical results are available in the literature for creeping motion and resistance coefficients of bodies having plane boundaries, e.g., short cylinders. However, these coefficients can be extracted from experimental data obtained from sedimentation rates of such particles [1].

3.4.4. The Hydrodynamic Resistance Tensors

The above results pertaining to single particle motion in an unbounded fluid under the creeping flow regime indicate that the hydrodynamic resistance force and torque can be expressed in the general form

$$F_h^{'} = -f(\eta,\eta_f)K_tU$$
$$To_h^{'} = -f(\eta,\eta_f)K_r\Omega \tag{348}$$

where f is a function of the viscosities of both the particle η_f and the fluid η, and K_t, K_r are functions of geometrical factors only (particle dimension and shape), independent of any properties of the fluid or the velocity of particle translation.

The separation of the geometrical and specific effects (stemming from properties of the suspending fluid) as well as the proportionality of the force to the velocity is a direct consequence of the linearity of the Stokes equation. For a solid spherical particle one has simply $f = \eta$, $K_t = 6\pi a$, $K_r = 8\pi a^3$. The parameter K_t is called the translation resistance coefficient and K_r the rotation resistance coefficient. However, for non-spherical bodies, i.e., ellipsoids or spheroids, the resistance coefficient, and consequently the resistance force, also become dependent on the direction of motion. For ellipsoids there are three different resistance coefficients depending solely on the length of the axes, and for spheroids there are two different coefficients (see Table 3.5).

A natural generalization of the Stokes law expressed by Eq. (348) involves replacing the resistance coefficients with a tensor defined in such a way that

$$\mathbf{F}_h' = -f(\eta,\eta_f)\mathbf{K}_t\cdot\mathbf{U} \tag{349}$$

where $\mathbf{K}_t$ is referred to as the translation resistance tensor [1]. In an analogous way, the rotational motion of a body can be characterized by the rotation tensor defined as

$$\mathbf{To}_h' = -f(\eta,\eta_f)\mathbf{K}_r\cdot\mathbf{\Omega} \tag{350}$$

By exploiting the energy dissipation theorem, it can be demonstrated that both the $\mathbf{K}_t$ and $\mathbf{K}_r$ tensors are symmetric and positively definite [1]. Hence, in the general case, for bodies of arbitrary shape, each of the tensors has six different components. The situation simplifies for bodies having symmetry planes or symmetry axes. Thus, for bodies having two or three symmetry planes (called orthotropic bodies) such as ellipsoids, these tensors can be expressed explicitly (in the coordinate system oriented along the symmetry planes or the symmetry axis) as

$$\mathbf{K}_t = \begin{vmatrix} K_{t_{11}} & & 0 \\ & K_{t_{22}} & \\ 0 & & K_{t_{33}} \end{vmatrix}$$

$$\mathbf{K}_r = \begin{vmatrix} K_{r_{11}} & & 0 \\ & K_{r_{22}} & \\ 0 & & K_{r_{33}} \end{vmatrix} \tag{351}$$

Table 3.5
Resistance force for various bodies under the Stokes (creeping flow) regime

Body	Resistance force	Averaged resistance force*
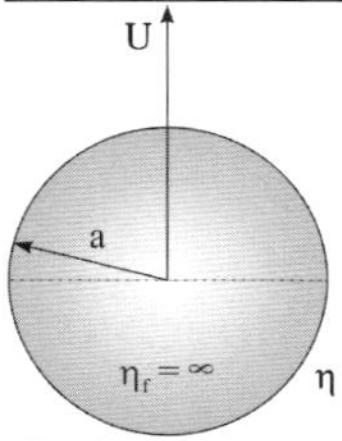 Solid sphere	$-6\pi\eta a U$	$-6\pi\eta a U$
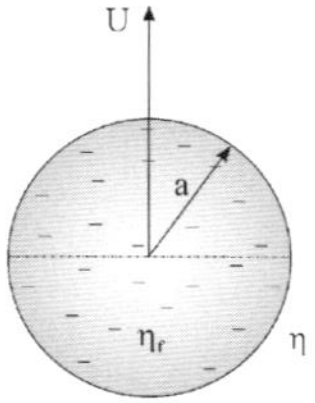 Liquid sphere	$-6\pi\eta a \frac{1+2\eta/3\eta_f}{1+\eta/\eta_f} U$	$-6\pi\eta a \frac{1+2\eta/3\eta_f}{1+\eta/\eta_f} U$
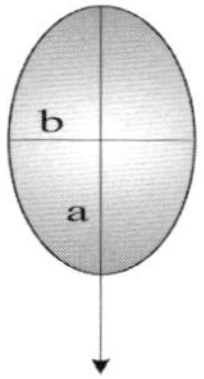	$-6\pi\eta a K_1 U$ $K_1 = \text{Eq. (336)}$	$-6\pi\eta a \langle K \rangle U$
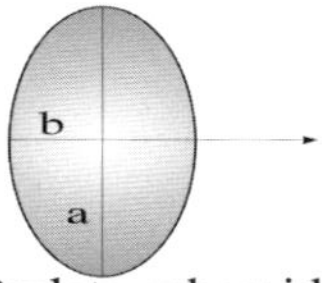 Prolate spheroids	$-6\pi\eta b K_2 U$ $K_2 = \text{Eq. (336)}$	$\langle K \rangle = \left(1 - \frac{b^2}{a^2}\right)^{\frac{1}{2}} / \cosh^{-1} \frac{a}{b}$

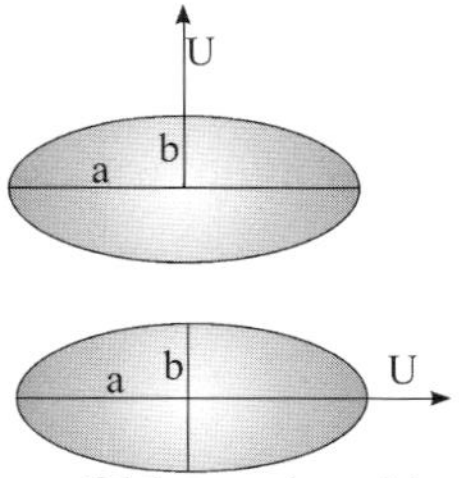 Oblate spheroids	$-6\pi\eta a K_1 U$ $K_1 = \text{Eq. (341)}$ $-6\pi\eta b K_2 U$ $K_2 = \text{Eq. (341)}$	$-6\pi\eta a\langle K\rangle U$ $\langle K\rangle = \left(1-\frac{b^2}{a^2}\right)^{\frac{1}{2}} / \cos^{-1}\frac{b}{a}$
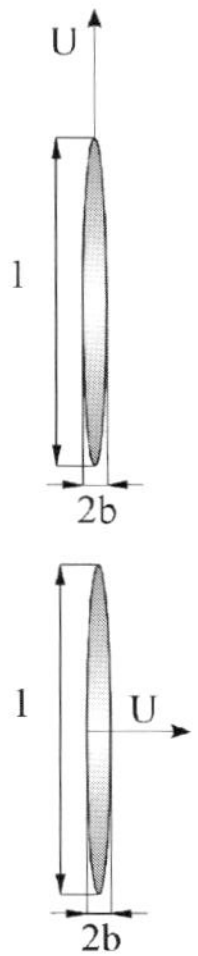 Slender bodies	$-3\pi\eta l K_1 U$ $K_1 = \frac{2}{3}\frac{1}{[\ln(l/b)-0.5]}$ (spheroid) $K_1 = \frac{2}{3}\frac{1}{[\ln(l/b)-0.72]}$ (cylinder) $-3\pi\eta l K_2 U$ $K_2 = \frac{4}{3}\frac{1}{\left[\ln\left(\frac{l}{b}\right)-0.5\right]}$ (spheroid)	$-3\pi\eta l\langle K\rangle U$ $\langle K\rangle = \frac{1}{\ln(l/b)}$ (spheroid) $\langle K\rangle = \frac{1}{\left[\ln\left(\frac{l}{b}-0.11\right)\right]}$ (cylinder)

Table 3.5 (continued)

Body	Resistance force	Averaged resistance force*
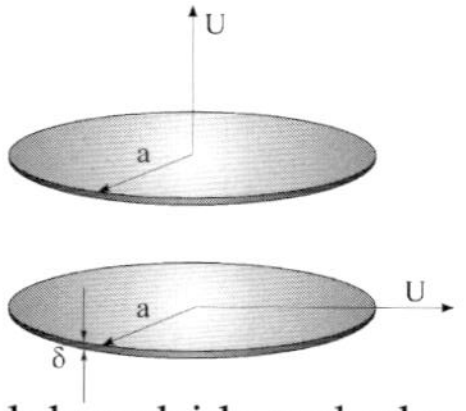 Disk broadside and edgewise	$-16\eta a \mathrm{U}$ $-\frac{32}{3}\eta a U$	$-12\eta a U$
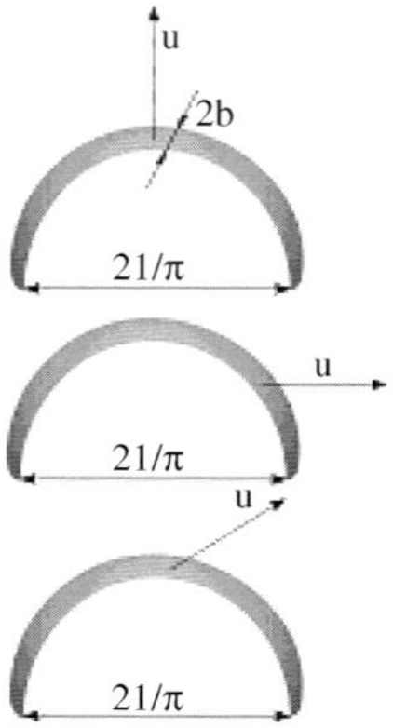 Half circle	$-\frac{3\pi\eta l}{\ln(l/b)-0.68}U$ $-\frac{3\pi\eta l}{\ln(l/b)-0.68}U$ $-\frac{4\pi\eta l}{\ln(l/b)+0.56}U$	$-\frac{3\pi\eta l}{\frac{11}{12}\ln(l/b)-0.31}U$

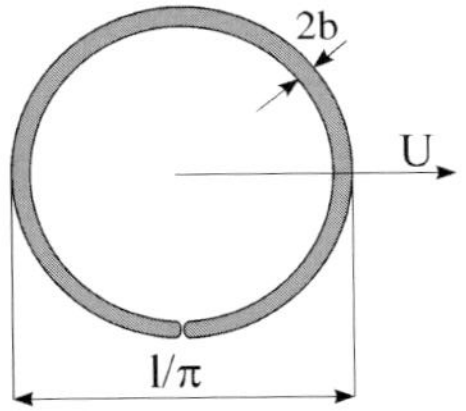	$-\dfrac{3\pi\eta l}{\ln(l/b)-2.09}U$	$-\dfrac{3\pi\eta l}{\dfrac{11}{12}\ln(l/b)-1.21}U$
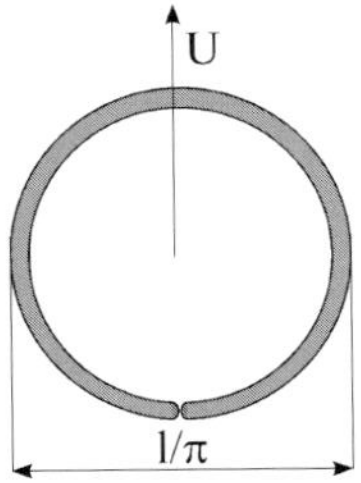 Circular ring	$-\dfrac{4\pi\eta l}{\ln(l/b)+0.75}U$	

$$* \ \frac{1}{\langle K\rangle}=\frac{1}{3}\left(\frac{1}{K_1}+\frac{2}{K_2}\right) \quad \langle K\rangle=\frac{3K_1K_2}{2K_1+K_2}$$

Note that each tensor has only three different diagonal components in this case. For ellipsoids they are given explicitly by

$$
\begin{aligned}
K_{t_{11}} &= 6\pi a\, K_1 = 16\pi \frac{abc}{\chi_0 + \alpha_1 a^2} \\
K_{t_{22}} &= 6\pi b\, K_2 = 16\pi \frac{abc}{\chi_0 + \alpha_2 b^2} \\
K_{t_{33}} &= 6\pi c\, K_3 = 16\pi \frac{abc}{\chi_0 + \alpha_3 c^2}
\end{aligned}
\tag{352}
$$

On the other hand, the components of the rotation tensor for ellipsoids can be expressed using Eq. (332) in the form

$$
\begin{aligned}
K_{r_{11}} &= \frac{4v_e(b^2+c^2)}{\alpha_2^2 b^2 + \alpha_3^2 c^2} \\
K_{r_{22}} &= \frac{4v_e(a^2+c^2)}{\alpha_1^2 a^2 + \alpha_3^2 c^2} \\
K_{r_{33}} &= \frac{4v_e(a^2+b^2)}{\alpha_1^2 a^2 + \alpha_2^2 b^2}
\end{aligned}
\tag{353}
$$

where $v_e = (4/3)\pi\, abc$ is the ellipsoid volume, and the functions $\alpha_1 \div \alpha_3$ are defined by Eq. (327) and Eq. (331).

For bodies of revolution, which have three perpendicular symmetry planes and a symmetry axis, the $\mathbf{K}_t$ and $\mathbf{K}_r$ tensors become

$$
\begin{aligned}
\mathbf{K}_t &= \begin{vmatrix} K_{t_{11}} & & 0 \\ & K_{t_{22}} & \\ 0 & & K_{t_{22}} \end{vmatrix} \\
\mathbf{K}_r &= \begin{vmatrix} K_{r_{11}} & & 0 \\ & K_{r_{22}} & \\ 0 & & K_{r_{22}} \end{vmatrix}
\end{aligned}
\tag{354}
$$

In this case, these tensors have only two different diagonal components as is the case of all the bodies listed in Table 3.5

$$K_{t_{11}} = 6\pi a K_1$$
$$K_{t_{22}} = 6\pi b K_2 \tag{355}$$

where $K_1 - K_2$ are defined by Eq. (336) (prolate spheroids) and Eq. (341) (oblate spheroids).

The components of the rotation tensor for spheroids are given by

$$K_{r_{11}} = \frac{16\pi ab^2}{3\alpha_2^2}$$
$$K_{r_{22}} = \frac{16\pi ab^2(a^2+b^2)}{3(\alpha_1^2 a^2 + \alpha_2^2 b^2)} \tag{356}$$

where α_1 and α_2 are defined by Eq. (331).

For spherically isotropic bodies the $\mathbf{K}_t$ and $\mathbf{K}_r$ tensors become simply

$$\mathbf{K}_t = K_t \mathbf{I}$$
$$\mathbf{K}_r = K_r \mathbf{I} \tag{357}$$

where K_t, K_r are scalars.

Naturally, spheres belong to this class of bodies and so do all particles of the polyhedra shape, like tetrahedra, cubes, octahedra, etc. In the case of spherically isotropic bodies, in contrast to other particle shapes, the direction of the resistance force vector is always parallel to the direction of the velocity vector. In particular, they will sedimentate parallel to the direction of gravity.

Knowledge of the $\mathbf{K}_t$ and $\mathbf{K}_r$ tensors is sufficient to calculate the resistance force using Eq. (349) for almost all particles of practical interest, such as colloids, proteins, polyelectrolytes, etc., in the bulk where there is no coupling between the translational and rotational motion. The coupling occurs only for artificially produced bodies of complex shapes, such as propellers, which are outside the interest of this book. However, the coupling between rotational and translational motion plays a role in spherical particles moving in the vicinity of boundaries, as discussed in the next section. When the coupling appears, Eq. (349) assumes a more general form [1]

$$\mathbf{F}_h' = -f(\eta,\eta_f)(\mathbf{K}_t \cdot \mathbf{U} + \mathbf{K}_c^{\dagger} \cdot \mathbf{\Omega})$$

$$\mathbf{To}_h' = -f(\eta,\eta_f)(\mathbf{K}_r \cdot \mathbf{\Omega} + \mathbf{K}_c \cdot \mathbf{U}) \tag{358}$$

where $\mathbf{K}_c$ is the coupling tensor and $\mathbf{K}_c^{\dagger}$ is the transposed coupling tensor.

For particles having two symmetry planes (like the screw propeller), the tensor $\mathbf{K}_c$ is explicitly given by [1]

$$\mathbf{K}_c = \begin{vmatrix} 0 & 0 & 0 \\ 0 & 0 & K_c \\ 0 & -K_c & 0 \end{vmatrix} \tag{359}$$

A similar shape of the $\mathbf{K}_c$ tensor occurs in the motion of a spherical particle near planar boundaries, as discussed later on.

Eq. (358) can also be written in a comprehensive form

$$\tilde{\mathbf{F}} = -f(\eta,\eta_f)\tilde{\mathbf{K}} \cdot \tilde{\mathbf{U}} \tag{360}$$

where $\tilde{\mathbf{K}}$, called the grand resistance matrix, is given by

$$\tilde{\mathbf{K}} = \begin{vmatrix} \mathbf{K}_t & \mathbf{K}_c^{\dagger} \\ \mathbf{K}_c & \mathbf{K}_r \end{vmatrix} \tag{361}$$

and $\tilde{\mathbf{F}}, \tilde{\mathbf{U}}$ are the generalized resistance force and velocity vectors

$$\tilde{\mathbf{F}} = \begin{vmatrix} \mathbf{F}_h' \\ \mathbf{To}_h' \end{vmatrix} \tag{362}$$

$$\tilde{\mathbf{U}} = \begin{vmatrix} \mathbf{U} \\ \mathbf{\Omega} \end{vmatrix}$$

Because of the linearity of the Stokes equation, the above relationships can be generalized to the case of particle motion in a fluid undergoing a linear flow having the velocity distribution

$$\mathbf{V} = \mathbf{V}_\infty + \boldsymbol{\omega}_0 \times \mathbf{x} + \boldsymbol{\Delta}_0 \cdot \mathbf{x} \tag{363}$$

where $\mathbf{V}_\infty$ is the velocity of the fluid at the particle center, $\mathbf{x}$ the distance measured from particle center, $\omega_0=(1/2)\nabla x\mathbf{V}$ the vorticity of the fluid element, Δ_0 the rate of the strain tensor given by Eq. (12).

It is interesting to note that the change in the uniform flow $\mathbf{V}_\infty$ occurring over distances comparable to particle size is of the order of $|\omega_0|\ a_p \cong a_p\, G_0$ (where G_0 is the fluid velocity gradient and a_p is the characteristic particle dimension). Because $G_0 \sim |\mathbf{V}_\infty|/L_{ch}$, where L_{ch} is the characteristic flow variation length scale, one can approximate Eq. (363) by the expression

$$\mathbf{V} \cong \mathbf{V}_\infty + \frac{a_p}{L_{ch}} \left| \mathbf{V}_\infty \right| \tag{364}$$

For a colloid particle, a_p is of the order of 10^{-7}m (100 nm), whereas the length scale of flow variations is 10^{-2}m. Therefore, the second term in Eq. (364) can be neglected in this case, so the particle can be treated as if it were immersed in a uniform fluid stream having the velocity $\mathbf{V}_\infty$. However, this is no longer true at interfaces when the length scale of flow variations becomes comparable to the particle dimensions.

It was shown [6] that for particles immersed in the linear fluid flow governed by Eq. (363) the net hydrodynamic force and torque on the particle are given by

$$\begin{aligned} \mathbf{F}_h &= -f(\eta,\eta_f)\left[\mathbf{K}_t\cdot(\mathbf{U}-\mathbf{V}_\infty)+\mathbf{K}_c^{\dagger}\cdot(\boldsymbol{\Omega}-\boldsymbol{\omega}_0)+\mathbf{K}_s\cdot\boldsymbol{\Delta}_0\right] \\ \mathbf{To}_h &= -f(\eta,\eta_f)\left[\mathbf{K}_c\cdot(\mathbf{U}-\mathbf{V}_\infty)+\mathbf{K}_r\cdot(\boldsymbol{\Omega}-\boldsymbol{\omega}_0)+\mathbf{K}'_s\cdot\boldsymbol{\Delta}_0\right] \end{aligned} \tag{365}$$

where $\mathbf{K}_s$, $\mathbf{K}'_s$, are third rank tensors.

For spherical particles, the tensors $\mathbf{K}_c$, $\mathbf{K}^{\dagger}_c$, $\mathbf{K}_s$, $\mathbf{K}'_s$, vanish, and Eq. (365) becomes

$$\begin{aligned} \mathbf{F}_h &= -f(\eta,\eta_f)\mathbf{K}_t\cdot(\mathbf{U}-\mathbf{V}_\infty) \\ \mathbf{To}_h &= -f(\eta,\eta_f)\mathbf{K}_r\cdot(\boldsymbol{\Omega}-\boldsymbol{\omega}_0) \end{aligned} \tag{366}$$

In the case of solid spherical particles Eq. (365) can be generalized to arbitrary, non-uniform flows. The hydrodynamic force and torque can be calculated from the modified equation called the Faxen law [5]

$$\mathbf{F}_h = 6\pi\eta a\left(1+\frac{a^2}{6}\nabla^2\right)\mathbf{V}_\infty - 6\pi\eta\mathbf{U}$$
$$\mathbf{To}_h = -8\pi\eta a^3(\mathbf{\Omega}-\boldsymbol{\omega}_0) \tag{367}$$

The Faxen law for fluid spheres and non-spherical particles (spheroids) becomes more complicated, as discussed in [5].

From Eqs. (366,367) it can be deduced that for a freely suspended particle when the hydrodynamic force and torque vanish

$$\mathbf{U} = \mathbf{V}_\infty$$
$$\mathbf{\Omega} = \boldsymbol{\omega}_0 \tag{368}$$

Hence the particle translates with a velocity equal to the free stream velocity and rotates with the angular velocity ω_0.

When there are no velocity gradients, or in the case when the flow variations over distances comparable to particle size are minor, Eq. (365) reduces to the form valid for particles of arbitrary shapes

$$\mathbf{F}_h = -f(\eta,\eta_f)\left[\mathbf{K}_t\cdot(\mathbf{U}-\mathbf{V}_\infty)+\mathbf{K}_c^\dagger\cdot\mathbf{\Omega}\right]$$
$$\mathbf{To}_h = -f(\eta,\eta_f)\left[\mathbf{K}_c\cdot(\mathbf{U}-\mathbf{V}_\infty)+\mathbf{K}_r\cdot\mathbf{\Omega}\right] \tag{369}$$

The above hydrodynamic resistance tensors, especially the translation tensor, are strongly modified by the presence of interfaces because of the reflection of velocity fields associated with moving particles. These effects, having profound consequences for particle transfer to interfaces, are discussed in the next section.

3.4.5. Motion of a Particle Near Interfaces – Wall Effects

Hydrodynamic forces and torques on moving particles are affected in many ways by the presence of boundary surfaces. These wall effects, whose spatial extension is comparable with the moving particle dimension manifest themselves through:

(i) Increase in the hydrodynamic resistance force and torque.
(ii) Coupling of the translational and rotational motion even for orthotropic particles.
(iii) Modification of the hydrodynamic drag force exerted by fluid streams on particles near interfaces or on those attached to interfaces.

The first effect, physically due to reflection of the velocity field associated with the moving particle, plays an especially important role in the perpendicular particle motion near the interface. For the parallel motion, the significance of this effect remains negligible until the gap width between the particle and the interface, h, becomes much smaller than the particle dimension. Because perpendicular and parallel particle motion near interfaces becomes affected in a different way, there appears an anisotropy of the resistance coefficient even for spherical particles. Thus, the particle translation motion can be characterized in terms of resistance tensor, which has a mathematical form similar to that of spheroidal particles in the bulk. Analogously, the rotational motion of a particle is affected in a different way depending on the orientation of the angular velocity vector relative to the interface. Moreover, the rotational motion near interfaces produces a translational motion and vice versa. However, the effects exerted by the interface on a rotating particle are much weaker than the analogous effects on a translating particle. Also, the modification of the hydrodynamic drag force on a particle immersed in simple flows near interfaces is not as significant as the modification of the resistance force of a translating particle.

Despite the apparent complexity of the wall effects, they can be quite efficiently analyzed for spherical particles because of the linearity of the governing Stokes equation. Obviously, the applicability of the quasi-stationary Stokes regime requires that the Reynolds number of the flow $Re = 2aU_p/\nu$ (where a is the characteristic particle dimension and U_p the characteristic particle velocity relative to the fluid) is much smaller than unity. The linearity of the Stokes equation significantly facilitates the analysis of complex flows near interfaces because the additivity principle can be applied. This also pertains to the net hydrodynamic force and torque on the particle, which are calculated as a vector sum of forces and torques derived from the elementary flows, such as

(i) perpendicular – and parallel to the interface translational motion of a particle in a quiescent fluid, treated separately;

(ii) rotational motion in a quiescent fluid with the angular velocity vector perpendicular and parallel to the interface; and
(iii) simple shear or stagnation-point flows past stationary particles near interfaces.

Accordingly, the translation motion of two spherical particles along their line of centers will be analyzed in this section, followed by the limiting case of a planar interface (solid or fluid). Next, the translational and rotational motion of a spherical particle close to a planar boundary will be considered, then simple flows past stationary particles near a planar boundary. Before performing a quantitative analysis, some useful estimates of the wall effects at large distances from interfaces will be given, derived by the method of reflections.

3.4.5.1. The Method of Reflections

The method of reflections was first developed by Lorentz [62] and Faxen [63] to analyze spherical particle motion near one or two planar interfaces. Smoluchowski [64] was the first to use this method for two different particle systems and for a system of n particles forming a cloud. He derived correction to the Stokes' resistance law in terms of a power series of the parameter a/l, where l is average distance between the particles.

The starting point of the method of reflections is the unperturbed velocity field $\mathbf{V}_1$ derived from the solution of the Stokes' equation for an isolated particle moving in a quiescent medium with a constant velocity $\mathbf{U}$. The translation resistance force is therefore $\mathbf{F}_\infty = -\mathbf{K}_t \cdot \mathbf{U}$ Obviously, the fluid velocity field produced by the moving particle fulfills the boundary conditions $\mathbf{V}_1 = \mathbf{U}$ at the particle surface and $\mathbf{V}_1 = 0$ at infinity. However, this field does not vanish at the interface located at a distance l from the particle, as required by the no-penetration boundary condition. To cancel the initial field an additional velocity field $\mathbf{V}_2$ must be added, which is referred to as the first reflection. Thus, $\mathbf{V}_1 + \mathbf{V}_2 = 0$ at the interface. The value of the reflected field $\mathbf{V}_2$ at the particle center is of the order of $\mathbf{V}_2\, a/l$. This field exerts an additional force on the particle, in addition to the original resistance force. Using the reciprocal theorem (a generalized Faxen law), one can calculate the net resistance force $\mathbf{F}_\infty$ on a particle as [1]

$$\mathbf{F}_1 = -\mathbf{K}\cdot\mathbf{V}_1(l) = \mathbf{F}_\infty\left[1 + \frac{|\mathbf{V}_2(l)|}{|\mathbf{U}|}\right] \qquad (370)$$

Eq. (370) represents the first-order correction to the resistance force.

However, the flow field $\mathbf{V}_2$ is further reflected from the particle, which requires that

$$\mathbf{V}_2 + \mathbf{V}_3 = 0 \tag{371}$$

where $\mathbf{V}_3$ is the third reflection field.

This reflection procedure is continued indefinitely producing a series of corrections to the resistance force, described by [1]

$$\mathbf{F}_h = \mathbf{F}_\infty \left[1 + \sum \left(\frac{|\mathbf{V}(l)|}{|\mathbf{U}|} \right)^n \right] \tag{372}$$

Because $\mathbf{F}_h$ decreases in a geometric series, one can express the net result of an infinite number of reflections in the form [1]

$$\mathbf{F}_h = \mathbf{F}_\infty \frac{1}{1 - |\mathbf{V}_2(l)||\mathbf{U}|} \tag{373}$$

By noting that $\mathbf{V}_2 = \mathbf{F}_\infty / 6\pi\eta l$ and $\mathbf{U} = -\mathbf{F}_\infty / 6\pi\eta a$, Eq.(373) can be expressed in the compact form

$$\mathbf{F}_h = \mathbf{F}_\infty \frac{1}{1 - K'(a/l)} \cong \mathbf{F}_\infty \left[1 + K' \frac{a}{l} + \left(K' \frac{a}{l} \right)^2 \cdots \right] \tag{374}$$

where the constant K' depends on the geometry of the interface (or interfaces), the shape of the particle and the direction of its motion.

In this case, for a solid spherical particle moving perpendicularly to a rigid wall the a/l coefficient equals 9/8, as was found by Lorentz [62]. Hence the correction to the Stokes law at large distances can be expressed as

$$\begin{aligned} \lambda_{f_\perp} &= F_h / F_\infty = F_h / 6\pi\eta a U = \frac{1}{1 - (9/8)(a/l) + (1/2)\left(\frac{a}{l} \right)^3} \\ &\cong 1 + \frac{9}{8}\frac{a}{l} + \frac{81}{64}\left(\frac{a}{l} \right)^2 - \frac{1}{2}\left(\frac{a}{l} \right)^3 \end{aligned} \tag{375}$$

where $\mathbf{F}_\infty$ is the resistance force on particle in an unbounded fluid.

In the case of a liquid/gas interface, Faxen and Dahl [65] showed that the correction is given by

$$\lambda_{f_\perp} = 1 + \frac{3}{4}\frac{a}{l} + \frac{9}{16}\left(\frac{a}{l}\right)^2 + \frac{19}{64}\left(\frac{a}{l}\right)^3 \tag{376}$$

Interestingly enough the leading term coefficient equal to 3/4 is the same as that for two equal-sized rigid particles, as first shown by Smoluchowski [64].

For a liquid/liquid interface (the viscosity of the liquid where the particle is immersed being η and the viscosity of the other liquid η_f), the correction calculated by Lee et al. [66] is

$$\lambda_{f_\perp} = 1 + \frac{3}{8}\left(\frac{2+3\bar{\eta}}{1+\bar{\eta}}\right)\left(\frac{a}{l}\right) + \frac{9}{64}\left(\frac{2+3\bar{\eta}}{1+\bar{\eta}}\right)^2\left(\frac{a}{l}\right)^2 \tag{377}$$

where $\bar{\eta} = \eta_f / \eta$.

For a sphere moving parallel to a solid interface, Faxen [63] derived the following expression for $\lambda_{f\text{II}}$

$$\lambda_{f_\text{II}} = \frac{1}{1-(9/16)(a/l)+(1/8)(a/l)^3-(45/256)(a/l)^4} \cong 1 + \frac{9}{16}\frac{a}{l} + \frac{81}{256}\left(\frac{a}{l}\right)^2 - \frac{1}{8}\left(\frac{a}{l}\right)^3 \tag{378}$$

On the other hand, for a liquid/liquid interface, the correction for the parallel motion of a solid sphere was calculated by Lee et al. [66]

$$\lambda_{f_\text{II}} = 1 - \frac{3}{16}\left(\frac{2-3\bar{\eta}}{1+\bar{\eta}}\right)\left(\frac{a}{l}\right) + \frac{9}{256}\left(\frac{2-3\bar{\eta}}{1+\bar{\eta}}\right)^2\left(\frac{a}{l}\right)^2 \tag{379}$$

Note that the correction factor $\lambda_{f_\parallel}$ becomes smaller than unity for viscosity ratio η_f/η smaller than 2/3. This mans that the resistance force on liquid particle moving parallel to a free interface becomes smaller than in the bulk.

The range of validity of the asymptotic solutions expressed by Eq. (377) and Eq. (379) was estimated by Lee et al. [67]. It was shown that for $= \eta_f/\eta \to \infty$ (a solid particle moving near a solid interface), these equations remain accurate for $a/l < 0.3$. For $\eta_f/\eta \to 0$ (solid particle moving near a liquid/gas interface), these equations remain accurate for $a/l < 0.5$.

Using the method of reflections the wall corrections to the rotational motion of a solid particle have also been calculated by Lee et al. [66]. In the case where the rotation axis is perpendicular to the wall, the correction is given by

$$\lambda_{r_{II}} = \frac{To_h}{8\pi\eta a^3} = 1 - \frac{1-\overline{\eta}}{8(1+\overline{\eta})}\left(\frac{a}{l}\right)^3 \tag{380}$$

For a solid/liquid interface, $\eta \rightarrow \infty$ one has

$$\lambda_{r_\perp} = 1 + \frac{1}{8}\left(\frac{a}{l}\right)^3 \tag{381}$$

If the rotation axis is parallel to the interface, the correction is given by

$$\lambda_{r_{II}} = 1 - \frac{1-5\overline{\eta}}{16(1+\overline{\eta})}\left(\frac{a}{l}\right)^3 \tag{382}$$

For a solid/liquid interface, one has

$$\lambda_{r_{II}} = 1 + \frac{5}{16}\left(\frac{a}{l}\right)^3 \tag{383}$$

Unfortunately, there are very few results for other particle shapes. Wakya [68] treated the case of spheroidal particles (both prolate and oblate) moving parallel with a constant velocity between two planar walls (the particle center was located at a distance l = (1/4) L from the nearest wall (where L is the gap between the walls). For smaller particle sizes, where, $a/l >> 1$, the presence of the other wall exerts little effect on the resistance force correction. It was found that for a/l = 10, $\lambda_{r_\parallel}$ = 1.06 for thin disks (oblate spheroids in the limit of negligible thickness) $\lambda_{r_\parallel}$ = 1.07 for spheres and $\lambda_{r_\parallel}$ = 1.11 for prolate spheroids having an axes ratio of 4. For a/l = 20, $\lambda_{r_\parallel}$= 1.03 for disks, $\lambda_{r_\parallel}$ = 1.03 for spheres and $\lambda_{r_\parallel}$= 1.05 for prolate spheroids having an axes ratio of 4. Note that the correction to the Stokes law due to the presence of the wall seems quite insensitive to particle shape.

It should be mentioned, however, that although the significance of these corrections has been confirmed experimentally [69], some doubt arises as to their validity for distances $l > a/Re$. This is so because the Stokes equation fails to correctly describe the fluid velocity fields at larger distances since it neglects the inertia effects.

3.4.5.2. Particle Motion in a Quiescent Fluid Near Interfaces

We now consider corrections to the resistance force and torque for particles moving near interfaces at arbitrary distances, including very small gaps, where the boundary effects play a dominant role. For the sake of generality, we discuss the case of two spherical particles of different radii a_1 and a_2 moving along their line of centers with constant velocities U_1 and U_2 (see Fig. 3.32) in a quiescent fluid. Since the particles are solid, their deformations upon approach are neglected. Obviously, in the case where $a_1/a_2 \rightarrow \infty$ and $U_2 = 0$ one can mimic the situation of a spherical particle approaching a planar interface. The fluid velocity field for the two-particle system can be calculated analytically using the stream function expressed in the bipolar coordinate system (ξ_s, η_s). The solution for equal sphere velocities was first obtained by Stimson and Jefferey [70] and

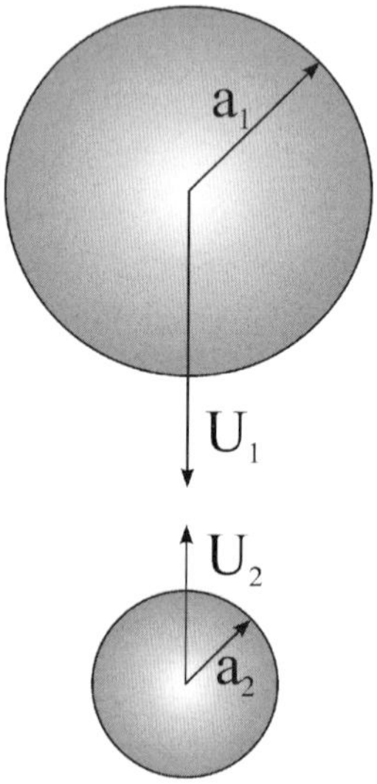

Fig. 3.32. A schematic view of motion of two particles along their line of centers.

generalized to arbitrary fluid velocities by Maude [71]. We follow hereafter the latter approach.

It was found that the stream function for this problem can be expressed, analogously as for the flow near spheroids, in the factorized form

$$\psi_s = \frac{1}{(\cosh\xi_s - \cos\eta_s)^{3/2}} \sum_{n=1}^{\infty} f_n(\xi_s) g_n(\eta_s) \tag{384}$$

where

$$\begin{aligned} f_n &= A_n \cosh\left(n - \frac{1}{2}\right)\xi_s + B_n \sinh\left(n - \frac{1}{2}\right)\xi_s \\ &\quad + C_n \cosh\left(n + \frac{3}{2}\right)\xi_s + D_n \sinh\left(n + \frac{3}{2}\right)\xi_s \\ g_n &= P_{n-1}(\cos\eta_s) - P_{n+1}(\cos\eta_s) \end{aligned} \tag{385}$$

and P_{n-1}, P_{n+1} are the Legendre polynomials.

The coefficients $A_n - D_n$ are determined from the boundary conditions usually employed on flow problems, i.e., no-penetration on the surfaces of the spheres and vanishing velocity field in the bulk.

From the symmetry of the problem it can be deduced that there is no torque on the spheres, and the hydrodynamic resistance force on each particle F_1 and F_2 is given by [72]

$$\begin{aligned} F_1 &= -\eta(K_{11}U_1 + K_{12}U_2) \\ F_2 &= -\eta(K_{21}U_1 + K_{22}U_2) \end{aligned} \tag{386}$$

This can also be written in the concise form

$$\mathbf{F}_h = -\eta \mathbf{K} \cdot \mathbf{U} \tag{387}$$

where **F** is the force vector, **K** is the resistance tensor and **U** is the particle velocity vector.

The force coefficients in Eq. (386) are given by

$$
\begin{aligned}
K_{11} &= F_0 \sum_{n=1}^{\infty} K_n (2n+1) \\
&\quad (-A_n + A'_n - B_n + B'_n - C_n + C'_n - D_n + D'_n) \\
K_{12} &= F_0 \sum_{n=1}^{\infty} K_n (2n+1) \\
&\quad (A_n + A'_n + B_n + B'_n + C_n + C'_n + D_n + D'_n) \\
K_{21} &= F_0 \sum_{n=1}^{\infty} K_n (2n+1) \\
&\quad (A_n - A'_n - B_n + B'_n + C_n - C'_n + D_n + D'_n) \\
K_{22} &= F_0 \sum_{n=1}^{\infty} K_n (2n+1) \\
&\quad (-A_n - A'_n + B_n + B'_n - C_n - C'_n + D_n + D'_n)
\end{aligned}
\tag{388}
$$

where $F_0 = \pi \bar{a}$

$$
K_n = \frac{n(n+1)\left[4\sinh^2\left(n+\frac{1}{2}\right)(\alpha_s' - \beta_s) - (2n+1)^2 \sinh^2(\alpha_s' - \beta_s)\right]^{-1}}{(2n-1)(2n+1)(2n+3)}
\tag{389}
$$

$$
\begin{aligned}
\bar{a} &= a_1 \sinh \alpha_s' = -a_2 \sinh \beta_s \\
\alpha_s' &= \cosh^{-1}\left[\frac{(r/a_1)^2 - (a_2/a_1)^2 + 1}{2r/a_1}\right] \\
\beta_s &= \cosh^{-1}\left[\frac{(r/a_1)^2 + (a_2/a_1)^2 - 1}{2ra_2/a_1^2}\right]
\end{aligned}
\tag{390}
$$

The coefficients $A_n \div D_n$ were first calculated by Stimson and Jefferey, whereas the remaining coefficients $A'_n - D'_n$ were given by Maude [71]. For the sake of convenience, they are given in the Appendix.

Eqs. (383,384) can be used for calculating the resistance coefficient for an arbitrary sphere size ratio and arbitrary sphere velocities. This can be exploited, for example, to determine the mutual diffusion coefficient of two spheres, as discussed later on. Also, interesting subcases can be derived from the above general solution by assuming, for example, that one of the spheres is stationary, so $U_1 = 0$. Then, the hydrodynamic resistance force on the moving sphere is given by

$$F_h' = -K_{22}U_2 \tag{391}$$

Moreover, when the size of the stationary sphere becomes very large, so $a_1/a_2 >> 1$, one can describe the case of a solid sphere moving toward a planar interface. In this case one has $\alpha_s' = 0$ and

$$\beta_s = \cosh^{-1}(H+1) = \ln\{H+1+[(H+1)^2-1]^{1/2}\} \tag{392}$$

where $H = h/a$ is the dimensionless gap width between the particle and the interface.

Consequently, the Stokes correction factor becomes in this case

$$\begin{aligned}\lambda_{f_\perp} &= K_{22}/6\pi\eta a_2 = \frac{1}{F_1(H)}\\ &= \frac{4}{3}\sinh\beta_s \sum_{n=1}^{\infty} \frac{n(n+1)}{(2n-1)(2n+3)}\left(\frac{Y_n}{T_n}-1\right)\end{aligned} \tag{393}$$

where

$$\begin{aligned}Y_n &= 2\sinh(2n+1)\beta_s + (2n+1)\sinh 2\beta_s\\ T_n &= 4\sinh^2\left(n+\frac{1}{2}\right)\beta_s - (2n+1)^2\sinh^2\beta_s\end{aligned} \tag{394}$$

and $F_1(H)$ is the universal hydrodynamic correction function calculated independently by Brenner [73].

For $H >> 1$, $\lambda_{f_\perp}$ approaches the value given by Eq. (375) whereas for small separations it can be approximated by

$$\lambda_{f_\perp} = \frac{1 + 0.974H - 0.21H \ln H}{H} \tag{395}$$

Eq. (395) indicates that for very small distances between a particle and the interface, the correction factor tends to infinity. This can be physically interpreted in terms of the lubrication theory [1] as a result of the increased force needed to drive the liquid out from the gap between the interface and the approaching particle.

It was found [14] that $\lambda_{f_\perp}$ can be well approximated for the entire range of H by the expression

$$\lambda_{f_\perp} = \frac{19H^2 + 26H + 4}{H(19H + 4)} \tag{396}$$

In many applications, $\lambda_{f_\perp}$ can be approximated by an even simpler interpolating function [28]

$$\lambda_{f_\perp} \cong 1 + \frac{1}{H} \tag{397}$$

The dependence of $\lambda_{f_\perp}(H)$ calculated from Eq. (391) and Eq. (393) for a system consisting of one stationary particle and another moving toward it along the line of centers is shown in Fig. 3.33. The theoretical results are compared with the experimental measurements of Adamczyk et al. [74] performed for nylon spheres with diameters 0.32–0.62 cm, immersed in a silicon oil of viscosity 9.83×10^{-4} m^2 s^{-1}. The correction was calculated by measuring the local particle translation velocity in the vicinity of the stationary particle, using the stroboscopic method (the multiple image obtained in this case is shown in the inset to Fig. 3.33).

It is interesting to note that for two equal particle system shown in Fig. 3.33, the correction factor can be well approximated by the following interpolation function [75]

$$\lambda_{f_\perp} = \frac{1}{H}\left(\frac{6H^2 + 13\mathrm{H} + 2}{6H + 4}\right) \tag{398}$$

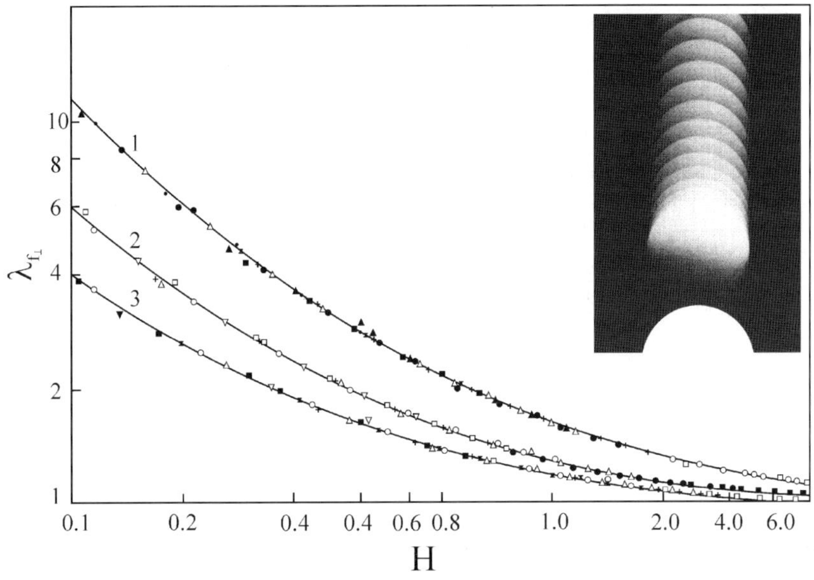

Fig. 3.33. The dependence of the Stokes correction factor $\lambda_{f_\perp}$ on the dimensionless distance H determined experimentally using the stroboscopic method for a system of a stationary and moving particle points. The solid lines denote theoretical results calculated from Eq. (391) and Eq. (393) for 1. $a_1/a_2 \to \infty$ (a sphere approaching a planar interface), 2. $a_1/a_2 = 2$ and 3. $a_1/a_2 = 1$. The inset shows the multiple images of the moving particle approaching the stationary particle [74].

The more general case of a fluid sphere (either a liquid drop or a gas bubble) approaching a free interface (liquid/liquid or liquid/gas) was studied by Bart [76]. The Stokes correction factor in this case was found to be

$$\lambda_{f_\perp} = \frac{4(\eta_f + \eta)}{(3\eta_f + 2\eta)} \sinh\beta_s \sum_{n-1}^{\infty} \frac{n(n+1)}{(2n-1)(2n+3)} \left[\frac{\eta\eta_f X_n + (\eta\eta_1 + \eta^2)Y_n + 2\eta\eta_1 Z_n}{\eta(\eta_f + \eta_1)V_n + \eta\eta_1 T_n + 2\eta^2 S_n} - 1 \right] \tag{399}$$

where η is the viscosity of the fluid in which the particle is immersed, η_1 is the viscosity of the second liquid, η_f is the viscosity of the fluid constituting the sphere, X_n, Y_n are given by Eq. (394) and the remaining functions are

$$Z_n = \cosh(2n+1)\beta_s + \cosh 2\beta_s$$

$$X_n = (2n+1)^2 \sin^2 \beta_s + 4\cosh^2\left(n+\frac{1}{2}\right)\beta_s$$

$$V_n = 2\sinh(2n+1)\beta_s - (2n+1)\sinh 2\beta_s \tag{400}$$

$$S_n = \cosh(2n+1)\beta_s - \cosh 2\beta_s$$

Several cases of practical significance can be derived from Eq. (399), e.g., for a gas bubble approaching a liquid/gas interface when $\eta_f = \eta_2 = 0$, the solution for $\lambda_{f_\perp}$ becomes

$$\lambda_{f_\perp} = 2\sinh\beta_s \sum_{n=1}^{\infty} \frac{n(n+1)}{(2n-1)(2n+3)}\left[\frac{Y_n}{2S_n} - 1\right] \tag{401}$$

For a gas bubble approaching a liquid/solid interface $\lambda_{f_\perp}$ is given by

$$\lambda_{f_\perp} = 2\sinh\beta_s \sum_{n=1}^{\infty} \frac{n(n+1)}{(2n-1)(2n+3)}\left[\frac{2Z_n}{V_n} - 1\right] \tag{402}$$

For a solid sphere moving toward a liquid/gas interface, when $\eta_l = 0$, $\eta_f \to \infty$, λ_f is given by

$$\lambda_{f_\perp} = (4/3)\sinh\beta_s \sum_{n=1}^{\infty} \frac{n(n+1)}{(2n-1)(2n+3)}\left[\frac{X_n}{V_n} - 1\right] \tag{403}$$

The latter case also was studied experimentally by Adamczyk et al [74]. The results shown in Fig. 3.34 clearly indicate that the theoretical results derived from Eq. (403) are in quantitative agreement with the experimental data for the entire range of the dimensionless distances studied (0.1 to 10).

It should be mentioned, however, that the experimental data shown in Figs. 3.33 and 3.34 have been obtained for low Reynolds number

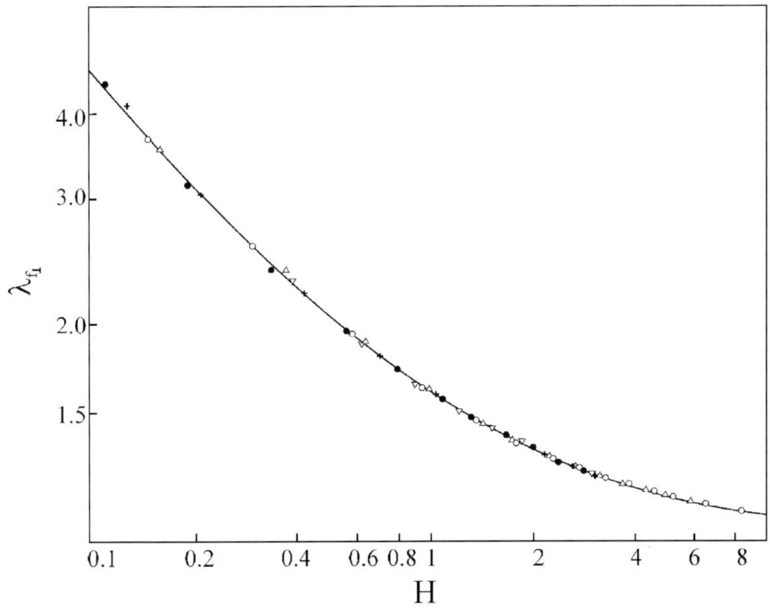

Fig. 3.34. The dependence of the Stokes correction factor $\lambda_{f_\perp}$ on the dimensionless distance H determined experimentally using the stroboscopic method for a solid particle approaching the liquid/air interface points. The solid line denotes the theoretical results calculated from Eq. (403) [74].

corresponding to motion of colloid particles in aqueous solutions. Hence, these data are of a special interest for colloid particle transport and deposition on surfaces. For other applications, it would be of interest to derive a correction to the resistance force for higher *Re* number, beyond the Stokes regime. This correction was calculated theoretically, by Cox and Brenner [77] for *Re* comparable to unity. Because of the approximate matching procedure the results obtained are not conclusive enough to be used to predict the behavior of a real system.

However, there exist interesting experimental data obtained by Małysa [78], who measured the correction to the Stokes law for spherical particles approaching a solid interface. The multiple image technique with stroboscopic illumination was used and the *Re* number range studied was 0.01–3.6. The results are shown in Fig. 3.35 in the form of the dependence of the correction factor $\lambda_{f_\perp}$ on the dimensionless gap width H between the particle and the interface. Note that for $Re > 1$, $\lambda_{f_\perp}$ decreases significantly, especially at larger

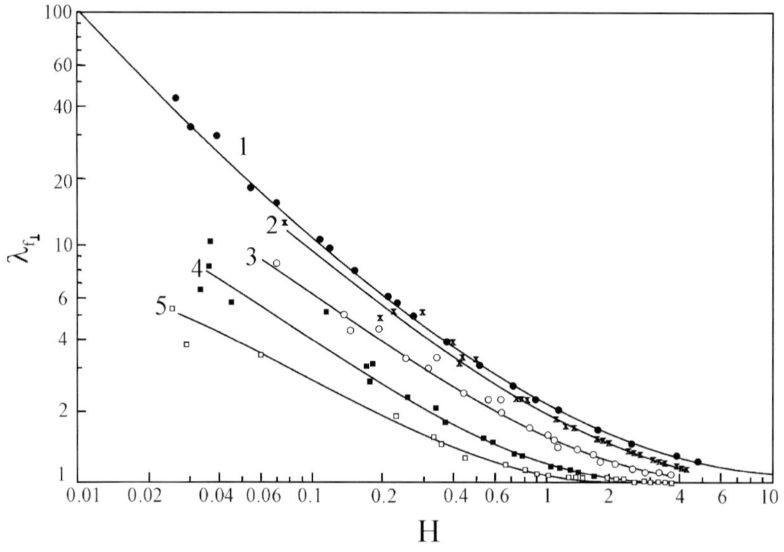

Fig. 3.35. The dependence of the Stokes correction factor $\lambda_{f_\perp}$ on the dimensionless distance H determined experimentally using the stroboscopic method for solid particles approaching the liquid/solid interface (points). (1) $Re = 10^{-2}$ (the solid line denotes the theoretical results calculated from Eq. (393)) (2) $Re = 0.14$; (3) $Re = 0.55$; (4) $Re = 1.57$; (5) $Re = 2.5$ (dashed lines denote interpolations) [78].

distances, when $H>1$. These data strongly suggest, therefore, that the wall effects are diminished for higher Reynolds flows, which is characteristic of particles of macroscopic dimensions, those larger than 10^{-4} m (100 μm). However, for colloid particles, of a size comparable to smaller than 1 μm, these effects are expected to play a significant role.

Other problems pertinent to calculations of resistance forces and torques on spheres moving in the vicinity of planar interfaces can also be solved in an exact way using bipolar coordinate system. Jefferey [79] and Stimson and Jefferey [59] first solved the axisymmetrical problem of a solid sphere rotating with a constant angular velocity about an axis perpendicular to a planar, solid interface. The expression for the hydrodynamic torque was found to be

$$To_h' = 8\pi a^3 \eta \sinh^3 \beta_s \sum_{n=1}^{\infty} \frac{1}{\sinh^3 n\beta_s} = 8\pi a^3 \eta \lambda_{r_\perp} \qquad (404)$$

where the correction factor $\lambda_{r_\perp}$ is given by

$$\lambda_{r_\perp} = \sinh^3\beta_s \sum_{n=1}^{\infty} \frac{1}{\sinh^3 n\beta_s} \tag{405}$$

For a liquid/gas interface the expression for $\lambda_{r_\perp}$ becomes

$$\lambda_{r_\perp} = \sinh^3\beta_{s_s} \sum_{n=1}^{\infty} \frac{(-1)^{n+1}}{\sinh^3 n\beta_s} \tag{406}$$

For a sphere translating parallel to the interface O'Neill [80] derived the following formulae describing the Stokes correction to the resistance force

$$\lambda_{f_{II}} = \frac{1}{F_4(H)} = \frac{\sqrt{2}}{6}\sinh\beta_s \sum_{n=0}^{\infty} \left[n(n+1)C_n + E_n\right] \tag{407}$$

where

$$\begin{aligned}
C_n &= -2k_n\left(\frac{n-1}{2n-1}A_{n-1} - A_n - \frac{n+2}{2n+3}A_{n+1}\right) \\
E_n &= \frac{2\sqrt{2}e^{-(n+(1/2))\beta_s}}{\sinh\mathrm{h}(n+(1/2))\beta_s} \\
&\quad + k_n\left[\frac{(n-1)n}{2n-1}A_{n-1} - \frac{(n+1)(n+2)}{2n+3}A_{n+1}\right] \\
k_n &= \left(n+\frac{1}{2}\right)\coth\left(n+\frac{1}{2}\right)\beta_s - \coth\beta_s
\end{aligned} \tag{408}$$

The A_n coefficients are given by the recurrence formula

$$\begin{aligned}&\left[(2n-1)k_{n-1}-(2n-3)k_n\right]\left[\frac{(n-1)A_{n-1}}{2n-1}-\frac{nA_n}{2n+1}\right]\\&-\left[(2n+5)k_n-(2n+3)k_{n+1}\right]\left[\frac{(n+1)A_n}{2n+1}-\frac{(n+2)A_{n+1}}{2n+3}\right]=\\&\sqrt{2}\left[2\coth\left(n+\frac{1}{2}\right)\beta_s-\coth\left(n-\frac{1}{2}\right)\beta_s-\coth\left(n+\frac{3}{2}\right)\beta_s\right]\end{aligned} \tag{409}$$

Because A_n converges to zero for every β_s, the set of equations represented by Eq. (409) can be solved backwards by assuming that for $n>>1$, the A_n coefficients equals zero [80].

The hydrodynamic torque on the sphere, necessary to prevent it from rotating, was found to be

$$To_h' = -8\pi\eta U a^2 \lambda_c \tag{410}$$

where λ_c, which can be regarded as the coupling coefficient, is given by

$$\begin{aligned}\lambda_c = &\frac{1}{9(2)^{1/2}\mathrm{cosech}^2\beta_s}\sum_{n=0}^{\infty}\left[2+e^{-(2n+1)\beta_s}\right]\left[n(n+1)\right.\\&\left.(2A_n+C_n\coth\beta_s)-(2n+1-\coth\beta_s)E_n\right]\\&+\sum_{n=0}^{\infty}\left[2-e^{-(2n+1)\beta_s}\right]\\&\left[n(n+1)B_n\coth\beta_s-(2n+1-\coth\beta_s)D_n\right]\end{aligned} \tag{411}$$

$$\begin{aligned}&B_n = (n-1)A_{n-1} - (2n+1)A_n + (n+2)A_{n+1}\\&D_n = -\frac{1}{2}(n-1)nA_{n-1} + \frac{1}{2}(n+1)(n+2)A_{n+1}\end{aligned} \tag{412}$$

The complementary problem of a sphere rotating in a quiescent fluid with constant angular velocity about the axis parallel to the interface was treated in a quite similar manner by Dean and O'Neill [81]. The force acting on the rotating sphere was found to be

$$F_h' = -8\pi\eta\Omega a^3 \lambda_{r_{II}} \tag{413}$$

where

$$\lambda_{r_{\text{II}}}=\frac{1}{3}\left\{1-\frac{1}{4(2)^{1/2}\operatorname{cosech}^3\beta_s}\sum_{n=0}^{\infty}\left[2+e^{-(2n+1)\beta_s}\right]\right.$$
$$\left[n(n+1)(2A_n+C_n\coth\beta_s)-(2n+1-\coth\beta_s)E_n\right]$$
$$\left.+\sum_{0}^{\infty}\left[2-e^{-(2n+1)\beta_s}\right]\left[n(n+1)B_n\coth\beta_s-(2n+1-\coth\beta_s)D_n\right]\right\}$$

and the coefficients A_n through E_n can be calculated as

$$\left[(2n-1)k_{n-1}-(2n-3)k_n\right]\left[\frac{(n-1)A_{n-1}}{2n-1}-\frac{nA_n}{2n+1}\right]$$
$$-\left[(2n+5)k_n-(2n+3)k_{n+1}\right]\left[\frac{(n+1)A_n}{2n+1}-\frac{(n+2)A_{n+1}}{2n+3}\right]$$
$$=\frac{2^{\frac{1}{2}}e^{-\left(n+\frac{1}{2}\right)\beta_s}}{(2n+1)\sinh\beta_s}\left[(2n+1)^2\left(\frac{e^{\beta_s}}{(2n-1)}+\frac{e^{-\beta_s}}{2n+3}\right)\operatorname{cosech}\left(n+\frac{1}{2}\right)\beta_s\right.$$
$$\left.-(2n-1)\operatorname{cosech}\left(n-\frac{1}{2}\right)\beta_s-(2n+3)\operatorname{cosech}\left(n+\frac{3}{2}\right)\beta_s\right]$$

$$K_n=(n+\tfrac{1}{2})\coth(n+\tfrac{1}{2})\beta_s-\coth\beta_s$$

$$B_n=(n-1)A_{n-1}-(2n+1)A_n+(n+2)A_{n+1}$$

$$C_n=4l_n\operatorname{cosech}\beta_s\operatorname{cosech}\left(n+\frac{1}{2}\right)\beta_s$$
$$-2k_n\left[\frac{(n-1)A_{n-1}}{2n-1}-A_n+\frac{(n+2)A_{n+1}}{2n+3}\right] \qquad (414)$$

$$l_n=-\frac{1}{(2)^{1/2}}\left[\frac{e^{-\left(n-\frac{1}{2}\right)\beta_s}}{2n-1}-\frac{e^{-\left(n+\frac{3}{2}\right)\beta_s}}{2n+3}\right]$$

$$D_n=-\frac{1}{2}(n-1)nA_{n-1}+\frac{1}{2}(n+1)(n+2)A_{n+1}$$

$$E_n = \left[(2)^{1/2}(2n+1)e^{-(n+1/2)\beta_s} - l_n \operatorname{cosech}\left(n+\frac{1}{2}\right)\beta_s\right]\operatorname{cosech}\left(n+\frac{1}{2}\right)\beta_s$$
$$+k_n\left[\frac{(n-1)nA_{n-1}}{2n-1} - \frac{(n+1)(n+2)nA_{n+1}}{2n+3}\right]$$

Since the analytical solutions converge quite poorly, especially for small separations, Goldmann et al. [82] derived analytical formulae for $\lambda_{r_\perp}$ and λ_c using a lubrication-type approximation with a semi-empirical correction to account for the exact data derived by O'Neill. These formulae are

$$\lambda_{f_{\text{II}}} = -\frac{8}{15}\ln H + 0.9588 \tag{415}$$

Note that, from Eq. (415), the correction to the resistance force for the parallel motion of a sphere diverges logarithmically for small gap widths. Thus, it is much smaller than for the perpendicular motion of the sphere when the $\lambda_{f_{\parallel}}$ correction factor diverges as H^{-1}. For example, at a distance H=0.1, for the parallel motion equals 2.19, whereas for the perpendicular motion, $\lambda_{r_\perp} \cong 11$ at the same distance. For H=0.01 one has $\lambda_{f_{\parallel}}$=3.4 and 101 for the parallel and perpendicular motion, respectively.

For large separations one can use the approximate expression derived by Faxen using the method of reflections, i.e., Eq. (378). The range of validity of these approximations can be estimated from the comparison of exact and approximate results shown in Fig. 3.36.

For many applications connected with particle deposition at interfaces it is more convenient to use the inverse of the Stokes correction factor denoted as the universal functions F_1 and F_4 for the perpendicular and parallel motions, respectively. These functions are shown in Fig. 3.37.

F_4 can be interpolated for small and large separations by [14,82]

$$F_4(H) = \frac{1}{\lambda_{f_{\parallel}}} \cong \frac{1}{-(8/15)\ln H + 0.9588} \quad \text{for} \quad H < 0.1$$

$$F_4(H) = \left(\frac{H}{2.639 + H}\right)^{1/4} \quad \text{for} \quad \text{H} > 0.1 \tag{416}$$

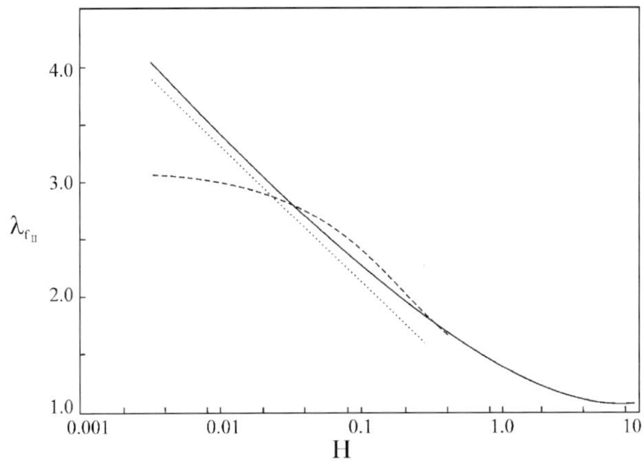

Fig. 3.36. The dependence of the Stokes correction factor $\lambda_{f_{||}}$ on the dimensionless distance H calculated theoretically for a solid sphere moving parallel to a solid interface. The solid lines denote the exact data derived from the analytical solutions of O'Neill, [80] the dotted line denotes the results derived from the corrected lubrication theory, Eq. (415) and the dashed line represents the large distance asymptotic results derived from Faxen's formula, Eq. (378).

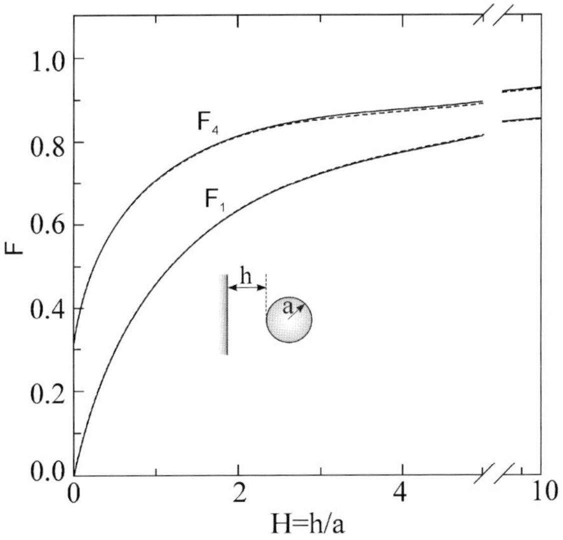

Fig. 3.37. The dependence of the $F_4 = 1/\lambda_{f_{||}}$ and $F_1 = 1/\lambda_{f_{\perp}}$ functions for a solid sphere translating parallel and perpendicular to a planar, solid interface. The dashed lines denote the approximate results calculated from Eq. (416) and Eq. (396), respectively. From Ref. [14].

The range of validity of theoretical predictions concerning the translation and rotation of a particle parallel to the interface has been estimated experimentally by Małysa and van de Ven [83] using, as previously, the stroboscopic technique.

Selected results showing the dependence of the reduced sphere mobility U_∞/U (where U is the sphere velocity far from the wall and U_∞ is the measured sphere velocity parallel to the wall at the distance H) are plotted in Fig. 3.38. A comparison with the lubrication-type approximation valid for small separations and the solution derived from the method of reflection is made. As can be seen, the data derived for $Re=4\times10^{-2}$, are in quantitative agreement with the theoretical predictions in the entire range of distances studied.

By exploiting the lubrication theory, corrections to the rotational particle motion about an axis parallel and perpendicular to a solid interface have also been calculated, as well as the expressions for the coupling tensor, described by Eq. (410) [82].

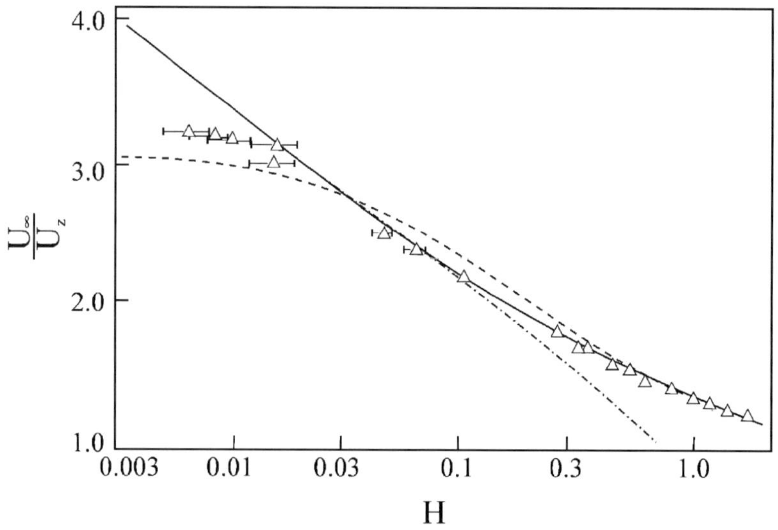

Fig. 3.38. The dependence of the reduced mobility U_∞ / U_z for a spherical particle moving parallel to a planar interface at the distance H. The solid line denotes the exact data derived from the analytical solutions, the dashed-dotted lines denote the results derived from the lubrication theory, Eq. (416) and the dashed lines represent the large distance asymptotic results derived from Faxen's formula, Eq. (378). From Ref [83].

Lee et al. [66,67] treated extensively the more general problem of a fluid particle of arbitrary viscosity moving without deformations in the vicinity of a solid interface. They considered all relevant cases, i.e., particle translation parallel and perpendicular to the interface as well as particle rotation and coupling between the translation and rotation motion. Extensive comparisons between the exact data and the approximate expression valid for large distances have been made.

All results discussed in this section that concern particle motion in a quiescent fluid near planar interfaces can be summarized as follows. The hydrodynamic resistance force and torque are described by the formulae involving the translation, rotation and coupling tensors

$$\mathbf{F}'_h = \eta\ (\mathbf{K}_t \cdot \mathbf{U} + \mathbf{K}^{\dagger}_c\ \Omega)$$

$$\mathbf{To}'_h = -\eta(\mathbf{K}_c \cdot \mathbf{U} + \mathbf{K}_r \cdot \mathbf{\Omega}) \tag{417}$$

where, explicitly

$$\mathbf{K}_t = \begin{vmatrix} K_{t_{\parallel}} & & \\ & K_{t_{\parallel}} & \\ & & K_{t_{\perp}} \end{vmatrix}$$

$$\mathbf{K}_r = \begin{vmatrix} K_{r_{\parallel}} & & \\ & K_{r_{\parallel}} & \\ & & K_{r_{\perp}} \end{vmatrix} \tag{418}$$

$$\mathbf{K}_c = \begin{vmatrix} 0 & K_c & 0 \\ -K_c & 0 & 0 \\ 0 & 0 & 0 \end{vmatrix}$$

In the bulk, one has $K_{t_{\perp}} = K_{t_{\parallel}} = 6\pi a$, $K_{r_{\perp}} = K_{r_{\parallel}} = 8\pi a^3$, $K_c = 0$.

For an arbitrary distance from the surface, the exact values of these coefficients are given by

$$K_{t_{\perp}} = 6\pi\, a \lambda_{f_{\perp}} = 6\pi\, a \frac{1}{F_1(H)}$$

$$K_{t_{\text{II}}} = 6\pi\, a \lambda_{f_{\text{II}}} = 6\pi\, a \frac{1}{F_4(H)}$$

$$K_{r_\perp} = 8\pi\, a^3 \lambda_{r_\perp} = 8\pi\, a^3 \frac{1}{F_5(H)}$$

$$K_{r_{\text{II}}} = 8\pi\, a^3 \lambda_{r_{\text{II}}} = 8\pi\, a^3 \frac{1}{F_6(H)} \tag{419}$$

$$K_c = 6\pi\, a^2 \lambda_c = 6\pi\, a^2 \frac{1}{F_7(H)}$$

where $F_1 = \lambda_{f\perp}^{-1}$ is given by Eq. (393) for solid spheres and Eq. (399) for fluid spheres. $F_4 = \lambda_{f_{||}}^{-1}$ is given by Eq. (407), $F_5 = \lambda_{r\perp}^{-1}$, by Eq. (405), $F_6 = \lambda_{r_{||}}^{-1}$ by Eq. (414) and $F_7 = \lambda_c^{-1}$ by Eq. (411).

On the other hand, from the lubrication theory [77], one can approximate the components of these tensors for solid particles as

$$K_{t_\perp} = 6\pi a \frac{1 + 0.974H - 0.21H \ln H}{H}$$

$$K_{t_{\text{II}}} = 6\pi a \left(-\frac{8}{15} \ln H + 0.9588 \right)$$

$$K_{r_\perp} = \begin{cases} 1.202\pi a^3, & \text{liquid/solid interface} \\ 0.9015\pi a^3, & \text{liquid/air interface} \end{cases} \tag{420}$$

$$K_{r_{\text{II}}} = 8\pi a^3 \left(-\frac{2}{5} \ln H + 0.3817 \right)$$

$$K_c = 6\pi a^2 \left(-\frac{2}{15} \ln H + 0.2526 \right)$$

For large separations, using the method of reflection combined with the singularity method, the approximate expressions for the components of the $\mathbf{K}_t$, $\mathbf{K}_r$ and $\mathbf{K}_c$ tensors for a solid sphere near a fluid/fluid interface of arbitrary viscosity ratio $\bar{\eta}$ are given by

$$K_{t_\perp} = 6\pi a\left[1+\frac{3}{8}\left(\frac{2+3\bar{\eta}}{1+\bar{\eta}}\right)\frac{1}{H+1}+\frac{9}{64}\left(\frac{2+3\bar{\eta}}{1+\bar{\eta}}\right)^2\frac{1}{(H+1)^2}\right]$$

$$K_{t_{\text{II}}} = 6\pi a\left[1-\frac{3}{16}\left(\frac{2-3\bar{\eta}}{1+\bar{\eta}}\right)\frac{1}{(H+1)}+\frac{9}{256}\left(\frac{2-3\bar{\eta}}{1+\bar{\eta}}\right)\frac{1}{(H+1)^2}\right]$$

$$K_{r_\perp} = 8\pi a^3\left[1-\frac{1}{8}\left(\frac{1-\bar{\eta}}{1+\bar{\eta}}\right)\frac{1}{(H+1)^3}\right] \qquad (421)$$

$$K_{r_{\text{II}}} = 8\pi a^3\left[1-\frac{1}{16}\left(\frac{1-5\bar{\eta}}{1+\bar{\eta}}\right)\frac{1}{(H+1)^3}\right]$$

$$K_c = 6\pi a^2\frac{1}{4(1+\bar{\eta})}\frac{1}{(H+1)^2}$$

where it was considered that $a/l = 1/(H+1)$.

Expressions for a solid sphere can be obtained from Eq. (421) by simply substituting $\bar{\eta} = \infty$. Alternatively, for large separations, one can use Eq. (375) for $K_{t_\perp}$ and Eq. (378) for $K_{t_{||}}$.

For the sake of convenience, the expressions for the components of the K_t, K_r and K_c tensors are collected in Table 3.6.

Exact results of Stokes' law correction due to the presence of interfaces discussed above are strictly valid only for spheres. No analytical solutions are known in the literature for the analogous problem of non-spherical particle motion near interfaces. This is so because of the lack of a proper orthogonal coordinate system reflecting exactly the shapes of two non-spherical particles or a particle and a plane. However, one can deduce from the analysis performed according to the method of reflections that at large separations the asymptotic results listed in Table 3.6 can be used as a reasonable approximation. As suggested by Brenner [1] the dimensionless gap width H should be calculated for non-spherical particles by taking their maximum dimension as the scaling factor. This conclusion seems to be further confirmed by the results of Wakya for spheroids [68] discussed above.

On the other hand, at separations much smaller than the local radius of curvature of a particle approaching the interface, the lubrication approximation seems reasonable for calculating the resistance force corrections. The functional dependence of the correction factor on the distance, i.e., H^{-1} for the perpendicular and -ln H for the parallel motion is expected to remain the same. Differences appear in the numerical coefficients, which is of secondary importance for predicting particle transfer rates to interfaces as proven later on. This conclusion is supported by the experimental data of Małysa et al. [84] shown in Fig. 3.39, which were obtained, as previously mentioned by the multiple image technique. The Stokes law correction factor $\lambda_{f_\perp}$ was determined for a two-particle aggregate approaching a solid–liquid interface. The experimental results were interpreted in terms of the approximate theoretical model using the following equation describing $\lambda_{f_\perp}$ for the doublet

$$\lambda^a_{f_\perp} \cong \lambda_f(1.4651+0.0852\cos 2\vartheta)\frac{1}{1+\cos 2\vartheta} \tag{422}$$

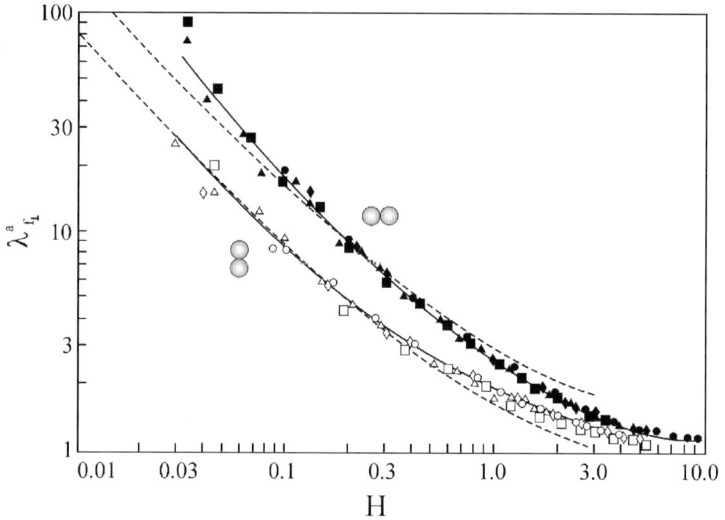

Fig. 3.39. The Stokes law correction factor $\lambda^a_{f\perp}$ for a two particle aggregate approaching a soild–liquid interface. The points denote the experimental results determined for the parallel orientation of the aggregate (upper curve) and perpendicular orientation (lower curve), the solid lines denote regression fits and the dashed line represent the theoretical results calculated from Eq. (422). From Ref [84]

where ϑ is the angle of the doublet axis relative to the direction of gravity (directed perpendicularly to the wall). As can be seen, Eq. (422) predicts that the resistance coefficient of the doublet can be estimated by multiplying for a single sphere by a factor independent of the distance. Hence, the large and small distance asymptotic solutions for the sphere are applicable for the doublet. This seems to be in a good agreement with the experimental data shown in Fig. 3.39 for the range of $0.05 < H < 1$. A quite natural tendency of the aggregate to align parallel to the interface upon approach (by rotation) was also observed experimentally.

This effect can be explained by the action of gravitational force on the sphere located further from the interface. These experiments suggest that the limiting solutions shown in Table 3.6 are applicable for non-spherical particles as well.

3.4.5.3. Particles at Interface in Simple Flows

All the results discussed in this section concern particle motion near interfaces in quiescent fluids. From the viewpoint of transport and deposition phenomena it is also relevant to calculate the forces and torques on particles at interfaces in simple flows such as the shearing and stagnation-point flow. This is so because according to previous analysis, see Section 3.4, all flows near macroscopic surfaces can be effectively decomposed into such simple flows.

The problem of determining the force and torque on a stationary particle held fixed in a simple shear (when the ambient fluid velocity is given as $\mathbf{V}_{sh} = Gz\mathbf{i}_z$) at an arbitrary distance from the interface was solved by Goldmann et al. [85]. They exploited the solutions of O'Neill [80] and Dean and O'Neill [81] in the form of the Legendre polynomials in bispherical coordinates. The same method of solution was applied later on by Goren and O'Neill [86].

It was shown that the hydrodynamic drag force and torque on the particle was described by

$$
\begin{aligned}
F_h &= 6\pi\eta a^2 G(H+1)F_8(H) \\
To_h &= 8\pi\eta a^3 G F_9(H)
\end{aligned}
\tag{423}
$$

where the functions $F_8(H)$ and $F_9(H)$ have been given in tabulated form only. It was demonstrated, however, using the method of reflections that for large separations the asymptotic expressions for $F_8(H)$ and $F_9(H)$ are

Table 3.6
The resistance force and torque on a solid sphere moving near planar interfaces

Flow configuration	Exact solutions	Large separations, $H >> 1$ free interfaces	Large separations, $H >> 1$ solid/fluid interfaces	Lubrication approximation solid/fluid interfaces
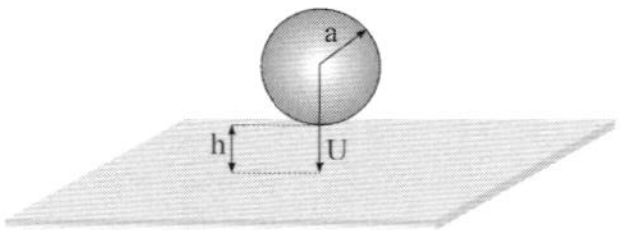 Translation perpendicular to planar interface	$F_h' = -\eta K_{t_\perp} U$ $= -6\pi\eta a \lambda_{f_\perp} U$ $= -6\pi\eta U \frac{1}{F_1(H)}$ $To_h' = 0$ $F_1(H)$ given by Eq. (393) for solid and Eq. (399) for fluid particles	$\lambda_{f_\perp} = 1 + \frac{3}{8}\left(\frac{2+3\bar{\eta}}{1+\bar{\eta}}\right) \frac{1}{H+1} + \frac{9}{64}\left(\frac{2+3\bar{\eta}}{1+\bar{\eta}}\right)^2 \frac{1}{(H+1)^2}$	$\lambda_{f_\perp} = 1 + \frac{9}{8(H+1)} + \frac{81}{64(H+1)^2}$	$\lambda_{f_\perp} = \frac{1+0.974H - 0.21H \ln H}{H}$
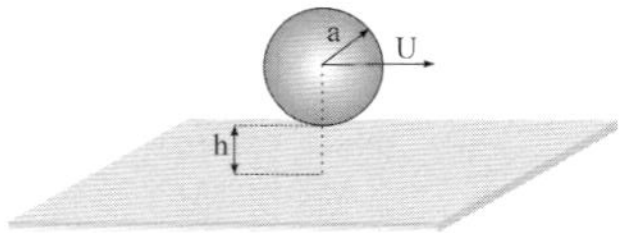 Translation parallel to planar interface	$F_h' = -6\eta K_{t_\parallel} U$ $U = -6\pi\eta a \lambda_{f_\parallel} U$ $= -6\pi\eta a U \frac{1}{F_4(H)}$ $To_h' = -\eta K_c U$ $= -6\pi\eta a^2 \lambda_c U$ $= -6\pi\eta a^2 U \frac{1}{F_7(H)}$ $F_1(H)$ given by Eq. (416) $F_7(H)$ given by Eq. (420)	$\lambda_{f_\parallel} = 1 - \frac{3}{16}\left(\frac{2-3\bar{\eta}}{1+\bar{\eta}}\right)\frac{1}{H+1} + \frac{9}{256}\left(\frac{2-3\bar{\eta}}{1+\bar{\eta}}\right)^2 \frac{1}{(H+1)^2}$ $\lambda_c = \frac{1}{4(1+\bar{\eta})}\frac{1}{(H+1)^2}$	$\lambda_{f_\parallel} = 1 + \frac{9}{16(H+1)} + \frac{81}{256(H+1)^2}$ $\lambda_c = \frac{1}{4(H+1)^2}$	$\lambda_{f_\parallel} = 2\frac{8}{15}\ln H + 0.9598$ $\lambda_c = -\frac{2}{15}\ln H + 0.2526$

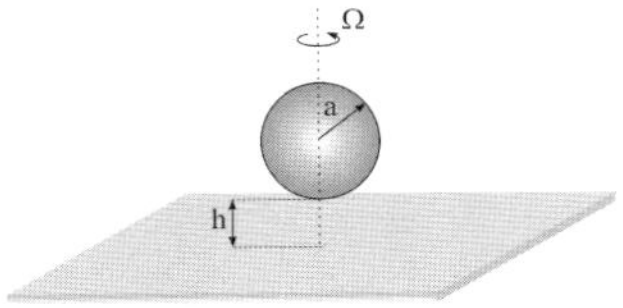 Rotation about an axis perpendicular to planar interface	$To'_h = -\eta K_{r_\parallel} \Omega$ $= -8\pi\eta a^3 \lambda_{r_\parallel} \Omega$ $= -8\pi\eta a^3 \Omega \frac{1}{F_5(H)}$ $F'_h = 0$ λ_r given by Eqs. (405, 406)	$\lambda_{r_\perp} = 1 - \frac{1}{8}\left(\frac{1-\bar{\eta}}{1+\bar{\eta}}\right) \frac{1}{(H+1)^3}$	$\lambda_{r_\perp} = 1 + \frac{1}{8(H+1)^3}$	$\lambda_{r_\perp} = 1.202$ liguid/solid interface $\lambda_{r_\perp} = 0.9015$ liquid/air interface
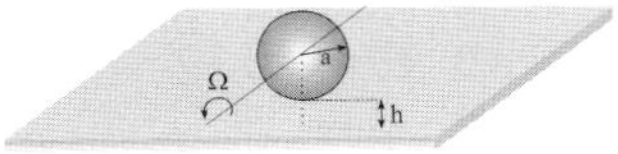 Rotation about an axis parallel to planar interface Rotation about an axis parallel to planar interface	$To'_h = -\eta K_{r_\parallel} \Omega$ $= -8\pi\eta a^3 \lambda_{r_\parallel} \Omega$ $= -8\pi\eta a^3 \Omega \frac{1}{F_6(H)}$ $\lambda_{f_{\parallel}}$ given by Eq. (413) $F'_h = -\eta K_c \Omega$ $= -6\pi\eta a^2 \Omega \lambda_c$ $= -6\pi\eta a^2 \Omega \frac{1}{F_7(H)}$ λ_c given by Eq. (411)	$\lambda_{r_\parallel} = 1 - \frac{1}{16}\left(\frac{1-5\bar{\eta}}{1+\bar{\eta}}\right) \frac{1}{(H+1)^3}$	$\lambda_{r_\parallel} = 1 + \frac{5}{16(H+1)^3}$	$\lambda_{r_\parallel} = -\frac{2}{15} \ln H + 0.3817$

η – viscosity of the fluid in which the particle is immersed; η_1 – viscosity of the second fluid (immiscible with the first one); $\bar{\eta} = \eta_1 / \eta$.
Useful interpolating functions for $F_1(H)$ valid for arbitrary range of H

$$F_1(H) \cong \frac{H}{H+1}, \quad F_1(H) \cong \frac{H(19H+4)}{19H^2+26H+4}$$

$$F_8 \cong 1 + \frac{9}{16}\frac{1}{(H+1)}$$

$$F_9 \cong 1 - \frac{3}{16}\frac{1}{(H+1)^3} \tag{424}$$

It also was shown that for $H \to 0$, $F_8(H)$ and $F_9(H)$ approach the limiting values

$$F_8 = 1.701$$

$$F_9 = 0.944 \tag{425}$$

Using Eq. (425), one can deduce that the maximum drag force on a particle attached to an interface when $H \to 0$ is given by

$$F_s = 10.21\pi\eta G a^2 \tag{426}$$

Eq. (426) indicates that the hydrodynamic force increases with the square of particle size. One can deduce, therefore, that there is a maximum particle size that can withstand the shearing hydrodynamic force because the dispersion and electrostatic force increase proportionally to the size of the particle, as demonstrated in the previous chapter. Eq. (426) has major significance in estimating the adhesion force of particles at surfaces measured experimentally by applying a simple shear, for example, by rotating the substrate surface [87].

O'Neill [88] calculated the force and torque on a particle attached to a solid interface immersed in a simple shear flow in terms of the Bessel function expansion. His approach allows one to evaluate fluid velocity components. Calculations of the fluid velocity field make it possible to demonstrate that the presence of a particle at the interface significantly reduces the ambient shearing flow intensity. This effect extends over distances considerably exceeding the particle diameter. This results in turn in diminishing the wall shear rate over large surface areas. This is clearly visible in Fig. 3.40 where the contours of the function G/G_0 are plotted (where G_0 is the unperturbed shear rate and G is the actual wall shear rate) around a solid particle attached to a solid interface. It is interesting to note that this perturbation is symmetric with respect to the flow direction, which is a consequence of the symmetry of

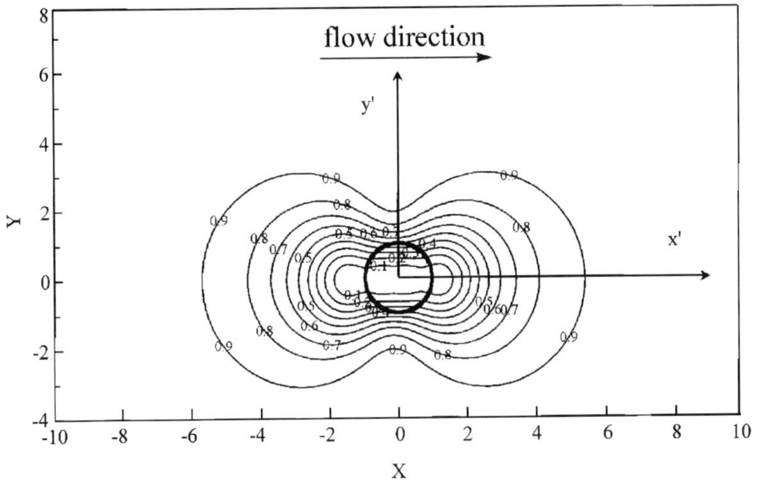

Fig. 3.40. The contours of the reduced wall shear rate G/G_0 (where G_0 is the unperturbed shear rate and G is the actual wall shear rate) near a solid particle attached to a solid interface.

the Stokes equation for this geometry. The decrease in the shear rate generated by the attached particle plays a decisive role in determining the streaming current and streaming potential for particle-covered (rough) surfaces. The same effect becomes clearly visible from the streamline pattern around a particle attached to the wall, placed in the shear flow, as shown in Fig. 3.41.

The problem of a particle at an interface immersed in a two-dimensional and axisymmetric stagnation flow with ambient velocity $\mathbf{V}_\infty = G_{st}(-z^2\mathbf{i}_z + zr\mathbf{i}_r)$ was solved by Goren [89] and Goren and O'Neill [86] using the bipolar coordinate system. It was shown that the hydrodynamic drag force and torque on the particle were described by

$$\begin{aligned} F_h &= 6\pi\eta a^3 G_{st} F_2(H) \\ To_h &= 0 \end{aligned} \tag{427}$$

where the function F_2(H) was given in tabulated form only.

It was also shown that in the limit of $H \to 0$, $F_2(H) = 3.23$ [89]. Thus, from from Eq. (416) one can deduce that the maximum drag force on a particle attached to an interface is given by

$$F_h = 19.4\pi\eta G_{st} a^3 \tag{428}$$

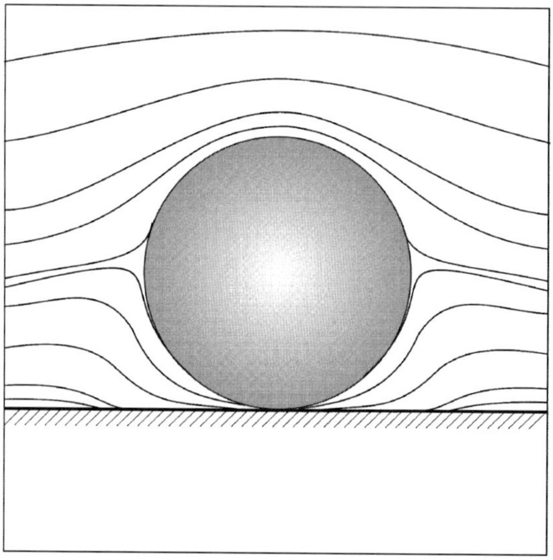

Fig. 3.41. Streamline pattern near a solid particle attached to a solid interface and immersed in a simple shear flow. From Ref [43]

As can be noted, F_h increases proportionally to a^3. This means that for larger particle sizes F_h will dominate over the hydrodynamic shearing force which increases proportionally to a^2, and over the dispersion and double-layer forces which increase proportionally to a.

The positive value of the drag force (directed toward the interface) occurs near the forward stagnation point of many macroscopic surfaces listed in Table 3.4, such as the sphere, spheroid and cylinder immersed in uniform flows. On the other hand, in the region close to the rear stagnation point of these interfaces (in the limit of low Re number), the drag force is directed outward from the surface. This is expected to significantly affect particle adhesion to surfaces in these regions.

For the entire range of H, the correction function $F_2(H)$ can be interpolated by [14]

$$F_2(H) = 1 + \frac{1.79}{(0.828 + H)^{1.167}} \tag{429}$$

For many applications, for example, for calculating particle deposition rates at surfaces, it is useful to know the product of $F_1(H)F_2(H)$. This can be interpolated by

$$F_1(H)F_2(H) = H\left(\frac{H+0.21}{H^3+1.37+0.21}\right)\left(1+\frac{1.79}{(0.828+H)^{1.167}}\right) \qquad (430)$$

The dependence of F_2 and the product $F_1(H)F_2(H)$ on particle–wall separation is also shown in Fig. 3.42. Unfortunately, there are no analytical solutions in the case of non-spherical particles in simple flows at interfaces.

In the case of non-spherical particles, more significant differences than for motion in quiescent fluid are expected in the case of shearing flows. In this case, just by simple geometrical constraints, the rotational motion of an elongated particle will be strongly influenced by the boundary. A definite

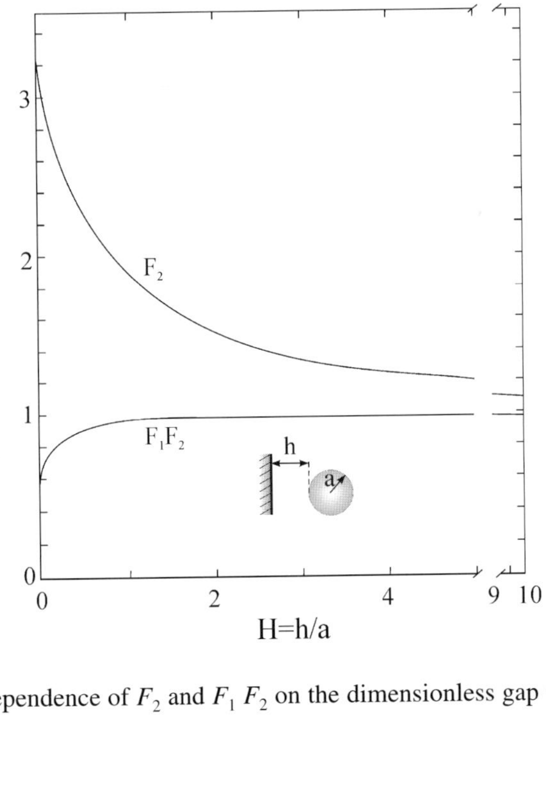

Fig. 3.42. The dependence of F_2 and F_1 F_2 on the dimensionless gap width H.

tendency to particle orientation parallel to the interface is expected, probably much far more than in the bulk [50]. Unfortunately, there are no analytical solutions to the problem of non-spherical particle motion near interfaces that could support this hypothesis.

In this chapter the so-called resistance problems [1,5] have been considered with the emphasis on calculating the hydrodynamic forces and torques on particles. They comprised

(i) the driving forces (and torques) exerted by fluid streams on particles, and
(ii) the resistance force which appears when particles move in a quiescent fluid.

A prerequisite for calculating the driving forces was a knowledge of macroscopic flows near various interfaces, which were, therefore, analyzed in detail as well. The related problem of determining particle velocity for a prescribed force field, pertinent to particle transport, will be treated in extensively in the next chapter.

APPENDIX

Expressions for the coefficients $A_n - D'_n$ in Eq. (388)

$$A_n = (2n+3)\left\{4E\left[-\left(n+\frac{1}{2}\right)y\right]S\left(n+\frac{1}{2}\right)y\right.$$
$$+(2n+1)^2 EySy + 2(2n-1)S\left(n+\frac{1}{2}\right)y$$
$$C\left(n+\frac{1}{2}\right)x - 2(2n+1)S\left(n+\frac{3}{2}\right)y$$
$$\left.C\left(n-\frac{1}{2}\right)x - (2n+1)(2n-1)SyCx\right\}$$

$$B_n = -(2n+3)\left[2(2n-1)S\left(n+\frac{1}{2}\right)yS\left(n+\frac{1}{2}\right)x\right.$$
$$\left.-2(2n-1)S\left(n+\frac{3}{2}\right)yS\left(n-\frac{1}{2}\right)x + (2n+1)(2n-1)Sy\,Sx\right]$$

$$C_n = -(2n-1)\left\{4E\left[-\left(n+\frac{1}{2}\right)y\right]S\left(n+\frac{1}{2}\right)y\right.$$
$$-(2n+1)^2 E(-y)Sy + 2(2n+1)$$
$$S\left(n-\frac{1}{2}\right)yC\left(n+\frac{3}{2}\right)x - 2(2n+3)S\left(n+\frac{1}{2}\right)y$$
$$\left.C\left(n+\frac{1}{2}\right)x + (2n+1)(2n+3)SyCx\right\}$$

$$D_n = (2n-1)\left[2(2n+1)S\left(n-\frac{1}{2}\right)yS\left(n+\frac{3}{2}\right)x - 2(2n+3)\right.$$
$$\left.S\left(n+\frac{1}{2}\right)yS\left(n+\frac{1}{2}\right)x + (2n+1)(2n+3)SySx\right]$$

$$A'_n = (2n+3)\left[2(2n-1)S\left(n+\frac{1}{2}\right)yS\left(n+\frac{1}{2}\right)x - 2(2n+1)\right.$$
$$\left.S\left(n+\frac{3}{2}\right)yS\left(n-\frac{1}{2}\right)x - (2n+1)(2n-1)SySx\right]$$

$$B'_n = -(2n+3)\left\{-4E\left[-\left(n+\frac{1}{2}\right)y\right]S\left(n+\frac{1}{2}\right)y\right.$$
$$-(2n+1)^2 EySy + 2(2n-1)S\left(n+\frac{1}{2}\right)y$$
$$C\left(n+\frac{1}{2}\right)x - 2(2n+1)S\left(n+\frac{3}{2}\right)y$$
$$\left.C\left(n-\frac{1}{2}\right)x + (2n-1)(2n+1)SyCx\right\}$$

$$C'_n = -(2n-1)\left[2(2n+1)S\left(n-\frac{1}{2}\right)yS\left(n+\frac{3}{2}\right)x\right.$$
$$\left.-2(2n+3)S\left(n+\frac{1}{2}\right)yS\left(n+\frac{1}{2}\right)y - (2n+1)(2n+3)SySx\right]$$

$$D_n' = (2n-1)\left\{-4E\left[-\left(n+\frac{1}{2}\right)y\right]S\left(n+\frac{1}{2}\right)x\right.$$
$$+(2n+1)^2 E(-y)Sy + 2(2n+1)S\left(n-\frac{1}{2}\right)y$$
$$C\left(n+\frac{3}{2}\right)x - 2(2n+3)S\left(n+\frac{1}{2}\right)y$$
$$\left.C\left(n+\frac{1}{2}\right)x - (2n+1)(2n+3)SyCx\right\}$$

where $x = \alpha_s' + \beta_s$, $y = \alpha_s' - \beta_s$ and S, C, and E denote sinh, cosh, and exp respectively (e.g., Sx = sinh) $(\alpha_s' - \beta_s)$) etc.

LIST OF SYMBOLS

Symbol	Definition	Character	Unit	Numerical value
$A(y)$, $A'(y')$	Amplitude of fluid velocity	S	$m\,s^{-1}$	
A_n, A_n'	Series expansion coefficients for two particle motion	S	[1]	
A_f	Flow parameter	S	[1]	
$\mathbf{a}$	Constant force vector	**V**	N	
a, a_p	Particle radius, spheroid semi-axis	S	m	
$B'(y')$	Function characterizing oscillatory flows	S	[1]	
B_n, B_n'	Series expansion coefficients for two particle motion	S	[1]	
$\mathbf{b}$	Constant force vector	**V**	N	
b	Shorter semi-axis of spheroids	S	m	
$2b$	Height of parallel-plate channel	S	m	
$\mathbf{C}$	Rotlet (couplet) tensor	**T**	$m^{-1}\,kg^{-1}\,s$	
$C(r/R)$	Universal function for radial impinging jet	S	[1]	
C_f	Flow intensity constant	S	$m^{-1}\,s^{-1}$	
C_0	Constant of integration	S	[1]	
C_0'	Constant of integration for streamlines	S	m^4	
C_s	Constant of integration for ellipsoid	S	$m^{-1}\,s^{-1}$	

List of Symbols (continued)

Symbol	Definition	Character	Unit	Numerical value
C_w	Constant describing velocity distribution for wall jet flows	S	$m^{11/4}\ s^{-1}$	
$\overline{C}_3$	Coefficient of wall jet velocity	S	[1]	
C_1, C_2, C_3, C_4	Coefficients of fluid velocity expansion outside sphere	S		
C_{1f}, C_{2f}, C_{3f} C_{4f}	Coefficients of fluid velocity expansion outside sphere	S		
$\mathbf{c}$	Constant force vector	**V**	N	
c	Ellipsoid semi-axis	S	m	
$2c$	Width of parallel-plate channel	S	m	
c_f	Focal point position for spheroids	S	m	
c_0, c_1, c_2, c_3, c_4	Dimensionless constants	S	[1]	
$\mathbf{D}$	Reflection tensor	**T**	$kg^{-1}\ s$	
$\mathbf{D}_p$	Potential dipole tensor	**T**	$kg^{-1}\ s$	
D_n, D'_n	Series expansion coefficients for two particles motion	S	[1]	
$2d$	Width of lot jet	S	m	
d_p	Width of plate in uniform flow	S	m	
E, E'	Electric fields	**V**	$V\ m^{-1}$	
$\overline{E}$	Dimensionless electric field	**V**	[1]	
E, E^0	Electric field (scalar)	S	$V\ m^{-1}$	
El	Dimensionless electrostatic number	S	[1]	
E_n	Coupling function expansion coefficients	S	[1]	
E_s	Stream function operator	S	m^{-2}	
e	Elementary charge	S	C	1.60218×10^{-19}
e_{ijk}	Components of alternating matrix	S	[1]	
$\mathrm{erf}\ (x)$	Error function	S	[1]	
$\mathbf{F}$	Force vector	**V**	$N = kg\ m\ s^{-2}$	
$\tilde{\mathbf{F}}$	Generalized driving force	**V**	$N\ m^{-3}$	
$\tilde{\mathbf{F}}'$	Generalized resistance force	**V**	N	
$\mathbf{F}_h, \boldsymbol{F}'_h$	Hydrodynamic force	**V**	N	
$\tilde{\boldsymbol{F}}_h(s)$	Laplace transform of hydrodynamic force	**V**	N	
F	Force of interactions (scalar)	S	N	
$<F>$	Average resistance force of ellipsoid	S	N	
$F(\xi)$	Universal expansion functions for oblique flow and rotating disk	S	[1]	

List of Symbols (continued)

Symbol	Definition	Character	Unit	Numerical value
$F_n'(\xi), F_n'(\xi)$	Universal expansion function for cylinder and sphere	S	[1]	
F_h'	Hydrodynamic force on channel walls	S	N	
F_h	Hydrodynamic resistance force	S	N	
F_o	Resistance force parameter	S	m	
Fr	Froude number	S	[1]	
F_s	Maximum drag force on particle	S	N	
F_w	Wall jet flow intensity function	S	$m^7 s^{-3}$	
F_1, F_2, F_3	Resistance force of ellipsoid	S	N	
$F_1(H) \div F_9(H)$	Universal hydrodynamic correction functions	S	[1]	
$\mathbf{f}_p$	Hydrodynamic point force	**V**	N	
$f(r)$	Function describing velocity near sphere	S	[1]	
$f(y)$	Function for oblique flow	S	s^{-1}	
$f(\lambda_s)$	Function describing flow distribution near spheroid	S	[1]	
$f(\xi)$	Universal function for plate	S	[1]	
$f(\eta, \eta_f)$	Fluid viscosity function	S	$kg\ m^{-1} s^{-1}$	1.2326
$f(0)$	Value of the $f(\xi)$ function at surface plate, stagnation flow intensity parameter	S	[1]	0.3321
$f_f(r)$	Function describing velocity inside sphere	S	[1]	
$f_n(\xi_s)$	Series expansion functions for two particle motion	S	[1]	
f_0	Dimensionless coefficients for cylinder in uniform flow	S	[1]	
f_0'	Dimensionless coefficient for spheroids	S	[1]	
$\mathbf{G}$	Oseen tensor	**T**	$m^3 kg^{-1} s$	
$\mathbf{G}_d$	Reflection tensor	**T**	$kg^{-1} s$	
$\mathbf{G}_{so}$	Stokeson tensor	**T**	$m^3 kg^{-1} s$	
$\mathbf{G}_w$	Modified Oseen tensor	**T**	$kg^{-1} m\ s$	
$G(\xi)$	Universal function for oblique flow and rotating disk	S	[1]	
$G(0), g(0)$	Value of the $G(\xi)$ function at disk surface	S	[1]	
G_f	Stagnation flow parameter for oblique flow	S	s^{-1}	

List of Symbols (continued)

Symbol	Definition	Character	Unit	Numerical value
G^0	Wall shear rate	S	s^{-1}	
G_{sh}, G^0	Simple shear flow rate	S	s^{-1}	
G_{St}	Stagnation flow shear rate	S	$m^{-1}\, s^{-1}$	
G_y^0, G_x^0	Wall shear rates in channel	S	s^{-1}	
$\mathbf{g}$	Gravity acceleration vector	**V**	$m\, s^{-2}$	
$\overline{\mathbf{g}}$	Reduced gravity vector	**V**	[1]	
$g(r)$	Functions describing velocity near sphere	S	[1]	
$g(y)$	Function of oblique flow	S	$m\, s^{-1}$	
$g(\lambda_s)$	Function describing flow distribution near spheroid	S	$m^3\, s^{-1}$	
$g_n(\xi_s)$	Series expansion function for two particle motion	S	[1]	
H	Dimensionless gap width	S	[1]	
$H(\xi)$	Universal function for rotating disk	S	[1]	
$H(0)$	Value of $H(\xi)$ function at disk surface	S	[1]	−1.0204
$H( )$	Value of $H(\xi)$ function at infinity	S	[1]	−0.886
h	Distance of particles from wall	S	m	
h_c	Distance of capillary from wall	S	M	
h_m	Distance between particle and surface	S	m	
h_1, h_2	Metric coefficients for spheroidal coordinates	S	m^{-1}	
$\mathbf{I}$	Unit tensor	**T**	[1]	
I	Ionic strength	S	m^{-3}	
$\mathbf{i}, \mathbf{i}_r, \mathbf{i}_H, \mathbf{i}_y, \mathbf{i}_z$	Unit vectors	**V**	[1]	
i	Imaginary unit	S	[1]	
J_0, J_l	Bessel functions of zeroth and first order	S	[1]	
$\mathbf{j}_f$	Flux of physical quantify **f**	**V**	arbitrary	
$\mathbf{K}$	Resistance tensor for two spheres	**T**	m	
$\tilde{\mathbf{K}}$	Grand resistance matrix	**T**	m	
$\mathbf{K}_c$, $\mathbf{K}_c^{\dagger}$	Coupling resistance tensor	**T**	m^3	
$\mathbf{K}_r$	Rotation resistance tensor	**T**	m^3	
$\mathbf{K}_s$, $\mathbf{K}'_s$	Third rank hydrodynamic tensors	**T**	m^3	
$\mathbf{K}_t$	Translation resistance tensor	**T**	m	
$<K>$	Average resistance force coefficient of ellipsoid	S	[1]	

List of Symbols (continued)

Symbol	Definition	Character	Unit	Numerical value
$K(\lambda_0), K'(\lambda_0)$	Functions describing resistance of ellipsoid	S	[1]	
K'	Coefficient of hydrodynamic force due to boundary	S	[1]	
K_c	Components of coupling resistance tensor	S	m^2	
K_n	Series expansion coefficients for two particle motion	S	[1]	
K_{Oseen}	Drag force coefficient	S	[1]	
$K_{r\parallel}, Kr_{\perp}, K_{r11}$ K_{r22}, K_{r33}	Components of rotation resistance tensor	S	m^3	
$K_{t11}, K_{t22},$ $K_{t33}, K_{t\perp}$	Components of translation resistance tensor	S	m^3	
$K_{11}, K_{22},$ K_{12}, K_{21}	Resistance tensor coefficients for two spheres	S	m	
k	Boltzmann constant	S	$J\,K^{-1}$	1.38065×10^{-23}
k_n	Series expansion coefficient for two sphere motion	S	[1]	
L	Length of channel, cylinder, plate	S	m	
L_c	Characteristic length of entrance region	S	m	
L_{ch}	Characteristic length	S	m	
Le	The diffuse double-layer thickness	S	m	
L_t	Flow transition length	S	m	
l	Virtual origin of wall jet, length of slender body	S	m	
$\bar{l}$	Entrance length parameter	S	[1]	
l_n	Coefficients for two sphere motion	S	[1]	
m	Fluid mass	S	kg	
m_f	Mass of displaced fluid	S	kg	
m_p	Particle mass	S	kg	
m'_p	Effective particle mass	S	kg	
$\hat{\mathbf{n}}$	Unit normal vector	**V**	[1]	
P	Pressure in channel	S	$N\,m^{-2}$	
P_n	Legendre polynomials	S	[1]	
$\mathbf{p}_m$	Linear momentum vector	**V**	$kg\,m\,s^{-1}$	
p, p', p_0	Hydrodynamic pressure	S	$N\,m^{-2}$	
$\tilde{p}(s)$	Laplace transform of pressure	S	$N\,m^{-2}$	

List of Symbols (continued)

Symbol	Definition	Character	Unit	Numerical value
p_e	Equilibrium pressure	*S*	N m^{-2}	
p_h	Non-equilibrium pressure	*S*	N m^{-2}	
p_1, p_2	Pressure in various phases	*S*	N m^{-2}	
$\mathbf{Q}_f$	Source term of quantity **f**	arbitrary	arbitrary	
Q	Volumetric flow rate	*S*	$m^3\ s^{-1}$	
q_h	Flow source strength	*S*	$m^3\ s^{-1}$	
R	Radius of cylindrical channel, cylindrical or spherical collector	*S*	m	
$\bar{R}$	Mean radius of curvature	*S*	m	
Re	Reynolds number	*S*	[1]	
R, R_1, R_2	Radii of curvature of interfaces	*S*	m	
R'_1, R''_1, R'_2, R''_2	Radii of curvature of bodies	*S*	m	
$\mathbf{r}, \mathbf{r}_1, \mathbf{r}_2$ $\mathbf{r}_i, \mathbf{r}_0$	Position vectors	**V**	m	
$\bar{\mathbf{r}}$	Reduced position vector	**V**	[1]	
(r_0, z_0)	Coordinates of point on streamline			
r_0	Critical radius for limiting trajectory	*S*	m^{-1}	
r_p	Radial position of particle in local flow	*S*	m	
$\bar{r}_1, \bar{r}_3, \bar{r}_5$	Expansion coefficients for sphere	*S*	m^{n-1}	
S	Surface area	*S*	m^2	
$S(r/R)$	Universal function for radial impinging jet	*S*	[1]	
$\delta S, dS$	Surface elements	*S*	m^2	
S_n	Coefficients of resistance force series expansion	*S*	[1]	
St	Stokes' number	*S*	[1]	
s	Laplace transformation parameter Integration parameter for ellipsoid	*S* *S*	[1] m^2	
To	Torque vector	**V**	N m	
$\mathbf{To}_{ext}$	External torque	**V**	N m	
$\mathbf{To}_h$	Hydrodynamic torque	**V**	N m	
T	Absolute temperature	*S*	K	
$\boldsymbol{t}$	Time	*S*	s	

List of Symbols (continued)

Symbol	Definition	Character	Unit	Numerical value
$\bar{\mathbf{t}}$	Dimensionless time	S	[1]	
$\mathbf{U}$	Velocity of particle	$\mathbf{V}$	m s^{-1}	
$\tilde{\mathbf{U}}$	Generalized velocity of particle	$\mathbf{V}$		
$\mathbf{U}$, $\mathbf{U}_o$	Velocity vectors of translating sphere	$\mathbf{V}$	m s^{-1}	
$\mathbf{U}$	Uniform translation velocity of sphere	S	m s^{-1}	
U', U^*, U_o	Particle velocities	S	m s^{-1}	
$\langle \mathbf{U} \rangle$	Averaged sedimentation velocity	S	m s^{-1}	
U_B	Apparent diffusion velocity of particle	S	m s^{-1}	
$\mathbf{V}, \mathbf{V}'$	Fluid velocity vectors	$\mathbf{V}$	m s^{-1}	
$\bar{\mathbf{V}}$	Reduced fluid velocity vector	$\mathbf{V}$	[1]	
$\tilde{\mathbf{V}}(s)$	Laplace transform of fluid velocity vector	$\mathbf{V}$	m s^{-1}	
$\mathbf{V}_{Sh}$	Shear flow	$\mathbf{V}$	m s^{-1}	
$\mathbf{V}_{St}$	Stagnation flow	$\mathbf{V}$	m s^{-1}	
$\mathbf{V}_h$, $\mathbf{V}_r$, $\mathbf{V}_x$, $\mathbf{V}_y$, $\mathbf{V}_z$	Simple flow and reflection fields	$\mathbf{V}$	m s^{-1}	
$\mathbf{V}_\infty$	Macroscopic flow at infinity	$\mathbf{V}$	m s^{-1}	
$<V>$	Mean fluid velocity in channel	S	m s^{-1}	
$\tilde{V}(s)$	Laplace transform of fluid velocity	S	m s^{-1}	
V_h, V_r, V_x V_y, V_z	Components of fluid velocity	S	m s^{-1}	
V_i	Characteristic velocity of jet	S	m s^{-1}	
V_{mx}	Maximum velocity in channels	S	m s^{-1}	
V_n	Velocity expansion coefficients	S	m s^{-1}	
V_n'	Coefficients of resistance force series expansion	S	[1]	
V_0	Wall velocity	$\mathbf{V}$	m s^{-1}	
V_s	Moving plate velocity	S	m s^{-1}	
V_{st}	Stationary fluid velocity distribution	S	m s^{-1}	
V_{ξ_s}, V_{η_s}	Components of fluid velocity near spheroids	S	m s^{-1}	
V_ϑ	Angular velocity component	S	m s^{-1}	
V_∞	Characteristic fluid velocity far from interface	S	m s^{-1}	
V_1, V_2, V_3	Components of fluid velocity near sphere	S	m s^{-1}	

List of Symbols (continued)

Symbol	Definition	Character	Unit	Numerical value
v	Volume	S	m^3	
δv	Control volume, volume element	S	m^3	
X_n	Coefficients of resistance force series expansion	S	[1]	
$\mathbf{x}$, $\mathbf{x}_0$	Position vectors	V	m	
$\bar{\mathbf{x}}$	Relative position vectors	**V**	m	
x	Distance	S	m	
$\bar{x}$	Dimensionless distance	S	[1]	
(x_0, y_0)	Coordinates of point on trajectory			
x_p	Temporary position of particle in local flow	S	m	
x_s	Stagnation point position	S	m	
Y_n	Coefficients of resistance force series expansion	S	[1]	
y	Distance from wall	S	m	
Z_n	Coefficients of resistance force series expansion	S	[1]	
z	Coordinate perpendicular to interface	S	m	
Greek				
α_r	Flow parameter for radial impinging jet	S	[1]	
α_s	Flow parameter for slot impinging jet	S	[1]	
α_s'	Dimensionless parameter for two particle motion	S	[1]	
α_1, α_2, α_3	Parameters describing ellipsoid resistance	S	[1]	
β	Oblique flow direction	S	[1]	
β_s	Dimensionless parameter for two-particle motion	S	[1]	
Δ, Δ_0	Rate of strain tensor	**T**	s^{-1}	
γ	Interfacial tension	S	$N\ m^{-1}=J\ m^{-2}$	
$\bar{\delta}$	Relative deformation of sphere	S	[1]	
δ_h	Hydrodynamic boundary layer thickness	S	m	
ζ	Zeta potential of interface	S	V	
η	Dynamic viscosity of fluid	S	$kg\ m^{-1}\ s^{-1}$	

List of Symbols (continued)

Symbol	Definition	Character	Unit	Numerical value
$\bar{\eta}_1, \bar{\eta}_2\ \bar{\eta}_3$	Viscosity ratios	S	[1]	
η_f	Dynamic viscosity of fluid particle	S	kg m^{-1} s^{-1}	
κ_h	Bulk viscosity	S	kg m^{-1} s^{-1}	
λ_c	Coupling coefficients	S	[1]	
$\lambda_{f_\perp}$	Stokes' law correction factor for perpendicular motion	S	[1]	
$\lambda_{f_\parallel}$	Stokes' law correction factor for parallel motion	S	[1]	
λ_{ft}^a	Stokes' law correction factor for translational motion of an aggregate	S	[1]	
λ_i	Dimensionless constant describing non-stationary motion of spheres	S	[1]	
λ_0	Dimensionless parameter for spheroid	S	[1]	
λ_s	Dimensionless parameter for spheroid	S	[1]	
$\lambda_{r_\parallel}$	Stokes' law correction factor for rotational motion parallel to interface	S	[1]	
$\lambda_{r_\perp}$	Stokes' law correction factor for rotational motion perpendicular to interface	S	[1]	
ν	Kinematic viscosity	S	N m^2	
ξ	Integration variable, dimensionless angular variable	S	[1]	
ξ_s	Dimensionless parameter of spheroidal coordinates of	S	V	
$\mathbf{\Pi}_d$, $\mathbf{\Pi}_s$	Hydrodynamic stress tensors	**T**	N m^2	
Π_0	Tangential stress at surface	S	m^2 s^{-1}	
$\Pi_0(x)$	Local stress on plate	S	N m^{-2}	
π	Pi number	S	[1]	3.141593
ρ	Fluid density	S	kg m^{-3}	
ρ_e	Electric charge density	S	C m^{-3}	
τ	Dimensionless time	S	[1]	
ϑ	Angular coordinate	S	[1]	
Φ_B	Bernoulli function	S	m^2 s^{-2}	
φ	Phase shift angle	S	[1]	

List of Symbols (continued)

Symbol	Definition	Character	Unit	Numerical value
φ_0	Angle parameter of oblique flow	S	[1]	
ψ	Electric potential	S	V	
$\overline{\psi}$	Dimensionless potential	S	[1]	
ψ_s	Stream function	S	$m^2\ s^{-1}$	
ψ_{sf}	Stream function for internal flow in sphere	S	$m^2\ s^{-1}$	
$\boldsymbol{\omega}$	Fluid vorticity	**V**	s^{-1}	
$\boldsymbol{\omega}_o$	Fluid vorticity at fixed point	**V**	s^{-1}	
ω	Angular velocity	S	s^{-1}	
$\bar{\omega}$	Reduced angular velocity	S	[1]	
$\boldsymbol{\Omega}$	Fluid angular velocity	**V**	s^{-1}	
Ω	Angular velocity of rotating disk and oscillating particle	S	s^{-1}	
Ω_1, Ω_2, Ω_3	Angular velocities of rotating sphere, ellipsoid	S	s^{-1}	

Note: S = scalar; **V** = vector; **T** = tensor (matrix); [1] = dimensionless.

REFERENCES

[1] J.Happel, and H.Brenner, Low Reynolds Number Hydrodynamics, Martinus Nijhoff, The Hague, 1983.
[2] C.Pozrikidis, Introduction to Theoretical and Computational Fluid Dynamics, Oxford University Press, 1997.
[3] L.D.Landau and E.M.Lifshitz, Fluid Mechanics, Pergamon Press, 1959.
[4] L.M.H.Navier, Mem.Acad.Sci., 6 (1827) 389.
[5] H. Schlichting, Boundary Layer Theory, 7th Edn. Mc Grow- Hill, 1968.
[6] S.Kim and S.J.Karilla, Microhydrodynamics, Principles and Selected Applications, Butterworth-Heinemann, Boston 1991.
[7] H.Lamb, Hydrodynamics, 6th Ed., Dover, New York, 1945.
[8] D.A.Edwards, H.Brenner, D.T.Wasan, Interfacial Transport Processes and Rheology, Butterworth-Heinemann, 1989.
[9] V. G. Levich, Physicochemical Hydrodynamics, Prentice-Hall, Englewood Cliffs, 1962.
[10] W. B. Russel, D.A. Saville, and W.R.Schowalter, Colloidal Suspensions, Cambridge University Press, 1993.
[11] P.J.Scales, F.Grieser, Th.W.Healy, L.R.White and D.Y.C.Chan, Langmuir, 8 (1992) 965.

[12] Z.Adamczyk, P.Warszyński and M.Zembala, Bull.Pol.Ac.Chem., 47 (1999) 239.
[13] R.Pal, Rheology of Emulsions Containing Polymeric Liquids, p. 161 in Encyclopedia of Emulsion Technology, V4, P. Becher edit, Marcel Dekker, 1996.
[14] P. Warszyński, Adv. Colloid Interface Sci., 84 (2000) 47.
[15] M. Smoluchowski, Phys. Zeit., 17 (1916) 557.
[16] D.Li, Electrokinetics in Microfluidics, Elsevier 2004.
[17] H.S.Carslow, J.C.Jeager, Conduction of Heat in Solids, Oxford University Press 1959.
[18] M.Abramowitz, I.E.Steguin, Handbook of Mathematical Functions, Dover, 1972.
[19] K.Himenz, Dinglers Polyt. J., 326 1911) 321.
[20] F.Homann, Z.Angew.Mech., 16 (1936) 153
[21] Th. von Karman Z.Angew.Math.Mech., 1 (1921) 233.
[22] W.G.Cochran, Proc. Camb.Philos.Soc., 30 (1934) 365.
[23] Z.Adamczyk, T.T.Dąbroś, J.Czarnecki, and T.G.M. van de Ven, Adv. Colloid Interface Sci,. 19 (1983) 183.
[24] T. T.Dąbroś , and T.G.M. van de Ven, Colloid Polym. Sci., 261 (1983) 694.
[25] Z.Adamczyk, B.Siwek, and Warszyński, and E. Musiał, J. Colloid Interface Sci., 242 (2001) 14.
[26] Z.Adamczyk, B.Siwek, P.Warszyński, J. Colloid Interface Sci., 248 (2002) 244.
[27] F.H.Harlow and.Welch, Phys.Fluids, 8 (1965) 2182.
[28] Z.Adamczyk, Kinetics of Particle and Protein Adsorption, Ch.5 in " Surface and Colloid Science V.17, E.Matijevic, M.Borkovec Edits. Kluwer Academic, New York 2004.
[29] Z.Adamczyk, L.Szyk, and P.Warszyński, J.Colloid Interface Sci., 209 (1999) 350.
[30] L.Rosenhead, Laminar Boundary Layers, Oxford 1963.
[31] H.Evans, Laminar Boundary Layers, Addison-Weseley, 1968.
[32] H.Blasius, Z.Math.Phys., 56 (1908) 1.
[33] L.Howarth, Proc.Royal Soc. London, 164 (1938) 547.
[34] S.Pai, Viscous Flow Theory, van Nostrand, New York, 1956.
[35] Z.Adamczyk and. T.G.M. van de Ven, Chem.Engng.Sci., 37 (1982) 869.
[36] B.C.Sakiadis, A.I. Che.E.J., 7 (1961) 26.
[37] B.C.Sakiadis, A.I. Che.E.J., 7 (1961) 221.
[38] Z.Adamczyk, T.Dąbroś and T.G.M. van de Ven, Chem.Engng.Sci., 37 (1982) 1513.
[39] W.Schwarz and B.Caswell, Chem.Engng.Sci., 16 (1961) 338.
[40] A.Goransson and Ch.Tragardh, J.Colloid Interface Sci., 231 (2000) 228.
[41] W.J. Albery, J.Electroanal.Chem., 191 (1985) 1.
[42] M.B.Glauert, J.Fluid.Mech., 1 (1956) 625.
[43] C.Pozrikidis , Little Book of Streamlines, Academic Press, 1999.
[44] M. E. Weber and, and D. Paddock, J.Colloid Interface Sci., 94 (1983) 328.
[45] J.R.Blake, Proc.Camb.Phil.Soc., 70 (1971) 303.
[46] R.G.Cox, I.Y.Z.Zia, S.G.Mason, J.Colloid Interfaces Sci., 27(1968) 7.
[47] G.G.Poe and A.Acrivos, J.Multiphase Flow, 2 (1976) 365.
[48] W.Rybczyński, Bull.Inst.Acad.Sci.Cracovie (ser. A), (1911) 40.
[49] J.S.Hadamard, Compt.Rend.Acad.Sci., (Paris) 152 (1911) 1735: 154,(1912) 109.
[50] T.G.M. van de Ven, Colloid Hydrodynamics, Academic Press (1984).
[51] K.Małysa, Adv.Colloid Interface Sci., 40 (1992) 37.
[52] C.W.Oseen, Ark.f.Mat.Astr.og Fys. 6 (1911) 7.

[53] C.W.Oseen, Neuere Methoden und Ergebnisse in der Hydrodynamik, Akademische Verlagengesellschaft, 1927.
[54] R.A.Simpson, Phil.Trans.Roy.Soc., A182 (1891) 449.
[55] D.R.Breach, J.Fluid Mech., 10 (1961) 306.
[56] J.H.Masliyah, N.Epstein, J.Fluid Mech., 44 (1970) 493.
[57] A.Oberbeek, J.Reine Angw. Math., 81 (1876) 62.
[58] G.B.Jefferey, Proc. Proc.Roy.Soc. (London), A102 (1922)161.
[59] M.Stimson and G.B.Jeffery, Proc.Roy.Soc. (London), A111 (1926) 110.
[60] H.Brenner, Int.J.Multiphase Flow, 1 (1974) 195.
[61] C.M.Tchen, J.Appl.Phys., 25 (1954) 463.
[62] H.A.Lorenz, Abhandl.theoret.Phys., 1 (1906) 23.
[63] H.Faxen, Arkiv.Mat.Astron.Fys., 17 (1923) 1.
[64] M.Smoluchowski, Bull.Inter.Acad.Sci.Cracovie Ser.A., 1A (1911) 28.
[65] H.Faxen, Arkiv.Mat.Astron.Fys,. 19A (1925) 21.
[66] S.H.Lee, R.S. Chadwick, L.G.Leal, J. Fluid Mech., 93 (1979) 705.
[67] S.H.Lee and L.G.Leal, J.Fluid.Mech., 98 (1980) 193.
[68] S.Wakya, J.Phys.Soc.Japan, 12 (1957) 1130.
[69] V.D.Hopper and A.M.Grant, Aust.J.Chem.Research, 1 (1948) 28.
[70] M.Stimson and G.B.Jeffery, Proc.Roy.Soc., A111, (1926) 110.
[71] A.D.Maude, Brit.J.Appl.Phys., 12 (1961) 293.
[72] L.A.Spielman, J.Colloid Interface Sci., 33 (1970) 562.
[73] H.Brenner, Chem.Eng.Sci., 16 (1961) 242.
[74] Z.Adamczyk M.Adamczyk, T.G.M. van de Ven, J.Colloid Interface Sci., 96 (1983) 204.
[75] M.Elimelech, J.Gregory, X.Jia, R.A.Williams, Particle Deposition and Aggregation, Butterworth -Heinemann, 1996.
[76] E.Bart, Chem.Eng.Sci. 23 (1968) 193.
[77] R.G.Cox and H.Brenner, Chem.Eng.Sci., 22 (1967) 1753.
[78] K.Małysa, Bull.Pol.Ac.Chem, 35 (1987) 441.
[79] G.B.Jeffery, Proc.London.Math.Soc. Ser. 2, 14 (1915) 327.
[80] M.E. O'Neill, Mathematika, 11 (1964) 67.
[81] W.R.Dean and M.E. O'Neill, Mathematika 10 (1963) 13.
[82] A.J.Goldman, R.G.Cox and H.Brenner, Chem.Eng.Sci., 22 (1967) 637.
[83] K.Małysa and T.G.M. van de Ven, Int.J.Multiphase Flow, 12 (1986) 459.
[84] K.Małysa, T.G.M. van de Ven, J.Colloid Interface Sci. 107 (1985) 477.
[85] A.J.Goldman, R.G.Cox and H.Brenner, Chem.Eng.Sci., 22 (1967) 653.
[86] S.L. Goren, M..E.O'Neill, Chem.Eng.Sci., 26 (1971) 325.
[87] J. Visser , J.Colloid Interface Sci. 55 (1976) 664.
[88] M.E.O'Neill, Chem.Eng.Sci,. 23 (1968) 1293.
[89] S.L. Goren, J.Fluid Mech., 41 (1970) 619.

Chapter 4

Transfer of Particles to Interfaces – Linear Problems

4.1. THE FORCE BALANCE AND THE MOBILITY OF PARTICLES

In previous chapters we were concerned with a detailed description of various forces exerted on particles. These included electrostatic and van der Waals forces appearing owing to the presence of fixed and fluctuating charges. They possessed a potential character, depending solely on space variables, but not on the particle velocity, its direction, acceleration, or the motion of the dispersing medium. Therefore, a quantitative analysis was made possible by introducing a scalar potential, identified with the interaction energy of a particle system, whose gradient produced the force vector directly. Moreover, the potential was an additive quantity, i.e., the contributions stemming from electrostatic and van der Waals interaction, as well as gravity, could simply be added to obtain the net interaction energy.

In contrast, the hydrodynamic interactions analyzed in the previous chapter were dependent not only on the instantaneous velocity of particles but also on their direction, presence of interfaces, acceleration and the history of particle motion. Consequently, hydrodynamic interactions are of a dissipative character, i.e., it is impossible to specify any potential function describing them, dependent on space variables only.

The simultaneous occurrence of potential and dissipative interactions complicate any theoretical analysis of particle motion near interfaces, which is needed to predict particle transport and deposition kinetics. In order to resolve this discrepancy, one has to perform the force and torque balance instead of the potential energy balance. This seems more complicated because the force balance is based on the assumption of simple additivity, which is not always obvious as can be realized just by considering the fact that fluid velocity fields perturb charge distributions and vice versa. Force balance based on the additivity rule is a powerful approach when dealing with dilute systems of particles of larger size, where the diffusion and interaction among particles can be ignored. One obtains in this case a set of differential equations of the second order, called the trajectory equation, whose integration can be used to calculate particle trajectory and particle deposition rates under the inertial transport regime. This approach belongs to the group of Lagrangian methods of analyzing particle deposition problems.

When the inertial effects are neglected, which is usually the case with colloid particles, the trajectory equations simplify to first order differential equations, which can be easily integrated for various flows and geometries of the interfaces, e.g., planar, spherical, cylindrical. In this way the mass transfer rate to these interfaces and particle deposition kinetics can efficiently be calculated. Moreover, for negligible inertia, the force balance equations can be inverted, which produces the so-called mobility relationship, i.e., the dependence of particle velocity on the specific and hydrodynamic forces acting on it [1].

Mobility relationships can be directly used to calculate particle-migration velocity under the action of external forces such as gravitational or electrostatic forces. This allows one, for example, to determine sedimentation rates of particles of various shapes as well as their migration velocity under the action of electrostatic forces in ion-free media, governed by the Henry's law.

In principle, the force balance equations can be extended by incorporating the random force, having the character of a white noise, exerted on particles by the thermal motion of the suspending medium molecules. This leads to the Langevine-type stochastic equations, whose integration can be exploited to describe Brownian motion and for calculating the diffusion coefficients of a single particle. By introducing the probability density function, which characterizes the chance of finding a particle somewhere in space, one can formulate the Fokker-Planck and Smoluchowski equations governing the temporal and spacial evolution of this quantity.

Moreover, using Einstein's ingenuous method, one can derive directly the expression for the diffusion matrix for particles of arbitrary shape and formulate the constitutive dependence of particle flux on the gradient of chemical potential (particle concentration). Then, by exploiting the flux concept, one can formulate the general mass balance equation both for one- and multi-component systems. This equation, often called the diffusion equation, assumes the form of the second-order partial differential equation expressed in terms of the continuous particle concentration field. The solution of this equation with appropriate initial and boundary conditions allows one to calculate particle transfer rates in various situations, including the non-stationary transport and the case of no convection. The latter could not be treated by the Lagrangian methods exploiting the particle trajectory concept. Thus, the use of the diffusion equation approach, belonging to the class of Eulerian methods of analyzing particle transfer and deposition on interfaces, seems more universal and will be extensively used later on in this chapter.

Accordingly, the organization of this chapter is as follows: in the first section the force and torque balance equation will be formulated and its limiting forms are derived, in particular the inertial and inertialess trajectory equations. In the latter case, the general mobility relationship will be derived, describing particle velocity in terms of the inverted hydrodynamic resistance matrix (called the mobility matrix) and the net force and torque field acting on particles. This universal mobility relationship will be used to calculate particle migration velocity under the action of external forces such as gravitational or electrostatic forces. This allows one to determine the sedimentation rates of particles of various shapes as well as their migration velocity under electrostatic forces in ion-free media. By further exploiting this method, an inertialess trajectory analysis of particles moving under various flows near interfaces will be presented, with the aim of calculating particle velocities, and consequently particle deposition rates. The trajectory analysis will be extended to the more complex system of two particles at a planar interface that allows one to analyze the hydrodynamic scattering effect, playing a significant role in particle deposition phenomena.

In the next section, the force balance equations will be extended by incorporating the random force, exerted on particles by the suspending medium molecules. The Langevine-type stochastic equations of motion will be derived as well as the Fokker-Planck and Smoluchowski equations for the case of particle motion in an external force field. Diffusion coefficients will be calculated, and the Brownian motion of a single particle in the bulk and near interfaces will be analyzed in detail.

The general diffusion equation for particle systems will be derived next by using the Einstein's method. Limiting forms and particle transport regimes, in particular, the convective diffusion regime, as well as the diffusion regime will be considered. Methods of analytical solutions of these equations are discussed with the emphasis focused on the Laplace transformation method. The one-dimensional transport of particles throughout the surface layer adjacent to interfaces is considered that allows one to formulate the foundations of the surface boundary layer (SFBL) theory and appropriate boundary conditions for the bulk transport equations. The boundary conditions in the form of the perfect sink and the linear adsorption isotherm will be discussed.

In the next section solutions to problems concerning linear transport of particles to interfaces are extensively discussed, in particular non-stationary, diffusional transport to spherical and planar interfaces, convective diffusion transport for a variety of flows and interfaces, transport to uniformly accessible surfaces and in the impinging-jet cells, etc. Analytical and numerical expressions for various transport conditions and interface shape obtained within the framework of the convective diffusion theory are tabulated. The influence of the electrostatic interactions and external forces on the limiting flux and mass transfer coefficients for various transport conditions is also discussed.

4.1.1. The Mobility Matrix

By assuming additivity of all forces affecting particle motion one can formulate the linear and angular momentum balance equations in the general form

$$\begin{aligned} \frac{d\mathbf{p}_m}{dt} &= m_p \frac{d\mathbf{U}}{dt} = \mathbf{F} \\ \frac{d\mathbf{p}_a}{dt} &= \mathbf{I_n} \cdot \frac{d\mathbf{\Omega}}{dt} = \mathbf{To} \end{aligned} \tag{1}$$

where $\mathbf{p}_m = m_p \mathbf{U}$ is the linear momentum vector of the particle, $\mathbf{p}_a = \mathbf{I_n} \cdot \mathbf{\Omega}$ is the angular momentum vector, $\mathbf{I_n}$ is the moment of inertia tensor and $\mathbf{F}$, $\mathbf{To}$ are the net force and torque on particle given by

$$\begin{aligned} \mathbf{F} &= \mathbf{F}_s + \mathbf{F}_{ext} + \mathbf{F}_h + \mathbf{F}^* \\ \mathbf{To} &= \mathbf{To}_s + \mathbf{To}_{ext} + \mathbf{To}_h + \mathbf{To}^* \end{aligned} \tag{2}$$

$\mathbf{F}_s$ and $\mathbf{To}_s$ are the specific force and torque (comprising the double-layer and van der Waals forces), $\mathbf{F}_{ext}$ and $\mathbf{To}_{ext}$ are the external forces and torques (gravitational or electrostatic), $\mathbf{F}_h$ and $\mathbf{To}_h$ are the net hydrodynamic forces and torques (comprising the resistance and drag hydrodynamic contributions) and $\mathbf{F}^*$ and $\mathbf{To}^*$ are the force and torque contributions stemming from the interactions with the suspending medium.

In principle, by formulating the force and torque balance, Eq. (2) one could consider the interaction among particles as well. However, this would complicate enormously the mathematical analysis of particle motion, because instead of two, one would have to deal with a system of $2N_p$ coupled differential equation (where N_p is the number of particles considered).

By using the previously derived expressions for the hydrodynamic resistance force and torque given by Eq. (358) in Chapter 3, one can convert Eq. (1) to the form

$$\begin{aligned} m_p \frac{d\mathbf{U}}{dt} &= m_p \frac{d^2\mathbf{r}}{dt^2} = -f(\eta,\eta_f)(\mathbf{K}_t \cdot \mathbf{U} + \mathbf{K}_c^{\dagger} \cdot \boldsymbol{\Omega}) + \mathbf{F} + \mathbf{F}^*(t) \\ \mathbf{In} \frac{d\boldsymbol{\Omega}}{dt} &= \mathbf{In} \frac{d^2\boldsymbol{\varphi}}{dt^2} = -f(\eta,\eta_f)(\mathbf{K}_r \cdot \boldsymbol{\Omega} + \mathbf{K}_c \cdot \mathbf{U}) + \mathbf{To} + \mathbf{To}^*(t) \end{aligned} \tag{3}$$

where $\mathbf{r}$ is the position vector of the particle, $\boldsymbol{\varphi}$ the angular position vector, $\mathbf{F}$ and $\mathbf{To}$ are the force and torque on particle consisting of specific, external and hydrodynamic contributions.

It should be mentioned that according to estimates done in Chapter 3, Eq.(3) remains valid for time exceeding L_{ch}^2/ν (where L_{ch} is the characteristic length scale of the motion, and ν is the kinematic viscosity of the medium).

Since the hydrodynamic force and torque are usually coupled, Eq. (3) becomes, in a general case, a system of two non-linear coupled differential equations of the second order in respect to the space variables. Hence, it requires four initial conditions, which can be specified in the general form

$$\left.\begin{aligned} \mathbf{U} &= \mathbf{U}_0 \\ \boldsymbol{\Omega} &= \boldsymbol{\Omega}_0 \\ \mathbf{r} &= \mathbf{r}_0 \\ \boldsymbol{\varphi} &= \boldsymbol{\varphi}_0 \end{aligned}\right\} \quad \text{for } t = 0 \tag{4}$$

where $\mathbf{U}_0$ and $\boldsymbol{\Omega}_0$ are the initial velocity and angular velocities of the particle and $\mathbf{r}_0$ and $\boldsymbol{\varphi}_0$ are the initial position and angular positions.

Eq. (3), describing the stochastic motion of particles of arbitrary shapes when the coupling between the translational and rotational motion occurs, is referred to as the generalized Langevine equation [2].

For solid spherical particles moving in a quiescent medium far from interfaces Eq. (3) reduces to the usual form of the Langevine equation [3]

$$\frac{d\mathbf{U}}{dt}=\frac{d^2\mathbf{r}}{dt^2}=-\frac{\eta K_t}{m_p'}\mathbf{U}+\frac{1}{m_p'}\mathbf{F}^*(t)=-\beta\mathbf{U}+\frac{1}{m_p'}\mathbf{F}^*(t) \tag{5}$$

where $m'_p = m_p + \frac{1}{2}m_f$ is the effective mass of the particle, m the mass of the fluid $K_t = 6\pi a$ and

$$\frac{1}{\beta}=\frac{m_p'}{\eta K_t}=\frac{2}{9}\frac{\left(\rho_p+\frac{1}{2}\rho\right)}{\eta}a^2 \tag{6}$$

the important parameter having the dimension of time. It characterizes the time of persistence of particle velocity (direction of motion) after sudden velocity change. This is discussed in more detail later on.

If the random forces due to the suspending medium are neglected, Eq. (3) reduces to the deterministic trajectory equation, having the form

$$\begin{aligned} m_p\frac{d\mathbf{U}}{dt}&=-f(\eta,\eta_f)\left(\mathbf{K}_t\cdot\mathbf{U}+\mathbf{K}_c^{\dagger}\cdot\mathbf{\Omega}\right)+\mathbf{F}\\ \mathbf{In}\frac{d\mathbf{\Omega}}{dt}&=-f(\eta,\eta_f)\left(\mathbf{K}_r\cdot\mathbf{\Omega}+\mathbf{K}_c\cdot\mathbf{U}\right)+\mathbf{To}\end{aligned} \tag{7}$$

If moreover, the particle inertia effects can be ignored, as is usually the case with colloids, Eq. (7) further simplifies to the inertia less trajectory equation

$$\begin{aligned} f(\eta,\eta_f)\left(\mathbf{K}_t\cdot\mathbf{U}+\mathbf{K}_c^{\dagger}\cdot\mathbf{\Omega}\right)&=\mathbf{F}\\ f(\eta,\eta_f)\left(\mathbf{K}_r\cdot\mathbf{\Omega}+\mathbf{K}_c\cdot\mathbf{U}\right)&=\mathbf{To}\end{aligned} \tag{8}$$

This can also be written in the concise form using the previously introduced notation (cf. Eqs.(360–362) in Chapter 3).

$$f(\eta,\eta_f)\tilde{\mathbf{K}}\cdot\tilde{\mathbf{U}} = \tilde{\mathbf{F}} = -\tilde{\mathbf{F}}' \tag{9}$$

where **U** is the generalized volecity

$$\tilde{\mathbf{F}} = \begin{vmatrix} \mathbf{F} \\ \mathbf{To} \end{vmatrix} \tag{10}$$

is the generalized force acting on a particle,

$$\tilde{\mathbf{F}}' = \begin{vmatrix} \mathbf{F}'_h \\ \mathbf{To}'_h \end{vmatrix} \tag{11}$$

is the generalized hydrodynamic resistance force and $\tilde{\mathbf{K}}$ is the grand resistance matrix defined by Eq. (361) in Chapter 3.

Eq. (9) can be formally inverted, which furnishes an important relationship

$$\tilde{\mathbf{U}} = \frac{1}{f(\eta,\eta_f)}\ \tilde{\mathbf{K}}^{-1}\cdot\tilde{\mathbf{F}} = \tilde{\mathbf{M}}\cdot\tilde{\mathbf{F}} \tag{12}$$

where $\tilde{\mathbf{M}}$ denotes the mobility matrix given by

$$\tilde{\mathbf{M}} = \frac{1}{f(\eta,\eta_f)}\ \tilde{\mathbf{K}}^{-1} \tag{13}$$

If one is interested in calculating the translation velocity only, for example for a spherical particle moving near a planar interface, one can derive from Eq. (8), by eliminating the torque, the following expression

$$\mathbf{U} = \frac{d\mathbf{r}}{dt} = \mathbf{M}\cdot\mathbf{F} + \mathbf{M}\cdot\mathbf{K}_c^{\dagger}\cdot\mathbf{K}_r^{-1}\cdot\mathbf{To} = \mathbf{M}\cdot\mathbf{F} + \mathbf{M}_c\cdot\mathbf{To} \tag{14}$$

where the translational and rotational mobility matrices are given by

$$\begin{aligned} \mathbf{M} &= \frac{1}{f(\eta,\eta_f)}(\mathbf{K}_t - \mathbf{K}_c^{\dagger}\cdot\mathbf{K}_r^{-1}\cdot\mathbf{K}_c)^{-1} \\ \mathbf{M}_c &= -\mathbf{M}\cdot\mathbf{K}_c^{\dagger}\cdot\mathbf{K}_r^{-1} \end{aligned} \tag{15}$$

By using the previously derived expressions for $\mathbf{K}_t$, $\mathbf{K}_c$, $\mathbf{K}_c^{\dagger}$ and $\mathbf{K}_r$ one can formulate the expression for the mobility matrix of solid spherical particles near planar interface in the form

$$\mathbf{M} = \begin{vmatrix} M_{||} & & 0 \\ & M_{||} & \\ 0 & & M_{\perp} \end{vmatrix}$$

$$\mathbf{M}_c = \begin{vmatrix} 0 & -K_c' & 0 \\ -K_c' & 0 & 0 \\ 0 & 0 & 0 \end{vmatrix} \tag{16}$$

where

$$M_{||} = \frac{1}{\eta\left(K_{t_{||}} - \dfrac{K_c^2}{K_{r_{||}}}\right)} = \frac{1}{6\pi\eta a}\frac{F_4}{\left[1-\dfrac{2}{3}\dfrac{F_4F_6}{F_7^2}\right]} = \frac{1}{6\pi\eta a}F_4'$$

$$M_{\perp} = \frac{1}{6\pi\eta K_{t_{\perp}}} = \frac{1}{6\pi\eta a}F_1$$

$$K_c' = \frac{K_c M_{||}}{K_{r_{||}}} = \frac{1}{8\pi\eta a^2}\frac{F_6 F_4'}{F_7} \tag{17}$$

$$F_4' = \frac{F_4}{1-\dfrac{2}{3}\dfrac{F_4F_6}{F_7^2}}$$

and F_1, F_4 , F_5 and F_7 are the universal hydrodynamic correction functions defined by Eq. (407) in Section 3.4.4.2.

On the other hand, for orthotropic bodies, e.g., ellipsoids, far from interfaces, $\mathbf{K}_c = 0$ and the expression for the translational mobility matrix assumes the form

$$\mathbf{M} = \begin{vmatrix} M_{11} & & 0 \\ & M_{22} & \\ 0 & & M_{33} \end{vmatrix} \tag{18}$$

where

$$M_{11} = \frac{\chi_0 + \alpha_1 a^2}{16\pi\eta bc} == \frac{1}{6\pi\eta K_1}$$

$$M_{22} = \frac{\chi_0 + \alpha_2 b^2}{16\pi\eta ac} = \frac{1}{6\pi\eta K_2} \tag{19}$$

$$M_{33} = \frac{\chi_0 + \alpha_3 c^2}{16\pi\eta ab} = \frac{1}{6\pi\eta K_3}$$

and K_1, K_2 and K_3 are the components of the resistance matrix, expressed by Eqs. (329–331) in Chapter 3.

For spheroids the expression for **M** simplifies to a form analogous to the case of a sphere near a planar interface, i.e.,

$$\mathbf{M} = \begin{vmatrix} M_{11} & & 0 \\ & M_{22} & \\ 0 & & M_{22} \end{vmatrix} \tag{20}$$

where

$$M_{11} = \frac{\chi_0 + \alpha_1 a^2}{16\pi\eta b^2} = \frac{1}{6\pi\eta a K_1}$$

$$M_{22} = \frac{\chi_0 + \alpha_2 b^2}{16\pi\eta a^2} = \frac{1}{6\pi\eta b K_2} \tag{21}$$

where the resistance coefficients are given by Eq. (336) for prolate and Eq. (341) for oblate spheroids, derived in Chapter 3.

Finally, for spherically isotropic bodies such as spheres, cubes octahedral, etc., the mobility matrix reduces to the simple form

$$\mathbf{M} = \begin{vmatrix} M_{11} & & 0 \\ & M_{11} & \\ 0 & & M_{11} \end{vmatrix} \tag{22}$$

where for spheres one has $M_{11} = 1/6\pi\eta a$.

4.2. MIGRATION OF PARTICLES IN EXTERNAL FIELDS

The general mobility relationships derived above, i.e., the dependencies of particle velocity on the forces applied, can be exploited for calculating several effects of basic practical significance for particle transfer and deposition. One of the most important of these is particle motion under the action of external forces, which is often referred to as the migration effect. In the case of the gravitational force, the migration effect is called sedimentation, which is playing an essential role in many large scale natural and industrial processes. Also particle migration in an external electric field plays an important role, especially in gaseous or ion-free media, where it is possible to apply large field strength without causing electrolysis. Both these effects will be analyzed in some detail in this section.

For the sake of simplicity we limit our discussion to dilute systems, where inter-particle interactions, both specific and hydrodynamic, can be ignored. Our attention is focused on orthotropic bodies of uniform density or uniform charge distribution where the center of mass and the center of buoyancy coincide.

Let us first consider orthotropic solid bodies moving in a quiescent fluid far from boundary surfaces. In this case $\mathbf{K}_c = 0$ and the mobility matrix has the diagonal form given by Eq. (18). The inertialess trajectory equation, governing particle velocity, derived from Eq. (14), has the form

$$\mathbf{U} = \frac{d\mathbf{r}}{dt} = \mathbf{M} \cdot \mathbf{F} \tag{23}$$

Let us assume that $\mathbf{F}$ consists of the gravitational force whose net effect on a particle of density ρ_p immersed in a medium of density ρ is described by

$$\mathbf{F} = m_p\,\mathbf{g} - m\mathbf{g} = v_p(\rho_p - \rho)\,\mathbf{g} \tag{24}$$

where the second term describes the buoyant force, m_p is the particle mass and m is the mass of the displaced fluid.

By substituting Eq. (24) into Eq. (23) one obtains the formula [1]

$$\mathbf{U} = v_p(\rho_p - \rho)\,\mathbf{M}\cdot\mathbf{g} = \frac{v_p\Delta\rho}{\eta}\mathbf{K}_t^{-1}\cdot\mathbf{g} \tag{25}$$

where $\Delta\rho = \rho_p - \rho$

Eq. (25), describing the stationary (terminal) velocity of particles in the gravitational field, is referred to as the Stokes' sedimentation law.

If **F** is due to the external electric field **E**, it can be described, in salt-free media, by the Lorenz equation, derived in Chapter 2

$$\mathbf{F} = q\mathbf{E} = C_e\,\psi^0\,\mathbf{E} \tag{26}$$

where q is the electric charge on the particle, C_e is the electric capacity of the particle and ψ^0 is the surface potential of the particle.

Using Eq. (26) the velocity of a particle in an electric field is given by

$$\mathbf{U} = q\mathbf{M}\cdot\mathbf{E} = C_e\psi^0\mathbf{M}\cdot\mathbf{E} \tag{27}$$

Eq. (27) is often referred to as Henry's law. It describes the migration velocity, often called electrophoretic motion, of charged bodies in electric fields when the free charge (ion) concentration is low, so the screening length *Le* becomes much larger than particle dimension. If the concentration of ions increases, the double-layer thickness becomes comparable to or smaller than particle dimensions, in this case, a proper calculation of the electrophoretic velocity of particles requires in addition the ion mass balance equation [2,3].

It is interesting to note that the particle translation velocity **U**, is expressed in the coordinate system associated with the particle rather than with a space-fixed coordinate system. This is so, because the mobility matrix **M** has a diagonal form only in the particle-fixed frame of references with axes parallel to the major axes of the body. If the major axes are not parallel to the direction of the external force, then it is necessary to transform the velocity from the particle- to the space-fixed coordinate system. This can be done by introducing the matrix **Tr** allowing for the transformation from the space-fixed (x,y,z) to the particle fixed (x',y',z') Cartesian coordinate systems. The external force is transformed as

$$\mathbf{F}' = \mathbf{Tr}\cdot\mathbf{F}$$
$$\mathbf{F} = \mathbf{Tr}^{-1}\cdot\mathbf{F}' \tag{28}$$

where $\mathbf{Tr}^{-1}$ is the inverse transformation matrix.

Both **Tr** and $\mathbf{Tr}^{-1}$ can be expressed via direction cosines of the angles formed by the corresponding axes of the two coordinate systems or in terms of the three Eulerian angles $(\phi_e,\psi_e,\vartheta_e)$ specifying the orientation of the particle relative to the direction of the external force [1].

Using Eq. (28) one can derive the following expression for the particle translation velocity in the coordinate system parallel to the direction of the gravitational force.

$$\mathbf{U} = \mathbf{Tr}^{-1}\cdot\mathbf{M}\cdot\mathbf{Tr}\cdot\mathbf{F} \tag{29}$$

Since **U** in the above equation depends on the orientation of the body relative to the direction of the external force, it is useful to calculate the average migration velocity. This can be done, in analogy to the calculations of the average dipole moment, using the general expression, Eq. (256) in Chapter 2. Thus, the average migration velocity is given by

$$\langle\mathbf{U}\rangle = \frac{1}{8\pi^2}\int_{\Omega_c} \mathbf{U}(\phi_e,\psi_e,\vartheta_e) f_p(\phi_e,\psi_e,\vartheta_e)\, d\Omega_c \tag{30}$$

where $f_p(\phi_e,\psi_e,v_e)$ is the probability density function of finding a particle orientation described by the Eulerian angles and Ω_c is the configurational space domain. Assuming that f_p is a uniform function of these angles and using Eq. (29) one can express Eq. (30) as

$$\langle\mathbf{U}\rangle = \frac{1}{8\pi^2}\int_0^{\pi}\int_0^{2\pi}\int_0^{2\pi} \mathbf{Tr}^{-1}(\phi_e,\psi_e,\vartheta_e)\cdot\mathbf{M}\cdot\mathbf{Tr}(\phi_e,\psi_e,\vartheta_e)\cdot\mathbf{F}\sin\vartheta\, d\vartheta_e d\psi_e d\phi_e \tag{31}$$

By integrating it can be shown that [1]

$$\langle \mathbf{U} \rangle = \langle M \rangle \, \mathbf{F} \tag{32}$$

where the average mobility, $\langle M \rangle$ being a scalar quantity, is given by

$$\langle M \rangle = \frac{1}{3}(M_{11} + M_{22} + M_{33}) = \frac{1}{\eta}\langle K \rangle^{-1} = \frac{1}{3\eta}\left(\frac{1}{K_1} + \frac{1}{K_2} + \frac{1}{K_3}\right)^{-1} \tag{33}$$

and $\langle K \rangle$ is the average resistance coefficient for bodies of various shape, listed in Table 3.5.

As can be noticed from Eq. (32), the average direction of particle motion coincides with the direction of the external force.

Knowing the translation velocity of a body, one can easily derive its trajectory by simple integration with respect to time t. In this way, by integrating Eq. (23), one obtains

$$\mathbf{r} = \mathbf{r}_0 + \mathbf{U}t \tag{34}$$

where $\mathbf{r}$ is the position vector in the space fixed coordinate system and $\mathbf{r}_0$ is the initial position of the particle at $t = 0$.

As can be noticed, Eq. (34) represents the parametric equation of a straight line in space. Thus, as expected, particle trajectories are described by lines parallel to the direction of the translation velocity $\mathbf{U}$. It is also interesting to note that Eq. (34) remains invariant upon the substitution of $\mathbf{U} = -\mathbf{U}$ and $t = -t$. This means that particle motion induced by external forces remains reversible in the sense that the change in the direction of forces reverses the direction of particle of motion and consequently its position in space.

Note that under the assumption of inertialess particle motion, Eqs. (29) and (34) remain valid if the mobility matrix and the external force become position dependent. This is the case, e.g., for spherical particles moving near planar interfaces when the mobility matrix, described by Eq. (16) is a function of the distance of the particle from the surface. Obviously, the *quasi-static* particle velocity becomes dependent on the distance in this case, so the particle trajectory equation, Eq. (34) attains the form

$$\mathbf{r} = \mathbf{r}_0 + \mathbf{U}(\mathbf{r})t \tag{35}$$

As can be noticed, Eq. (35) is implicit, which means that in general, particle trajectories are curvilinear, except for some limiting cases analyzed below.

Let us apply the above analysis to concrete situations of practical significance. In the case of spherically isotropic bodies, such as spheres, cubes, tetrahedrons, octahedrons, etc. the mobility matrix assumes the simple form given by Eq. (22) $\mathbf{M} = \mathrm{M}_{11}\,\mathbf{I}$. The single mobility coefficient M_{11} can be expressed for the spherically isotropic bodies in the general form [1]

$$M_{11} = \frac{1}{\eta} K_t^{-1} = \frac{2}{9}\frac{K_s\, a_s^2}{\eta v_p} \tag{36}$$

where K_s is the sphericity coefficient given by [1]

$$K_s = \frac{\log(\psi_i/0.065)}{\log(1/0.065)} = 0.842 \log\left(\frac{\psi_i}{0.065}\right) \tag{37}$$

where the shape parameter

$$\psi_i = \frac{4\pi a_s^2}{S_p} \tag{38}$$

is the ratio of the surface area of the equivalent sphere to the actual surface area of the particle S_p. The equivalent sphere is defined as having the same volume as the particle v_p. Thus, the radius a_s is given by

$$a_s = \left(\frac{3v_p}{4\pi}\right)^{1/3} \tag{39}$$

By considering the above definitions one obtains from Eq. (25) the following sedimentation law for spherically isotropic bodies

$$\mathbf{U} = \frac{2}{9} K_s \frac{\Delta\rho\, a_s^2}{\eta}\mathbf{g} \tag{40}$$

Analogously, by using Eq. (27) one obtains the migration velocity under electric fields

$$\mathbf{U} = \frac{2}{9} K_s \frac{a_s^2}{\eta v_p} q \mathbf{E} \tag{41}$$

For solid spheres $K_s = 1$ and Eq. (40) becomes

$$\mathbf{U} = \frac{2}{9} \frac{\Delta \rho a^2}{\eta} \mathbf{g} \tag{42}$$

Eq. (42) is the well-known Stokes' formula indicating that the sedimentation velocity of spherical particles is proportional to the square of the particle radius and is inversely proportionally to the fluid viscosity.

In the case of fluid spheres, where the mobility coefficient $M_{11}=(1+\eta/\eta_f)\,/6\pi\eta a(1+2\eta/3\eta_f)$, Eq. (42) becomes

$$\mathbf{U} = \frac{2}{9} \frac{(1+\eta/\eta_f)}{(1+2\eta/3\eta_f)} \frac{\Delta \rho a^2}{\eta} \mathbf{g} \tag{43}$$

Interestingly enough, the sedimentation velocity becomes negative if the particle density is smaller than the fluid density. This means that the particle will rise in the direction opposite to that of the gravitational force. This is often the case with liquid drops (emulsions) and gas bubbles.

Analogously, the migration velocity for a sphere under an electric field is

$$\mathbf{U} = \frac{1}{6\pi\eta a} q \mathbf{E} = \frac{2}{3} \frac{\varepsilon \psi^0}{\eta} \mathbf{E} \tag{44}$$

The sedimentation and migration velocities of solid spherical particles of various sizes are compiled in Table 4.1.

In the case of fluid spheres, one has

$$\mathbf{U} = \frac{2}{3} \frac{(1+\eta/\eta_f)}{(1+2\eta/3\eta_f)} \frac{\varepsilon \psi^0}{\eta} \mathbf{E} \tag{45}$$

Note that Eq. (45) remains accurate as long as drop deformations remain small, i.e., when the parameter $\Delta_d = \frac{9\varepsilon}{16\gamma} |\mathbf{E}|^2 a < 1$ (where γ is the interfacial tension). By taking $\varepsilon = 7.1\times10^{-10}$ F m^{-1}, $a = 10^{-6}$ m, $|\mathbf{E}| = 10^4$ V m^{-1}, $\gamma = 30$ m N m^{-1} one has $\Delta_d = 1.33\times10^{-6}$.

Eq. (44) and (45) are the well-known Henry formula describing the electrophoretic motion of spherical particles in an electric field. A unique feature of these formulae is that in contrast to sedimentation, the migration velocity in the electric field does not depend on particle size. This has profound practical significance indicating that the separation of colloid and smaller sized particles, which cannot be produced by sedimentation, may quite feasibly happen in the electric field. This is evident from the data compiled in Table 4.1.

In the case of cubes, where $a_s = (3/4\pi)^{1/3} d$ (where d is the size of the cube), $\psi_i = 0.806$ and the coefficient $K_s = 0.921$. By considering this, one obtains from Eq. (40) the formula for the sedimentation rate of cube-shaped particles

$$\mathbf{U} = 0.0788 \frac{\Delta\rho d^2}{\eta} \mathbf{g} \tag{46}$$

As can be noticed, the sedimentation velocity is again proportional to the square of the particle dimension, in a similar way to spheres. This is also the case for all spherically isotropic bodies.

Analogously, using Eq. (41), one obtains the migration velocity of the cube-shaped particles in electric fields

$$\mathbf{U} = 0.0788 \frac{C_e \psi^0}{\eta d} \mathbf{E} \tag{47}$$

Table 4.1
The sedimentation and migration velocities of spherical particles of various sizes and surface potential in water and in air

a (nm)	Sedimentation velocity (m s^{-1})		ψ^0 (mV)	Migration velocity (m s^{-1})	
	Water	Air		Water	Air
1	2.18×10^{-12}	2.41×10^{-10}	1	4.72×10^{-8}	3.26×10^{-6}
10	2.18×10^{-10}	2.41×10^{-8}	10	4.72×10^{-7}	3.26×10^{-5}
100	2.18×10^{-8}	2.41×10^{-6}	100	4.72×10^{-6}	3.26×10^{-4}
10^3	2.18×10^{-6}	2.41×10^{-4}	10^3	4.72×10^{-5}	3.26×10^{-3}
10^4	2.18×10^{-4}	2.41×10^{-2}	10^4	4.72×10^{-4}	3.26×10^{-2}

Notes: T = 293 K, $\Delta\rho = 10^3$ kg m^{-3}, $\eta = 10^{-3}$ kg m^{-1} s^{-1}, $|\mathbf{E}| = 10^2$ V m^{-1} (water), $\Delta\rho = 2(10^3$ kg m^{-3}, $\eta = 1.81\times10^{-5}$ kg (m s)$^{-1}$, $|\mathbf{E}| = 10^4$ V m^{-1} (air)

In the case of orthotropic bodies, such as ellipsoids and spheroids, the situation becomes more complicated because the migration velocity depends on the orientation of the particle with respect to the external force. As mentioned, the orientation can be expressed in terms of the three Eulerian angles ($\phi_e, \psi_e, \vartheta_e$). By setting $\phi_e = 0$, $\psi_e = 0$ and $\vartheta_p = 0$, the three basic modes of particle motion can be expressed, i.e., when the major axes of the moving ellipsoid coincide with the direction of the external force. The sedimentation velocities along these axes are given by

$$
\begin{aligned}
U_1 &= \frac{4}{3}\pi abc\,\Delta\rho\, M_{11}\,|\mathbf{g}| = \frac{\Delta\rho\, a}{12\eta}(\chi_0 + \alpha_1 a^2)|\mathbf{g}| \\
U_2 &= \frac{4}{3}\pi abc\,\Delta\rho\, M_{22}\,|\mathbf{g}| = \frac{\Delta\rho\, b}{12\eta}(\chi_0 + \alpha_2 b^2)|\mathbf{g}| \\
U_3 &= \frac{4}{3}\pi abc\,\Delta\rho\, M_{33}\,|\mathbf{g}| = \frac{\Delta\rho\, c}{12\eta}(\chi_0 + \alpha_3 c^2)|\mathbf{g}|
\end{aligned}
\tag{48}
$$

In the case of prolate spheroids (where $b = c < a$), one can formulate Eq. (48) explicitly as

$$
\begin{aligned}
U_1 &= \frac{2}{9}\frac{\Delta\rho b^2}{\eta K_1}|\mathbf{g}| \\
U_2 = U_3 &= \frac{2}{9}\frac{\Delta\rho ab}{\eta K_2}|\mathbf{g}|
\end{aligned}
\tag{49}
$$

where the resistance coefficients $K_1(a/b)$, $K_2(a/b)$ are given by Eqs. (336–341) in Chapter 3. The average sedimentation velocity for prolate spheroids is

$$
\langle \mathbf{U} \rangle = \frac{\Delta\rho\, 2ab\left(2K_1 + \frac{b}{a}K_2\right)}{27\eta K_1 K_2}\mathbf{g} = \frac{2}{9}\frac{\Delta\rho\, b^2 \cosh^{-1}(a/b)}{\eta(1 - b^2/a^2)^{1/2}}\mathbf{g} \tag{50}
$$

Analogously, in the case of an electric field, the average migration velocity of prolate spheroids is given by

$$\langle \mathbf{U} \rangle = \frac{\left(2K_1 + \frac{b}{a} K_2\right)}{18\pi b K_1 K_2} q\,\mathbf{E} = \frac{\cosh^{-1}(a/b)}{6\pi\eta a(1-b^2/a^2)^{1/2}} q\mathbf{E} \tag{51}$$

In the limit of very elongated spheroids, when $a/b \gg 1$, the average sedimentation velocity becomes

$$\langle \mathbf{U} \rangle = \frac{2}{9} \frac{\Delta\rho \ln(2a/b) b^2}{\eta} \mathbf{g} \tag{52}$$

Eq. (52) can be used for predicting sedimentation of slender bodies of length $l = 2a$.

As can be noticed from Eq. (52), the sedimentation velocity is a more complicated function of particle dimensions than it is for spheres. It increases proportionally to $b^2 \ln(2a/b)$, which means that an increasing in the length of a particle will increase its sedimentation velocity in a logarithmic way only.

Analogously, in the case of an electric field, the average migration velocity of slender bodies is given by

$$\langle \mathbf{U} \rangle = \frac{\ln(2a/b)}{3\pi\eta l} q\,\mathbf{E} \tag{53}$$

For oblate spheroids (where $b < a = c$), the averaged sedimentation velocity is

$$\langle \mathbf{U} \rangle = \frac{2}{27} \frac{\left(2K_1 + \frac{b}{a} K_2\right) a^2}{\eta K_1 K_2} \mathbf{g} = \frac{2}{9} \frac{\Delta\rho\, ab \cos^{-1}(b/a)}{\eta(1-b^2/a^2)^{1/2}} \mathbf{g} \tag{54}$$

The averaged migration velocity of oblate spheroids is

$$\langle \mathbf{U} \rangle = \frac{2K_1 + \frac{b}{a} K_2}{18\pi\eta a K_1 K_2} q\,\mathbf{E} = \frac{\cos^{-1}(b/a)}{6\pi\eta a(1-b^2/a^2)^{1/2}} q\mathbf{E} \tag{55}$$

In the limit of very flat spheroids (disks), where $b/a \ll 1$, the average sedimentation velocity is

$$\langle \mathbf{U} \rangle = \frac{\Delta\rho\ \pi ab}{9\eta}\mathbf{g} \tag{56}$$

The average migration velocity for disks in the electric field is

$$\langle \mathbf{U} \rangle = \frac{1}{12\eta a} q\,\mathbf{E} \tag{57}$$

For the sake of convenience particle sedimentation and migration velocities for bodies of practical significance are compiled in Table 4.2.

Note that the formulae derived in this section describing sedimentation and migration velocities are strictly valid only for isolated bodies, i.e., in the limit of an infinitely low concentration of particles. For particle systems of low but finite concentration, there occur additional effects stemming from two-particle hydrodynamic and specific interactions. This results in a decrease in the sedimentation and migration velocities of particles in suspensions. In the case of spheres, the first- order correction was determined by Batchelor [4] who neglected all specific interactions and assumed that particle positions in the suspension are not correlated. The expression for the average sedimentation or migration velocity of particles in a suspension $\langle \mathbf{U} \rangle$ has the form

$$\langle \mathbf{U} \rangle = \mathbf{U}(1 - 6.55\Phi_v) \tag{58}$$

where $\langle \mathbf{U} \rangle$ is the averaged single-particle velocity calculated above and

$$\Phi_v = \frac{4}{3}\pi a^3 n_b \tag{59}$$

is the volume fraction of the spheres in the suspension having the number concentration n_b.

As can be noticed from Eq. (58), the sedimentation velocity of suspensions decreases proportionally to the volume fraction. The main contribution ($-5.5\ \Phi_v$) to this effect stems from the counter-flow of the fluid induced by

Table 4.2
Sedimentation and migration velocity in the electric field E under the Stokes regime for orthotropic bodies of various shapes

Body	Sedimentation velocity $\mathbf{U}_s$	Averaged $\langle \mathbf{U}_s \rangle$	Migration velocity $\mathbf{U}_m$	Averaged $\langle \mathbf{U}_m \rangle$
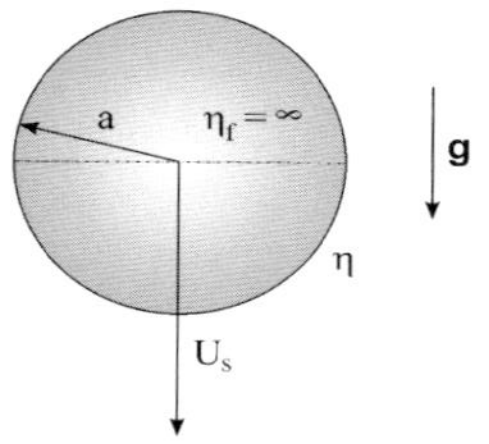 Solid sphere	$\frac{2}{9}\frac{\Delta\rho a^2}{\eta}\mathbf{g}$	$\frac{2}{9}\frac{\Delta\rho a^2}{\eta}\mathbf{g}$	$\frac{q}{6\pi\eta a}\mathbf{E}$ $\frac{2}{3}\frac{\varepsilon\psi^0}{\eta}\mathbf{E}$	$\frac{2}{3}\frac{\varepsilon\,\psi^0}{\eta}\mathbf{E}$
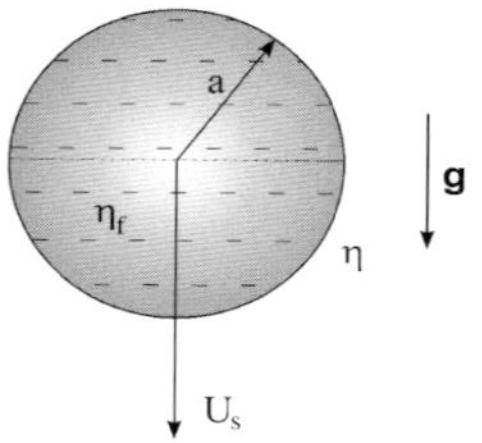 Liquid sphere	$\frac{2}{9}\tilde{\eta}\frac{\Delta\rho a^2}{\eta}\mathbf{g}$	$\frac{2}{9}\tilde{\eta}\frac{\Delta\rho a^2}{\eta}\mathbf{g}$	$\frac{2}{3}\tilde{\eta}\frac{\varepsilon\psi^0}{\eta}\mathbf{E}$	$\frac{2}{3}\tilde{\eta}\frac{\varepsilon\psi^0}{\eta}\mathbf{E}$
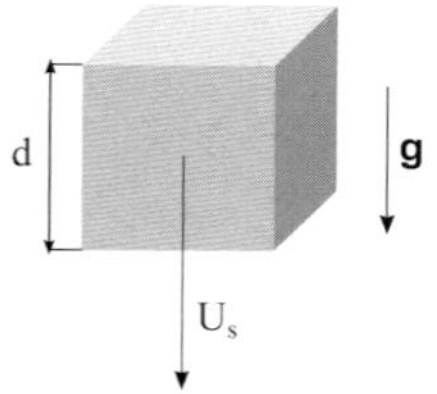 Cube	$0.0788\frac{\Delta\rho d^2}{\eta}\mathbf{g}$	$0.0788\frac{\Delta\rho d^2}{\eta}\mathbf{g}$	$0.0788\frac{q}{\eta d}\mathbf{E}$	$0.0788\frac{q}{\eta d}\mathbf{E}$

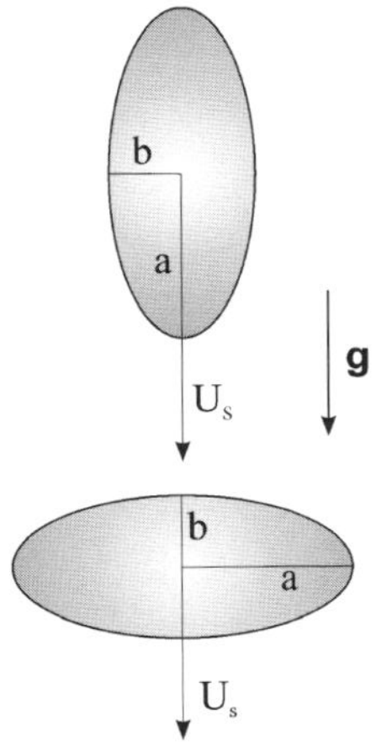	$\frac{2}{9}\Delta\rho\frac{b^2}{\eta K_1}\lvert\mathbf{g}\rvert$		$\frac{q}{6\pi\eta a K_1}\lvert\mathbf{E}\rvert$	
		$\frac{2}{9}\frac{\Delta\rho b^2\cosh^{-1}(a/b)}{\eta(1-b^2/a^2)^{1/2}}$		$\frac{\cosh^{-1}(b/a)q\,\mathbf{E}}{6\pi\eta a(1-b^2/a^2)^{1/2}}$
	$\frac{2}{9}\Delta\rho\frac{ab}{\eta K_2}\lvert\mathbf{g}\rvert$		$\frac{q}{6\pi\eta b K_2}\lvert\mathbf{E}\rvert$	
Prolate spheroids				
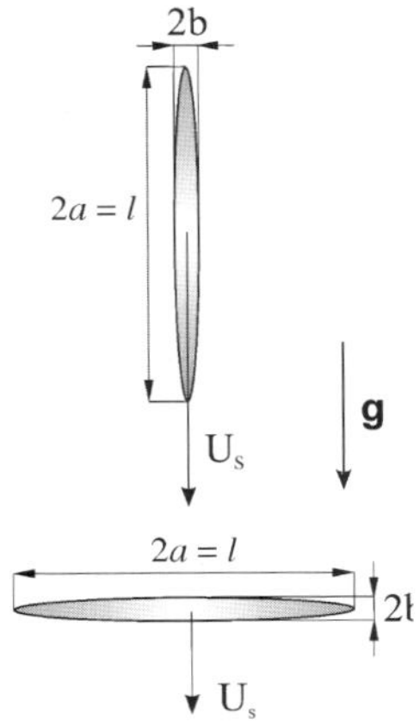	$\frac{1}{3}\Delta\rho\frac{b^2[\mathrm{In}(l/b)-0.5]}{\eta}\lvert\mathbf{g}\rvert$		$\frac{[\mathrm{In}(l/b)-0.5]\,q}{2\pi\eta l}\lvert\mathbf{E}\rvert$	
		$\frac{2}{9}\Delta\rho\frac{b^2\ln(l/b)}{\eta}\mathbf{g}$		$\frac{\ln(l/b)\,q}{3\pi\eta l}\mathbf{E}$
	$\frac{1}{6}\Delta\rho\frac{ab[\mathrm{In}(l/b)+0.5]}{\eta}\lvert\mathbf{g}\rvert$		$\frac{[\mathrm{In}(l/b)+0.5]\,q}{4\pi\eta l}\lvert\mathbf{E}\rvert$	
Slender bodies				

Table 4.2 (continued)

Body	Sedimentation velocity $\mathbf{U}_s$	Averaged $\langle \mathbf{U}_s \rangle$	Migration velocity $\mathbf{U}_m$	Averaged $\langle \mathbf{U}_m \rangle$
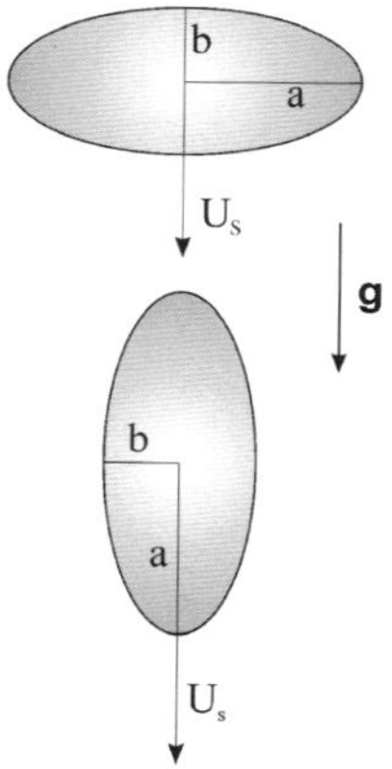	$\frac{2}{9}\frac{ab\Delta\rho}{\eta K_1}\lvert\mathbf{g}\rvert$		$\frac{q}{6\pi\eta a K_1}\lvert\mathbf{E}\rvert$	
		$\frac{2}{9}\frac{\Delta\rho\, ab\cos^{-1}(b/a)}{\eta(1-b^2/a^2)^{1/2}}$		$\frac{\cos^{-1}(b/a) q\,\mathbf{E}}{6\pi\eta a(1-b^2/a^2)^{1/2}}$
Oblate spheroids	$\frac{2}{9}\frac{a^2\Delta\rho}{\eta K_2}\lvert\mathbf{g}\rvert$		$\frac{q}{6\pi\eta a K_2}\lvert\mathbf{E}\rvert$	

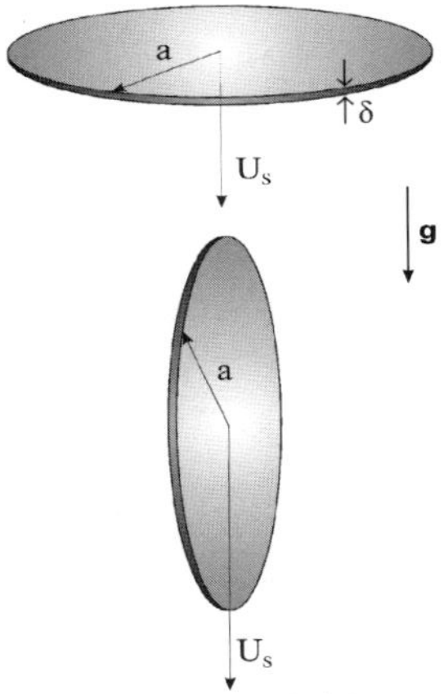 Disk broadside and disk edgewise	$\frac{\pi}{12}\frac{a\delta\,\Delta\rho}{\eta}\lvert\mathbf{g}\rvert$	$\frac{\pi}{9}\frac{a\delta\,\Delta\rho}{\eta}\mathbf{g}$	$\frac{q}{16\eta a}\lvert\mathbf{E}\rvert$	$\frac{q}{12\eta a}\mathbf{E}$
	$\frac{\pi}{8}\frac{a\delta\,\Delta\rho}{\eta}\lvert\mathbf{g}\rvert$		$\frac{3}{12}\frac{q}{\eta a}\lvert\mathbf{E}\rvert$	

Notes:

$\tilde{\eta} = (1+\eta/\eta_f)/(1+2\eta/3\eta_f)$

Prolate spheroids

$$K_1 = \frac{4}{3}(b/a)\left[\frac{2(a^2/b^2)-1}{\left[a^2/b^2-1\right]^{3/2}}\cosh^{-1}(a/b)-\frac{a/b}{a^2/b^2-1}\right]^{-1} \qquad K_2 = \frac{8}{3}\left[\frac{2(a^2/b^2)-3}{\left[a^2/b^2-1\right]^{3/2}}\cosh^{-1}(a/b)+\frac{a/b}{a^2/b^2-1}\right]^{-1}$$

Oblate spheroids

$$K_1 = \frac{4}{3}\left[\frac{1-2b^2/a^2}{(1-b^2/a^2)^{3/2}}\cos^{-1}(b/a)+\frac{(b/a)}{(1-b^2/a^2)}\right]^{-1} \qquad K_2 = \frac{8}{3}(a/b)\left[\frac{3-2(b/a)^2}{(1-b^2/a^2)^{1/2}}\cos^{-1}(b/a)+\frac{(b/a)}{(1-b^2/a^2)^{3/2}}\right]^{-1}$$

moving particles and from the reflection of their flow field from other particles ($-1.55\ \phi_v$). There is also a small contribution ($0.5\ \Phi_v$), which is the effect of the decreased hydrodynamic pressure of the counter-flow of the fluid.

For typical colloid deposition processes, $\Phi_v < 0.01$, so the correction to the sedimentation velocity calculated from Eq. (58) is of about 7 per cent or less.

The validity of Eq. (58) has been confirmed experimentally [2–3], although, in the case of colloid particles, significantly smaller values of the correction constant were found. This can be attributed to the specific interaction between particles, mainly the repulsive electrostatic double-layer interactions, which induce negative correlations between particle positions at short distances.

Knowing the averaged sedimentation or migration velocity, one can calculate the flux of particles, which is a quantity of primary practical interest, from the constitutive dependence

$$\mathbf{j} = n\langle \mathbf{U} \rangle \tag{60}$$

where n is the local concentration of particles in the suspension. As can be noticed from Eq. (60), the particle flux $\mathbf{j}$ is a vector quantity. It is of major significance in formulating mass conservation equations.

Accordingly, the number of particles crossing an arbitrary surface per unit of time (rate of particle transfer) is

$$\frac{dN}{dt} = \int_S \mathbf{j}(t)\cdot\hat{\mathbf{n}}dS \tag{61}$$

where $\hat{\mathbf{n}}$ is the unit normal vector.
Because, in the case of sedimentation or migration, $\mathbf{j}$ remains constant in time, and $n = n_b$ is postulated to be uniform throughout the suspension, the number of particles, which have passed through the surface S after the time t is given by

$$N(t) = \left(\int_S \mathbf{j}(t)\cdot\hat{\mathbf{n}}dS\right) n_b\, t \tag{62}$$

If S is chosen as a flat boundary perpendicular to the averaged velocity vector, for example the container's bottom, Eq. (62) simplifies to

$$N = USn_bt \tag{63}$$

where $U = |\langle \mathbf{U} \rangle|$.
Eq. (63), which indicates that the amount of matter accumulated at the interface, is a linear function of time and the particle bulk concentration, of major practical significance in predicting particle transfer to boundary surfaces under external force regimes.

4.3. PARTICLE MOTION NEAR BOUNDARY SURFACES – TRAJECTORY ANALYSIS

As was shown above, the sedimentation rates decrease significantly with particle dimensions (in the case of spherically isotropic bodies, proportionally to the square of particle size). Hence, for particles the size of a micrometer, sedimentation plays a minor role in transferring them to boundary surfaces, especially for liquids, where the densities of the particles and the medium are similar. In order to enhance these rates one often applies forced convection, discussed in Chapter 3. Because of fluid motion particles are brought into the immediate vicinity of interfaces, often called collectors, where the specific force field, composed of electrostatic and van der Waals forces, induces their attachment and deposition. These effects can also be studied in terms of the trajectory approach, provided that the specific and hydrodynamic force fields are well defined. Attention will be focused on spherical particles, for which the hydrodynamic resistance tensors near boundaries as well as specific interactions are known.

The starting point for the trajectory analysis is Eqs. (14),(15) with the mobility matrices $\mathbf{M}$, $\mathbf{M_r}$ defined explicitly by Eq. (16). Although these tensors are strictly valid only for solid spherical particles in the vicinity of planar interfaces, they can be used as a good approximation for curved interfaces as well, e.g., for spherical or cylindrical interfaces. This is so because the size of transported particles is usually much smaller than the local radius of curvature of the interface R, so $a/R \ll 1$. This means that the interface can be treated as locally planar from the viewpoint of hydrodynamics. By accepting this postulate, one can introduce the local Cartesian coordinate system with the axes (x',y',z') tangential and the axis z' perpendicular to the interface, see Fig. 4.1. The coordinate system has its origin at the point $P_p(\xi_{1p}, \xi_{2p}, \xi_c)$ defining the local position of the particle at the interface and (ξ_1, ξ_2, ξ_3) is an arbitrary curvilinear coordinate system appropriately chosen to reflect the shape

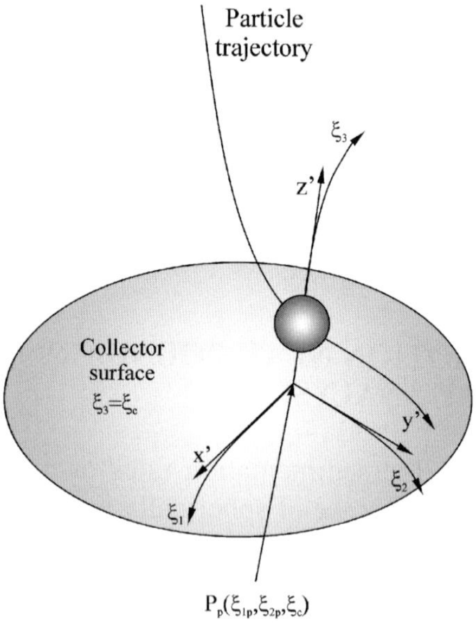

Fig. 4.1. A schematic view of coordinate systems used for formulating the trajectory concept for spherical particles.

of the collector. Because the interface can be treated as locally planar, all the components of the resistance tensors are dependent on the z' coordinate only. Moreover, using the method described in Chapter 3, the macroscopic flows near interfaces can be decomposed into a sum of elementary flows such as the stagnation-point flow or the simple shear flow. This allows one to calculate in a quite simple manner the hydrodynamic forces and torques on particles as the sum of contributions stemming from these simple flows, i.e.,

$$\begin{aligned} \mathbf{F}_h &= \mathbf{F}_{St} + \sum \mathbf{F}_{Sh} \\ \mathbf{To}_h &= \sum \mathbf{To}_{Sh} \end{aligned} \tag{64}$$

where

$$\mathbf{F}_{St} = -6\pi\eta a G_{St}(P_p) F_2(z') z'^2 \mathbf{i}_z \tag{65}$$

is the force due to the stagnation point flow and

$$\mathbf{F}_{Sh_1} = 6\pi\eta a G_{Sh_x}(P_p) F_8(z') z' \mathbf{i}_x \tag{66}$$

$$\mathbf{F}_{Sh_2} = 6\pi\eta a G_{Sh_y}(P_p) F_8(z') z' \mathbf{i}_y$$

is the force due to the shear flow in the x' and y' directions (parallel to the interface).

G_{St}, G_{Sh} in Eqs. (64),(65) are the shear rates at the wall, which are listed for various collectors in Table 3.4.

The hydrodynamic torque has only tangential components because the stagnation point flows do not produce any torque on a spherical particle, so

$$\mathbf{To}_{x'} = 4\pi\eta a^3 G_{Sh_y} F_9(z')$$

$$\mathbf{To}_{y'} = 4\pi\eta a^3 G_{Sh_x} F_9(z') \tag{67}$$

The net force on a particle is the sum of the hydrodynamic force given by Eqs. (65),(66) and the specific $\mathbf{F}_s$ and external $\mathbf{F}_{ext}$ contribution, hence

$$\mathbf{F} = \mathbf{F}_h + \mathbf{F}_s + \mathbf{F}_{ext} \tag{68}$$

For spherical particles, with homogeneous surface and bulk properties, the external and specific torques vanish, so the net torque equals the hydrodynamic torque $\mathbf{To}_h$.

Considering Eqs. (65), (66) the trajectory equations, Eqs.(14),(15) can be expressed in the local coordinate system (x, y, z), in the form

$$U_{x'} = \frac{dx'}{dt} = M_{||} F_{x'} - K_c' To_{y'} = \frac{F_4'}{6\pi\eta a} F_{x'} + G_{Sh_x} z' F_3$$

$$U_{y'} = \frac{dy'}{dt} = M_{||} F_{y'} - K_c' To_{x'} = \frac{F_4'}{6\pi\eta a} F_{y'} + G_{Sh_y} z' F_3 \tag{69}$$

$$U_{z'} = \frac{dz'}{dt} = M_{\perp} F_{z'} - F_1 F_2 G_{St} z'^2 = \frac{F_1}{6\pi\eta a} F_{z'} - F_1 F_2 G_{St} z'^2$$

where

$$F_3(z') = F_4'(z')F_8(z') - \frac{2}{3}K_c' F_9(z')\frac{a}{z} \tag{70}$$

is the universal correction function for the parallel motion of a spherical particle in the simple shear flow [5].

It can be deduced from Eq. (69) that in the absence of external forces, the tangential velocity of a particle in the shear flow is

$$U_x' = G_{Sh_x} z' F_3(z') \tag{71}$$

Eq. (71) has a practical significance enabling one to calculate the separation distance of a particle moving parallel to a wall in the shear flow from microscopic observations of its motion. This is especially important for colloid particles [6].

The $F_3(H)$ function (where $H = z/a - 1$) is shown in Fig. 4.2. As discussed in Ref. 6 it can be approximated by the following formulae

$$\begin{aligned} F_3(H) &\cong \frac{1}{0.754 - 0.256\ln H} \quad \text{for} \quad H < 0.15 \\ F_3(H) &\cong 1 - \frac{0.304}{(1+H)^3} \quad \text{for} \quad H > 0.15 \end{aligned} \tag{72}$$

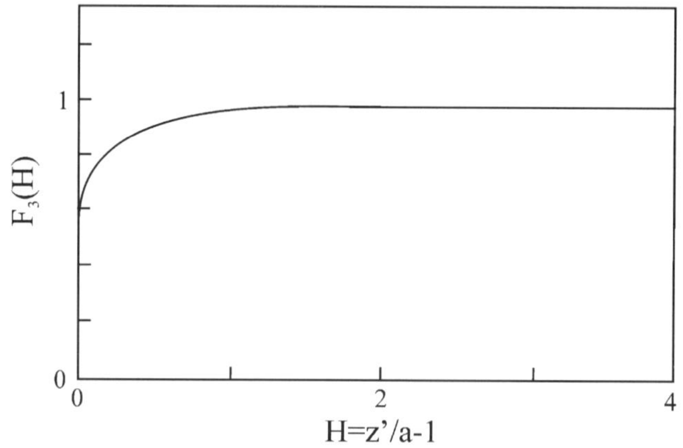

Fig. 4.2. The hydrodynamic correction function $F_3(H)$.

On the other hand, Goldman *et al.* [5] showed that in the limit $H \to 0$ the F_3 function can be approximated by

$$F_3(H) = \frac{0.743}{0.6376 - 0.20 \ln H} \tag{73}$$

It can be calculated from Eqs.(72),(73) that for $H = 1$ the correction equals 0.96, for $H = 0.1$, $F_3 = 0.75$ and for $H = 0.001$, $F_3 = 0.397$. As can be seen, the correction remains rather insignificant except for very small distances, where $H < 0.01$.

For calculating particle transfer rates it is advantageous to formulate Eq. (69) in an appropriate curvilinear coordinate system fixed in space. In this way one obtains

$$\begin{aligned} U_{\xi_1} &= f_1(\xi_1, \xi_2, \xi_3) \\ U_{\xi_2} &= f_2(\xi_1, \xi_2, \xi_3) \\ U_{\xi_3} &= f_3(\xi_1, \xi_2, \xi_3) \end{aligned} \tag{74}$$

where the functions $f_1 - f_3$ depend in a non-linear way on the curvilinear coordinates, (ξ_1, ξ_2, ξ_3), describing the actual particle position near the interface.

As can be noticed, Eqs. (69) and (74) represent a system of three non-linear ordinary differential equations of the first order. Therefore, they should be complemented with the initial condition, i.e., the position of the particle for the time $t = 0$. In this way an initial-value problem arises whose solution in the general case, although possible, can only be achieved by numerical methods.

It is useful to formulate Eq. (74) for collectors of practical significance. For example, in the case of the rotating disk, using the lowest order terms of the fluid velocity expansion, Eq. (129) in Chapter 3, one can formulate Eq. (74) in the explicit form

$$\begin{aligned} U_r &= \frac{dr}{dt} = F_4'(H)\frac{F_r}{6\pi\eta a} + f(0)\frac{\Omega^{3/2} a}{\nu^{1/2}} F_3(H)(H+1)r \\ U_\varphi &= a\frac{d\varphi}{dt} = F_4'(H)\frac{F_\varphi}{6\pi\eta a} + \Omega\left[1 - g(0)\left(\frac{\Omega}{\nu}\right)^{1/2} a F_3(H)(H+1)\right] r \\ U_z &= \frac{dz}{dt} = a\frac{dH}{dt} = F_1(H)\frac{F_z}{6\pi\eta a} - F_1(H)F_2(H)(H+1)^2 f(0)\frac{\Omega^{3/2} a}{\nu^{1/2}} \end{aligned} \tag{75}$$

where (r,φ,z) are coordinates of the cylindrical system having their origin at the center of the disk and $H=z\,/a^{-1}$ is the dimensionless gap width, $f(0) = 0.5102$, $g(0) = 0.616$.

As can be seen, particle trajectories are truly three-dimensional in this case. However, by assuming that the external and specific force is directed along the H axis, Eq. (75) can be reduced to the following 2D form because the change in the φ coordinate does not affect particle transfer to the interface

$$\frac{dr}{dt} = f(0)\frac{\Omega^{3/2}a}{\nu^{1/2}}F_3(H)(H+1)r$$
$$a\frac{dH}{dt} = F_1(H)\frac{F_H}{6\pi\eta a} - F_1(H)F_2(H)(H+1)^2 f(0)\frac{\Omega^{3/2}a^2}{\nu^{1/2}} \tag{76}$$

where $F_H = F_z$ is the H component of the external and specific forces.

Analogously, 2D trajectory equations can be formulated for the impinging jet flows, where the external force acts perpendicular to the surface

$$U_r = \frac{dr}{dt} = \alpha_r\frac{V_\infty a}{R}S(r/R)(H+1)F_3$$
$$U_H = a\frac{dH}{dt} = F_1(H)\frac{F_H}{6\pi\eta a} - F_1(H)F_2(H)(H+1)^2\alpha_r\frac{V_\infty a^2}{R^2}C(r/R) \tag{77}$$

where α_r is the flow intensity parameter shown in Fig. 3.12 and $S(r/R)$, $C(r/R)$ are the correction functions shown in Fig. 3.10. For $r/R < 1$ they can be approximated by $2/\pi\sin(\pi r/2R)$ and $\cos(\pi r/2R)$, respectively.

For spherical and cylindrical collectors, assuming that the external force acts parallel to the uniform flow direction at infinity, Eq.(74) becomes

$$U_\vartheta = R\frac{d\vartheta}{dt} = F_4'(H)\frac{F_{ext}\sin\vartheta}{6\pi\eta a} + G_{Sh}^0 aF_3(H)\,(H+1)S(\vartheta) \tag{78}$$

$$U_H = a\frac{dH}{dt} = F_1(H)\frac{F_{sH}+F_{ext}\cos\vartheta}{6\pi\eta a} - F_1(H)\,F_2(H)\,(H+1)^2G_{St}^0a^2C(\vartheta)$$

where text is the extended force parallel to the flow direction, F_s is the specific force (G^0_{Sh} and G^0_{St} for these collectors are given in Table 3.4), $\vartheta = x/R$ is the tangential coordinate measured from the forward stagnation point to the actual particle position, $S(\vartheta)$, $C(\vartheta)$ are the functions of ϑ. As discussed in Chapter 3, for low *Re* number flows, $Re < 1$, these functions for both spherical and cylindrical collectors become simply $S = \sin\vartheta$ and $C = \cos\vartheta$. For higher *Re* number, as discussed in Chapter 3, flow separation at the rear of these collectors occurs, when the angle assumes the critical value of ~109°. In this case the trajectory equation remains accurate for the front part of the collector only.

The above trajectory equations can be further simplified in the case where the flow vanishes and particle transport to the interface occurs under the external and specific forces only, acting toward the collector (opposite to the *H* axis direction). For any planar interface (e.g., the rotating disk), one obtains the one-dimensional trajectory equation in the form

$$a\frac{dH}{dt} = F_1(H)\frac{F_{ext} + F_s(H)}{6\pi\eta a} = F_1(H)\left[-U + \frac{F_s(H)}{6\pi\eta a}\right] \tag{79}$$

where *U* is the previously determined sedimentation or migration velocity of the particle directed to the surface.

Eq. (79) can be integrated analytically which results in the following implicit dependence of the distance from the interface vs. the time *t*

$$\int_{H_0}^{H} \frac{dH'}{F_1(H')\left[U - \dfrac{F_s(H')}{6\pi\eta a}\right]} = -\frac{t}{a} \tag{80}$$

where H_0 is the separation of the particle from the interface at $t = 0$.

Assuming that the net force consists of the external force only (gravitational or electrostatic) and using the approximation $F_1(H)=H/(H+1)$ Eq. (80) can be expressed as

$$-U\frac{t}{a} = H - H_0 + \ln\frac{H}{H_0} \tag{81}$$

For small distances $H << 1$, Eq. (81) simplifies to

$$H = H_0 e^{-\frac{U}{a}t} \tag{82}$$

Note that the final position at the interface is approached exponentially. By taking $U = 10^{-6}$ m s^{-1} and $a = 10^{-6}$ m, one can calculate from Eq. (82) that the sphere will approach the interface at the distance of $\delta_m = 0.5$ nm (the initial minimum distance) starting from a H_0 = a = 10^{-6} m after a time of 7.6s., instead of 1 s if it had moved with a constant velocity. This indicates that the hydrodynamic effects for the perpendicular particle motion are quite significant.

However, the hydrodynamic effects will be decreased by the attractive van der Waals interactions described by the formula (cf. Eq. (306) in Chapter 2).

$$F_s = -\frac{A_{132}\,a}{6h^2} = -\frac{A_{132}}{6aH^2} \tag{83}$$

In this case, integration of Eq. (80) with $F_1 = H/(H+1)$ gives for small distance,

$$H = \left[\left(H_0^2 + \frac{U_s}{U}\right)e^{-\frac{2Ut}{a}} - \frac{U_s}{U}\right]^{1/2} \tag{84}$$

where, $U_s = A_{132}/36\pi\eta a$.

From Eq. (84), one can derive the expression for the characteristic time of particle transfer from H to the initial minimum distance δ_m under the external force

$$t = \frac{1}{2}\frac{a}{U}\ln\frac{H_0^2 + \frac{U_s}{U}}{\bar{\delta}_m^2 + \frac{U_s}{U}} \tag{85}$$

were $\bar{\delta}_m = \delta_m/a$

By taking the same data as before, i.e., $U = 10^{-6}$ m s^{-1}, $a = 10^{-6}$ m and the Hamaker constant $A_{132} = 5\times10^{-21}$ J, one can calculate from Eq. (85) that the sphere will approach the interface at the distance of 0.5 nm starting from $H_0 = a = 10^{-6}$ m after a time of about 1.6 s. This estimation confirms that the attractive van der Waals forces (or electric double-layer forces) can play quite a significant role in particle deposition. It can be, expected, therefore that by measuring the transfer time of particles under gravitational force, one determine the specific force field due to the van der Waals attraction, or the electrostatic double-layer forces.

When the flow occurs, the trajectory equations cannot be solved analytically. However, they can be converted to a more useful form by eliminating the time variable. In this way one obtains a non-linear differential equation involving the normal and tangential coordinates. For the rotating disk, one can reduce the equation system, Eq. (76) to the form

$$r\frac{dH}{dr} = \frac{F_1(H)}{F_3(H)(H+1)}\left[\frac{\nu^{1/2}}{6\pi\eta a^3 f(0)\Omega^{3/2}}F_H - F_2(H)(H+1)^2\right] \tag{86}$$

Because the r and H variables are separated, Eq.(86) can be integrated to the form

$$r(H) = C_d\, e^{-I_d(H)} \tag{87}$$

where C_d is the integration constant and

$$I_d(H) = \int \frac{F_3(H)(H+1)\,dH}{F_1(H)\left[F_2(H)(H+1)^2 - f_H F_H\right]} \tag{88}$$

and $f_H = \nu^{1/2}/6\pi\eta a^3 f_0\, \Omega^{3/2}$

For a sphere and a cylinder, Eq. (78) can be expressed as

$$\frac{dH}{d\vartheta} = \frac{R}{a}F_1(H)\frac{(F_{sH}+F_{ext}\cos\vartheta)/6\pi\eta a^2 - G^0_{St}(\vartheta)aF_2(H)\,(H+1)^2C(\vartheta)}{\dfrac{F_4'(\mathrm{H})F_{ext}\sin\vartheta}{6\pi\eta a^2} + G^0_{Sh}(\vartheta)F_3(H)\,(H+1)S(\vartheta)} \tag{89}$$

Eqs. (86) and (89) are exploited in the trajectory approach shown schematically in Fig. 4.3 [7–10]. The method is based on integration of these equations starting from a point at the collector surface (described by a fixed value of the coordinate tangential to the collector) where the particle touches its surface, i.e., when $H = \delta_m$. Such a trajectory is called the

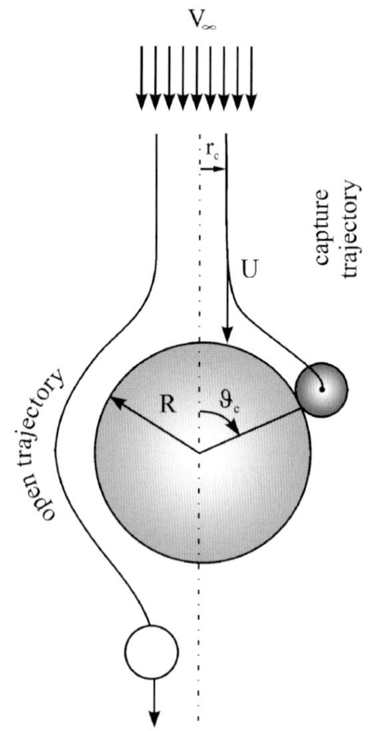

Fig. 4.3. A schematic representation of the limiting trajectory approach for spherical particles, showing the open and capture trajectories, respectively.

capture trajectory (see Fig. 4.3). By proceeding with the integration procedure in such a way that H tends to infinity (in practice to a large number, $H >> 1$), one obtains the value of the critical radius r_c depending obviously on the location of the starting point at the collector. All particles whose centers are separated from the axis of symmetry by a distance smaller than r_c will be deposited within this area. Using Eq. (61) the rate of particle deposition can be expressed as

$$\frac{dN}{dt} = \pi r_c^2 U(H) n_b \tag{90}$$

where $U(H)$ is the normal component of particle velocity far from the interface.

For the rotating disk, one can calculate r_c from Eq. (87) by postulating, in accordance with the trajectory approach, that $H = \delta_m$ for $r = r_p$, where r_p is the radial position of the captured particle.

$$C_d = (\bar{\delta}_m + 1) r_p e^{I_d(\bar{\delta})} \cong r_p e^{I_d(\bar{\delta}_m)} \tag{91}$$

By substituting the expression for C_d into Eq. (87) and assuming that particle velocity at large separations coincides with the fluid velocity, the rate of particle deposition can be expressed as

$$\frac{dN}{dt} = \pi r_c^2 (H)(H+1)^2 f(0) \frac{\Omega^{3/2}}{\nu^{1/2}} a^2 n_b I_d' = \pi r_p^2 f(0) \frac{\Omega^{3/2}}{\nu^{1/2}} a^2 n_b I_d' \tag{92}$$

where

$$I_d' = e^{-2 \int_{\bar{\delta}_m}^{\infty} \left[\frac{F_3(H)(H+1)}{F_1(H)\left[F_2(H)(H+1)^2 - f_H F_H\right]} - \frac{1}{H+1} \right] dH} \tag{93}$$

Note that the upper integration limit in Eq. (93) was extended to infinity because the integral is convergent. It is also interesting to note that I_d', and consequently particle deposition rate depends in a complex way on particle size, viscosity, disk angular velocity, the Hamaker constant and other parameters characterizing the electrostatic interactions.

The local flux of particles can be calculated from Eq. (92) using the relationship

$$-j = \frac{dN}{dS} = \frac{1}{2\pi r_c} \frac{dN}{dr_c} = f(0) \frac{\Omega^{3/2}}{\nu^{1/2}} a^2 n_b I_d' \tag{94}$$

(the negative sign appears because the flux is directed opposite to the H axis)

As can be deduced from Eq. (94), the flux, being independent of the radial coordinate, is uniform over the entire disk surface. Collector surfaces exhibiting such a property are called uniformly accessible surfaces [11].

Although it is not possible to calculate the integral of Eq. (93) analytically in the general case, a useful limiting form can be derived by assuming that $F_1 = F_2 = F_3 = 1$ and by neglecting all external and specific forces. In this case $I_d' = 1$ and Eq. (93) becomes

$$-j = f(0)\frac{\Omega^{3/2}}{\nu^{1/2}}a^2 n_b \tag{95}$$

Eq. (95) is expected to be a valid approximation for particles of micrometer size range, in the absence of electrostatic interactions.

The flux described by Eq. (95) is often called the interception flux. [8–10]. It is physically due to the fact that particles of finite dimensions that follow the fluid streamlines must collide with the collector at a given distance r_p. A characteristic feature of the interception flux is that it is proportional to the square of particle size, so this mode of transportation becomes particularly efficient for larger sized particles.

For the impinging-jet, spherical and cylindrical collectors, calculation of the particle flux becomes feasible via numerical solutions of the trajectory equation, which is a standard and simple procedure. Extensive calculations have been reported in [8–10,12,13] for the rotating disk, the spherical and cylindrical collectors, both isolated and forming a packed bed column.

However, as is the case of the rotating disk, the interception flux for these collectors can be calculated analytically by substituting into Eq. (89) $F_1 = F_2 = F_3 = 1$ and by neglecting all external and specific forces. In this case the trajectory equation for the impinging-jet and the spherical collector is

$$\frac{dH}{d\vartheta} = -(H+1)\frac{C(\vartheta)}{S(\vartheta)} \tag{96}$$

$\vartheta = \pi r/2R$ for the impinging-jet collector.

For the cylindrical collector and the two slot impinging- jet collectors, one has

$$\frac{dH}{d\vartheta} = -\frac{1}{2}(H+1)\frac{C(\vartheta)}{S(\vartheta)} \tag{97}$$

$\vartheta = \pi d/2R$ for the slot impinging-jet collector.

Eqs. (96),(97) can be integrated analytically when $C(\vartheta)=\frac{dS(\vartheta)}{d\vartheta}$. This is a valid approximation for the spherical and cylindrical collectors of low *Re* number and for the impinging-jet collector of $r/R<1$ (see Eq. (137) in Chapter 3). By exploiting additionally the condition $H = \delta_m$ for $\vartheta=\vartheta_p$ where ϑ_p is the captured particle position one obtains

$$S(\vartheta)=\frac{\overline{\delta}_m+1}{H+1}\,S(\vartheta_p)\cong\frac{S(\vartheta_p)}{H+1}\,; \qquad \text{sphere, impinging - jet} \tag{98}$$

$$S(\vartheta)=\frac{\overline{\delta}_m+1}{H+1}\,S(\vartheta_p)\cong\frac{S(\vartheta_p)}{(H+1)^2}\,; \qquad \text{cylinder, slot impinging - jet}$$

Considering that $S(\vartheta)\rightarrow 0$ for $\vartheta\rightarrow 0$ and $r_c=R\vartheta$, the overall deposition rate for the spherical collector can be expressed as

$$\frac{dN}{dt}=\frac{3}{2}\pi A_f V_\infty a^2 S^2(\vartheta_p)n_b \tag{99}$$

Analogously, one obtains for the impinging-jet

$$\frac{dN}{dt}=\pi\alpha_r V_\infty a^2 S^2(\vartheta_p)n_b \tag{100}$$

For the cylindrical collector and the slot impinging-jet the expression for N becomes

$$\frac{dN}{dt}=\frac{1}{R}A_f V_\infty a^2 S(\vartheta_p)\,n_b; \quad \text{cylinder} \tag{101}$$

$$\frac{dN}{dt}=\frac{1}{d}\alpha_s V_\infty a^2 S(\vartheta_p)n_b\,; \quad \text{slot impinging-jet}$$

The local particle flux for these collectors is given by

$$-j=\frac{3}{2}\,A_f V_\infty\frac{a^2}{R^2}\frac{S(\vartheta_p)C(\vartheta_p)}{\sin\vartheta_p}\,n_b; \quad \text{sphere}$$

$$-j=\alpha_r V_\infty\frac{a^2}{Rr_p}\,S(\vartheta_p)C(\vartheta_p)n_b; \qquad \text{impinging - jet} \tag{102}$$

$$-j=A_f V_\infty\frac{a^2}{R^2}\,C(\vartheta_p)\,n_b; \qquad \text{cylinder}$$

$$-j=\alpha_s V_\infty\frac{a^2}{d^2}\,C(\vartheta_p)\,n_b; \qquad \text{slot impinging - jet}$$

As can be deduced from Eq. (102), the interception flux in the case of these collectors depends on the position over the interface, which is governed by the ϑ variable, and decreases to zero for $\vartheta \to \pi/2$, because $C(\pi/2) = 0$. For $\vartheta > \frac{\pi}{2}$ the interception flux vanishes because the normal component of the fluid velocity is directed outward from the collector surface. Hence, particle deposition in this region only becomes feasible if external or specific forces acting toward the collector surface appear (opposite to the flow direction).

For $\vartheta \to 0$, near the forward stagnation point, Eq. (102) reduces to

$$\begin{aligned}
-j &= \frac{3}{2} A_f V_\infty \frac{a^2}{R^2} n_b; && \text{sphere} \\
-j &= \alpha_r V_\infty \frac{a^2}{R^2} n_b; && \text{impinging - jet} \\
-j &= A_f V_\infty \frac{a^2}{R^2} n_b; && \text{cylinder} \\
-j &= \alpha_s V_\infty \frac{a^2}{d^2} n_b; && \text{slot impinging - jet}
\end{aligned} \tag{103}$$

These equations, analogous to those previously obtained for the rotating disk, Eq. (95), indicate that the flux becomes uniform in the region near the forward stagnation point, hence the surface of these collectors is uniformly accessible.

However, in spite of its simplicity and a direct physical interpretation, the limiting trajectory approach has many disadvantages, such as

(i) The inability to properly describe deposition processes of smaller particles under no flow conditions when Brownian motion become the dominating transport mechanism as discussed later;
(ii) The difficulty in treating transient, non-stationary deposition phenomena, non-stationary (oscillating), rectilinear and turbulent flows; and
(iii) The inability to treat the reversible deposition of particles and their detachment from surfaces.

However, the limiting trajectory approach remains an useful, and often the only approach available, for more complex situations, such as multilayer particle deposition from gaseous phases studied in Ref. 13 and for predicting particle trajectories near other particles adsorbed on the surfaces. In the latter case there is no orthogonal coordinate system, that could reflects exactly this many-body geometry, so no analytical results can be derived. The moving particle trajectories can only be determined by applying elaborate numerical techniques. Usually, an integral form of the Stokes equation is exploited for this purpose [14–15].

Dąbroś [16] and Dąbroś and van de Ven [17–18] developed two convenient methods for the numerical solution of the Stokes equation for multiparticle systems. Both are based on the singularity representation exploiting the Oseen and the Blake wall tensors discussed in Chapter 3. The integrals over entire particle surfaces are replaced by sums of integrals evaluated over smaller surface elements having the curvilinear triangle shape. This method is referred to as the boundary element method (BEM), or over subunits of quite arbitrary shapes (the multisubunit, MS, method). The moving particle velocity is obtained either directly by solving a linear system of equations in the BEM method, or in a more indirect way by determining the components of the grand mobility matrix $\tilde{\mathbf{M}}$ (defined by Eq. (13)). The latter approach is more flexible because it allows the use of lubrication approximation for small separations between particles [19]. This is necessary because the MS method gives inaccurate results at distances much smaller than particle size. Most of the calculations have been done for a two-particle system, one attached firmly to the wall and the other moving under gravity or driven by simple flows such as the shearing or the stagnation point flow. All the specific interactions between particles, such as van der Waals interactions, were neglected except for the short-range interactions of the Born type. The trajectory of the particle was determined by numerical integration of its velocity using the predictor-corrector method.

Examples of such calculations confronted with experimental measurements of Malysa et al. [19] made using the stroboscopic method using nylon spheres of the diameter $2a=6.3\times10^{-3}$ m are shown in Fig. 4.4. As can be noticed, for larger initial separation from the wall ($H = x/a = 1.9$, external trajectory), the calculated trajectory of the moving particle, which reflects the experimental points very well, remains symmetric as is expected for Stokes flows. This was so because the shorter distance between the moving and attached particles was much larger than the range of the Born interactions. On the other hand, for small initial distances of the particle from the wall ($H = 0.36$, internal trajectory), the moving particle approaches the attached particle at very small distances, so the Born forces come into operation. This results in a considerable asymmetry of particle trajectory. Analogous results have been obtained using laser interferometry, and a macroscopic

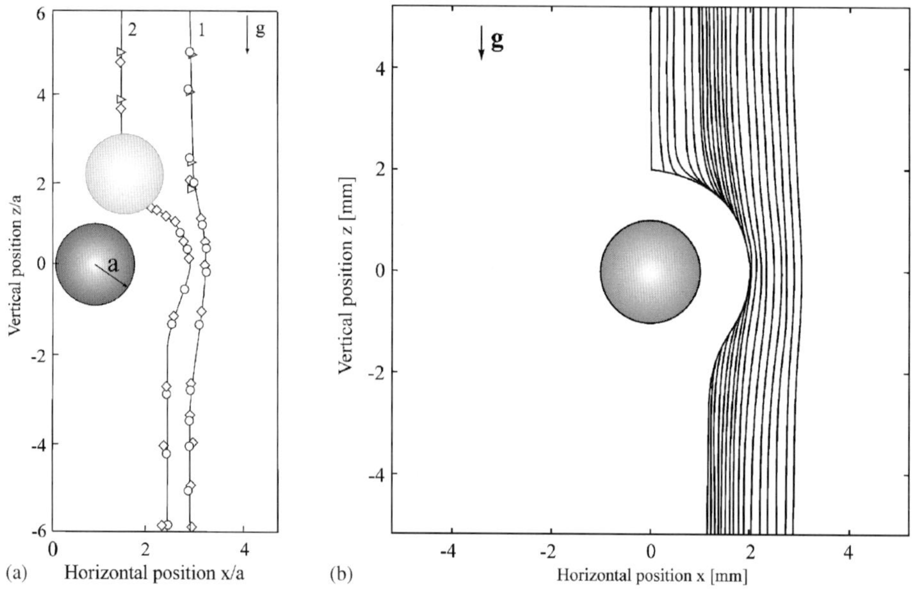

Fig. 4.4. (a) shows the experimentally determined trajectories of a particle sedimentating in a quiescent fluid near another particle attached to the wall (points) [19]. Solid lines represent the trajectories derived numerically using the MS method for two different initial separations of the moving particle from the wall $H = x/a - 1 = 1.9$ (external trajectory) and $H = 0.36$ (internal trajectory). (b) shows the results obtained by using laser interferometry for a polyacetate sphere of 6.3×10^{-3} m of diameter in silicon oil. From Ref. [20].

sphere 6.3×10^{-3} m in diameter sedimentating in silicon oil near a sphere attached to the wall [20] (see part "b" in Fig. 4.4). The receding part is significantly shifted away from the surface.

This effect also occurs if the moving particle is driven by fluid flows rather than by external force. This can be observed in Fig. 4.5 where the family of particle trajectories around a stationary particle attached to the wall is shown in the case of the simple shear flow directed along the Y axis. As can be seen, the receding trajectories of the particles form a semi-circular scattering pattern whose radius is dependent on the magnitude and the range of the specific particle-particle interactions. This effect can be exploited to determine these interactions as was done in the experimental technique by van de Ven et al. [21], referred to as the colloidal particle collider method.

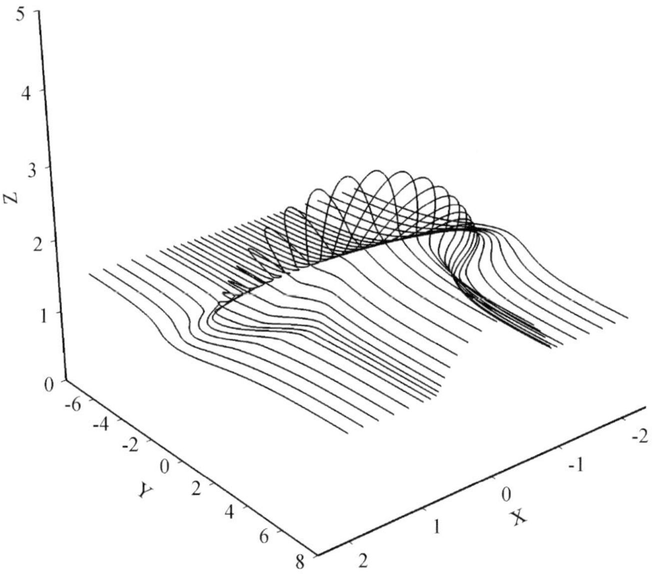

Fig. 4.5. The collision pattern formed by the trajectories of particles moving in the simple shear at a rigid wall in the proximity of a particle attached to it. The initial position of the particle was $H = 0.6$ (vertical) and –2 to 2 (tangential). The calculations were performed using the MS method. From Ref. [6].

Analogous results obtained for the stagnation-point flow are shown in Fig. 4.6. As can again be observed, the particle, after the collision with the attached particle, moves at a much larger distance from the interface than initially. This phenomenon, often referred to as the hydrodynamic scattering (HS), plays a significant role in the convection driven deposition of larger particles at surfaces [22,23]. Due to the HS effect, the surface area behind adsorbed particles becomes inaccessible for moving particles which leads to enhanced surface blocking effects [6,11,22–23]. The length of this hydrodynamic "shadow" depends on the flow configuration (either simple shear or stagnation point flow), specific interactions between the particle and the interface, and most significantly on the local flow rate governed by the G_{sh} or G_{st} parameters. This can be observed in Fig. 4.7, where the shapes of the shadow are shown for various values of the dimensionless parameter $6\pi\eta a^3 G_{Sh} / kT$.

The limitation of the deterministic trajectory method discussed in this section is that it does not allow for the consideration of the Brownian motion, which plays an increasingly important role for colloid particles

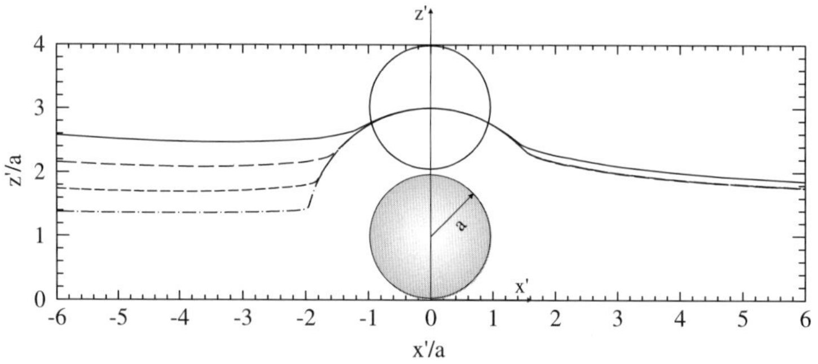

Fig. 4.6. The trajectories of particles moving in the stagnation point flow at a rigid wall in the proximity of a particle attached to it. The initial position of the particle was varied between $H = 0.4$ (vertical) for the lowest trajectory and $H = 1.6$ for the highest trajectory. The calculations were performed using the MS method. From Ref. [6].

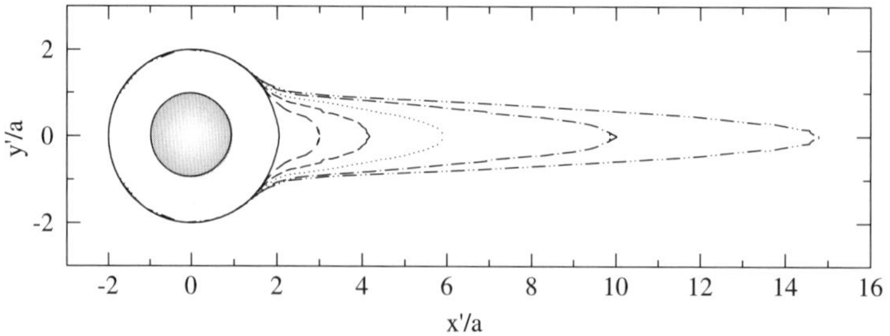

Fig. 4.7.The size of the hydrodynamic shadow behind a particle attached to the interface and immersed in a stagnation flow, various contours corresponding to the dimensionless parameter $6\pi\eta a^3 G_{Sh}/kT$ equal 0.01, 0.05, 0.2 and 0.5, respectively. The calculations were performed using the MS method. From Ref. [6].

smaller then a micrometer. If Brownian motion is taken into consideration, particle trajectories become stochastic in nature and their quantification requires more sophisticated methods as discussed next.

4.4. BROWNIAN MOTION AND DIFFUSION

For particles of a submicrometer size range, in addition to hydrodynamic and specific forces, Brownian motion significantly affects their trajectories and transport to boundary surfaces. Particles undergo Brownian motion because of collisions with molecules of the suspending medium. Since the collision events are uncorrelated, a random (fluctuating) force occurs, which has the character of white noise, i.e., zero mean value and a strength independent of the previous history of the motion. The perpetual particle motion induced by the fluctuating force is completely unpredictable as far as its direction and velocity is concerned. Thus, an analysis of a single particle trajectory is meaningless because one cannot predict the consecutive particle position, no matter how long it has been observed before. The only things that matter in this case are the statistical averages extended over ensembles of particles or over long times. In this way one can only determine the probability of a displacement of a particle from its initial position as function of the time of observation rather than its instantaneous position or velocity.

The discovery of this phenomenon is attributed to the botanist Robert Brown who made the first observations involving colloid particle (pollens). Gouy excluded any extraneous causes of Brownian motion such as light, mechanical vibrations, convective currents, shaking, evaporation, etc., therefore the molecular origin of this motion was the only one left. However, the representatives of the phenomenological school of Ostwald and Mach refused to accept the very existence of molecules till the end of the 19th century. The major source of controversy was that the velocity of a colloid particle predicted from the Maxwell distribution of velocities of a gas was more than three orders of magnitude smaller than 'velocity' observed using, e.g., microscope methods.

A proper explanation of Brownian motion was furnished by Einsten in 1905–1906 [24,25], who developed a general theory based on the statistical mechanic approach and independently by Smoluchowski [26] who used the more mechanistic approach. Both of them treated the Brownian motion of particles as a special case of the general thermodynamic fluctuation phenomena originated from collisions of the particle with molecules of the suspending medium. Most of the momentum impulses stemming from these collisions cancel each other out and only a small fraction effectively propels the particle. As a result, particle displacements observed in the macroscale remain much smaller than predicted by the kinetic theory of gases. Moreover, these displacements are governed by the square root of time rather than by linear dependence observed for deterministic motions. The situation is quite analogous to a walk with completely random directions, changed unpredictably in time. Hence, Brownian motion is often referred to as a random walk.

The Einstein–Smoluchowski theory has been confirmed quantitatively in the experiments of Perrin [27,28] who used the microscope observation method to measure displacements of particles. However, the instantaneous velocity of the particle could not be measured or calculated by differentiating particle trajectory with respect to time.

The Brownian motion of particles can be effectively described by the Langevine equation specified above with the random force evaluated by assuming that the kinetic energy of particles is equally partitioned among the three translational degrees of freedom. In order to formulate laws governing the temporal and spatial evolution of the probability density of finding a particle, Enistein's method that exploits the Markoff equation is used. Furthermore, Einstein's concept of the osmotic force balancing the resistance force (this corresponds to the thermodynamic Gibbs–Duhem relationship) will be used to describe the particle diffusion process, which is the net result of the Brownian motion of assemblages of particles.

4.4.1. Isolated Particle Motion – Langevine Equation

For the sake of clarity, we consider the one-dimensional Langevine equation, which has the form

$$m_p' \frac{dU'}{dt} = m_p' \frac{d^2 x'}{dt^2} = -M_{11}^{-1}\, U' + F + F^*(t) \tag{104}$$

where x' is the particle displacement from its initial position x_0, U' is the x component of the particle velocity, M_{11} is the component of the particle mobility tensor in this direction, and F, F^*(t) are the components of the deterministic and random force in the x direction, respectively. Because of its white noise character, the random force has the following basic properties [3]:

$$\left\langle F^*(t) \right\rangle = \frac{1}{N_p} \sum_{i}^{N_p} F^*(t) = 0$$

$$\left\langle F^*(t) F(t+\Delta t) \right\rangle = \frac{1}{N_p} \sum_{i}^{N_p} F^*(t) F(t+\Delta t) = F_0^2 \delta_D(t) \tag{105}$$

where $\langle\ \rangle$ means the ensemble average extended over a large system of N_p identical particles, Δt the time interval, δ_D the Dirac delta function and F_0 the constant to be calculated from the thermodynamic equilibrium condition written as

$$\frac{1}{2}m_p'\left\langle U^*(t)U^*(t)\right\rangle = \frac{1}{2}kT \tag{106}$$

From Eq. (106) it follows that the absolute value of the averaged particle velocity is

$$\left\langle U^2\right\rangle^{1/2} = U_0 = \left(\frac{kT}{m_p'}\right)^{1/2} \tag{107}$$

Eq. (104) can be split into a system of two independent equations by assuming that the net velocity is the sum of systematic and random contributions, i.e., $U' = U + U^*$. In this way one has for the systematic component

$$m_p'\frac{dU}{dt} = -M_{11}^{-1}U + F \tag{108}$$

By neglecting the inertia effects as done before, Eq. (108) becomes

$$U = M_{11}F \tag{109}$$

Eq. (109) is the one-dimensional analog of the previously analyzed trajectory equation, Eq. (12).

If both the net force and the mobility coefficient remain constant (independent of x'), then Eq. (109) becomes simply

$$x' = M_{11}Ft \tag{110}$$

On the other hand, after subtracting the solution for the systematic velocity from Eq. (104), one obtains the relationship describing the effect of the random force

$$\frac{dU^*}{dt} = \frac{dx'^2}{dt^2} = -\frac{1}{m_p'}M_{11}^{-1}U^* + \frac{1}{m_p'}F^*(t) \tag{111}$$

Eq. (111) can be formally integrated with respect to the time variable, which results in the following expression for the instantaneous particle velocity

$$U^{*}(t) = U_0 e^{-\beta t} + \frac{1}{m_p'} e^{-\beta t} \int_0^t F^{*}(t') e^{\beta t'} dt' \tag{112}$$

where U_0' is the initial particle velocity at $t = 0$ and β the inverse time parameter defined previously, cf. Eq. (6).

Eq. (112) is of limited applicability because the explicit time dependence of $F^{*}(t)$ cannot be specified. Also the initial velocity U_0' cannot be exactly determined. However, Eq. (112) can be used to estimate the relaxation time of particle velocity change (this can be referred to as the persistence time of particle motion) and the free path of the particle. The velocity relaxation time is simply

$$t_{ch} = 1/\beta = m_p' M_{11} \tag{113}$$

and the free path length l_{ch} is given by

$$l_{ch} = U_0 t_{ch} \cong U_0 m_p' M_{11} = (kTm_p')^{1/2} M_{11} \tag{114}$$

where U_0' was approximated by the averaged value of the velocity calculated from Eq. (107).

Typical values of U_0, t_{ch} and l_{ch} calculated from Eqs.(107, 113, 114) for a spherical particle are compiled in Table 4.3. As can be noticed, the relaxation time t_{ch} for a particle of the radius of a 10^3 nm (1 μm) at room temperature is 4.4×10^{-7} s and $l_{ch} \cong 0.31$ nm. For a particle of a radius of 10nm, one has, $t_{ch} = 4.4 \times 10^{-11}$ s and $l_{ch} = 0.031$ nm, respectively. This simple estimate indicates that particle velocity changes from 10^7 to 10^{10} times per second for the typical colloid size range, which means that this quantity cannot be determined by microscope observation as it has the duration time of the order of seconds.

Eq. (112) can be exploited to calculate the velocity autocorrelation function, which has an essential significance in determining particle size using the photon correlation spectroscopy [2,3]. The function is defined as

Table 4.3
The characteristic relaxation time t_{ch} [s], the characteristic velocity U_0 [nm s^{-1}] and the persistence length l_{ch} [nm] for particles of various sizes

a [nm]	t_{ch} [s]	U_0 [nm s^{-1}]	l_{ch} [nm]
1	4.4×10^{-13}	2.2×10^{10}	9.8×10^{-3}
10	4.4×10^{-11}	7.0×10^{8}	0.031
100	4.4×10^{-9}	2.2×10^{7}	0.098
10^3	4.4×10^{-7}	7.0×10^{5}	0.31
10^4	4.4×10^{-5}	2.2×10^{4}	0.98

Notes: T = 293 K, $\eta = 10^{-3}$ kg(m s)$^{-1}$, $\rho' = 2.10$ kg m^{-3}

$$t_{ch} = \frac{2\rho'}{9\eta}a^2; \quad U_0 = \frac{1}{2}\left[\frac{3kT}{\pi a^3\rho'}\right]^{1/2}; \quad l_{ch} = t_{ch}\,U_0$$

$$A(\Delta t) = \left\langle U^*(t)U^*(t+\Delta t)\right\rangle \tag{115}$$

Using Eq. (112) the $A(\Delta t)$ function can be expressed as

$$A(\Delta t) = \left\langle U_0^2\, e^{-2\beta t}\right\rangle + \frac{U_0^2}{m_p'} e^{-2\beta t}\int_0^t \left\langle F^*(t')\right\rangle e^{\beta t'}dt' + \frac{e^{-\beta(2t+\Delta t)}}{m_p'^2} \int_0^t e^{\beta t'}dt' \int_0^{t+\Delta t} \left\langle F(t')F(t'')dt''\right\rangle \tag{116}$$

By considering the properties of the random force expressed by Eq. (105), one can integrate Eq. (116) analytically, with the result

$$A(\Delta t) = V_0^2\, e^{-2\beta t} + \frac{F_0^2}{2m_p'^2} e^{-\beta\Delta t}(1 - e^{-2\beta t}) \tag{117}$$

For $t >> 1/\beta$, Eq. (117) becomes

$$A(\Delta t) = \frac{F_0^2}{2m_p'^2\,\beta} e^{-\beta\Delta t} = \frac{A_0\, M_{11}}{2m_p'} = A_0\, e^{-\beta\Delta t} \tag{118}$$

where

$$A_0 = \frac{F_0^2 M_{11}}{2m_p^{'}} \tag{119}$$

By noting that

$$A_0 = \left\langle U^*(t)U^*(t)\right\rangle = U_0^2 = \frac{kT}{m_p^{'}} \tag{120}$$

one obtains explicitly

$$F_0^2 = 2kT\, M_{11}^{-1} \tag{121}$$

By integrating Eq. (112) once more, one obtains the expression for the displacement

$$x^* = \frac{U_0^{'}}{\beta}(1-e^{-\beta t}) + \frac{1}{\beta m_p^{'}}\int_0^t F^*(t')dt' - \frac{e^{-\beta t}}{\beta m_p^{'}}\int_0^t F^*(t')e^{\beta t'}dt' \tag{122}$$

Eq. (122) as it stands is of limited applicability. However, it can be converted to a useful form by squaring and calculating the time average [2]. In this way one obtains

$$x^{*2} = \frac{\left\langle U_0^{'2}\right\rangle}{\beta^2}(1-e^{-\beta t})^2 + kT\, m_p^{'} M_{11}^2 (4e^{-\beta t} - e^{-2\beta t} - 3) + 2kT\, M_{11}\, t \tag{123}$$

In the limit of $t = 1/\beta >> t_{ch}$, Eq. (123) becomes

$$\left\langle x^{*2}\right\rangle = 2kT\, M_{11}\, t = 2D_{11}\, t \tag{124}$$

where the quantity

$$D_{11} = kT\, M_{11} \tag{125}$$

is defined as the translational diffusion coefficient of the particle.

Eq. (125) is usually referred to as the Einstein relationship.

As can be deduced from Eq. (124), the mean square displacement increases proportionally to the temperature and inversely proportionally to the medium viscosity and particle size. Contrary to intuitive expectations, $\langle x^{*2} \rangle$, and consequently the particle diffusion coefficient, does not depend on particle mass. This is so because with increasing mass the average velocity decreases but the persistence length increases in the same way, so both effects cancel each other out.

Using Eq. (36), one can formulate the explicit expression for D_{11} in the case of spherically symmetric bodies in the form

$$D_{11} = D = kT\,M_{11} = \frac{2}{9} kT \frac{K_s\, a_s^2}{\eta v_p} \tag{126}$$

where a_s is the equivalent radius of the body and K_s the shape parameter, defined by Eqs. (37),(38). For a sphere $a_s = a$, $K_s = 1$, and Eq. (126) becomes

$$D_{11} = D = \frac{kT}{6\pi\eta a} \tag{127}$$

For a fluid sphere one has

$$D = \frac{kT}{6\pi\eta a}\left(\frac{1+\eta/\eta_f}{1+2\eta/3\eta_f}\right) \tag{128}$$

On the other hand, for a cube having the size d, $K_s = 0.921$ as previously calculated, and Eq. (126) becomes

$$D = 0.0788\frac{kT}{\eta d} \tag{129}$$

As can be noticed from Eqs. (126–129), in the case of spherically isotropic bodies, the diffusion coefficient is inversely proportional to particle size. For spherically shaped bodies the diffusion coefficient varies from $2.15\times10^{-9}\,m^2\,s^{-1}$ for a 0.1nm diameter particle (the typical size of ions) in water to 2.15×10^{-14} m^2 s^{-1} for the particle of the size of 10^4 nm (10 μm), which is typical range of particle forming suspensions. Values of D for spherical particles of various size in water and in air at T = 293 K are given in Table 4.4. Note that in the case of aqueous solutions the diffusion coefficient increases quite significantly with temperature because of exponential decrease in the viscosity. On the other hand, for gases the diffusion coefficient remains much less sensitive to temperature because the viscosity increases in a linear manner with the temperature.

Once the diffusion coefficients are known, one can calculate the mean displacement of the particle, defining the characteristic length scale of Brownian motion transport. From Eq. (124) one can deduce that

$$l_B = \left\langle x^{*2} \right\rangle^{1/2} = (2D_{11}\,t)^{1/2} \tag{130}$$

For spherical particles one has

$$l_B = \left(\frac{kT}{3\pi\eta a} t \right)^{1/2} \tag{131}$$

Values of l_B calculated from Eq. (130) for a spherical particle in an aqueous medium are compiled in Table 4.5. For the colloid particle size range,

Table 4.4
The translational and rotational diffusion coefficients of spherical particles in water and air

a [nm]	Translational diffusion coefficient [m^2 s^{-1}]		Rotational diffusion coefficient [s^{-1}]	
	water	Air	water	Air
0.1 (ions)	2.15×10^{-9}	1.19×10^{-7}	1.61×10^{11}	8.89×10^{12}
1(surfactants)	2.15×10^{-10}	1.19×10^{-8}	1.61×10^{8}	8.89×10^{9}
10 (proteins)	2.15×10^{-11}	1.19×10^{-9}	1.61×10^{5}	8.89×10^{6}
100 (colloids, aerosols, viruses)	2.15×10^{-12}	1.19×10^{-10}	1.61×10^{2}	8.89×10^{3}
10^3 (bacteria)	2.15×10^{-13}	1.19×10^{-11}	1.61×10^{-1}	8.89
10^4 (suspensions)	2.15×10^{-14}	1.19×10^{-12}	1.61×10^{-4}	8.89×10^{-3}

Notes: T = 293 K, $\eta = 10^{-3}$ kg (m s)$^{-1}$ – water; $\eta = 1.81\times10^{-5}$ kg (m s)$^{-1}$ – air i.e., 100–1000 nm, l_b is of the order of 10^3 nm (for a time of one second). The displacement increases proportionally the square root of time, becoming of the order of 10^4 nm, for $t = 1000$ s. This means that Brownian motion is an inefficient mode of transportation for long times in comparison with external force and convective transport when particle displacement remains a linear function of time analyzed above. This becomes evident from the ratio of Brownian l_B to external force $l_l = Ut$ length scales

$$\frac{l_B}{l_l} = \left(\frac{2D_{11}}{U^2 t} \right)^{1/2} \tag{132}$$

Thus, the transport due to an external force always dominates over Brownian motion for long times when $t > 2D_{11}/U^2$.

By formally differentiating Eq. (130) with respect to time t, one obtains the expression for the apparent particle velocity

$$U_B = \frac{d}{dt} \left\langle x^{*2} \right\rangle^{1/2} = \left(\frac{D_{11}}{2t} \right)^{1/2} \tag{133}$$

For spherical particles one has

$$U_B = \left(\frac{kT}{12\pi\eta a t} \right)^{1/2} \tag{134}$$

Table 4.5
The characteristic velocity U_0, the apparent diffusion velocity U_B and the length l_B for spherical particles in water

a [nm]	U_0 [nm s^{-1}]	U_B [nm s^{-1}] t = 1 s	l_B [nm] t = 1 s	l_B[nm] t = 10^3 s
1	2.2×10^{10}	1.04×10^4	2.07×10^4	6.5×10^5
10	7×10^8	3.3×10^3	6.5×10^3	2.07×10^5
100	2.2×10^7	1.04×10^3	2.07×10^3	6.5×10^4
10^3	7×10^5	3.3×10^2	6.5×10^2	2.07×10^4
10^4	2.2×10^4	1.04×10^2	2.07×10^2	6.5×10^3

Notes: $T = 293$ K, $\eta = 10^{-3}$ kg (m s)$^{-1}$ $\rho'_p = 2\times10^3$ kg m^{-3} $U_B=(kT/12\pi\eta at)^{1/2}$ $l_B = (kTt/3\pi\eta a)^{1/2}$

Obviously, Eq. (134) is only valid for $t >> t_{ch}$ (which was estimated to be of the order of 10^{-7}– 10^{-11} s) for colloid particle. As can be seen, the apparent particle velocity decreases with the square root of time, so for experimental observation times of the order of seconds, it is a few orders of magnitude smaller than U_0. This is evident from the data compiled in Table 4.5, containing the values of U_0 calculated for spherical particles in an aqueous medium.

The above results obtained for particle translational motion in one direction can be generalized to particle motion in space (in three directions) and for rotational motion as well. This is so because the random force is uncorrelated as far as the space directions are concerned. Therefore, for spherically symmetric bodies, by repeating the integration procedure of the Langevine equation for other directions in space, the mean square displacement can be expressed in the form

$$\left\langle y^{*2} \right\rangle = 2D_{22}\,t \qquad \left\langle z^{*2} \right\rangle = 2D_{33}\,t \tag{135}$$

where, D_{22}, D_{33} are the diffusion coefficients for the y and z direction, respectively, assumed to be different for sake of generality.

Accordingly, the translation diffusion coefficient now becomes a diagonal matrix given by

$$\mathbf{D} = \begin{vmatrix} D_{11} & & 0 \\ & D_{22} & \\ 0 & & D_{33} \end{vmatrix} \tag{136}$$

The mean square displacement in this case (in three dimensions) is given by

$$\left\langle r^{*2} \right\rangle = \left\langle x^{*2} \right\rangle + \left\langle y^{*2} \right\rangle + \left\langle z^{*2} \right\rangle = 2(D_{11} + D_{22} + D_{33})\,t = 6\langle D \rangle t \tag{137}$$

where $\langle D \rangle = (1/3)\,(D_{11}+D_{22}+D_{33})$ is the averaged diffusion coefficient.

The matrix defined by Eq. (136) describes the diffusion of spherically symmetric particles in an anisotropic medium. In ordinary fluids, all the components of the diffusion matrix are equal, so

$$\mathbf{D} = D\,\mathbf{I} \tag{138}$$

The three dimensional mean square displacement in this case is given by

$$\left\langle r^{*2} \right\rangle = 6Dt \tag{139}$$

The anisotropy of the translational diffusion coefficient occurs for particle motion near interfaces because of the hydrodynamic boundary effects discussed extensively in Chapter 3. Particle mobility in this case becomes dependent on direction and is different for parallel or perpendicular motions relative to the interface. Exact expressions for the mobility matrix exist only for spherical particles moving near planar interfaces. It was shown by Brenner [29] that in this case the Einstein relationship expressed by Eq. (125) remains valid with the components of the hydrodynamic resistance tensors depending on the particle position. Accordingly, the translational diffusion tensor of a spherical particle near interfaces can be expressed as

$$\mathbf{D} = kT\,\mathbf{M} = \begin{vmatrix} D_{\|} & & 0 \\ & D_{\|} & \\ 0 & & D_{\perp} \end{vmatrix} = D \begin{vmatrix} F_4'(H) & & 0 \\ & F_4'(H) & \\ 0 & & F_1(H) \end{vmatrix} \mathbf{I} \tag{140}$$

where $D_{\perp}$, $D_{\|}$ are the diffusion coefficients for perpendicular and parallel particle motion given explicitly by

$$\begin{aligned} D_{\perp} &= \frac{kT}{6\pi\eta a} F_1(H) \\ D_{\|} &= \frac{kT}{6\pi\eta a} F_4'(H) \end{aligned} \tag{141}$$

where $F_1(H)$ and $F_4'(H)$ are the universal correction functions defined before.

As can be seen from Eq. (141), the mobility of the particle depends on the dimension less distance from the interface only.

The mean square displacement in this case is given by

$$\left\langle r^{*2} \right\rangle = 2D_{\perp}\, t + 4D_{\parallel}\, t \tag{142}$$

However, it should be mentioned, that in this case, because diffusion coefficients depend on the separation distance from the interface, the mean square displacement remains a well-defined quantity in a local sense only, i.e., it remains valid if $\langle r^{*2}\rangle^{1/2} \ll aH$, when the displacement remains much smaller than the particle distance from the interface.

In an analogous way, using the Langevine equation, one can analyze the rotational Brownian motion of particles. By considering the one-dimensional case, one has

$$In\frac{d\Omega}{dt} = In\frac{d^2\varphi}{dt^2} = -Mr_{11}^{-1}\,\Omega + To + To^{*}(t) \tag{143}$$

where I_n is the moment of inertia for rotation about this axis (a scalar), φ is one of the Eulerian angles and Mr_{11} is the rotational mobility matrix component. For spheres $I_n = (4/5)ma^2$.

Performing an analogous analysis as before for the translational motion and considering that

$$\frac{1}{2} In\left\langle \Omega^{*}(t)\Omega^{*}(t) \right\rangle = \frac{1}{2}kT \tag{144}$$

it can be shown that after a short transition time of the order of $t_{ch} = I_n\, Mr_{11}$ the mean square angular displacement of the particle is described by

$$\left\langle \varphi^{*2} \right\rangle = 2kT\, Mr_{11}\, t = 2Dr_{11}\, t \tag{145}$$

where

$$Dr_{11} = kT\, Mr_{11} \tag{146}$$

is defined as the rotational diffusion coefficient of a particle. For spheres one has $Mr_{11} = 1/8\pi\eta a^3$ and the rotational diffusion coefficient is given by

$$Dr_{11} = Dr = \frac{kT}{8\pi\eta a^3} \tag{147}$$

Values of Dr calculated for a spherical particle immersed in water and air are collected in Table 4.4.

From Eq. (145) one can deduce that for spheres

$$\varphi_B = \left\langle \varphi^{*2} \right\rangle^{1/2} = \left(\frac{kT\,t}{4\pi\eta a^3} \right)^{1/2} \tag{148}$$

Therefore, the apparent angular velocity of the rotational Brownian motion of a sphere is

$$\Omega_B = \frac{d}{dt}\varphi_B(t) = \left(\frac{Dr}{2t} \right)^{1/2} = \left(\frac{kT\,t}{16\pi\eta a^3} \right)^{1/2} \tag{149}$$

Using the data given in Table 4.4 one can estimate that for an aqueous medium, assuming $a = 1\text{nm}$, and $t = 1$ s, $Dr = 1.61 \times 10^8\,\text{s}^{-1}$, so $\varphi_B = 1.79 \times 10^4$. This rotational displacement corresponds to $\varphi_B / 2\pi = 2.9 \times 10^3$ rotations of the particle per second. For a colloid particle of radius $a = 1000$ nm, and for $t = 1$ s, $Dr = 0.161\ \text{s}^{-1}$, so $\varphi_B = 0.57$. This rotational displacement corresponds to 9×10^{-2} rotations per second.

The anisotropy of the rotational diffusion coefficient appears for particle motion near interfaces because of the hydrodynamic boundary effects similar to the case of translational motion. Particle rotational mobility in this case becomes dependent on the direction of the angular velocity vector, which is different for the parallel and perpendicular rotation relative to the interface. Analogous to previous cases, the rotational diffusion tensor of a spherical particle near interfaces can be expressed as

$$\mathbf{Dr} = \begin{vmatrix} Dr_{\|} & & 0 \\ & Dr_{\|} & \\ 0 & & Dr_{\perp} \end{vmatrix} = Dr \begin{vmatrix} F_{11}(H) & & 0 \\ & F_{11}(H) & \\ 0 & & F_{10}(H) \end{vmatrix} \tag{150}$$

where $D_{\perp}, Dr_{\|}$ are the rotational diffusion coefficients for the perpendicular and parallel particle motion and $F_{10}(H) = F_6/(1-(3/4)((F_4F_6)/F_7^2))$, $F_{11}(H) = F_5/(1-(3/4)((F_1F_5)/F_7^2))$ are the universal correction functions.

Note that, the rotational mobility of a particle depends on the dimensionless distance from the interface only as is the case with the translational motion.

4.4.2. Brownian Motion of Non-Spherical Particles

In the case of non-spherical particles the Langevine equation can be, in principle, used to derive the translational and rotational displacements and to define diffusion coefficients. However, the analysis becomes rather complicated because of the coupling between these two modes of particle motion. There are two types of coupling: (i) orientation coupling and (ii) hydrodynamic coupling. The former appears because the resistance coefficient of a translating non-spherical particle depends on its orientation with respect to a space-fixed coordinate system. Thus, rotational Brownian motion induces changes in particle orientation that affect its translation. However, particle rotation is not coupled with the translational motion because the rotational mobility matrix does not depend on particle position (except for in the region close to boundary surfaces). In principle, orientational coupling can be eliminated when considering translational motion in a particle-fixed coordinate system. However, such results are physically meaningful in the short-time limit only, when particle translational displacement remains so small (*quasi*-static) that its orientation can be assumed fixed. In order to calculate a real long-time displacement, one has to solve the multidimensional diffusion equations as discussed in [2]. Our further discussion in this section is based on the assumption of the quasi-static translational motion postulate.

By accepting this assumption, one can derive general expressions for the diffusion matrix of non-spherical particles in the case, where hydrodynamic coupling between the rotational and the translational motion occurs. It was postulated by Brenner [30,31] and later on rigorously proven by Condiff and Dahler [32] that in this case the diffusion matrix is given by

$$\tilde{\mathbf{D}} = kT\tilde{\mathbf{M}} = kT\left|\tilde{\boldsymbol{K}}^{-1}\right| \tag{151}$$

where $\tilde{\mathbf{M}}$ is the grand mobility matrix defined previously.

The relation expressed by Eq. (151) is often called the generalized Einstein relationship.

For orthotropic bodies far from boundary surfaces, where the there is no hydrodynamic coupling, the grand diffusion matrix can be decomposed into two parts

$$\begin{aligned}
\mathbf{D} = kT\,\mathbf{M} = kT \begin{vmatrix} M_{11} & & 0 \\ & M_{22} & \\ 0 & & M_{33} \end{vmatrix} \\
\mathbf{D}\mathrm{r} = kT\,\mathbf{M} = kT \begin{vmatrix} Mr_{11} & & 0 \\ & Mr_{22} & \\ 0 & & Mr_{33} \end{vmatrix}
\end{aligned} \tag{152}$$

The translational mean square displacement in this case (in three dimensions) is given by

$$\left\langle r^{*2} \right\rangle = 2(D_{11} + D_{22} + D_{33}) = 6\langle D \rangle t \tag{153}$$

where the average diffusion coefficient is

$$\langle D \rangle = kT\langle M \rangle = \frac{1}{3}kT(M_{11} + M_{22} + M_{33}) \tag{154}$$

and $\langle M \rangle = (1/3)\ (M_{11}+M_{22}+M_{33})$ is the average mobility of the orthotropic bodies (a scalar quantity) defined previously by Eq. (32).

The averaged diffusion coefficient is a quantity of major practical significance because it can be used to approximate dynamic phenomena occurring in non-spherical particle suspensions such as sedimentation or transport to interfaces when Brownian motion dominates over convection [2]. Then, because of rotary Brownian motion, particle orientations become nearly random, so a particle can be treated as a sphere, characterized by the effective diffusion coefficient $\langle D \rangle$.

Using the previously derived expressions for $\langle M \rangle$ one can derive formulae for the averaged diffusion coefficient of bodies of various shape. For the sake of convenience they are compiled in Table 4.6. It is interesting to note that for spherically isotropic bodies the average diffusion coefficient is inversely proportional to particle size. For particles of other shapes $\langle D \rangle$ is a complex function of particle dimensions, e.g., in the case of slender bodies of length l, one has

$$\langle D \rangle = \frac{kT}{6\pi\eta l}\ln(l/b) \tag{155}$$

Eq. (155) indicates that the value of the average diffusion coefficient of very elongated (needle-like) particles is controlled by their length rather than by their diameter. This becomes more obvious when we realize that for $l/b = 100$, a tenfold increase in the length of the particle causes a decrease in the diffusion coefficient by a factor of 6.7, whereas a tenfold increase in the diameter of the particle cause a decrease in the diffusion coefficient by a factor of 1.5 only.

Table 4.6
Translation diffusion coefficients of orthotropic bodies of various shapes

Body	Diffusion coefficient	Averaged diffusion coefficient
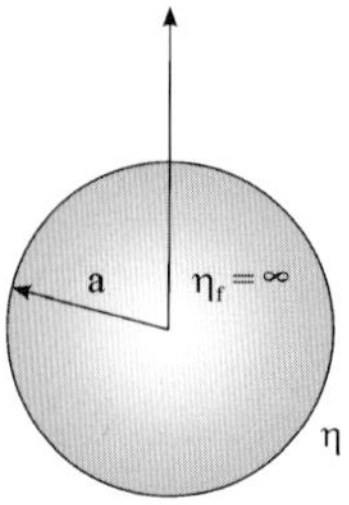 Solid sphere	$\frac{kT}{6\pi\eta a}$	$\frac{kT}{6\pi\eta a}$
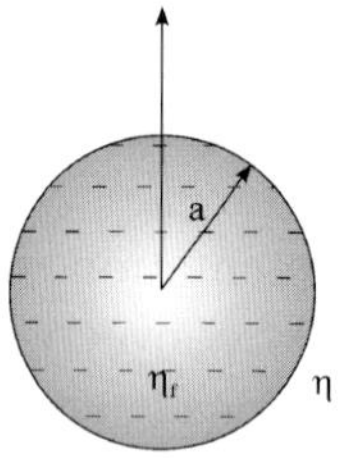 Liquid sphere	$\frac{kT}{6\pi\eta a}\left(\frac{1+\eta/\eta_f}{1+2\eta/3\eta_f}\right)$	$\frac{kT}{6\pi\eta a}\left(\frac{1+\eta/\eta_f}{1+2\eta/3\eta_f}\right)$
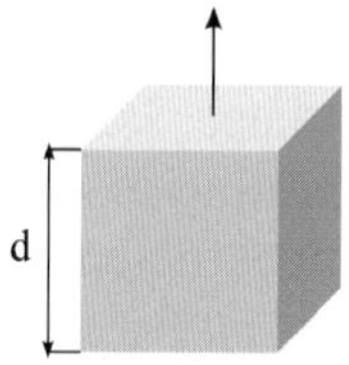 Solid cube	$0.0788\frac{kT}{6\pi\eta d}$	$0.0788\frac{kT}{6\pi\eta d}$
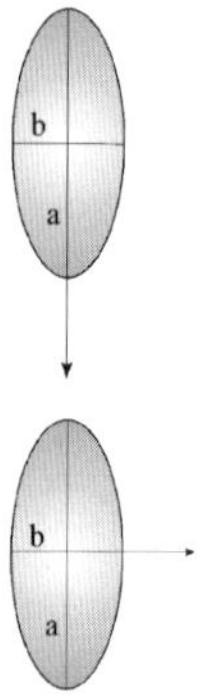 Prolate spheroids	$\frac{kT}{6\pi\eta a K_1}$ $\frac{kT}{6\pi\eta b K_2}$	$\frac{kT \cosh^{-1}(a/b)}{6\pi\eta a[1-(b/a)^2]^{1/2}}$

Table 4.6 (continued)

Body	Diffusion coefficient	Averaged diffusion coefficient
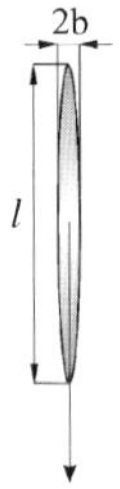	$\frac{kT}{2\pi\eta l}[\mathrm{In}(l/b) - 0.5]$	
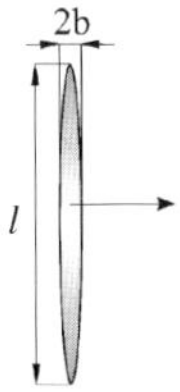	$\frac{kT}{4\pi\eta l}[\mathrm{In}(l/b) + 0.5]$	$\frac{kT}{3\pi\eta l}\ln(l/b)$
Slender bodies		
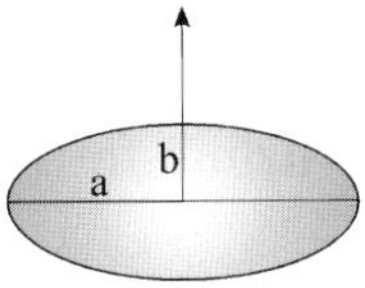	$\frac{kT}{6\pi\eta a K_1}$	
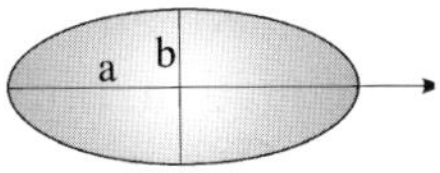	$\frac{kT}{6\pi\eta b K_2}$	$\frac{kT}{6\pi\eta a}\frac{\cos^{-1}(b/a)}{\left[1-(b/a)^2\right]^{1/2}}$
Oblate spheroids		
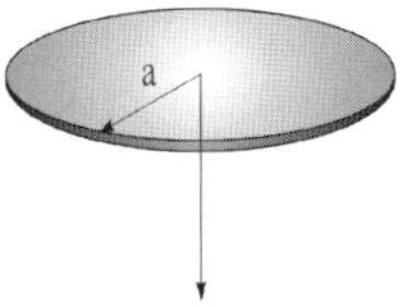	$\frac{kT}{16\eta a}$	
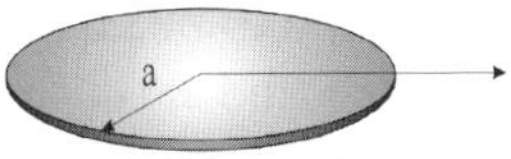	$\frac{3kT}{32\eta a}$	$\frac{kT}{12\eta a}$
Disk broadside and disk edgewise		

Expressions for the K_1 and K_2 constants for prolate and oblate spheroids are given in Table 4.2.

It is also interesting to note that in the case of disks

$$\langle D \rangle = \frac{kT}{12\eta a} \tag{156}$$

regardless of the thickness of the disk. Thus, the diffusion coefficient of a disk is only $\pi/2 = 1.571$ times larger that of a sphere of the same radius a. This means that the shape exerts quite an insignificant influence on diffusion coefficients of oblate particles.

4.4.3. Diffusion of Isolated Particles – the Fokker-Planck and Smoluchowski Equations

All the above considerations allow one to calculate the mean displacements of Brownian particles as a function of their diffusion coefficient and time. However, no expressions were formulated for the flux of particle probability density, that could make it possible to formulate the general diffusion equation (conservation law) of this quantity. Once the diffusion equation is known, various effects of practical interest can be quantitatively analyzed, e.g., the role of the initial conditions, the geometry of the system confining the particle, the coupling with external, specific or hydrodynamic forces leading to particle dispersion in various flows, etc.

The constitutive equation for the diffusion flux can be derived using the method originally employed by Einstein [24,25] and extended later on by Chandrasekhar [33]. The method is based on the Markoff equation, often referred to as the master equation, describing probability evolution in time for uncorrelated stochastic processes.

Beforc doing this, it is useful to derive, as was done by Smoluchowski [34], the relationship governing the distribution of the displacement probability as a function of time. Let us assume the one-dimensional case when a Brownian particle undergoes a large number N_B of elementary displacements (jumps) of equal length l_b in two opposite directions. The probability of making an elementary positive or negative jump are assumed to be equal. Denote by n_+ the number of jumps in the positive direction and n_- in the negative direction, respectively. Obviously, on average they are equal, but due to the stochastic nature of this process the difference $n_B = n_+ - n_-$, although being very small in comparison with N_B, remains non-zero. The probability of finding any n_B (which can be treated as a measure of success in the Brownian motion game) is governed by the well-known Bernoulli distribution [34]

$$P_B(N_B, n_B) = \frac{N_B!}{\left[\frac{1}{2}(N_B + n_B)\right]!\left[\frac{1}{2}(N_B - n_B)\right]!} \tag{157}$$

For $N_B >> 1$ and $n_B/N_B << 1$, which is always the case in any real situation involving Brownian motion, P_B can be transformed to a more useful form by taking the logarithm and using Stirling's formula $N_B \cong (N_B + (1/2))\ln N_B - N_B + (1/2)\ln 2\pi$. In this way one obtains

$$P_B = \left(\frac{2}{\pi N_B}\right)^{1/2} e^{-\frac{n_B^2}{2N_B}} \tag{158}$$

By defining a *quasi*-continuous variable

$$x = (2N_B c_B)^{1/2} n_B \tag{159}$$

where c_B is an arbitrary constant to be determined, one can covert Eq. (158) to the useful form

$$P_B(x)dx = \left(\frac{c_B}{\pi}\right)^{1/2} e^{-c_B x^2} dx \tag{160}$$

Eq. (160) representing the general distribution law of particle displacements has a form analogous to the Gauss law of error distributions. Indeed, the normalization condition is fulfilled because

$$\int_{-\infty}^{\infty} P_B(x)dx = 1 \tag{161}$$

Also the first moment (averaged displacement) vanishes since

$$\langle x \rangle = \int_{-\infty}^{\infty} x P_B(x)dx = \left(\frac{c_B}{\pi}\right)^{1/2} \int_{-\infty}^{\infty} x e^{-c_B x^2} dx = \frac{1}{2(\pi c_B)^{1/2}} e^{-c_B x^2} \Big|_{-\infty}^{\infty} = 0 \tag{162}$$

The mean square displacement (the second moment of the distribution P_B) is

$$\langle x^2 \rangle = \int_{-\infty}^{\infty} x^2 P_B(x)\, dx = \frac{1}{2c_B} \tag{163}$$

Using Eq. (124) one can calculate the constant c_B, which is given by

$$c_B = \frac{1}{4D_{11}t} \tag{164}$$

Hence, the explicit expression for the displacement probability is

$$P_B(x)\,dx = \frac{1}{2(\pi D_{11} t)^{1/2}} e^{-\frac{x^2}{4D_{11}t}} dx \tag{165}$$

Particle motion along the y- and z-axes (of the coordinate system fixed at the particle) can be considered in a similar way. The overall displacement in this case is given by

$$P_B\, dx\, dy\, dz = \frac{1}{8(D_{11}D_{22}D_{33})^{1/2}(\pi t)^{3/2}} e^{-\frac{1}{4t}\left(\frac{x^2}{D_{11}}+\frac{y^2}{D_{22}}+\frac{z^2}{D_{33}}\right)} dx\, dy\, dz \tag{166}$$

For spherically isotropic bodies, Eq. (166) simplifies to

$$P_B\, dv = \frac{\pi r^2\, dr}{2(\pi D t)^{3/2}} e^{-\frac{r^2}{4Dt}} dv \tag{167}$$

where $dv = dx\, dy\, dz = 4\pi r^2 dr$, $r^2 = x^2 + y^2 + z^2$.
For two-dimensional Brownian motion one has, respectively

$$P_B\, dS = \frac{r\, dr}{2Dt} e^{-\frac{r^2}{4Dt}} dS \tag{168}$$

where $r^2 = x^2 + y^2$ and $dS = 2\pi r dr$.

Eqs. (165–168) are useful for predicting the evolution of the probability of finding a Brownian particle in a space after a given time provided that its initial position is known. However, they still do not furnish information on the diffusion flux and its relation to external and specific forces. This can be done using the Einstein-Chandresakhar approach [33]. The starting point for this approach is the Markoffian equation governing the time evolution of the probability of occurrence of any state of a system undergoing a statistical process. $f_p(\tilde{\mathbf{R}},t)$ denotes the probability density of finding a particle in the state $\tilde{\mathbf{R}}$ at the time t, where $\tilde{\mathbf{R}}$ is a generalized vector describing the phase point of a system in a generalized space of position and orientation. By defining further $K_p(\tilde{\mathbf{R}} - \Delta\tilde{\mathbf{R}}|\Delta\tilde{\mathbf{R}},\tau)$ as the probability that the particle state changes from $\tilde{\mathbf{R}} - \Delta\tilde{\mathbf{R}}$ to $\tilde{\mathbf{R}}$ during the time interval τ, one can formulate the Markoffian equation [32]

$$f_p(\tilde{\mathbf{R}},t+\tau) = \int K_p(\tilde{\mathbf{R}} - \Delta\tilde{\mathbf{R}}|\Delta\tilde{\mathbf{R}},\tau)\, f_p(\tilde{\mathbf{R}} - \Delta\tilde{\mathbf{R}},\tau)\, d(\Delta\tilde{\mathbf{R}}) \tag{169}$$

Eq. (169) is independent of any particular mechanism of particle transfer from the state $\tilde{\mathbf{R}} - \Delta\tilde{\mathbf{R}}$ to $\tilde{\mathbf{R}}$, which can be convection, external force, Brownian motion, etc.

It is usually assumed that the K_p function in Eq. (169) (kernal) falls off rapidly with $\tilde{\mathbf{R}}$, so the absolute value of state increment $\tilde{\mathbf{R}}$ during the time τ remains much smaller than the absolute value (length) of the vector $\tilde{\mathbf{R}}$. By virtue of this assumption, one can expand the K_p function into the Taylor series (by retaining terms of the second order only), which results in the following expression

$$f_p(\tilde{\mathbf{R}},t+\tau) = f_p(\tilde{\mathbf{R}},t) + \tau\frac{\partial f_p}{\partial t}$$

$$= \int\left[1 - \Delta\tilde{\mathbf{R}}\frac{\partial}{\partial\tilde{\mathbf{R}}} + \frac{1}{2}\Delta\tilde{\mathbf{R}}\Delta\tilde{\mathbf{R}}:\frac{\partial}{\partial\tilde{\mathbf{R}}}\frac{\partial}{\partial\tilde{\mathbf{R}}} + \cdots\right]$$

$$K_p(\tilde{\mathbf{R}}/\Delta\tilde{\mathbf{R}},\tau)\, f_p(\tilde{\mathbf{R}},t)\, d(\Delta\tilde{\mathbf{R}})$$

$$= f_p(t,\tilde{\mathbf{R}}) - \frac{\partial}{\partial\tilde{\mathbf{R}}}\cdot\langle\Delta\tilde{\mathbf{R}}\rangle f_p + \frac{1}{2}\frac{\partial}{\partial\tilde{\mathbf{R}}}\frac{\partial}{\partial\tilde{\mathbf{R}}}:\langle\Delta\tilde{\mathbf{R}}\Delta\tilde{\mathbf{R}}\rangle f_p \tag{170}$$

where $\langle(...)\rangle = \int(...)K_p d(\Delta\tilde{\mathbf{R}})$ means the phase space averaging, in particular the probability normalization condition

$$\langle 1\rangle = \int K_p d\tilde{\mathbf{R}} = 1 \tag{171}$$

This considered, Eq. (170) can be finally rearranged to the form of the generalized Fokker-Planck equation [32]

$$\frac{\partial f_p}{\partial t} = \frac{\partial}{\partial\tilde{\mathbf{R}}}\frac{\partial}{\partial\tilde{\mathbf{R}}}:\left[\frac{1}{2\tau}\langle\Delta\tilde{\mathbf{R}}\Delta\tilde{\mathbf{R}}\rangle f_p\right]-\frac{\partial}{\partial\tilde{\mathbf{R}}}\cdot\left[\frac{1}{\tau}\langle\Delta\tilde{\mathbf{R}}\rangle f_p\right] \tag{172}$$

By choosing the simplest case of $\tilde{\mathbf{R}} = x$ (one-dimensional translation of a particle without rotation) one can formulate Eq. (172) as

$$\frac{\partial f_p}{\partial t} = \frac{\partial}{\partial x}\left[\frac{\partial}{\partial x}\frac{\langle\Delta x^2\rangle}{2\tau}f_p - \frac{\Delta x}{\tau}f_p\right] = \frac{\partial}{\partial x}\left[\frac{\partial}{\partial x}(D f_p) - U f_p\right] \tag{173}$$

where

$$D = \frac{\langle\Delta x^2\rangle}{2\tau} \tag{174}$$

and $U=\Delta x/\tau$ is the particle velocity due to the net deterministic force acting on the particle.

Eq. (173) is the general diffusion equation in one dimension incorporating in an exact manner the effects stemming from the stochastic Brownian force and from all other deterministic forces. The quantity in brackets in Eq. (173) can be identified as the flux of the probability density f_p, which is given by

$$-j_{f_p} = \frac{\partial}{\partial x}(D_{11} f_p) - U f_p = \frac{\partial}{\partial x}(D_{11} f_p) - D_{11}\frac{F}{kT}f_p \tag{175}$$

Eq. (175), indicates that the effects stemming from diffusion and the deterministic forces F are additive. Moreover, the diffusion flux is proportional to the gradient of the product $D_{11} f_p$ or to the gradient of f_p in the case when D_{11} is position independent. Using Eq. (173) and Eq. (175)

it is possible to formulate the probability conservation law (balance equation of the f_p variable)

$$\frac{\partial f_p}{\partial t}+\frac{\partial}{\partial x}j_{f_p}=0 \tag{176}$$

Several cases of major practical significance can be derived from Eq. (176), e.g., for position-independent diffusion coefficient and no external force, one has

$$\frac{\partial f_p}{\partial t}=D\frac{\partial^2 f_p}{\partial x^2} \tag{177}$$

Eq. (177) is the basic diffusion equation in one dimension, having an analogous form to the heat diffusion equation [35].

On the other hand, if the diffusion coefficient remains constant but an external force appears, then Eq. (176) becomes

$$\frac{\partial f_p}{\partial t}=D\frac{\partial^2 f_p}{\partial x^2}-D\frac{\partial}{\partial x}\left(\frac{F}{kT}f_p\right) \tag{178}$$

Eq. (178) is widely known as the Smoluchowski equation [2].

If moreover F is position-independent, as is the case for example with the gravitational force, Eq. (178) becomes

$$\frac{\partial f_p}{\partial t}=D\frac{\partial^2 f_p}{\partial x^2}-\frac{DF}{kT}\frac{\partial f_p}{\partial x}=D\frac{\partial^2 f_p}{\partial x^2}-U\frac{\partial f_p}{\partial x} \tag{179}$$

In a more general case when diffusion proceeds in three dimensions and a coupling between the translational and rotational motions occurs, the expression for the flux of f_p can be derived from Eq. (172). Condiff and Dahler [32] showed that the generalized flux is given by

$$\tilde{\mathbf{j}}=-\tilde{\mathbf{D}}\cdot\nabla f_p+\tilde{\mathbf{U}}f_p=-\tilde{\mathbf{D}}\cdot\nabla f_p+\frac{1}{kT}\tilde{\mathbf{D}}\cdot\tilde{\mathbf{F}} \tag{180}$$

where $\tilde{\mathbf{F}}$ is the generalized force vector incorporating the net force and torque on a particle.

The conservation equation for f_p becomes

$$\frac{\partial f_p}{\partial t} + \nabla \cdot \tilde{\mathbf{j}} = \frac{\partial f_p}{\partial t} + \nabla \cdot (-\tilde{\mathbf{D}} \nabla f_p + \frac{1}{kT} \tilde{\mathbf{D}} \cdot \tilde{\mathbf{F}}) = 0 \tag{181}$$

Eq. (181) is expressed in a particlefixed coordinate system. When formulating in the space-fixed coordinate system it becomes, in the general case of non-spherical particles, a sixth-order partial differential equation, whose solution can be found for some limiting cases only [3], e.g., for spherical particles.
In the case of a particle moving near planar interfaces, Eq. (181) becomes

$$\frac{\partial f_p}{\partial t} + \mathbf{\nabla} \cdot (-\mathbf{D} \cdot \mathbf{\nabla} f_p + \mathbf{U} f_p) = 0 \tag{182}$$

where the flux of f_p is given by

$$\tilde{\mathbf{j}} = \tilde{\mathbf{D}} \cdot \nabla f_b + \mathbf{U} f_p - \tilde{\mathbf{D}} \cdot \nabla f_p + (\mathbf{M} \cdot \mathbf{F} + \mathbf{M}_c \cdot \mathbf{To}) f_p \tag{183}$$

An analogous equation can be formulated for rotational flux, which is not relevant here, however, because particle orientations are indistinguishable.

On the other hand, for a spherical particle in an isotropic medium far from interfaces, where the wall effects are negligible and $\mathbf{D} = D\,\mathbf{I}$, the diffusion equation assumes thc form

$$\frac{\partial f_p}{\partial t} = D \mathbf{\nabla}^2 f_p - \frac{D}{kT} \mathbf{\nabla} \cdot (\mathbf{F} f_p) \tag{184}$$

In the absence of forces, one has

$$\frac{\partial f_p}{\partial t} = D\, \mathbf{\nabla}^2 f_p \tag{185}$$

Eq. (185) is the general diffusion equation of spherical particles in three dimensions.

4.4.4. Limiting Solutions of the Diffusion Equation for an Isolated Particle

Consider the one-dimensional motion of a spherically symmetric particle, placed at $t = 0$ in an infinite medium at the point (x_0, y_0, z_0). Assume that there is no external force. The three-dimensional diffusion equation now becomes

$$\frac{\partial f_p}{\partial t} = D_{11}\frac{\partial^2 f_p}{\partial x^2} + D_{22}\frac{\partial^2 f_p}{\partial y^2} + D_{33}\frac{\partial^2 f_p}{\partial z^2} \tag{186}$$

The initial condition for Eq. (186) has the form

$$f_p = \delta_D(x_0, y_0, z_0), \quad \text{for} \quad t = 0 \tag{187}$$

where $\delta_D(x_0, y_0, z_0)$ is the Dirac delta function.
This corresponds to a unit source placed initially at the point (x_0, y_0, z_0) [35]. The boundary condition from Eq. (186) is the postulate that f_p vanishes at all times at a large distance from the origin (x_0,y_0,z_0). The solution of Eq. (187) with these initial and boundary conditions is [35]

$$f_p = \frac{1}{8(D_{11}D_{22}D_{33})^{1/2}(\pi t)^{3/2}} e^{-\frac{1}{4t}\left[\frac{(x-x_0)^2}{D_{11}}\frac{(y-y_0)^2}{D_{22}}\frac{(z-z_0)^2}{D_{33}}\right]} \tag{188}$$

Eq. (188) having analogous form as previously derived, cf. Eq. (166), describes the diffusion of spherically symmetric particles in an anisotropic medium.

In the case of spheres and an isotropic medium, the probability of finding a particle in the shell $4\pi r^2 dr\, f_p$, where $r=(x-x_0)^2+(y-y_0)^2+(z-z_0)^2$, is given by

$$P_B = \frac{\pi r^2}{2(\pi D t)^{3/2}} e^{-\frac{r^2}{4Dt}} dr = \frac{4}{\pi^{1/2}} \xi^2 e^{-\xi^2} d\xi \tag{189}$$

where $\xi = r/2(Dt)^{1/2}$ is the dimensionless variable.

As can be noticed, P_B vanishes at the origin, for $t > 0$. Because $f_p \rightarrow 0$ in the limit of $r \rightarrow \infty$ a maximum of the $P_B(r)$ function appears. This can be seen in Fig. 4.8 where the dependence of P_B on ξ is shown.

In the two-dimensional case, the probability of finding a particle in the ring $2\pi r^2 dr\, f_p$, where $r^2=(x-x_0)^2+(y-y_0)^2$, is given by

$$P_B = \frac{r}{2Dt} e^{-\frac{r^2}{4Dt}} dr = 2\xi e^{-\xi^2} d\xi \tag{190}$$

Finally, in the one-dimensional case, the probability of finding a particle within dx is given by

$$P_B = \frac{1}{(\pi D t)^{1/2}} e^{-\frac{(x-x_0)^2}{4Dt}} dx = \frac{1}{2(\pi)^{1/2}} e^{-\xi^2} d\xi \tag{191}$$

where, in this case, $\xi = (x-x_0)/2(Dt)^{1/2}$.

The dependence of the $P_B/d\xi$ function on ξ for the two-dimensional and one-dimensional case is also plotted in Fig. 4.8.

When a spherical particle is placed initially at a distance z above an impenetrable wall (see Fig. 4.9), the solution of the diffusion equation, Eq. (185), was first found by Smoluchowski [36] using the method of reflections. It has the form

$$f_p = \frac{1}{8(\pi D t)^{3/2}} e^{-\frac{r^2}{4Dt}} \left[e^{-\frac{(z-z_0)^2}{4Dt}} + e^{-\frac{(z+z_0)^2}{4Dt}} \right] \tag{192}$$

where $r^2=x^2+y^2$ and the vertical position of the particle z is measured from its surface, rather than from its center as done by Smoluchowski.

Note that by deriving Eq. (192), the change in the diffusion coefficient with the distance from the wall was neglected, so it is inaccurate for $z/a < 1$.

Other analytical solutions of quite general validity can be derived in the case, where the Brownian motion of a particle is affected by an external force. Let us consider the case, first analyzed by Smoluchowski [36], of a Brownian particle moving in the vicinity of a planar wall under the action

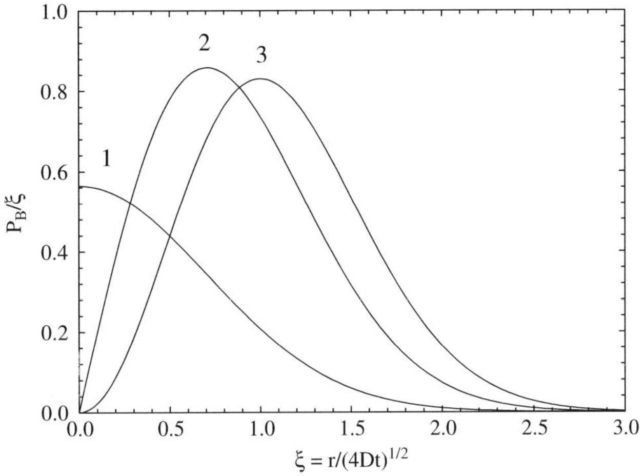

Fig. 4.8. The dependence of the $P_B\,/d\xi$ function on $\xi = r/(4D\ t)^{1/2}$ for the one-dimensional Brownian motion of a spherical particle (curve 1), two-dimensional (curve 2) and three-dimensional (curve 3).

of an external force **F** directed to the interface along the z axis (see Fig. 4.9). The governing equation now assumes the form

$$\frac{\partial f_p}{\partial t} = \nabla\cdot\left(\mathbf{D}\cdot\nabla f_p - \frac{1}{kT}\mathbf{D}\cdot\mathbf{F}\ f_p\right) = \nabla\cdot(\mathbf{D}\cdot\nabla f_p - \mathbf{U} f_p) \tag{193}$$

The initial conditions are identical to those expressed by Eq. (187), i.e., the particle was placed initially at a distance z_0 above the plane defined by $z = 0$. In the general case, there can be two different boundary conditions, at the plane,

$$\begin{aligned} &f_p = 0 && \text{at} \quad z = 0 \\ &or && \\ &j_z = \left(-D\frac{\partial f_p}{\partial z} + U_z f_p\right) = 0 && \text{at} \quad z = 0 \end{aligned} \tag{194}$$

where U_z is the z component of the particle velocity. The first condition, Eq. (194), corresponds to perfect sink model, where a particle touching the plane becomes irreversibly attached, and the second condition corresponds to the impenetrable wall assumption used above to derive the solution for no external

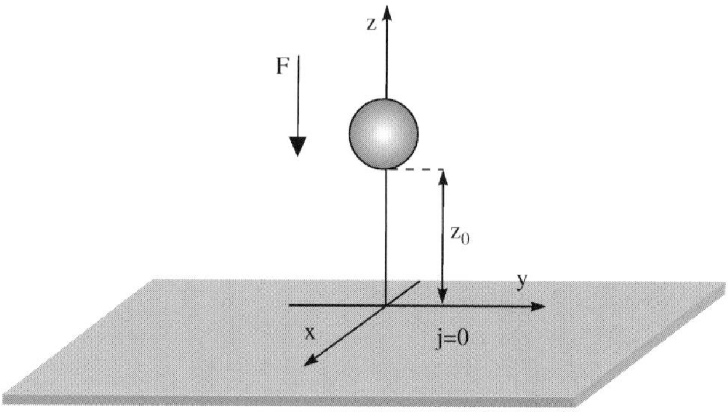

Fig. 4.9. Brownian motion of a spherical particle under the action of an external force F near a solid planar wall.

force. We will consider the latter possibility, because the perfect sink model is discussed extensively later on. Moreover, the stationary solutions derived with the condition $j = 0$ correspond to the thermodynamic equilibrium postulate.

Unfortunately, no analytical solution of the non-stationary Eq. (193) can be found in the general case of a variable diffusion coefficient and the force depending on the position. However, a stationary solution of practical significance can be found by assuming that equilibrium is established, when $\frac{\partial f_p}{\partial t} = 0$ and $j_z = 0$. By integrating Eq. (193) one obtains in this case

$$f_p = \frac{e^{-\phi(z)/kT}}{\int_0^\infty e^{-\phi(z)/kT} dz} \tag{195}$$

where

$$\phi(z) = -\int_0^z F(z')dz' \tag{196}$$

is the potential energy due to the force F.

Eq. (195) is the well-known Boltzmann law, which is valid for particles of arbitrary sizes and for an arbitrary interaction energy, e.g., the electrostatic energy, which was exploited before to express ion concentration distribution in the double-layer. For a constant force, e.g., gravitational or electric force discussed in Section 4.2, the Boltzmann distribution, Eq. (195), assumes the form

$$f_p = \frac{e^{Fz/kT}}{\int\limits_0^\infty e^{Fz/kT}\, dz} = f_0\, e^{Fz/kT} = f_0\, e^{-\phi/kT} = \frac{U}{D} e^{-\frac{U}{D}z} \tag{197}$$

where $f_0 = F/kT$ is the probability of finding the particle at $z = 0$, and U the steady migration velocity given in Table 4.2. In the case of the gravitational force, Eq. (197) becomes

$$f_p = \frac{m'|\mathbf{g}|}{kT}\, e^{-m'|\mathbf{g}|\, z/kT} = f_0\, e^{-\frac{\frac{4}{3}\pi a^3 \Delta\rho |\mathbf{g}| N_{Av}}{R_g} z} \tag{198}$$

where R_g is the gas constant, and N_{Av} the Avogadro constant (see Table 2.1).

Eq. (198) indicates that by measuring the probability of finding a particle subject to gravitational force at a given height from the interface, one can determine Avogadro number N_{Av} when the particle size, and the density and viscosity of the medium are known. This approach was used by Perrin [27,28] who determined $N_{Av} = 6.83 \times 10^{23}$ for emulsions of latex particles. The deviation from the exact value of N_{Av} was probably due to errors in estimating particle density, which depends critically on particle porosity. This method of determining N_{Av} is of historical interest only because there are many other much more accurate methods, for example, at this same time Millikan determined $N_{Av} = 6.06 \times 10^{23}$ by measuring the elementary charge). The possibility of determining the real density of suspended particles or the viscosity of the dispersing medium by this method seems to be more interesting.

Using Eq. (197) one can calculate the averaged distance of the particle from the interface which is given by

$$\langle z \rangle = \int\limits_0^\infty z\, f_p\, dz = \frac{D}{U} = \frac{kT}{F} \tag{199}$$

As can be noticed, the displacement assumes a finite value, which indicates that, owing to the random force exerted by the medium, particle position deviates from the equilibrium position, $z = 0$, by the increment $\langle z \rangle$. Hence, the deviation of the particle energy from the thermodynamic equilibrium equals $\langle \phi \rangle = F\langle z \rangle = kT$. Although the deviation in terms of energy is equal for particles of arbitrary sizes, the deviation in terms of the distance increases with decreasing particle size (mass) according to the formula

$$\langle z \rangle = \frac{3kT}{4\pi \Delta\rho |\mathbf{g}|} \frac{1}{a^3} \tag{200}$$

Hence, in the limit of particle mass (particle size) tending to zero the value of $\langle z \rangle$ tends to infinity. This effect is observed, e.g., for smaller planets when the gravitational force becomes too weak to keep gas molecules close to the surface.

It should be noted, however, that the Boltzmann distribution is established after a relaxation time, which can be precisely estimated by solving the non-stationary equation, Eq. (193) for the limiting case of a constant force and position-independent diffusion coefficient. The general solution of Eq. (193) in this case can be expressed as

$$f_p = \frac{1}{4\pi Dt} e^{-\frac{r^2}{4Dt}} f_p'(z) \tag{201}$$

where $r^2 = x^2 + y^2$ and the function f_p' fulfills the one-dimensional Smoluchowski equation for the z variable, Eq. (178)

$$\frac{\partial f_p'}{\partial z} = D\left(\frac{\partial^2 f_p'}{\partial z^2} - \frac{F}{kT} \frac{\partial f_p'}{\partial z} \right) \tag{202}$$

with the boundary condition

$$j = -D\left(\frac{\partial f_p'}{\partial z} + \frac{F}{kT} f_p' \right) = 0, \quad \text{at} \quad z = 0 \tag{203}$$

By making these assumptions Smoluchowski [36] analytically solved Eq. (202) by introducing the substitution

$$f_p' = f_p''(z,t)e^{-\frac{U}{2D}(z-z_0)-\frac{U^2t}{4D}} \tag{204}$$

The solution has the form

$$\begin{aligned} f_p' = {} & \frac{1}{2(\pi Dt)^{1/2}}\left[e^{-\frac{(z-z_0)^2}{4Dt}} + e^{\frac{(z+z_0)^2}{4Dt}}\right]e^{-\frac{U}{2D}(z-z_0)-\frac{U^2t}{4D}} \\ & + \frac{zU}{\pi^{1/2}D}e^{-\frac{Uz}{D}}\operatorname{erfc}\left(\frac{z+z_0-Ut}{2(Dt)^{1/2}}\right) \end{aligned} \tag{205}$$

$$\operatorname{erfc} = \frac{2}{\pi^{1/2}}\int_{\xi}^{\infty} e^{-\xi^2}d\xi \tag{206}$$

The first term on the right hand side of Eq. (205) is the non-stationary (transient) term, which vanishes if

$$t \gg \frac{D}{U^2} \tag{207}$$

On the other hand, the second term attains the equilibrium Boltzmann distribution when

$$t > (z+z_0)/U \tag{208}$$

The latter condition suggests that the equilibrium distribution is first established in the region close to the surface. Note that the relaxation time increases proportionally to the distance from the interface.

Another limiting case can be derived from Eq. (205) if the inequality $\frac{z^2}{Dt} \gg 1$ is met, i.e., either in the case of large initial distance of the particle from the interface or for short times. Then, the solution derived by Smoluchowski [36] has the form

$$f_p' = \frac{1}{2(\pi D t)^{1/2}} e^{-\frac{z'^2}{4Dt}} \tag{209}$$

where

$$z' = z - z_0 - Ut \tag{210}$$

Hence, Eq. (210) describes 'wave' of particle probability density, which traveling with the velocity U toward the interface undergoes diffusion at the same time.

The mean square displacement in this case is

$$\left\langle (z - z_0)^2 \right\rangle = 2Dt + U^2 t^2 \tag{211}$$

As can be noticed, for $t < 2D/U^2$ the displacement of the particle is mainly controlled by the Brownian motion, and for $t > 2D/U$ by the migration effect.

Smoluchowski [36] also considered another interesting case of particle motion in the elastic force field described by

$$F = -k_e z \tag{212}$$

where k_e is the elasticity constant.

The elastic interaction energy is

$$\phi(z) = \frac{1}{2} k_e z^2 \tag{213}$$

Hence, the energy distribution given by Eq. (213) can well approximate the case of a particle undergoing Brownian motion around an energy minimum. For colloid particles such a minimum can be produced by the superposition of gravitational forces acting toward the interface and repulsive electric double-layer forces.

The probability density function for a particle initially placed at $z = z_0$ was found to be [36]

$$f_p = \left[\frac{k_e}{2\pi kT(1 - e^{-2k_e M t})} \right]^{1/2} e^{-\frac{k_e (z - z_0 e^{-k_e M t})^2}{2kT(1 - e^{-2k_e M t})}} \tag{214}$$

where $M = \dfrac{1}{6\pi\eta a}$

The mean square displacement is

$$\left\langle (z-z_0)^2 \right\rangle = \frac{kT}{k_e}\left[1-e^{-2k_e M t}\right] + z_0^2\, e^{-2k_e M t} \tag{215}$$

For $t > 1/Mk_e$, the steady state is attained with f_p given by the expression

$$f_p(z) = \left(\frac{k_e}{2\pi kT}\right)^{1/2} e^{-\frac{k_e z^2}{2kT}} = f_0\, e^{-\phi(z)/kT} \tag{216}$$

Hence, Eq. (216) is in accordance with the general Boltzmann distribution, given by Eq. (195). The averaged displacement under the steady state is

$$\left\langle z^2 \right\rangle^{1/2} = \left(\frac{kT}{k_e}\right)^{1/2} \tag{217}$$

Accordingly, the averaged energy of the particle is

$$\langle \phi \rangle = \frac{1}{2} k_e \left\langle z^2 \right\rangle = \frac{1}{2} kT \tag{218}$$

Note that the above non-stationary results derived by Smoluchowski, in particular, particle motion to the interface under external force, are strictly valid for spherically symmetric particles, characterized by hydrodynamic mobility independent of orientation. Other solutions derived for non-spherical particles, e.g., spheroids or ellipsoids in quiescent media and in various shearing flows were extensively discussed in the monograph by van de Ven [2]. The rotary Brownian motion in shearing flows is of major practical significance for determining the viscosity of non-spherical particle suspensions. As discussed by van de Ven [2], rotary Brownian motion under shear is not coupled with translational motion. However, the governing diffusion equation under shearing conditions can only be solved analytically for limiting cases of low and high values of the rotary Peclet number, defined as G/D_r (where G is the characteristic flow shear rate. In the case of $Pe = 0$, an isotropic distribution of particle orientations was predicted.

It was also shown that the translational motion of non-spherical particles is coupled with their rotary motion when expressed in a space-fixed coordinate system. This makes it necessary to consider simultaneously both modes of Brownian motion in the general case, governed by a sixth-dimensional diffusion equation. No analytical solution of this equation was found except for low and large values of the translation *Pe* number defined as $U\ a/D$ (where U is the characteristic translation velocity), which becomes much smaller than unity. Then, the above diffusion equations, in particular, the Smoluchowski equation, can be effectively used to analyze the dynamics of particle suspensions, e.g., sedimentation in the gravitational field, provided that the diffusion coefficient is replaced with its averaged counterpart $\langle D \rangle$ listed in Table 4.2. This is a valid approximation for small colloid particles when $Pe \ll 1$, when the rotary Brownian motion ensures randomization of particle orientations.

Other complications to the Brownian motion appear in the case of charged particles, surrounded by electric double-layers discussed in Chapter 2. The friction coefficient increases in this case because of the deformation of the double-layer induced by Brownian motion. As discussed by van de Ven [2] this results in a decrease in the diffusion coefficient of the particle. This effect can be evaluated analytically for spherical particles only. As shown in [2], however, for the typical surface potential occurring under experimental conditions, the maximum correction to the diffusion coefficient occurring at κa is of the order of a few per cent only.

Other corrections to the Brownian motion arise because of the finite concentration of particles in suspensions occurring under real conditions. These effects will be discussed in further sections of this chapter concerned with the transport of suspended particles to various interfaces.

4.5. PHENOMENOLOGICAL TRANSPORT EQUATIONS

The results discussed in the previous section, in particular, the constitutive equation defining the probability density flux, Eq. (180), were derived for an isolated Brownian particle. In this section these results will be generalized to a more relevant situation of ensembles of particle forming suspensions. In this case, one is interested in determining particle concentration distribution and mass transfer rates to various interfaces, rather than in the time evolution of the probability density. The most natural concentration unit is the number concentration, i.e., the number of particles per unit volume, which can be connected to the f_p function discussed previously through the volume integral

$$n = \frac{1}{\delta v} \int_{\delta v} f_p \, dv \tag{219}$$

where δv is an arbitrary volume element.

Considering this definition one can convert the constitutive relationship for the flux, Eq. (180) to the form involving the number concentration of particles

$$\tilde{\mathbf{j}} = -\tilde{\mathbf{D}}\cdot\nabla n + (\tilde{\mathbf{M}}\cdot\mathbf{F})n = -\tilde{\mathbf{D}}\cdot\nabla n + \frac{1}{kT}(\tilde{\mathbf{D}}\cdot\tilde{\mathbf{F}})n \tag{220}$$

where $\tilde{\mathbf{j}}$ is the generalized flux vector incorporating the translational and the rotary fluxes, $\tilde{\mathbf{D}}$ is the generalized diffusion coefficient and $\tilde{\mathbf{F}}$ is the generalized force.

Consequently, the continuity (particle mass balance) equation becomes

$$\frac{\partial n}{\partial t} = \nabla\cdot\tilde{\mathbf{j}} = 0 \tag{221}$$

As mentioned, the translational and rotary fluxes are usually coupled in the case of non-spherical particles, so in the general case the mass balance equation, Eq. (221), becomes a system of partial differential equations of the parabolic type expressed in three spatial and three orientation coordinates.

In the case of a spherical particle having homogeneous properties, the rotary flux becomes irrelevant and the translation flux becomes

$$j = -\mathbf{D}\cdot\nabla n + \frac{1}{kT}(\mathbf{D}\cdot\mathbf{F})n = -\mathbf{D}\cdot\nabla n - \frac{1}{kT}(\mathbf{D}\cdot\nabla\phi)n + \mathbf{U}_h n \tag{222}$$

where ϕ is the interaction potential of the net conservative force $\mathbf{F} = -\nabla\phi$ incorporating the specific interaction potential ϕ_s with the interface, external force potential, etc., and $\mathbf{U}_h = \mathbf{M}\cdot\mathbf{F}_h + \mathbf{M}_c\cdot\mathbf{To}_h$ is the particle velocity resulting from hydrodynamic forces $\mathbf{F}_h$ and torques $\mathbf{To}_h$.

Using the expression for the flux, Eq. (222), one can formulate the mass conservation as

$$\frac{\partial n}{\partial t} = \nabla\cdot\left[\mathbf{D}\cdot\nabla n + \frac{1}{kT}(\mathbf{D}\cdot\nabla\phi)n - \mathbf{U}_h n\right] \tag{223}$$

It should be remembered, however, that by formulating Eq. (223) all hydrodynamic and specific interactions among particles were neglected and no coupling was assumed between hydrodynamic and specific interactions. Moreover, the diffusion coefficient occurring in Eq. (223) was assumed to be independent of the particle concentration n and its gradient. Therefore, the diffusion coefficient is often referred to as the self diffusion coefficient [2,3]. As a consequence of these simplifying assumptions Eq. (223) remains strictly valid for diluted suspensions of non-interacting particles only.

However, it still may be a useful approximation for calculating the transfer and deposition rates of particles because the volume fraction of particles in these processes $\Phi_v = n\, v_1$ (where v_1 is the volume of a single particle) is usually of the order of 0.01 or lower. Thus, the average distance between particles in the suspension l_s normalized to the particle characteristic dimension $d \cong v_1^{1/3}$ is of the order of $\Phi_v^{1/3}$, i.e., about 5–10. This means that the particles in the bulk interact quite weakly with each other, so Eq. (223) can be used upon introducing corrections depending solely on the volume fraction of particles. The first type of these corrections is due to multiparticle hydrodynamic interactions, which lead to a decrease in particle hydrodynamic mobility, and consequently a decrease in the diffusion coefficient. This effect can be described formally by the expansion

$$\mathbf{D}' = kT\,\mathbf{M}(\Phi_v) = \left(1 + \sum_l C_l\, \Phi_v^l\right)\mathbf{D} \tag{224}$$

where $\mathbf{D}'$ is the corrected diffusion tensor and C_l are the dimensionless constants to be determined from the multibody hydrodynamics.

In the case of spheres for short times, when particle displacements remain comparable with their dimensions, the constant C_1 was found to be -1.83 [3], so the first order correction is

$$\mathbf{D}' = (1 - 1.83\Phi_v)\mathbf{D} \tag{225}$$

As can be noticed, the correction is much smaller than that previously found for particle sedimentation, described by Eq. (58), because in the short-time limit the back-flow processes remain negligible. For longer times, when the displacement exceeds particle dimensions, the first order correction is expected to be the same as that given by Eq. (58), i.e.,

$$\mathbf{D}' = (1 - 6.55\Phi_v)\mathbf{D} \tag{226}$$

Unfortunately, no analytical results are available for higher order terms and particle motion near boundary surfaces. There are no results for non-spherical particles either.

In addition to these hydrodynamic corrections, in the case of finite particle concentration, there appear thermodynamic effects stemming from specific interactions between particles including hard-body interactions. This increases the local osmotic pressure of the particles and their chemical potential μ, which in turn increases the particle concentration gradient. The diffusion process in this case, which is referred to as the gradient diffusion, can be described using the method of Einstein [24,25], who postulated that the force on a particle stemming from the osmotic pressure gradient $\mathbf{F}=(1/n)\nabla\mathbf{\Pi}$ is compensated, under local equilibrium conditions, by the hydrodynamic resistance force. In the general case one has, therefore

$$\frac{1}{n}\nabla\mathbf{\Pi} + \tilde{\mathbf{K}}\cdot\mathbf{U} = 0 \tag{227}$$

Thus, the generalized diffusion flux is given by

$$\tilde{\mathbf{j}}_d = -\tilde{\mathbf{U}}n = -\tilde{\mathbf{K}}^{-1}\cdot\nabla\mathbf{\Pi} = -\frac{1}{kT}\tilde{\mathbf{D}}'\cdot\nabla\mathbf{\Pi} \tag{228}$$

where $\tilde{\mathbf{D}}' = kT\tilde{\mathbf{M}}'$ is the grand diffusion matrix defined previously.

Because of the Gibbs-Duhem thermodynamic relationship, $\nabla\mathbf{\Pi}=-n\nabla\mu$, one can alternatively express Eq. (228) as

$$\tilde{\mathbf{j}}_d = -\frac{1}{kT}(\tilde{\mathbf{D}}'\cdot\nabla\mu)n \tag{229}$$

For spheres the translational flux is given by

$$\mathbf{j}_d = -\frac{1}{kT}(\mathbf{D}'\cdot\nabla\mu)n \tag{230}$$

As can be noticed, the diffusion flux is proportional to the gradient of the chemical potential or osmotic pressure rather than to particle concentration.

Using Eq. (230) one can formulate the mass conservation equation as

$$\frac{\partial n}{\partial t} = \nabla\cdot\left[\frac{1}{kT}\mathbf{D}'\cdot(\nabla\mu+\nabla\phi)n - \mathbf{U}_h n\right] \tag{231}$$

Eq. (231) is the generalized version of Eq. (223) valid for arbitrary particle concentration. It can be expressed as an explicit function if the dependence of the chemical potential on the concentration is known. An alternative, more efficient method relies on the use of the equation of state, i.e., the dependence of $\mathbf{\Pi}$ on Φ_v. For hard spheres the equation of state is well approximated by [37]

$$\mathbf{\Pi} = kT\,v_1\,\Phi_v\,\frac{1+2\Phi_v+\Phi_v^2}{(1-\Phi_v)^3}\,\mathbf{I} \tag{232}$$

For lower Φ_v the range of the equation of state can be expressed in terms of the virial expressions for the osmotic pressure, given by

$$\mathbf{\Pi} = kT\,v_1\,\Phi_v\left[1+\sum_{l\geq 2}\bar{B}_l\,\Phi_v^{l-1}\right]\mathbf{I} \tag{233}$$

where $\bar{\mathrm{B}}_l$ are the reduced virial coefficients to calculated from multiple configurational integrals involving two, three, etc., particles.

For hard spheres in three dimensions, the virial coefficients up to the order of six are known [38], in particular

$$\bar{B}_2 = \frac{B_2}{v_1} = 4$$

$$\bar{B}_3 = \frac{B_3}{v_1^2} = 10 \tag{234}$$

$$\bar{B}_4 = \frac{B_4}{v_1^3} = 18.36$$

In the two-dimensional case, these virial coefficients have been derived for non-spherical particles of various shapes [39]. They will be discussed later on in relation to particle blocking effects occurring during particle adsorption.

Using the virial expansion given by Eq. (233) the diffusion flux for hard spheres can be expressed in the form

$$\mathbf{j}_d = -\left[1+\sum_{l\geq 2}\bar{B}_l\,(l)\Phi_v^{l-1}\right]\mathbf{D}'\cdot\nabla n = -\mathbf{D}_g\,\nabla n \tag{235}$$

where

$$\mathbf{D}_g = \left[1+\sum_{l\geq 2}\bar{B}_l(l)\Phi_v^{l-1}\right]\left[1+\sum_{l\geq 1}C_l\,\Phi_v^l\right]\mathbf{D} \tag{236}$$

is often defined as the gradient diffusion coefficient [4].

For hard spheres, by noting that $C_1=-6.55$ and $\bar{B}_2=4$, Eq. (236) becomes

$$\mathbf{D}_g \cong (1+1.45\Phi_v)\mathbf{D} \tag{237}$$

As can be noticed, the gradient diffusion coefficient decreases linearly with the volume fraction because the thermodynamic excluded volume correction is smaller than the hydrodynamic correction, thus decreasing particle mobility. For the typical range of volume fraction met in particle deposition processes, the correction to the diffusion coefficient is of the order of one per cent and can be neglected in comparison with, e.g., temperature variations affecting diffusion coefficients in an exponential way. Measurable effects of the diffusion coefficient changes are only expected in the case of particle sedimentation in a concentrated suspension [3]. However, the gradient diffusion coefficient is significantly affected by the volume fraction of particles in the case of non-spherical, elongated particles, interacting electrostatically, e.g., polyelectrolytes, because the second virial coefficient attains values much higher than four [3].

4.5.1. Limiting Forms – Transport Regimes

If there is more than one class of particles present in the suspension, e.g., in the case of polydisperse systems, one can formulate the mass conservation equation for any particle class in the form

$$\frac{\partial n_l}{\partial t}+\nabla\cdot\mathbf{j}_l = Q_l^{'} \tag{238}$$

where $\mathbf{j}_l$ is the flux of the l-th type of particle, and $Q_l^{'}$ the source term describing the annihilation or creation of the l-th particle class in the process of coalescence or aggregation. Because the source terms usually depends on the concentration of particles of various classes, Eq. (238) becomes a set of partial differential equations of an non-linear type.

The initial conditions for the above mass conservation equations are usually specified in the form

$$n = n_0(\mathbf{r}), \quad \text{for} \quad t = 0 \tag{239}$$

where n_0 is the initial concentration of particles in the volume close to the collector. Most often one assumes that the initial concentration in the suspension is uniform, i.e., $n = n_b$ where n_b is the bulk concentration of particles. However, this simplification may induce inaccurate predictions of mass transfer rates because the initial concentration in flowing suspensions influence the particle deposition rate over macroscopic distances as discussed later on.

Formulating appropriate boundary conditions for the mass conservation equations, Eqs. (223) and (231) is not trivial either, especially at the collector surface. The most natural way of formulating this boundary condition would be to accept the no-penetration postulate, analogous to that of the Navier-Stokes, hydrodynamic equation, i.e.,

$$\mathbf{j} \cdot d\mathbf{S} = 0 \tag{240}$$

at the collector surface.

However, by postulating this, problems with specifying the interaction potential at short distances of the particle from the interface arice. In order to obtain any meaningful solutions, it is necessary to postulate that this potential also has a tangential component, which ensures particle immobilization. At present, however, there is not general theory allowing for calculating such interaction potentials.

It is therefore often postulated, in accordance with the Smoluchowski approach [40], that the boundary condition at the interface has the form

$$n = 0, \quad \text{at } z = \delta_m \tag{241}$$

where $z = \delta_m$ is the primary minimum distance.

This corresponds to the perfect sink model introduced by Smoluchowski, cf. Eq. (194). This boundary condition simplifies numerical handling of conservation equations. However, the postulate $n = 0$ is physically incorrect since particle concentration at the interface remains always finite and increases with time in deposition processes. An appropriate formulation of the boundary condition at the collector surface can only be attained when considering the specific energy profile at the surface. This leads to kinetic boundary conditions, expressed in terms of the surface concentration of particles rather than bulk concentration. This aspect of particle deposition will be discussed in detail later on.

It is useful to derive the limiting forms of Eq. (223) or Eq. (231) because their analytical solutions are impractical in the general case owing to a complex dependence of the specific interaction potential on the distance from the interface.

One of these limiting forms is derived when the flow vanishes and particle transport can be assumed to be one-dimensional. This situation occurs for colloid particles in the thin region adjacent to the interface of thickness δ_a, often called the surface boundary layer [41–44]. If, moreover, the dependence of the chemical potential on particle concentration is neglected (*quasi*-dilute limit), Eq. (231) reduces to

$$\frac{\partial n}{\partial t} = \frac{\partial}{\partial z}\left[D(z)\left(\frac{\partial n}{\partial z} - \frac{F}{kT}n\right)\right] = \frac{\partial n}{\partial z}\left[D(z)\left(\frac{\partial n}{\partial z} + \frac{\nabla\phi}{kT}n\right)\right] \tag{242}$$

where $D(z) = F_1(z)kT/6\pi\eta a$ is the local value of the diffusion coefficient of the particle, and ϕ is the net interaction potential stemming from the particle surface interactions, incorporating also the part originating from pre-adsorbed particles [45]. Eq. (242) is important because its solutions can be used to formulate boundary conditions for the bulk transport equation.

If all specific interactions are neglected together with the hydrodynamic boundary effects, Eq. (223) simplifies to the form called the Smoluchowski–Levich equation [3]

$$\frac{\partial n}{\partial t} = D\,\nabla^2 n - \frac{D}{kT}\nabla\cdot(\mathbf{F}n) - \mathbf{V}\cdot\nabla n \tag{243}$$

where $\boldsymbol{V}$ is the unperturbed (macroscopic) fluid velocity vector.

On the other hand, under the no-convection conditions, when $\mathbf{V} = 0$, Eq.(243) reduces to the form called the Smoluchowski equation, analogous to Eq. (184) derived previously for the single particle case

$$\frac{\partial n}{\partial t} = D\,\nabla^2 n - \frac{D}{kT}\nabla\cdot(\mathbf{F}n) = D\,\nabla^2 n - \nabla\cdot(\mathbf{U}n) \tag{244}$$

If the force **F** remains constant, Eq. (244) becomes

$$\frac{\partial n}{\partial t} = D\left(\nabla^2 n - \frac{F}{kT}\nabla n\right) = D\,\nabla^2 n - U\,\nabla n \tag{245}$$

On the other hand, when the external force **F** vanishes, the Smoluchowski equation is reduced to the following form, which is in principle, a mathematical formulation of Ficks' second law

$$\frac{\partial n}{\partial t} = D\,\nabla^2 n \tag{246}$$

The non-stationary Eq. (246), being of a linear, parabolic type, can be solved analytically, using for example the Laplace transformation method in many situations of practical interest, e.g., adsorption on a spherical interface from a finite or infinite volume [46,47].

The boundary conditions for Eqs. (243–246) are usually specified in the form of the perfect sink model, expressed by Eq. (241)

Unfortunately, in the general case of multidimensional problems, analytical solutions of Eq. (243) under transient conditions are prohibitive. Therefore, it is instructive to perform a dimensional analysis of these equations in order to formulate criteria of their reduction to simpler, stationary forms. This can be done by introducing the characteristic length scale L_{ch}, time scale t_a, convection velocity V_{ch} and migration velocity U_{ch}. Then, Eq. (243) can be expressed as

$$T_a \frac{\partial n}{\partial \tau} = \bar{\nabla}^2 n - Pe\,\bar{\mathbf{V}}\cdot\bar{\nabla} n - Ex\,\bar{\mathbf{U}}\cdot\bar{\nabla} n \tag{247}$$

where the dimensionless variables are defined as

$$\tau = t/t_a, \qquad \bar{\nabla} = L_{ch}\,\nabla, \qquad \bar{\mathbf{V}} = \frac{1}{V_{ch}}\mathbf{V}, \qquad \bar{\mathbf{U}} = \frac{1}{U_{ch}}\mathbf{U} \tag{248}$$

Consequently, the dimensionless criterion numbers (analogous to the Reynolds number) are defined as

$$T_a = \frac{L_{ch}^2}{D t_a} \qquad \text{(relaxation number)}$$

$$Pe = \frac{V_{ch} L_{ch}}{D} \qquad \text{(Peclet number)} \tag{249}$$

$$Ex = \frac{U_{ch} L_{ch}}{D} \qquad \text{(external force number)}$$

The T_a number can be treated as the dimensionless relaxation time.

Useful limiting regimes can be derived from Eq. (247) by exploiting these definitions. When $T_a \ll 1$, which corresponds to the condition $t_a \gg L_{ch}^2/D$, steady-state transport conditions are established. This significantly simplifies the mathematical handling of Eq. (247). For colloid particles, assuming the typical values of the diffusion coefficient $D = 10^{-12}$ m^2 s^{-1} and $L_{ch} = 0.1$ μm (particle dimension), one obtains $t_a \gg 10^{-2}$ s as a criterion for attaining the steady-state conditions. For $L_{ch} = 1$ μm, this characteristic time becomes one second. Assuming the steady-state and $Ex \ll 1$, Eq. (247) reduces to the simple form

$$\bar{\nabla}^2 n - Pe\, \bar{V} \cdot \bar{\nabla} n = 0 \tag{250}$$

Eq. (250), often called the convective diffusion equation, was exploited widely for describing the transfer of particles to collectors of a simple geometry [3,11,12,48]. However, this equation becomes inappropriate for larger colloidal particles when the specific interactions and hydrodynamic wall effects play an important role. In this case one has to use either numerical techniques to solve the exact transport equation, Eq. (223), or use the approximate approaches combining the bulk transport with the surface boundary layer transport as discussed in the next section.

4.5.2. The Near Surface Transport

Let us consider particle transport through a thin layer adjacent to the interface where convection effects can be neglected. The thickness of this layer δ_a is comparable to the range of specific interactions, being usually much smaller than particle dimensions and the collector dimensions. Therefore the local curvature effects can be neglected and particle transport can be considered one-dimensional, governed by Eq. (242), i.e.,

$$\frac{\partial n}{\partial t} = \frac{\partial}{\partial z}\left[D(z) e^{-\frac{\phi(z)}{kT}} \frac{\partial}{\partial z}\left(n e^{\frac{\phi(z)}{kT}} \right)\right] \tag{251}$$

where $\phi(z) = -\int_0^z F dz'$ is the interaction potential.

In order to derive exact results for the non-stationary case, we assume that the diffusion coefficient and the interaction force remain constant within the boundary layer δ_a, denoted by D and F, respectively. This means that the interaction potential ϕ is a linear function of z, i.e.,

$$\phi = -F\int_0^{z'} dz' = -Fz + \phi_0 \tag{252}$$

where ϕ_0 is the reference energy of interaction. By assuming that $\phi = \phi_b$ at $z = 0$ and $\phi = 0$ at $h = \delta_a$ one can describe the effect of a triangular energy barrier of the width δ_b and the height $\phi_b = F\delta_a$ (see Fig. 4.10), characterized by the energy distribution

$$\phi = -\phi_b \frac{z}{\delta_a} + \phi_b \tag{253}$$

By virtue of these assumptions, Eq. (251) simplifies to

$$\frac{\partial n}{\partial t} = D\frac{\partial^2 n}{\partial z^2} - U\frac{\partial n}{\partial z} \tag{254}$$

Where $U - D\phi_b/\delta_a kT$

Consider now a non-stationary transport of particles through the barrier described by Eq. (253) in the case where a constant concentration of particles is maintained, equal to n_1 at $z = \delta_b$ and $n = n_2$ at $z = 0$. Thus, the boundary and initial conditions for the Smoluchowski equation, Eq. (254), take the form

$$\begin{aligned} n &= n_1, & \text{at } z &= \delta_a, & t &> 0 \\ n &= n_2, & \text{at } z &= 0, & t &> 0 \\ n &= 0, & \text{for } 0 < z &< \delta_a, & t &= 0 \end{aligned} \tag{255}$$

Eq. (254) with the boundary and initial conditions Eq. (255), can be solved by applying the Laplace transformation method. In this way, the flux of particles crossing the barrier was found to be [49]

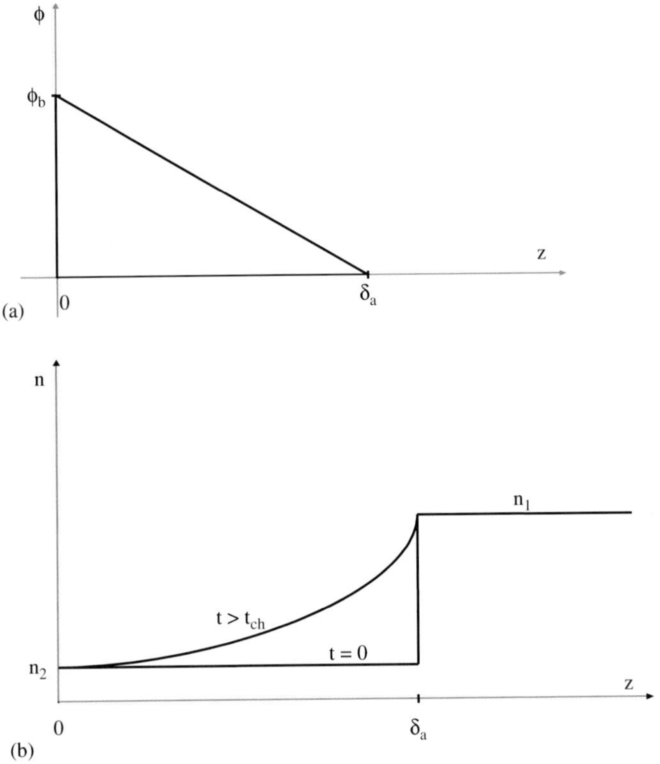

Fig. 4 .10. (a) shows schematic view of the triangular energy barrier, and (b) shows the particle concentration distribution in the boundary layer δ_a.

$$j = j_{st} + j_t(t) \tag{256}$$

where j_{st} is the stationary (time-independent) flux given by

$$-j_{st} = \frac{D(n_1 - n_2)}{\delta_a} \frac{1}{\frac{kT}{\phi_b}\left(e^{\phi_b/kT} - 1\right)} = \frac{D(n_1 - n_2)}{R_{bar}} \tag{257}$$

and

$$R_{bar} = \int_0^{\delta_a} e^{\phi/kT}\, dz = \frac{kT\delta_a}{\phi_b}\left[e^{\phi_b/kT} - 1\right]$$

can be treated as the barrier resistance.

It is interesting to note that in the limit of a negligible barrier height when $\phi_b \to 0$, $R_{bar} \to \delta_a$, Eq. (257) reduces to the simple form

$$-j_{st} = D\left(\frac{n_1 - n_2}{\delta_b}\right) \tag{258}$$

Eq. (258) represents Fick's first law of diffusion stating that the diffusion flux of particles is proportional to the diffusion coefficient and the gradient of particle concentration $(n_1 - n_2)/\delta_a$.

The transient flux j_t in Eq. (256) is given by

$$-j_t = \frac{2D(n_1 - n_2)}{\delta_a} e^{-\phi_b/2kT} \sum_{m=1}^{\infty} \frac{1}{\left(1 + \frac{\phi_b}{2kT\pi^2 m^2}\right)} e^{-\frac{D}{\delta_a^2}\left[\left(\frac{\phi_b}{2kT}\right)^2 + m^2\pi^2\right]t} \tag{259}$$

One can notice that the transient flux vanishes for the adsorption time fulfilling the inequality $t > t_{ch}$, where t_{ch} is the characteristic relaxation time given by

$$t_{ch} = \frac{\delta_a^2}{D\left[\left(\frac{\phi_b}{2kT}\right)^2 + \pi^2\right]} \tag{260}$$

For colloid particles, assuming $D = 10^{-12}\ \mathrm{m^2\,s^{-1}}$, and $\delta_a = 10$ nm (0.01 μm), one can estimate that $t_{ch} < 10^{-5}$ s even for ϕ_b as low as 1 kT. For higher barriers the relaxation time becomes even smaller. This is a negligible value from an experimental viewpoint. One can predict, therefore, that after the short transition time, a steady flux of particles is established that assumes a constant value through the entire domain $0 < z < \delta_b$.

Assume now that after attaining the steady-state conditions, particle concentration at $z = \delta_a$ (the edge of the barrier) is varied abruptly, attaining the new value n_1'. It can be easily proved by the Laplace transformation technique that the new steady-state flux value is established after exactly the same relaxation time. This observation allows one to draw the important conclusion that Eq. (257) properly describes a *quasi*-stationary transport of particles through the barrier if the characteristic time of the concentration variations remains less than t_{ch}. In view of the above estimation, it seems that this condition is always valid for colloid particle transport.

By virtue of this *quasi*-static transport postulate, one can neglect the time derivative in Eq. (251) and derive explicit results for arbitrary potential distributions (barrier shape) discussed in Chapter 2. The stationary form of Eq. (251) is

$$\frac{\partial}{\partial z}\left[D(z)e^{-\frac{\phi(z)}{kT}}\frac{\partial}{\partial z}\left(ne^{\frac{\phi(z)}{kT}}\right)\right]=-\frac{\partial}{\partial z}j=0 \tag{261}$$

Assuming that the interaction potential is of type 2, analyzed previously (see Fig. 2.33), i.e., an energy barrier of arbitrary high and a primary minimum of the depth ϕ_m at the distance $z = \delta_m$ appears. Particle concentration at $z = \delta_m$ equals $n(\delta_m)$ and at $z = \delta_a$ (adsorption layer thickness) equals $n(\delta_a)$. Under these assumptions Eq. (261) can be integrated twice, which results in the following expression for particle concentration distribution

$$n(z)=\left[n(\delta_m)e^{\phi(\delta_m)/kT}-j_a\int_{\delta_m}^{z}\frac{e^{\phi(z')/kT}}{D(z')}dz'\right]e^{-\phi(z)/kT}$$
$$=n(\delta_m)e^{\phi(\delta_m)/kT}e^{-\phi(z)/kT}+n^*(z) \tag{262}$$

where

$$n^*(z)=-j_a\left(\int_{\delta_m}^{z}\frac{e^{\phi(z')/kT}}{D(z')}dz'\right)e^{-\phi(z)/kT} \tag{263}$$

is the perturbing term and j_a is the constant flux through the boundary layer given by the expression,

$$-j_a=\frac{n(\delta_a)e^{\phi(\delta_a)/kT}-n(\delta_m)e^{\phi(\delta_m)/kT}}{\int_{\delta_m}^{\delta_a}\frac{e^{\phi(z')/kT}dz'}{D(z')}}=k_a'\,n(\delta_a)-k_a'\,n(\delta_m) \tag{264}$$

and

$$k_a' = \frac{e^{\phi(\delta_a)/kT}}{\int_{\delta_m}^{\delta_a} \frac{e^{\phi/kT}}{D(z')} dz'} = \frac{e^{\phi(\delta_a)/kT}}{R_{bar}} \tag{265}$$

$k_d' = k_a' \, e^{[\phi_m - \phi(\delta a)]/kT}$

are the adsorption and desorption rate constants, and $R_{bar} = \int_{\delta_m}^{\delta_a} e^{\phi/kT} dz'/D(z')$.

It is interesting to note that particle concentration distribution around the region of the primary minimum is well approximated by a *quasi*-Boltzmann distribution

$$n(z) \cong n_m \, e^{-\phi(z)/kT} \tag{266}$$

where $n_m = n(\delta_m) e^{\phi(\delta_m)/kT}$
This is so because the term

$$n^* \cong n(\delta_m) e^{\phi(\delta_m)/kT} \, e^{-\phi_b/kT} \ll n(\delta_m),$$

$$\text{if } \phi_b >> 1 \tag{267}$$

Eq. (266) has a major theoretical significance indicating that under non-equilibrium conditions, when a stationary value of particle flux is maintained through a layer, particle concentration remains locally equilibrated governed by the Boltzmann distribution with the appropriate normalization constant.

One can convert Eq. (264) to a more convenient form by introducing the surface concentration, defined as

$$N = \int_{\delta_m}^{\delta_a} n \, dz \tag{268}$$

by using Eq. (266) this can be expressed as

$$N = n_m \int_{\delta_m}^{\delta_a} e^{-\phi(z')/kT} \, dz' = n_m I_m \tag{269}$$

By substituting this relationship into Eq. (264) and assuming $\phi(\delta_a)=0$ one obtains

$$-j_a = k_a\, n(\delta_a) - k_d\, N \tag{270}$$

where

$$k_a = \frac{1}{\displaystyle\int_{\delta_m}^{\delta_a} \frac{e^{\phi(z)/kT}}{D(z')}\, dz'}$$

$$k_d = \frac{k_a}{I_m} = \frac{k_a}{\displaystyle\int_{\delta_m}^{\delta_a} e^{-\phi(z')/kT}\, dz'} \tag{271}$$

Instead of the surface concentration N one oft en introduces the dimensionless coverage Θ defined as

$$\Theta = S_g\, N \tag{272}$$

where S_g is the characteristic crosssection of the particle, equal to πa^2 for spherical particles.

Using this definition, Eq. (270) becomes

$$-j_a = k_a\, n(\delta_a) - \frac{k_d}{S_g}\Theta \tag{273}$$

Eq. (273) represents a general expression describing the reversible adsorption of particles under linear conditions (in the case of negligible surface blocking effects). It is often used as the "kinetic" boundary condition for bulk transport problems [41,42,46,47,49].

Under equilibrium conditions when $j = 0$, Eq. (273) becomes

$$\Theta = S_g\, K_a\, n(\delta_a) \tag{274}$$

where

$$K_a = k_a/k_d = \int_{\delta_m}^{\delta_a} e^{-\phi/kT} dz = I_m$$

Eq. (274) expresses the Henry's adsorption law, which says that the amount adsorbed is proportional to the bulk concentration.

On the other hand, in the case when $|\phi_m| >> kT$, the constant k_d vanishes, and particle adsorption becomes practically irreversible. The expression for particle flux Eq. (273) becomes

$$-j_a = k_a n(\delta_a) \tag{275}$$

When $k_a \rightarrow \infty$, which is the case with energy profile of type Ib (see Fig. 2.32), Eq. (275) simplifies to

$$n(\delta_a) \rightarrow 0, \quad at \quad z = \delta_a \tag{276}$$

In this way one obtains the perfect-sink boundary condition introduced by Smoluchowski.

One can evaluate analytically k_a, k_d, K_a for simple shapes of the specific interaction energy profile. For example, in the case of the triangular barrier considered previously ($\phi_b >> kT$) and the primary minimum of the depth ϕ_m occurring δ_m

$$k_a = \frac{D}{a}\left(\frac{\phi_b}{kT}\right) e^{-\phi_b/kT}$$

$$k_d = \frac{k_a}{\delta_m} e^{\phi_m/kT} \tag{277}$$

$$K_a = \delta_m e^{-\phi_m/kT}$$

In a more general case, the energy distributions around the primary minimum and the barrier region can be approximated by a parabolic distribution. Then, these constants are given by [49]

$$k_a = D(\delta_b)\left(\frac{\gamma_b}{2\pi kT}\right)^{1/2} e^{-\phi_b/kT} \cong \frac{D}{a}\left(\frac{\phi_b}{\pi kT}\right)^{1/2} e^{-\phi_b/kT}$$

$$k_d = k_a\left(\frac{\gamma_m}{2\pi kT}\right)^{1/2} e^{\phi_m/kT} \cong \frac{k_a}{\delta_m}\left(\frac{\phi_m}{\pi kT}\right)^{1/2} e^{\phi_m/kT} \tag{278}$$

$$K_a = \left(\frac{2\pi kT}{\gamma_m}\right)^{1/2} e^{-\phi_m/kT} \cong \left(\frac{\pi kT}{\phi_m}\right)^{1/2} \delta_m e^{-\phi_m/kT}$$

where

$$\gamma_b = -\left(\frac{d^2\phi}{dh^2}\right)_{\delta_b} \cong \frac{2\phi_b}{\delta_b^2}, \quad \gamma_m = \left(\frac{d^2\phi}{dh^2}\right)_{\delta_m} \cong \frac{2\phi_m}{\delta_m^2}$$

and $D(\delta_b)$ is the value of the diffusion coefficient in the barrier region, approximated according to Eq. (141) by $D(\delta_b)=(\delta_b/a)D$.

Eq. (270) and its irreversible counterparts Eqs. (275,276) have a large significance because they can be used as boundary condition for bulk transport problems governed by the Smoluchowski–Levich (SL) equation. This is justified by the fact that the thickness of the surface boundary layer δ_a is generally much smaller than the diffusion boundary layer defined as the region where particle concentration changes occur. Hence, except for particle size above micrometer, the surface and bulk transport steps can be decoupled.

4.6. SOLVED PROBLEMS OF LINEAR TRANSPORT TO INTERFACES

In the general case, the kinetics of particle transfer to interfaces can be predicted by solving the governing mass-balance equation, Eq. (231), which incorporates the effects of specific, external and hydrodynamic forces in an exact manner. However, this is only possible by numerical techniques, e.g., the finite difference method, which become quite complicated for non-stationary transport and multidimensional problems. This is so, because of the disparity between the range scale of the specific force and the diffusion boundary layer thickness, which differ by orders of magnitude, thus making it necessary to use special functions for transforming the grid. Numerical solutions become quite efficient for the one-dimensional case only, e.g., for collectors of simple geometry and stagnation point flows as discussed later on.

When dealing with transport problems concerning particles of submicrometer size, it is considerably more efficient to use approximate approaches providing one with analytical solutions for a broad range of applications. These methods constitute in splitting the entire transport path of particles from the bulk to the interface into two separate regions

(i) The bulk diffusion boundary-layer (see Fig. 4.11), where specific forces and hydrodynamic wall effects are neglected; and
(ii) The surface boundary layer, where fluid convection effects are neglected but the specific interactions, as well as hydrodynamic wall corrections are taken into account.

The dividing plane of these two regions is located at the distance $z = \delta_a$ from the collector where the specific surface forces vanish. Usually, the value of δ_a, referred to as the surface force boundary, is much smaller than particle dimensions. At this plane the boundary conditions for the bulk transport equation are specified by exploiting the near surface transport

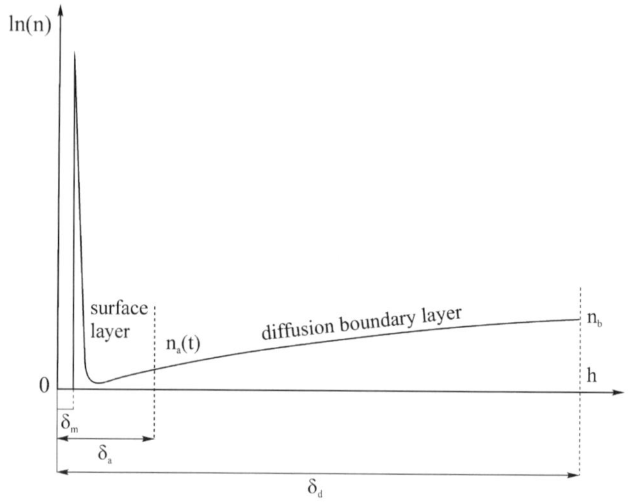

Fig. 4.11. A schematic view of the particle concentration distribution in the diffusion δ_d and surface force boundary layer δ_a.

results discussed above. In the general case these are the kinetic boundary conditions, given by Eq. (273), which are most appropriate for describing surfactant and molecule adsorption. For protein and colloid particles, one usually assumes irreversible adsorption, so the boundary condition is given by Eq. (276). The approach exploiting this type of boundary condition is referred to in the literature as the surface force boundary layer (SFBL) method [42–44]. This methods becomes quite involved, however, in the case of collectors of a more complicated geometry when the bulk transport equation becomes two-dimensional. Useful analytical results can be derived in this case by assuming the perfect sink boundary conditions.

Using these approaches we will analyze in this section the non-stationary transport of particles to a spherical interface under the diffusion-controlled transport conditions, transport under a uniform external force and finally the convective transport under *quasi*-stationary conditions for interfaces of various shapes.

4.6.1. Diffusion Transport to Spherical and Planar Interfaces

Let us consider particle transfer to a spherically shaped homogeneous interface (often called a collector or an adsorber) of radius R_D (see Fig. 4.12), which may represent either a liquid drop, a gas bubble or a solid particle in contact with a dispersion of particles of arbitrary size. Neglecting the specific force, convection and external forces are neglected, the governing equation, Eq. (246), for the spherical geometry assumes the form

$$\frac{\partial n}{\partial t} = D\frac{1}{r^2}\frac{\partial}{\partial r}\left(r^2\frac{\partial n}{\partial r}\right) \tag{279}$$

In accordance with Eq. (270), the boundary conditions at the edge of the surface boundary layer are expressed as [46]

$$-j_a = \left(D\frac{\partial n}{\partial r}\right)_{\delta_a} = k_a\, n_a - k_d\, N \tag{280}$$

where $n_a = n(\delta_a)$ is the particle concentration at $r = \delta_a$, often called the subsurface concentration and j_a is the flux of particles reaching the adsorption layer.

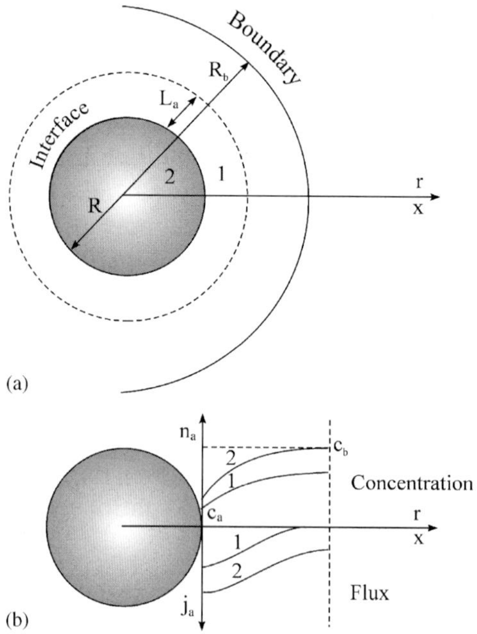

Fig. 4.12. (a) shows a schematic diagram of particle adsorption onto a spherical interface (collector) from a finite volume and (b) shows the concentration and particle flux distributions; case (i): perfectly reflecting boundary and case (ii): contact with a well-stirred reservoir.

When the flux is known, the coverage of adsorbed particles as a function of time can be calculated, from the surface mass-balance equation

$$\frac{dN}{dt} = -j_a(t) \tag{281}$$

The boundary condition in the bulk, at the outer interface of a spherical shape and radius R_b (see Fig. 4.12), is specified as

$$-j_a = D\frac{\partial c}{\partial r} = 0, \quad \text{at } r = R_b \text{ case (i)} \tag{282}$$

This boundary condition, called a symmetry boundary condition, corresponds directly to the case of particle adsorption from a reservoir of radius

R_b having non-adsorbing walls. It can also approximate well the situation of particle adsorption on an assemblage of spherical collectors immersed in a suspension, as is the case, e.g.,in deep bed filtration processes. The value of R_b, which depends on the volume fraction of collectors, can be treated as the averaged size of the cell.

Another external boundary condition for the particle adsorption problem is formulated as

$$n = n_b, \quad \text{at } r = R_b \text{ case (ii)} \tag{283}$$

where n_b is the bulk concentration of particles. This boundary condition, approximates the situation where the collector is in contact with a well-stirred suspension of particles, so at the distance $r = R_b$ a uniform concentration of particles is maintained during the entire adsorption run. Then, R_b can be treated as the effective mixing length depending on whether the mode and intensity of mixing are known [50].

The initial conditions for this problem are usually postulated to initially have the form of a uniform particle distribution, i.e.,

$$\left.\begin{array}{ll} n = n_b, & \text{for} \quad R \le r \le R_b \\ N = N_i & \end{array}\right\} \quad \text{at } t = 0 \tag{284}$$

where N_i is the initial surface concentration.

Melik [50] considered a more complicated situation where the initial distribution of particles was non-uniform (segregated initial condition).

It is useful to covert the boundary value problem represented by Eq. (279) and the above boundary and initial conditions to a dimensionless form by introducing the transformations reduced variables defined as

$$\begin{gathered} \bar{n} = n/n_b, \qquad \bar{r} = (r - R - a)/L_a \\ \tau = t/t_a, \qquad \Theta = N/N_m \\ \bar{k}_a = k_a \frac{t_a n_b}{N_m}, \qquad \bar{k}_d = k_d t_a, \qquad \bar{K}_a = \bar{k}_a/\bar{k}_d \\ t_{ch} = L_a^2/D, \qquad Ar = L_a/R \end{gathered} \tag{285}$$

where N_m is the characteristic surface concentration of the monolayer, $L_a = N_m/n_b$ the characteristic adsorption length scale and t_a the characteristic adsorption time. It is interesting to note that L_a can be treated as the characteristic depletion length. On the other hand, the adsorption time t_a characterizes the duration of the non-stationary conditions. Note that this time increases proportionally to n_b^{-2}, which means that it may assume large values for dilute particle suspensions.

For the sake of convenience, values of N_m, L_a and t_a calculated for particles of various sizes, ranging from ions to colloids, are collected in Table 4.7. The bulk concentration of particles expressed in the volume fraction Φ_v was assumed to be equal to 10^{-3} and the monolayer surface concentration $N_m = 0.55/\pi a^2$ (this corresponds to a random monolayer as discussed later on). As can be noticed, for particles of molecular dimensions, e.g., surfactants, the characteristic length scale L_a is of the order of 10^{-6} m and the characteristic adsorption time t_a is of the order of 10^{-3} s. On the other hand, for colloid particle solutions the adsorption time t_a assumes values of the order of hours, which suggests that non-stationary transport may be easily measured experimentally.

Using the scaling variables expressed by Eq. (285) one can convert the boundary value problem, given by Eqs. (279–283) to the concise form

$$\frac{\partial \bar{n}'}{\partial \tau} = \frac{\partial^2 \bar{n}'}{\partial \bar{r}^2}$$
$$\frac{d\Theta}{d\tau} = \bar{k}_a \bar{n}_a - \bar{k}_d H \qquad (286)$$

Table 4.7
The adsorption length scale L_a and the diffusion relaxation time t_a calculated for spherical particles of various size

Particle radius [nm]	Diffusion coefficient [m^2 s^{-1}]	N_m [m^{-2}]	L_a [m]	t_a [s]
0.1 (ions)	2.15×10^{-9}	1.75×10^{19}	7.3×10^{-8}	2.48×10^{-6}
1 (surfactants)	2.15×10^{-10}	1.75×10^{17}	7.3×10^{-7}	2.48×10^{-3}
10 (proteins)	2.15×10^{-11}	1.75×10^{15}	7.3×10^{-6}	2.48
10^2 (colloids)	2.15×10^{-12}	1.75×10^{13}	7.3×10^{-5}	2.48×10^{3}
10^3 (bacteria)	2.15×10^{-13}	1.75×10^{11}	7.3×10^{-4}	2.48×10^{6}

Notes: T = 293 K, $\eta = 10^{-3}$ kg (m s)$^{-1}$ (aqueous solutions), $\Phi_v = 10^{-3}$, $n_b = 3/(4\pi a^3)\Phi_v$ $N_m = 0.55/\pi a^2$, $L_a = N_m/n_b$, $t_a = L_a^2/D$

with the boundary and initial conditions

$$\frac{\partial \bar{n}'}{\partial \bar{r}} = \bar{k}_a \bar{n}_a - \bar{k}_d \Theta, \quad at \quad \bar{r} = (R+a)/L_a$$

$$\frac{\partial \bar{n}}{\partial \bar{r}} = 0, \quad \text{at} \quad \bar{r} = R_b/L_a \qquad \text{case (i)}$$

$$\bar{n} = 1, \qquad \text{at} \quad \bar{r} = R_b/L_a \qquad \text{case (ii)}$$

where $\bar{n}' = \bar{n}\,\bar{r}$.

The boundary value problem expressed by Eq. (286) describes non-stationary adsorption kinetics for the Henry isotherm. It can be solved analytically in the limiting case of major practical significance, when $\bar{k}_a >> 1$ and $L_a << R_b$ so adsorption can be treated as irreversible and proceeding from an infinite volume. The solution describing the flux of particles in this case was first derived by Smoluchowski in relation to particle aggregation kinetics [40]. By reverting to the dimensional units it assumes the form

$$-j_a = \left[\frac{D_{12}}{a+R} + \left(\frac{D_{12}}{\pi t} \right)^{1/2} \right] n_b \tag{287}$$

where $D_{12} = D + D_c$ is the relative diffusion coefficient of the particle relative to the collector, which may undergo diffusion in the general case and D_c the collector diffusion coefficient.

As one can deduced from Eq. (287), for longer times, when, $(a + R)^2 / D_{12} t < 1$, the flux attains a steady-state value equal to

$$-j_a = \frac{D_{12}}{a+R} n_b \tag{288}$$

It can be estimated by assuming $a = R = 100$ nm $D_{12} = 10^{-12}\,\mathrm{m^2\,s^{-1}}$ that the critical relaxation time equals 4×10^{-2} s. For $a = 1000$ nm the relaxation time becomes 1s.

Eq. (288) has major significance since it can be used to calculate the coagulation rate constants, as done originally by Smoluchowski [40]. The rate of coagulation r_{12} of two particles assumed to have a spherical shape and radii a_1 and a_2 is given by the formula

$$r_{12} = -k_{12}\, n_1\, n_2 \tag{289}$$

where n_1, n_2 are the number concentrations of the particles in the bulk.

The coagulation rate constant k_{12} is given by

$$k_{12} = -4\pi(a_1 + a_2)^2\, j_a/n_b = 4\pi(a_1 + a_2)(D_1 + D_2) \\ = \frac{2}{3}\frac{kT}{\eta}\left(2 + \frac{a_1}{a_2} + \frac{a_2}{a_1}\right) = \frac{2}{3}\frac{kT}{\eta} \tag{290}$$

where D_1, D_2 are the diffusion coefficients of particle 1 and particle 2, respectively. The function $f_{12}=2+a_1/a_2+a_2/a_1$ occurring in Eq. (290) characterizes the geometrical effect, which is quite independent of particle dimension. For example, a twofold increase in particle size ratio causes a 12.5 per cent increase in the value of the function f_{12} only. For $a_1/R_2 = 4$, the increase equals 56.2 per cent.

From Eq. (287) one can also deduce that for $R_b \to \infty$ (adsorption on a planar interface), the flux becomes unsteady for all times and is given by the well-known expression

$$-j_a = \left(\frac{D}{\pi t}\right)^{1/2} \tag{291}$$

As can be noticed, the flux vanishes with time proportionally to $t^{-1/2}$, which means that particle adsorption on planar interfaces driven by diffusion alone becomes an inefficient mode of transport for long times.

By integrating Eq. (291), one obtains the expression for surface coverage

$$N = 2\left(\frac{Dt}{\pi}\right)^{1/2} n_b \tag{292}$$

This equation can be exploited to determine the particle diffusion coefficient by experimentally measuring the surface concentration N of an irreversibly adsorbed particle as a function of time [11,49,51].

In the general case of reversible adsorption from a finite volume, Eq. (286) was solved by the Laplace transformation method [46], assuming that the size of the particle a is much smaller than the adsorbed radius R.

This is a valid approximation for all problems involving colloid and smaller particles because the collector size usually exceeds 10 μm. The general expression for the coverage of particles is given by

$$\Theta = \Theta_i e^{-\bar{k}_d \tau} + \Theta_\infty (1 - e^{-\bar{k}_d \tau}) + \bar{k}_a \bar{K} \sum_{l=1}^{\infty} Q_l \left(e^{-\alpha_l^2 \tau} - e^{-\bar{k}_d \tau}\right) \tag{293}$$

where $\Theta_\infty = \bar{k}_a \bar{n}_b / \bar{k}_d = \bar{K}_a \bar{n}_b$ is the final (equilibrium) coverage, $\bar{K} = \bar{k}_d \Theta_i - k_a$

$$Q_l = \frac{\alpha_l^2}{P_l(\bar{k}_d - \alpha_l^2)} \left[1 - \frac{Ar(c_3 \alpha_l^2 - c_1)}{\bar{R}(\alpha_l^4 - c_2 \alpha_l^2 + c_0)}\right]$$

$$\begin{aligned} P_l = \frac{1}{(\alpha_l^4 - c_2 \alpha_l^2 + c_0)} \Big[& C_4 \alpha_l^8 + (C_3 c_3 - C_4 c_2 + C_2)\alpha_l^6 \\ & + (C_0 - C_2 c_2 - C_3 c_1 + C_4 c_0)\alpha_l^4 \\ & + C_1 c_3 \alpha_l^3 + (C_2 c_0 - C_0 c_2)\alpha_l^2 - C_1 c_1 \alpha_l + C_0 c_0 \Big] \end{aligned} \tag{294}$$

α_l are the positive roots of the non-linear trigonometric equation,

$$tg\, \alpha_l \bar{H} = \frac{\alpha_l (c_3 \alpha_l^2 - c_1)}{(\alpha_l^4 - c_2 \alpha_l^2 + c_o)}$$

$$c_0 = -\bar{k}_d \frac{Ar^2}{\bar{R}}, \quad c_1 = \bar{k}_d Ar^2 \frac{\bar{H}}{\bar{R}},$$

$$c_2 = \bar{k}_d - (\bar{k}_a + Ar)\frac{Ar}{\bar{R}}, \quad c_3 = \frac{1}{\bar{k}_a} + Ar - \frac{Ar}{\bar{R}} \tag{295}$$

$$C_0 = -\frac{1}{2}(c_o \bar{H} + c_1), \qquad C_1 = c_2 + \frac{1}{2} c_1 \bar{H}$$

$$C_2 = \frac{1}{2} c_2 \bar{H} + \frac{3}{2} c_3, \qquad C_3 = -\left(\frac{1}{2} c_3 \bar{H} + 2\right), \qquad C_4 = -\frac{1}{2} \bar{H}$$

$$\bar{H} = (R_b - R)/L_a$$
$$\bar{R} = Ar\,\bar{H} + 1 \tag{296}$$

$$\bar{n}_\infty = 1 + \frac{\bar{K}}{\bar{R}\left(\frac{1}{6}c_0\bar{H}^3 + \frac{1}{2}c_1\bar{H}^2 + c_2\bar{H} + c_3\right)} \tag{297}$$

From Eq. (293) one can see that for $\tau >> 1(L_a^2/D >> 1)$

$$\Theta = \Theta_\infty = \bar{K}_a\, n_b \tag{298}$$

In case (ii) (contact with a well-stirred suspension) one has analogously [46]

$$\Theta = \Theta_i\, e^{-\bar{k}_d\tau} + \Theta_\infty\left(1 - e^{-\bar{k}_d\tau}\right) + \bar{k}_a\bar{K}\sum_{l=1}^{\infty} Q_l\left(e^{-\alpha_l^2\tau} - e^{-\bar{k}_d\tau}\right) \tag{299}$$

where $\Theta_\infty = \bar{K}_a$

$$Q_l = \frac{1}{P_l\left(\bar{k}_d - \alpha_l^2\right)}$$

$$P_l = \frac{1}{\alpha_l^2\left(c_1 - \alpha_l^2\right)}\Big[-C_3\alpha_l^6 + \left(C_2c_2 + C_3c_1 - C_1c_1\right)\alpha_l^4$$
$$+\left(C_1c_1 + C_0c_2 + C_2c_0\right)\alpha_l^2 + C_0c_0\Big] \tag{300}$$

$$C_0 = -\frac{1}{2}(c_0\bar{H} + c_1), \qquad C_1 = \frac{1}{2}c_1\,\bar{H} + c_2$$

$$C_2 = \frac{1}{2}c_2\bar{H} + \frac{3}{2}, \qquad C_3 = -\frac{1}{2}\bar{H}$$

and α_l are the positive roots of the non-linear trigonometric equation

$$ctg\,\alpha_l\,\bar{H} = \frac{c_2\,\alpha_l^2 + c_0}{\alpha_l\,(c_1 - \alpha_l^2)} \tag{301}$$

The general solutions, Eqs. (293) and (299) are rather cumbersome for direct use. It is, therefore, advantageous to derive some limiting solutions, which are useful for interpreting experimental data concerning adsorption kinetics. We first consider the short time limit where the system is close to its initial state described by the uniform concentration in the bulk $\overline{n} = 1$ and $\Theta = \Theta_i$. This adsorption regime prevails if $\tau << \overline{H}^2$ or, in dimensional units when $t < (R_b - R)^2 / D$. When moreover, $c_0\,\tau^2 << 1$, $c_1\,\tau^{3/2} << 1$, $c_2\,\tau << 1$, and $c_3^2\,\tau << 1$, one obtains from Eqs. (293) and (299) the same limiting solution for the coverage

$$\Theta = \Theta_i e^{-\overline{k}_d\tau} + \Theta_\infty\left(1 - e^{-\overline{k}_d\tau}\right) + \frac{\overline{k}_a\overline{K}}{\pi^{1/2}\overline{k}_d^{\,3/2}} e^{-\overline{k}_d\tau} \int_0^{(\overline{k}_d\tau)^{1/2}} \xi^2 e^{\xi^2}\, d\xi \qquad (302)$$

If $\overline{k}_d\,\tau << 1$, Eq. (302) further reduces to

$$\Theta = \Theta_i + \left(\overline{k}_a - \Theta_i\,\overline{k}_d\right)\tau + \overline{k}_a\overline{K}\frac{4}{3(\pi)^{1/2}}\tau^{3/2} \qquad (303)$$

The particle flux to the collector is given by

$$-j_a = \overline{k}_a - \overline{k}_d\,\Theta_i + 2\overline{k}_a\,\overline{K}\left(\frac{\tau}{\pi}\right)^{1/2} - \overline{k}_a\,(\overline{k}_a - \overline{k}_d\,\Theta_i)\tau \qquad (304)$$

It is interesting to note that the surface coverage always increases linearly for short adsorption times, and the flux remains finite in this limit, equal to $\overline{k}_a$.

Particle concentration distribution in the suspension is given by [46]

$$\overline{n} = 1 + \frac{\overline{K}}{(Ar\,\overline{z} + 1)}\left[2\left(\frac{\tau}{\pi}\right)^{1/2} e^{-\frac{\overline{z}^2}{4\tau}} - \overline{z}\,\mathrm{erfc}\frac{\overline{z}}{2\tau^{1/2}}\right] \qquad (305)$$

where $\overline{z} = (r - R)/L_a$

The subsurface concentration at $\overline{r} = (R+a)/L_a$ is

$$\overline{n}_\infty = 1 + 2\overline{K}\left(\frac{\tau}{\pi}\right)^{1/2} \qquad (306)$$

Another limiting case of practical interest arises when $\overline{H} = (R_b - R)/L_a \ll 1$, which corresponds to the thin film adsorption regime. The solution for the coverage in case (i) is

$$\Theta = \Theta_\infty + \overline{H}\,\frac{\Theta_i - \overline{K}_a}{\overline{K}_a + \overline{H}}\,e^{-\left(\frac{\overline{k}_a}{\overline{H}}+\overline{k}_d\right)\tau} \tag{307}$$

where $\Theta_\infty = \dfrac{\overline{k}_a}{\overline{k}_d}\,\dfrac{\overline{H} + \Theta_i}{\overline{H} + \overline{K}_a} = \overline{K}_a\,\overline{n}_\infty$

As can be noticed from this equation Θ approaches the final value Θ_∞ exponentially with the rate constant $\overline{k}_a/\overline{H} + \overline{k}_d$, which is inversely proportional to the thickness of the film.

Particle concentration distribution in the film is given by the expression [46]

$$\overline{n} = \overline{n}_\infty - \frac{\overline{K}}{\overline{k}_a + \overline{k}_d\overline{H}}\,e^{-\left(\frac{\overline{k}_a}{H}+\overline{k}_d\right)\tau} \tag{308}$$

As can be seen the concentration distribution is uniform and attains the final concentration distribution in an exponential manner.

In case (ii) (contact with a well-stirred suspension) the solution for the coverage is

$$\Theta = \Theta_i\,e^{-\overline{k}_d\,\tau} + \Theta_\infty\left(1 - e^{-\overline{k}_d\,\tau}\right) + \left(\Theta_i - K_a\right)\left(e^{-\left(\frac{\overline{k}_d}{1+\overline{k}_a\overline{H}}\right)\tau} - e^{-\overline{k}_d\,\tau}\right) \tag{309}$$

If $\overline{k}_a\overline{H} \ll 1$, Eq. (309) further simplifies to

$$\overline{\Theta} = \frac{\Theta - \Theta_\infty}{\Theta_i - \Theta_\infty} = e^{-\overline{k}_d\left(1-\overline{k}_a\,\overline{k}_d\,\overline{H}\right)\tau} \tag{310}$$

If the adsorption volume is very large in comparison with the collector volume, so that $R_b \gg 1$, Eqs. (293) and (299) reduce to the same limiting expression for the coverage

$$\bar{\Theta} = \Theta_i e^{-\bar{k}_d \tau} + \Theta_\infty \left(1 - e^{-\bar{k}_d \tau}\right) + \bar{k}_a \bar{K} \sum_{l=1}^{3} \frac{C_l' r_{o_l}}{r_{o_l}^2 + \bar{k}_d} \left[e^{r_{o_l}^2 \tau} \operatorname{erfc} r_{o_l} \tau^{1/2} + \left(1 + r_{o_l} I_1\right) e^{-\bar{k}_d \tau} \right] \tag{311}$$

where

$$I_1 = \frac{2}{(\pi \bar{k}_d)^{1/2}} \int_0^{(\bar{k}_d \tau)^{1/2}} e^{\xi^2} d\xi \tag{312}$$

$$C_1' = \frac{1}{\left(r_{o_1} - r_{o_2}\right)\left(r_{o_3} - r_{o_1}\right)}$$
$$C_2' = \frac{1}{\left(r_{o_1} - r_{o_2}\right)\left(r_{o_2} - r_{o_3}\right)} \tag{313}$$
$$C_3' = \frac{1}{\left(r_{o_3} - r_{o_1}\right)\left(r_{o_2} - r_{o_3}\right)}$$

and $r_{0_l} = -z_{0_l}$ are the real roots of the third-order equation

$$z_o^3 + \left(\bar{k}_a + Ar\right) z_o^2 + \bar{k}_d z_o + \bar{k}_d Ar = 0 \tag{314}$$

If one of the roots is real and the two others complex conjugate the solution for Θ is given in Ref. [46].

If the adsorption and desorption rate constants are large, so that the inequalities $\bar{k}_a^2 \tau \gg 1$ and $\bar{k}_d \tau \gg 1$ are met simultaneously (this is referred to as the diffusion-controlled adsorption regime), Eq. (311) reduces to the form [52]

$$\bar{\Theta} = \frac{\Theta(\tau) - \Theta_i}{\Theta_i - \Theta_\infty} = \frac{1}{r_{o_2} - r_{o_1}} \left[r_{o_2} e^{r_{o2}^2 \tau} \operatorname{erfc}(r_{o_2} \tau^{1/2}) - r_{o_1} e^{r_{o1}^2 \tau} \operatorname{erfc}\left(r_{o_1} \tau^{1/2}\right) \right] \tag{315}$$

where

$$r_{o_1} = \frac{1}{2\bar{K}_a} \left[1 + (1 - 4Ar)^{1/2} \right]$$
$$r_{o_2} = \frac{1}{2\bar{K}_a} \left[1 - (1 - 4Ar)^{1/2} \right] \tag{316}$$

If $4\,Ar > 1$, the solution is

$$\Theta = \left[\left(\frac{b_1}{b_2}\cos b_3\,\tau + \sin b_3\,\tau\right) I_2 + \left(\cos b_3\,\tau - \frac{b_1}{b_2}\sin b_3\,\tau\right) I_1'\right] \tag{317}$$

where $b_1 = 1/(2\overline{K}_a)$, $b_2 = 4(Ar - 1)/2\overline{K}_a$, $b_3 = 2b_1 b_2$

$$I_1' = e^{-b_2^2\,\tau} - \frac{2}{\pi^{1/2}} \int_0^{b_1\tau^{1/2}} \cos(2b_2\,\tau^{1/2}\,t)\,e^{-t^2}\,dt$$

$$I_2 = e^{-b_2^2\,\tau}\,\mathrm{erf}\,(b_2\,\tau^{1/2}) - \frac{2}{\pi^{1/2}} \int_0^{b_1\tau^{1/2}} \sin(2b_2\,\tau^{1/2} t) e^{-t^2}\,dt$$

$$\mathrm{erf}\,z = \int_0^{z} e^{-\xi^2}\,d\xi \tag{318}$$

For a planar geometry the general solution for θ under the kinetic boundary condition is [52]

$$\bar{\Theta} = \frac{1}{r_{o_2} - r_{o_1}}\left[r_{o_2}\,e^{r_{o2}^2\,\tau}\,\mathrm{erfc}\left(r_{o_2}\,\tau^{1/2}\right) - r_{o_1}\,e^{r_{o1}^2\,\tau}\,\mathrm{erfc}\left(r_{o_1}\,\tau^{1/2}\right)\right] \tag{319}$$

where

$$r_{o_1} = \frac{\bar{k}_a}{2}\left[1 + \left(1 - \frac{4\bar{k}_d}{\bar{k}_a^2}\right)^{1/2}\right]$$

$$r_{o_2} = \frac{\bar{k}_a}{2}\left[1 - \left(1 - \frac{4\bar{k}_d}{\bar{k}_a^2}\right)^{1/2}\right]$$

When $4\bar{k}_d / \bar{k}_a^2 > 1$, the solution is

$$\bar{\Theta} = e^{b_1^2 \tau}\left[\frac{b_1}{b_2}(\cos b_3\tau + \sin b_3\tau) I_2 + \left(\cos b_3\tau - \frac{b_1}{b_2}\sin b_3\tau\right) I_1'\right]$$
$$b_1 = \frac{\bar{k}_a}{2}, \quad b_2 = \frac{\bar{k}_a}{2}\left(\frac{4\bar{k}_d}{\bar{k}_a^2} - 1\right)^{1/2}, \quad b_3 = 2b_1 b_2 \tag{320}$$

Under the diffusion-controlled adsorption regime when $\bar{k}_a \gg 1$ and $\bar{k}_d \gg 1$ and desorption, Eq. (315) reduces to the form

$$\bar{\Theta} = e^{\frac{\tau}{\bar{K}_a^2}} \operatorname{erfc}\left(\frac{\tau^{1/2}}{\bar{K}_a}\right) \tag{321}$$

It is interesting to note that the limiting form of Eq. (321) for short times, when $\tau/\bar{K}_a^2 \ll 1$, is

$$\Theta = \Theta_i - \left(\frac{\Theta_i}{\bar{K}_a} - 1\right)\left[2\left(\frac{\tau}{\pi}\right)^{1/2} + Ar\,\tau\right] \tag{322}$$

When $\bar{K}_a \gg 1$, which is tantamount to the perfect sink model, Eq. (322) simplifies to

$$\Theta = \Theta_i + 2\left(\frac{\tau}{\pi}\right)^{1/2} + Ar\,\tau \tag{323}$$

By revering to the dimensional units and assuming $\Theta_i = 0$ this reduces to

$$N = \left[2\left(\frac{Dt}{\pi}\right)^{1/2} + \frac{D}{R}t\right] n_b \tag{324}$$

Hence the flux of particles is

$$-j_a = \frac{dN}{dt} = \left[\left(\frac{D}{\pi t}\right)^{1/2} + \frac{D}{R}\right] n_b \tag{325}$$

Eq. (325) agrees with the previously derived expression, Eq. (287), when substituting $a = 0$.

For longer times, when $R^2/Dt \ll 1$, Eq. (325) becomes

$$N = \frac{D}{R} n_b \, t \tag{326}$$

The flux has a constant value

$$-j_a = \frac{D}{R} n_b \tag{327}$$

On the other hand, for a planar geometry, when $R_b \rightarrow \infty$, Eqs. (325,326) reduce to the previously derived Eqs. (291,292).

These results have been generalized to multicomponent particle systems, adsorption from two phases (inside of the adsorber) and deforming interfaces, as is the case for liquid/gas or liquid/liquid interfaces [47,52]. In the latter case the adsorption or desorption of surfactants change the interfacial tension. This indices a force that resists the interfacial area changes, which is referred to as the surface elasticity force. It plays a significant role in emulsion dynamic and also determines foam stability. As shown in Ref. 47 in the general case of multi-component adsorption from two phases on deformable surfaces, the coverage in the Laplace transformation space is given by

$$\tilde{\boldsymbol{\Theta}}(s) = -\tilde{\mathbf{S}}(s) \cdot \tilde{\mathbf{E}}(s) \cdot \tilde{\boldsymbol{\Theta}}_e \tag{328}$$

where $\tilde{\boldsymbol{\Theta}}(s)$ is the vector having N_a components (where N_a is the number of adsorbing species), $\tilde{\mathbf{S}}(s)$ is the Laplace transform of the surface deformation function and $\tilde{\mathbf{E}}(s)$ is the surface elasticity, which can be defined in the Laplace transformation space as the complex operator (transfer function), being a square matrix of the dimension N_a and $\boldsymbol{\Theta}_e$ is the equilibrium surface concentration vector.

Eq. (328) indicates that by knowing $\tilde{\mathbf{E}}(s)$, one can determine $\overline{\Theta}(s)$ for arbitrary surface deformation (including a sudden expansion corresponding to the adsorption case studied above). The dependence of the surface coverage vector $\overline{\Theta}(\tau)$ on the adsorption time can be obtained by exploiting the inversion theorem of the Laplace transformation [47]

$$\Theta(\tau) = -\frac{1}{2\pi i}\int_{\lambda-i\infty}^{\lambda+i\infty} \tilde{\boldsymbol{\Theta}}(s)e^{s\tau}ds = -\frac{1}{2\pi i}\int_{\lambda-i\infty}^{\lambda+i\infty} \tilde{\mathbf{S}}(s)\cdot\,\tilde{\mathbf{E}}(s)\cdot\tilde{\boldsymbol{\Theta}}_e \tag{329}$$

The expressions for the $\tilde{\mathbf{E}}(s)$ operator for multi-component systems are given in Ref. 47,52.

Here we present a simpler expression for an one-component system and adsorption from one phase. Then, $\tilde{\mathbf{E}}(s)$ becomes a scalar (one component matrix), which has the following form for case (i) [52]

$$\begin{aligned}\tilde{\mathbf{E}}(s) = s[Ths + (Ar + k_a^* - \gamma_a)s^{1/2} - \gamma_a(k_a^* + Ar)Th]/\\ \{Ths^2 + (k_a^* + Ar - \gamma_a)s^{3/2} + Th[k_d^* - \gamma_a(k_a^* + Ar)]s\\ + k_d^*(Ar - \gamma_a)s^{1/2} - k_d^*\gamma_a ArTh\}\end{aligned} \tag{330}$$

where $Th = \tanh(s^{1/2}\overline{H})$, $\gamma_a = L_d/R_b$, $k_a^* = k_a f_e$, $k_d^* = \overline{k}_a + \overline{k}_d$ and f_e is the function describing the adsorption isotherm.

In case (ii) one has

$$\tilde{E}(s) = s\frac{s^{1/2} + (k_a^* + Ar)Th}{s^{3/2} + Th(k_a^* + Ar)s + k_d^* s^{1/2} + k_d^* ArTh} \tag{331}$$

For planar adsorption $Ar = 0$ and $\gamma_a = 0$, thus Eqs. (330) and (331) become

$$\tilde{E}(s) = s^{1/2}\frac{Th\,s^{1/2} + k_a^*}{Th\,s + k_a^* s^{1/2} + k_d^* Th} \tag{332}$$

In case (ii), one has

$$\tilde{E}(s) = s^{1/2}\frac{s^{1/2} + k_a^* Th}{s + k_a^* Th\,s^{1/2} + k_d^*} \tag{333}$$

For the most widely studied case of diffusion-controlled adsorption on a planar interface when $K_a^* \gg 1$ and $k_d \gg 1$ the solutions in case (i) and (ii) become

$$\tilde{E}(s) = s^{1/2}\frac{K_a^*}{K_a^*\,s^{1/2} + Th} \tag{334}$$

In case (ii), one has

$$E(s) = s^{1/2}\frac{K_a^* Th}{K_a^* Th\, s^{1/2} + 1} \tag{335}$$

where $K_a^* = k_a^* / k_d^*$
The limiting expressions for the infinite volume adsorption can be directly derived from Eqs. (330–335) by substituting $Th = 1$

4.6.2. Particle Adsorption Driven by External Forces

As can be noticed from Eq. (291), adsorption kinetics on planar interfaces, even in the most favorable case when $\bar{k}_a \gg 1$ (infinite adsorption rate constant), becomes inefficient for long times since the flux becomes proportional to $t^{-1/2}$. Because of the sluggishness of diffusion-controlled adsorption for longer times, the external force (gravitational) and convection effects start to play a more significant role. The external force effects have been analyzed before by neglecting diffusion, cf. Table 4.2. On the other hand, the pure convective transport of particles has been considered before the limiting trajectory approach, cf. Eqs. (102), (103). These solutions are most appropriate for particles larger than a micrometer. In this section we consider the case, more relevant to colloid particles transport where the effect of diffusion and the external force or diffusion and convection are combined.

Let us first consider the case of non-stationary external force-driven adsorption on a planar interface. The governing Smoluchowski equation now has the form

$$\frac{\partial n}{\partial t} = D\frac{\partial^2 n}{\partial z^2} - U\frac{\partial n}{\partial z} = D\left(\frac{\partial^2 n}{\partial z^2} - \frac{F}{kT}\frac{\partial n}{\partial z}\right) \tag{336}$$

In terms of dimensionless variables, Eq. (336) can be written as

$$\frac{\partial \bar{n}}{\partial \tau} = \frac{\partial^2 \bar{n}}{\partial H^2} - Ex\frac{\partial \bar{n}}{\partial H} \tag{337}$$

where $\tau = (D/a^2)t$, $H = z/a - 1$, $\bar{n} = \bar{n}/n_b$ and $Ex = Fa/kT$ is the external force parameter.

For the gravitational force, it is given explicitly by the expression

$$Ex = Gr = \frac{(m_p - m)a}{kT}|\mathbf{g}| = \frac{v_p(\rho_p - \rho)}{kT}|\mathbf{g}| \tag{338}$$

For the uniform electric force and spherical particles one has

$$Ex = \frac{qa}{kT}|\mathbf{E}| \tag{339}$$

The positive value of *Ex* in Eqs. (338,339) denotes the external force acting toward the interface, and the negative value of *Ex*, the opposite direction.

As discussed before, Eqs. (336,337) describe precisly the case of spherically isotropic bodies. They can also provide a good approximation of the adsorption of non-spherical particles if the inequality is fulfilled

$$U/Dr\ a << 1 \tag{340}$$

Considering that the rotary diffusion coefficient is approximately proportional to a^{-3} and the translational diffusion coefficient to a^{-1} (where a is the characteristic dimension of the particle), one can express Eq. (340) as

$$Ex << 1 \tag{341}$$

For the sake of simplicity the boundary condition for Eq. (337) is expressed the form of the perfect sink , i.e.,

$$\bar{n} = 0, \quad \text{at } H = 0 \tag{342}$$

and adsorption is assumed to proceed from an infinite medium, thus

$$\bar{n} \to 1, \quad \text{for } H \to \infty \tag{343}$$

The initial condition is

$$\bar{n} = 1, \quad \text{for } 0 < H < \infty, \quad t = 0 \tag{344}$$

By using the Laplace transformation method one can show that the solution of Eq. (337) with the boundary conditions Eqs. (342–344) has the form [53]

$$\bar{n} = 1 - \frac{1}{2}\left[e^{-Ex\,H}\operatorname{erfc}\left(\frac{H - Ex\,\tau}{2\tau^{1/2}}\right) + \operatorname{erfc}\left(\frac{H + Ex\,\tau}{2\tau^{1/2}}\right)\right] \tag{345}$$

The flux of particles at the solid surface is

$$j_a = -\left(\frac{D}{\pi t}\right)^{1/2} n_b\, e^{-(Ex^2/4a^2)Dt} - \frac{1}{2}\frac{D n_b}{a} Ex \operatorname{erfc} - \left(\frac{Ex(Dt)^{1/2}}{2a}\right) \tag{346}$$

One can deduce from Eq. (346) that for $Ex > 0$ after the transition time $t_{ch} \cong 4\,a^2\,/\,D\,Ex^2$, equal in the case of gravity to $4(kT/m_p')^2/D$ where $m_p' = (4/3)\pi a^3(\rho_p - \rho)$, the flux attains the stationary value

$$j_a = -\frac{D\,Ex}{a} n_b = -U\,n_b \tag{347}$$

where U is the steady sedimentation or migration velocity of particles listed in Table 4.2.

On the other hand, if $Ex < 0$ (the gravity force acting opposite to the interface), the particle flux vanishes asymptotically for long times according to the formula

$$j_a = -2\left(\frac{D}{\pi t}\right)^{1/2} e^{-\frac{Ex^2\,Dt}{4a^2}}\, D\,t\,n_b \tag{348}$$

By integrating Eq. (346) with respect to adsorption time one obtains the expression for the surface concentration of particles [53]

$$N = \frac{2a n_b}{Ex}\operatorname{erf}\frac{Ex}{2a}(Dt)^{1/2} + \frac{1}{2}\frac{D n_b t}{a} Ex \operatorname{erfc}\left[-\frac{Ex}{2a}(Dt)^{1/2}\right] - \frac{4a n_b}{Ex}\frac{1}{\pi^{1/2}}\int_0^{\frac{Ex}{2a}(Dt)^{1/2}} \xi^2 e^{-\xi^2}\,d\xi \tag{349}$$

For $t > t_{ch}$ and $Ex > 0$, Eq. (349) simplifies to

$$N = \frac{D\,Ex}{a} n_b\,t + \frac{a\,n_b}{Ex} \cong U\,t \tag{350}$$

In the case of $Ex < 0$, the long-time solution is

$$N = \left(\frac{Dt}{\pi}\right)^{1/2} n_b\, e^{-\frac{Ex^2 Dt}{4a^2}} - \frac{a n_b}{Ex} \tag{351}$$

4.6.3. Convective Diffusion Transport to Various Interfaces

In addition to external forces, fluid convection is another mechanism of enhancing particle transfer to interfaces. Because of flows, particles are brought to the region close to interfaces, which equalizes their concentration, except for a thin layer at the collector called the diffusion boundary layer δ_d (see Fig. 4.11). Its thickness remains usually much smaller than the hydrodynamic boundary layer thickness analyzed in Chapter 3 and the characteristic collector dimension R. Accordingly, the characteristic relaxation time δ_d^2/D of establishing steady-state conditions of transport is of the order of seconds for colloid particles (quantitative estimations of the relaxation time derived by numerical solutions of the exact transport will be presented later on). This is a negligible value in comparison with typical experimental times ranging from 10^2 to 10^5 s [11,49]. Therefore, for describing particle transfer rates one can use the stationary convective diffusion equation, Eq. (250). Note that in deriving this equation, the hydrodynamic wall corrections, as well as the specific force field have been neglected. The latter can be considered via the boundary conditions specified in Section 4.6.1

Eq. (250) can be further simplified for many situations of practical interest by performing the dimensional analysis analogous to that of the hydrodynamic boundary layer theory (see Chapter 3). This is feasible because $\delta_d/R \ll 1$, so the terms describing the diffusion tangential to the interface become negligible in comparison with the terms describing the diffusion normal to the interface. By assuming additionally that the flows near collectors can be expressed in terms of two spatial coordinates (z, x), which is the case for all flows listed in Table 3.4, one can reduce Eq. (250) to

$$\frac{\partial^2 \bar{n}}{\partial \bar{z}^2} = Pe\left(\bar{V}_\perp \frac{\partial \bar{n}}{\partial \bar{z}} + \bar{V}_\parallel \frac{\partial \bar{n}}{\partial \bar{x}}\right) \tag{352}$$

where $\bar{V}_\perp = V_\perp/V_{ch}$ and $\bar{V}_\parallel = V_\parallel/V_{ch}$ are the dimensionless components of fluid velocity, perpendicular and parallel to the interface, $\bar{z} = (z + a)/L_{ch}$ the dimensionless coordinate perpendicular to the interface, $\bar{x} = x/L_{ch}$ the coordinate parallel to the interface and V_{ch} the characteristic velocity.

Eq. (352) can be further simplified if the normal component of the flow remains independent of the tangential coordinate, and when the tangential flow component vanishes. In this case there is no gradient of particle concentration parallel to the interface (if it is homogeneous). As discussed previously, such a property is characteristic of the flow in the vicinity of the stagnation point, near the rotating disk, the impinging-jet cells, a sphere or a cylinder placed in a uniform flow in the region close to the stagnation point, etc. The interface exposed to this type of flow is called the uniformly accessible surface. Eq. (352) becomes in this case

$$\frac{\partial^2 \bar{n}}{\partial \bar{z}^2} = Pe \bar{V}_\perp \frac{\partial \bar{n}}{\partial \bar{z}} \tag{353}$$

The thickness of the diffusion boundary layer δ_d can be estimated now using Eq. (353) by considering that $\frac{\partial^2 \bar{n}}{\partial \bar{z}^2} \sim \frac{1}{\delta_a^2}, \frac{\partial \bar{n}}{\partial \bar{z}} \sim \frac{1}{\delta_d}$ and by assuming that $\bar{V}_\perp \sim \delta_h^2$; thus

$$\frac{\delta_d}{L_{ch}} \sim Pe^{-1/3} = \left(\frac{D}{V_{ch} L_{ch}} \right)^{1/3} \tag{354}$$

Taking the typical data: $D = 10^{-12}\,\mathrm{m^2\,s^{-1}}$, $V_{ch} = 10^{-2}\,\mathrm{m\,s^{-1}}$, $L_{ch} = 10^{-3}\,\mathrm{m}$ one obtains from Eq. (354) $\delta_d / L_{ch} = 4.6\mathrm{x}10^{-3}$. As can be seen, the thickness at the diffusion boundary layer is indeed much smaller than L_{ch}.

Also note that the thickness of the diffusion boundary layer decreases with fluid velocity proportionally to $V_{ch}^{-1/3}$.

The boundary conditions for Eq. (353) are usually specified by assuming reversibility in the form

$$-\bar{j}_a = -j_a \frac{L_{ch}}{D n_b} = \bar{k}_a \left(\frac{\partial \bar{n}}{\partial \bar{z}} \right), \quad \text{at } \bar{z} = \bar{z}_0 \tag{355}$$

where $\bar{z}_0 = (\delta_a + a) / L_{ch}$ and the adsorption constant $\bar{k}_a$, depending on the shape of the specific interaction potential, is given explicitly by Eq. (271). Eq. (355) can also be expressed in an alternative form

$$\bar{n} = \bar{n}^*, \quad \text{at } \bar{z} = \bar{z}_0 \tag{356}$$

where $\bar{n}^*$ remains an arbitrary concentration to be determined. In the case where the adsorption rate constant $\bar{k}_a$ tends to infinity, one arrives at the perfect sink model, where the boundary condition at $\bar{z} = \bar{z}_0$ becomes

$$\bar{n} = 0, \quad \text{at } \bar{z} = \bar{z}_0 \tag{357}$$

Eq. (353) with the boundary condition Eq. (356) can be solved analytically by considering that the normal velocity component in the diffusion boundary layer is described by the formula

$$\bar{V}_{\perp} = -C_{\perp}\bar{z}^2 \tag{358}$$

Values of the constant $C_{\perp}$ for various collectors are listed in Table 3.4.

By subsisting Eq. (358) into Eq. (353) and performing the integration one obtains the expression for particle concentration distribution

$$\bar{n}(z) = \bar{n}^* + \frac{(1-n^*)}{I_{\perp}(\bar{z}_0)}\left(\frac{C_{\perp}Pe}{3}\right)^{1/3}\int_{\bar{z}_0}^{\bar{z}} e^{-\xi^3}\,d\xi \tag{359}$$

where

$$I_{\perp}(\bar{z}_0) = \int_{\bar{z}_0}^{\infty} e^{-\xi^3}\,d\xi \tag{360}$$

The flux of particles at $\bar{z} = \bar{z}_0$ is given by

$$-j = \frac{Dn_b}{L_{ch}}\left(\frac{C_{\perp}Pe}{3}\right)^{1/3}\frac{1-\bar{n}^*}{I_{\perp}} = k_c'(n_b - n^*) \tag{361}$$

where

$$k_c' = \frac{D}{L_{ch}I_{\perp}}\left(\frac{C_{\perp}Pe}{3}\right)^{1/3} \tag{362}$$

is the mass transfer coefficient, often called the reduced flux, characterizing the rate of particle transport from the bulk to the edge of the adsorption layer.

In practice, because $\delta_a < \delta_d$, one can assume that the lower integration limit in Eq. (360) equals zero, so the mass transfer coefficient k_c is given by

$$k_c^{'} = k_c = \frac{D}{L_{ch}\Gamma\left(\frac{4}{3}\right)}\left(\frac{C_\perp Pe}{3}\right)^{1/3} \tag{363}$$

where $\Gamma(4/3) = I_\perp(0) = 0.893$ is the Euler gamma function value for 4/3.

The unknown concentration n^* can be eliminated from Eq. (361) by postulating that the flux from the bulk to the adsorption layer equals the adsorption flux

$$-k_c(n_b - n^*) = j_a = -k_a n^* \tag{364}$$

where k_a is the modified adsorption constant given by the expression

$$k_a^{''} = \frac{1}{\displaystyle\int_{\delta_m}^{\delta_a} \frac{e^{\phi/kT} - 1}{D(z)} dz} \tag{365}$$

From Eq. (364) one obtains

$$n^* = \frac{k_c}{k_c + k_a^{''}} n_b \tag{366}$$

Thus, the overall flux, $k_a^{''} n^*$ is given by

$$-j = \frac{k_c}{1 + \dfrac{k_c}{k_a^{''}}} n_b = \frac{j_0}{1 + \dfrac{j_0}{n_b k_a^{''}}} \tag{367}$$

where

$$-j_0 = k_c n_b = \frac{D n_b}{L_{ch}} Sh_0 \tag{368}$$

is the stationary flux from the bulk to the primary minimum called the limiting flux and $Sh_0 = -j_0 L_{ch}/D\, n_b$ is the dimensionless Sherwood number characterizing the mass transfer rate from the bulk to the primary minimum in the case of no barrier.

The definite integral occurring in Eq. (365) can be evaluated analytically for a triangular or parabolic barrier whose height exceeds a few kT units, as done previously (cf. Eqs. 277,278). By doing so one can express Eq. (367) for a triangular barrier, by assuming $L_{ch} = a$, in the form

$$j = j_0 \frac{1}{1 + Sh_0 \dfrac{kT}{\phi_b} e^{\phi_b/kT}} \tag{369}$$

For a parabolic barrier one has

$$j = j_0 \frac{1}{1 + Sh_0 \left(\dfrac{2\pi kT}{\phi_b} \right)^{1/2} e^{\phi_b/kT}} \tag{370}$$

The dependence of the reduced particle flux j/j_0 on the reduced barrier height ϕ_b/kT calculated from Eq. (369) for various Sh_0 is shown in Fig. 4.13.

It can be deduced from Eqs. (369), (370) that for a high energy barrier, the role of the bulk transport becomes negligible and the net flux is given by

$$-j = k_a n_b \tag{371}$$

Eq. (371) for triangular and parabolic barriers becomes

$$\begin{aligned} -j &= \frac{Dn_b}{a} \frac{\phi_b}{kT} e^{-\phi_b/kT} \\ -j &= \frac{Dn_b}{a} \left(\frac{\phi_b}{2\pi kT} \right)^{1/2} e^{-\phi_b/kT} \end{aligned} \tag{372}$$

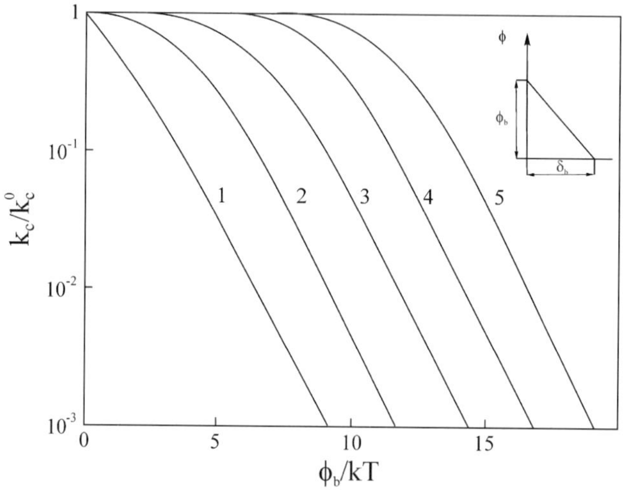

Fig. 4.13. The dependence of the reduced particle flux $k_c/k_c^0 = j/j_0$ (where j_0 = - $k_c\, n_b$ is the flux for no barrier) on the reduced barrier height ϕ_b/kT calculated from Eq. (367), (1) Sh_0=1, (2) Sh_0=0.1, (3) $Sh_0=10^{-2}$, (4) $Sh_0=10^{-3}$, (5) $Sh_0=10^{-4}$.

Eqs. (371), (372), which indicate that the flux decreases exponentially with the barrier height and is independent of the flow intensity, describe the so-called activated or barrier-controlled transport of particles to interfaces.

On the other hand, in the case when $k_c/k_a' \ll 1$ (fast transfer rate through the adsorption layer), Eq. (367) simplifies to

$$-\frac{j_0}{n_b} = k_c = \frac{D}{\Gamma(4/3)}\left(\frac{C_\perp Pe}{3}\right)^{1/3} \tag{373}$$

In this case particle adsorption is governed by bulk transport alone, which is referred to as the barrierless adsorption regime.

Analytical expressions for k_c are collected in Table 4.8 for uniformly accessible surfaces. They have been calculated from Eq. (363) using the near-surface fluid velocity fields given previously for various collectors in Table 3.4. Also the definitions of the Peclet number are given calculated by assuming $L_{ch} = a$.

Selected solutions of practical significance are discussed below. For the rotating disk whose entire surface is uniformly accessible for particle transport, one has $Pe = 1.02\,\dfrac{\Omega^{3/2}a^3}{\nu^{1/2}D}$, $C_\perp = 1/2$, and the expression for k_c becomes

$$k_c = 0.620\frac{\Omega^{1/2}D^{2/3}}{\nu^{1/6}} \tag{374}$$

Eq. (374), first derived by Levich [48], has a fundamental significance in calculating the limiting current to the rotating disk electrode used widely in electrochemistry.

Analogously, for a sphere in an uniform flow, in the region of the forward stagnation point, the surface becomes uniformly accessible and the limiting flux is given by

$$k_c = 0.889\,A_f^{1/3}\frac{V_\infty^{1/3}D^{2/3}}{R^{2/3}} \tag{375}$$

where A_f is the flow model parameter given in Table 3.4.

In the case of oblate spheroids Eq. (375) can be used with the flow parameter A_f can be calculated from Eq. (305) in Chapter 3. It is explicitly given by

$$A_f = \frac{4}{3}\frac{\lambda_0^3}{(\lambda_0^2+1)^2\left[\lambda_0 - (\lambda_0^2-1)\cot^{-1}\lambda_0\right]} \tag{376}$$

where

$$\lambda_0 = R_{b_s}/(R_{a_s}^2 - R_{b_s}^2)^{1/2} \tag{377}$$

R_{a_s} is the longer and R_{b_s} the shorter semi-axis of the spheroid.

Another solution of practical importance can be derived for the cylinder in a uniform flow, in the region near the forward stagnation point

$$k_c = 0.776\,A_f^{1/3}\frac{V_\infty^{1/3}D^{2/3}}{R^{2/3}} \tag{378}$$

where A_f is the flow model parameter given in Table 4.8.

Table 4.8
Pe definitions and bulk transfer rate constants (reduced flux) k_c for uniformly accessible surfaces

Collector and flow configuration	*Pe* definition microscopic $L_{ch} = a$	*Pe* definition macroscopic	$k_c = -j/n_b$
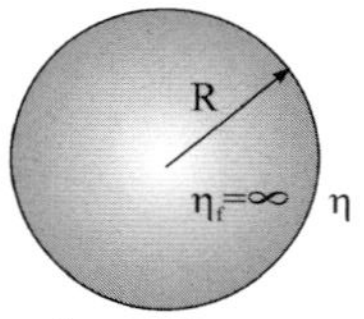 Sphere in quiescent fluid	0	0	$\dfrac{D}{R}$
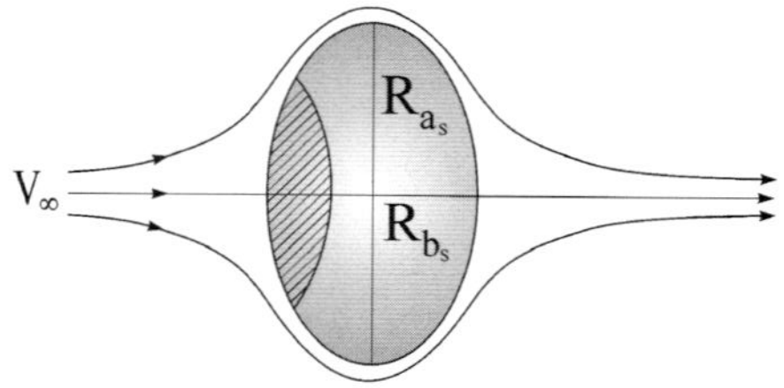 Oblate spheroid in uniform flow (near stagnation point)*	$3A_f\left(\dfrac{R_{a_s}}{R_{b_s}}\right)\dfrac{V_\infty a^3}{R_{b_s}^2 D}$	$2A_f\left(\dfrac{R_{a_s}}{R_{b_s}}\right)\dfrac{V_\infty R_{b_s}}{D}$	$0.888\dfrac{A_f^{1/3}V_\infty^{1/3}D^{2/3}}{R_{b_s}^{2/3}}$
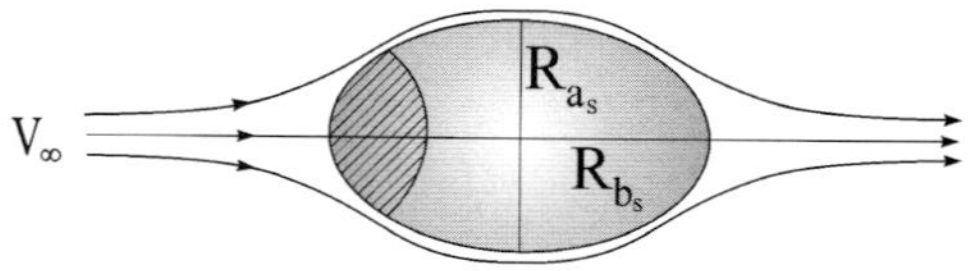 Prolate spheroid in uniform flow (near stagnation point)	$3A_f\left(\dfrac{R_{a_s}}{R_{b_s}}\right)\dfrac{V_\infty a^3}{R_{a_s}^2 D}$	$2A_f\left(\dfrac{R_{a_s}}{R_{b_s}}\right)\dfrac{V_\infty R_{a_s}}{D}$	$0.888\dfrac{A_f^{1/3}V_\infty^{1/3}D^{2/3}}{R_{a_s}^{2/3}}$

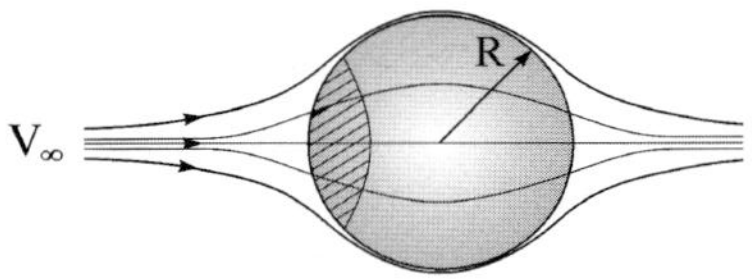 Sphere in uniform flow (near stagnation point)**	$\frac{3A_f(Re)V_\infty}{R^2}\frac{a^3}{D}$	$2A_f(Re)V_\infty\frac{R}{D}$	$0.888\frac{A_f^{1/3}V_\infty^{1/3}D^{2/3}}{R^{2/3}}$
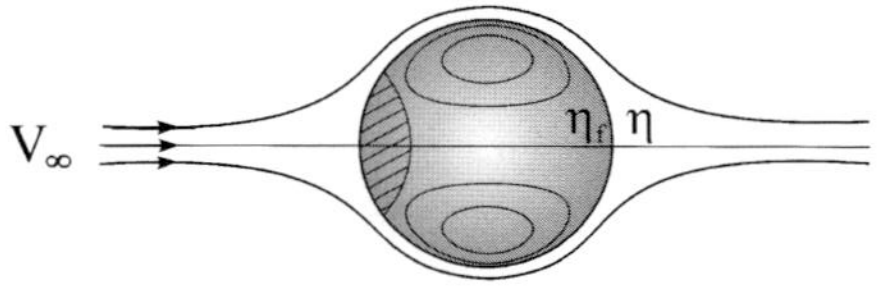 Liquid sphere in uniform flow	$2\left(\frac{\eta}{\eta_f+\eta}\right)\frac{V_\infty a^2}{DR}$	$\frac{2V_\infty R}{D}$	$\left[\frac{\eta}{2(\eta_f+\eta)}\frac{DV_\infty}{\pi R}\right]^{1/2}$
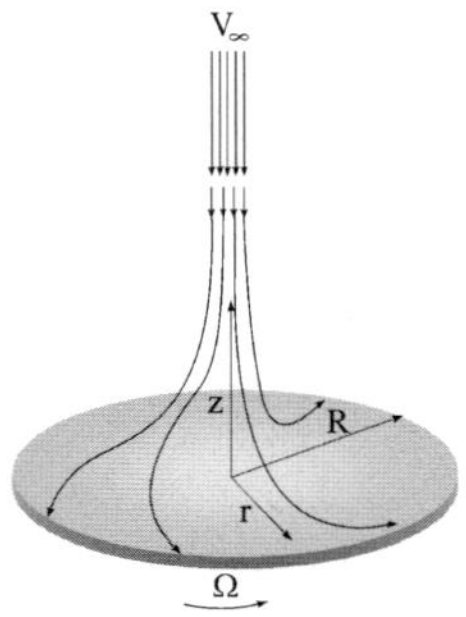 The rotating disk	$\frac{1.021\Omega^{3/2}}{\nu^{1/2}}\frac{a^3}{D}$	$\frac{2V_\infty\delta_h}{D}=8.86\frac{\nu}{D}$	$0.620\frac{\Omega^{1/2}D^{2/3}}{\nu^{1/6}}$

Table 4.8 (continued)

Collector and flow configuration	*Pe* definition microscopic $L_{ch} = a$	*Pe* definition macroscopic	$k_c = -j/n_b$
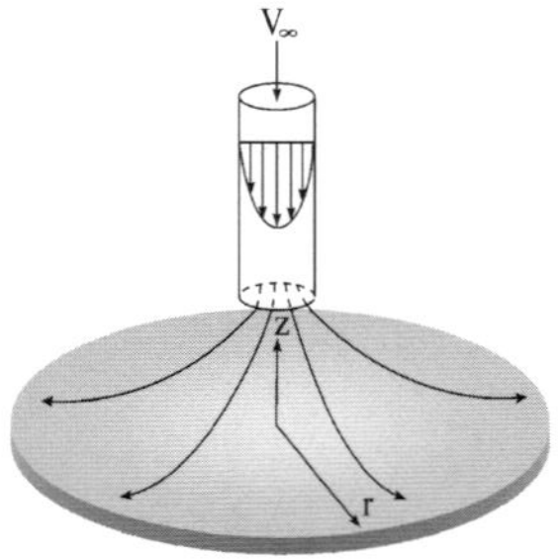 Radial impinging-jet RIJ (near stagnation point)***	$\frac{2\alpha_r V_\infty}{R^2}\frac{a^3}{D}$	$\frac{2V_\infty R}{D}$	$0.776\frac{\alpha_r^{1/3} V_\infty^{1/3} D^{2/3}}{R^{2/3}}$ $0.530\frac{\alpha_r^{1/3} Q^{1/3} D^{2/3}}{R^{4/3}}$
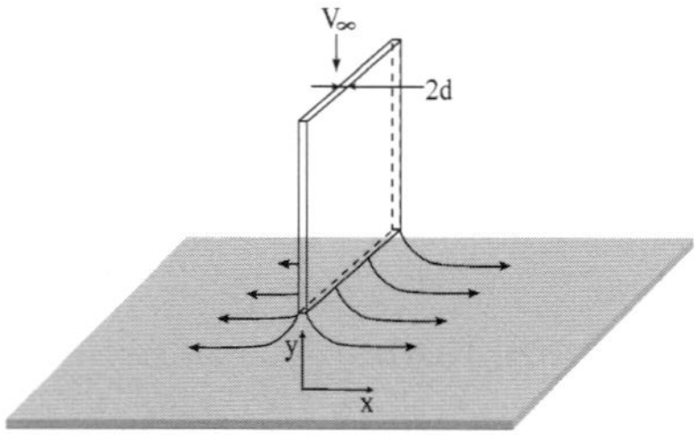 Slot impinging-jet cell (near stagnation point)****	$\frac{2\alpha_s V_\infty a^3}{d^2 D}$	$\frac{2V_\infty d}{D}$	$0.776\frac{\alpha_s^{1/3} V_\infty^{1/3} D^{2/3}}{d^{2/3}}$ $0.616\frac{\alpha_s^{1/3} Q^{1/3} D^{2/3}}{d\, l^{1/3}}$

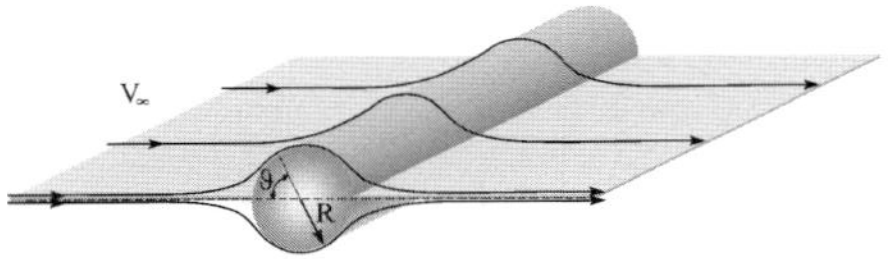 Cylinder in uniform flow (near stagnation point)*****	$\frac{2A_f V_\infty}{R^2}\frac{a^3}{D}$	$\frac{A_f V_\infty R}{D}$	$0.776\frac{A_f^{1/3} v_\infty^{1/3} D^{2/3}}{R^{2/3}}$

* Oblate spheroids

$$A_f\left(\frac{R_{a_s}}{R_{b_s}}\right)=\frac{4}{3}\frac{\lambda_0^3}{(\lambda_0^2+1)^2[\lambda_0-(\lambda_0^2-1)\cot^{-1}\lambda_0]},\quad \lambda_0=R_{b_s}/(R_{a_s}-R_{b_s})^{1/2},$$

Prolate spheroids

$$A_f\left(\frac{R_{a_s}}{R_{b_s}}\right)=\frac{4}{3}\frac{\lambda_0^3}{(\lambda_0^2-1)^2\left[\frac{1}{2}(\lambda_0^2+1)\ln\frac{\lambda_0+1}{\lambda_0-1}-\lambda_0\right]}$$

** $A_f(Re)=\frac{3}{2}\left(1+\frac{\frac{3}{16}Re}{1+0.249\,Re^{0.56}}\right)$ valid for $Re=\frac{2RV_\infty}{v}<300$

*** α_r for various (h/R) and Re are given in Section 3.3
$V_\infty = Q/\pi R^2$ Q – volumetric flow rate
$Re = 2V_\infty R/v$

**** α_s for various d and Re are given in Section 3.3
$V_\infty = Q/2dl$ Q – volumetric flow rate
$Re = 2V_\infty d/v$

***** $A_f = (\beta_c - 0.48\beta_c^3)$, $\beta_c = (2 - ln\ Re)^{-1}$, for $Re<1$
$A_f = 0.44\ Re^{0.52}$ $Re = 2V_\infty R/v$ for $1<Re<200$

As can be noticed from the results collected in Table 4.8, in all cases k_c increases proportionally to $D^{2/3}$ rather than to D is as intuitively expected. Because D is inversely proportional to particle size, this means that the convective flux decreases as $a^{-2/3}$ with particle radius. It is also interesting to observe that k_c is rather insensitive to the fluid velocity V_∞ increasing in all cases proportionally to $V_\infty^{1/3}$ except for the case of the rotating disk where k_c remains proportional to the $V_\infty = 0.886\ (\Omega\nu)^{1/2}$. For the fluid sphere $k_c \sim V_\infty^{1/2}$ as discussed later on.

In the case of non-uniformly accessible surfaces, expressions for k_c can be derived using Eq. (352) by assuming that the velocity components near the surface are given by [54]

$$\begin{aligned} \bar{V}_\perp &= f_1(\bar{x})\bar{z}^2 \\ \bar{V}_{||} &= f_2(\bar{x})\bar{z} \end{aligned} \tag{379}$$

where $f_1(x)$ and $f_2(x)$ are functions of the tangential coordinate alone, which can be extracted from velocity fields given in Table 3.4.

For a fluid drop the velocity components are

$$\begin{aligned} \bar{V}_\perp &= f_1'(\bar{x})\bar{z} \\ \bar{V}_{||} &= f_2'(\bar{x}) \end{aligned} \tag{380}$$

By considering Eq. (379) and introducing the new variable

$$\xi = \bar{z}\, Pe^{1/3} \tag{381}$$

one can convert Eq. (352) to the form

$$f_1 \xi^2 \frac{\partial \bar{n}}{\partial \xi} + f_2 \xi \frac{\partial \bar{n}}{\partial \bar{z}} = \frac{\partial^2 \bar{n}}{\partial \xi^2} \tag{382}$$

Introducing the similarity transformation

$$Z = \frac{\xi}{g(\bar{x})} = \frac{\bar{z}\, Pe^{1/3}}{g(\bar{x})} \tag{383}$$

one can reduce Eq. (382) to the ordinary differential equation

$$-3Z^2 \frac{\partial \bar{n}}{\partial Z} = \frac{\partial^2 \bar{n}}{\partial Z^2} \tag{384}$$

where $g(\bar{x})$ is the unknown function to be found by solving the differential equation

$$f_1 g^3 - f_2 g^2 g' = f_1 G_3 - \frac{1}{3} f_2 G_3' = -3 \tag{385}$$

where $G_3(x) = g^3(\bar{x})$.

For a fluid sphere, upon substituting $Z = \dfrac{\bar{z}\, Pe^{1/2}}{g'(\vartheta)}$, Eq. (352) can be converted to the form

$$-2Z \frac{\partial \bar{n}}{\partial Z} = \frac{\partial^2 \bar{n}}{\partial Z^2} \tag{386}$$

where $g(\vartheta)$ is the function to be calculated from the differential equation

$$f_1 G_2 - \frac{1}{2} f_2 G_2' = -2 \tag{387}$$

where $G_2 = g^2(\vartheta)$

The general kinetic boundary condition for Eq. (384) is given by

$$\frac{Pe^{1/3}}{g(\bar{x})} \frac{\partial \bar{n}}{\partial Z} = \frac{L_{ch}}{D n_b} k_a n^* = \bar{k}_a \bar{n}^* \quad \text{at } Z = Z_0 = \frac{\bar{z}_0\, Pe^{1/3}}{g(\bar{x})} \tag{388}$$

where $\bar{k}_a = k_a L_{ch} / D n_b$; $\bar{n}^* = n^*/n_b$

In the case of the perfect sink model, $\bar{n}^* = 0$ and the boundary condition becomes

$$\bar{n}^* = 0 \quad \text{at} \quad Z = Z_0 \tag{389}$$

The boundary condition for Eq. (385), being in principle, the *quasi*-initial condition, has the form

$$f_1(0)G_3(0) - \frac{1}{3}f_2(0)G_3'(0) = -3 \tag{390}$$

where $f_1(0), f_2(0)$ are the values of the functions $f_1(\bar{x}), f_2(\bar{x})$ at $\bar{x} = 0$.

In the case of symmetric flows, e.g., near sphere or cylinder in uniform flow $f_2(0)=0$ and Eq. (390) simplifies to

$$G_3(0) = -\frac{3}{f_1(0)} \tag{391}$$

On the other hand, for unidirectional flows occurring, e.g., in the parallel-plate channel or a cylindrical channel $f_1(0)=0$ and the initial condition for the G_3 function is

$$G_3'(0) = \frac{9}{f_2(0)} \tag{392}$$

For the fluid sphere one has the condition

$$G_2(0) = -\frac{2}{f_1(0)} \tag{393}$$

Since Eq. (385) is linear its general solution can be expressed as

$$G_3(\bar{x}) = C_3\, e^{3F(\bar{x})} \int \frac{e^{-3F(\bar{x})}}{f_2(\bar{x})}\, d\bar{x} \tag{394}$$

where

$$F(\bar{x}) = \int \frac{f_1(\bar{x})}{f_2(\bar{x})}\, d\bar{x} \tag{395}$$

and C_3 is the constant of integration to be determined from the boundary condition, Eq. (390).

For the fluid sphere one has

$$G_2(\bar{x}) = C_2 e^{2F} \int \frac{e^{-2F(\bar{x})}}{f_2(\bar{x})}\, d\bar{x} \tag{396}$$

Eq. (394) can be integrated analytically for many simple flows of practical interest, e.g., in the case of a sphere in uniform flow one has $f_1 = -\frac{1}{2}\cos\vartheta$, $f_2 = \frac{1}{2}\sin\vartheta$, therefore $F(\vartheta) = -\ln\sin\vartheta$ and

$$g(\bar{x}) = G_3^{1/3}(\bar{x}) = 9^{1/3}\,\frac{\left(\vartheta - \frac{1}{2}\sin 2\vartheta\right)^{1/3}}{\sin\vartheta} \tag{397}$$

where $\vartheta = \bar{x} = x/R$ and $C_3=9$ were derived from the boundary condition, Eq. (391).

For the fluid sphere $f_1 = -\frac{1}{2}\cos\vartheta$, $f_2 = \frac{1}{4}\sin\vartheta$, therefore

$$g(\vartheta) = G_3^{1/2} = \left[\frac{4}{3}\right]^{1/2}\frac{(\cos 3\vartheta - 9\cos\vartheta + 8)^{1/2}}{\sin^2\vartheta} \tag{398}$$

For the cylinder one has $f_1 = -\frac{1}{2}\cos\vartheta$, $f_2 = \sin\vartheta$, therefore

$$g(\bar{x}) = G_3^{1/3}(\bar{x}) = 9^{1/3}\,\frac{\left[\int_0^{\vartheta}\sin^{1/2}\vartheta'\,d\vartheta'\right]^{1/3}}{\sin^{1/2}\vartheta} \tag{399}$$

$\bar{x} = x/R = \vartheta$

Analogously, for the parallel-plate and cylindrical channel one has $f_1 = 0$, $f_2 = \frac{1}{2}$. By considering the boundary condition, Eq. (392), one obtains for $g(\bar{x})$ the expression

$$g(\bar{x}) = G_3^{1/3}(\bar{x}) = (18\bar{x})^{1/3} \tag{400}$$

For the radial wall-jet one has $f_1 = \frac{1}{2}\bar{x}^{-15/4}$, $f_2 = \frac{4}{7}\bar{x}^{-11/4}$, $\bar{x} = r/R$.

From the boundary condition, Eq. (385), one has $G_3 = 14$, therefore the function is

$$g(\bar{x}) = (14)^{1/3}\bar{x}^{5/4} \tag{401}$$

The $g(\bar{x})$ function for other collectors and flow configurations can be derived in an analogous manner. In the general case, however, for arbitrary flows, Eq. (395) can only be integrated numerically as done by Dabros et al. [54] for a cylinder placed in a combined uniform and shearing flow.

Once the function $g(\bar{x})$ is known, one can integrate Eq. (384) with the perfect sink boundary condition obtaining the expression for the particle concentration distribution in the form

$$\bar{n}(Z) = \frac{\int_{Z_0}^{Z} e^{-\xi^3}\, d\xi}{\int_{Z_0}^{\infty} e^{-\xi^3}\, d\xi} = \frac{1}{I_\perp(Z_0)} \int_{Z_0}^{Z} e^{-\xi^3}\, d\xi \tag{402}$$

The mass transfer rate constant at the surface is

$$k_c(\bar{x}) = \frac{D}{L_{ch}} \frac{Pe^{1/3}}{g(\bar{x})} \frac{1}{I_\perp(Z_0)} \tag{403}$$

For small colloid particles when the interception effect can be neglected, $Z_0 \ll 1$ and Eq. (403) simplifies to

$$k_c(\bar{x}) \cong \frac{D}{L_{ch}} \frac{Pe^{1/3}}{\Gamma\left(\frac{4}{3}\right) g(\bar{x})} \tag{404}$$

For the fluid sphere, one has

$$k_c(\bar{x}) = \frac{D}{L_{ch}} \frac{Pe^{1/2}}{\Gamma\left(\frac{1}{2}\right) g(\bar{x})} \tag{405}$$

Eq. (404) represents the general solution valid for any flow configuration except for the fluid sphere. As can be noticed, the flux is always proportional to $Pe^{1/3}$ and depends on the tangential coordinate $\bar{x}$. For the fluid sphere $k_c \sim Pe^{1/2}$.

The integration of the local flux over the entire collector surface gives the averaged flux,

$$\langle k_c \rangle = \frac{1}{S_c}\int_{S_c} k_c(\bar{x})\,dS(\bar{x}) = C_c \frac{D}{L_{ch}}\frac{Pe^{1/3}}{\Gamma\left(\frac{4}{3}\right)} = \frac{D}{L_{ch}}\langle Sh \rangle \tag{406}$$

For the fluid sphere

$$\langle k_c \rangle = C_c \frac{D}{L_{ch}}\frac{Pe^{\frac{1}{2}}}{\Gamma\left(\frac{1}{2}\right)} \tag{407}$$

where $C_c = (1/S_c)\int_{Sc} dS/g(\bar{x})$ and $\langle Sh \rangle = \langle k_c \rangle L_{ch}/D$ is the averaged Sherwood number. As can be deduced from Eq. (406), the averaged flux (independent of the tangential coordinate) also increases proportionally to $Pe^{1/3}$ for all collectors, except for the fluid sphere.

By combining Eq. (403) with the previously derived expression for the $g(\bar{x})$ function, Eq. (397), one obtains for the sphere, the explicit expression for the local transfer rate (see Table 4.9)

$$k_c(\vartheta) = 0.776 \frac{\sin\vartheta}{\left(\vartheta - \dfrac{1}{2}\sin 2\vartheta\right)^{1/3}} \frac{A_f^{1/3} V_\infty^{1/3} D^{2/3}}{R^{2/3}} \tag{408}$$

Eq. (408) was first derived by Levich [48] using a different similarity transformation.

For the fluid sphere one has

$$k_c(\vartheta) = \left(\frac{3\eta}{2(\eta_f + \eta}\right)^{1/2} \left(\frac{DV_\infty}{\pi R}\right)^{1/2} \frac{\sin^2\vartheta}{(\cos 3\vartheta - 9\cos\vartheta + 8)^{1/2}} \tag{409}$$

It can be deduced from Eq. (408) that the flux attains the maximum value: $0.889(A_f^{1/3} V_\infty D^{2/3})/R^{2/3}$ at the forward stagnation point and vanishes at the rear stagnation point. The average flux is

$$\langle k_c \rangle = 0.624 \frac{A_f^{1/3} V_\infty^{1/3} D^{2/3}}{R^{2/3}} \tag{410}$$

The ratio of the local to the averaged flux is

$$J = \frac{k_c}{\langle k_c \rangle} = 1.24 \frac{\sin\vartheta}{\left(\vartheta - \frac{1}{2}\sin 2\vartheta\right)^{1/3}} \tag{411}$$

In Fig. 4.14, the dependence of the reduced flux calculated from Eq. (410) for the spherical collector is plotted as a function of the reduced tangential distance $x' = x/2\pi R$.

It is interesting to compare the averaged flux at the sphere due to convection, given by Eq. (410) with the diffusion-controlled flux (given by Eq. (327)) . The ratio of both is

$$\frac{j_c}{j_d} = 0.624\left(\frac{A_f V_\infty R}{D}\right)^{1/3} \tag{412}$$

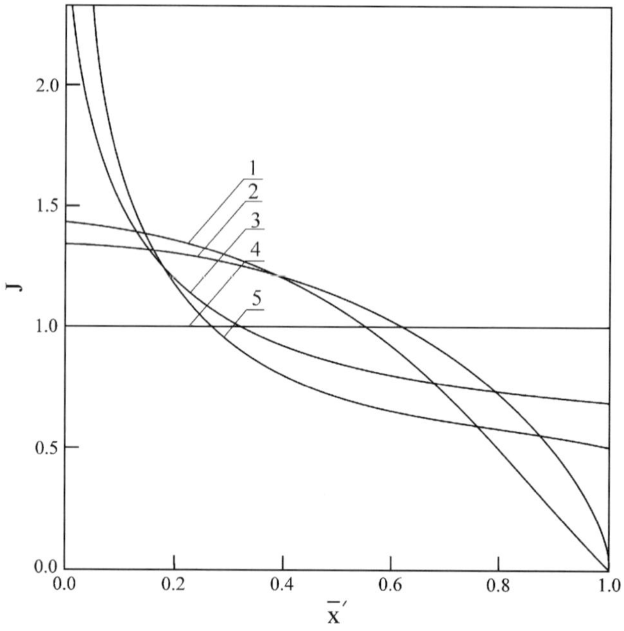

Fig. 4.14. The dependence of the normalized flux $J=k_c/\langle k_c \rangle$ on the reduced tangential coordinate $\bar{x}'$, for various collectors (1) a sphere in uniform flow, (2) a cylinder in uniform flow, (3) a parallel-plate or cylindrical channel, (4) a rotating disk and (5) a plate in uniform flow.

As expected, this ratio increases with the uniform flow velocity V_∞, the collector radius R and particle size. Taking into account the typical data pertinent to deep bed filtration experiments [12], i.e., $A_f = 3/2$, $V_\infty = 10^{-2}$ m s^{-1}, $R = 10^{-3}$ m, $D = 10^{-12}$ m^2 s^{-1} one can calculate from Eq. (412) that the convective flux is 154 times greater than, the diffusion flux.

In the case of a cylinder in uniform flow the local mass transfer rate is

$$k_c = 0.678 \frac{\sin^{1/2}\vartheta}{\left[\int_0^{\vartheta} \sin^{1/2}\vartheta' d\vartheta'\right]} \frac{A_f^{1/3} V_\infty^{1/3} D^{2/3}}{R^{2/3}} \tag{413}$$

Similarly as for the sphere, the flux attains the maximum value, $0.776 \dfrac{A_f^{1/3} V_\infty^{1/3} D^{2/3}}{R^{2/3}}$ at the forward stagnation point and vanishes at the rear stagnation point. The average mass transfer rate is

$$k_c = 0.582 \frac{A_f^{1/3} V_\infty^{1/3} D^{2/3}}{R^{2/3}} \tag{414}$$

Eq. (414) was first derived by Natanson [55]

In the case of a cylinder placed in a simple shear flow of the shear rate, G the velocity components are

$$\overline{V}_\perp = \frac{2G}{R} \sin 2\vartheta \bar{z}^2 \tag{415}$$

$$\overline{V}_\parallel = \frac{G}{R} (2\cos 2\vartheta - 1) z$$

thus $f_1 = -\frac{1}{2} \sin 2\vartheta$, $f_2 = \frac{1}{4} (2\cos 2\vartheta' - 1)$.

Consequently $g(\vartheta) = \dfrac{\left[36\int_0^{\vartheta}(2\cos 2\vartheta' - 1)^{1/2}\, d\vartheta'\right]^{1/3}}{(2\cos 2\vartheta - 1)^{1/2}}$ and the local flux is given by

$$k_c = 0.538 \frac{(2\cos 2\vartheta - 1)^{1/2}}{\left[36 \int_0^{\vartheta} (2\cos 2\vartheta' - 1)\, d\vartheta'\right]^{1/3}} \left(\frac{G}{R}\right)^{1/3} D^{2/3} \tag{416}$$

Table 4.9
Pe definitions and bulk transfer rates (local and averaged) for non-uniformly accessible surfaces

Collector and flow configuration	*Pe* definition microscopic $L_{ch} = 2a$	Local rate constant $k_c = \langle \bar{x} \rangle$	Averaged rate constant $\langle \boldsymbol{k}_c \rangle$
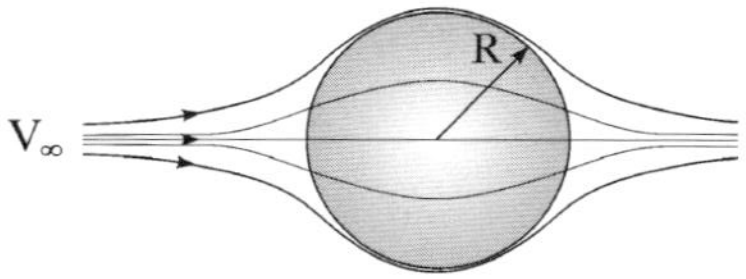 Sphere in uniform flow	$\frac{3A_f(Re)V_\infty}{R^2}\frac{a^3}{D}$	$0.776\, f_1(\bar{x})\, \frac{A_f^{1/3}V_\infty^{1/3}D^{2/3}}{R^{2/3}}$ $f_1 = \frac{\sin\vartheta}{\left(\vartheta - \frac{1}{2}\sin 2\vartheta\right)^{1/3}}$ $\vartheta = \bar{x} = x/R$	$0.624\frac{A_f^{1/3}V_\infty^{1/3}D^{2/3}}{R^{2/3}}$
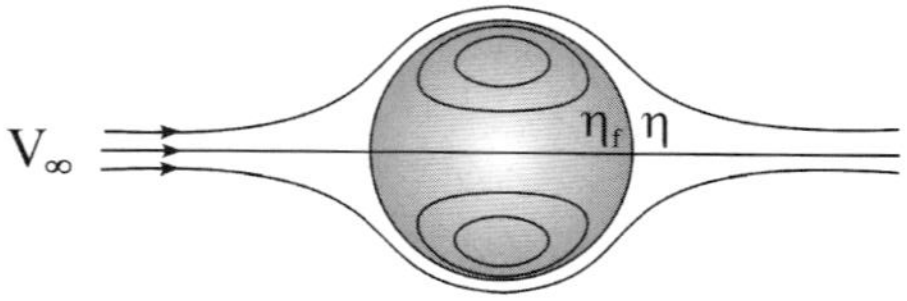 Liquid sphere in uniform flow	$2\left(\frac{\eta}{\eta_f + \eta}\right)\frac{V_\infty a^2}{DR}$	$\left[\frac{3\eta}{2(\eta_f+\eta)}\right]^{1/2}\left(\frac{DV_\infty}{\pi R}\right)^{1/2}$ $\frac{\sin^2\vartheta}{(\cos 3\vartheta - 9\cos\vartheta + 8)^{1/2}}$	$\left[\frac{2}{3}\frac{\eta D V_\infty}{(\eta_f + \eta)}\right]^{1/2}$
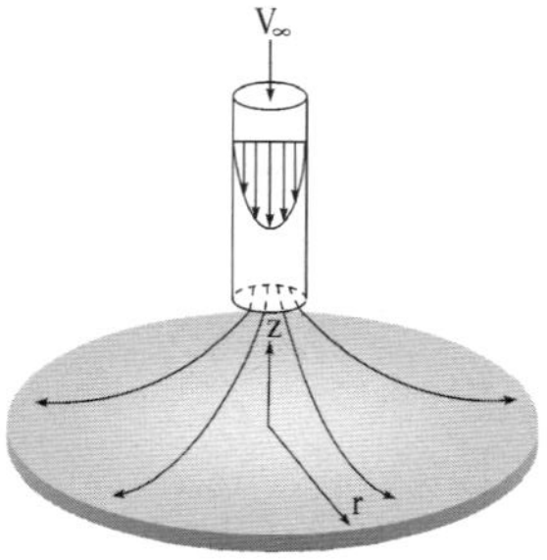 Radial impinging-jet RIJ	$\frac{2\alpha_r V_\infty}{R^2}\frac{a^3}{D}$	$0.678 f_1(\bar{x})\frac{\alpha_r^{1/3}V_\infty^{1/3}D^{2/3}}{R^{2/3}}$ $\bar{x} = \frac{\pi r}{2R}$	$0.339\frac{\alpha_r^{1/3}V_\infty^{1/3}D^{2/3}}{R^{2/3}}$

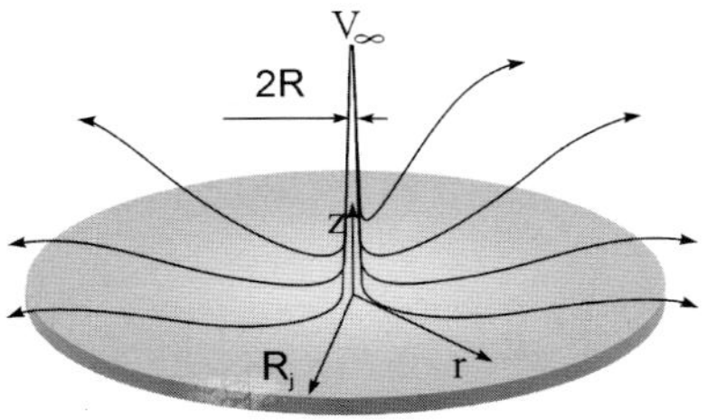 Wall-jet (axisymmetric) *	$\frac{7}{4}\frac{C_w}{R^{15/4}}\frac{a^3}{D}$	$0.560\frac{C_w^{1/3}D^{2/3}}{x^{5/4}}$	$1.493\frac{C_w^{1/3}D^{2/3}}{R_j^{5/4}}$
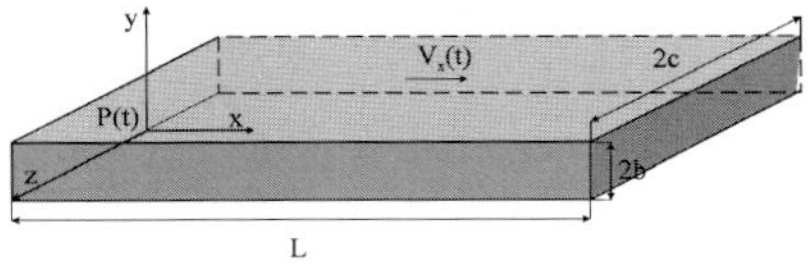 Parallel-plate channel **	$\frac{6\langle V\rangle_m}{b^2}\frac{a^3}{D}$	$0.776\frac{\langle V\rangle^{1/3}D^{2/3}}{b^{2/3}\bar{x}^{1/3}}$ $\bar{x}=x/b$ $\langle V\rangle=\frac{2}{3}V_m=\frac{Pb^2}{3\eta L}$	$1.165\frac{\langle V\rangle^{1/3}D^{2/3}}{b^{1/3}L^{1/3}}$
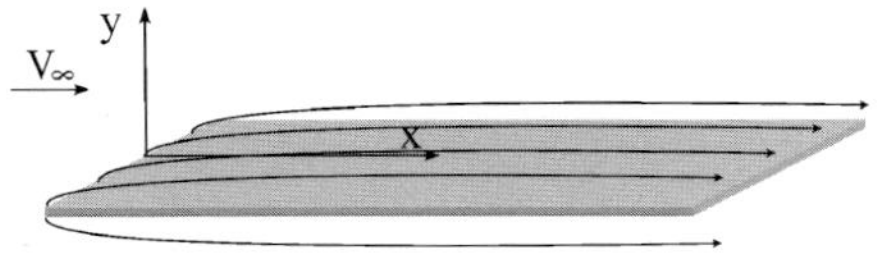 Plate in uniform flow	$0.332\frac{V_\infty^{3/2}}{2\nu^{1/2}L^{3/2}}\frac{a^3}{D}$	$0.3395\frac{V_\infty^{1/2}D^{2/3}}{\nu^{1/6}L^{1/2}\bar{x}^{1/2}}$ $\bar{x}=x/L$	$0.679\frac{V_\infty^{1/2}D^{2/3}}{\nu^{1/6}L^{1/2}}$

Table 4.9 (continued)

Collector and flow configuration	Pe definition microscopic $L_{ch} = 2a$	Local rate constant $k_c = \langle \bar{x} \rangle$	Averaged rate constant $\langle \boldsymbol{k}_c \rangle$
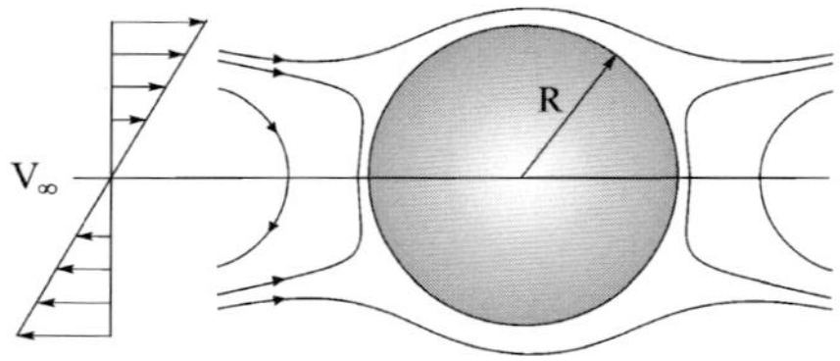 Cylinder in simple shear flow	$4\frac{G}{R}\frac{a^3}{D}$	$f_3(\vartheta)\left(\frac{G}{R}\right)^{1/3} D^{2/3}$ $f_3(\vartheta)=\frac{(2\cos\vartheta - 1)^{1/2}}{\left[\int_0^{\vartheta}(2\cos 2\vartheta' - 1)^{1/2}\,d\vartheta'\right]^{1/3}}$	$0.724\left(\frac{G}{R}\right)^{1/3} D^{2/3}$
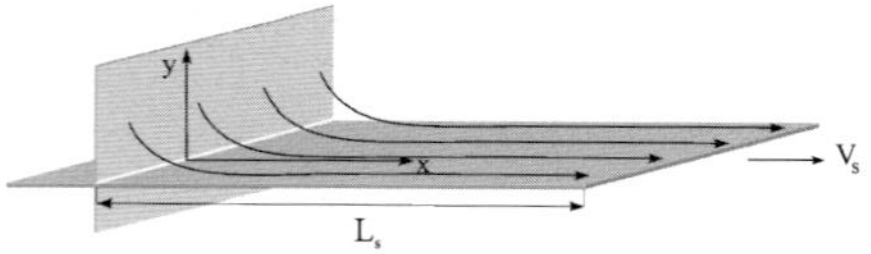 Continuous moving plate	$0.222\frac{V_s^{3/2}}{\nu^{1/2}L^{3/2}}\frac{a^3}{D}$	$0.564\left(\frac{V_s D}{L\,\bar{x}}\right)^{1/2}$ $\bar{x} = x/L$	$1.128\left(\frac{V_s D}{L}\right)^{1/2}$
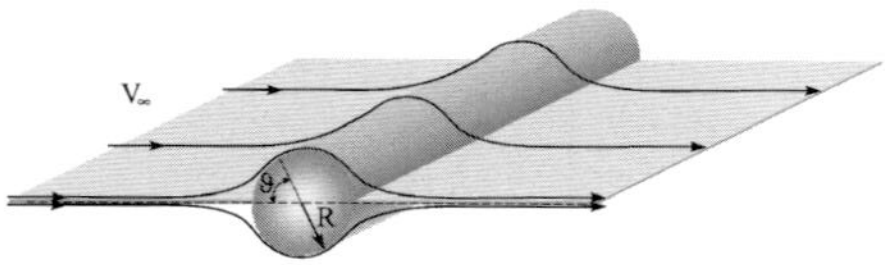 Cylinder in uniform flow	$\frac{2A_f(Re)V_\infty}{R^2}\frac{a^3}{D}$	$0.678\, f_2(\vartheta)\quad \frac{A_f^{1/3}V_\infty^{1/3}D^{2/3}}{R^{2/3}}$ $f_2(\vartheta)=\frac{\sin^{1/2}\vartheta}{\left[\int_0^{\vartheta}\sin^{1/2}\vartheta'\,d\vartheta'\right]^{1/3}}$ $\vartheta = \bar{x} = x/R$	$0.582\frac{A_f^{1/3}V_\infty^{1/3}D^{2/3}}{R^{2/3}}$

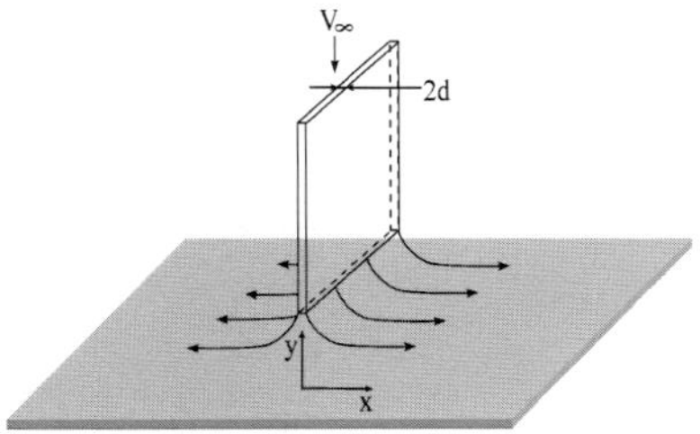 Slot impinging-jet cell RIJ	$\frac{2\alpha_s V_\infty a^3}{d^2 D}$	$0.678\, f_2(\bar{x})\frac{\alpha_s^{1/3} V_\infty^{1/3} D^{2/3}}{d^{2/3}}$ $\bar{x} = \frac{\pi x}{2d}$	$\frac{\alpha_s^{1/3} V_\infty^{1/3} D^{2/3}}{d^{2/3}}$
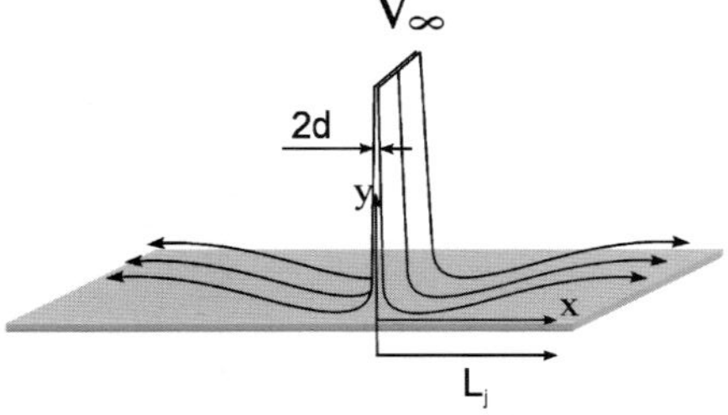 Wall-jet 2D **	$\frac{5}{4}\frac{C_w}{d^{9/4}\bar{x}^{9/4}}\frac{a^3}{D}$	$0.388\frac{C_w^{1/3} D^{2/3}}{x^{3/4}}$	$1.55\frac{C_w^{1/3} D^{2/3}}{L_j^{3/4}}$
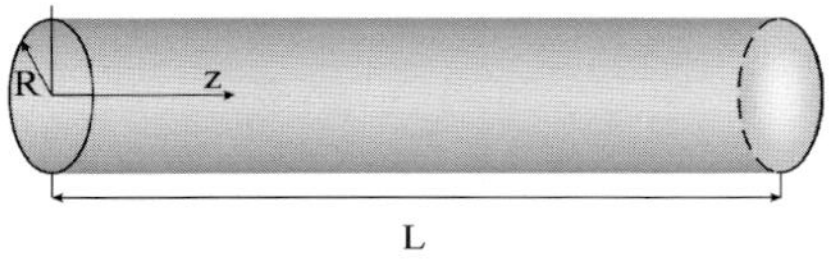 Cylindrical channel	$\frac{8\langle V\rangle}{R^2}\frac{a^3}{D}$	$0.854\frac{\langle V\rangle^{1/3} D^{2/3}}{R^{2/3}\bar{x}^{1/3}}$ $\bar{x} = x/R$ $\langle V\rangle = \frac{1}{2}V_{m^x} = \frac{P R^2}{8\eta L}$	$1.282\frac{\langle V\rangle^{1/3} D^{2/3}}{R^{1/3} L^{1/3}}$

Notes:

* $C_w = 0.131\left(\frac{V_\infty^3 Q^6}{\nu^5}\right)^{1/4}$ ** $C_w = 0.0329\left(\frac{V_\infty^3 Q^6}{\nu^5}\right)^{1/4}$

The fluid stream lines and the distribution of the local flux around a cylinder are shown schematically in Fig. 4.15 (the arrows are proportional to the absolute value of the local flux).

The averaged flux for the cylinder in the shear flow is [57]

$$k_c = 0.724\left(\frac{G}{R}\right)^{1/3} D^{2/3} \tag{417}$$

It is interesting to analyze the local flux for the impinging-jet collectors. As previously discussed in Chapter 3, for distances from the center smaller than the characteristic dimension of the jet R_j and not too small Re number values ($Re > 4$), the normal and tangential velocity components for the radial jet can be approximated by

$$V_{\perp} = -\alpha_r V_{\infty}\left(\frac{z}{R}\right)^2 \cos\bar{x} = -\alpha_r V_{\infty}\left(\frac{z}{R}\right)^2 f_1(\bar{x})$$

$$V_{||} = \alpha_r V_{\infty}\left(\frac{z}{R}\right)\frac{2}{\pi}\sin\bar{x} = \alpha_r V_{\infty}\left(\frac{z}{R}\right) f_2(\bar{x}) \tag{418}$$

valid for $\bar{x} = \pi r/2\,R < \pi/2$

As can be noticed, this flow pattern is similar to that of the sphere in uniform flow. Therefore, the $g(\bar{x})$ function for the impinging-jet is identical to that of the sphere, so the local flux is given by

$$k_c = 0.678\,\frac{\sin\vartheta}{\left(\vartheta - \frac{1}{2}\sin\vartheta\right)^{1/3}}\;\frac{\alpha_r^{1/3} V_{\infty}^{1/3} D^{2/3}}{R^{2/3}} \tag{419}$$

The averaged flux is

$$k_c = 0.339\frac{\alpha_r^{1/3} V_{\infty}^{1/3} D^{2/3}}{R^{2/3}} \tag{420}$$

However, for $\bar{x} = \pi r/2\,R > \pi/2$ the flow pattern near the impinging-jet deviates from that of the sphere (see Fig. 3.10) since in this region there appears the wall-jet whose velocity components are given by Eq. (179) in Chapter 3. The $g(\bar{x})$ function for this region is

$$g(\bar{x}) = (14)^{1/3}\left(\frac{r}{R}\right)^{5/4} \tag{421}$$

where $\bar{x} = r/R$.

The local mass transfer rate is given by

$$k_c = 0.560\frac{C_w^{1/3}D^{2/3}}{r^{5/4}} \tag{422}$$

where C_w is the flow intensity function given by Eq. (182) in Chapter 3 and R_j the outer radius of the jet (see Table 4.9). Eq. (422) was first derived by Albery [56].

The averaged flux is

$$k_c = 1.493\frac{C_w^{1/3}D^{2/3}}{R_j^{5/4}} \tag{423}$$

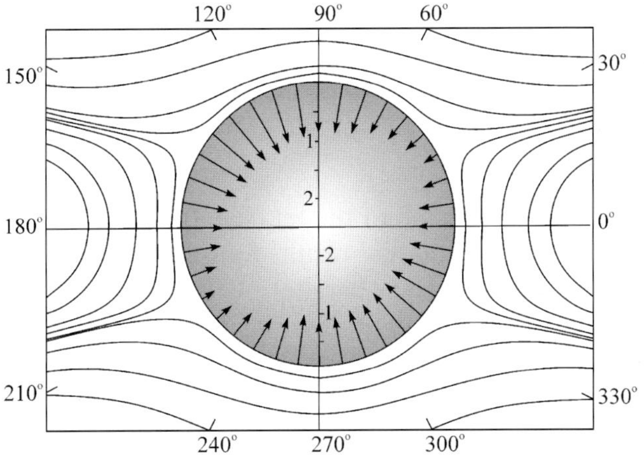

Fig. 4.15. Streamline pattern around a cylinder in a simple shear flow and the distribution of the local flux (reduced) depicted by arrows whose length is proportional to the function 1/g.

Analogously, for the slot impinging-jet, the $g(\bar{x})$ function for the wall-jet is the same as for the cylinder in uniform if $R_j >> R$ flow (with $\bar{x} = \pi x / 2\pi d < \pi/2$) and the local flux is given by

$$k_c = 0.678 \frac{\sin^{1/2}\vartheta}{\left[\int_0^{\vartheta} \sin^{1/2}\vartheta' d\vartheta'\right]} \frac{\alpha_s^{1/3} V_\infty^{1/3} D^{2/3}}{d^{2/3}} \tag{424}$$

The averaged flux in this region for $\bar{x} \leq \pi/2$ is

$$k_c = 0.5797 \frac{\alpha_s^{1/3} V_\infty^{1/3} D^{2/3}}{d^{2/3}} \tag{425}$$

In the wall-jet region (two-dimensional), for , $\bar{x} = x/d > 1$ the $g(\bar{x})$ function is

$$g(\bar{x}) = 30^{1/3} \bar{x}^{3/4} \tag{426}$$

The local transfer rate is

$$k_c = 0.388 \frac{C_w^{1/3} D^{2/3}}{x^{3/4}} \tag{427}$$

where C_w is the flow intensity function given by Eq. (185) in Chapter 3.

The averaged flux for the wall-jet region is

$$k_c \cong 1.55 \frac{C_w^{1/3} D^{2/3}}{L_j^{3/4}} \tag{428}$$

valid for $L_j >> d$

where L_j is the outer dimension of the jet (see Table 4.9).

It is also interesting to discuss results for other collectors, which do not have the stagnation point, such as parallel-plate and cylindrical channels. In this case $g(\bar{x}) = (18\bar{x})^{1/3}$ and the expressions for the local flux are [57]

$$k_c = 0.776 \frac{\langle V \rangle^{1/3} D^{2/3}}{b^{2/3} \bar{x}^{1/3}} = 0.678 \frac{V_m^{1/3} D^{2/3}}{b^{2/3} \bar{x}^{1/3}} \quad \text{(parallel - plate channel)}$$

(429)

$$k_c = 0.854 \frac{\langle V \rangle^{1/3} D^{2/3}}{R^{2/3} \bar{x}^{1/3}} = 0.678 \frac{V_m^{1/3} D^{2/3}}{R^{2/3} \bar{x}^{1/3}} \quad \text{(cylindrical channel)}$$

where $\bar{x} = x / L$ is the dimensionless distance from the inlet to the channel and

$$\langle V \rangle = \frac{2}{3} V_{mx} = \frac{2}{3} \frac{P R^2}{\eta L} \tag{430}$$

is the averaged fluid velocity in the parallel-plate channel.
In the case of the cylindrical channel one has

$$\langle V \rangle = \frac{1}{2} V_{mx} = \frac{P R^2}{8 \eta L} \tag{431}$$

Note that the local flux for both parallel-plate and cylindrical channel diverges for $\bar{x} \to 0$ (entrance to channels) proportionally to $\bar{x}^{-1/3}$. This non-physical behavior is caused by the postulate that the initial particle concentration distribution in the channel was uniform up to the interface. In reality, at the entrance area, the suspension is devoid of particles in the thin layer adjacent to the interface. By considering this, the singularity occurring in Eq. (421) is eliminated [58].

The flux averaged over the entire channel area is

$$k_c = \langle k_c \rangle = 1.1646 \frac{\langle V \rangle^{1/3} D^{2/3}}{b^{1/3} L^{1/3}} \tag{432}$$

In the case of the cylindrical channel one has

$$k_c = \langle k_c \rangle = 1.281 \frac{\langle V \rangle^{1/3} D^{2/3}}{R^{1/3} L^{1/3}} \tag{433}$$

It is interesting to note that local flux in this channel attains a value equal to the average flux at the distance $x = 8/27 = 0.296$ from the entrance.

The dependence of the local flux on the distance for these channels is also plotted in Fig. 4.14.

Analogously, in the case of a plate in the uniform flow, the function $g(\bar{x}) = 6^{1/3} x^{1/2}$ $(\bar{x} = x / L)$, and the local flux is given by

$$k_c = 0.339 \frac{V_\infty^{1/3} D^{2/3}}{\nu^{1/3} L^{1/3} \bar{x}^{1/2}} \tag{434}$$

As can be seen, the local flux diverges for $\bar{x} \to 0$ (entrance to channels) proportionally to $x^{-1/2}$ owing to the same reasons as for the channel flows. The flux averaged over the entire plate is given by

$$\langle k_c \rangle = 0.678 \frac{V_\infty^{1/3} D^{2/3}}{\nu^{1/3} L^{1/2}} \tag{435}$$

It is interesting to note that in the case of the plate, the local and average fluxes increase with the fluid velocity proportionally to $V_\infty^{1/2}$, rather than to $V_\infty^{1/3}$ as was the case for other collectors.

For the sake of convenience the local and average flux expressions for various collectors are given in Table 4.9.

All the above expressions for the flux to non-uniform surfaces were derived by assuming the perfect sink model, where $k_a' \to \infty$. Unfortunately, there is no solution in the more general case of an arbitrary value of k_a', valid for all collectors. However, several specific analytical solutions were derived for parallel-plate and the cylindrical channels. In this case the solution of Eq. (352) with the boundary condition, Eq. (355), has the form [42]

$$\bar{n} = \frac{\int_0^Z e^{-\xi^3} d\xi + \frac{1}{k_a'} \left(\frac{\langle V \rangle D^2}{3b\,x} \right)^{1/3}}{\Gamma\left(\frac{4}{3}\right) + \frac{1}{k_a'} \left(\frac{\langle V \rangle D^2}{3b\,x} \right)^{1/3}} \tag{436}$$

The local mass transfer is

$$k = k_c(x) = \frac{\frac{2}{3}\langle k_c \rangle}{\left(\frac{x}{L} \right)^{1/3} + \frac{2}{3}\bar{K}_c} \tag{437}$$

where $\langle k_c \rangle$ is the average mass transfer rate constant for the channel given in Table 4.9 and $\bar{K}_c = \langle k_c \rangle / k_a''$ An identical expression for the flux is obtained for a circular channel.

The dependence of the reduced flux $J=k_c(x)/\langle k_c\rangle$ on x/L calculated from Eq. (437) for various $\bar{K}_c$ values is plotted in Fig. 4.16. It can be observed that the influence of the surface force barrier, characterized by the k_a' constant, is most significant at the entrance region of the channel, when $x/L \ll 1$. Consequently, for $\bar{K}_c \ll 1$ (the barrier-controlled deposition regime) the flux in the channel remains practically uniform over macroscopic distances comparable with the channel length.

By integrating Eq. (437) one obtains the following expression for the averaged flux in a parallel-plate or cylindrical channel

$$k_c = \langle k_c\rangle \left[1 - \frac{4}{3}\bar{K}_c + \frac{8}{9}\bar{K}_c^2 \ln \frac{\frac{2}{3}\bar{K}_c + 1}{\frac{2}{3}\bar{K}_c}\right] \tag{438}$$

For $\bar{K}_c = 0$ Eq. (438) reduces to Eq. (432) or Eq. (433) derived previously for parallel-plate and cylindrical channels, respectively, in the case of the perfect-sink model (no barrier). On the other hand, for $\bar{K}_c \gg 1$, Eq. (438) simplifies to

$$\langle k'_c\rangle \cong \langle k_c\rangle \frac{1}{\bar{K}_c} = k_a'' \tag{439}$$

Other solutions for the barrier-controlled transport regime have been derived for the spherical and cylindrical collectors. In the former case the local flux was found to be [41]

$$k_c = k_c(\vartheta)\frac{1}{1+\bar{K}_c}\left[1 + \frac{3\Gamma\left(\frac{4}{3}\right)}{2^{1/3}} F(\xi)\right] \tag{440}$$

$$\xi = \frac{1}{2}\left(\vartheta - \frac{1}{2}\sin^2 2\vartheta\right)$$

where k_c is the mass transfer rate constant at the forward stagnation point given in Table 4.8.

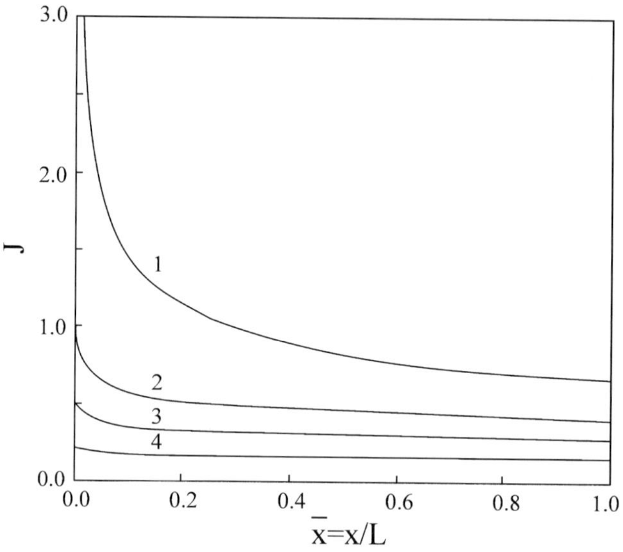

Fig. 4.16. The dependence of the reduced flux $J = k(\bar{x})/\langle k_c \rangle$ on $\bar{x} = x / L$ calculated from Eq. (437). (1) $\bar{K}_c = 0$ (barrierless transport), (2) $\bar{K}_c = 1$, (3) $\bar{K}_c = 2$, (4) $\bar{K}_c = 5$.

The function $F(\xi)$ can be calculated from the linear Volterra equation of the second kind

$$F(\xi) + \frac{3^{1/3}}{3\Gamma\left(\frac{4}{3}\right)\Gamma\left(\frac{2}{3}\right)\bar{K}_c \sin\vartheta} \int_0^{\xi} \frac{F(\xi')d\xi'}{(\xi-\xi')^{2/3}} = -\frac{2^{1/3}}{3\Gamma\left(\frac{4}{3}\right)\xi^{1/3}} + \frac{6^{1/3}}{3\Gamma\left(\frac{4}{3}\right)\sin\vartheta} \qquad (441)$$

where $\Gamma\left(\frac{2}{3}\right) = 1.3539$ is the Gamma Euler function value for $x = \frac{2}{3}$.

The flux averaged over the entire collector surface is given by

$$k_c = \frac{\langle k_c \rangle}{1+\bar{K}_c} F_c(\bar{K}_c) \qquad (442)$$

where $\langle k_c \rangle$ is the averaged mass transfer rate constant given in Table 4.9, and the function $F_s(\bar{K}_c)$ was calculated by solving numerical solution of Eq. (441) and was given in a tabulated form in Ref. 41. It varies slightly with $\bar{K}_c$

parameter assuming the value of 1.423 for $\bar{K}_c = \infty$ (barrier-controlled) deposition, 1.3446 for $\bar{K}_c = 5$, 1.197 for $\bar{K}_c = 1$ and 1.0621 for $\bar{K}_c = 0.2$.

An analogous expression for the averaged flux in the case of the cylindrical collector is [41]

$$k_c = \frac{\langle k_c \rangle}{1+\bar{K}_c} F_c(\bar{K}_c) \tag{443}$$

where $\langle k'_c \rangle$ is the averaged mass transfer rate constant for the cylinder in uniform flow given in Table 4.9 and the function $F_c(\bar{K}_c)$ is given in a tabulated form in Ref. 41. It assumes the value of 1.339 for $\bar{K}_c = \infty$ (barrier-controlled) deposition, 1.279 for $\bar{K}_c = 5$, 1.164 for $\bar{K}_c = 1$ and 1.053 for $\bar{K}_c = 0.2$.

As can be deduced from Eqs. (437–443) the effect of the surface force barrier on particle transport to non-uniformly accessible surfaces is quite similar to uniformly accessible surface case. Therefore, it seems advisable to use the much simpler Eqs.(367–370) for predicting the mass transfer rate coefficients in the general case of bulk transport and barrier controlled regime.

Because of approximations, the results discussed in the section are appropriate for particles smaller than 500 nm, where the range of hydrodynamic wall effects and specific surface interactions remains negligible in comparison with bulk transfer distances.

4.6.4 Exact Transport Equations

For larger particles, the hydrodynamic as well as external and specific forces are coupled in a complex way. This makes it necessary to use the exact transport equation, Eq. (223), for calculating mass transfer rates of particles of arbitrary size under the simultaneous action of specific and external forces. This equation assumes the explicit form in the (x,y,z) Cartesian coordinate system

$$\frac{\partial n}{\partial t} = \frac{\partial}{\partial x}\left[D_{\parallel}\left(\frac{\partial n}{\partial x} + \frac{1}{kT}\frac{\partial \phi}{\partial x} n\right) - U_{h_x} n\right] + \frac{\partial}{\partial y}\left[D_{\parallel}\left(\frac{\partial n}{\partial y} + \frac{1}{kT}\frac{\partial \phi}{\partial y} n\right) - U_{h_y} n\right]$$
$$+ \frac{\partial}{\partial x}\left[D_{\perp}\left(\frac{\partial n}{\partial z} + \frac{1}{kT}\frac{\partial \phi}{\partial z} n\right) - U_{h_z} n\right] \tag{444}$$

where (x,y) are coordinates parallel to the collector (assumed to have a planar shape), z is the coordinate perpendicular to the collector, $D_{\parallel}$, $D_{\perp}$ are the parallel and perpendicular components of the translation diffusion tensor discussed above, $-\frac{\partial \phi}{\partial x} = F_x$, $-\frac{\partial \phi}{\partial y} = F_y$ and $-\frac{\partial \phi}{\partial z} = F_z$ are the components of the net interaction potential gradient (force) and U_{h_x}, U_{h_y}, and U_{hz} are the components of the particle velocity vector owing to fluid motion.

Eq. (444) is appropriate for *D* stagnation point flows (orthogonal or oblique), impinging-jet, wall-jets, parallel-plate channels, a plate in uniform flow, a continuous moving plate and the rotating disk.

An analogous transport equation can be derived for the cylindrical and spherical coordinate systems, which are appropriate for a cylindrical channel, a cylinder in uniform and simple shear flows, and for a spherical collector in uniform flow.

The major difficulty in dealing with the exact transport equation, Eq. (444), is the lack of appropriate analytical methods for their solution because of complex dependence of the net potential ϕ on the distance from the collector. Therefore, exact solutions can only be derived by rather sophisticated numerical techniques, which become quite inefficient for multidimensional problems.

However, for most cases involving nano-sized particles, the governing transport equation can be simplified considerably by applying dimensional analysis, similar to the derivation of Eq. (352). Because the particle concentration gradient in the direction perpendicular to the collector is of the order of n_b/δ_h, and for the direction parallel to the surface gradient is n_b/R, one can deduce that the ratio of the tangential to normal diffusion terms $(\partial/\partial x)D_{\parallel}(\partial n/\partial x)/(\partial/\partial z)D_{\perp}(\partial n/\partial z) \sim \delta_h^2/R^2$.

Therefore, by considering that $\delta_h \ll R$, the tangential diffusion terms can be neglected. Moreover, the collector surface can be treated as locally planar, which allows one to use the known expressions for the mobility matrix in terms of the hydrodynamic correction functions $F_1 \div F_4$ discussed above. By virtue of these assumptions, one can express the particle velocity components near the interface in the form similar to those obtained by the trajectory analysis

$$
\begin{aligned}
U_{h_x} &= V_1 (H+1) F_3(H) S(x,y) + V_1' \\
U_{h_y} &= V_2 (H+1) F_3(H) S'(x,y) + V_2' \\
U_{h_z} &= V_3 (H+1)^2 F_1(H) F_2(H) C(x,y)
\end{aligned}
\tag{445}
$$

where V_1, V_1', V_2, V_2' and V_3 are constants with the dimension of velocity, and S, S', C are functions of the tangential coordinates alone.

For 2D flows, Eq. (445) simplifies to

$$U_{h_x} = V_1 (H+1) F_3(H) S(x,y) + V_1' \tag{446}$$

$$U_{h_z} = V_3 (H+1)^2 F_1(H) F_2(H) C(x,y)$$

where the functions S and C are connected through the continuity relationship

$$S = -\frac{2V_3}{V_1}\int C(x)dx \tag{447}$$

By neglecting the tangential diffusion terms and considering Eq. (445), the transport equation, Eq. (444), can be formulated in the dimensionless form

$$\begin{aligned}\frac{\partial \bar{n}}{\partial \tau} &= \frac{\partial}{\partial H}\left[F_1(H)\left(\frac{\partial \bar{n}}{\partial H} + \frac{\partial \bar{\phi}}{\partial H}\bar{n}\right) - \frac{1}{2} Pe\, F_1(H) F_2(H)(H+1)^2 C(\bar{x},\bar{y})\bar{n}\right] \\ &+ \frac{\partial}{\partial \bar{x}} A\left\{F_4(H)\frac{\partial \bar{\phi}}{\partial \bar{x}}\bar{n} - \left[V_1' + \frac{V_1}{2V_3} Pe\, F_3(H)(H+1) S(\bar{x},\bar{y})\bar{n}\right]\right\} \\ &+ \frac{\partial}{\partial \bar{y}} A\left\{F_4(H)\frac{\partial \bar{\phi}}{\partial \bar{y}}\bar{n} - \left[V_1 + \frac{V_2}{2V_3} Pe\, F_3(H)(H+1) S'(\bar{x},\bar{y})\bar{n}\right]\right\}\end{aligned} \tag{448}$$

where $\tau=(D/a^2)t$ is the dimensionless time, $\bar{x} = x / L_{ch}$, $A = a/R$ is the interception parameter $\bar{\phi} = \phi/kT$ is the dimensionless potential and $Pe=2|V_3|a/kT$ is the Peclet number.

The external and specific interaction term (the H-component) appearing in Eq. (448) can be evaluated as

$$\frac{\partial \bar{\phi}}{\partial H} = -\frac{a}{kT}F_H = -\left(\bar{F}_{ext_H} + \bar{F}_{el_H} + \bar{F}_{s_H}\right) \tag{449}$$

where F_{ext_H} is the external force consisting of the gravitational force and the Lorenz force, leading to migration effects discussed in Section 4.2,

$$\bar{F}_{ext_H} = \frac{a}{kT}(m_p' \mathbf{g}\cdot\mathbf{i}_H + q_0 \mathbf{E}\cdot\mathbf{i}_H) = \frac{Gr}{|\mathbf{g}|}\mathbf{g}\cdot\mathbf{i}_H + \frac{El}{|\mathbf{E}|}\mathbf{E}\cdot\mathbf{i}_H \tag{450}$$

where $Gr = m_p'|\mathbf{g}|a / kT = (m_p' - m_f)|\mathbf{g}|a / kT$ is the gravitational number, and $El = q_0|\mathbf{E}|a / kT$ is the electrostatic number characterizing the effect of a uniform electric field $\mathbf{E}$ on a particle bearing the charge q_o (for the sake of convenience these dimensionless numbers have been compiled in Table 4.10). When there is no electric field, the force acting on a charged particle consists of the image force, described by the expression (see Table 2.6)

$$\bar{F}_{in_H} = -\frac{a}{kT}\frac{\varepsilon_2 - \varepsilon}{4\pi\varepsilon}\frac{q^{0^2}}{(\varepsilon_2 + \varepsilon)a^2}\frac{1}{4(H+1)^2} = -Ei\frac{1}{(H+1)^2} \tag{451}$$

where $Ei = (\varepsilon_2 - \varepsilon)q^{0^2}/16\pi\varepsilon(\varepsilon_2 + \varepsilon)a\ kT$ is the dimensionless image electrostatic number also given in Table 4.10. For a metallic plate when $\varepsilon_2 \to \infty$, $Ei = q^{0^2}/8\pi\varepsilon akT$.

If the particle is uncharged, a polarization force appears in the vicinity of metallic electrodes, described by the formula (see Table 2.6).

$$\bar{F}_{p_H} = -\frac{a}{kT}\frac{3}{2}\frac{\pi a^2(\varepsilon_p - \varepsilon)^2}{(\varepsilon_p + 2\varepsilon)^2}\frac{|\mathbf{E}|^2}{a}\frac{1}{(H+1)^2} = -Ep\frac{1}{(H+1)^4} \tag{452}$$

where $Ep = 3\pi a^3(\varepsilon_p - \varepsilon)^2|\mathbf{E}|^2/2\ (\varepsilon_p + 2\varepsilon)^2\ kT$ is the dimensionless polarization electrostatic number (Table 4.10).

In polar media, owing to the formation of the electric double-layer, the electrostatic forces are modified and described by various formulae given in Table 2.10. They can be approximated for most situations of practical interest by the LSA model, which furnishes the two-parametric expression

$$\bar{F} = \frac{\kappa a}{kT}\phi_0 e^{-\kappa a H} = \kappa a Dl e^{-\kappa a H} \tag{453}$$

where $Dl = \phi_0/kT$ is the electrostatic double-layer number, $\kappa a = a/Le$ is the parameter describing the ratio of the particle radius to the double-layer thickness and $Le = (\varepsilon kT/2e^2 I)^{1/2}$.

The van der Waals interactions can be described by

$$\bar{F} = -\frac{A_{132}}{kT} f_a(z,\lambda_r) = -Ad\,\bar{f}_a(H,\bar{\lambda}_r) \tag{454}$$

where $Ad = A_{132}/6kT$ is the dimensionless adhesion number, $\bar{\lambda}_r = \lambda_r/a$ the dimensionless parameter describing the retardation effect and $f_a(H,\bar{\lambda}_r)$ the dimensionless function to be found in Table 2.10. For distances from the interface smaller than particle dimensions, the function $\bar{f}_a(H,\lambda_r)$ can well be approximated for spherical particles by the Derjaguin expression

$$\bar{f}_a = \frac{1}{H^2} \tag{455}$$

By considering the above expressions, the normal component of the dimensionless interaction force can be expressed as

$$\bar{F}_H = \frac{Gr}{|\mathbf{g}|}\mathbf{g}\cdot\mathbf{i}_H + \frac{El}{|\mathbf{E}|}\mathbf{E}\cdot\mathbf{i}_H - Ei\frac{1}{(H+1)^2} - Ep\frac{1}{(H+1)^4} + Dl\,e^{-\kappa a H} - Ad\,\bar{f}_a(H,\bar{\lambda}_r) \tag{456}$$

The tangential force components can be evaluated in the same way. They consist, however, of the gravitational force only because there are no specific and electrostatic contributions acting parallel to the surface. Therefore, the $\bar{F}_x$ and $\bar{F}_y$ force components are

$$\bar{F}_x = \frac{Gr}{|\mathbf{g}|}\mathbf{g}\cdot\mathbf{i}_x$$
$$\bar{F}_y = \frac{Gr}{|\mathbf{g}|}\mathbf{g}\cdot\mathbf{i}_y \tag{457}$$

Considering the above expressions for external and specific forces one can formulate Eq. (448) in the explicit form

$$\begin{aligned}\frac{\partial \bar{n}}{\partial \tau} = {} & \frac{\partial}{\partial H}\left[F_1(H)\left(\frac{\partial \bar{n}}{\partial H} - \bar{F}_H\,\bar{n}\right) - \frac{1}{2}Pe\,F_1(H)F_2(H)(H+1)^2\,C(\bar{x},\bar{y})\bar{n}\right] \\ & + \frac{\partial}{\partial \bar{x}}A\left\{-F_4(H)\,\bar{F}_x\,\bar{n} - \left[V_1' + \frac{V_1}{2V_3}Pe\,F_3(H)(H+1)S(\bar{x},\bar{y})\right]\bar{n}\right\} \\ & + \frac{\partial}{\partial \bar{y}}A\left\{-F_4(H)\bar{F}_y\,\bar{n} - \left[V_2' + \frac{V_2}{2V_3}Pe\,F_3(H)(H+1)S'(\bar{x},\bar{y})\right]\bar{n}\right\}\end{aligned} \tag{458}$$

Table 4.10
Dimensionless numbers pertinent to particle deposition and their physical significance

Dimensionless number	Name	Definition	Physical interpretation
* Pe	Peclet number	$\frac{V_{ch}L_{ch}}{D}$	Ratio of convection to diffusion rates
** Re	Reynolds number	$\frac{V_{ch}L_{ch}}{\nu}$	Ratio of inertia to viscous forces
Sc	Schmidt number	$\frac{\nu}{D}$	Ratio of mass to momentum transport relaxation time
A	Interception Parameter	$\frac{a}{R}$	Particle to collector size ratio
Ex	External force number	$\frac{\phi_{\text{ext}}}{kT}=\frac{F_{ext}L_{ch}}{kT}$	Ratio of characteristic external energy to thermal energy
Gr	Gravitational number	$\frac{m'_p\lvert\mathbf{g}\rvert a}{kT}$	Ratio of gravitational energy to thermal energy
El	Electrostatic number	$\frac{q^{0}\lvert\mathbf{E}\rvert a}{kT}$	Ratio of electrostatic Lorenz energy to thermal energy
E_i	Electrostatic image number	$\frac{(\varepsilon_2-\varepsilon)q^{0^2}}{16\pi\varepsilon(\varepsilon_2+\varepsilon)a\,kT}$	Ratio of electrostatic image energy to thermal energy
Ep	Electrostatic polarization number	$\frac{3}{2}\frac{\pi a^3(\varepsilon_p-\varepsilon)^2\lvert\mathbf{E}\rvert^2}{(\varepsilon_p+2\varepsilon)^2kT}$	Ratio of electrostatic polorization energy to thermal energy
Dl	Electrostatic double-layer number	$\frac{\phi_0}{kT}$	Ratio of electrostatic double-layer interaction energy to thermal energy
κa	Reciprocal double-layer thickness	$\frac{a}{Le}$	Ratio of particle radius to electric double-layer thickness

Table 4.10 (continued)

Dimensionless number	Name	Definition	Physical interpretation
Ad	Dispersion interaction parameter (adhesion number)	$\frac{A_{132}}{6kT}$	Ratio of the van der Waals energy to thermal energy
$\bar{\lambda}r$	Retardation parameter	$\frac{\lambda_r}{a}$	Ratio of the characteristic wavelength to particle radius
$Sh(\bar{x})$	Local Sherwood number	$\frac{k_c L_{ch}}{D}$	Dimensionless mass transfer rate (local) Ratio of local flux to characteristic flux
$\langle Sh \rangle$	Averaged Sherwood number	$\frac{\langle k_c \rangle L_{ch}}{D}$	Dimensionless mass transfer rate (averaged)

* L_{ch} characteristic length scale of the mass transport problem usually equal to particle radius a (microscopic Pe number) or collector dimension R (macroscopic Pe number). V_{ch} characteristic fluid velocity in the diffusion boundary-layer.

** L_{ch} characteristic length scale of the flow problem (usually particle or collector diameter). V_{ch} characteristic velocity of the flow, e.g., the fluid approach velocity V_∞.

Eq. (458) represents a general formulation of the particle transport equation valid for all collectors, if the diffusion boundary-layer is thin in comparison with the characteristic dimension of the collector. Several cases of practical interest can be derived from Eq. (458), if the external force is directed parallel to the symmetry axis (coinciding usually with the flow direction at infinity). Then Eq. (458) can be reduced to a two-dimensional form,

$$\frac{\partial \bar{n}}{\partial \tau} = \frac{\partial}{\partial H}\left[F_1(H)\left(\frac{\partial \bar{n}}{\partial H} - \bar{F}_H\,\bar{n}\right) - \frac{1}{2}Pe\,F_1(H)F_2(H)(H+1)^2\,C(\bar{x})\bar{n}\right]$$
$$+\frac{\partial}{\partial \bar{x}}A\{-F_4(H)\,\bar{F}_x\,\bar{n} - \left[V_1' + Pe\,F_3(H)(H+1)S(\bar{x})\right]\bar{n}\} \tag{459}$$

Eq. (459) is valid in the case of 2D stagnation point flows (orthogonal or oblique), impinging-jet and wall-jets, cylinders and spheres in uniform flows, parallel-plate channels, plates in uniform flows and continuous moving plates. Expressions for the S and C functions for these flows are given in Table 4.11.

For flows of radial symmetry, e.g., for radial impinging-jet flows and wall-jet Eq. (448) can be expressed as

$$\frac{\partial \bar{n}}{\partial \tau}=\frac{\partial}{\partial H}\left[F_1(H)\left(\frac{\partial \bar{n}}{\partial H}-\bar{F}_H\,\bar{n}\right)-\frac{1}{2}Pe\,F_1(H)F_2(H)(H+1)^2\,C(\bar{x},\bar{y})\bar{n}\right]$$
$$+\frac{1}{\bar{x}}\frac{\partial}{\partial \bar{x}}A\,\bar{x}[-F_4(H)\,\bar{F}_x\,\bar{n}-Pe\,F_3(H)(H+1)S(\bar{x})\bar{n}] \quad (460)$$

where $\bar{x} = r / R$ and r is the radial distance from the center. Expressions for the S and C functions for these flows are also given in Table 4.11.

For uniformly accessible, homogeneous surfaces, where the specific interaction potential and the C function do not depend on position, Eq. (460) reduces to the one-dimensional form

$$\frac{\partial \bar{n}}{\partial \tau}=\frac{\partial}{\partial H}\left[F_1(H)\left(\frac{\partial \bar{n}}{\partial H}-\bar{F}_H\,\bar{n}\right)+\frac{1}{2}Pe\,F_1(H)F_2(H)(H+1)^2\,\bar{n}\right]$$
$$-Pe\,F_3(H)(H+1)\,\bar{n} \quad (461)$$

It is interesting to observe that Eq. (461) is a parabolic partial differential equation, which simplifies its numerical analysis.

The boundary and initial conditions for Eqs. (456–461) are the same as previously discussed in section 3.5.

Stationary forms are obtained from Eqs. (458–461) by putting $\partial\bar{n}/\partial\tau=0$. The 2D transport equation, Eq. (459), then becomes

$$A\left[F_4(H)\bar{F}_x+\bar{V}_1+Pe\,F_3(H)(H+1)S(\bar{x})\right]\frac{\partial \bar{n}}{\partial \bar{x}}$$
$$=\frac{\partial}{\partial H}\left[F_1(H)\left(\frac{\partial \bar{n}}{\partial H}-\bar{F}_H\,\bar{n}\right)\right.$$
$$\left.-\frac{1}{2}Pe\,F_1(H)F_2(H)(H+1)^2\,C(\bar{x})\bar{n}\right]-Pe\,F_3(H)\,\bar{n} \quad (462)$$

It is interesting to observe that Eq. (462) can be converted to a *quasi*-parabolic partial differential equation, analogous to the non-stationary Eq. (461)

$$\frac{\partial \bar{n}}{\partial \tau'}=\frac{\partial}{\partial H}\left[F_1(H)\left(\frac{\partial \bar{n}}{\partial H}-\bar{F}_H\,\bar{n}\right)-\frac{1}{2}Pe\,F_1(H)F_2(H)(H+1)^2\,C(\bar{x})\bar{n}\right]$$
$$-Pe\,F_3(H)(H+1)C(\bar{x})\bar{n} \quad (463)$$

Table 4.11
Expressions for the functions appearing in Eq. (459) for various collectors

Collector	$C(\bar{x})$	$S(\bar{x})$	$\bar{C}_1'$	$(\bar{x})$
Stagnation point-flow (2D)	1	$\bar{x}$	0	$\bar{x}$
Slot impinging-jet	$-\cos\bar{x}$	$\sin\bar{x}$	0	$\frac{\pi x}{2d}$
Wall-jet (2D)	$\bar{x}^{-9/4}$	$\bar{x}^{-5/4}$	0	$\frac{x}{d}$
Parallel-plate channel	0	1	0	$\frac{x}{b}$
Plate in uniform flow	$-\bar{x}^{-3/2}$	$\bar{x}^{-1/2}$	0	$\frac{x}{L}$
Continuous moving plate	$-\bar{x}^{-3/2}$	$\bar{x}^{-1/2}$	$\frac{V_s a}{D}$	$\frac{x}{L}$
Cylinder in uniform flow	$-\cos\bar{x}$	$\sin\bar{x}$	0	$\frac{x}{R}$
Sphere in uniform flow	$-\cos\bar{x}$	$\sin\bar{x}$	0	$\frac{x}{R}$
Radial impinging-jet	$-\cos\bar{x}$	$\sin\bar{x}$	0	$\frac{\pi r}{2R}$
Axisymmetric wall-jet	$\bar{x}^{-15/4}$	$\bar{x}^{-11/4}$	0	$\frac{r}{R}$
Rotating-disk	-1	1	0	$\frac{r}{R}$
Cylindrical channel	0	1	0	$\frac{r}{R}$

where

$$\tau' = A\{[F_4(H)\bar{F}_x + \bar{V}_1]\,\bar{x} + Pe\,F_3(H)(H+1)C(\bar{x})\} \tag{464}$$

The $\bar{x}$ variable can be eliminated from Eq. (463) by inverting the non-linear relationship, Eq. (464). In the usual case where $\bar{F}_x = \bar{V}_1 = 0$, this can be done analytically, so

$$C(\bar{x}) = \frac{\tau'}{A\,Pe\,F_3(H)(H+1)} \tag{465}$$

Consequently

$$\bar{x} = C^{-1}(\tau') \tag{466}$$

where C^{-1} means the inverse function.

On the other hand, the one-dimensional Eq. (461) under the steady-state is converted to the ordinary differential equation

$$\frac{\partial}{\partial H}\left[F_1(H)\left(\frac{\partial \bar{n}}{\partial H} - \bar{F}_H\,\bar{n}\right) + \frac{1}{2}Pe\,F_1(H)F_2(H)(H+1)^2\,\bar{n}\right]$$

$$= -\frac{\partial \bar{j}_H}{\partial H} = Pe\,F_3(H)(H+1)\bar{n} \tag{467}$$

where

$$-\bar{j}_H = -\frac{a}{Dn_b}\,j_H = Sh \tag{468}$$

is the component of the dimensionless flux perpendicular to the surface.

The boundary condition at the surface for Eqs. (462,463,467) is usually expressed in the form of the perfect sink, i.e., $\bar{n} = 0$ at $H_m = \delta_m / a$ (where δ_m is the primary minimum distance).

Most of the above transport equations can be effectively solved by using the finite-difference methods discussed in some detail elsewhere [12]. For the one-dimensional non-stationary transport equations pertinent to uniformly accessible surfaces, one can effectively use the implicit Crank–Nicholson scheme with appropriate net transformation functions [57,58,59,61]. The set of linear algebraic equations originating from the discretization procedure is solved directly by the Gauss elimination procedure. This allows one to treat problems, which have a large number of mesh points.

Calculations for both transient and stationary conditions have been performed for the rotating disk [62,63], and impinging-jet collectors [64,65]. The same method can be effectively used to solve the stationary transport equations for a spherical [66] or cylindrical [59] collector, the parallel-plate and cylindrical channels [58], a plate in uniform flow [60] and a continuous moving plate [61]. This is so because in all these cases, the governing transport equation under stationary conditions can be reduced to the *quasi*-one-dimensional form, as shown above.

The numerical calculations allow one to calculate the particle concentration distribution as a function of time and spatial variables. If $\bar{n}(\bar{x},H,\tau)$ is known, one can calculate the local mass transfer rate (the flux component perpendicular to the interface) by using the constitutive dependence

$$Sh = -\bar{j}_H = F_1(H)\left(\frac{\partial \bar{n}}{\partial H} - \bar{F}_H \bar{n}\right) + \frac{1}{2} Pe\, F_1(H) F_2(H)(H+1)^2 \bar{n}$$
$$\text{for} \quad H \to \delta_m / a \tag{469}$$

Because j_H depends on the set of dimensionless parameters listed in Table 4.10, the Sh number can be expressed by

$$Sh = Sh(\bar{x}, \tau, Pe, Re, Sc, A, Gr, El, Ei, Ep, Dl, \kappa a, Ad, \lambda_r) \tag{470}$$

Obviously, under the steady-state conditions, when $\partial \bar{n}/\partial \tau = 0$, Sh does not depend on the time.
The averaged Sherwood number, defined previously by Eq. (406), is given by

$$\langle Sh \rangle = \int_0^1 Sh \, d\bar{x} \tag{471}$$

For uniformly accessible surfaces $Sh = \langle Sh \rangle$.

As can be noticed from Eqs. (460,461), the Sh and $\langle Sh \rangle$ numbers depend on so many dimensionless parameters that a systematic analysis of their significance seems prohibitive. In numerical calculations one is usually concentrated on the most significant parameters, which are the Peclet number *Pe*, the gravitational number *Gr*, and the parameters describing specific interactions, i.e., *Dl*, κa and *Ad*. In the next section, representative numerical results, illustrating the significance of these parameters are discussed. These results can also be exploited to estimate the range of validity of previously discussed limiting deposition regimes.

4.6.4.1. Exact numerical calculations of particle deposition rates

Most of the calculations discussed in this section were obtained for uniformly accessible surfaces, using the non-stationary, Eq. (461) and stationary, Eq. (467) forms of the transport equation. The range of validity of these results can be extended to non-uniformly accessible surfaces, especially for spherical and cylindrical collectors, because a significant part of their surfaces in the region close to the forward stagnation point is uniformly accessible.

The non-stationary effects, especially the duration of the transient deposition regime, were determined in Ref. 63. The effect of flow and particle size (governed by the *Pe* number), as well as the particle density, governed by the gravitational number *Gr*, have been studied systematically for the rotating disk case. In these calculations the dependence of the non-stationary dimensionless flux vs. the dimensionless time $\tau=(D/a^2)t$ was obtained. For the sake of convenience the results were presented as the dependence of the reduced flux $J=j(\tau)/j(\infty)=k_c(\tau)/k_c(\infty)=Sh(\tau)/Sh(\infty)$ (where $j(\infty)$, $k_c(\infty)$ and $Sh(\infty)$ are the corresponding values of these functions after infinite time) on τ.

With this definition on hand the transition time τ_r can be defined by postulating that $J = 0.99$ for $\tau=\tau_r$, i.e., the particle flux attains 99% of its final value. The results plotted in Fig. 4.17 indicate that the transition time increases with decreasing *Pe* number, roughly proportionally to $Pe^{-2/3}$.

This behavior can be predicted by exploiting the usual formula for the diffusion relaxation time

$$t_r = \frac{\delta^2}{D} \tag{472}$$

Thus, the dimensionless relaxation time is given by

$$\tau_r = \frac{D}{a^2} t_r = \left(\frac{\delta_d}{a} \right)^2 = \bar{\delta}_d^2 \tag{473}$$

where δ_d is the diffusion-boundary layer thickness, and $\bar{\delta}_d = \delta_d/a$ is the thickness of this boundary layer can be most directly calculated from the formula

$$\bar{\delta}_d = \frac{1}{Sh} = \frac{1}{C\,Pe^{1/3}} \tag{474}$$

Therefore, the dimensionless relaxation time is

$$\tau_r = C'\,Pe^{-2/3} \tag{475}$$

where $C' = C^{-2/3}$.
Eq. (475) is in agreement with the results shown in Fig. 4.17. Dąbroś et al. [54] confirmed also the validity of Eq. (475) for the case of a cylinder placed in uniform and shearing flow.

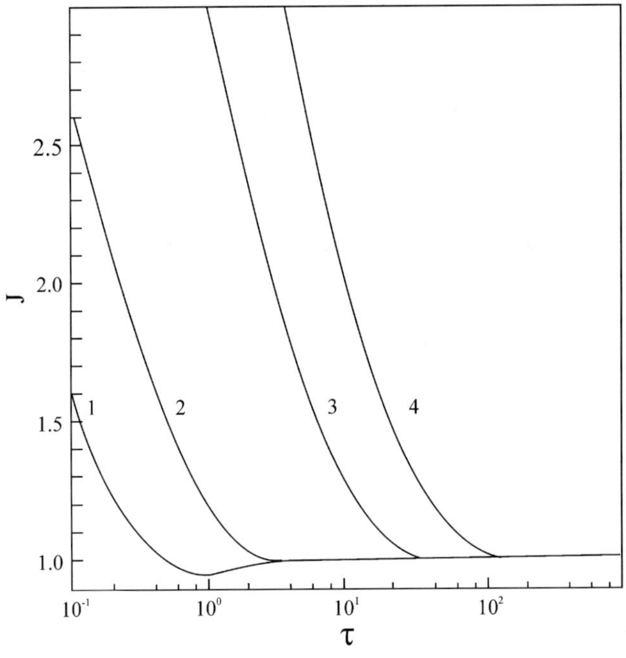

Fig. 4.17. The normalized transient flux J vs. the dimensionless deposition time $\tau=(D/a^2)t$ calculated numerically for the rotating disk (Dl = 0, Ad = 0.41). (1) Pe = 0.64, (2) Pe = 0.16, (3) Pe = 10^{-2}, and (4) Pe = 2×10^{-3}. From Ref [63]

For the rotating disk, using the definition of the Pe number (see Table 4.8) and considering that $C = 0.616$, one obtains the expression for the dimensional relaxation time in the form

$$t_r = \frac{a^2}{D}\tau_r = \frac{1}{(0.616)^{2/3}}\frac{a^2}{D}\left(\frac{\nu^{1/2} D}{1.021 a^3 \Omega^{3/2}}\right)^{2/3}$$
$$= 1.36\left(\frac{\nu}{D}\right)^{1/3}\frac{1}{\Omega} \qquad (476)$$

As can be noticed, the relaxation time increases with particle size proportionally to $a^{1/3}$ and decreases inversely proportionally to the angular velocity of the disk Ω. By taking the typical data $a = 100$ nm, $D = 2.15\times10^{-12}\ \mathrm{m^2\,s^{-1}}$, $\nu = 10^{-6}\ \mathrm{m^2\,s^{-1}}$ $\Omega = 25\ \mathrm{s^{-1}}$ one obtains $t_r = 4.21$ s. For $a = 10$ nm and the same angular velocity one has $t_r = 1.95$ s and for $a = 1000$ nm, $t_r = 9.1$ s.

Analogously, for the impinging-jet cell the relaxation time is given by the expression

$$t_r = 1.38\left(\frac{R}{2\alpha_r V_\infty}\right)^{2/3} D^{-1/3} \qquad (477)$$

As can be noticed, the relaxation time increases with particle size proportionally to $a^{1/3}$ similarly as for the rotating disk and decreases proportionally to $V_\infty^{-2/3}$ (the mean velocity in the inlet capillary).

By again taking typical experimental data: $a = 100$ nm, $D = 2.15\times 10^{-12}\ \mathrm{m^2\,s^{-1}}$, $V_\infty = 0.1\ \mathrm{m\,s^{-1}}$, $R = 10^{-3}$ m (Re $=10$, $\alpha_r = 4$), one obtains $t = 1.24$ s. For $a = 10$ nm and the same velocity one has $t_r = 0.57$ s and for $a = 1000$ nm, $t_r = 2.63$ s. These estimates suggest that for typical particle deposition experiments, lasting usually for more than 10^3 s, the transition deposition regime is of minor significance. However, the role of the transition deposition regime may become appreciable for deposition experiments carried out for $Re < 1$.

The duration of the transition regime increases significantly in the case of particle sedimentation in the presence of an energy barrier, when $Dl \gg 1$ [67]. This is so because of the formation of a particle concentration peak in the local energy minimum appearing before the energy barrier. When the flow is applied, the concentration peak attains only a limited high value, strictly related to the flow intensity (characterized by the *Pe* number value). Particle deposition under such conditions was studied theoretically [68,69]. For the sake of generality, instead of the perfect sink model, the more realistic no-penetration boundary condition was applied, by postulating that the normal component of the flux vanishes for $H \to 0$. This approach represents, in principle, the extension of the adsorption model discussed in Section 3, taking into account the additional process of particle removal (desorption) from the energy minimum induced by the flow. Hence, the particles that accumulate at the primary energy minimum, occurring because of attractive electrostatic interactions, retained their ability to move tangentially to the fluid stream. However, they cannot penetrate the collector surface. The rate of particle accumulation is the net result of the normal flux, which becomes steady after the short transition time estimated above, and the tangential flux, proportional to the concentration (coverage) of particles in the energy minimum and to their tangential velocity.

Using this model, extensive theoretical studies of particle accumulation have been carried out with the aim of estimating the range of validity of the perfect sink boundary condition and the adsorption approach. One of the major findings was that particle concentration profiles under transient times, shown in Fig. 4.18, are well described by the *quasi*-Boltzmann distribution given by

$$\bar{n} = \bar{N}(\tau)\frac{e^{-\phi/kT}}{\bar{K}_a} + \bar{n}^*(H,\tau) \tag{478}$$

where $\bar{N} = \int \bar{n}\, dH$ is the dimensionless surface concentration of particles accumulated in the energy minimum, $\bar{K}_a = \int_{\min} \mathrm{e}^{-\phi/kT}\, dH$ is the dimensionless adsorption constant and $\bar{n}^*$ is perturbing term, given by the expression

$$\bar{n}^* = Sh\frac{e^{-\phi/kT}}{\bar{K}_a}\int_{H_m}^{H}\left(\int_{H_m}^{H} e^{\phi/kT}\, dH\right)\frac{e^{\phi/kT}}{F_1(H)}\, dH \tag{479}$$

where $H_m = \delta_m / a$.

The perturbing concentration term remains practically negligible in comparison with the *quasi*-Boltzmann particle concentration peak in the region of the energy minimum, which can be well seen in Fig. 4.19. This is in full agreement with the analytical predictions derived previously for

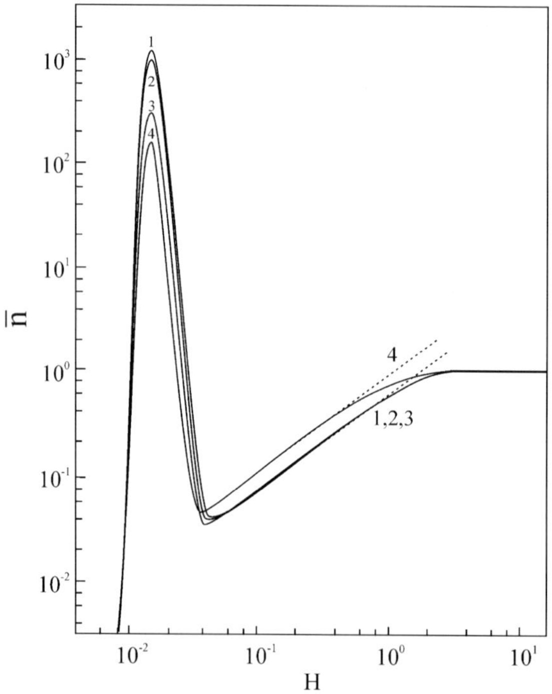

Fig. 4.18. The dimensionless particle concentration profiles for various times calculated numerically for the rotating disk, $Pe = 0.1$, $Gr = 0$. (1) $\tau = \infty$ (steady-state), (2) $\tau = 30$, (3) $\tau = 3$, and (4) $\tau = 1$. The dashed lines show the approximate analytical results calculated from Eq. (478). From Ref. [69].

diffusion-controlled transport through the energy barrier, expressed by Eq. (264). This observation seems to validate the adsorption model, in particular the concept of particle coverage adopted when deriving Eq. (283).

Another interesting observation, evident from the results shown in Fig. 4.19 (part "b"), is that particle flux remains constant over a wide range of distances from the interface ($2\times10^{-2} < H < 2$) and equal to the flux calculated using the perfect-sink model. This means that in order to apply the perfect-sink model, one does not need to know the exact distance, where the condition $n = 0$ is postulated. The distance can, therefore, be chosen quite arbitrarily.

However, as a result of convection, the amount of particles accumulated at the energy minimum under the steady state is considerably smaller than is expected under equilibrium conditions, using Eq. (274). This becomes evident using the formula derived in Ref. 68, for uniformly accessible surfaces by performing the mass balance for the minimum

$$\Theta = \pi a^3 n_b \frac{C\, Pe^{-2/3}}{\bar{f}(H_m)} \tag{480}$$

where $\Theta = \pi a^2 \bar{N}$ is the coverage of particles accumulated at the collector and $\bar{f} = (1+H_m)\, F_3(H_m)$.

For the rotating disk one has explicitly

$$\Theta = \pi \frac{0.607}{f(H_m)} a\, n_b \frac{\nu^{1/3} D^{2/3}}{\Omega} \tag{481}$$

By taking the typical data: $a = 100$ nm, $D = 2.15 \times 10^{-12}\ \mathrm{m^2\,s^{-1}}$, $\Omega = 25\ \mathrm{s^{-1}}$, $n_b = 10^{15}\ \mathrm{m^{-3}}$, $\bar{f} = 0.42$ one obtains $\Theta = 0.003$, which is only 0.3 per cent of the surface area of the collector. For $a = 10$ nm and the same parameters, one has $\Theta = 0.00652$, which is 0.65 per cent of the surface area. For comparison, the coverage calculated from Eq. (274) for $\phi = -20\ kT$ is 60 (obviously such a large value can only be treated as hypothetical, because the surface blocking effects discussed next make it impossible to attain such coverage).

As these estimates indicate, the predicted coverage of particles accumulated dynamically in the region at the energy minimum is very small, well below 1 per cent of a monolayer (characterized by a surface concentration of $1/\pi a^2$). Moreover, the particles accumulated in the energy minimum are expected to move tangentially to the fluid stream.

In reality, all experiments performed for colloid particles indicate that there is no lateral motion of deposited particles and their coverage attains significant fraction of a monolayer [11,49]. These facts suggest that appear specific interactions acting tangentially to the interface, strong enough to prevent particle motion. As yet, no comprehensive model of these interactions has been formulated. For example, in Refs. 68,69 an immobilization reaction model was proposed, which could explain quantitatively the high coverage of particles. In essence, this model approach was an extension of the perfect-sink with an allowance made for the finite rate of particle immobilization. However, the model gave no hints as to how the immobilization rate constant could be calculated for any real situation. It seems that even in the future this will be prohibitive because the specific interactions leading to particle immobilization on surfaces depend on many ill-defined parameters.

This suggests that the perfect sink model still remains the most useful one for colloidal particle deposition phenomena. Hence, in further discussion we will focus our attention on results derived by using solely the perfect sink model. We first discuss the results obtained for negligible

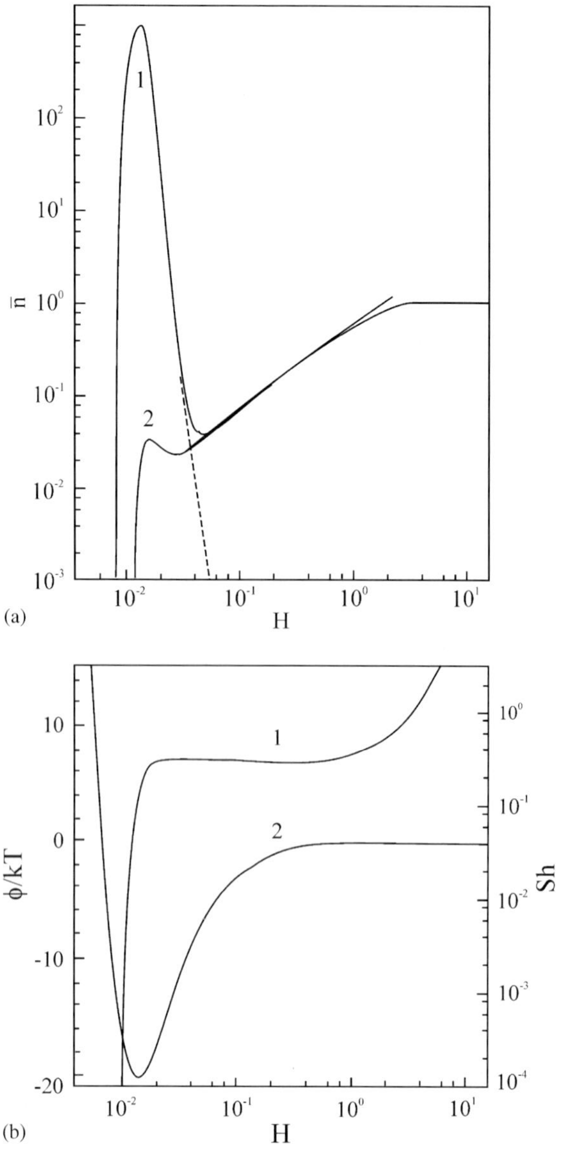

Fig. 4.19. (a) shows the stationary concentration profile calculated numerically for the rotating disk, $Pe = 0.1$, $Gr = 0$, (solid line 1), line 2 denotes the perturbation concentration distribution calculated from Eq. (479) and the dashed line represents the *quasi*-Botzmann distribution $\bar{n} = \bar{N}\ (\tau)e^{-\phi/kT}/\bar{K}_a$. (b) shows the dependence of the normal component of particle flux on the the dimensionless distance from the surface H (curve 1) and the specific interaction energy profile ϕ/kT (curve 2). From Ref. [69].

double-layer forces ($Dl = 0$) illustrating the role of particle size, dispersion interactions and external force. Then, the effect of attractive double-layer interactions is illustrated, as well as repulsive interactions leading to the formation of an energy barrier.

In Fig. 4.20 the theoretical results are shown, obtained for the rotating disk by solving numerically the stationary one-dimensional transport equation, Eq. (465), with the perfect sink boundary condition. The mass transfer rate constant (reduced flux) k_c is plotted vs. particle size in micrometers. The comparison of the exact results with the limiting analytical solution of Levich, i.e., $k_c = 0.620\ (\Omega^{1/2}D^{2/3}/\nu^{1/6})$ shows that the latter can well be used for particles with a radius below 0.2 μm. For increasing particle size, Levich's formula overestimates the exact flux values, which can be attributed to the effect of increased hydrodynamic resistance of a particle moving perpendicularly to the disk (this effect is described by the $F_1(H)$ correction function). For particle sizes above 1μm, however, the exact data become larger than Levich's formula predictions owing to the interception effect, described by Eq. (95). Indeed, this equation reflects quite well the characteristic feature of theoretical predictions, i.e., the parabolic dependence of k_c on particle radius. The positive deviation of the exact numerical data from the interception formula is caused by the attractive van der Waals forces characterized by the parameter $Ad=A_{132}/6kT$. For example, the increase in the Hamaker constant from 2.5×10^{-21} J to 2×10^{-20} J resulted in a 15 per cent increase in the calculated mass transfer rate for particles of radius $a = 0.5\ \mu$m. This effect becomes slightly more pronounced for larger particles, since for $a = 2\ \mu$m the mass transfer rate is expected to increase by 20 per cent for the same increase in the Hamaker constant.

As a result of diffusion and interception (governed by convection velocity and the Hamaker constant) there appears a characteristic minimum on the k_c vs. particle size curve, located approximately at $a = 0.75\ \mu$m (for $\Omega = 25\ \mathrm{s}^{-1}$). Similar theoretical results have been reported by Yao et al. [70] in the case of a spherical collector. This occurrence of the minimum has implications for particle filtration processes, e.g., deep bed filtration, indicating that collection of particle of the size of around a micrometer may be the least efficient.

It is interesting to observe that the location and depth of the minimum can be to some extent regulated by changing the flow intensity. In the case of the rotating disk this can be achieved by changing the angular velocity, because the flow approach velocity equals to $0.886\ (\Omega\nu)^{1/2}$. The results shown in Fig. 4.21 indicate quite unequivocally that the minimum is shifted

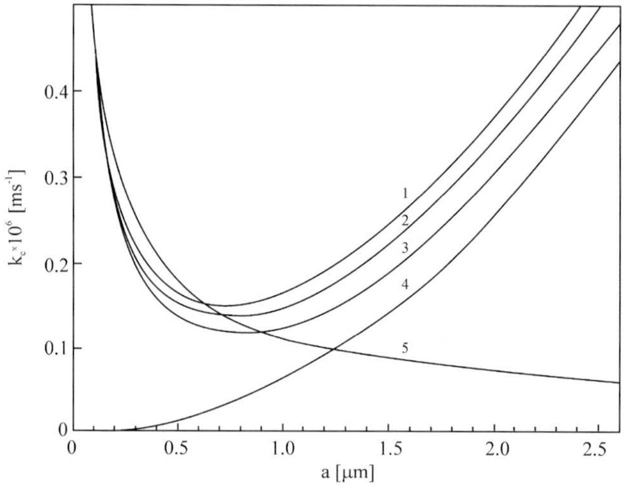

Fig. 4.20. The mass transfer rate constant k_c vs. particle radius a calculated numerically for the rotating disk, $\Omega = 25\ s^{-1}$, T =293K, $\Delta\rho = 0$ (neutrally buoyant particles), $Dl = 0$ (no electrostatic interactions). (1) $A_{132} = 2\times10^{-20}$ J ($Ad = 0.8$), (2) $A_{132} = 10^{-20}$ J ($Ad = 0.4$), (3) $A_{132} = 2.5\times10^{-21}$ J($Ad = 0.1$), (4) analytical results calculated for interception alone, k_c=0.5102($\Omega^{3/2}a^2/\nu^{1/2}$), and (5) analytical results calculated from Levich's formula k_c=0.620($\Omega^{3/2}\ D^{2/3}/\nu^{1/6}$). From Ref. [62].

to smaller particle sizes, occurring at $a = 0.5$ μm for $\Omega = 63\ s^{-1}$. At the same time the minimum value of k_c increases more than twice from 1.7×10^{-6} m s^{-1} to 3.7×10^{-6} m s^{-1} when the disk angular velocity increases from 25 s^{-1} to 63 s^{-1}.

In order to make the numerical calculation universally valid for other uniformly accessible surfaces it is useful to present the results of numerical calculations in the *Sh* vs. *Pe* coordinate system as shown in Fig. 4.22. The results obtained in Ref. 71 for a wide range of *Pe* number defined as 1.021 ($\Omega^{3/2}a^3/\nu^{1/2}D$), being therefore two times smaller than the usual definition given in Table 4.8, and the *Ad* parameter are plotted in Fig. 4.22.

The double-layer interactions have been neglected in these calculations.

The same trend as previously can be observed, i.e., for low values of the Peclet number, $Pe < 10^{-5}$, the computed *Sh* number approaches the Levich's prediction $Sh = 0.776\ (Pe')^{1/3} = 0.616\ Pe^{1/3}$ independently of the value of the *Ad* parameter. This is so because the diffusion boundary layer thickness, equal to 30 for $a = 10^{-5}$ as previously estimated, becomes considerably larger than the range of the van der Waals forces. As the *Pe* number increases, the diffusion boundary layer thickness becomes comparable with particle dimensions, so the van der Waals attractive forces start to play a significant role. For lower values of the *Ad* parameter they are, however, compensated by the increased hydrodynamic resistance of particles described by the correction function F_1. The net result of these two effects is that for $Ad < 17$ and $Pe' < 10^2$ the calculated flux becomes smaller than Levich's formula prediction. This is in full agreement with the results shown in Fig. 4.22.

On the other hand, for $Pe' > 10^2$ van der Waals attraction plays a dominant role since the diffusion boundary layer becomes much thinner than particle dimensions. For this deposition regime, diffusion is negligible and mass transfer rates can be predicted, in principle, from the trajectory analysis discussed previously.

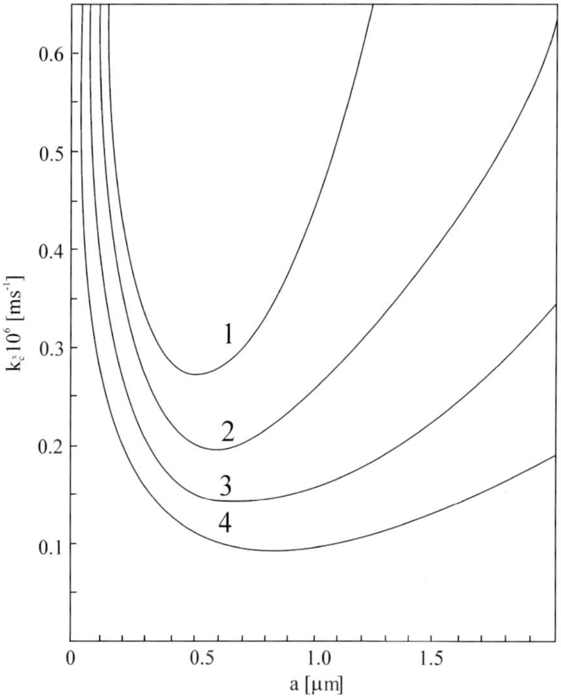

Fig. 4.21. The mass transfer rate constant k_c vs. particle radius a calculated numerically for the rotating disk, $T = 293$ K, $A_{132} = 10^{-20}$ J ($Ad = 0.4$), $\Delta\rho = 0$, $Dl = 0$ (no electrostatic interactions). The effect of flow intensity (angular velocity of the disk). (1) $\Omega = 63$ s^{-1}, (2) $\Omega = 40$ s^{-1}, (3) $\Omega = 25$ s^{-1}, and (4) $\Omega = 15.75$ s^{-1}. From Ref. [62].

It can be then predicted that the *Sh* number in the limit of $Pe' \gg 1$and $Ad \gg 1$ is [71]

$$Sh = 1.59 Ad^{1/3} Pe'^{2/3} = 1.00\ Ad^{1/3} Pe^{2/3} \tag{482}$$

Note, however, that the limiting form, Eq. (482) seems valid for $Ad > 17$, which corresponds to a Hamaker constant greater than 4×10^{-19} J. This value is rather too high from a practical viewpoint, corresponding to interactions of metal particles with metallic interfaces. For most situations of practical interest, the value of *Ad* is comparable to or smaller than unity. As can be seen in Fig. 4.22, under this intermediate deposition regime, for $Ad \ll 1$, and $Pe \gg 1$, the *Sh* number is better described by the limiting law

$$Sh \sim Pe \tag{483}$$

All the calculations discussed above were done for neutrally buoyant particles having the same density as the suspending medium, so $Gr = 0$, and the sedimentation effect was effectively eliminated. For most situations of practical interest, however, the particle density is larger than the density of

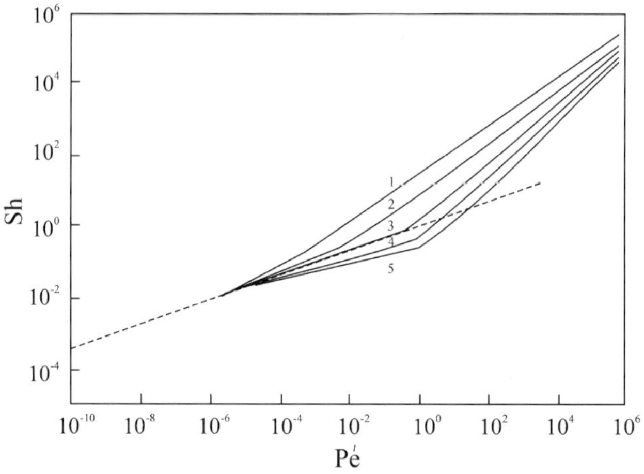

Fig. 4.22. The *Sh* number (dimensionless flux) vs. the *Pe* number ($Pe' = (1/2)\ Pe = 0.5102\ (\Omega^{3/2} a^3/\nu^{1/2} D)$), calculated numerically for uniformly accessible surfaces (solid lines). The dashed line denotes the analytical results calculated from the Levich formula $Sh = 0.776\ Pe' = 0.616\ Pe$. (1) $Ad = 1.67\times10^3$, (2) $Ad = 1.67\times10^1$, (3) $Ad = 0.167$, (4) $Ad = 1.67\times10^{-1}$, and (5) $Ad = 1.67\times10^{-3}$. From Ref. [71].

the medium, especially for a medium composed of particles of mineral origin, which induces significant migration. Depending on the orientation of the collector surface, either toward or outward from the direction of the gravitational force, one obtains $Gr=(2/9)(\Delta\rho|\mathbf{g}|a^2/\eta)>0$ or $Gr<0$, respectively. Calculations performed in this case for the rotating disk over a wide range of particle sizes are shown in Fig. 4.23 [72]. The exact numerical data are compared with the limiting analytical expressions derived for the diffusion $k_c=0.62(\Omega^{1/2}/\nu^{1/6})D^{2/3}$ interception $k_c = 0.5102\ (\Omega^{3/2}a^2/\nu^{1/2})$ and sedimention $k_c = \left(\frac{2}{9}\right)\frac{\Delta\rho|\mathrm{g}|\mathrm{a}^2}{\eta}$

As can be noticed, for $a < 0.2\ \mu$m the role of the interception and migration effects becomes negligible and the net deposition rate is dominated by convective diffusion with $k_c{\sim}a^{-2/3}$.

On the other hand, for $a > 0.5\ \mu$m the migration effect plays a decisive role, which significantly enhances particle deposition for $Gr > 0$, or reduces the deposition rate for $Gr < 0$. The high sensitivity of the particle deposition rate toward $\Delta\rho$ predicted for $Gr < 0$, implies a possibility of a very accurate experimental measurement of the apparent particle density in diluted suspensions if particle size is known.

In addition to van der Waals and gravitational interactions, the particle deposition rate can be changed to a significant extent changed by the electrostatic double-layer interactions. As discussed in Chapter 2, the magnitude and the range of these interactions is governed by two major parameters $\kappa a = a/Le = (2\,e^2 I\,a^2/\,\varepsilon\,kT)^{1/2}$ and Dl, which for the LSA model can be expressed as

$$Dl = 4\pi\varepsilon\frac{kT}{e^2}a16\tanh\left(\frac{\zeta_1 e}{4kT}\right)\tanh\left(\frac{\zeta_2 e}{4kT}\right) \tag{484}$$

where ζ_1 is the ζ potential of the collector and ζ_2 is the ζ potential of the particle.

Because the ionic strength of the electrolyte can be varied in wide limits (for aqueous suspension practically between 10^{-6} and 5M), κa can be varied between 0.1 and 10^3. Also the sign and the magnitude of the Dl parameter can be regulated within wide limits by the change of the ionic strength, pH and adsorption of surfactants. For $Dl < 0$ the zeta potentials of the particle and the interface are opposite, which produces attractive interactions, and for $Dl > 0$, one has to deal with repulsive interactions leading to an energy barrier. In the following figures, theoretical results obtained for both cases are presented.

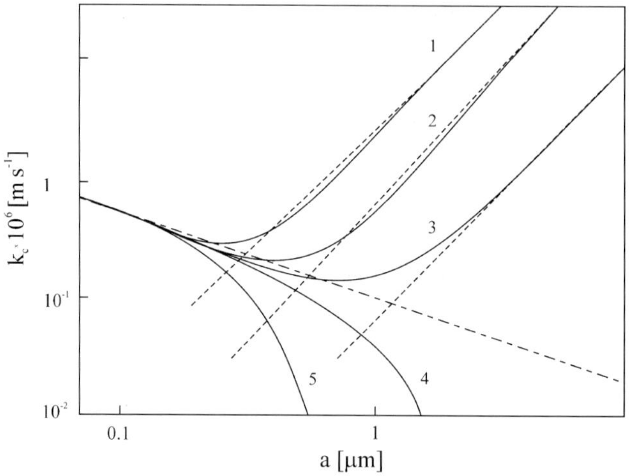

Fig. 4.23. The mass transfer rate constant k_c vs. particle radius a calculated numerically for the rotating disk, $\Omega = 25\ \mathrm{s}^{-1}$, $T = 293$ K, $A_{132} = 10^{-20}$ J ($Ad = 0.4$), $Dl = 0$ (no electrostatic interactions). The effect of gravitational force. (1) $\Delta\rho = 6\times10^2\ \mathrm{kg\,m^{-3}}$ (gravitational force directed to the disk surface), (2) $\Delta\rho = 3\times10^2$ kg m$^{(3}$ (gravitational force directed to the disk surface), (3) $\Delta\rho = 0$ (no sedimentation), (4) $\Delta\rho = -3\times10^2$ kg m^{-3} (gravitational force directed outwards from the disk surface), and (5) $\Delta\rho = -6\times10^2$ kg m^{-3}. The dashed lines denote the limiting deposition regimes, i.e., the diffusion controlled flux k_c=0.62($\Omega^{1/2}/\nu^{1/6}$)$D^{2/3}$, the sedimentation flux k_c=(2/9)($\Delta\rho\,|\mathbf{g}|\,a^2/\eta$) and the interception flux k_c=0.5102 ($\Omega^{3/2}a^2/\nu^{1/2}$) (curve 3). From Ref.[72].

In Fig. 4.24 the dependence of *Sh* on the *Pe* number calculated numerically in the case, where attractive electrostatic interactions appeared (type Ib energy profile discussed in Chapter 2), is present. These results have been obtained for uniformly accessible surfaces for $Gr = 0$ and $Ad = 0.2$ [73]. One can observe that the effect of electrostatic interactions starts to plays a significant role already for $Pe > 10^{-2}$. For $Pe > 1$ the *Sh* number increases almost proportionally to *Pe*, which is caused by the interception effect and enhanced by the attractive electrostatic interactions. This can be well reflected by the modified interception formula [73]

$$Sh = \frac{1}{2}(1 + H^*)^2 Pe \tag{485}$$

where $H^* = h^*/a$ is the effective range of the attractive double-layer interactions.

By equating the electrostatic forces and the hydrodynamic forces due to fluid convection it was shown in Ref. 73 that H^* can well be approximated by the expression

$$H^* = \frac{h^*}{a} = \frac{1}{\kappa a}\left[\ln\frac{2\left|Dl\right|\kappa a}{Pe} - 2\ln\left(1+\frac{1}{\kappa a}\ln\frac{2\left|Dl\right|\kappa a}{\kappa a\,Pe}\right)\right] \tag{486}$$

As can be seen in Fig. 4.24, Eq. (485) reflects quite well the exact numerical data for $Pe > 1$ and all values of a study. On the other hand, for $Pe < 10^{-2}$ the role of electrostatic interaction becomes negligible and the dimensionless flux is given by Levich's limit formula.

Since the effect of the electrostatic interactions depends to a critical extent on the κa parameter, it is useful to consider the dependence of Sh on κa. Calculations of this type, done in Ref. 73 are shown in Fig. 4.25. As can be seen, the analytical limiting formula, Eq. (485) describes satisfactorily the numerical data for $Pe > 1$. These results also suggest that the initial deposition rate of colloidal particles can be increased by more than an order of magnitude for large Pe (large particles or high flow intensities) by decreasing the ionic strength of the suspension, when $\kappa a \to 1$.

Analogous deposition rate enhancement was predicted for spherical collectors forming a packed bed having a porosity of 0.36 [74]. In order to make this effect more visible, the deposition rate was expressed in the reduced form $k_c/k_c^0 = Sh/Sh^0$ (where k_c is the mass transfer rate for negligible electrostatic interactions) as a function of the ionic strength I of a 1:1 electrolyte. The calculations were done for $R = 1.5\times10^{-4}$ m, $V = 10^{-3}$ m s^{-1}, $T = 298$ K, $Ad = 0.4$ ($A_{132} = 10^{-20}$ J) the zeta potentials of particles and collectors were 40 mV and –30 mV, respectively. As can be observed in Fig. 4.26, for particles with a diameter of 0.8 μm, the deposition rate is predicted to increased more than five times by reducing the ionic strength from 10^{-3} to 10^{-6} M. For a particle size of 0.1 μm, the deposition rate enhancement is limited to a factor 1.8.

The deposition rate increase due to the electrostatic interactions is also predicted for other collector geometries, such as the cylindrical collector and the parallel-plate channel, widely used in practice and in experiments aimed at particle and protein deposition. Examples of numerical calculations performed in this case are shown in Figs. 4.27–4.28. In the former, the dependence of the local deposition Sh number (dimensionless deposition rate) on the angle measured against the uniform flow direction is plotted.

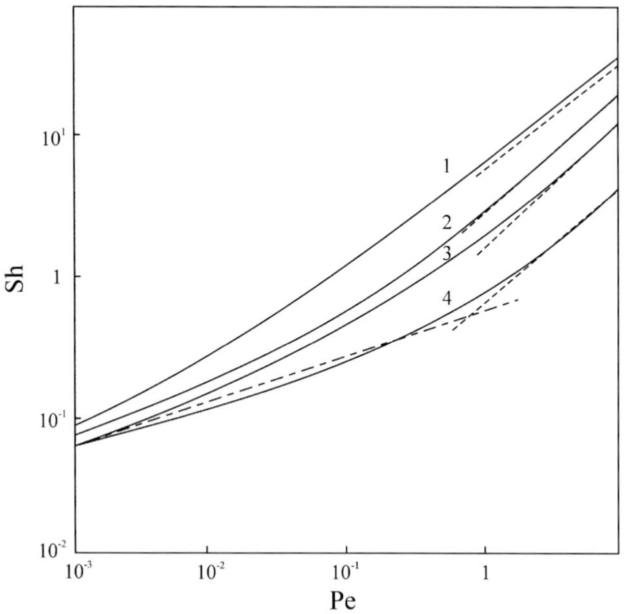

Fig. 4.24. The dependence of the *Sh* number (reduced bulk transfer rate $k_c a/D$ on the Peclet number *Pe* calculated numerically for uniformly accessible surfaces in the case of no external force, $Ad = 0.2$, $\phi_0 = -10^3\,kT$: (1) $a/Le = 2$, (2) $a/Le = 5$, (3) $a/Le = 10$ and (4) $a/Le = 100$. The limiting analytical results have been calculated from $Sh = 0.616\,Pe^{1/3}$ (diffusion-controlled transport) and denoted by the dashed-dotted line. Those calculated from Eq. (485) (interception-controlled flux) are denoted by the dashed line. From Ref. [73].

It can be observed in Fig. 4.27 that the effect of the electrostatic interactions plays an appreciable role in the region close to the stagnation point for $Pe>1$. For example, at $\vartheta = 0$ (forward stagnation point) the *Sh* number becomes almost an order of magnitude larger when electrostatic interactions occur ($Dl = -4\times10^3$, $\kappa a = 5$) compared to the case of no electrostatic interactions. The effect of electrostatic interactions plays a significant role for $\vartheta<120°$.

As the value of *Pe* decreases, i.e., for small particles and low approach velocities, the influence of the attractive double layer interactions becomes quite minor. For example, for $Pe = 10^{-5}$, the increase in the *Sh* number at the stagnation point is limited to a few percent only. In this case the local *Sh* number for the entire range of angle ϑ is well reflected by the analytical formula, Eq. (413) (these results coincide with the solid line in Fig. 4.27). Note also that for such small *Pe* values, the particle deposition rate (governed by the *Sh* number)

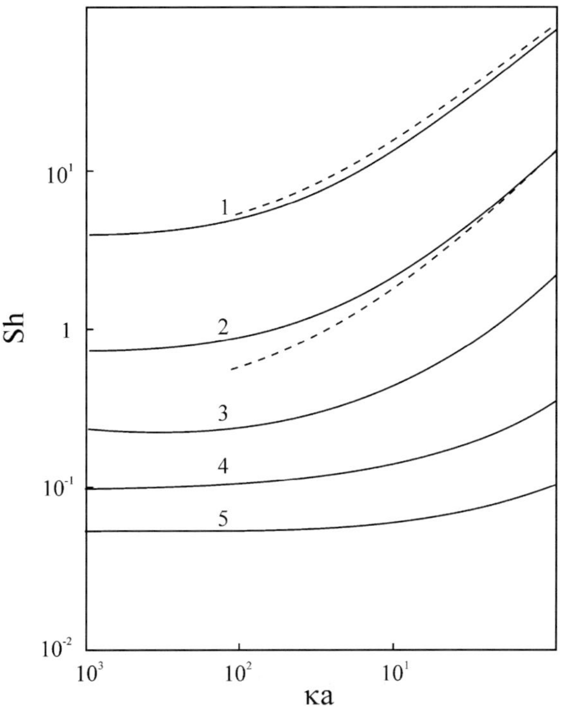

Fig. 4.25. The dependence of the Sherwood number on the $\kappa a = a/Le$ parameter (inverse double-layer thickness) calculated numerically for uniformly accessible surfaces in the case of no external force, $Ad = 0.2$, $Dl = -10^3\,kT$ (HHF double-layer model)(1) $Pe = 10$, (2) $Pe = 1$, (3) $Pe = 10^{-1}$, (4) $Pe = 10^{-2}$, and (5) $Pe = 10^{-3}$. Dashed lines denote the limiting analytical results calculated for the interception effect from Eq. (485). From Ref. [73].

is much more uniform. Thus, for $Pe = 10$, the deposition rate at $\vartheta \cong 150°$ is more than an order of magnitude smaller than at the stagnation point, whereas for $Pe = 10^{-2}$ the difference is only 20 percent. Because the local flux decreases more abruptly in the region near the rear stagnation point, the influence of electrostatic interactions on the averaged *Sh* number is expected to be less pronounced than for uniformly accessible surfaces. This trend is indeed observed in Fig. 4.28, which shows the dependence of $\langle Sh \rangle$ (averaged over the entire cylinder surface) on *Pe* for the same parameters as in Fig. 4.27.

The attractive electrostatic interaction also play a role in particle deposition in channel flows, especially at the entrance region, as demonstrated in Ref. 58. Examples of such calculation performed for a parallel-plate channel ($Pe = 1$, $Gr = 0$, $b/L = 10^{-3}$, $\kappa a = 5$) are shown in Fig. 4.29.

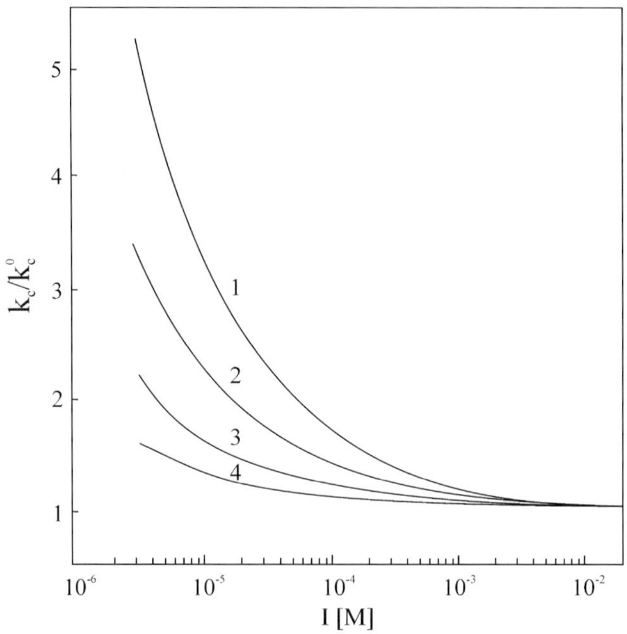

Fig. 4.26. The dependence of the normalized mass transfer rate $k_c/k_c^0 = Sh/Sh^0$ (where k_c^0 is the mass transfer rate for negligible electrostatic interactions) vs. the ionic strength I of a 1:1 electrolyte calculated numerically (solid lines) for a packed bed of spherical collectors. $R = 1.5 \times 10^{-4}$ m, $V_\infty = 10^{-3}$ m s^{-1}, bed porosity 0.36, $T = 298$ K, $Ad = 0.4$ ($A_{132} = 10^{-20}$ J), zeta potentials of particles and collectors were 40mV and –30 mV, respectively. (1) $a = 0.8$ μm, (2) $a = 0.4$ μm, (3) $a = 0.2$ μm, and (4) $a = 0.1$ μm. From Ref. [74].

As can be observed, the attractive electrostatic interactions play a significant role in the region $x/b < 10$. This corresponds to $x/L < 10^{-2}$, i.e., to 1 per cent of the entire channel surface only, because the b/L parameter in these calculations was 10^{-3}. For negligible electrostatic interactions the local deposition rate was predicted to be considerably smaller in this region. Concluding one can state that deposition rate enhancement in channel flows is much less pronounced than for collectors that have a region of uniform accessibility to transport (near stagnation points).

The electrostatic interactions can also be of repulsive character, when the zeta potential of the particle and the collector are of opposite sign. The energy barrier, which then arises, considerably reduces the deposition rate of particles. The significance of this effect is shown in Fig. 4.30, where the results of numerical calculations performed for uniformly accessible surfaces are shown (for $Gr = 0$, $Ad = 0.2$, $\kappa a = 5$).

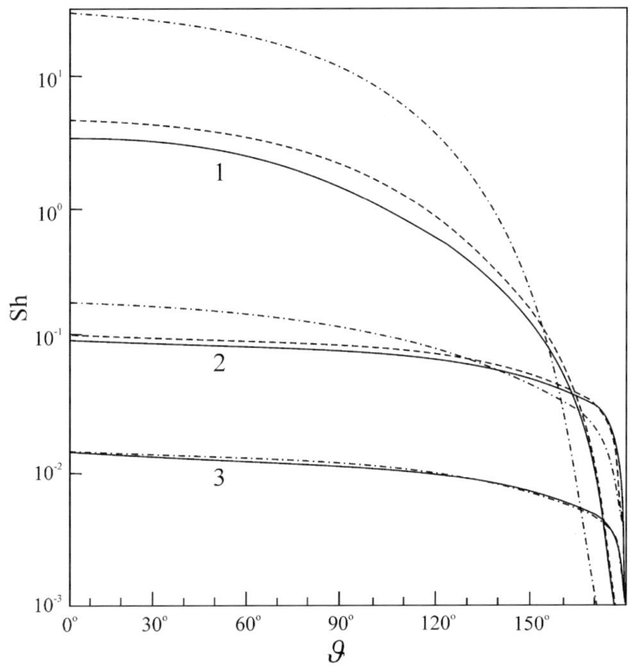

Fig. 4.27. The dependence of the local mass transfer (*Sh* number, dimensionless deposition rate) on the angle measured against the incident flow direction), calculated numerically for the cylindrical collector in the absence of external forces. (1) $Pe = 10$, (2) $Pe = 10^{-2}$, and (3) $Pe = 10^{-5}$. The solid lines denote $Ad = 0.04$ ($A_{132} = 10^{-21}$ J), $Dl = 0$, the dashed lines $Ad = 0.4$ ($A_{132} = 10^{-20}$ J), $Dl = 0$, the dashed-dotted lines $Ad = 0.4$ ($A_{132} = 10^{-20}$ J), $Dl = -4\text{x}10^3$, $\kappa a = 5$. From Ref. [59].

The electrostatic double-layer interaction force was calculated from the linear HHF formula

$$\bar{F}_H = Dl\,\kappa a\left[\frac{e^{-\kappa a H}}{1 \pm e^{-\kappa a H}} \mp \frac{1}{2}\frac{(\zeta_1 - \zeta_2)^2}{\zeta_1\zeta_2}\frac{e^{-2\kappa a H}}{1 - e^{-2\kappa a H}}\right]$$
$$Dl = 4\pi\varepsilon a\,\zeta_1\zeta_2 / kT \tag{487}$$

where the plus and minus sign correspond to the constant potential and constant charge model.

As can be seen in Fig. 4.30, for the c.p. model, the *Sh* number for the positive branch of the *Dl* parameter (repulsion) decreases exponentially with the slope independently of the *Pe* number. This behavior is in accordance with the analytical formula derived for the barrier-controlled deposition regime i.e.,

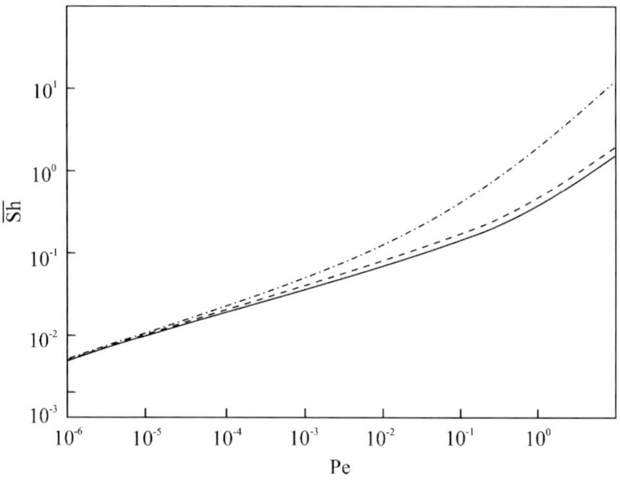

Fig. 4.28. The dependence of the averaged mass transfer number $\langle Sh \rangle$ number on the Peclet number Pe calculated numerically for a cylindrical collector in the absence of external forces, the solid lines denote $Ad = 0.04$ ($A_{132} = 10^{-21}$ J), $Dl = 0$, the dashed lines $Ad = 0.4$ ($A_{132} = 10^{-20}$ J), $Dl = 0$, the dashed-dotted lines $Ad = 0.4$ ($A_{132} = 10^{-20}$ J), $Dl = -4\times10^3$, $\kappa a = 5$. From Ref. [59].

$$Sh = K_a' \frac{a}{D} = \left(\frac{\phi_b}{kT}\right) e^{-\phi_b/kT} \tag{488}$$

Eq. (488) is valid for a triangular barrier. Qualitatively the same type of dependence of Sh on Dl is observed for the constant charge (c.c.) model if $Dl > 0$. However, in the case of negative Dl values (attraction) the c.c. model produces results, which seem entirely unphysical, i.e., an increase in the zeta potential of the particle and the collector surface diminishes the calculated flux values. Hence, these calculations suggest that the linear c.c. model is rather inadequate for interpreting particle deposition phenomena.

A comparison of barrier-controlled deposition results obtained for various models of electrostatic interactions is shown in Fig. 4.31. These calculations have been performed in Ref. 75 for $Pe = 0.1$, $Gr = 0$, $Ad = 0.4$, and $\kappa a = 11.52$. One can observe that the constant potential, the mixed (constant potential, constant charge) and the LSA models produce results that are quite similar, whereas the c.c. model predictions are much lower.

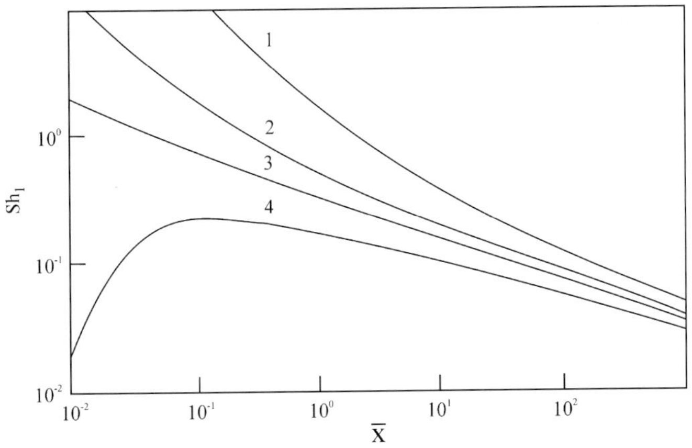

Fig. 4.29. The dependence of the local mass transfer number *Sh* number on the distance from the inlet point calculated numerically for a cylindrical channel for $Pe = 1$, $Gr = 0$, $b/L = 10^{-3}$. (1) $Ad = 0.4$, $Dl = -4\text{x}10^3$, $\kappa a = 5$; (2) $Ad = 0.4$, $Dl = 0$; (3) analytical approximation $Sh = 0.427\bar{x}^{1/3}$; and (4) $Ad = 0.04$, $Dl = 0$. From Ref. [58].

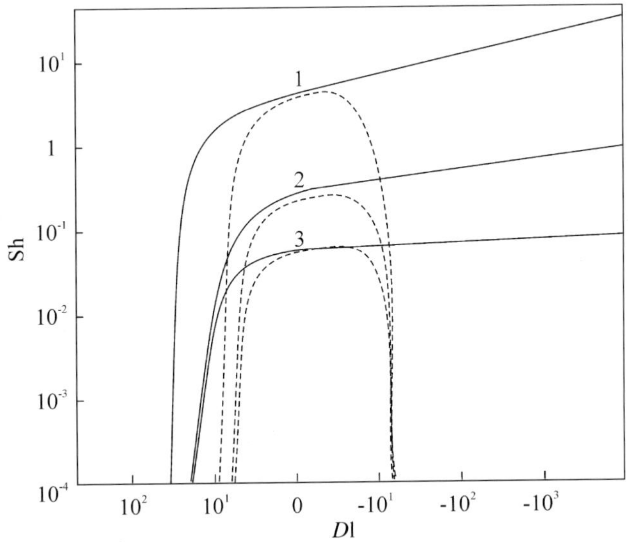

Fig. 4.30. The dependence of the *Sh* number on the double-layer parameter *Dl* calculated numerically for uniformly accessible surfaces in the case of no external force, $Ad = 0.2$, κa=5. (1) $Pe = 10$, (2) $Pe = 10^{-1}$, and (3) $Pe = 10^{-3}$. The solid lines denote the constant-potential HHF model of electrostatic interactions described by Eq. (487) and the dashed lines denote the constant-charge double-layer model. From Ref. [72].

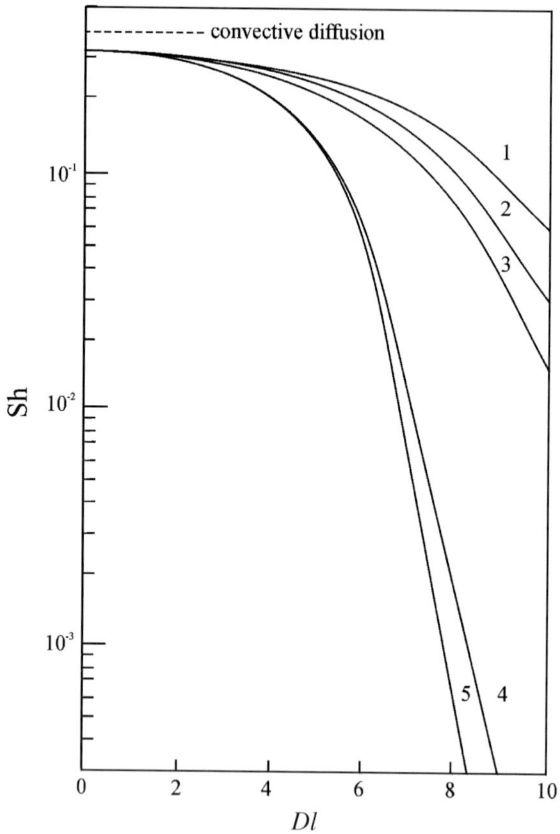

Fig. 4.31. The dependence of the Sherwood number on the double layer parameter $Dl=4\pi\ \varepsilon\ a\ \zeta_1\ \zeta_2 / kT$ calculated numerically for uniformly accessible surfaces a nd various double-layer models. $Pe = 0.1$, $Gr = 0$, $Ad = 0.4$, $\kappa a = 11.52$ (1) constant potential approximation, (2) mixed case, (3) LSA approximation (linear superposition approximation), (4) constant charge (non-linear), and (5) constant charge (linear). From Ref. [75].

The numerical calculations reported in this section allow one, by using the dimensionless parameters, to formulate the areas of various deposition regimes, governed by simple limiting laws. For the sake of convenience they have been collected in Table 4.12.

It should be remembered that under the irreversible adsorption (deposition) regimes (perfect-sink model) the flux remains independent of particle coverage. Therefore, the coverage of particles can be calculated from the

Table 4.12
Limiting laws for calculating the local mass transfer rate k_c (for uniformly accessible surfaces)

Limiting law	Range of parameters	Remarks
$Sh = 0.616\, Pe^{1/3}$	$Pe \ll 1, \quad Ex \ll Pe$ $Dl = Ad = 0$	Diffusion controlled regime
$Sh = (1/2)\, Pe$	$Pe \gg 1, \quad Ex \ll Pe$ $Dl = Ad = 0$	Geometrical interception regime
$Sh = Ex$ $Sh = Gr$ (gravitational force)	$Ex \gg Pe$ $Dl = 0$	External force regime
$Sh = Ad^{1/3} Pe^{2/3}$	$Pe \gg 1, \quad Ad \gg 1$ $Dl = 0$	van der Waals force regime
$Sh = (1/2)\, Pe\, (1+H^*)^2$	$Pe \gg 1$ $Dl = 0$	Attractive double-layer force regime $H^* \cong \frac{1}{\kappa a}\left[\ln \frac{2\lvert Dl\rvert \kappa a}{Pe} - 2\ln\left(1+\frac{1}{\kappa a}\ln\frac{2\lvert Dl\rvert \kappa a}{Pe}\right)\right]$
$Sh = \bar{k}_a$	$Sh \ll 1$ $\phi_b > 5kT$	Barrier controlled regime $\bar{k}_a = \frac{a}{\int_{\delta_m} \frac{e^{\phi_b/kT} - 1}{F_1(z')} dz'} \cong \frac{\phi_b}{kT} e^{-\phi_b/kT}$ for a triangular barrier

surface mass balance equation by noting that the rate of particle deposition equals the rate of their arrival at the primary minimum, i.e., the limiting flux -j. This can be expressed as

$$\frac{1}{S_g}\frac{d\Theta}{dt} = \frac{dN}{dt} = -j(t) = k_c\, n^b \tag{489}$$

By integrating Eq. (489) with the assumption that there were no particles initially one obtains the expression

$$\Theta = S_g N = -S_g \int_0^t j(t')\, dt' \tag{490}$$

On the other hand, when the flux becomes stationary, under the convection-controlled transport, the change in Θ is given by the linear relationship

$$\Theta = S_g\,(\Theta_0 + k_c n^b t) \tag{491}$$

where Θ_0 is the surface coverage at the time when the flux becomes steady, equal to the above-discussed limiting flux $j_0 = -k_c n^b$.

The results discussed in this section concerning the particle deposition rate under linear regimes have a practical significance. For example, they can be exploited for verifying experimental results concerning colloid or protein-adsorption kinetics by indicating artifacts that may appear when indirect methods are used. If the experimentally measured transfer rate is larger than the maximum value k_c one should suspect that the transport mechanism has been modified, e.g., by natural convection. On the other hand, too small values of the measured transfer rate suggest surface contamination or aggregation of the suspension. Hence, the initial flux measurements in conjuncture with theoretical predictions given above can be used as a sensitive tool for detecting local particle/surface interactions. In this way, by using larger colloidal particles as markers, surface heterogeneity can be detected, e.g., the presence of nano-sized colloid contaminants.

LIST OF SYMBOLS

Symbol	Definition	Character	Unit	Numerical value
A	Interception parameter (particle to collector size ratio)	S	[1]	
$A(\Delta t)$	Velocity autocorrelation function	S	$m^2\ s^{-2}$	
Ad	Dispersion interaction parameter	S	[1]	

List of Symbols (continued)

Symbol	Definition	Character	Unit	Numerical value
A_f	Flow parameter	*S*	[1]	
A_0	Velocity autocorrelation function (initial value)	*S*	$m^2\ s^{-2}$	
Ar	Dimensionless aspect ratio parameter	*S*	[1]	
A_{132}	Hamaker constants for interactions of two different particles in medium	*S*	J	
a, a_1, a_2	Radii of particles, longer axis of oblate spheroid	*S*	m	
$\bar{a}$	Mean radius of spheres	*S*	m	
a_p	Characteristic particle dimension	*S*	m	
a_s	Radius of equivalent sphere	*S*	m	
$\bar{B}_2, \bar{B}_3, \bar{B}_4, \bar{B}_n,$	Reduced virial coefficients	*S*	[1]	
b	Spheroid semi-axis	*S*	m	
$2b$	Height of parallel-plate channel	*S*	m	
b_1, b_2, b_3	Dimensionless coefficients	*S*	[1]	
C, C'	Dimensionless constants for rotating disk	*S*	[1]	
$C(r/R), C(\vartheta)$	Universal function for radial impinging jet	*S*	[1]	
C_c	Dimensionless constant	*S*	[1]	
C_e	Electric capacity	*S*	$F=C\ V^{-1}$	
C_l	Coefficients of series expansion	*S*	[1]	
C_w	Constant describing velocity distribution for wall jet flows	*S*	$m^{11/4}\ s^{-1}$	
$C_\perp$	Dimensionless constant characterizing perpendicular flow component	*S*	[1]	
$C_1\ C_2, C_3, C_4$ $C'_1\ C'_2, C'_3, C'_4$	Dimensionless constants	*S*	[1]	
c	Ellipsoid semi-axis	*S*	m	
c_B	Constant characterizing probability distribution	*S*	m^{-2}	
c_0, c_1, c_2, c_3, c_4	Dimensionless constants	*S*	[1]	
$\mathbf{D}, \mathbf{D}'$	Translation diffusion tensors of particles	**T**	$m^2\ s^{-1}$	
$\tilde{\mathbf{D}}$	Grand diffusion matrix	**T**	$m^2\ s^{-1}$	
$\boldsymbol{D}_g$	Gradient diffusion tensor of particles	**T**	$m^2\ s^{-1}$	

List of Symbols (continued)

Symbol	Definition	Character	Unit	Numerical value
Dr	Rotation diffusion tensor of particles	**T**	s^{-1}	
$D, D_{11}, D_{22}, D_{33}$	Translation diffusion coefficients of particles	*S*	$m^2\ s^{-1}$	
$\langle D \rangle$	Averaged diffusion coefficient	*S*	$m^2\ s^{-1}$	
D_c	Diffusion coefficient of collector	*S*	$m^2\ s^{-1}$	
Dl	Electrostatic double-layer number	*S*	[1]	
$Dr, Dr_{11}, Dr_{22}, Dr_{33}$	Rotation diffusion coefficients of particles	*S*	s^{-1}	
D_{12}	Relative diffusion coefficient of two particles	*S*	$m^2\ s^{-1}$	
$D_\rho, D_\perp$	Particle diffusion coefficients perpendicular and parallel to interfaces	*S*	$m^2\ s^{-1}$	
d	Size of cube	*S*	m	
$2d$	Width of slot-jet	*S*	m	
d_p	Width of plate	*S*	m	
E, **E**′	Electric field	**V**	$V\ m^{-1}$	
$\tilde{\mathbf{E}}(s)$	Complex surface elasticity operator (matrix)	**T**	[1]	
E, E^0	Electric field (scalar)	*S*	$V\ m^{-1}$	
$\tilde{E}(s)$	Complex surface elasticity operator (scalar)	*S*	[1]	
Ei	Electrostatic image force number	*S*	[1]	
El	Electrostatic number	*S*	[1]	
Ep	Electrostatic polarization number	*S*	[1]	
Ex	External force number	*S*	[1]	
e	Elementary charge	*S*	C	1.60218×10^{-19}
erf (x)	Error function	*S*	[1]	
erfc (x)	Subsidiary error function	*S*	[1]	1- erf (x)
F	Force vector	**V**	$N = kg\ m\ s^{-2}$	
$\tilde{\mathbf{F}}$	Generalized driving force	**V**	N	
F′	Generalized resistance force	**V**	N	
F*	Random force	**V**	N	
$\mathbf{F}_{ext}$	External force	**V**	N	
$\mathbf{F}_h$	Hydrodynamic force	**V**	N	

List of Symbols (Continued)

Symbol	Definition	Character	Unit	Numerical value
$\mathbf{F}_s$	Specific force	**V**	N	
$\mathbf{F}_{Sh}$	Force exerted by simple shear flow	**V**	N	
$\mathbf{F}_{St}$	Force exerted by stagnation-point flow	**V**	N	
F	Force on particle	S	N	
$F(\bar{x})$	Function describing velocity distribution near collectors	S	[1]	
$F^*(t)$	Random force	S	N	
$F(\xi)$	Universal function for spherical and cylindrical collectors	S	[1]	
F_{ext}	External force	S	N	
F_H	Component of net force on particle	S	N	
F_0	Constant force	S	N	
Fr	Froude number	S	[1]	
F_s	Specific force of interactions (scalar)	S	N	
F_x, F_y, F_z	Components of net force on particle	S	N	
$F_1(H) \div F_{11}(H)$	Universal hydrodynamic correction functions	S	[1]	
$f(0)$	Dimensionless coefficient for rotating disk	S	[1]	0.5102
$f(\lambda_s)$	Function describing flow distribution near spheroid	S	[1]	
f_i	Activity coefficient of ion	S	[1]	
f_e	Dimensionless adsorption equilibrium parameter	S	[1]	
$\bar{f}_0(H_m)$	Dimensionless coefficient for rotating disk	S	[1]	
$f_0(H,\lambda_r)$	Function describing van der Waals interactions	S	[1]	
f_p	Probability density	S	m^{-1}	
$f_p\ (k,t)$	Probability density of particle state	S	[1]	
$f_1(0)$, $f_2(0)$	Dimensionless constants	S	[1]	
$f_1(x)$, $f_2(x)$, $f_3(x)$ $f_1'(x)$, $f_2'(x)$	Functions characterizing flows near and inside collectors	S	[1]	

List of Symbols (continued)

Symbol	Definition	Character	Unit	Numerical value
f_{12}	Dimensionless geometrical parameter	S	[1]	
G	Shear rate of flow near cylinder	S	s^{-1}	
G_D	Derjaguin factor	S	m	
Gr	Dimensionless gravity number	S	[1]	
G_{Sn}	Simple shear flow rate	S	s^{-1}	
G_{St}	Stagnation flow shear rate	S	$m^{-1}\ s^{-1}$	
G_2, G_3	Functions describing velocity distribution near collectors	S	[1]	
$\mathbf{g}$	Gravity acceleration vector	**V**	$m\ s^{-2}$	
$\overline{\mathbf{g}}$	Reduced gravity vector	**V**	[1]	
$g(x)$, $g(\vartheta)$	Functions describing velocity distribution near collectors	S	[1]	
$g(\lambda_s)$	Function describing flow distribution near spheroid	S	$m^3\ s^{-1}$	
$g(0)$	Dimensionless coefficient for rotating disk	S	[1]	0.616
H	Dimensionless gap width	S	[1]	
$\overline{H}$	Dimensionless adsorption parameter	S	[1]	
H^*	Dimensionless effective range of interactions	S	[1]	
H_0	Dimensionless initial separation from surface	S	[1]	
h	Distance from wall	S	m	
h^*	Effective range of interactions	S	m	
h_m	Distance between particle and surface	S	m	
$\mathbf{I}$	Unit tensor	**T**	[1]	
$\mathbf{In}$	Moment of inertia of particle	**T**	$kg\ m^2$	
I	Ionic strength	S	m^{-3}	
$\overline{I}$	Dimensionless ionic strength	S	[1]	
I_d, I'_d, I_1	Definite integrals	S	[1]	
Im	Configurational integral	S	m	
In	Moment of inertia	S	$kg\ m^2$	
$I_{\perp}, I(0)$	Definite integrals	S	[1]	
$\mathbf{i}$, $\mathbf{i}_H$, $\mathbf{i}_r$, $\mathbf{i}_y$	Unit vectors	**V**	[1]	
i	Imaginary unit	S	[1]	

List of Symbols (continued)

Symbol	Definition	Character	Unit	Numerical value
J	Relative adsorption flux of particles	S	[1]	
$\mathbf{j}$	Particle flux	**V**	$m^{-2}\,s^{-1}$	
$\tilde{j}$	Generalized particle flux	**V**	$m^{-2}\,s^{-1}$	
$\mathbf{j}_d$	Generalized diffusion flux	**V**	$m^{-2}\,s^{-1}$	
j	Particle flux (scalar)	S	$m^{-2}\,s^{-1}$	
j_a	Adsorption flux of particles	S	$m^{-2}\,s^{-1}$	
$\bar{j}_a$	Reduced adsorption flux of particles	S	[1]	
j_H	Normal component of flux of particles	S	$m^{-2}\,s^{-1}$	
$\overline{j_H}$	Reduced normal component of flux of particles	S	[1]	
j_0	Limiting flux of particles	S	$m^{-2}\,s^{-1}$	
j_{st}	Stationary particle flux	S	$m^{-2}\,s^{-1}$	
j_t	Transient particle flux	S	$m^{-2}\,s^{-1}$	
$\mathbf{K}$	Resistance tensor for two spheres	**T**	m	
$\tilde{\mathbf{K}}$	Grand resistance matrix	**T**	m	
$\mathbf{K}_c$, $\tilde{\mathbf{K}}_c$	Coupling resistance tensors	**T**	m^2	
$\mathbf{K}_r$	Rotation resistance tensor	**T**	m^3	
$\mathbf{K}_t$	Translation resistance tensor	**T**	m	
$\overline{K}$	Dimensionless adsorption constant	S	[1]	
K_a	Equilibrium adsorption constant	S	m	
$\overline{K_a}$	Dimensionless equilibrium adsorption constant	S	[1]	
K_a^*	Dimensionless adsorption parameter	S	[1]	
$\overline{K_c}$	Reduced mass transfer rate constant for channel flows	S	[1]	
K_c'	Components of coupling resistance tensor	S	m^2	
K_p	Probability of particle state change	S	[1]	
K_r	Components of rotation resistance tensor	S	m^3	
K_s	Sphericity coefficient	S	[1]	

List of Symbols (continued)

Symbol	Definition	Character	Unit	Numerical value
$K_{t\parallel}$, $K_{t\perp}$	Components of translation resistance tensor	S	m	
K_1, K_2, K_3	Functions describing resistance of ellipsoid	S	[1]	
k	Boltzmann constant	S	J K^{-1}	1.38065×10^{-23}
$<k_c>$	Averaged mass transfer rate	S	m s^{-1}	
k_a, k_a'	Adsorption rate constants	S	m s^{-1}	
$\bar{k}_a, \bar{k}_a'$	Dimensionless adsorption constants	S	[1]	
K_a^*	Dimensionless adsorption rate parameter	S	[1]	
k_c, k_c'	Mass transfer coefficients	S	m s^{-1}	
k_d	Desorption rate constant	S	s^{-1}	
$\bar{k}_d$	Dimensionless desorption constant	S	[1]	
k_e	Elasticity constant	S	kg s^{-2}	
k_{12}	Coagulation rate constant	S	m^3 s^{-1}	
L	Length of channel, cylinder, plate	S	m	
L_a	Characteristic adsorption length scale	S	m	
Le	The diffuse double-layer thickness	S	m	
L_c	Characteristic length of entrance region	S	m	
L_{ch}	Characteristic adsorption length	S	m	
L_t	Flow transition length	S	m	
l	Length of slender body	S	m	
l_B	Characteristic length of Brownian motion	S	m	
l_{ch}	Length of free path of particle	S	m	
M	Translational mobility matrix of particle	**T**	kg^{-1} s	
$\mathrm{M_r}$	Mobility matrix of particle	**T**	m^{-1} kg^{-1} s	
M	Particle mobility in one dimension	S	kg^{-1} s	
$<M>$	Averaged particle mobility	S	kg^{-1} s	
Mr_{11}, Mr_{22}, Mr_{33}	Diagonal components of rotation mobility matrix	S	J^{-1} s^{-1}	
M_{11}, M_{22}, M_{33}	Diagonal components of translational mobility matrix	S	kg^{-1} s	

List of Symbols (continued)

Symbol	Definition	Character	Unit	Numerical value
m, m_f	Fluid mass	S	kg	
m_p	Particle mass	S	kg	
m'_p	Effective particle mass	S	kg	
N	Surface concentration of particles	S	m^{-2}	
$\overline{N}(\tau)$	Dimensionless coverage of particles	S	[1]	
$N(t)$	Number of sedimentating particles	S	[1]	
N_{Av}	Avogadro constant	S	[1]	6.02214×10^{23}
N_B	Total number of elementary displacements of Brownian particle	S	[1]	
N_i	Initial surface concentration of particles	S	m^{-2}	
N_p	Number of particles	S	[1]	
N_m	Characteristic monolayer concentration of particles	S	m^{-2}	
$\hat{\mathbf{n}}$	Unit normal vector	**V**	[1]	
$\overline{n}$	Dimensionless particle concentration	S	[1]	
n, n_a, n_m, n_l, n_1, n_2, n^*	Number concentrations of particles	S	m^{-3}	
n_b	Number concentration of particles in bulk	S	m^{-3}	
n_+, n_-, n_B	Numbers of displacements of particle	S	[1]	
n_0(r)	Initial particle concentration distribution	S	m^{-3}	
n_b	Dimensionless equilibrium concentration of particles	S	[1]	
P	Osmotic pressure	**T**	N m^{-2}	
Pe	Peclet number	S	[1]	
P_1	Dimensionless functions characterizing adsorption	S	[1]	
$\mathbf{p}_a$	Angular momentum vector	**V**	kg m^2 s^{-1}	
$\mathbf{p}_m$	Linear momentum vector	**V**	kg m s^{-1}	
p_B	Probability of particle displacement	S	[1]	
$\mathbf{Q}_f$	Source term of quantity f	Arbitrary	Arbitrary	
Q	Volumetric flow rate	S	m^3 s^{-1}	

List of Symbols (continued)

Symbol	Definition	Character	Unit	Numerical value
Q_1	Dimensionless functions characterizing adsorption	S	[1]	
Q_1'	Particle concentration sources	S	$m^{-3}\ s^{-1}$	
$q,\ q^0$	Electric charge	S	C	
$\tilde{R}$	Particle state vector	**V**		
R	Radius of interface, cylindrical channel, cylindrical or spherical collectors, rotating disk	S	m	
$\bar{R}$	Adsorption parameter	S	[1]	
$R_{as},\ R_{bs}$	Radii of spheroidal collector	S	m	
R_b	Radius of outer boundary	S	m	
R_{bar}	Energy barrier resistance	S	m	
Re	Reynolds number	S	[1]	
R_g	Gas constant	S	J	
$\mathbf{r},\mathbf{r}_1,\mathbf{r}_2,\mathbf{r}_i$	Position vectors	**V**	m	
r_0	Initial position vector	**V**	m	
r	Radial coordinate	S	m	
$(r,\ \vartheta,\ z)$	Cylindrical coordinates	S		
r_p	Radial position of particle	S	m	
r_c	Critical radius for limiting trajectory	S	m	
$\mathrm{r}_{o_1},\ \mathrm{r}_{o_2},\ \mathrm{r}_{o_3}$	Roots of third-order equation	S	[1]	
r_p	Radial position of particle in local flow	S	m	
r_{12}	Coagulation rate of particles	S	$m^{-3}\ s^{-1}$	
$\delta\mathbf{S}$	Surface element vector	**V**	m^2	
S	Surface area	S	m^2	
$S(r/R),\ S(\vartheta)$	Universal function for radial impinging jet	S	[1]	
$\tilde{S}(s)$	Laplace transformation of surface deformation function	S	[1]	
$S(\vartheta)$	Universal function for impinging-jet flows	S	[1]	
$\delta S, dS$	Surface elements	S	m^2	
S_c	Surface area of collector	S	m^2	
Sc	Schmidt number	S	[1]	
S_g	Characteristic cross-section of particle	S	m^2	
$Sh,\ Sh_0$	Dimensionless mass transfer Sherwood numbers	S	[1]	

List of Symbols (continued)

Symbol	Definition	Character	Unit	Numerical value
$Sh(x)$	Local Sherwood number	S	[1]	
$\langle Sh \rangle$	Averaged Sherwood number	S	[1]	
S_p	Surface area of particle	S	m^2	
St	Stokes number	S	[1]	
s	Laplace transformation parameter	S	[1]	
$\mathbf{To}$	Torque	**V**	N m	
$\mathbf{To}^*$	Random torque	**V**	N m	
$\mathbf{To}_{ext}$	External torque	**V**	N m	
$\mathbf{To}_h$	Hydrodynamic torque	**V**	N m	
$\mathbf{To}_s$	Specific torque	**V**	N m	
$\mathbf{T}_r$	Transformation matrix	**T**	[1]	
T	Absolute temperature	S	K	
T_a	Relaxation number	S	[1]	
T_{ch}	Characteristic time	S	s	
Th	Dimensionless parameter	S	[1]	
To_x, To_y, To_z	Components of hydrodynamic torque vector	S	N m	
t	Time	S	s	
$\bar{t}$	Dimensionless time	S	[1]	
t_a	Characteristic time scale	S	s	
t_{ch}	Velocity relaxation time	S	s	
t_r	Relaxation time of particle accumulation	S	s	
$\mathbf{U}_h$	Velocity of particle	**V**	$m\ s^{-1}$	
$\mathbf{U}$	Generalized velocity of particle	**V**		
$\mathbf{V}$	Averaged sedimentation velocity	**V**	$m\ s^{-1}$	
U', U^*, U_0	Particle velocities	S	$m\ s^{-1}$	
U_B	Apparent diffusion velocity of particle	S	$m\ s^{-1}$	
U_{ch}, U_s	Characteristic particle migration velocity	S	$m\ s^{-1}$	
U_h, U_ϑ, U_r U_x, U_y, U_z $U_{\xi 1}, U_{\xi 2}, U_{\xi 3}$	Components of particle velocity vector	S	$m\ s^{-1}$	
$U_{h_x}, U_{h_y}, U_{h_z}$	Components of particle velocity vector due to flow	S	$m\ s^{-1}$	
U_1, U_2, U_3	Sedimentation velocities of particle	S	$m\ s^{-1}$	

List of Symbols (continued)

Symbol	Definition	Character	Unit	Numerical value
$\mathbf{V},\mathbf{V}',\mathbf{V}_0$	Fluid velocity vectors	**V**	m s^{-1}	
$\overline{\mathrm{V}}$	Reduced fluid velocity vector	**V**	[1]	
$\mathbf{V}_1,\mathbf{V}_2,\mathbf{V}_3,\mathbf{V}_4$	Simple flow fields	**V**	m s^{-1}	
$\langle \mathrm{V} \rangle$	Mean fluid velocity in channels	*S*	m s^{-1}	
V_{ch}	Characteristic velocity of fluid	*S*	m s^{-1}	
V_{mx}	Maximum velocity in channels	*S*	m s^{-1}	
V_r	Radial component of fluid velocity	*S*	m s^{-1}	
V_s	Moving plate velocity	*S*	m s^{-1}	
V_x	Fluid velocity component in *x* direction	*S*	m s^{-1}	
V_y	Fluid velocity in *y* direction	*S*	m s^{-1}	
V_∞	Macroscopic fluid velocity far from interface	*S*	m s^{-1}	
V_1,V_1',V_2',V_2,V_3	Characteristic fluid velocities	*S*	m s^{-1}	
$V_\perp,V_{\parallel}$	Perpendicular and parallel components of fluid velocity near interfaces	*S*	m s^{-1}	
$\overline{V}_\perp,\overline{V}_{\parallel}$	Dimensionless perpendicular and parallel components of fluid velocity	*S*	[1]	
v	Volume	*S*	m^3	
v_p	Volume of particle	*S*	m^3	
δv	Control volume, volume element	*S*	m^3	
$\mathbf{x}, \mathbf{x}_0$	Position vectors	**V**	m	
x	Distance	*S*	m	
$\overline{x}$	Dimensionless distance from interface	*S*	[1]	
x^*, y^*, z^*	Particle displacements	*S*	m	
Y^0, Y_1^0, Y_2^0	Effective surface potentials	*S*	[1]	
Z	Similarity variable	*S*	[1]	
z	Coordinate perpendicular to interface	*S*	m	
$\overline{z}, \overline{z}_0$	Reduced coordinate perpendicular to interface	*S*	[1]	

List of Symbols (continued)

Symbol	Definition	Character	Unit	Numerical value
Greek				
α_l	Roots of trigonometric equation	S	[1]	
α_r	Flow parameter for radial impinging jet	S	[1]	
α_s	Flow parameter for slot impinging jet	S	[1]	
α_1, α_2, α_3	Parameters describing ellipsoid resistance	S	[1]	
β	Parameter characterizing particle inertia effect	S	s^{-1}	
β_c	Parameter characterizing flow near cylinder	S	[1]	
δ_a	Thickness of adsorption layer	S	m	
δ_b	Energy barrier distance	S	m	
δ_h	Hydrodynamic boundary layer thickness	S	m	
δ_D	Dirac delta function	S	[1]	
δ_d	Diffusion boundary layer thickness	S	m	
δ_m	Primary minimum distance	S	m	
$\bar{\delta}_m$	Dimensionless primary minimum distance	S	[1]	
$\Gamma(1/2)$	Gamma Euler function	S	[1]	1.772454
$\Gamma(4/3)$	Gamma Euler function	S	[1]	0.8930
γ	Interfacial tension	S	$N\ m^{-1}$	
γ_a	Dimensionless parameter	S	[1]	
γ_b, γ_m	Parameters of adsorption	S	$J\ m^{-2}$	
ε	Permittivity of medium	S	$F\ m^{-1}$	
ε_0	Permittivity of vacuum	S	$F\ m^{-1}$	8.85419×10^{-12}
$\varepsilon_1, \varepsilon_2$	Permittivities of particles	S	m^{-1}	
ζ_1	Zeta potential of interface	S	V	
ζ_2	Zeta potential of particle	S	V	
η	Dynamic viscosity of fluid	S	$kg\ m^{-1}\ s^{-1}$	
$\bar{\eta}_1, \bar{\eta}_2$	Fluid viscosity ratios	S	[1]	
η_f	Dynamic viscosity of particle	S	$kg\ m^{-1}\ s^{-1}$	
$\tilde{\mathbf{\Theta}}_e$	Laplace transform of surface coverage vector	**V**	[1]	

List of Symbols (continued)

Symbol	Definition	Character	Unit	Numerical value
$\boldsymbol{\Theta}_e$	Equilibrium surface coverage vector	**V**	[1]	
Θ	Surface coverage of particles	S	[1]	
$\overline{\Theta}$	Reduced coverage of particles	S	[1]	
Θ_i	Initial coverage of particles	S	[1]	
Θ_∞	Final coverage of particles	S	[1]	
$\kappa = Le^{-1}$	Reciprocal double-layer thickness	S	m^{-1}	
λ_0	Dimensionless parameter for prolate spheroid	S	[1]	
λ_r	Dimensionless retardation parameter	S	[1]	
λ_c	Coupling coefficient	S	[1]	
μ	Chemical potential of particles	S	J	
μ_i	Electrochemical potential of ion	S	J	
ν	Kinematic viscosity of fluids	S	$m^2\ s^{-1}$	
ξ	Integration variable, dimensionless angular variable	S	[1]	
(ξ_1, ξ_2, ξ_3)	Curvilinear coordinate system	S		
$(\xi_{1p}, \xi_{2p}, \xi_{3p})$	Local curvilinear collector coordinates	S		
Π, Δ**Π**	Osmotic pressure tensors	**T**	Nm^{-2}	
ρ	Fluid density	S	$kg\ m^{-3}$	
ρ_e	Electric charge density	S	$C\ m^{-3}$	
ρ_p, ρ'	Particle density	S	$kg\ m^{-3}$	
$\Delta\rho$	Apparent density of particle	S	$kg\ m^{-3}$	
$\sigma, \sigma(r_s)$	Surface charge	S	$C\ m^{-2}$	
τ	Dimensionless time	S	[1]	
τ_r	Dimensionless relaxation time	S	[1]	
$\boldsymbol{\Phi}_v$	Volume fraction of particles	S	[1]	
ϕ	Net interaction energy of particle	S	J	
ϕ_b	Energy barrier height	S	J	
$(\phi_e, \psi_e, \vartheta_e)$	Eulerian angles	S		
ϕ_o	Reference energy of interaction	S	J	
ϕ_m	Depth of the energy minimum	S	J	
	Angular position of particle	**V**	[1]	

List of Symbols (continued)

Symbol	Definition	Character	Unit	Numerical value
φ_0	Initial angular position of particle	**V**	[1]	
φ_B	Angular displacement of particle due to Brownian motion	*S*	[1]	
ϑ	Angular coordinate	*S*	[1]	
χ_o	Parameter describing ellipsoid resistance	*S*	m^2	
ψ	Electric potential	*S*	V	
$\overline{\psi}$	Dimensionless potential	*S*	[1]	
ψ_i	Shape parameter	*S*	[1]	
$\psi^0, \psi_1^0, \psi_2^0$	Surface potentials	*S*	V	
$\mathbf{\Omega}$	Angular velocity of particle	**V**	s^{-1}	
Ω	Angular velocity of rotating disk	*S*	s^{-1}	
Ω_B	Velocity of rotational Brownian motion	*S*	s^{-1}	
Ω_C	Configurational space domain	*S*	[1]	
ω	Angular velocity	*S*	s^{-1}	

Note: *S* = scalar; **V** = vector; **T** = tensor (matrix); [1] = dimensionless.

REFERENCES

[1] J.Happel and H.Brenner, Low Reynolds Number Hydrodynamics, Martinus Nijhoff, The Hague, 1986.
[2] T.G.M.van de Ven, Colloidal Hydrodynamics, Academic Press, New York 1989.
[3] W.B.Russel, D.A.Saville and W.R.Schowalter, Colloidal Suspensions, Cambridge University Press, Cambridge, 1993.
[4] G.K.Batchelor, J.Fluid Mech., 52 (1972) 245.
[5] A.J.Goldman, R.G.Cox and H. Brenner, Chem. Eng. Sci., 22 (1967) 653.
[6] P. Warszyński, Adv. Colloid Interface Sci., 84 (2000) 47.
[7] G.M.Hidy and J.R.Brock, The Dynamics of Aerocolloidal Systems Vol. 1, Pergamon Press, Oxford, 1970.
[8] L.A.Spielman and J.A.Fitzpatrick, J.Colloid Interface Sci., 42 (1973) 607.
[9] L.A.Spielman and P.M.Cukor, J.Colloid Interface Sci., 43 (1973) 51.
[10] M.E.Weber and D.Paddock, J.Colloid Interface Sci., 94 (1983) 328.
[11] Z. Adamczyk, B. Siwek, M. Zembala and P. Belouschek. Adv. Colloid Interface Sci., 48 (1994) 151.
[12] M.Elimelech, J.Gregory, X.Jia and R.A.Williams, Particle Deposition and Aggregation, Buttereworth–Heinemann, Oxford, 1996.

[13] A.C.Payatakes, Chi-Tien and R.M. Turian, AIChEJ, 20 (1974) 890.
[14] C.Pozrikidis, Introduction to Theoretical and Computational Fluid Dynamics, Oxford University Press, New York, Oxford, 1997.
[15] S.Kim and S.J.Karilla, Microhydrodynamics, Principles and Selected Applications, Butterworth-Heinemann, Oxford, 1991.
[16] T.Dąbroś, Coll. Surf., 39 (1989) 127.
[17] T.Dąbroś, and T.G.M.van de Ven, J.Colloid Interface Sci., 149 (1992) 493.
[18] T.Dąbroś, and T.G.M.van de Ven, Int. J.Multiphase Flow, 18 (1992) 751.
[19] K.Małysa, T.Dąbroś, and T.G.M.van de Ven, J.Fluid Mech., 162 (1986) 157.
[20] M.L.Ekiel-Jeżewska, F.Feuillbois, N.Lecoq, M.Masmoudi, R.Anthore, F.Bostel and W.Wainryb, Contact Friction in Hydrodynamic Interactions between Two Spheres, Materials: Third Conference on Multiphase Flow pp. 1–8, ICMF 98, Lyon, France, 1998,
[21] T.G.M van de Ven, P.Warszyński, X.Wu and T.Dąbroś, Langmuir, 10 (1994) 3046
[22] Z.Adamczyk, P.Warszyński, L.Szyk-Warszyńska and P.Weroński, Coll. Surf. A, 165 (2000) 157.
[23] Z.Adamczyk, B.Siwek and P. Warszyński, J. Colloid Interface Sci., 248 (2002) 244
[24] A.Einstein, Ann. Phys., 17 (1905) 549.
[25] A.Einstein, Ann. Phys., 19 (1906) 371.
[26] M.Smoluchowski, Ann. Phys., 21 (1906) 756.
[27] J.Perrin, Kolloid. Chem. Beihefte, 1 (1910) 221.
[28] J.Perrin, Ann. Chim. Phys., 18 (1909) 5–114; idem., Brownian Motion and Molecular Reality, Taylor and Francis, London, 1910.
[29] H.Brenner and L.J. Gayados, J. Colloid Interface Sci., 58 (1977) 372.
[30] H.Brenner, J.Colloid Interface Sci., 20 (1965) 104.
[31] H.Brenner, J.Colloid Interface Sci., 23 (1967) 407.
[32] D.W.Condiff and J.S.Dahler, J. Chem. Phys., 44 (1966) 3988.
[33] S.Chandrasekhar, Rev. Mod. Phys., 15 (1943) 1
[34] M.Smoluchowski, Bull. Int. Acad. Sci. Cracovie, Ser. A, (1913) 428.
[35] H.S.Carslow and J.C.Jeager, Conduction of Heat in Solids, Oxford, 1959.
[36] M. Smoluchowski, Ann. Phys., 48 (1915) 1103.
[37] D.Henderson and S.G.Davison, Equilibrium Theory of Liquids and Liquid Mixtures, in "Physical Chemistry, Advanced Treatise" (H. Eyring ed.), Vol. 2, p.339, Academic Press, New York, 1967.
[38] F.H.Ree and W.G.Hoover, J. Chem. Phys., 40 (1964) 939.
[39] S.M.Ricci, J.Talbot, G.Tarjus and P. Viot, J. Chem. Phys., 97 (1992) 5219.
[40] M.Smoluchowski, Phys. Z., 17 (1916) 557.
[41] L.A.Spielman and S.K.Friedlander, J. Colloid Interface Sci., 46 (1974) 22.
[42] B.D.Bowen, S.Levine and N.Epstein, J. Colloid Interface Sci., 54 (1976) 375.
[43] E.Ruckenstein, J. Colloid Interface Sci., 66 (1978) 531.
[44] D.C.Prieve and M.M.J.Lin, J. Colloid Interface Sci., 76 (1980) 32.
[45] Z.Adamczyk and P.Weroński, Adv. Colloid Interface Sci., 83 (1999) 137.
[46] Z.Adamczyk and J. Petlicki: J. Colloid Interface Sci., 118 (1987) 20.
[47] Z.Adamczyk, J. Colloid Interface Sci., 133 (1989) 23.
[48] V.G.Levich, Physicochemical Hydrodynamics, Prentice-Hall, Englewood Cliffs, NJ, 1962.

[49] Z.Adamczyk, Kinetics of Particle and Protein Adsorption. "Surface and Colloid Science" (E. Matijevic and M. Borkovec eds.), Vol. 17, Ch. 5, pp. 300–462, Kluwer Academic, New York, 2004.
[50] D.H.Melik, J. Colloid Interface Sci., 138 (1990) 397.
[51] Z.Adamczyk, J. Colloid Interface Sci., 229 (2000) 477.
[52] Z.Adamczyk, Bull. Pol. Acad. Sci. Chem., 35 (1987) 417.
[53] Z.Adamczyk and L. Szyk, Langmuir, 16 (2000) 5730.
[54] T.Dąbroś, Z.Adamczyk and T.G.M. van de Ven, PCH Physicochem. Hydrodynamics, 5 (1984) 67.
[55] G.L.Natanson, Dokl. Akad. Nauk SSSR, 112 (1957) 696.
[56] W.J.Albery, J. Electroanal. Chem., 191 (1985) 1.
[57] M.A.Leveque, Ann. Mines, 13 (1928) 201.
[58] Z.Adamczyk and T.G.M. van de Ven, J. Colloid Interface Sci., 80 (1981) 340.
[59] Z.Adamczyk and T.G.M. van de Ven, J. Colloid Interface Sci., 84 (1981) 497.
[60] Z.Adamczyk and T.G.M. van de Ven, Chem. Eng. Sci., 37 (1982) 869.
[61] Z.Adamczyk, T. Dąbroś and T.G.M. van de Ven, Chem. Eng. Sci., 37 (1982) 1513.
[62] T.Dąbroś Z. Adamczyk and J. Czarnecki, J. Colloid Interface Sci., 62 (1977) 529.
[63] Z.Adamczyk, J. Colloid Interface Sci., 78 (1980) 559.
[64] Z.Adamczyk, L. Szyk and P. Warszyński, Coll. Surf. A, 75 (1993) 185.
[65] Z.Adamczyk, B. Siwek, P. Warszyński and E. Musiał, J. Colloid Interface Sci., 242 (2001) 14.
[66] D.C.Prieve and E. Ruckenstein, AIChEJ, 20 (1974) 117.
[67] Z.Adamczyk, J. Colloid Interface Sci., 79 (1981) 381.
[68] Z.Adamczyk and T.G.M. van de Ven, J. Colloid Interface Sci., 97 (1984) 68.
[69] Z.Adamczyk, T.Dąbroś, J.Czarnecki and T.G.M. van de Ven, J. Colloid Interface Sci., 97 (1984) 91.
[70] K.-M.Yao, M.T.Habibian and C.R. O'Melia, Environ. Sci. Technol., 5 (1971) 1105.
[71] D.C.Prieve and M.M.J. Lin, J. Colloid Interface Sci., 76 (1980) 32.
[72] Z.Adamczyk, Colloids Surf., 39 (1989) 1.
[73] Z.Adamczyk, B.Siwek, M.Zembala and P.Warszyńki, J. Colloid Interface Sci., 130 (1989) 578.
[74] M.Elimelech, J.Colloid Interface Sci., 164 (1994) 190.
[75] R.Rajagopalan and J.S.Kim, J. Colloid Interface Sci., 83 (1981) 428.

Chapter 5

Non–linear Transport of Particles

5.1. INTRODUCTION

The linear transport conditions discussed previously are relatively short lasting, especially for small colloid particles and protein suspensions at higher volume fractions. As shown in Table 4.7, the characteristic time of completing a monolayer under the diffusion controlled transport for a 10 nm radius particle is only 2.5 s (for $\Phi_v = 10^{-3}$). Under convection-controlled transport this relaxation time becomes even smaller.

With the progress of particle adsorption, there appear deviations from linearity, with respect to time, caused mainly by

(i) particle desorption phenomena; and
(ii) volume exclusion effects induced by preadsorbed particles.

The former effect is important for reversible systems when the energy minimum depth is not too high. It can be accounted for to some extent by the kinetic boundary conditions discussed above. In contrast, a proper description of the volume exclusion effect, called often the surface blocking effect, is more complicated. The direct cause of the volume exclusion effects is the dynamic interactions, both hydrodynamic and specific, between adsorbed particles and those moving in their vicinity. These interactions depend not only on particle coverage but also on their distribution over interfaces, which is in turn dependent on mechanisms of particle transfer like diffusion, flow, migration, etc. As a result of this non-linear coupling, a rigorous analysis of surface exclusion phenomena becomes prohibitive. Therefore no theory of general validity has emerged yet. There are, however, several approximate models, which can be successfully used for predicting basic features of particle adsorption under non-linear regimes.

For a lower coverage range one can use the statistical mechanical approach, in particular the scaled particle theory (SPT) [1–5], which allows one to evaluate the chemical potential of particles, the structure of the adsorbed layers, fluctuations, adsorption isotherms, etc. This can be done for two-dimensional adsorption of convex particles, e.g., monodisperse and polydisperse disks, ellipses, squares, rectangles, etc. Also the available surface function (ASF) can be determined, which can be connected with adsorption kinetics. However, theories based on the statistical–mechanical approach fail in the case of polydisperse spheres, spheroids adsorbing under various orientations, for interacting particles and for adsorption on centers of finite dimension. In all these cases particle adsorption is a truly three-dimensional (3D) process. In particular, the jamming limit of particles cannot be predicted, a quantity of primary interest from an experimental viewpoint.

The jamming coverage, the structure of adsorbed layers and to some extent adsorption kinetics can be evaluated by using other approaches, most often various brands of the random sequential adsorption (RSA) model [6–13]. Although they do not provide one with analytical results for two dimensions, the RSA configurations of particles can efficiently be generated by using Monte Carlo type simulation schemes. Results are available for monodisperse and polydisperse spherical particles, interacting spherical particles, prolate and oblate spheroids cylinders, spherocylinders, etc. By modifying the basic algorithm one can simulate adsorption on a heterogeneous surface, e.g., pre-covered by adsorption centers of various shape and concentrations. Also, layer by layer adsorption processes leading to multilayer coverage can be modeled. Except for the jamming coverage, the RSA simulations enable one to determine the ASF (the surface blocking function) at any coverage of particles. This function can then be used as a non-linear kinetic boundary condition for the bulk transport equation.

Accordingly, the organization of this chapter is as follows: in the first section the 2D equilibrium systems of particles will be considered. Using the virial expansion and the SPT, the chemical potential of particles, the available surface function, adsorption isotherms, density fluctuations, and structure of the adsorbed layers will be analyzed for 2D convex particles: monodisperse and polydisperse disks, ellipses, squares, rectangles, etc. Analytical expression, useful for testing the accuracy of numerical approaches, will be derived.

In the next section, various RSA models are considered together with methods of simulations of the particle configurations. Monolayers of various density of monodisperse and polydisperse spherical particles, prolate and oblate spheroids interacting spherical cylinders, spherocylinders, etc.

are discussed. The jamming coverage results are widely analyzed and interpolating functions for calculating them are given. The important case of particle adsorption on heterogeneous surface, e.g., pre-covered by adsorption centers of various shape and concentration is discussed in detail. Except for the jamming coverage, the RSA simulations enable one to determine the blocking function at any coverage of particles. Limiting analytical and interpolating expressions are derived, which can be exploited as non-linear kinetic boundary conditions for the bulk transport equation.

The last section is devoted to the problem of adsorption kinetics under the non-linear transport conditions, in particular the non-stationary, diffusion-controlled transport to planar interfaces, quasi-stationary transport under convective diffusion conditions and transport to heterogeneous surfaces. Implications for protein adsorption are pointed out.

5.2. REVERSIBLE, TWO-DIMENSIONAL PARTICLE SYSTEMS

Because of the lack of an exact theory of irreversible adsorption of particles, it is useful to consider first reversible systems under thermodynamic equilibrium, where one can derive analytical expressions of practical interest. The equilibrium systems are uniquely defined in terms of the absolute temperature T and the 2D particle density (coverage), defined as $\Theta = NS_g$, where N is the surface concentration of particles, S_g is the characteristic cross-section of adsorbing particles. The advantage of this definition of coverage is that it is absolute, i.e., no empirical parameters are needed, in contrast to the often used empirical definition $\Theta = N/N_{\mathrm{mx}}$, where N_{mx} is the maximum concentration, derived *a posteriori* from experimental measurements.

Let us consider a one-component system of uncharged particles forming a 2D adsorption phase. The particles are assumed to have arbitrary shape but they remain oriented parallel to the surface (lying side-on). In the case of spheres this condition is fulfilled automatically, but for spheroids, cylinders and other axis-symmetric particles the side-on orientation implies that there are specific forces between the particle and interface that are strong enough to maintain the flat orientation.

The quantity of basic importance for characterizing the state of the particle monolayer is the chemical potential given by

$$\mu = \mu^0 + kT \ln f(\Theta)\Theta \tag{1}$$

where μ_0 is the reference potential, and $f(\Theta)$ the activity coefficient assumed to depend on particle coverage alone.

The chemical potential can be calculated from the Gibbs–Duhem thermodynamic relationship used before for the 3D case

$$\mu(\Theta) = S_g \int \frac{d\pi_s(\Theta)}{\Theta} \tag{2}$$

where $(\pi_s(\Theta)$ is the 2D pressure of the adsorbed particle layer.

By comparing Eqs. (1,2) one obtains the expression for the activity coefficient in the form

$$f(\Theta) = e^{\mu_r/kT} = e^{\left(S_g \int \frac{d\Pi_s(\Theta)}{\Theta} - \mu^0 - kT\ln\Theta)/kT\right)} \tag{3}$$

where $\mu_r = \mu - \mu^0 - kT\ln\Theta$ is the excess (residual) chemical potential characterizing the deviation from the ideal system behavior.

As demonstrated by Widom [14,15], the available surface function, which is of primary importance for predicting deposition kinetics, is connected with μ_r through the relationship

$$ASF(\Theta) = B(\Theta) = e^{-\mu_r/kT} = \frac{1}{f(\Theta)} \tag{4}$$

where $B_0(\Theta)$ is referred to, less precisely, as the surface blocking function.

Thc ASF can be interpreted as an averaged Boltzmann factor for a particle wandering within the 2D adsorbing particle system frozen in a given configuration. This can be expressed as

$$ASF(\Theta) = \left\langle e^{-\phi_p/kT} \right\rangle \tag{5}$$

where ϕ_p is the interaction energy of the adsorbing (wandering) particle with all other particles and $\langle\ \rangle$ means the value averaged over the system large enough to neglect the boundary effects.

In the case of noninteracting (hard) particles, the ASF has a simple geometrical interpretation as the area available to the wandering particle

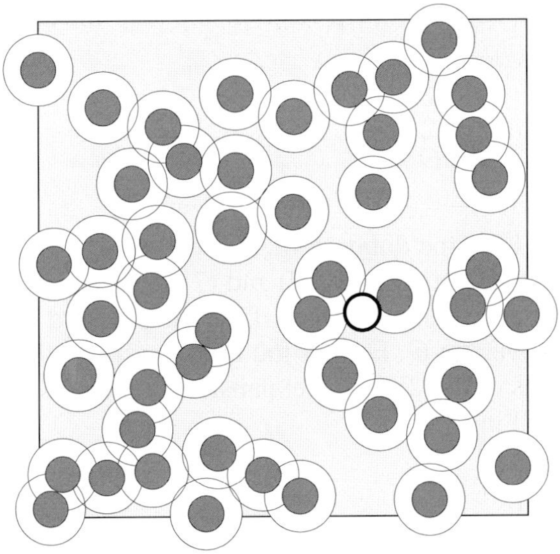

Fig. 5.1. A typical configuration of particles adsorbed at an interface (black disks); the white disks show the exclusion areas, whereas the shadowed zones represent the areas (targets) available for the center of the wandering particle depicted by the white disk.

(see Fig. 5.1). The topology of this available surface area can be obtained by drawing a circle of radius $2a$ around each adsorbed particle. Then, the ratio of the area lying outside these circles to the overall area is just the ASF, which depends not only on particle coverage Θ (the fraction of the surface area occupied by particles) but also on particle distribution over the surface.

Another quantity of major significance for characterizing particle monolayers is the variance of the coverage fluctuation, which is given in the grand canonical ensemble by the known thermodynamic relationship [16]

$$\frac{\sigma_v^2}{\langle N_p \rangle} = \bar{\sigma}_v^2 = \frac{kT}{\Theta}\left(\frac{\partial \mu}{\partial \Theta}\right)_{S,T}^{-1} \tag{6}$$

where $\langle N_p \rangle$ is the averaged number of particles over a fixed surface area ΔS.

By using the Gibbs–Duhem relationship, Eq. (6) can be converted to the useful form

$$\frac{1}{\bar{\sigma}_v^2} = \frac{1}{kT}\left(\frac{\partial \pi_s}{\partial N}\right)_{S,T} = \left(\frac{\partial \bar{\pi}}{\partial \Theta}\right)_{S,T} \tag{7}$$

where $\bar{\pi} = \pi_s S_g / kT$ is the dimensionless pressure.

As can be deduced from Eqs. (2) and (7) the chemical potential of particles and the variance can be derived if the 2D pressure is know as a function of particle coverage Θ. For not too high coverage, the pressure can be expressed in terms of the 2D virial expansion, analogous to that previously used, cf. Eq. (233) in Chapter 4

$$\frac{\bar{\pi}}{\Theta} = 1 + \sum_{l \geq 2} \bar{B}_l \Theta^{l-1} \tag{8}$$

where $\bar{B}_l$ is the 2D virial expansion coefficients to be derived from the configurational integrals [16–18]

$$\bar{B}_l = \frac{1}{S_g^l}\frac{1-l}{l!}\frac{1}{S}\int \cdots \int F_l d\mathbf{r}_1 \cdots d\mathbf{r}_l \tag{9}$$

where $F_l = \sum \prod_{i>j}^{l} f_{ij}, f_{ij} = e^{-\phi_{ij}/kT} - 1$ are Mayer functions, $\mathbf{r}_l$ are the position vectors (2D) of each particle in the cluster and ϕ_{ij} the interaction energies of the i and j particles in the cluster, derived by assuming additivity.

The summation in Eq. (9) extends over all physically accessible (irreducible) cluster integrals [16–18].

The second virial coefficient, which is a quantity of major significance for predicting the first order correction to linear adsorption regimes, involves a cluster composed of two particles only. In this case Eq. (9) for spherical particles reduces to the simple form

$$\bar{B}_2 = \frac{1}{2}\frac{1}{S_g}\frac{1}{S}\iint \left(e^{-\phi_{12}/kT} - 1\right) d\mathbf{r}_1\, d\mathbf{r}_2 \tag{10}$$

where ϕ_{12} is the energy of interaction of two particles.

By substituting $r = |\mathbf{r}_1 - \mathbf{r}_2|$, $d\mathbf{r}_1 = 2\pi r dr$, $\int d\mathbf{r}_2 = S$, one can convert Eq. (10) into the usual form

$$\bar{B}_2 = \frac{1}{2}\int_0^2 \left(1 - e^{-\phi_{12}/kT}\right)\bar{r}\, d\bar{r} \tag{11}$$

where $\bar{r} = r/a$, a is the particle radius.

For hard (non-interacting) particles, where $\phi_{12} = 0$ for $r > a$, $\phi = \infty$ for $r \le a$, one can calculate from Eq. (11) that $\bar{B}_2 = 2$. On the other hand, for the triangular interaction potential (see Fig. 4.10), $\bar{B}_2$ becomes

$$\bar{B}_2 = 2(1+\bar{\delta}) + \frac{\bar{\delta}^2}{2} - \frac{2\bar{\delta}}{\bar{\phi}_b}\left(1 - e^{-\bar{\phi}_b}\right) - \frac{\bar{\delta}^2}{\bar{\phi}_b} + \frac{\bar{\delta}^2}{\bar{\phi}_b^2}\left(1 - e^{-\bar{\phi}_b}\right) \tag{12}$$

where $\bar{\phi}_b$ is the height of the energy barrier $\bar{\phi}_b = \phi_b/kT$, δ the thickness of the barrier and $\bar{\delta} = \delta/a$.

If $\bar{\phi}_b > 10$, Eq. (12) simplifies to

$$\bar{B}_2 = 2 + 2\bar{\delta} + \frac{\bar{\delta}^2}{2} = 2\left(1 + \frac{1}{2}\bar{\delta}\right)^2 \tag{13}$$

The triangular barrier reflects the situation arising due to superposition of van der Waals attraction and double-layer repulsion.

However, for more complicated potentials, an analytical evaluation of $\bar{B}_2$ is not possible.

The third and fourth virial coefficient were evaluated analytically for hard particles [17,18]

$$\bar{B}_3 = 4\left(\frac{4}{3} - \frac{3^{1/2}}{\pi}\right) = 3.128 \tag{14}$$

$$\bar{B}_4 = 8\left(2 - \frac{9}{2}\frac{3^{1/2}}{\pi} + \frac{10}{\pi^2}\right) = 4.258$$

Because of the occurrence of multiple integrals, the fifth and higher order virial coefficients can only be calculated numerically, using Monte Carlo methods. They are [17,18]

$$\begin{aligned} \bar{B}_5 &= 5.341 \\ \bar{B}_6 &= 6.374 \\ \bar{B}_7 &= 7.302 \end{aligned} \tag{15}$$

Bublik [5] showed that in the case of convex particles the second virial coefficient is given by the analytical formula

$$\bar{B}_2 = 1 + \frac{P^2}{4\pi S_g} = 1 + \gamma_p \tag{16}$$

where $\gamma_p = \dfrac{P^2}{4\pi S_g}$ is the particle shape parameter and P the perimeter of the particle. For spheres $\gamma_p = 1$.

The $\bar{B}_3$ and $\bar{B}_4$ coefficients were calculated numerically for spheroids (ellipses), cylinders (rectangles) and spherocylinders (diskorectangles) [10,19].

Using the expansion, Eq. (8) one obtains from Eq. (2) the expression for the chemical potential

$$\mu = \mu^0 + kT \ln\Theta + kT \sum_{l \geq 2} \frac{1}{l-1} \Theta^{l-1} \tag{17}$$

Accordingly, the ASF can be expressed using Eq. (4) as

$$ASF(\Theta) = e^{-\mu_r/kT} = e^{-\sum\limits_{l \geq 2} \frac{l}{l-1} \Theta^{l-1}} \cong 1 - C_1\Theta + C_2\Theta^2 + C_3\Theta^3 + 0(\Theta)^4 \tag{18}$$

where the first coefficients of the expansion are

$$\begin{aligned} C_1 &= 2\bar{B}_2 = 4 \\ C_2 &= \left(2\bar{B}_2^2 - \frac{3}{2}\bar{B}_3\right) = \frac{6(3)^{1/2}}{\pi} = 3.308 \\ C_3 &= 3\bar{B}_2\bar{B}_3 - \frac{4}{3}\left(\bar{B}_2^2 + \bar{B}_4\right) = 2.424 \end{aligned} \tag{19}$$

On the other hand, the reduced variance of the coverage fluctuation can be calculated from Eq. (7)

$$\bar{\sigma}_v^2 = \frac{1}{1+\sum_{l\geq 2} l\bar{B}_l \Theta^{l-1}} \tag{20}$$

Eq. (20) can be expanded to

$$\begin{aligned}\bar{\sigma}_v^2 &= 1-2\bar{B}_2\Theta+(4\bar{B}_2^2-3\bar{B}_3)\Theta^2 \\ &\quad +\left[12\bar{B}_2\bar{B}_3-4\left(2\bar{B}_2^3+\bar{B}_4\right)\right]\Theta^3+0(\Theta^4) \\ &= 1-4\Theta+6.616\Theta^2-5.960\Theta^3+0(\Theta^4)\end{aligned} \tag{21}$$

Although the virial expansion for the pressure, Eq. (8), gives fairly exact results for coverages as high as 0.5 [18] it is rather awkward to use. More convenient approximations can be derived from the SPT [1–5], which is based on the constitutive thermodynamic relationship

$$\frac{\bar{\pi}}{\Theta} = 1+2\Theta g(2a,\Theta) \tag{22}$$

where $g\,(2a,\Theta)$ is the contact radial distribution function [3,4] directly related to the reversible work $W(R_c)$ of creating a spherical cavity in the fluid of radius R_c,

$$g(2a,\Theta) = \frac{a^2}{2\Theta R_c kT}\frac{\partial W}{\partial R_c} \tag{23}$$

Then the work $W\,(R_c)$ is expanded in polynomial series of R_c up to the order of three with the leading term representing the volumetric work against the pressure. The remaining coefficients of the expansion are calculated by matching the work with the analytical expressions and its derivative, known for $R_c<a$. This analysis leads to the following expression for the pressure

$$\frac{\bar{\pi}}{\Theta} = \frac{1}{(1-\Theta)^2} \tag{24}$$

Boublik [5] used the SPT approach to analyze the behavior of convex particles (in the side-on orientation as mentioned above). The expression for the pressure derived by him has the form

$$\frac{\bar{\pi}}{\Theta} = \frac{1+(\gamma_p - 1)\Theta}{(1-\Theta)^2} \tag{25}$$

The virial expansion of Eq. (25) is

$$\frac{\bar{\pi}}{\Theta} = 1 + \sum_{l \geq 1} (1 + l\gamma_p)\Theta^l \tag{26}$$

hence, $\bar{B}_l = 1 + l\gamma_p$.

In the case of spheres

$$\frac{\bar{\pi}}{\Theta} = 1 + \sum_{l \geq 1} (1 + l)\Theta^l \tag{27}$$

hence, $\bar{B}_l = 1 + l$. By comparing this result with the exact values given by Eqs. (14,15) one can notice that the SPT approximation is a reasonable one for a broad range of coverages.

Using Eq. (25) one obtains the following expression for the chemical potential of convex particles

$$\frac{\mu}{kT} = \frac{\mu^0}{kT} + \ln\frac{\Theta}{1-\Theta} + (1+2\gamma_p)\frac{\Theta}{1-\Theta} + \gamma_p\left(\frac{\Theta}{1-\Theta}\right)^2 \tag{28}$$

For spheres Eq. (28) becomes

$$\frac{\mu}{kT} = \frac{\mu^0}{kT} + \ln\frac{\Theta}{1-\Theta} + \frac{3\Theta}{1-\Theta} + \left(\frac{\Theta}{1-\Theta}\right)^2 \tag{29}$$

This SPT has been extended to convex particle mixtures [4,20]. The general expression for the chemical potential of the ith component of a convex particle mixture, having the area S_i, is given by

$$\frac{\mu}{kT}=\frac{\mu_i^0}{kT}+\ln\frac{\Theta_i}{1-\Theta}+\frac{\sum_j(\lambda_{ij}^2+\lambda_{ij}^{'})\Theta_j}{1-\Theta}+\frac{\left(\sum_j\lambda_{ij}^{''}\Theta_j\right)^2}{(1-\Theta)^2} \tag{30}$$

where $\Theta=\sum_j \Theta_j$ and

$$\lambda_{ij}^2=\frac{S_i}{S_j};\quad \lambda_{ij}^{'}=\frac{P_{r_i}P_{r_j}}{2\pi S_j};\quad \lambda_{ij}^{''}=\left(\frac{S_i}{4\pi}\right)^{1/2}\frac{P_{r_j}}{S_j} \tag{31}$$

where P_{ri}, P_{rj} are the perimeters of the particles.

For polydisperse disks, Eq. (30) reduces to the form derived by Lebowitz et al. [4]

$$\frac{\mu_i}{kT}=\frac{\mu_i^0}{kT}+\ln\frac{\Theta_i}{1-\Theta}+\frac{\sum_j(\lambda_{ij}^2+2\lambda_{ij})\Theta_j}{1-\Theta}+\frac{\left(\sum_j\lambda_{ij}\Theta_j\right)^2}{(1-\Theta)^2} \tag{32}$$

where $\lambda_{ij}=a_i/a_j$ are the disk size ratios.

For a bimodal mixture, Eq. (32) simplifies to

$$\frac{\mu_1}{kT}=\frac{\mu_1^0}{kT}+\ln\frac{\Theta_1}{1-\Theta}+\frac{3\Theta_1+\lambda(\lambda+2)\Theta_2}{1-\Theta}+\frac{(\Theta_1+\lambda\Theta_2)^2}{(1-\Theta)^2} \tag{33}$$

$$\frac{\mu_2}{kT}=\frac{\mu_2^0}{kT}+\ln\frac{\Theta_2}{1-\Theta}+\frac{3\Theta_2+\frac{1}{\lambda}\left(\frac{1}{\lambda}+2\right)\Theta_1}{1-\Theta}+\frac{\left(\Theta_2+\frac{1}{\lambda}\Theta_1\right)^2}{(1-\Theta)^2}$$

where $\Theta=\Theta_1+\Theta_2$ and $\lambda=a_1/a_2$ is the disk size ratio.

Knowing the chemical potential, the ASF can be calculated from Eq. (18). For a monodisperse convex particle system one has

$$ASF = B(\Theta) = (1-\Theta)e^{-(1+2\gamma_p)\frac{\Theta}{1-\Theta}-\gamma_p\left(\frac{\Theta}{1-\Theta}\right)^2} \tag{34}$$

For spheres one has

$$ASF = B(\Theta) = (1-\Theta)e^{-\frac{3\Theta}{1-\Theta}-\left(\frac{\Theta}{1-\Theta}\right)^2} \tag{35}$$

It can be deduced from Eq. (34) that in the limit of low coverage where $\Theta << 1$, the ASF is given by

$$ASF = 1 - C_1\Theta + \left(2\gamma_p^2 + \gamma_p + \frac{1}{2}\right)\Theta^2 + \left(4\gamma_p^2 + \gamma_p + \frac{3}{2}\right)\Theta^3 + 0(\Theta^4) \tag{36}$$

where $C_1 = 2\overline{B}_2 = 2(\gamma_p + 1)$.

The C_1 constant for circles equals 4. On the other hand, for ellipses with axis ratio 2:1, $C_1 = 4.38$, for axis ratio 5:1, $C_1 = 6.47$ [10]. Very similar values are obtained for rectangles (cylinders) and diskorectangles (spherocylinders). The C_1S_g coefficient has the geometrical interpretation as the area blocked by one adsorbed particle. Obviously, for spheres this area equals to $4\pi a^2$.

For a bimodal mixture of disks having radii a_1 and a_2, respectively, the ASFs are given by the expressions

$$ASF_1 = (1-\Theta)e^{-\frac{3\Theta_1+\lambda(\lambda+2)\Theta_2}{1-\Theta}-\frac{(\Theta_1+\lambda\Theta_2)^2}{(1-\Theta)^2}}$$

$$ASF_2 = (1-\Theta)e^{-\frac{3\Theta_2+1/\lambda(1/\lambda+2)\Theta_1}{1-\Theta}-\frac{(\Theta_2+1/\lambda\Theta_1)^2}{(1-\Theta)^2}} \tag{37}$$

Using the expression for the pressure, Eq. (25) one can express the reduced variance for monodisperse convex particles in the form

$$\overline{\sigma}_v^2 = \frac{(1-\Theta)^3}{1+(2\gamma_p - 1)\Theta} \tag{38}$$

The low coverage expansion of Eq. (38) is

$$\bar{\sigma}_v^2 = 1 - C_1\Theta + (8\gamma_p - 1)\Theta^2 + (6 - 14\gamma_p)\Theta^3 + 0(\Theta^4) \quad (39)$$

As discussed in Refs. [21,22] Boublik's expression for the pressure, Eq. (25), becomes rather inaccurate for elongated particles when $\gamma_p >> 1$. In this case an alternative expression was proposed [21]

$$\frac{\bar{\pi}}{\Theta} = \frac{1 + (\bar{B}_2 - 2)\Theta + (1 - \bar{B}_2\gamma_1)\Theta^2 + B_2\gamma_2\Theta^3}{(1-\Theta)^2} \quad (40)$$

where

$$\gamma_1 = 2 - \frac{\bar{B}_3}{\bar{B}_2}$$
$$\gamma_2 = 1 - 2\frac{\bar{B}_3}{\bar{B}_2} + \frac{\bar{B}_4}{\bar{B}_2} \quad (41)$$

Using Eq. (40), the reduced variance can by calculated from Eq. (7)

$$\bar{\sigma}_v^2 = \frac{(1-\Theta)^3}{1 + (2\gamma_p - 1)\Theta + c_2\Theta^2 + c_3\Theta^3 + c_4\Theta^4} \quad (42)$$

where

$$c_2 = 3\left[1 - 2(1+\gamma_p) + \bar{B}_3\right]$$
$$c_3 = -1 + 6(1+\gamma_p) - 9\bar{B}_3 + 4\bar{B}_4 \quad (43)$$
$$c_4 = -2(1+\gamma_p) + 4\bar{B}_3 - 2\bar{B}_4$$

The disadvantage of the SPT, in particular Boublik's expression for the pressure, Eq. (25), as well as the improved formula, Eq. (40), is that the interactions among particles are not considered. This is so because considering attractive interaction is inconsistent with the postulate of a strictly 2D adsorption of a monolayer type.

A heuristic improvement of the theory is often sought by adopting the van der Waals model, developed originally to account for the behavior of real gases. The two-parametric equation for the pressure has the form

$$\left(\bar{\pi}+\frac{A_w}{\Theta^2}\right)\left(\frac{1}{\Theta}-B_w\right)=1 \tag{44}$$

where the first parameter A_w accounts for the attractive interactions between particles and B_w coefficient accounts for the excluded volume effect. Eq. (44) can be solved for the pressure

$$\bar{\pi}=\frac{\Theta}{1-B_w\Theta}-A_w\Theta^2 \tag{45}$$

Eq. (45) is often used to interpret the behavior of Langmuir–Blodgett films of insoluble surfactants. An expansion of Eq. (45) results in the virial series

$$\frac{\bar{\pi}}{\Theta}=1+(B_w-A_w)\Theta+\sum_{l>2}(B_w\Theta)^l \tag{46}$$

Therefore, the second virial coefficient in the van der Waals model is

$$\bar{B}_2=B_w-A_w \tag{47}$$

Note that in order to match the exact data for hard particles (when $A_w = 0$), the B_w coefficient in the van der Waals expression should be equal 2.

A combination of Eq. (45) with Eq. (2) gives the chemical potential the expression

$$\frac{\mu}{kT}=\frac{\mu^0}{kT}+\ln\frac{\Theta}{1-B_w\Theta}+\frac{\Theta}{1-B_w\Theta}-2A_w\Theta \tag{48}$$

Hence, ASF and the variance in particle coverage are given by

$$ASF = \frac{\Theta}{1 - B_w\Theta} e^{-\frac{\Theta}{1-B_w\Theta}+2A_w\Theta} \tag{49}$$

$$\bar{\sigma}_v^2 = \frac{(1 - B_w\Theta)^2}{1 - 2A_w\Theta(1 - B_w\Theta)^2} \tag{50}$$

Knowing the chemical potential one can derive the adsorption isotherms, which are of primary interest for equilibrium systems, especially for predicting adsorption of surfactants, proteins and smaller colloid particles. These isotherm equations are the extension of the Henry isotherm derived previously (cf. Eq. (274) in Chapter 4) to the non-linear adsorption conditions. They can be most directly derived by equating the chemical potential of the surface phase with the bulk chemical potential of the particle $\mu = \mu^0 + kT \ln f_b n_b$ (where f_b is the bulk activity coefficient of particles).

In this way one obtains for monodisperse convex particles the following expression

$$K_a f_b n_b S_g = \frac{\Theta}{1 - \Theta} e^{(1+2\gamma_p)\frac{\Theta}{1-\Theta}+\gamma_p\left(\frac{\Theta}{1-\Theta}\right)^2} \tag{51}$$

where K_a is the equilibrium adsorption constant, which remains an adjustable parameter, however, and cannot be calculated within the framework of the SPT .

Eq. (51) can also be expressed in the form

$$K_a^{'}(\Theta) n_b S_g = \frac{\Theta}{1 - \Theta} \tag{52}$$

where $K'_a = K_a f_b e^{-(1+2\gamma_p)\frac{\Theta}{1-\Theta}-\gamma_p\left(\frac{\Theta}{1-\Theta}\right)^2}$.

Eq. (52) resembles the widely used Langmuir isotherm with the adsorption constant decreasing with the coverage.

For the van der Waals model, the adsorption isotherm is

$$K_a f_b n_b S_g = \frac{\Theta}{1 - B_w\Theta} e^{\frac{\Theta}{1-B_w\Theta}-2A_w\Theta} \tag{53}$$

Eq. (53) can also be expressed in the form

$$K_a^{'}(\Theta) f_b n_b S_g = \frac{\Theta}{1 - B_w \Theta} \tag{54}$$

where $K_a' = K_a e^{-\frac{\Theta}{1-B_w\Theta} - +A_w\Theta}$

The equilibrium results presented in this section can be exploited as useful reference states for non-equilibrium (irreversible) systems of adsorbed particle monolayers, especially for predicting their structure. Although particle adsorption kinetics cannot be directly predicted using equilibrium theories, one often uses the ASF specified above to formulate the kinetic equation in the form [7,13,20]

$$\frac{d\Theta}{d\tau} = \tilde{k}_a ASF(\Theta) - \tilde{k}_d \Theta \tag{55}$$

where τ is the dimensionless time.

However, the adsorption $\tilde{k}_a$ and desorption $\tilde{k}_a$ rate constants appearing in Eq. (55) cannot be evaluated in terms of the equilibrium theories, remaining adjustable parameters only. Obviously, these constants are expected to depend not only on the particle wall interaction potential but also on the bulk transport rate of particles governed by the k_c value, characterized by various transport mechanisms analyzed in the previous chapter.

Also, in the case of irreversible adsorption, the above expressions for ASF become inaccurate for higher coverage. This is so because the distribution of particles differs usually from the equilibrium configurations, especially when Θ approaches the jamming limit, which is inaccessible from equilibrium theories. In order to determine the jamming limit for various particle shapes and properly reflect the kinetic aspects of particle adsorption, other approaches are required. One such efficient model is the RSA approach discussed in the next section. An important feature of this model is that it allows one to generate not only irreversible particle populations at arbitrary coverage but also to determine the ASF and the jamming coverage. The extension of the RSA to 3D situations enables one to connect the surface layer transport, governed by the $\tilde{k}_a$, $\tilde{k}_d$ rate constants with the bulk transport, governed by the k_c constant.

5.3. THE RANDOM SEQUENTIAL ADSORPTION RSA MODEL

The basic rules of the RSA simulation scheme shown schematically in Fig. 5.2 for spheroidal particles adsorbing side-on are [6–13]

(i) An adsorbing (virtual) particle is created whose position and orientation is described by the set of coordinates selected at random within prescribed limits defining the adsorption domain (position and orientation).
(ii) If the virtual particle fulfils the adsorption criteria, it is adsorbed with unit probability and its position remains unchanged during the entire process.
(iii) If the adsorption criteria are violated, a new adsorption attempt is made that is fully uncorrelated with previous attempts.

The dimensionality of the RSA process depends on the number of spacial coordinates chosen. One coordinate defines the 1D problem, i.e., adsorption of linear particles on a line, two coordinates define adsorption of various particles on a plane, and three coordinates define random addition

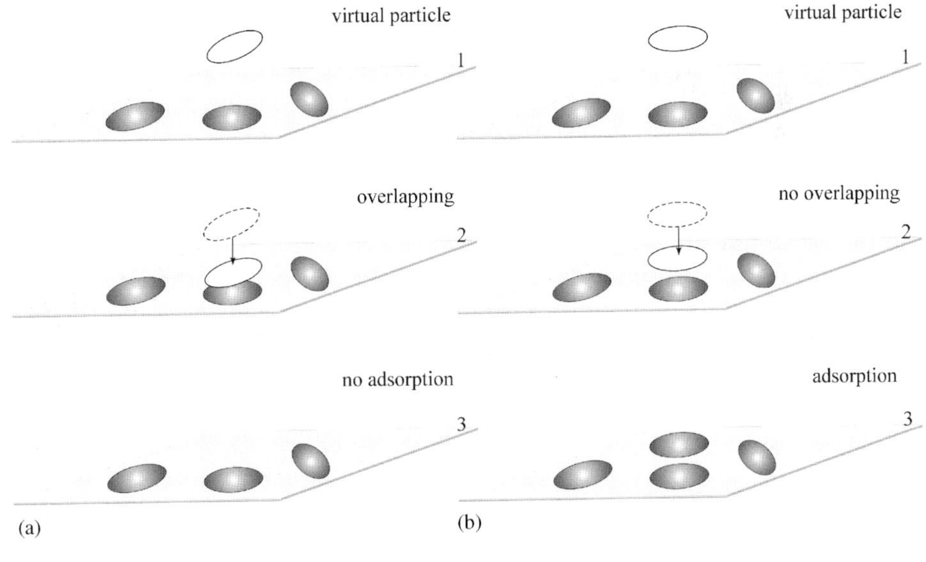

Fig. 5.2. A schematic view of the standard RSA method of simulations.

of particles to a space. In the latter cases, we have the possibility of selecting the orientation of particles if they are non-spherical.

There are usually two major adsorption criteria: (i) no overlapping with any previously adsorbed particles, and (ii) contact with adsorption centers, whose distribution can be either discrete, as is the case for lattice and heterogeneous models, or quasi-continuous where the average distance between adsorption centers and their size is much smaller than the particle dimension. Obviously, in the case of the 3D random addition, there is only the no overlapping adsorption criterion. It should be noted that efficient performance of the overlapping test is critical for developing a robust simulation scheme.

Despite the simplicity of the governing rules, the RSA is very powerful for producing particle configuration of a very large size. The density of the configuration is characterized in terms of the generalized coverage defined as

$$\Theta \sim N_p \left(\frac{a}{L} \right)^{d_{RSA}} \tag{56}$$

where N_p is the number of particles, a the characteristic particle dimension, L the characteristic dimension of the domain and d_{RSA} the dimensionality of the RSA problem.

Once a particle configuration or an ensemble of configurations at fixed Θ is created, it can be analyzed in many ways, for example its structure can be described by the pair correlation function. Also, the variance of the local coverage fluctuations can be evaluated. On the other hand, the available domain function (where the domain means line, surface or volume, depending on the dimensionality of the RSA process) can be directly derived via the Monte-Carlo type integration. This consists in performing, for a fixed coverage Θ, a large number of virtual adsorption attempts N_{att} out of which N_{suc} is potentially successful. Then, the available space function, called for sake of convenience the blocking function in two dimensions, is calculated from the expression

$$B(\Theta) = \frac{N_{suc}}{N_{att}} \tag{57}$$

for $N_{att} \to \infty$

In terms of the $B(\Theta)$ function, one can define implicitly the jamming coverage Θ_∞ of the RSA process

$$B(\Theta_\infty) = 0 \tag{58}$$

Thus, by definition, at the jamming state there is no space available for adsorption of particles of a given size. However, deriving the jamming coverage by exploiting this definition is impractical, because of the infinite simulation time needed to attain the jammed state. A more efficient procedure consists in extrapolating the apparent kinetic runs of the RSA process, i.e., the dependence of the coverage Θ on the dimensionless adsorption time τ defined as

$$\tau \sim N_{att}\left(\frac{a}{L}\right)^{d_{RSA}} \tag{59}$$

where N_{att} is the number of elementary adsorption events.

By defining τ and knowing the $B(\Theta)$ function, one can express the apparent kinetics of the RSA process in the form

$$\frac{d\Theta}{d\tau} = B(\Theta) \tag{60}$$

A formal integration of Eq. (60) gives

$$\Theta = F^{-1}(\tau) \tag{61}$$

where $F = \int_0^{\Theta} \frac{d\Theta'}{B(\Theta')}$.

Usually, for large but finite times, the dependence of Θ on τ derived from Eq. (61) is described by simple power laws, allowing one to perform efficient extrapolation to the infinite adsorption time.

Unfortunately, there are no analytical solutions describing the kinetics, blocking function or jamming coverage of RSA processes, except for the 1D case, i.e., adsorption on a line, often referred to as the "car parking problem" [12]. This case was first treated by Renyi [23], Gonzales et al. [24] and others [25]. The 1D RSA process consists in attempting to place on a line of length L, a linear object (line segment) of much smaller length d. If there is empty space large enough to accommodate the object, it is irreversibly adsorbed, if not, the adsorption attempt is repeated. The adsorption time is now $\tau = N_{att}d/L$ and the coverage $\Theta = N_p d/L$ (where N_p is the overall number of adsorbed objects). It was found that the kinetics of this process is described by the integral [25]

$$\Theta(\tau) = \int_0^\tau \left(e^{-2\int_0^{t'} \frac{1-e^{-\xi}}{\xi} d\xi} \right) dt' \tag{62}$$

For short adsorption times, when $\tau < 1$, the limiting form of Eq. (62) is

$$\Theta = \frac{1}{2}(1 - e^{-2\tau}) \tag{63}$$

From Eq. (63) one can deduce that $B(\Theta)$ in the limit of short times is given by

$$B(\Theta) = \frac{d\Theta}{d\tau} \cong 1 - 2\Theta \tag{64}$$

Hence, Eq. (64) reflects the fact that one adsorbed particle blocks, on average, a section two times larger than its length.

On the other hand, for $\tau >> 1$ one can derive from Eq. (62) the limiting expression

$$\Theta = \Theta_\infty - \frac{e^{-2\gamma_E}}{\tau} \tag{65}$$

where $\gamma_E = 0.5772$ is Euler's constant and Θ_∞ the jamming coverage in 1D equal to

$$\Theta_\infty = 0.7476 \tag{66}$$

Therefore, the blocking function in the limit of long times, when $\Theta \to \Theta_\infty$, is given by

$$B(\Theta) \cong \frac{(\Theta_\infty - \Theta)^2}{e^{-2\gamma_E}} \tag{67}$$

5.3.1. The 2D random sequential adsorption of spherical particles

In the case of 2D and 3D RSA processes, results have been derived in terms of Monte Carlo simulations, described in some detail elsewhere [6,7,13]. The process involves placing on a square plane of surface area ΔS, spherical particles (disks) of the radius a and the surface area $S_g = \pi a^2$. It is assumed that particle adsorption occurs on an uncovered surface only upon touching the interface. This is tantamount to the postulate of a continuous distribution of adsorption centers of size much smaller than particle dimensions.

The coverage is defined as $\Theta = N_p \pi a^2/\Delta S$ (where N_p is the overall number of adsorbed particles) and the adsorption time as $\tau = N_{att} \pi a^2/\Delta S$. The influence of the finite size of the adsorption domain is eliminated by superimposing the periodic boundary conditions over the perimeter of the adsorption domain. Typical configurations of particles produced in these RSA simulations are shown in Fig. 5.3. Also the available surface area is shown. The topology of available surface is obtained by drawing a circle with radius $2a$ around each adsorbed particle. Thus, the exclusion area of one particle, i.e., the place where there can be no center of another particle, equals $S_l = 4\pi a^2$. As can be seen in Fig. 5.3, for low coverage there is very little overlap of the exclusion zones, because the particles are separated on average by a distance larger than $4a$. One can therefore deduce that in the limit of low coverages the total area excluded is $N_p S_1$.

Thus the available surface area is $1 - N_p S_1$, so the ASF or blocking function is given by

$$B(\Theta) \cong 1 - \frac{N_p S_1}{\Delta S} = 1 - 4\Theta \tag{68}$$

As can be seen, this agrees with the expression derived for the equilibrium system, Eq. (19), in respect to the first order term.

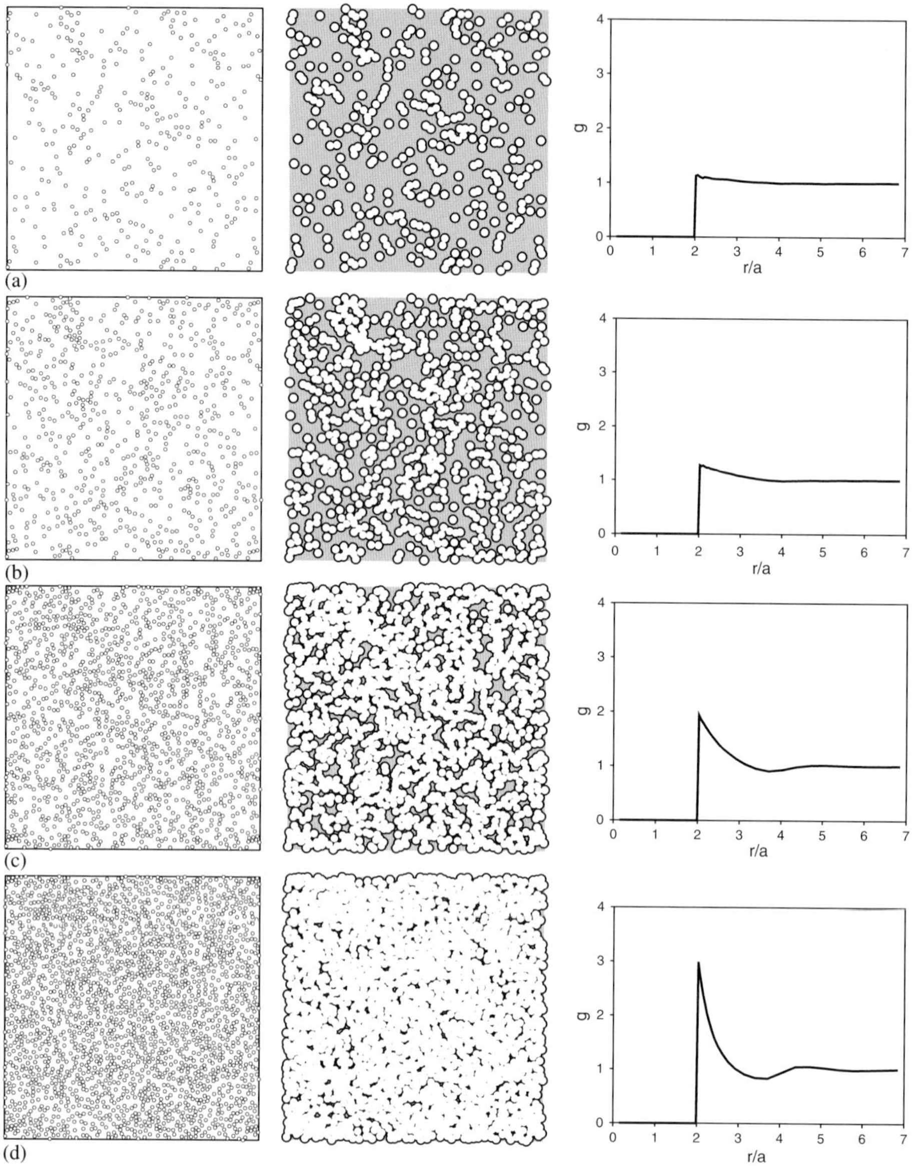

Fig. 5.3. Typical configurations of adsorbed particles generated in the RSA simulations of spheres (disks) and the corresponding available surface distributions. The white areas show the exclusion zones around adsorbed particles and the shadowed areas represent the available surface area (holes). The last column shows the pair correlation function of particles (a) $\Theta = 0.1$, (b) $\Theta = 0.2$, (c) $\Theta = 0.3$, (d) $\Theta = 0.5$.

However, for higher coverages the exclusion areas start to overlap (see Fig. 5.3), which means that the total blocked area of a particle pair separated by the distance r is smaller than S_1 by the increment $S_2(r)$, where $S_2(r)$ is the common area of the exclusion zones around two particles forming the pair (see Fig. 5.4). From simple geometry, see Fig. 5.4, the common area is the following function of the distance between particle pairs [8]

$$S_2(r) = 8a^2 \cos^{-1}\left(\frac{r}{4a}\right) - r\left(4a^2 - \frac{r^2}{4}\right)^{1/2} \tag{69}$$

Thus,

$$S_2(2a) = \pi a^2 \left(\frac{16}{3} - \frac{2(3)^{1/2}}{\pi}\right)$$

$$S_2(4a) = 0$$

By exploiting Eq. (69) one can calculate the averaged common area from the integral extended over all pairs present at the surface

$$I_2 = \frac{1}{2(\Delta S)^2} \int_{2a}^{4a} N_2(r) S_2(r) 2\pi r dr \tag{70}$$

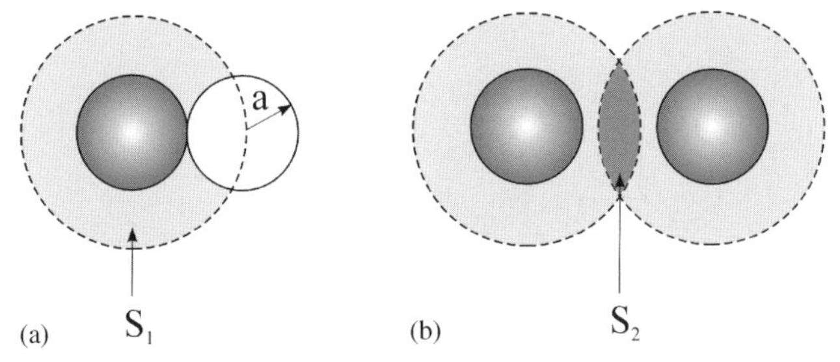

Fig. 5.4. (a) Exclusion area for one particle (shadowed), (b) Exclusion areas for a particle pair with the common part S_2.

where $N_2(r)$ is the number of pairs separated by the distance r. The number of pairs can be expressed in terms of particle coverage by exploiting the definition [16]

$$N_2(r) = N^2 g(r) \tag{71}$$

where $g(r)$ is the pair correlation function, also called the radial distribution function. Its value reflects the density of pairs separated by the distance r averaged over the entire population and normalized to the uniform bulk density (see Fig. 5.5). The $g(r)$ function can be defined for both reversible and irreversible systems, provided that there are no microscopic gradients in particle density (coverage).

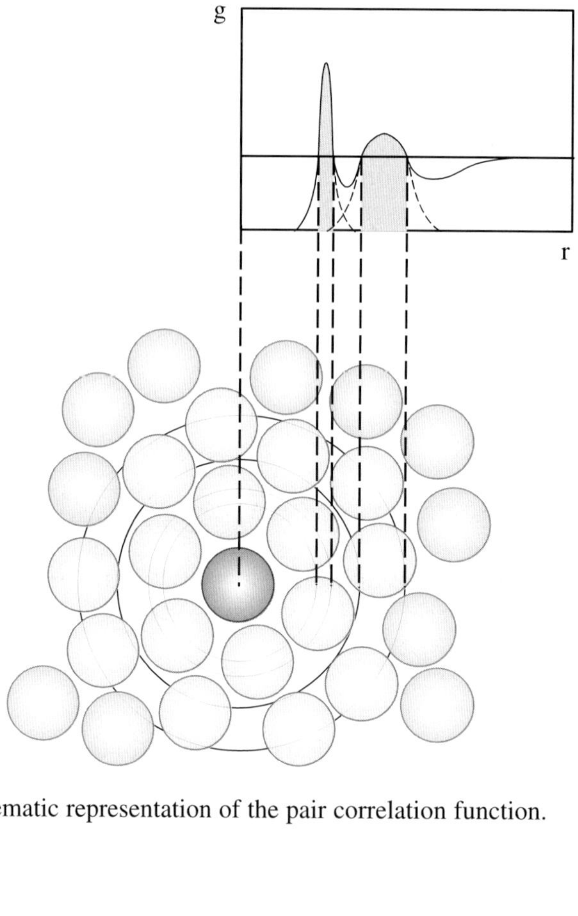

Fig. 5.5. A schematic representation of the pair correlation function.

In the general case, the pair correlation function depends both on the distance r and the coverage Θ. For not too high coverage it can be expressed in terms of the series expansion

$$g(r) = 1 + g_1(r)\Theta + \cdots \tag{72}$$

where $g_1(r)$ is a function of the distance r.

By inserting Eq. (72) into Eq. (70) and performing the integration, one obtains

$$I_2 = \frac{N^2}{2(\Delta S)^2}\int_{2a}^{4a} S_2(r)2\pi r dr + \frac{N^2}{2\,\Delta\,S^2}\Theta\int_{2a}^{4a} g_1(r)S_2(r)2\pi r dr \cong -C_2\Theta^2 + 0(\Theta^3)$$

$$\text{where } C_2 = \frac{1}{2(\pi a^2)^2}\int_{2a}^{4a} S_2(r)2\pi r dr = \frac{6(3)^{1/2}}{\pi} = 3.308 \tag{73}$$

Using Eq. (73) the second order expression for the available surface function becomes

$$B(\Theta) = 1 - 4\Theta + \frac{6(3)^{1/2}}{\pi}\Theta^2 + 0(\Theta^3) \tag{74}$$

Note that, the second order term is again identical to the equilibrium case (cf. Eq. (19)).

By considering the common exclusion area for these particles, Schaaf, and Talbot [7] showed that the third order correction, which deviates from the equilibrium one is given by

$$B(\Theta) = 1 - 4\Theta + \frac{6(3)^{1/2}}{\pi}\Theta^2 + \left(\frac{40}{\pi(3)^{1/2}} - \frac{176}{3\pi^2}\right)\Theta^3 + 0(\Theta^4) \tag{75}$$

Note that the coefficient in Eq. (75) at Θ^3 equals 1.407, which differs from the equilibrium expansion coefficient, equal to 2.424. As can be seen in Fig. 5.6, the results calculated from Eq. (75) reflect well the exact data derived from simulations for particle coverage $\Theta < 0.4$. However, higher order corrections were not calculated because of difficulties in expressing the four and higher particle correlation function.

A more efficient procedure for obtaining expressions for $B(\Theta)$ in the long time limit consists of analyzing the topology of available surface areas when particle coverage approaches the jamming limit. In this case, the areas consist exclusively of isolated spots having mostly a curvilinear triangle shape, characterized by the the linear dimension h, much smaller than the particle radius a. This can be clearly visible in Fig. 5.3 for higher coverage, $\Theta > 0.3$. As a result, one target area can accommodate only one adsorbing

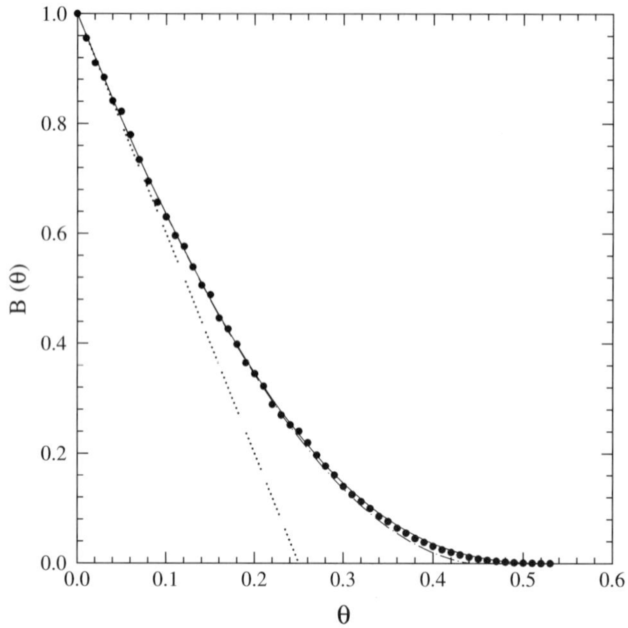

Fig. 5.6. The surface blocking function $B(\Theta)$, (ASF), for hard spherical particles. The points denote the exact numerical Monte Carlo simulations, the dotted and dashed-dotted lines denote the analytical results calculated from the low coverage expansion, Eq. (75) (using one and two terms respectively). The continuous line represents the results calculated from the fitting function, Eq. (86)

particle. The probability of particle adsorption over the target is proportional to its area,

$$p_t \sim S_t = C_p h_t^2 \tag{76}$$

where C_p is a constant depending on the shape of the target.

Denoting the surface concentration of targets of size h_t at time τ by $N_t(h, \tau)$ one can express the kinetics of their disappearance by [6]

$$\frac{d}{d\tau} N_t(h_t, \tau) = -C_p h^2 N_t(h_t, \tau) \tag{77}$$

After integration one has

$$N_t(h_t, \tau) = N_t(h_t, \tau_0) e^{-C_p h_t^2 \tau} = N_0(h_t) e^{-C_p h_t^2 \tau} \tag{78}$$

where $N_0(h_t)$ is the target distribution at the time $\tau = \tau_0$, when the asymptotic regime started.

The total concentration of targets of any size can be calculated from Eq. (78) by integrating with respect to h_t

$$\langle N_t(\tau) \rangle = \int_0^\infty N_0(h_t) e^{-C_p h_t^2 \tau} d\tau \tag{79}$$

where the integration limit was extended to infinity. This produces a small error because the contribution stemming from large targets is negligible (they disappear fast with the progress of adsorption).

By assuming that $N_0(h_t)$ is uniform [6], Eq. (79) can be evaluated as

$$\langle N_t(\tau) \rangle = C_p' \tau^{-1/2} \tag{80}$$

where C_p' is the dimensionless constant.

As the number of all available targets at the time τ (each of them able to accommodate one particle) is known, the kinetic equation describing the change in particle coverage with time can be formulated as

$$\Theta = \Theta_\infty - \frac{\pi a^2}{\Delta S} \langle N_t(\tau) \rangle = \Theta_\infty - C_p' \tau^{-1/2} \tag{81}$$

Using Eq. (81) the blocking function can be expressed as

$$B(\Theta) = \frac{d\Theta}{d\tau} = \frac{1}{2C_p'^2}(\Theta_\infty - \Theta)^3 \tag{82}$$

As noted from Eq. (81), the jamming coverage Θ_∞ can be determined by extrapolating the dependence of Θ on $\tau^{-1/2}$ obtained from numerical simulations to $\tau^{-1/2} \to \bar{0}$ (infinite adsorption time). Alternatively, according to Eq. (82) one can perform the extrapolation by plotting $B(\Theta)^{1/3}$ vs. Θ, obtained from numerical simulations. The numerical results plotted in this way are shown in Fig. 5.7a, b.

It was found by extrapolation that the jamming limit of spheres in two dimensions is 0.5467. The random monolayer at Θ close to this value is shown in Fig. 5.8e. It is interesting to compare the jamming coverage for random sphere adsorption with analytical results predicted for various monolayers of regular type. Obviously, the highest density monolayer is the hexagonal one, reflecting a 2D crystal phase with the jamming coverage $\Theta_\infty = \dfrac{\pi}{2(3)^{1/2}} = 0.9069$ and the smallest distance between particle centers equal to $2a$ (see Fig. 5.8).

On the other hand, the regular monolayer has the density $\pi/4 = 0.7854$ and the same distance between particles (Fig. 5.8). By expanding the distance between particles to $4a$ one can achieve the jammed state for the hexagonal monolayer at $\Theta_\infty = \pi/6(3)^{1/2} = 0.3023$. This is the minimum jamming coverage of spherical particles because for an expanded regular monolayer the jammed coverage is $\Theta_\infty = \pi/8 = 0.3927$. By taking an average from all these values one obtains $\Theta_\infty = 0.5968$, which is pretty close to the random monolayer value.

The pair correlation function obtained for the above monolayers is shown in Fig. 5.9. For the crystalline phases there are discrete peaks of infinite height appearing at a distance corresponding to the consecutive coordination shells. On the other hand, for the random monolayer the pair correlation function remains continuous except for the point $r/a=2$, where it diverges logarithmically as predicted theoretically [6]. In this limit the pair correlation function can be well approximated by the relationship

$$g(r) = -1.25 \ln \frac{r-2a}{a} = -1.25 \ln H \tag{83}$$

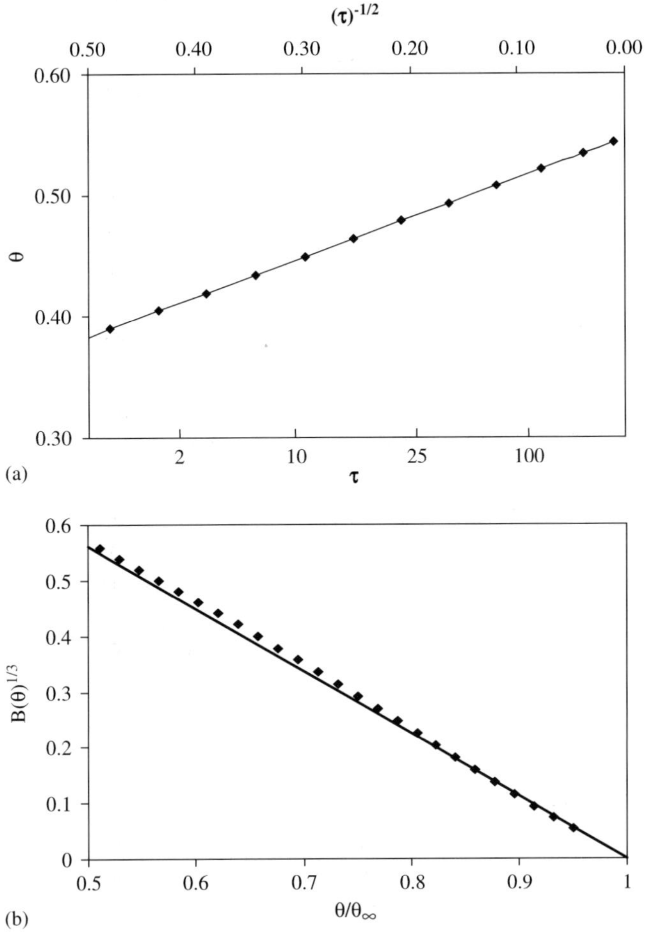

Fig. 5.7. Part "a". The dependence of Θ on $\tau^{-1/2}$ determined from Monte-Carlo simulations for hard spheres, performed according to the RSA model. Part "b". The dependence of $B(\theta)^{1/3}$ on Θ determined from the same simulations.

For $H=(r\text{-}2a)/a=0.1$, one obtains from Eq. (83) $g = 2.88$. For $H = 0.01$, $g = 5.76$ and for $H = 0.001$, $g = 8.63$. As can be seen, g function diverges if the gap width H tends to zero. In this respect, the contact correlation function for irreversible adsorption deviates from the equilibrium contact correlation function given by the expression [2–4]

$$g(2a,\Theta) = \frac{1}{2\Theta}\left(\frac{\overline{\pi}}{\Theta} - 1\right) \tag{84}$$

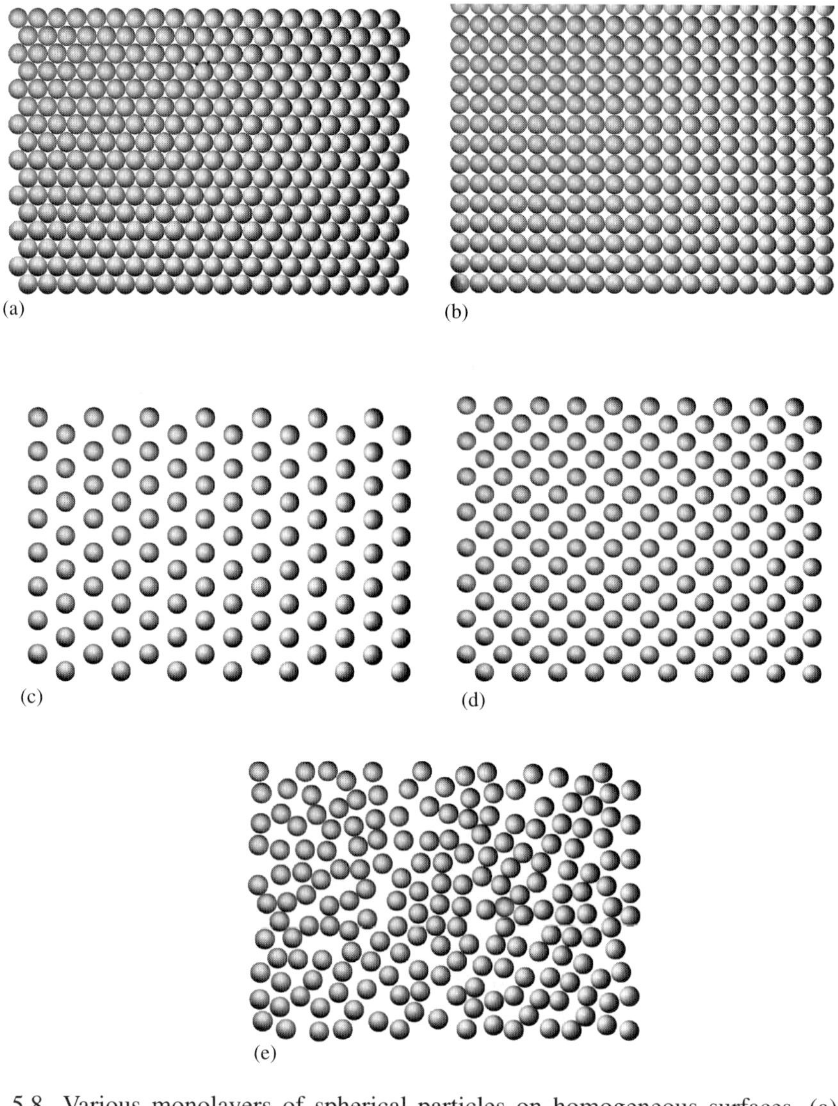

Fig. 5.8. Various monolayers of spherical particles on homogeneous surfaces, (a) the hexagonal monolayer with $\Theta = 0.9069$; (b) the regular monolayer with $\Theta = 0.7854$; (c) the extended hexagonal monolayer with $\Theta = 0.3023$; (d) the extended regular monolayer with $\Theta = 0.3927$; (e) the random monolayer with $\Theta = 0.5467$.

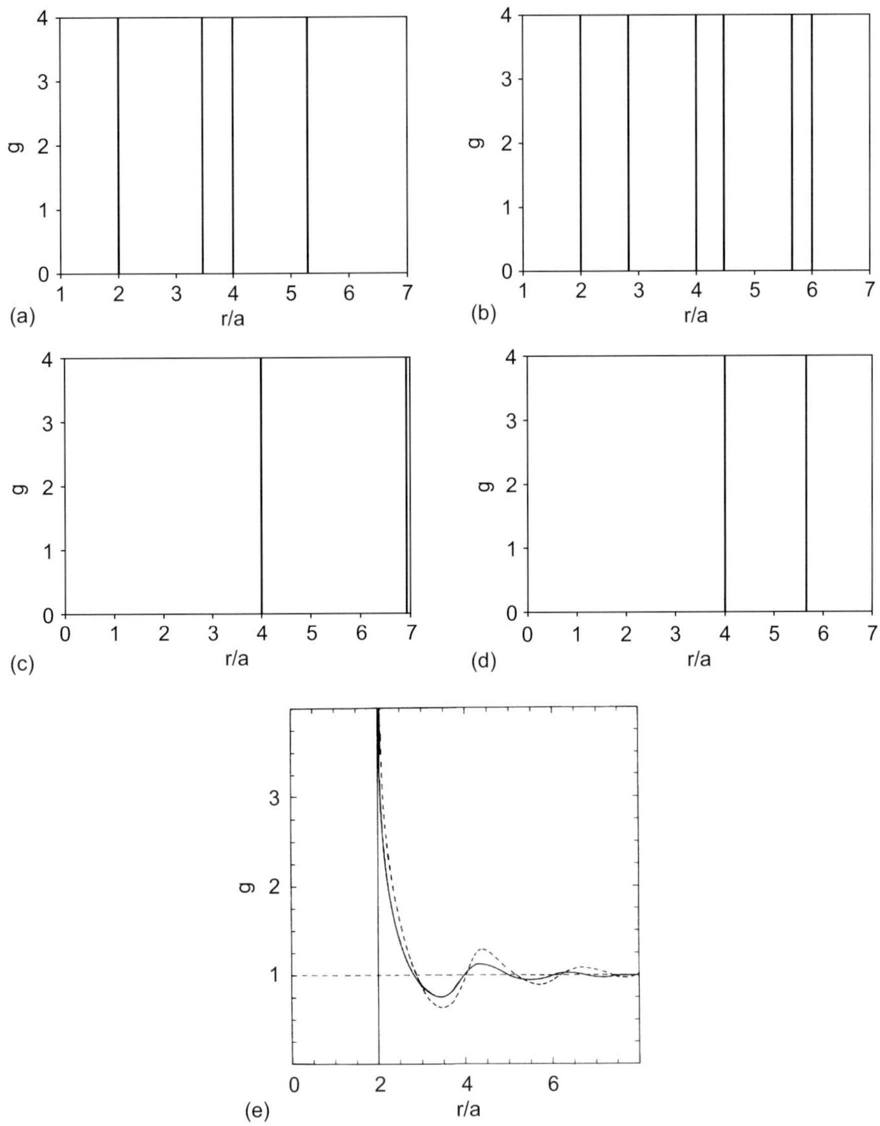

Fig. 5.9. The pair correlation function for various monolayers (a) the hexagonal monolayer with $\Theta = 0.9069$, (b) the regular monolayer with $\Theta = 0.7854$, (c) the extended hexagonal monolayer $\Theta = 0.3023$, (d) the extended regular monolayer $\Theta = 0.3927$, (e) the random monolayer with $\Theta = 0.547$ (the continuous line shows the smoothened numerical results and the dashed line shows the theoretical results for a 2D hard sphere equilibrium fluid for $\Theta = 0.547$).

Using the SPT result for the pressure, Eq. (25), the contact correlation function becomes

$$g(2a,\Theta)=\frac{1}{2\Theta}\left[\frac{1+(\gamma_p-1)\Theta}{(1-\Theta)^2}-1\right] \tag{85}$$

Substituting $\gamma_p = 1$ (spheres), and $\Theta = 0.547$ one obtains $g = 3.540$. The equilibrium pair correlation function for a 2D fluid at $\Theta = 0.547$ is plotted in Fig. 5.8.

Knowing the jamming coverage one can formulate various interpolating functions, which describe well $B(\Theta)$ for a broad range of coverages, especially close to the jamming state. One of the most convenient expressions given by Schaaf and Talbot [7] has the form

$$B(\Theta)=\left[1+0.812\frac{\Theta}{\Theta_{mx}}+0.426\left(\frac{\Theta}{\Theta_{mx}}\right)^2+0.0716\left(\frac{\Theta}{\Theta_{mx}}\right)^3\right]\left(1-\frac{\Theta}{\Theta_{mx}}\right)^3 \tag{86}$$

where Θ_{mx}=0.553.

One can deduce from Eq. (86) that for $\Theta \rightarrow \Theta_{mx}$ the blocking function is given by the asymptotic formula

$$B(\Theta)=2.310\left(1-\frac{\Theta}{\Theta_{mx}}\right)^3 \tag{87}$$

Note that the blocking function in the limit of high coverages close to the jamming state deviates from the Langmuir model often used in the literature, where it is postulated that $B(\Theta)=1-\Theta/\Theta_{mx}$, where Θ_{mx} is an empirical parameter identified with the coverage attained for a long adsorption time. The blocking effects predicted by the RSA model are, therefore, considerably more pronounced.

All the results discussed in this section were obtained for the case of planar interfaces. In many practical situations, however, particle adsorption occurs on curved interfaces, for example, on the spherical or cylindrical collectors, used widely in filtration. RSA type simulations in this case were done in Ref. [26]. Adsorption kinetics and jamming coverage of spherical particles were determined. It was shown that the jamming coverage can accurately be predicted by the formula

$$\Theta'_{\infty} = 0.5467\left(1 \pm \frac{a}{R}\right)^{l} \qquad (88)$$

where the plus sign means particle adsorption on the external side of the collector surface and the minus sign means adsorption on the internal side. The exponent $l = 1$ for adsorption on the cylinder and $l = 2$ for adsorption on the spherical collector.

The results expressed by Eq. (88) indicate that the jamming coverage calculated for the effective surface is the same as for the planar interface. However, the effective surface differs from the geometrical one by the increment $\left(1 \pm \frac{a}{R}\right)^{l}$, which is reflected in Eq. (88). One can calculate from Eq. (88) that for the spherical collector (adsorption of smaller spheres on larger spheres) and $a/R = 0.1$, the jamming coverage equals 0.6615. Remember that in this case the jamming coverage should be interpreted as an average taken for many collectors. For a cylindrical collector with the length much larger than its radius, which corresponds to adsorption of particles on a fiber, the jamming coverage for $a/R = 0.1$ equals 0.6014.

Other corrections to the jamming coverage arise in the case of polydisperse particles. This is an essential issue because under practical situations all particle suspensions are characterized by a degree of polydisperisty. The polydisperisty is expressed in terms of the relative standard deviation of the particle size distribution, which is often approximated by a Gaussian distribution

$$\delta_p(a) = \frac{1}{\sigma_p (2\pi)^{1/2}} e^{-\frac{1}{2}\left(\frac{\langle a \rangle - a}{\sigma_p}\right)^2} \qquad (89)$$

where δ_p is the probability of finding the particle of the size a, $\langle a \rangle$ the average particle size in the bulk and σ_p the standard deviation.

For example, polystyrene latex suspensions obey quite well the Gaussian size distribution with relative standard deviation $\overline{\sigma}_p = \sigma_p / a$ equal to 5–10% [27,28].

However, the use of the Gaussian distribution in RSA simulations is not advantageous, due to the finite probability of generating an arbitrary small particle over a long simulation time, which makes the entire simulation procedure ill-defined. This effect is eliminated by truncating the Gaussian distribution [29] or by using a uniform size distribution having the same standard deviation [28]

$$\delta_p(a) = \frac{\langle a \rangle}{2(3)^{1/2}\,\sigma_p} \tag{90}$$

for $a_{min} < a < a_{mx}$ (where a_{min} is the minimum and a_{mx} the maximum particle radius). The width of the distribution, i.e., $a_{mx} - a_{min}$, equals $2(3)^{1/2}\sigma_p$.

The coverage of particle monolayers produced in RSA simulations can be defined as

$$\Theta = \sum_{l=1} S_{g_l} N_l \tag{91}$$

where $S_{gl} = \pi a_l^2$ is the actual cross-section area of particles having the size a_l and N_l their surface concentration.

Although this definition is unequivocal physically, it is difficult for implementation, because measuring sizes of each adsorbing particle is impractical. Therefore, in experimental work one uses exclusively another definition of Θ

$$\Theta_{exp} = \pi \langle a \rangle^2 N_t \tag{92}$$

where N_t is the total number of particles per unit area.

This coverage definition is charged with a significant error, however, especially for suspensions with $\bar{\sigma}_p > 0.1$ because particle size distribution on the surface is quite different from the bulk distribution. This is clearly visible in Fig. 5.10 where the particle monolayer derived from RSA simulations is shown together with the size distribution in the bulk (uniform) and on the surface. Note that the size distribution of particles became highly non-uniform with a much higher contribution of smaller particle than in the bulk. As a result, the averaged particle size on the surface was reduced to 0.83 $\langle a \rangle$.

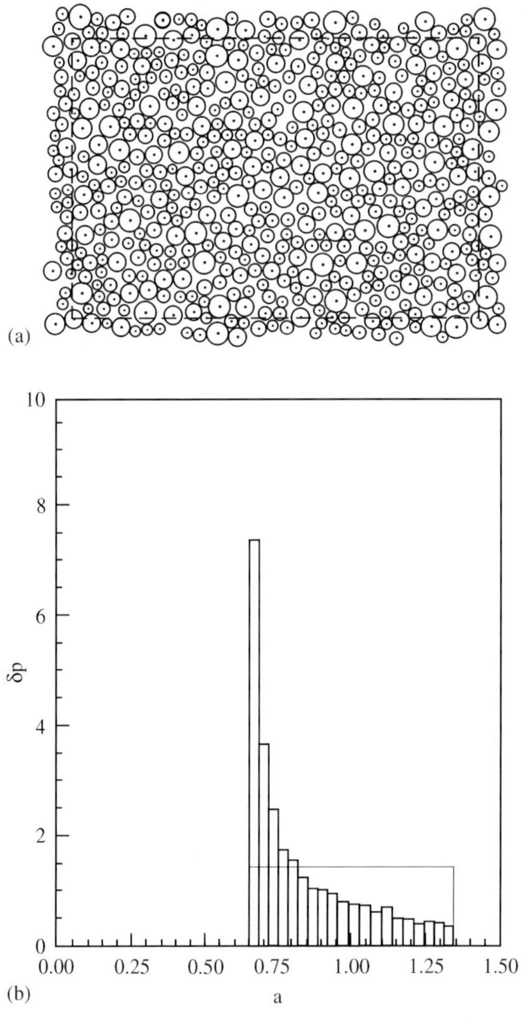

Fig. 5.10. (a) The monolayer of polydisperse particles derived from RSA simulations, $\Theta = 0.61$. (b) The histograms showing the particle size distributions, uniform in the bulk characterized with relative standard deviation of 0.2, and non-uniform on the surface with the averaged particle size 0.83 of the mean value in the bulk. From Ref. [28].

This difference in the particle size distribution in the bulk and on the surface is reflected in a large difference in coverage when calculated from Eq. (91) or Eq. (92). This is clearly seen in Fig. 5.11 showing the dependence of Θ expressed in these two ways on the polydispersity parameter of the particle suspension in the bulk [28]. It was found that the exact RSA numerical results can be well fitted by the linear dependence

$$\Theta_\infty = 0.547 + 0.53\,\bar{\sigma}_p \tag{93}$$

Similar simulations were performed by Meakin and Jullien for polydisperse disks [29]. They determined that Θ_∞ increased proportionally to $\bar{\sigma}_p^{0.86}$.

However, if the coverage is calculated as is usually done in experiments, i.e., from the dependence $\Theta_\infty = \pi\langle a\rangle^2 N$, then jamming coverage

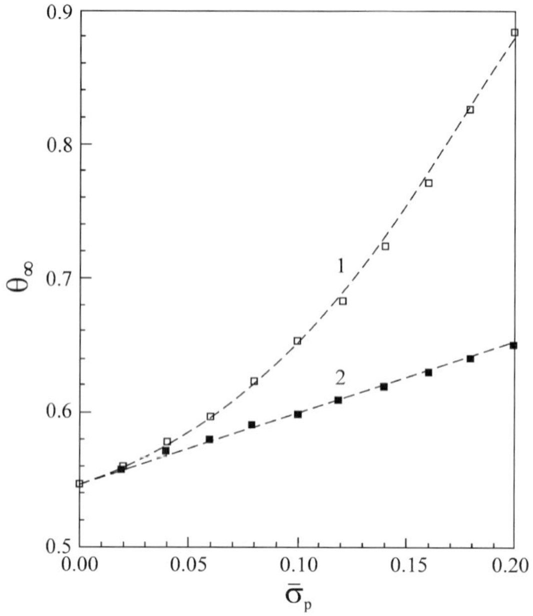

Fig. 5.11. The dependence of the jamming coverage Θ_∞ on the polydispersity parameter for particles characterized by a uniform size distribution in the bulk. (1) The coverage calculated from Eq. (92) as $\pi\langle a\rangle^2 N_t$ (definition used in experiments), (2) The coverage calculated from Eq. (91) as $\pi \sum_{l=1} a_l^2 N_l$ (proper physical definition). The points denote RSA simulations and the dashed lines the results calculated from fitting functions given by Eq. (94) and Eq. (93), respectively. From Ref. [28].

increased more rapidly with $\bar{\sigma}_p$. This dependence was fitted well by the parabolic function

$$\Theta_\infty = 0.547 + 0.458\,\bar{\sigma}_p + 6.055\,\bar{\sigma}_p^2 \qquad (94)$$

The above RSA simulation scheme for polydisperse mixtures can be adopted to the practically relevant situation of bimodal mixture adsorption of particles widely differing in size. Calculations for such a system aimed at determining kinetics of this process were performed [30]. It was found that smaller particles adsorbing more rapidly efficiently blocked the surface, decreasing significantly larger particle jamming coverage. This is in accordance with previous analytical estimations obtained for the bimodal disk system, done by Talbot and Schaaf [31]. However, for spheres finding any analytical expression for adsorption kinetics was not possible, because the problem was ill-posed. This is so, because adsorption of smaller particles is a 3D process depending on the topology of larger particles distribution, not only on their coverage. It is, therefore, more effective to study a similar problem of larger particle adsorption on surfaces pre-covered with smaller sized particles, which is uniquely posed from a mathematical point of view. It can also be treated as one of the limiting cases of particle adsorption on heterogeneous surfaces analyzed in the next section.

5.3.2. Adsorption on heterogeneous surfaces

We consider in some detail spherical particle adsorption on heterogeneous (rough) surfaces for two complementary cases:

(i) pre-covered surfaces, where larger particles can only adsorb upon touching the surface (they interact with the pre-adsorbed particles via the hard-body potential);

(ii) random site surfaces, where larger particles can only adsorb upon touching the adsorbing site but not on the surface.

Adsorption on pre-covered surfaces often occurs under experimental situations for polymer/protein, protein/surfactant or protein/colloid mixtures. On the other hand, adsorption on random-site surfaces is met in various separation processes, especially in selective removal of proteins, bacteria, cancerous cells and other bio-particles.

5.3.2.1. Adsorption on pre-covered surfaces

Let us consider the first case, i.e., adsorption of larger particles of radius a_p on a rough surface produced by attaching irreversibly N_s smaller sized particles of the radius a_s to a homogeneous surface. Assume for sake of simplicity that the particles are of spherical shape (see Fig. 5.12), so their coverage is $\Theta = \pi a_s^2 N_s$ and that they have been generated in a standard RSA process described above.

As a consequence, their distribution can be quantitatively characterized in terms of the RSA pair correlation function g, which tends to unity for $\Theta_s \rightarrow 0$ and $r \geq 2$. In practice, for $\Theta_s < 0.1$ one can assume that the particle distribution is statistically uniform. As mentioned, the adsorbing particles interact with the pre-adsorbed ones according to the hard-particle type, i.e., the interaction energy becomes infinity when they overlap and zero otherwise. This situation can be realized experimentally for higher ionic strength of the suspension if the smaller and larger particle zeta potentials are of the same sign. From a simple geometry, see Fig. 5.12, one can deduce that the exclusion area for this system, being the function of the distance h (measured from the primary energy distance), is [32,33]

$$S_1(h) = \pi\left[4a_s a_p + (2a_s - 2a_p - h)\,h\right] \tag{95}$$

where a_p is the adsorbing particle radius.

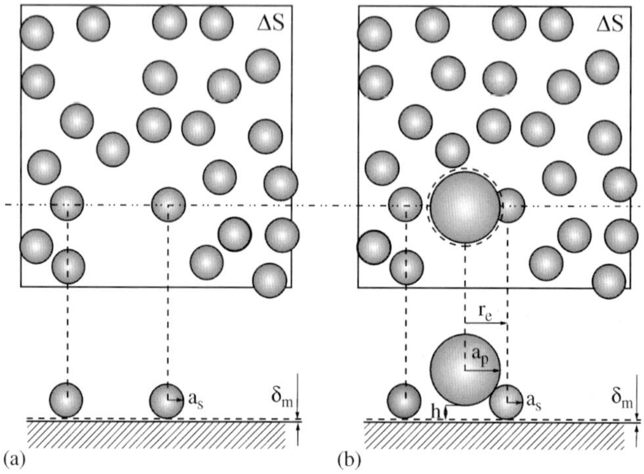

Fig. 5.12. (a) A random surface produced in RSA process of smaller particles of radius a_s. (b) A schematic view of larger particle adsorption showing the exclusion areas at height h.

For $h = 0$ the exclusion area is

$$S_1(0) = 4\pi a_s a_p = 4\pi a_s^2 \lambda' \tag{96}$$

where $\lambda' = a_p/a_s$ is the larger to smaller particle size ratio.

From Eq. (96) one can deduce that the blocking function of larger particles in the limit of low coverage Θ_s and negligible coverage of particles $\Theta_p = \pi a_p^2 N_p \to 0$, is given by

$$B^0(\Theta_l) \cong 1 - \frac{S_1(0)}{\pi a_s^2}\Theta_p \cong 1 - 4\lambda'\Theta_s \tag{97}$$

For larger coverage of pre-adsorbed particles, where their exclusion areas start to overlap, useful analytical expressions for the ASF can be derived from the binomial distribution

$$B^0(\Theta_p) \cong \left(1 - \frac{S_1}{S}\right)^{N_s} = \left(1 - \frac{S_1}{S}\right)^{\frac{S}{S_1}\langle n_s \rangle} \tag{98}$$

where $\langle n_s \rangle = N_s S_1/S$ is the averaged number of smaller particles expected within the area S_1 and S is the overall area of the interface.

Eq. (98) describes the probability of finding a surface area (hole) of the area S_1 devoted to all smaller particles. Because for most situations of practical interest $S_1/S \ll 1$, Eq. (98) simplifies to

$$B^0(\Theta) = e^{-\langle n_s \rangle} = e^{-4\lambda'\Theta_s} \tag{99}$$

Eq. (99) gives the probability of finding no particle over the area S_1, where the distribution of particles is governed by the Poisson distribution.

As can be deduced from Eq. (99), the available surface area of larger particles decreases exponentially with the surface coverage of smaller particles, proportionally to the λ' parameter. This means that at fixed Θ_s, the initial adsorption rate of larger particles, which is a parameter of vital practical significance, becomes negligible for higher values of particle size ratio λ'. In other words, the smaller the pre-adsorbed particles, compared to the adsorbing ones, the more efficient is the surface blocking effect. The abrupt decrease in the surface available surface area can be well observed in Fig. 5.13, where the distribution of holes is shown for $\lambda' = 1$, $\lambda' = 2.2$ and $\lambda' = 5$.

However, Eq. (99) is strictly valid for low Θ_s, where the smaller particle distribution remains uniform. For higher coverage, a more accurate expression can be derived from the SPT, using Eq. (37) for bimodal disk adsorption. The validity of this expression was extended to bimodal sphere adsorption by matching the low coverage expansion for the blocking function, Eq. (99), with the expansion derived from Eq. (37). It was then shown that these asymptotic solutions can be matched if [33]

$$\lambda = 2(\lambda')^{1/2} - 1 \tag{100}$$

Using Eq. (100) one obtains from Eq. (37) the following expressions for the blocking function of larger particles

$$B(\Theta_p, \Theta_s) = (1-\Theta) e^{-\frac{3\Theta_p + (4\lambda'-1)\Theta_s}{1-\Theta} - \left[\frac{\Theta_p + \left[2(\lambda'^{1/2}-1)\Theta_s\right]}{1-\Theta}\right]^2} \tag{101}$$

where $\Theta = 1 - \Theta_p - \Theta_s$.

The blocking function of larger particles can also be expressed in the useful form [33]

$$B(\Theta_p, \Theta_s) = B^0(\Theta_s)\left(1 - \frac{\Theta_p}{1-\Theta_s}\right) \times e^{-\frac{[(4\lambda'-1)\Theta_s + 3]\Theta_p}{(1-\Theta_s)(1-\Theta)} - \left[\Theta_p - \frac{(2\lambda'^{1/2}-1)^2 \Theta_s^2 \Theta_p}{(1-\Theta_s)^2} + \frac{2(2\lambda'^{1/2}-1)^2 \Theta_s^2}{1-\Theta_s} + 2(2\lambda'^{1/2}-1)^2 \Theta_s\right] \frac{\Theta_p}{(1-\Theta)^2}} \tag{102}$$

where

$$B^0(\Theta_s) = (1-\Theta_s) e^{-\frac{(4\lambda'-1)\Theta_s}{1-\Theta_s} - \left[\frac{(2\lambda'^{1/2}-1)\Theta_s}{1-\Theta_s}\right]^2} \tag{103}$$

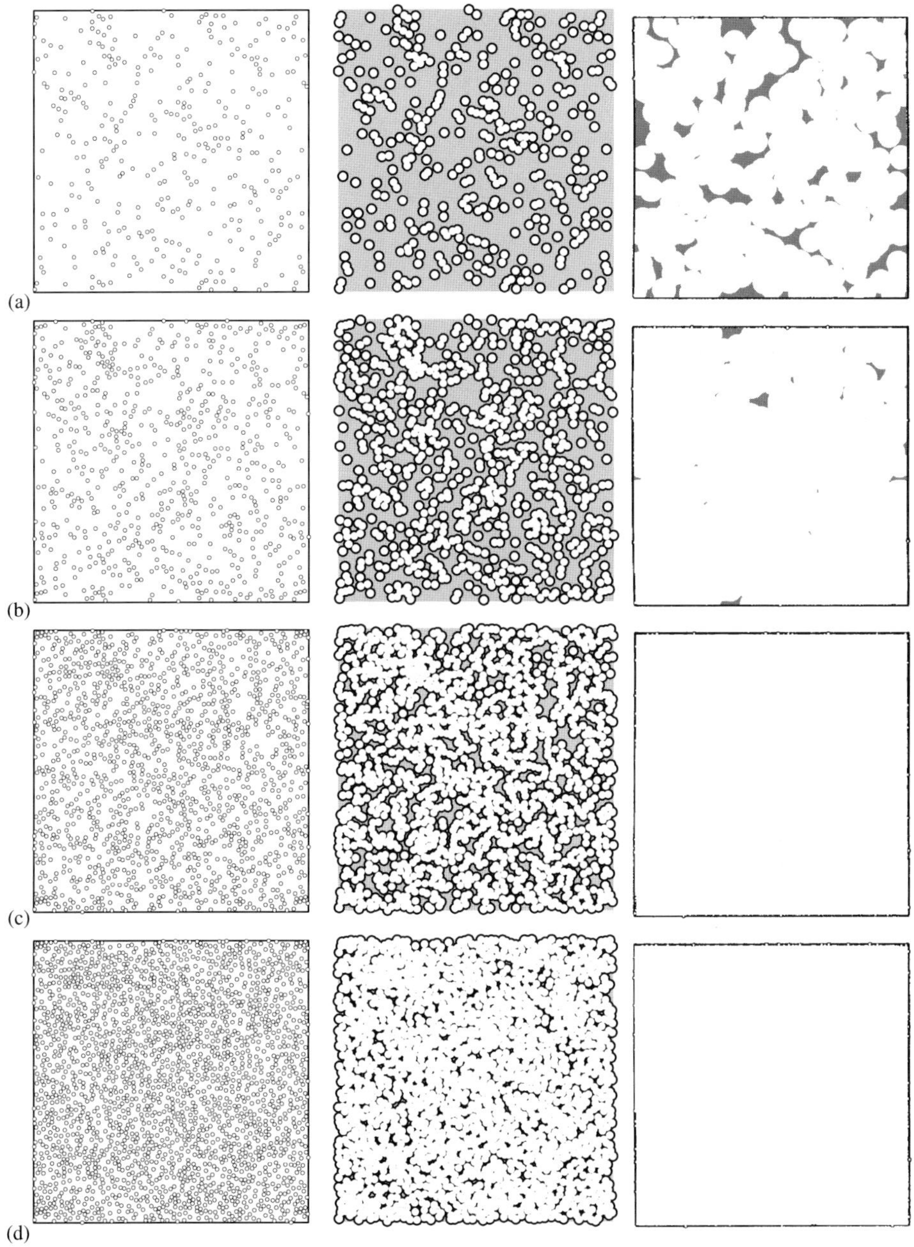

Fig. 5.13. Typical configurations of pre-adsorbed particles generated in RSA process. The available surface area for $\lambda' = 2.2$ and $\lambda' = 5$, is shown in columns 2 and 3, respectively. (a) $\Theta = 0.1$, (b) $\Theta = 0.2$, (c) $\Theta = 0.3$, (d) $\Theta = 0.5$.

is the function describing initial deposition rate of larger particles representing the extension of Eq. (99).

The second order expansion of Eq. (102) is [33]

$$B = B^0\left[1 - C_1'(\Theta_s)\Theta_p + C_2'(\Theta_s)\Theta_p^2\right] + 0\left(\Theta_p^3\right) \tag{104}$$

where

$$C_1' = \frac{4 + \left(4\lambda' + 4\lambda'^{1/2} - 7\right)\Theta_s + 2\left(2\lambda'^{1/2} - 1\right)^2 \dfrac{\Theta_s}{1 - \Theta_s}}{(1 - \Theta_s)^2}$$

$$C_2' = \frac{C_3'(\Theta_s)}{(1 - \Theta_s)^3} + \frac{C_3'^2(\Theta_s)}{2(1 - \Theta_s)^4} + \frac{C_4'(\Theta_s)}{(1 - \Theta_s)^2}$$

$$C_3'(\Theta_s) = 3 + 2\left(2\lambda' + 2\lambda'^{1/2} - 3\right)\Theta_s + 2\left(2\lambda'^{1/2} - 1\right)^2 \frac{\Theta_s^2}{1 - \Theta_s}$$

$$C_4'(\Theta_s) = -4 - \Theta_s \frac{4\lambda' + 8\lambda'^{1/2} - 5}{1 - \Theta_s} - 3\left[\frac{\left(2\lambda'^{1/2} - 1\right)\Theta_s}{1 - \Theta_s}\right]^2$$

The useful, first order expansion of Eq. (104) is

$$B = 1 - \frac{\Theta_p}{\Theta_{mx}} \tag{105}$$

where

$$\Theta_{mx} = \frac{1}{4 + \left(4\lambda' + 4\lambda'^{1/2} - 7\right)\Theta_s} \tag{106}$$

As can be seen, Eq. (105) resembles formally the Langmuir model with the saturation coverage equal to Θ_{mx}.

The theoretical predictions calculated from Eqs. (103) and (99) are compared in Fig. 5.14 with the exact numerical simulations performed according to the RSA model [32]. As can be observed, the exact numerical data are well reflected for the entire range of smaller particle coverage by Eq. (103). On the other hand, Eq. (99) remains an accurate approximation for Θ_s<0.1 only.

The theoretical predictions shown in Fig. 5.14 suggest, therefore, that the presence of trace amounts of nano-sized particles on the surface, invisible under an optical microscope, can completely block the interface significantly reducing the adsorption rate of particles of micrometer size.

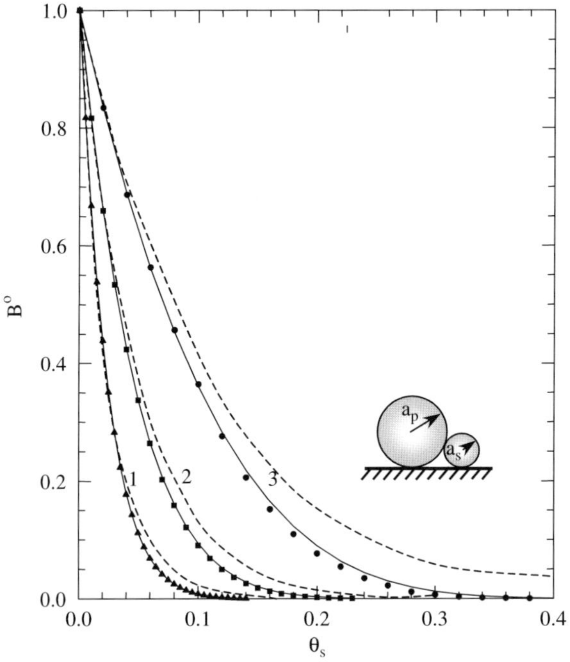

Fig. 5.14. The blocking function B^0 for adsorption at surfaces precovered by smaller sized particles (characterized by the coverage Θ_s), the points denote the exact numerical simulations performed according to the RSA method for: 1. a_p/a_s = 10, 2. a_p/a_s = 5, 3. a_p/a_s = 2.2. The continuous lines denote the equilibrium SPT results calculated from Eq. (103), the dashed lines show the results calculated from Eq. (99). From Ref. [32].

This can be deduced from the kinetic equation derived by integrating Eq. (60) with the blocking function given by Eq (105).

$$\Theta_p = \Theta_{mx}\left[1 - e^{-\frac{B^0(\Theta_s)}{\Theta_{mx}}\tau}\right] \tag{107}$$

where $\tau = \pi a_p^2 N_{att}$ is the dimensionless time τ.

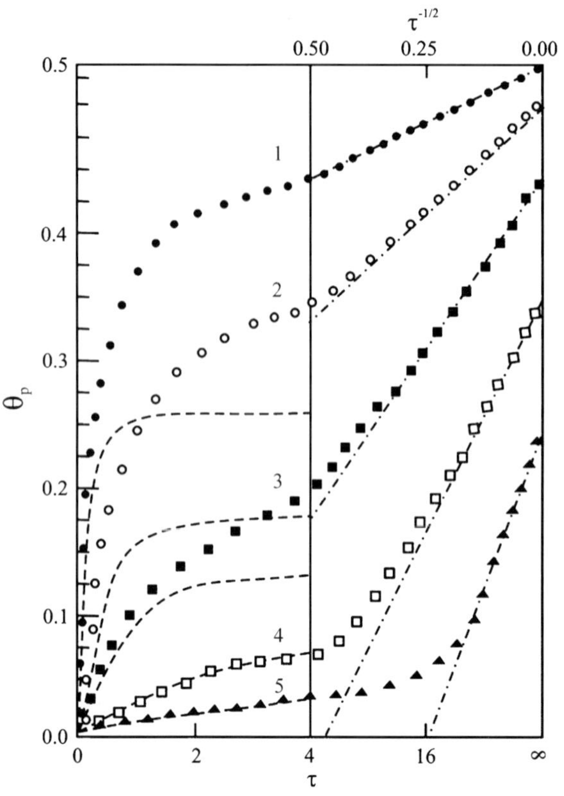

Fig. 5.15. The kinetics of larger particle adsorption on surfaces pre-covered with smaller particles determined in RSA simulations, $\lambda' = 2.2$; 1. $\Theta_s = 0$, 2. $\Theta_s = 0.05$, 3. $\Theta_s = 0.1$, 4. $\Theta_s = 0.15$, 5. $\Theta_s = 0.2$. The dashed lines denote the approximate results calculated from Eq. (107), the dashed dotted lines denote the linear fits calculated from Eq [81]. Reused with permission from Z. Adamczyk, P. Weroński, J. Chem. Phys 108 (1998) 9857. Copyright 1998, American Institute of Physics [33].

As can be observed in Fig. 5.15, the analytical results calculated from Eq. (107) can be used as a reasonable estimate of the numerical simulations for adsorption times $\tau < 1$. For longer times, when $\Theta_p > 0.1$ one can observe a systematic positive deviation of the numerical results from this formula. This was so because the apparent jamming limit is underestimated by Eq. (106).

The true jamming limit of larger particles as a function of Θ_s can be obtained in a precise way by extrapolating the kinetic runs, shown in the r.h.s. of Fig. 5.15, using the Θ_p vs. $\tau^{-1/2}$ coordinate system. As can be seen, all the kinetic runs shown in Fig. 5.15 for various Θ_s became linear in this coordinate system, which validates this extrapolation procedure. From this behavior one can conclude that the blocking function of larger particles in the limit of coverage approaching the jamming coverage can be well approximated by the expression

$$B_p \cong \left(\Theta_p^{\infty} - \Theta_p\right)^3 \tag{108}$$

where the jamming coverage Θ_p^{∞} (Θ_s) is dependent in this case on the initial coverage of smaller particles as shown in Fig. 5.16. As seen in Fig. 5.16, the jamming coverage Θ_p^{∞} decreases abruptly with Θ_s, especially for larger particle size ratio λ'. This suggests that the heterogeneity of an interface (the presence of adsorbed particles) could be detected by adsorption of larger particles serving as markers. Another interesting feature of adsorption at pre-covered surfaces is that the net surface coverage $\Theta_s + \Theta_p^{\infty}$ passes through a minimum whose depth increases considerably with λ'. Thus, the minimum net coverage was found equal to 0.378 for $\lambda' = 2.2$ ($\Theta_s = 0.34$), 0.261 for $\lambda' = 5$ ($\Theta_s = 0.24$) and 0.181 for $\lambda' = 10$ ($\Theta_s = 0.16$). These results represent a spectacular manifestation of the irreversibility effect since the composition and density of adsorbed particle monolayers is dependent on the path of adsorption. Such a process can be physically realized by replacing the smaller particle suspension after attaining a desired Θ_s, by a larger particle suspension. The exact data shown in Fig. 5.16 can well be interpolated by the fitting function [33]

$$\Theta_p^{\infty} = \Theta_{\infty}^{[c/(c-\Theta_s)]^2} \tag{109}$$

where $c = 0.596$ for $\lambda' = 2.2$, $c = 0.404$ for $\lambda' = 5$ and $c = 0.274$ for $\lambda' = 10$.

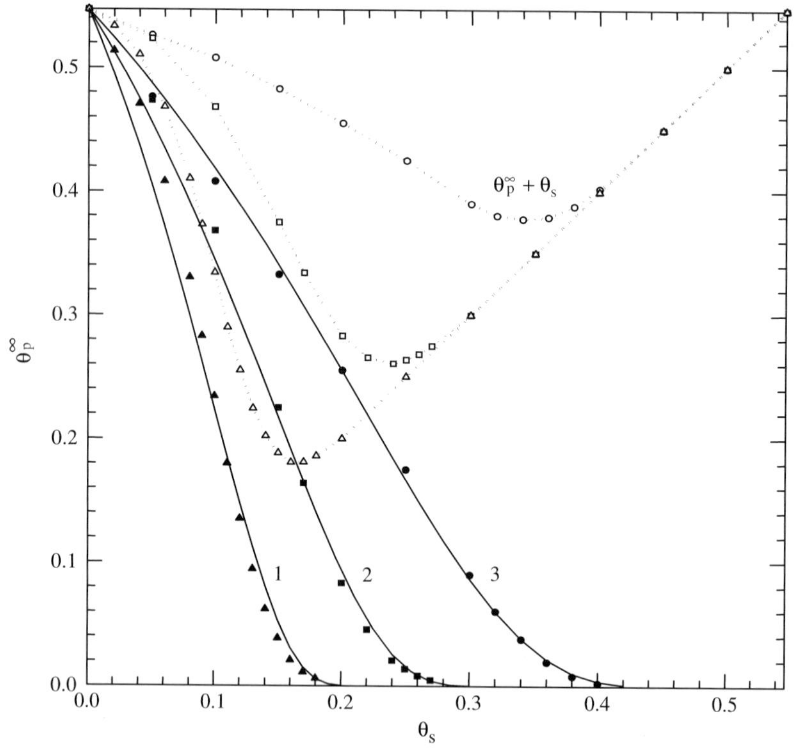

Fig. 5.16. The dependence of the jamming coverage of larger particles Θ_p^∞ on the smaller particle coverage Θ_s; the points denote the numerical RSA simulations performed for: 1. $\lambda' = 10$, 2. $\lambda' = 5$, 3. $\lambda' = 2.2$, the solid lines represent the fitting functions calculated from Eq.(109) [33]. Reused with permission from Z. Adamczyk, P. Weroński, J. Chem. Phys 108 (1998) 9851. Copyright 1998, American Institute of Physics.

In the low coverage limit, where Θ_p^∞, a more appropriate fitting function was [33]

$$\Theta_p^\infty = c\lambda' B^0(\Theta_s) \tag{110}$$

where $c = 3.78$ for $\lambda' = 2.2$, $c = 6.67$ for $\lambda' = 5$ and $c = 9.3$ for $\lambda' = 10$.

The monolayers of larger particles at the jamming state generated in the RSA simulations for pre-covered surfaces are shown in Fig. 5.17. The results shown in Figs. 5.16 and 5.17 suggest that the presence of nano-sized particles, invisible under an optical microscope, can be detected by determining experimentally the jamming coverage of micro-particles serving as markers.

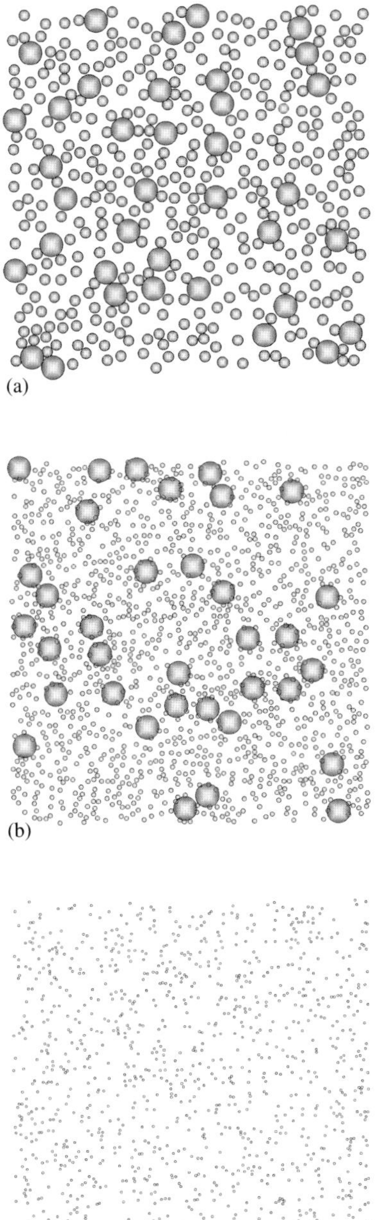

Fig. 5.17. The large particle monolayers at the jamming state on surfaces pre-covered by smaller particles, derived from RSA simulations: (a) $\lambda' = 2.2$, $\Theta_p^\infty = 0.1$, $\Theta_s = 0.3$, (b) $\lambda' = 5$, $\Theta_p^\infty = 0.1$, $\Theta_s = 0.2$, (c) $\lambda' = 10$, $\Theta_p^\infty = 0.1$, $\Theta_s = 0.14$. From Ref. [32].

5.3.2.2. Adsorption on random-site surfaces

Adsorption on surfaces bearing randomly distributed sites, which can irreversibly bind particles, is of vital significance in many practical and natural processes such as filtration, paper-making, chromatography, separation of proteins, viruses, bacteria, pathological cells, etc. The effectiveness of these processes is often enhanced by the use of coupling agents bound to interfaces (called often precursors), e.g., polyelectrolytes [34–37]. In biomedical applications special proteins (antibodies) attached to the surface are used for a selective binding of a desired ligand from protein mixtures as is the case of affinity chromatography [38], recognition processes (biosensors) [39,40], immunological assays [41,42], etc. Similarly, many experimental studies on colloid particle adsorption were carried out for surfaces modified by adsorption of surfactants, polyvalent ions, or chemical coupling agents (silanes), which change the natural surface charge of substrate surfaces [27,43]. This often leads to a discrete distribution of binding sites as is the case for packed bed filtration experiments [44,45]. The appearance of such heterogeneous interfaces exerts an important influence on transport and distribution of colloid contaminants in aqueous porous media, e.g., in soils [46].

A quantitative analysis of the RSA on random site surfaces, referred to as a RSS process, was performed by Jin et al. [47,48] who considered the simplest possibility of point-like sites perfectly distributed randomly over a homogeneous surface. A correspondence (mapping function) between the adsorption process at the RSS surfaces and the widely studied continuous surface RSA model was found.

Later on, more elaborate models were developed [49,50] by considering finite dimensions of sites having the form of either hard disks of the diameter $2a_s$ incorporated into the substrate see (Fig. 5.18, part "a") or hard spheres attached to the surface, (see Fig. 5.18b). The configuration of the sites was produced by performing the RSA simulations described above. However, the range of applicability of the disk model is limited by the fact that such a site configuration is rather specific and difficult to realize in practice.

More relevant for practical applications is the sphere-shaped adsorption site model. The basic assumption of this model is that the colloid particle can only be adsorbed upon touching the site (see Fig. 5.18b). Otherwise, at bare interfaces, the particle will not adsorb. Physically, this corresponds to the situation where the particles are irreversibly bound to the

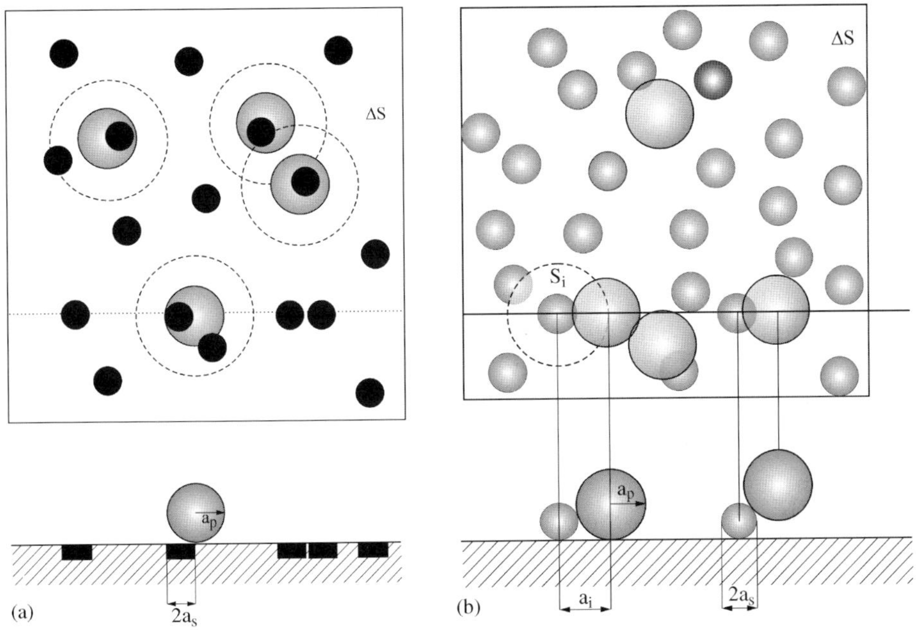

Fig. 5.18. A schematic representation of particle adsorption on heterogeneous surfaces bearing (a) disk-shaped adsorption sites (b), and spherically shaped sites.

sites due to short-ranged attractive interactions of an electrostatic or chemical nature. Furthermore, particle adsorption was assumed irreversible and localized, which means that all particle positions remained fixed during the entire simulation run.

Note that particle adsorption at heterogeneous surfaces covered by spherically shaped sites is a three dimensional process, opposite to adsorption at disk-shaped sites or adsorption at pre-covered surfaces. In the latter case, particle adsorption occurred at a bare substrate surface only, hence all particle centers were located in one plane.

In accordance with these assumptions, the modeling of RSA processes on site-covered surfaces was carried out according to a similar procedure as for the pre-covered surfaces. The first step was identical and consisted in deposition of sites according to the classical RSA of desired surface density N_s and the coverage $\Theta_s = \pi a_s^2 N_s$.

Then, the modeling of RSS process for the site-covered surface was carried out according to the algorithm whose main steps were [49,50]

(i) An adsorbing (virtual) particle of diameter $2a_p$ was generated at random within the simulation area ΔS with periodic boundary conditions on its perimeter; if it did not touch any of the sites, the particle was rejected and another virtual particle was generated (the number of attempts N_{att} was increased by one),

(ii) Otherwise, if the particle touched any of the sites, the overlapping test was performed according to the usual RSA rules, i.e., it was checked whether if there was any previously adsorbed particle within the exclusion volume; if there was overlapping the simulation loop was repeated (the number of attempts was increased by one),

(iii) If there was no overlapping, the virtual particle was assumed to be irreversibly adsorbed at a given position.

Analogously as in the classical RSA simulations, the coverage of particles at heterogeneous surfaces was expressed as $\Theta_p = \pi a_p^2 N_p$. Accordingly, the dimensionless adsorption time was defined as [49,50]

$$\bar{t} = \pi a_p^2 \frac{N_{att}}{\Delta S} = \pi a_p^2 N_{att} \qquad (111)$$

for sake of simplicity ΔS was assumed unity.

This definition, Eq. (111), allows one to derive kinetics of particle adsorption, i.e., the dependence of Θ_p on the adsorption time $\bar{t}$ if the available surface function is known. For site-covered surfaces the ASF_{RSS} function was defined as the probability p_a of adsorbing the virtual particle for a fixed coverage and configuration of sites and particles

$$ASF_{RSS} = B_{RSS} = p_a = p_0(\Theta_s)\bar{B}_{RSS}(\Theta_s, \Theta_p, \lambda') \qquad (112)$$

where $p_0(\Theta_s)$ is the initial adsorption probability if there are only sites of a given coverage and distribution on the surface.

Similarly, as in the classical RSA simulations, the $\bar{B}_{RSS}$ function was evaluated by performing, at fixed Θ_s, Θ_p, a large number of adsorption trials N_{att}, N_{succ} of them being potentially successful. Then, the ASF is defined as the limit of N_{succ}/N_{att} when $N_{att} \rightarrow \infty$. Especially important is the value of ASF in the limit of negligible coverage, $\Theta_p \rightarrow 0$, since it characterizes the initial adsorption rate for heterogeneous surfaces. This quantity is of primary interest from an experimental viewpoint.

Knowing the blocking function, one can calculate particle adsorption kinetics by integrating the constitutive dependence [49,50].

$$\frac{d\Theta_p}{d\bar{t}} = B_{RSS}(\Theta_s, \Theta_p, \lambda') \tag{113}$$

Despite the complexity of the RSS processes, several analytical expressions can be derived for low coverage of adsorbed particles. Because for disks the interaction areas $S_i = \pi a_s^2$ do not overlap, see Fig. 5.18, the initial probability is described by the formula that is valid for arbitrary Θ_s

$$p_0 = \Theta_s \tag{114}$$

On the other hand, for spherical sites, the interaction areas $S_i = 4\pi a_s a_p$ overlap for higher coverage Θ_s, therefore p_0 can be calculated from the Poisson distribution [50]

$$p_0 = 1 - e^{-S_i N_s} = 1 - e^{-4\lambda' \Theta_s} \tag{115}$$

In the limit of low site coverage, if $\Theta_s \ll 1$, Eq. (115) becomes

$$p_0 = 4\lambda' \Theta_s \tag{116}$$

As can be noted from Eqs.(115,116), the initial adsorption probability on site-covered surfaces equals $1 - B^0$, where B^0 is the initial probability of adsorption on pre-covered surfaces. Therefore, adsorption of particles on site-covered surfaces is in a sense complementary to adsorption on the pre-covered surfaces analyzed above.

More accurate expressions can be derived in the limit of higher Θ_s by exploiting the modified SPT results, analogously as for the adsorption on pre-covered surfaces.

For spherical sites one has simply $p_0 = 1 - B^0$, thus

$$p_0 = 1 - B^0 = 1 - (1 - \Theta_s) e^{-\frac{(4\lambda' - 1)\Theta_s}{(1 - \Theta_s)} - \left[\frac{(2\lambda'^{1/2} - 1)\Theta_s}{1 - \Theta_s}\right]^2} \tag{117}$$

The p_0 function is of primary practical interest because it represents the averaged probability of adsorbing the particle at surfaces covered by a given number of sites. Hence, by knowing p_0 one can calculate the initial flux of solute (particles) to heterogeneous surfaces. The dependence of p_0 on Θ_s calculated for $\lambda' = 2$, 5 and 10 is plotted in Fig. 5.19 [50].

Note that the adsorption probability of particles increases abruptly with Θ_s, especially for larger λ' values. For $\lambda' = 10$, the probability of adsorption reaches unity (the value pertinent to homogeneous surfaces) for Θ_s as low as 0.1. This behavior is well reflected by Eq. (117), being in quantitative agreement with the numerical data for the entire range of Θ_s and λ' studied. It is also interesting to note that adsorption probability at spherical sites becomes considerably larger than for disk-shaped sites (see the dashed line in Fig. 5.19). For example, at $\lambda' = 2$, p_0 for spherical sites increases proportionally to $8\Theta_s$ (for $\Theta_s<0.1$), whereas for the disk sites proportionally to Θ_s. For $\lambda' = 10$, this difference

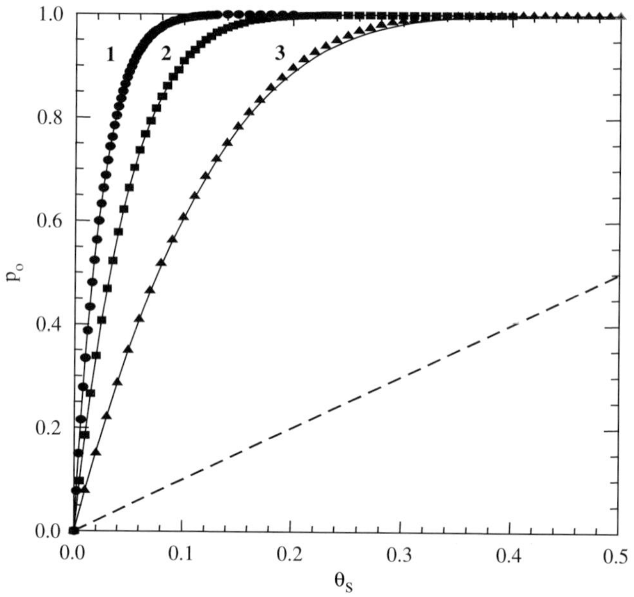

Fig. 5.19. The dependence of the initial adsorption probability p_0 at surfaces covered by sites on their coverage Θ_s. The points denote the numerical simulations performed for spherically shaped sites and 1. $\lambda' = 10$, 2. $\lambda' = 5$, 3. $\lambda' = 2$, the solid lines represent the analytical results calculated from Eq.(117) and the dashed line shows the analytical predictions for disk-shaped sites, i.e., $p_0=\Theta_s$. From Ref. [50].

becomes more than two orders of magnitude because for spherical sites $p_0 \sim 400\ \Theta_s$ initially, whereas for disks $p_0 \sim \Theta_s$. This observation has practical implications showing that the geometry of the active sites plays a more decisive role than their surface concentration.

The results shown in Fig. 5.19 imply that by measuring experimentally the flux of larger colloid particles (which can easily be done by direct microscope observations) one can detect the presence of nano-scale surface heterogeneities, that are invisible under the microscope. If the surface concentration of the sites can be estimated, one can determine their size or shape from the particle deposition experiments.

It should be mentioned that the results shown in Fig. 5.19 describe the particle adsorption rate at heterogeneous surfaces in the limit when particle accumulation is negligible, i.e., for $\Theta_p \to 0$ only. If Θ_p becomes finite, the probability of particle adsorption decreases as a result of the volume exclusion effects analyzed above. By considering the exclusion effects between adsorbed particles, the first order expressions for the ASF in the case of disk-shaped sites becomes simply

$$
\begin{aligned}
B_{RSS} &= \pi a_s^2 (N_s - N_p)(1 - 4\Theta_p) \\
&= \Theta_s \left(1 - \frac{\Theta_p}{\alpha_p}\right)(1 - 4\Theta_p) \cong \Theta_s \left(1 - \frac{\Theta_p}{\Theta_p^{mx}}\right)
\end{aligned}
\tag{118}
$$

where $\Theta_p^{mx} = \dfrac{\alpha_p}{1+4\alpha_p}$, $\alpha_p = \Theta_s \lambda'^2$

By substituting Eq. (118) into Eq. (113) upon integration one obtains the expression for particle adsorption kinetics

$$
\Theta_p = \Theta_p^{mx}\left(1 - e^{-\bar{\tau}/\Theta_p^{mx}}\right) \tag{119}
$$

For spherical sites, one has analogously

$$
\begin{aligned}
B_{RSS} &= 4\pi a_s a_p (N_s - N_p/l_s)(l - C_1' \Theta_p) \\
&= p_0 \left(1 - \frac{\Theta_s}{\alpha_p l_s}\right)(1 - C_1' \Theta_p) \cong p_0 \left(1 - \frac{\Theta_p}{\Theta_p^{mx}}\right)
\end{aligned}
\tag{120}
$$

where l_s is the site coordination number (multiplicity), C_1' the dimensionless accounting for the 3D exclusions effect of adsorbed particles and

$$\Theta_p^{mx} = \lambda'^2 l_s \Theta / \left(1 + C_1' \lambda'^2 l_s \Theta_s\right) \tag{121}$$

From a simple geometry one can deduce that for $\lambda' > 4$ one site can coordinate only one particle, so $l_s = 1$. For $\lambda' < 4$, the site coordination number becomes larger than unity. However its exact value can only be determined from numerical simulations as discussed later on.

An integration of Eq. (120) gives the kinetic dependence

$$\Theta_p = \Theta_p^{mx}\left(1 - e^{-\bar{t}/\Theta_p^{mx}}\right) \tag{122}$$

In the case of disk-shaped sites a useful analytical expression can also be derived in the case when the size of the particle becomes much larger than the site size, which can therefore be treated as point-like objects. Then the ASF becomes proportional to the number of sites N_s' lying outside exclusion areas of adsorbed spheres [49] and the ASF can be expressed as

$$B_{RSS} = \pi a_s^2 N_s' = \Theta_s \frac{N_s'}{N_s} \tag{123}$$

By postulating further a uniform distribution of sites, one can demonstrate that

$$N_s' = \left(N_s - N_{att}\right) B(\Theta_p) \tag{124}$$

where $B(\Theta_p)$ is the blocking function for the adsorption process over homogeneous surfaces governed by the classical RSA model. This is so because particle configuration at a heterogeneous surface bearing N_s sites becomes identical to the RSA configuration over continuous surfaces after N_{att} adsorption attempts [49].

Using Eq. (124) the ASF for the RSS process can be expressed as

$$B_{RSS} = \Theta_s \frac{\alpha_p - \tau}{\alpha_p} B(\Theta_p) \tag{125}$$

where $\tau = \pi a_p^2 N_{att}$ is the time of the RSA process.

This relationship indicates that an unequivocal mapping between the RSA process on homogeneous surfaces and the RSS on surfaces bearing point-like sites. The time-transforming function between them can be derived from Eq. (125) by noting that for equal coverage Θ_p

$$\frac{d\tau}{d\bar{t}} = \Theta_s \frac{\alpha_p - \tau}{\alpha_p} \tag{126}$$

By integrating Eq. (126) one obtains

$$\tau = \alpha_p\left(1 - e^{-\bar{t}\Theta_s/\alpha_p}\right) \tag{127}$$

As can be seen from Eq. (127) for $\bar{t}/\alpha_p \ll 1$, $\tau = \bar{t}\,\Theta_s$, which means that the RSS process becomes identical to the RSA process on homogeneous surfaces. On the other hand, for $\bar{t} \to \infty$, $\tau = \alpha_p$. This indicates that the jamming coverage of the RSS process can be obtained from the RSA kinetic by substituting $\tau = \alpha_p$.

By eliminating τ from Eq. (125), we obtain the following expression for the B_{RSS}

$$B_{RSS} = \frac{d\Theta_p}{d\bar{t}} = \Theta_s\, e^{-\Theta_s \bar{t}/\alpha_p} B\left(\Theta_p\right) \tag{128}$$

Eq. (128) can be further rearranged by integration to a more useful form

$$\int_0^{\Theta_p} \frac{d\Theta'}{B(\Theta')} = \alpha_p\left(1 - e^{-\Theta_s \bar{t}/\alpha_p}\right) = \tau \tag{129}$$

By substituting the low coverage limiting expression for $B(\Theta)=1-4\Theta$ one obtains from Eq. (129) the kinetic equation

$$\Theta_p = \frac{1}{4}\left[1 - e^{-4\alpha_p\left(1 - e^{-\Theta_s \bar{t}/\alpha_p}\right)}\right] \tag{130}$$

On the other hand, for the long-time regime, when $B(\Theta) = (1/2C_{RSA})(\Theta_\infty - \Theta_p)^3$ (where $C_{RSA} = 0.245$, $\Theta_\infty = 0.5467$) the limiting kinetic expression becomes

$$\Theta_p = \Theta_\infty - \frac{C_{RSA}}{\left[\alpha_p\left(1-e^{-\Theta_s \bar{t}/\alpha_p}\right)\right]^{1/2}} = \Theta_\infty - \frac{C_{RSA}}{\tau^{1/2}} \tag{131}$$

Hence, the ASF for the RSS process is

$$B_{RSA} = \Theta_s\left[1-\frac{C_{RSA}^2}{\alpha_p\left(\Theta_\infty-\Theta_p\right)^2}\right]B(\Theta_p) \tag{132}$$

For an arbitrary time span the kinetics of the RSS process can be described by the interpolating function proposed by Jin et al. [47]

$$\Theta_p = \Theta_\infty\left(1-\frac{1+0.314\tau^2+0.45\tau^3}{1+1.83\tau+0.66\tau^3+\tau^{7/2}}\right) \tag{133}$$

where $\tau = \alpha_p(1 - e^{-\Theta_s \bar{t}\alpha_p})$.

By substituting $\tau = \alpha_p = \Theta_s \lambda'^2$, one can calculate from Eq. (133) the jamming coverage for the RSS process, which is given explicitly by

$$\Theta_p^\infty(\Theta_s) = \Theta_\infty\left(1-\frac{1+0.314\Theta_s^2\lambda'^4+0.45\Theta_s^3\lambda'^6}{1+1.83\Theta_s\lambda'^2+0.66\Theta_s^3\lambda'^6+\Theta_s^{7/2}\lambda'^7}\right) \tag{134}$$

For not too large Θ_s, Eq. (134) can well be interpolated by a much simpler function

$$\Theta_p^\infty(\Theta_s) = \Theta_\infty\frac{\Theta_s\lambda'^2}{\Theta_\infty+\Theta_s\lambda'^2} \tag{135}$$

In the limit of $\Theta_s \lambda'^2 << 1$ this simplifies to

$$\Theta_p^\infty = \Theta_s\lambda'^2 \tag{136}$$

On the other hand, for $\Theta_s\lambda'^2 >> 1$one can predict, using Eq. (134), the asymptotic relationship between the jamming coverage and a

$$\Theta_p^{\infty}(\Theta_s) = \Theta_{\infty} - \frac{C_{RSA}}{\Theta_s^{1/2}\lambda'} = \Theta_{\infty} - \frac{C'_{RSA}}{aN_s^{1/2}} \tag{137}$$

where $C'_{RSA} = C_{RSA}/\pi^{1/2}$.

Note that the analytical results expressed by Eqs.(131–136) are strictly valid in the limit of negligible site dimension compared to adsorbing particle size, i.e., for $\lambda' >> 1$. Extensive simulations for a more realistic situation of finite site dimension were performed [49] by applying the RSS algorithm described above. The available surface function, the jamming coverage and pair correlation function were calculated for $\lambda' = 2$, 5 and 10. It was found that the blocking function became identical to the standard RSA model for $\lambda' = 10$. Also the pair correlation function was practically identical for $\lambda' = 10$ and $\Theta_s > 0.1$ [50]. On the other hand, the results of the jamming coverage Θ_p^{∞}, calculated in Ref. [49] are shown in Figs. 5.20 and 5.21. In the former figure, the dependence of Θ_p^{∞} vs. Θ_s (in logarithmic scale) is plotted in order to cover a broad range of site coverage.

As can be seen, the low coverage asymptotic formula, Eq. (126), $\Theta_p^{\infty} = \lambda'^2\Theta_s$ describes well the exact results for $\Theta_p^{\infty} < 0.2$ only. The comparison of the numerical with the analytical formula, Eq. (134), is done in Fig. 5.21 by using the universal coordinate system Θ_p^{∞} vs. $\alpha_p = \lambda'^2\,\Theta_s$. It can be seen that the exact numerical data are in agreement with the analytical results derived from Eq. (134) for $\alpha_p < 2$, only. For $\alpha_p > 2$, a positive deviation of numerical results from the analytical curve derived from Eq. (134) is observed for all λ'. This effect can be attributed to the fact that due to finite site dimensions, a fraction of their surface sticks out from the exclusion zones of adsorbed particles (see Fig. 5.18). This effect is completely absent if the sites are point-like. It was found that the exact results can be well interpolated by the function analogous to Eq. (135)

$$\Theta_p^{\infty} = \Theta_{\infty}\frac{C'_0\alpha_s^l}{1 + C'_0\alpha_s^l} \tag{138}$$

where C'_0, l equals to 1.75, 1 for $\lambda' = 5$ and 1, 1.5 for $\lambda' = 2$.

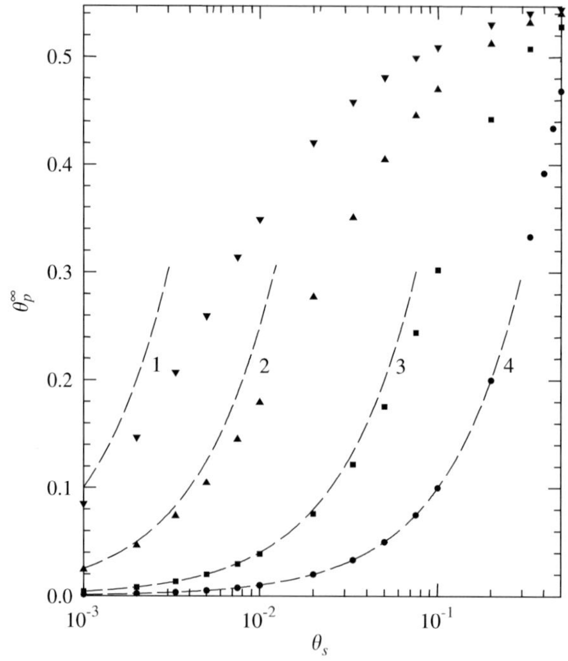

Fig. 5.20. The dependence of the jamming coverage of particles Θ_p^∞ on the coverage of the disk-shaped adsorption sites Θ_s; the points denote the results of numerical simulations, performed for 1. $\lambda' = 10$, 2. $\lambda' = 5$, 3. $\lambda' = 2$ 4. $\lambda' = 1$, the dashed lines show the results derived from the equation $\Theta_p^\infty = \lambda'^2 \Theta_s$ [49]. Reused with permission from Z. Adamczyk, P. Weroński, E. Musiał J. Chem. Phys 116 (2002) 167. Copyright 2002, American Institute of Physics.

It is also interesting to observe that the jamming coverage Θ_p^∞ in the case of disk-shaped sites never exceeds the jamming limit in the RSA process for homogeneous surfaces, i.e., $\Theta_\infty = 0.547$.

Interesting RSS processes occur in the case of spherical sites. As already observed in Fig. 5.19 the initial adsorption probability is much higher in this case than for the disk sites. Also, in contrast to disks, there appears a possibility of multiple coordination of one site, which is expected to have profound implications for the jamming limit and pair correlation function. However, for spherical sites, because of the overlapping of the interaction areas, no simple analytical relationships describing the jamming limit, analogous to Eq. (134), can be derived. These data are accessible from numerical simulations only, obtained by applying the RSS algorithm [51].

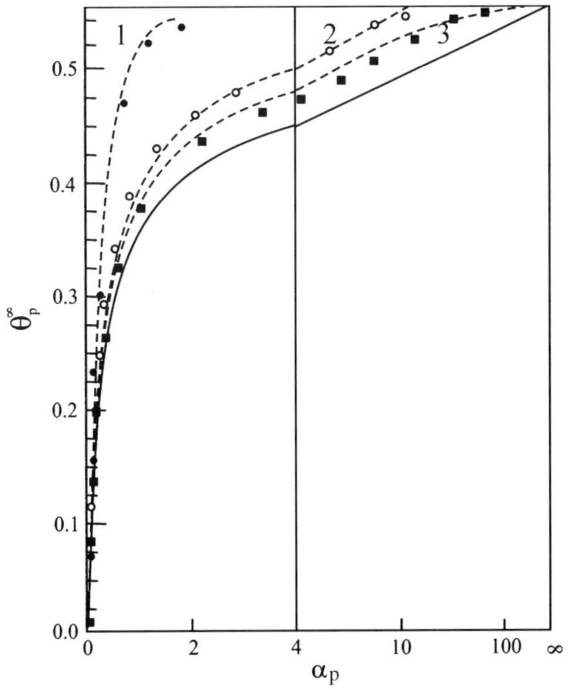

Fig. 5.21. The universal plot showing the jamming coverage of particles Θ_p^∞ vs. $\alpha_p \lambda'^2 \Theta_s$ for the disk-shaped adsorption sites; the points denote the results of numerical simulations, performed for 1. $\lambda' = 2$, 2. $\lambda' = 5$, 3. $\lambda' = 10$, the solid line denotes the results calculated from Eq.(134) and the dashed lines show the results calculated from the interpolating function, given by Eq.(135). Reused with permission from Z. Adamczyk, P. Weroński, E. Musiał, J. Chem. Phys 116 (2002) 67. Copyright 2002, American Institute of Physics [49].

Except for the jamming coverage, the available surface function, the site coordination number and the pair correlation function were calculated for $\lambda' = 1, 2, 5$ and 10 [51].

The coordination number can be determined most directly by plotting the dependence of the jamming coverage Θ_p^∞ on $\lambda'^2 \Theta_s$. The slope of this dependence in the limit of Θ_s tending to zero, when the volume exclusion effects between adsorbed particles remain negligible, gives directly the site coordination number l_s. Examples of such calculations performed for various λ', ranging from 1 to 4, are shown in Fig. 5.22. As was found from linear regression fits, for $\lambda' = 1$, $l_s = 5.5$ for $\lambda' = 2$, $l_s = 2.4$, and for $\lambda' = 4$, $l_s = 1$ as expected from geometrical considerations. The theoretical dependence of l_s on λ' plotted in Fig. 5.22 can well be fitted by the simple interpolating function

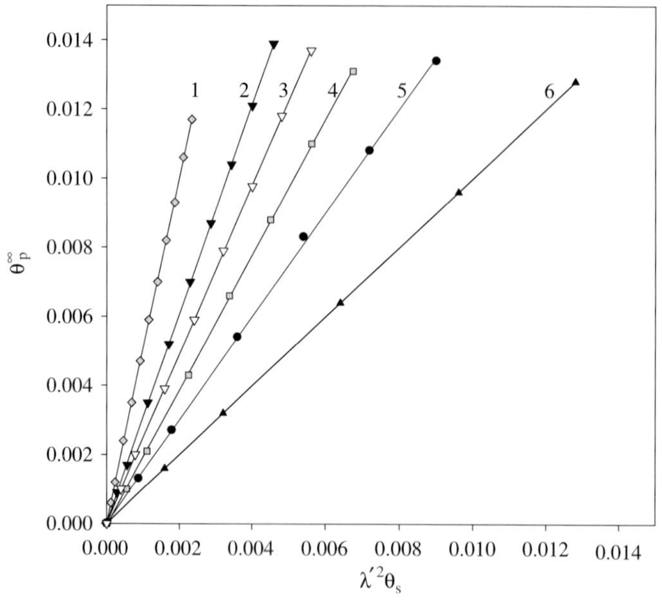

Fig. 5.22. The dependence of the jamming coverage of particles Θ_p^∞ on the $\lambda'^2\Theta_s$ parameter (spherically shaped adsorption sites); the points denote the results of numerical simulations, performed for 1. $\lambda' = 1$, 2. $\lambda' = 1.7$, 3. $\lambda' = 2$, 4. $\lambda' = 2.37$, 5. $\lambda' = 3$, 6. $\lambda' = 4$, The solid lines represent linear regression. Reprinted with permission from Z. Adamczyk, K. Jaszczółt, B. Siwek, P. Weroński, Langmuir 21 (2005), 8952, Copyright (2005) ACS [51].

$$l_s = 5.967/\lambda' - 0.517 \tag{139}$$

Obviously, Eq. (139) is valid for $\lambda' < 4$, otherwise for $\lambda' \geq 4$, $l_s = 1$.

As shown in Fig. 5.23, l_s increases abruptly when the size of the particle approaches the site dimension, that is for $\lambda' \to 1$. From the results shown in Figs. 5.22 and 5.23, one can deduce that the jamming coverage in the RSS process on spherical sites can well be approximated for small values of $\lambda'^2\Theta_s$ by the interpolating function

$$\Theta_p^\infty = \lambda'^2 l_s = \left(\frac{5.967}{\lambda'} - 0.517\right)\lambda'^2\Theta_s \tag{140}$$

The range of validity of Eq. (140) can be determined from Fig. 5.24 where the results of numerical calculations are presented in the form of the dependence of Θ_p^∞ on Θ_∞ plotted in logarithmic scale to cover a broad range

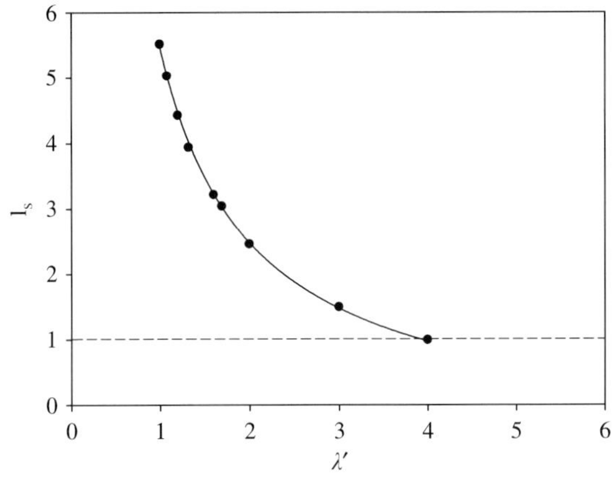

Fig. 5.23. The dependence of the site coordination number l_s on particle to site size ratio λ' (spherically shaped sites). The solid line represents the hyperbolic fitting function: l_s=5.967/λ'−0.517 Reprinted with permission from Z. Adamczyk, K. Jaszczółt, B. Siwek, P. Weroński, Langmuir 21 (2005), 8952, Copyright (2005) ACS [51].

of site coverage. As seen in Fig. 5.24 the limiting analytical results predicted from Eq. (140) reflect well the exact numerical data if $\Theta_p^\infty < 0.2$.

For broader range of jamming coverage and $\lambda' > 2$ the numerical results can be well interpolated by the simple analytical function [52]

$$\Theta_p^\infty = \Theta_\infty \left(1 - e^{-\frac{\lambda'^2 \Theta_s}{\Theta_\infty}}\right) \tag{141}$$

Eq. (141) gives a satisfactory accuracy for the entire range of Θ_s studied. However, it breaks down for $\lambda' = 2$ and 1 when a maximum on the Θ_p vs. Θ_s dependence is observed for Θ_s about 0.25. This maximum jamming coverage for $\lambda' = 2$ attained the value of 0.57, being slightly larger than $\Theta = 0.547$. The maximum appears because, for Θ_s about 0.25 the area accessible for adsorption becomes larger than the geometrical interface area. This is so because particles can touch the sites at various distances from the interface. When Θ_s increases further above this critical value, the accessible area again becomes very close to the geometrical interface area. Although this effect is interesting from a theoretical point of view it does not have significant experimental implications. This is so because in practice it is very difficult to measure surface coverage with a relative accuracy

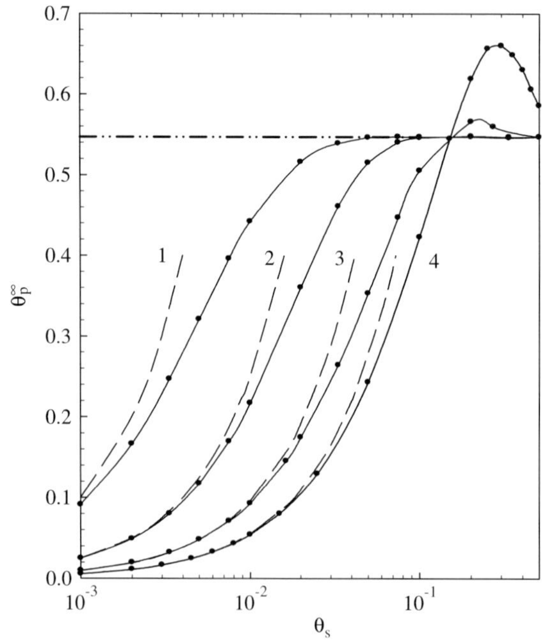

Fig. 5.24. The dependence of the jamming coverage of particles Θ_p^∞ on the coverage of the spherically shaped desorption sites Θ_s; the points denote the results of numerical simulations, performed for 1. $\lambda' = 10$, 2. $\lambda' = 5$, 3. $\lambda' = 2$, 4. $\lambda' = 1$, the solid lines represent the results calculated from the fitting function given by Eq.(141) (for $\lambda' < 2$) and the dashed lines show the results derived from Eq.(136), $\Theta_p = \lambda'^2\Theta_s$. From Ref. [52].

better than 5% due to the polydispersity of real particle suspensions [28]. This hypothesis is confirmed by the fact that in the case of disk-shaped sites, where all particles are adsorbed at the interface only in one plane, no such maximum appeared (see Fig. 5.20). It is also interesting to note that because of the site coordination larger than unity (for $\lambda' < 4$) the jamming coverage in the case of spherical adsorption sites increases with Θ_s more abruptly than for disk-shaped sites.

The results shown in Figs. 5.22 and 5.23 suggest that particle clusters of targeted composition, containing between two and six particles coordinated at one site, can be produced by exploiting the RSS process at spherically shaped sites. For example, on the basis of Fig. 5.22, one can predict that $l_s = 2$ for $\lambda' = 2.37$, $l_s = 3$ for $\lambda' = 1.7$, $l_s = 4$ for $\lambda' = 1.4$ and $l_s = 5$ for $\lambda' = 1.1$.

These predictions were confirmed by the simulations of particle monolayers shown in Fig. 5.25 for $\lambda' = 2.37$, 1.7 and 1.1. The site coverage in these calculations was kept constant at $\Theta_s = 0.005$, and the corresponding particle coverage Θ_p^∞ was 0.053, 0.0428 and 0.0294, respectively. The simulations provided the actual values of $l_s = 1.9$ for $\lambda' = 2.37$, $l_s = 3$ for $\lambda' = 1.7$, and $l_s = 4.9$ for $\lambda' = 1.1$. Note that the number of particles attached to one site, for most of the clusters, was close to the above coordination numbers. The structure of particles forming the surface clusters is reflected by the pair correlation function, which is also shown in Fig. 5.25. Note that for the RSS process there are three various pair correlation functions (i) site–site, given by the homogeneous RSA model, (ii) particle–site and (iii) particle–particle. Only the latter is shown in Fig. 5.25. It was calculated from the constitutive relationship

$$g(r) = \frac{\pi a_p^2}{\Theta_p} \left\langle \frac{\Delta N_p}{2\pi r \, \Delta r} \right\rangle \tag{142}$$

where $\langle \; \rangle$ means the ensemble average and N_p is the number of particles adsorbed within the ring $2\pi r \Delta r$ drawn around a central particle (see Fig. 5.5). For convenience the distance r was normalized by using the particle radius a_p as a scaling variable. Note that the distance r appearing in Eq. (142) was measured between the projection of the adsorbed particle centers on the adsorption plane. Obviously, all particles located close to the perimeter of the simulation area were discarded from the averaging procedure. For calculating $g(r)$, particle populations reaching 10^5 were considered.

As seen in Fig. 5.25, the pair correlation function exhibited a maximum of a considerable height at the distance $r \sim 2a$ that varies from 22 (for $\lambda' = 2.37$) to 35 (for $\lambda' = 1.1$). Interestingly, for $\lambda' = 1.1$ there was a secondary peak at the distance $r = 3.5\, a$, which reflects particles located on opposite sides of the cluster.

The ordering observed in the clusters was dependent on the coverage of particles as seen in Fig. 5.26. In this figure, the monolayers of particles at the jamming state with the corresponding pair correlation function are presented for $\lambda' = 1.1$ and Θ_s increasing progressively from 0.01 to 0.1. The jamming coverage was $\Theta_p = 0.058$, 0.113 and 0.255, respectively. Note that the primary peak height decreased from 18 to ~ 3 when Θ_p increased from 0.058 to 0.437. Hence, these results indicate that the degree

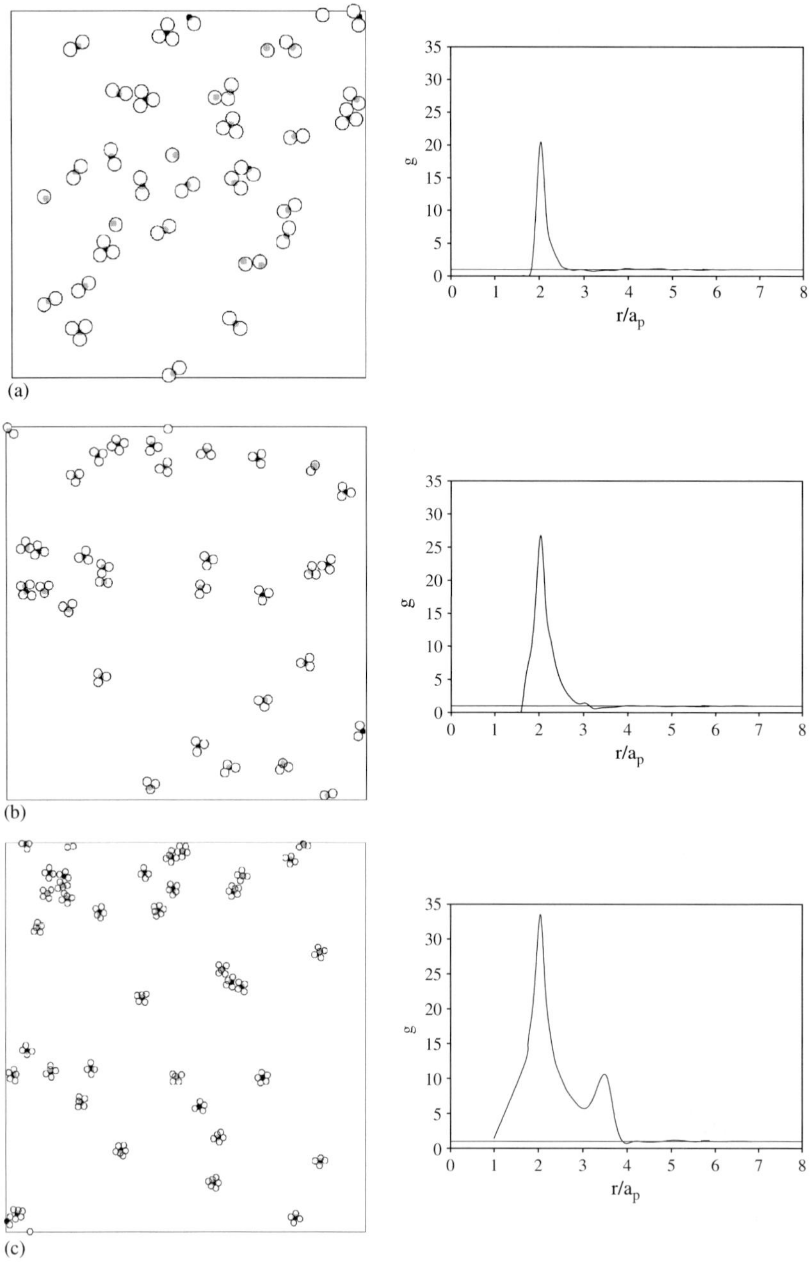

Fig. 5.25. Monolayers of the particles at site-covered surfaces (spherically shaped) with the corresponding pair correlation function derived from simulations $\Theta_s = 0.005$. (a) $\lambda' = 2.37$, $\Theta_p = 0.053$. (b) $\lambda' = 1.7$, $\Theta_p = 0.0428$. (c) $\lambda' = 1.1$, $\Theta_p = 0.0294$. Reprinted with permission from Z. Adamczyk, K. Jaszczółt, B. Siwek, P. Weroński, Langmuir 21 (2005), 8952, Copyright (2005) ACS [51]

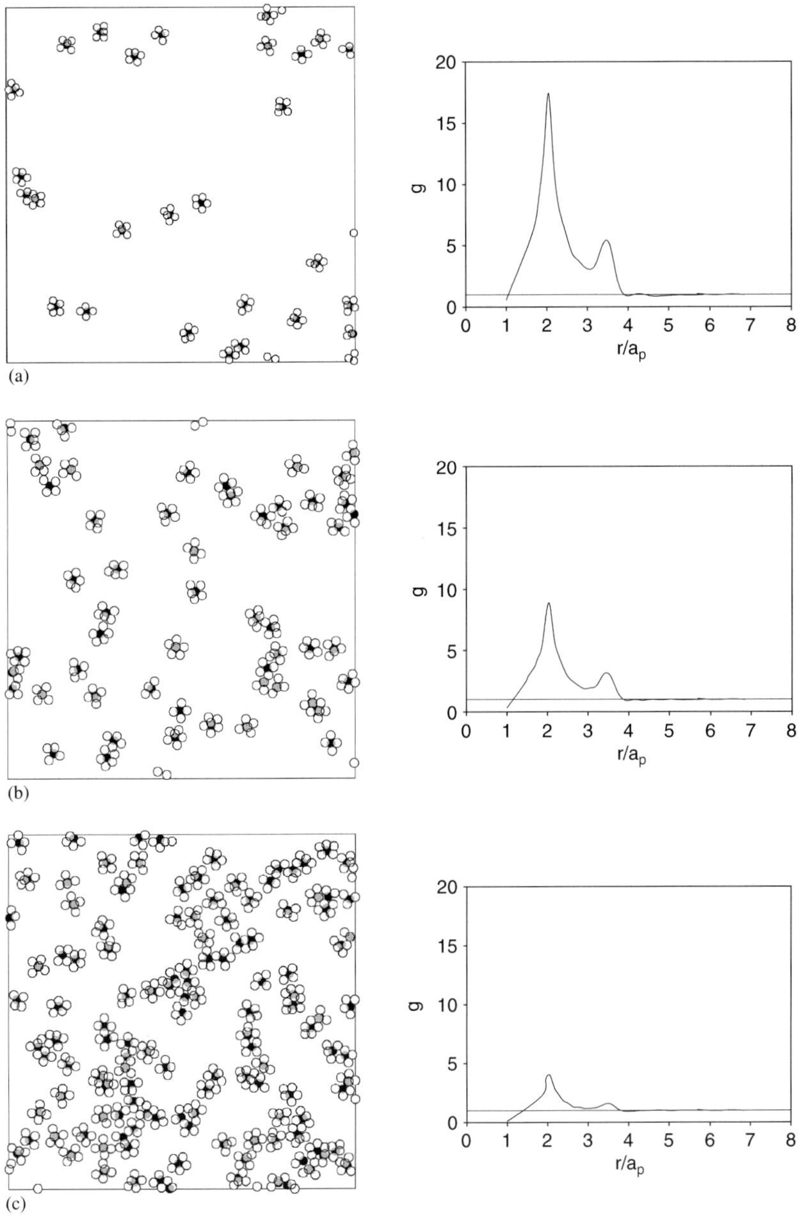

Fig. 5.26. Monolayers of the particles at the jamming state at site-covered surfaces (spherically shaped) with the corresponding pair correlation function derived from simulations for $\lambda' = 1.1$. (a) $\Theta_s = 0.01$, $\Theta_p = 0.058$. (b) $\Theta_s = 0.02$, $\Theta_p = 0.113$. (c) $\Theta_s = 0.05$, $\Theta_s = 0.255$. Reprinted with permission from Z. Adamczyk, K. Jaszczółt, B. Siwek, P. Weroński, Langmuir 21 (2005), 8952, Copyright (2005) ACS [51].

of ordering of particles largely vanished because of the overlapping of the adjacent clusters. One can deduce from the results shown in Figs. 5.22 – 5.26 and others, discussed in Refs.[51,52] that surface clusters of desired composition can be produced in RSS processes on surfaces covered by spherical sites in the limit of low coverage, $\Theta_p < 0.01$.

For convenience, jamming coverage of spherical particles occurring under various conditions is collected in Table 5.1.

The RSS simulations for spherically shaped sites allow one to determine the jamming coverage, the structure of the monolayers and the blocking function B, which is of importance for predicting particle adsorption kinetics. As can be deduced, in the general case the reduced $\bar{B} = \bar{B}/p_0$, which exhibits a more universal character, is dependent on Θ_s, Θ_p and λ'. This makes a systematic analysis of its behavior quite complicated. Therefore, the main goal of the calculations performed in Ref. [50] was to find links between the blocking function for the RSS process with the well known B function for the RSA over homogeneous surfaces.

In Fig. 5.27 the dependence of the reduced $\bar{B}$ function on the particle coverage Θ_p is shown, derived from RSS simulations for $\lambda' = 2$ and various Θ_s [50]. As can be seen, for lower site density, $\Theta_p = 0.1$ ($\alpha_s < 0.4$), and $\Theta_s < 0.2$, the B function can be well described by the *quasi*-Langmuir model, expressed by Eq. (120). On the other hand, for higher site density, $\Theta_s = 0.5$ ($\alpha_p = 2$) the B function in the RSS process, for the entire range of Θ_p, can be well approximated by the B pertinent to homogeneous surfaces, calculated from Eq. (86). Similar results were obtained for higher λ' values as well, see Fig. 5.28. These results indicate that in the RSS process on heterogeneous surfaces, for $\alpha_p = \lambda'^2 \Theta_s > 1$, the B function is described well by the formula

$$p = p_0\, B(\Theta_p) = (1-\Theta_s)\, e^{-\frac{(4\lambda'-1)\Theta_s}{1-\Theta_s} - \left[\frac{(2\lambda'^{1/2}-1)\Theta_s}{1-\Theta_s}\right]^2} \left(1+\bar{C}_1\bar{\Theta}_p + \bar{C}_2\bar{\Theta}_p^2 + \bar{C}_3\bar{\Theta}_p^2\right)(1-\Theta_p)^3 \tag{143}$$

where $\bar{C}_1 = 0.812$, $\bar{C}_2 = 0.426$, $\bar{C}_3 = 0.0716$

The jamming coverage occurring in Eq. (143) can be calculated from Eqs.(140,141). In the limit when $\bar{\Theta}_p \to 1$, $B(\bar{\Theta})$ for homogeneous surfaces assumes the form $B(\bar{\Theta}) \cong 2.31(1-\bar{\Theta})^3$, as previously derived. Therefore, in the limit of $\Theta_p \to \Theta_\infty$, the $B(\Theta)^{1/3}$ function determined for the RSS process is expected to depend linearly on the reduced particle coverage Θ_p/Θ_p^∞. The

Table 5.1
Jamming coverages of spherical particles

Monolayer configuration	Θ_{mx}	Remarks
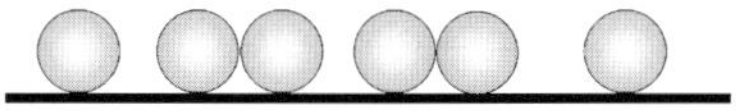 1D, random	0.7476	Parking problem, analytical solution
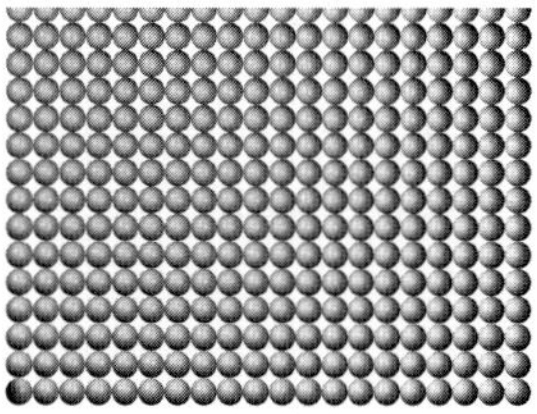 2D, regular	$\frac{\pi}{4} = 0.7854$	From geometry
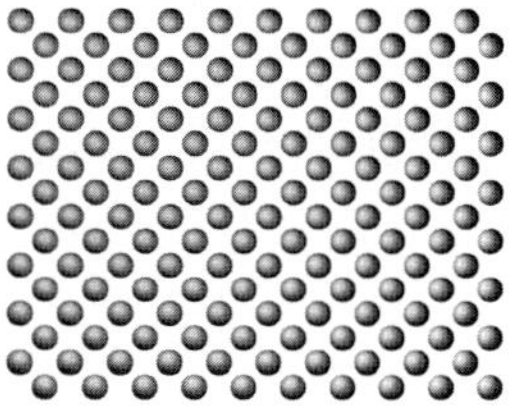 2D, regular-extended	$\frac{\pi}{8} = 0.3927$	From geometry
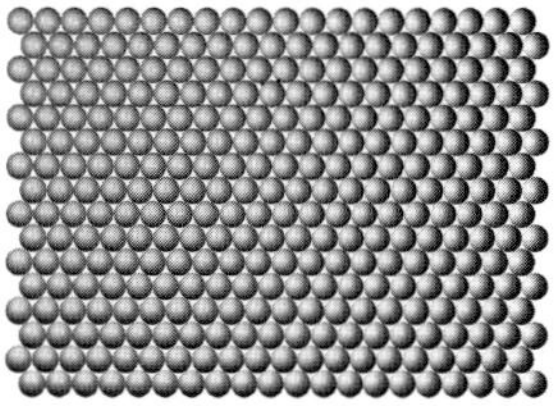 2D, hexagonal	$\frac{\pi}{2(3)^{1/2}} = 0.9069$	From geometry

Table 5.1 (continued)

Monolayer configuration	Θ_{mx}	Remarks
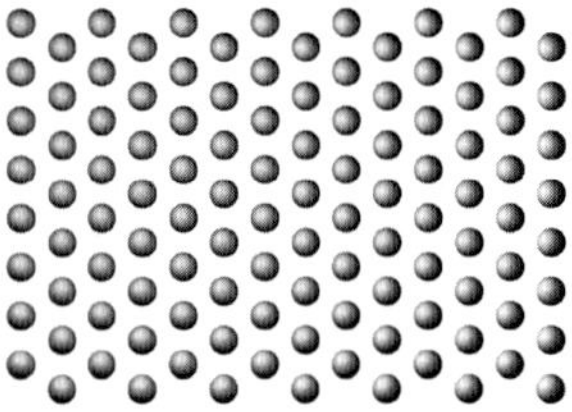 2D, hexagonal-extended	$\frac{\pi}{6(3)^{1/2}} = 0.3023$	From geometry
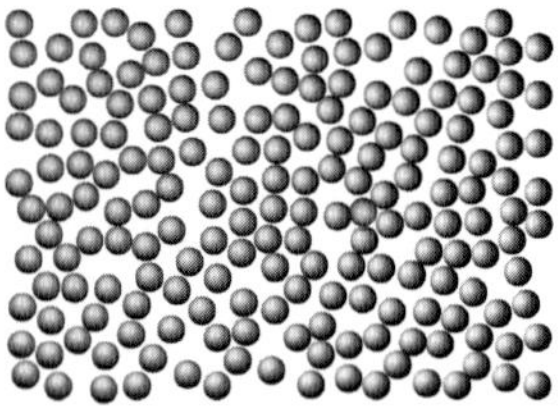 2D, random	0.5467	Derived from RSA simulation
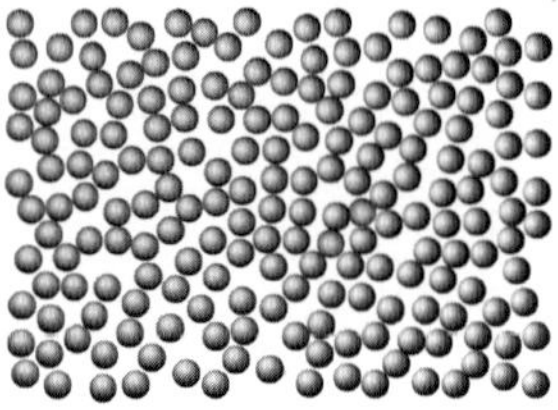 2D, correlated	0.6105	Derived from the ballistic model
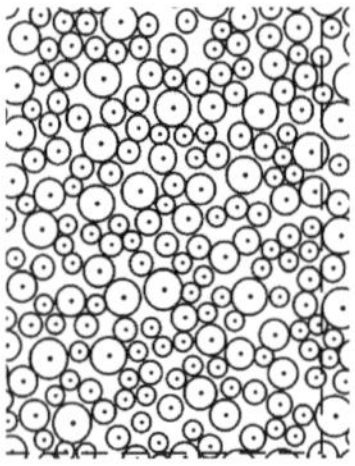 2D, random polydisperse	$\Theta_{\infty} + 0.53\bar{\sigma}_p$ for $\Theta = \pi\langle a\rangle^2 N$ $\Theta_{\infty} + 0.458\bar{\sigma}_p + 6.055\bar{\sigma}_p^2$ for $\Theta = \sum_l \pi a_l^2 N_l$	Derived from RSA simulations for uniform particle size distribution having the standard deviation $\bar{\sigma}_p$

Table 5.1 (continued)

Monolayer configuration	Θ_{mx}	Remarks
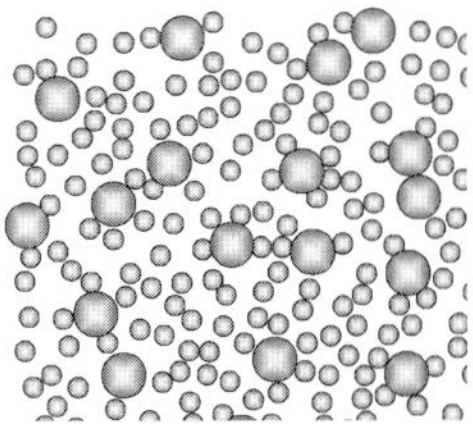 2D, precovered surfaces $\lambda'=2.2$	$\Theta_\infty\left(\frac{C_{RSS}}{C_{RSS}-\Theta_s}\right)^{-2}$ $C_{RSS}=0.596$ (for $\lambda'=2.2$) $C_{RSS}=0.404$ (for $\lambda'=5$) $C_{RSS}=0.274$ (for $\lambda'=10$)	Derived from RSA simulations, Θ_s = smaller particle coverage $\lambda'=a_p/a_5$
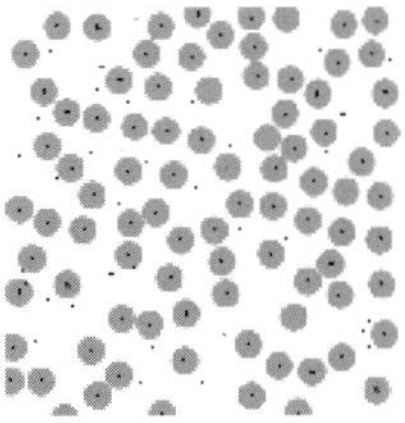 2D, heterogeneous, point-like sites	$1-\frac{1+0.314\alpha_p+0.45\alpha_p^2}{1+1.83\alpha_p+0.66\alpha_p^2+\alpha_p^{7/2}}$ $\alpha_p=\Theta_s\,\lambda'^2$	Interpolating function valid for arbitrary α_p
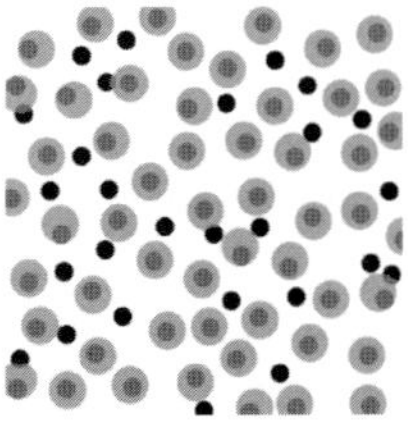 2D, heterogeneous disk-shaped sites, $\lambda'=2.2$	$\Theta_\infty\frac{C_0'\alpha_p^l}{1+C_0'\alpha_p^l}$ $\alpha_p=\Theta_s\,\lambda^2$	Derieved from RSS simulations, Θ_s – initial disk coverage
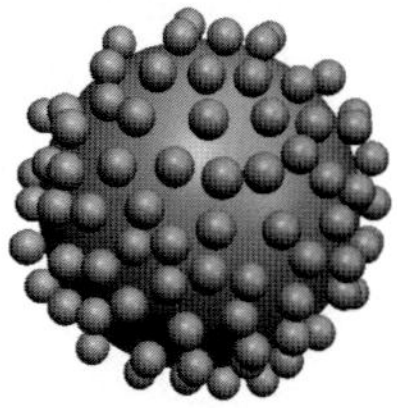 2D, random, spheres on spherical interface $a/R=10$	$\Theta_\infty\left(1\pm\frac{a}{R}\right)^2$	Derived from RSA simulations, upper sign – adsorption on outside surface lower sign – adsorption on inside surface

Table 5.1 (continued)

Monolayer configuration	Θ_{mx}	Remarks
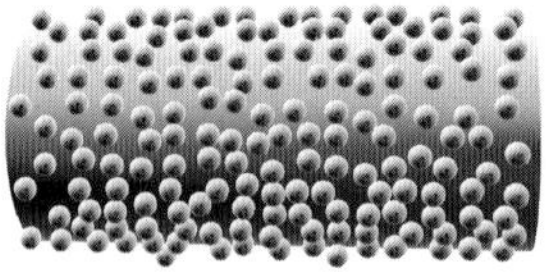 2D, random, spheres on cylindrical interface $a/R = 10$	$\Theta_\infty\left(1 \pm \frac{a}{R}\right)$	Derived from RSA simulations, upper sign – adsorption on outside surface lower sign – adsorption on inside surface
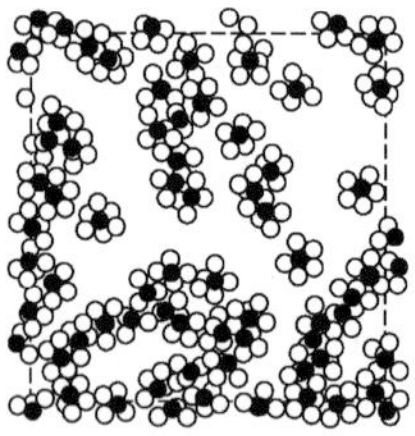 3D, heterogeneous, spherical sites, $\lambda' = 1$	$\Theta_\infty\left(1-e^{-\frac{\lambda'^2\Theta_s}{\Theta_\infty}}\right)$ $\Theta_s\lambda'^2\left(\frac{5.967}{\lambda'}-0.517\right)$	Interpolating function derived from RSA simulations valid for $\lambda'>2$ valid for $\lambda'< 4$ and $\Theta_s\lambda'^2<<1$
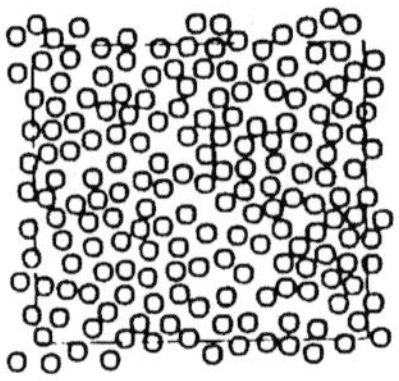 2D, random, soft particles $h^*/a = 0.1$	$\frac{\Theta_\infty}{\left(1+\frac{h^*}{a}\right)^2}$ $\frac{h^*}{a} \cong \frac{1}{2\kappa a}\ln(\phi_0/kT)$	Analytical approximation derived from RSA simulations

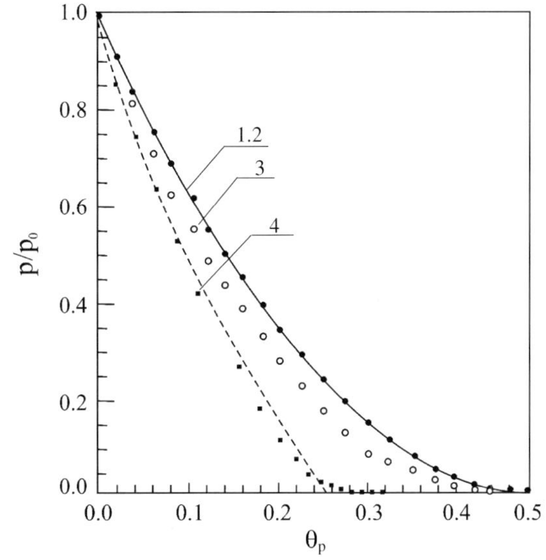

Fig. 5.27. The dependence of the reduced blocking function $\bar{B} = p/p_0$ on the particle coverage Θ_p for spherically shaped sites. The points denote numerical results obtained for $\lambda' = 2$. 1. $\Theta_s = 0.5$ ($\alpha_p = 2$), 2. $\Theta_s = 0.2$ ($\alpha_p = 0.8$), 3. $\Theta_s = 0.1$ ($\alpha_p = 0.4$), 4. $\Theta_s = 0.05$ ($\alpha_p = 0.2$), the solid line represent the result derived from the RSA model for homogeneous surfaces and the dashed line shows the analytical results calculated from Eq. (120). From Ref. [50].

results of numerical calculations for various site coverage and λ' values plotted in Fig. 5.29 using this transformation confirm fully this hypothesis and the validity of Eq. (143). This equation has a major significance because it allows one to transfer results derived for homogeneous surfaces to heterogeneous surface adsorption. In particular, by substituting the expression for the $B(\Theta)$ given by Eq. (143) into the constitutive dependence, Eq. (60) one obtains upon integration the kinetic expression

$$\int_0^{\Theta} \frac{d\bar{\Theta}'}{B(\bar{\Theta}')} = p_0\tau/\Theta_p^{\infty} = \tau' \tag{144}$$

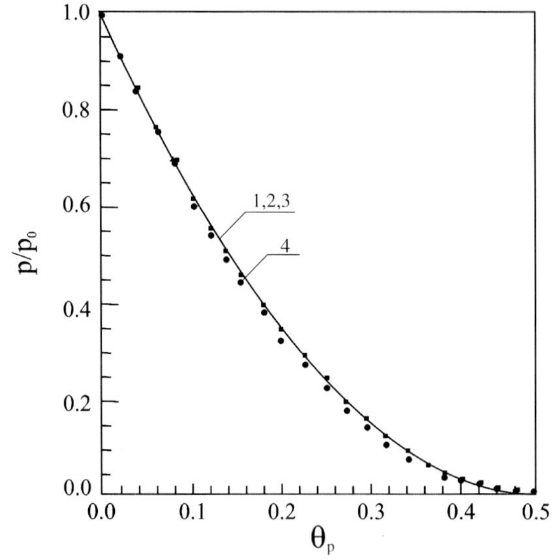

Fig. 5.28. The dependence of the reduced blocking function $\bar{B} = p/p_0$ on the particle coverage Θ_p for spherically shaped sites. The points denote numerical results obtained for $\lambda' = 5$. (1) $\Theta_s = 0.5$ ($\alpha_p = 12.5$), (2) $\Theta_s = 0.2$ ($\alpha_p = 5$), (3) $\Theta_s = 0.1$ ($\alpha_p = 2.5$), (4) $\Theta_s = 0.05$ ($\alpha_p = 1.25$), the solid line represents the result derived from the RSA model for homogeneous surfaces. From Ref. [50].

Eq. (144) shows that all the kinetic results known previously for the continuous surfaces can be directly transferred to heterogeneous surfaces by introducing the transformed adsorption time $\tau' = p_0(\Theta_s, \lambda')\,\tau/\Theta_p^\infty$. In particular, by substituting $B(\bar{\Theta}) \cong 2.31/(1-\bar{\Theta}_p)^3$ into Eq. (144) one obtains upon integration the limiting result

$$\Theta(\tau) = \Theta_p^\infty\left(1 - \frac{C'}{(\tau')^{1/2}}\right) \tag{145}$$

where

$$C' = \left(\frac{\Theta_p^\infty}{4.62\,p_0}\right)^{1/2} \tag{146}$$

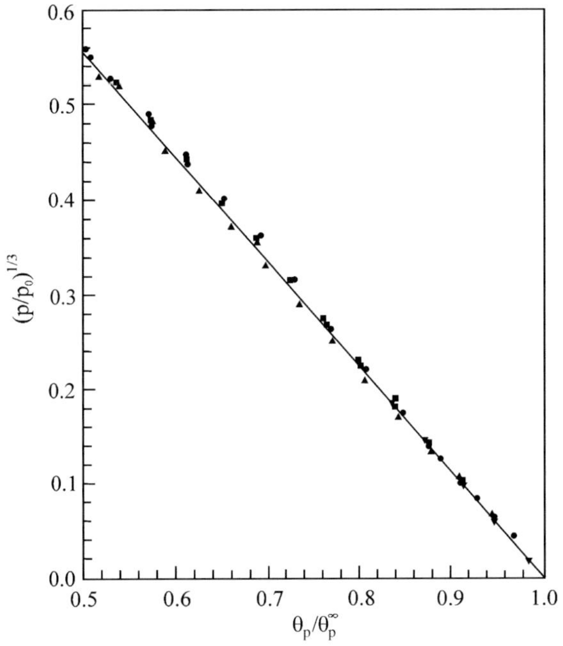

Fig. 5.29. The dependence of $(p/p_0)^{1/3} = \bar{B}^{1/3}$ on the reduced coverage of particles Θ_p/Θ_p^∞ for spherically shaped sites. The points denote numerical results obtained for various λ' and spherical site coverage Θ_s. From Ref. [50].

However, in the general case, with $B(\bar{\Theta}_p)$ given by Eq. (86), the integration procedure is rather awkward.

5.4. THE RSA MODEL OF NON-SPHERICAL PARTICLES

The RSA model can also be effectively used for deriving information on non-spherical (anisotropic) particle adsorption, especially the blocking function, structure of monolayers, fluctuations, the jamming coverage, etc. Studying adsorption of anisotropic particles is of vital significance because the shape of most surfactant molecules, polyelectrolytes bacteria and proteins deviates from the spherical shape, often resembling prolate spheroids. For example bovine serum albumin (BSA) as well as human serum albumin (HSA) can be approximated by a prolate spheroid of the dimensions 14 × 4 × 4 nm [53]. On the other hand, fibrinogen (important for blood clogging protein) has an elongated, spheroidal shape and has the of dimensions 45 × 6 × 6 (nm. [54,55]. The small protein lysosyme has also a spheroidal shape and the dimensions 4.5 × 3 × 3 nm [56]. Other examples of highly anisotropic particles are the red blood cells, blood platelets, silicates, synthetic colloids, gold, silver, silver iodide, barium sulfate, (poly-tetra fluoro-ethylene) PTFE.

Using the Derjaguin model discussed in Chapter 2 the interaction energy of an oblate spheroid with the interface in the side-on orientation is given by

$$\phi_{\parallel} = As\,\phi_0 \tag{147}$$

where As $= a/b$ is the spheroid ax is ratio, ϕ_0 is the interaction energy of a spherical particle of radius a_s (equal to the major axis of the spheroid). For the perpendicular (edge-on) orientation one has

$$\phi_{\perp} = \frac{\phi_0}{As} \tag{148}$$

Hence, the interaction energy in the perpendicular orientation is As^2 times smaller.

For prolate spheroids one has

$$\begin{aligned} \phi_{\parallel} &= \phi_0 \\ \phi_{\perp} &= \frac{1}{As^2}\phi_0 \end{aligned} \tag{149}$$

for the side-on and edge-on orientations, respectively.

By taking the typical data $A_{132} = 10^{-20}$ J, $\delta_m = 0.5$ nm , one can calculate that the van der Waals interaction energy $\phi_0 = -8.2\,kT$ for $a_s = 10$ nm (a typical value for proteins). Thus, for oblate spheroids, characterized by $As = 3$, one has $\phi_{\parallel} = -24\,kT$ in the side-on orientation and $\phi_{\perp} = -2.7$ for the edge-on orientation. For prolate spheroids with the same parameters one has $\phi_{\parallel} = -8.2$ kT for the side-on orientation and $\phi_{\perp} = -0.9\,kT$ for the edge-on orientation. These estimates suggest that the van der Waals interactions are not sufficient to ensure an irreversible deposition of particles of this size in the edge-on orientation. However, one can calculate in an analogous way that in the energy of electrostatic interactions, for typical values of zeta potential, equal to 25 mV for particles and –50 mV for the substrate surface, is three times lower than the van der Waals interaction energy [57].

From these estimations one can predict that for particles of elongated shape (like prolate spheroids) smaller than 10 nm the side-on adsorption regime will be the only one favored energetically.

The unoriented adsorption regime will be possible for particles larger than 50 nm. On the other hand, for the oblate spheroids the unoriented regime may appear already for particle size about 10 nm. In this section, we consider these two regimes separately by assuming that the size of the adsorption sites is much smaller than the particles and that their surface coverage is large enough, so the parameter $\lambda'^2\ \Theta_s >> 1$. It is further assumed that after adsorption, particle positions and orientations are time independent, which corresponds to the postulate of localized and irreversible adsorption. Under this assumption, the side-on and unoriented adsorption regimes can be effectively studied by applying the RSA model used above for spheres. Let the RSA process of particles, of axis-symmetric shape, proceed for some time. Then, there are N_p adsorbed particles over the simulation plane whose positions are described uniquely by the set of surface position vectors $\mathbf{r}_1$ to $\mathbf{r}_n$ and the orientation versors $\hat{\mathbf{e}}_1$ to $\hat{\mathbf{e}}_n$ depending on the angles $\alpha_n E\ \langle 0;\pi\rangle$ measured between the x- axis of the Cartesian reference system and the projection of the symmetry axis on the adsorption plane. The angles $\beta_n E\ \langle 0;\pi/2\rangle$ are measured between the z- axis and the particle symmetry axis (see Fig. 5.30).

It was shown in Refs.[58,59] that the $B(\Theta)$ function, defined as the probability of adsorption of the $N+1$ particle (virtual particle), can be expressed in terms of the series expansion, analogous to the sphere adsorption case

$$B(\Theta) = 1 - C_1\Theta + C_2\Theta^2 + 0(\Theta)^3 \tag{150}$$

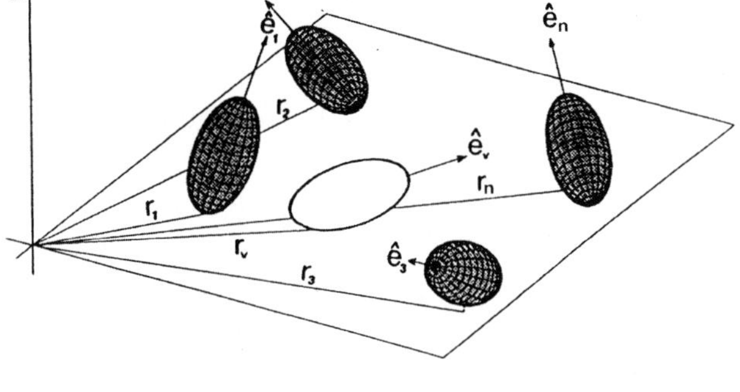

Fig. 5.30. A schematic representation of unoriented adsorption of axis-symmetric particles on a planar interface.

where the coverage is defined as $S_g N_p$, with S_g being the characteristic cross-section area of the particle.

The C_1 constant can be evaluated from the multiple integral originating from the geometry of the two-particle system: the adsorbed particle and the virtual particle contacting the surface, see Fig. 5.31 a [58,59]

$$C_1 = -\frac{4}{\pi^3 S_g} \int_0^{\pi} d\alpha_{1v} \int_0^{\pi/2} d\beta_1 \int_0^{\pi/2} d\beta_v \int_s f_{1v} d\mathbf{r}_v \tag{151}$$

where $\mathbf{r}_v$ is the relative position vector of the virtual particle, α_v the relative orientation angle, $f_{1v}=\rho_{pv}-1$ the Mayer function and ρ_{pv} the probability density of adsorbing the virtual particle, which is dependent on the energy of the pair interaction.

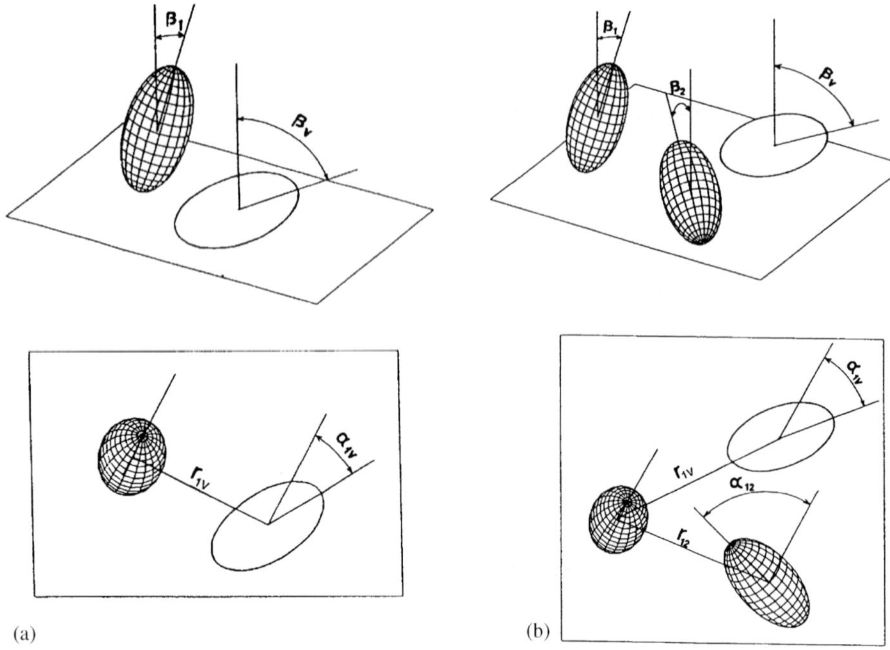

Fig. 5.31. A perspective view (upper part) and the projection (lower part) of the two particle configuration occurring in calculations of the C_1 constant. (b) A perspective view (upper part) and the projection (lower part) of the three particle configuration occurring in calculations of the C_2 constant.

For hard particles the interaction energy becomes infinitely large if particles overlap ($f_{1v} = -1$) and zero otherwise ($f_{1v} = 0$). In this case the C_1 constant can be interpreted, analogously to spheres, as the excluded area averaged over all orientations of the adsorbed and virtual particle.

In an analogous way the C_2 constant can be evaluated from the multiple integral originating from the geometry of the three particle system: two adsorbed particles forming a pair and the virtual particle contacting the surface, see Fig. 5.30 part "b" [59]

$$C_2 = \frac{4}{\pi^5 S_g^2} \int_0^{\pi} g d\alpha_{12} \int_0^{\pi} d\alpha_{1v} \int_0^{\pi/2} d\beta_1 \int_0^{\pi/2} d\beta_2 \int_0^{\pi/2} d\beta_v \iint_{ss} \left[2C_1 + f_{2v}\right] d\mathbf{r}_{12} d\mathbf{r}_v \quad (152)$$

where $g(\mathbf{r}_{12},\alpha_{12},\beta_1,\beta_2)$ is the pair correlation function, depending on the energy of interaction of the particle pair, $\mathbf{r}_{12}=\mathbf{r}_1-\mathbf{r}_2$ the relative position vector of the two adsorbed particles, $\alpha_{12}=\alpha_1-\alpha_2$ the relative orientation angle, $f_{2v}=\rho_{pv}-1$ the Mayer function for the cluster and ρ_{pv} the probability density of adsorbing the virtual particle near the pair of particles, which is dependent on the energy of the pair interaction.

For hard particles, the C_2 constant can be interpreted, analogously to spheres, as the common area of the two excluded areas near the pair of adsorbed particles, averaged over all orientations and positions.

Eqs.(151),(152) are valid for both reversible (thermodynamic) and irreversible systems. Under thermodynamic equilibrium conditions one usually assumes that the probability density is described by the Boltzmann law

$$\rho_{pv} = e^{-\phi_v / kT} \quad (153)$$

where ϕ_v is the interaction energy of the particle with the clusters (composed of one or two particles). Usually, the additivity principle is applied to calculate the energy of interaction.

The Bolzmann postulate is often extended to RSA processes as discussed later on.

5.4.1. The side-on adsorption of non-spherical particles

For the side on adsorption, all the β_n angles become $\pi/2$. This greatly simplifies the expressions for the C_1 constant, which is becomes

$$C_1 = -\frac{1}{\pi S_g} \int_0^{\pi} d\alpha_v \int f_{1v} d\mathbf{r}_v \tag{154}$$

For hard particles, f_{1v} = -1 when the particles overlap, so the C_1 constant equals the exclusion area averaged on all virtual particle orientations.

Considering that $C_1 = 2\bar{B}_2$, one does not need to evaluate the integral, Eq.(154), because the second virial coefficient $\bar{B}_2$ is known in an analytical form for hard convex bodies from Boublik's theory, being $1 + \gamma_p$. Therefore, in order to evaluate C_1 one has to calculate the particle shape parameter given by the expression

$$\gamma_p = \frac{P_r^2}{4\pi S_g} \tag{155}$$

Explicitly, for cylinders adsorbing side on, whose characteristic cross-section area is a rectangle, $S_g = 2bL$ (see Table 5.2), and γ_p is given by

$$\gamma_p = \frac{(As+1)^2}{\pi As} \tag{156}$$

where $As = L/2b$.
If $L = 2b$ (squares), $\gamma_p = 4/\pi$, and for needles, $\gamma_p = As/\pi$.

For spherocylinders of length L and width $2b$ (see Table 5.2)

$$\gamma_p = \frac{(2As+\pi-2)^2}{4\pi\left(As-1+\frac{\pi}{4}\right)} \tag{157}$$

For prolate spheroids (ellipses) of length $L = 2a_s$ and width $2b_s$ (see Table 5.2) the γ_p parameter is given by [19]

$$\gamma_p = \frac{4As}{\pi^2} E^2\left(\frac{As^2-1}{As^2}\right) \tag{158}$$

where

$$E(x) = \int_0^{\pi/2} (1 - x\sin^2\xi)d\xi \tag{159}$$

is the complete elliptic integral of the second kind.

One can calculate from the above equations that for $As = 2$, the C_1 constant equals 4.86 for cylinders, 4.38 for spheroids and 4.356 for spherocylinders. For very elongated particles, where $As = 15$, the C_1 constant equals 12.86 for cylinders, 14.35 for spheroids, and 12.44 for spherocylinders. In the limit of needle-like objects, the C_1 constant becomes $2As/\pi$ [19]. For convenience the C_1 values for various particle shapes are compiled in Table 5.2. The C_2 constant, given by Eq.(152), for the side- on adsorption, can be calculated from the integral [59]

$$C_2 = \frac{1}{2\pi^2 S_g^2} \int_0^{\pi} g d\alpha_{1v} \int_0^{\pi} d\alpha_{1v} \iint_{ss} \left[2C_2 + f_{2v}\right] d\mathbf{r}_{12} d\mathbf{r}_v \tag{160}$$

Unfortunately the integral, Eq.(160), cannot be evaluated analytically, not even for hard particles. The most efficient calculation method is the Monte Carlo type integration [60], which is a rather tedious procedure. However, useful estimates of C_2 can be derived by using the SPT expansion, Eq.(36)

$$C_2 = 2\gamma_p^2 + \gamma_p + \frac{1}{2} \tag{161}$$

For rectangles Eq.(161) becomes explicitly

$$C_2 = 2\frac{(As+1)^4}{\pi^2 As^2} + \frac{(As+1)^2}{\pi As} + \frac{1}{2} \tag{162}$$

Table 5.2
Values of γ_p, $\overline{B}_3/\overline{B}_2^2$, C_1, C_2, and Θ_∞ for side-on adsorption of convex particles of various shape

Particle	Axis ratio $As = a_s/b_s = L/2b_s$	γ_p	$\overline{B}_3/\overline{B}_2^2$,	$C_1 = 2\overline{B}_2$	C_2	Θ_∞
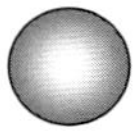 Sphere (disk)	1:1	1	0.782	4.00 (3.50)[a]	3.31	0.547
 Cylinder (square)	1:1	1.273	0.770	4.55 (5.01)[a]	4.37	0.530
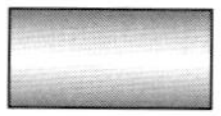 Cylinder (rectangle)	2:1	1.432	0.750	4.86 (6.03)[a]	5.17	0.548
 Spheroid (ellipse)	2:1	1.189	0.750	4.38 (4.52)[a]	4.20	0.583

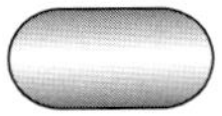 Spherocylinder (diskorectangle)	2:1	1.178	0.756	4.36	4.11 (4.45)[a]	0.581
 Cylinder (rectangle)	5:1	2.292	0.675	6.58	10.7 (13.3)[a]	0.510
 Spheroid (ellipse)	5:1	2.236	0.651	6.47	10.71 (12.7)[a]	0.536
 Spherocylinder (diskorectangle)	6:1	2.375	0.655	6.75	11.6 (14.2)[a]	0.524
 Cylinder (rectangle)	15:1	5.432	0.588	12.9	46.3 (64.9)[a]	0.483

Table 5.2 (continued)

Particle	Axis ratio $As = a_s/b_s = L/2b_s$	γ_p	$\overline{B}_3/\overline{B}_2^2$,	$C_1 = 2\overline{B}_2$	C_2	Θ_∞
 Spherocylinder	15:1	5.22	0.582	12.4	43.6 (60.3)[a]	0.445
Needle[b]	∞	$\frac{2}{\pi}$	0.514	0.637	0.124	–

General expressions:

$$C_1 = 2\overline{B}_2$$

$$C_2 = \frac{1}{2} C_1^2 \left(1 - \frac{3\overline{B}_3}{4B_2^2}\right)$$

[a]SPT results for equilibrium systems [9, 19]: $C_1 = 2(1+\gamma_p)$; $C_2 = 2+\gamma_p^2+\gamma_p+\frac{1}{2}$, $\gamma_p = \frac{P_r^2}{4\pi S_g}$

[b]By defining $S_g = L^2, \Theta = L^2 N_p$.

Special cases:

$$\gamma_p = \frac{(As+1)^2}{\pi As} \quad \text{cylinders} \; ; \quad \gamma_p = \frac{(\pi + 2As - 2)^2}{4\pi\left(As - 1 + \frac{\pi}{4}\right)} \quad \text{spherocylinders}$$

$$\gamma_p = \frac{4As}{\pi^2}\left[\int_0^{\pi/2}\left(1 - \frac{As^2 - 1}{As^2}\sin^2\xi\right)d\xi\right]^2 \quad \text{spheroids}$$

$$C_1 = \frac{4(\pi^2 - 4)}{\pi^2} + \frac{8}{\pi^2}\left(\frac{1}{As} + As\right)$$

$$C_2 = q_s(As)C_1^2, \quad q_s(As) = 0.126 + 0.181\left(\frac{\frac{1}{As} + As}{\frac{1}{As} + As + 0.448}\right)^4$$

Exact results for $As \to \infty$ (needles) [19]

$\overline{B}_2 = 1 + As/\pi$

$C_1 = 2(1 + As/\pi)$

$C_2 = 1.229\,(1 + As/\pi)^2$

For squares

$$C_2 = \frac{32}{\pi^2} + \frac{4}{\pi} + \frac{1}{2} = 5.015 \tag{163}$$

For spherocylinders one has accordingly

$$C_2 = \frac{(2As + \pi - 2)^4}{8\pi^2\left(As - 1 + \frac{\pi}{4}\right)^2} + \frac{(2As + \pi - 2)^2}{4\pi\left(As - 1 + \frac{\pi}{4}\right)} + \frac{1}{2} \tag{164}$$

For needles of unit length

$$C_2 = \frac{2A_S^2}{\pi^2} \tag{165}$$

It was demonstrated in Ref.[60] that in the case of spheroids (ellipses), the C_1 and C_2 constants can be well approximated by the interpolating functions

$$C_1 = \frac{4(\pi^2 - 4)}{\pi^2} + \frac{8}{\pi^2}\left(\frac{1}{As} + As\right)$$

$$C_2 = q_s(As)C_1^2 \tag{166}$$

$$q_s(As) = 0.126 + 0.181\left(\frac{As + \frac{1}{As}}{As + \frac{1}{As} + 0.448}\right)^4$$

Eq. (166) ensures a few percent accuracy for $1 < As < 15$.

These approximate results are compared in Table 5.2 with the exact data derived by numerical integration of Eq. (160).

However, evaluation of higher order expansion coefficients for anisotropic particles becomes inefficient, therefore no exact numerical data are known. A good estimate of the blocking function for the RSA of anisotropic particles for a broad range of coverage can be gained by using the equilibrium ASF, expressed by Eq. (18). However, this formula becomes inaccurate at higher coverage, close to the jamming state, which can only be derived from numerical simulations.

On the other hand the asymptotic form of the ASF for the side-on adsorption regime close to the jamming state can be deduced analogously as for spheres, by analyzing the kinetics of target disappearance. In this case, the isolated targets are 3D domains consisting of two spacial and one orientational coordinate. Thus the adsorption probability is proportional to this domain size [9]

$$p_t \sim \delta_{x_v} \delta_{y_v} \delta_{\alpha_v} \cong C_p h_t^3 \tag{167}$$

where h_t is the size of the target and $\delta\alpha_v \sim h_t$.

Accordingly, upon integration of the kinetic equation analogous to Eq. (77) derived previously for spheres, the total concentration of targets of any size is given by

$$N_t\,(h_t,\tau) = N_0\, e^{-C_p h_t^3 \tau} \tag{168}$$

Assuming that N_0 is uniform [6], Eq. (168) can be integrated with respect to h_t, which results in the formula

$$\Theta = \Theta_\infty - C_p' \tau^{-1/3} \tag{169}$$

where C_p' is the dimensionless constant.

Thus, the $B(\Theta)$ function becomes

$$B(\Theta) = \frac{d\Theta}{d\tau} = \frac{1}{3C_p'}(\Theta_\infty - \Theta)^4 \tag{170}$$

As can be observed from Eq. (169), the jamming coverage Θ_∞ can be determined by extrapolating the dependence of Θ on $\tau^{-1/3}$ obtained from numerical simulations to $\tau^{-1/3} \to 0$ (infinite time). It is also interesting to mention that Eqs.(168),(169) are valid for particles of arbitrary shape.

By exploiting the asymptotic form of the blocking function, given by Eq. (170), Ricci et al. [10] proposed a convenient interpolating expression

$$B(\Theta) = \left(1 + C_1' \bar{\Theta} + C_2' \bar{\Theta}^2\right)\left(1 - \bar{\Theta}\right)^4 \tag{171}$$

where $\bar{\Theta} = \Theta/\Theta'_\infty$ and C'_1, C'_2,Θ'_∞ are the fitting parameters to be derived by matching Eq. (171) with the low coverage expansion, Eq. (150). For spheres and spheroids it was found [10]

$$\begin{aligned} C_1' &= 0.788, \quad C_2' = 0.375, \quad \Theta_\infty' = 0.553 \text{ for } As = 1 \text{(spheres)} \\ C_1' &= 1.346, \quad C_2' = 0.928, \quad \Theta_\infty' = 0.606 \text{ for } As = 2 \\ C_1' &= 1.111, \quad C_2' = 0.580, \quad \Theta_\infty' = 0.446 \text{ for } As = 5 \end{aligned} \tag{172}$$

Although Eq. (171) reflects well the $\bar{B}(\Theta)$ function, it does not predict properly the jamming coverage of particles, which was determined from numerical simulations performed according to the RSA algorithm, similar to the spherical particle case. The jamming coverage for spheroids, spherocylinders and cylinders was determined in Ref. [9] and for squares in Ref. [61]. These data are compiled in Table 5.2 for various *As*. It is interesting to note that the jamming coverage for all particles passes through a maximum, located around $As = 2$ [9]. For spheroids one determined $\Theta'_\infty = 0.580$ for $As = 1.5$, $\Theta'_\infty = 0.583$ for $As = 2$ and $\Theta'_\infty = 0.536$ for $As = 5$ [9]. For spherocylinders $\Theta'_\infty = 0.580$ for $As = 1.5$, $\Theta'_\infty = 0.581$ for $As = 2$, $\Theta'_\infty = 0.524$ for $As = 6$ and $\Theta'_\infty = 0.445$ for $As = 15$. For cylinders $\Theta'_\infty = 0.530$ for $As = 1$ (squares), $\Theta'_\infty = 0.552$ for $As = 1.5$, $\Theta'_\infty = 0.548$ for $As = 2$, $\Theta'_\infty = 0.510$ for $As = 5$ and $\Theta'_\infty = 0.483$ for $As = 15$ [9]. No theoretical arguments were provided which would predict such behavior for the side-on adsorption of these particles.

Examples of spheroid and spherocylinder monolayers obtained from simulations for the coverage $\Theta = 0.5$, close to jamming [10] are shown in Fig. 5.32.

Knowing Θ_∞, more accurate fitting functions can be derived by exploiting Eq. (170). The unknown constant C'_p can be determined using the equilibrium expression for the blocking function given by Eq. (34), which was found to be quite accurate for a broad range of coverage and various

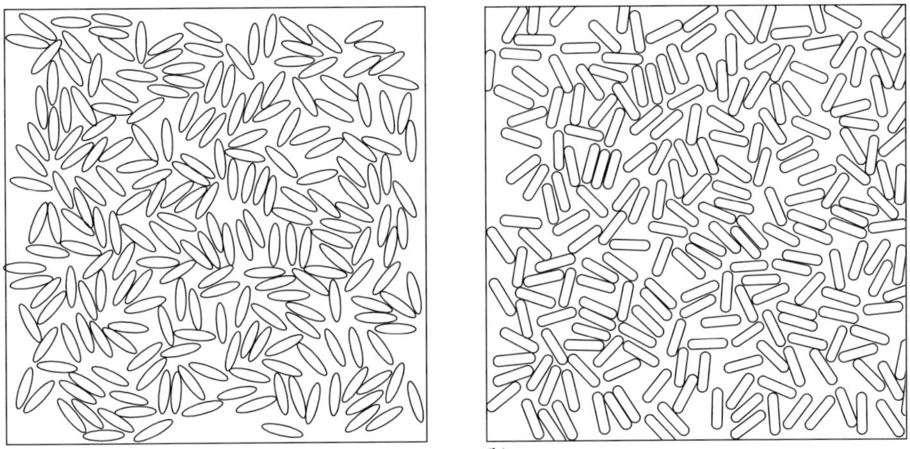

Fig. 5.32. Typical configurations of adsorbed particles generated in the RSA simulations of (a) spheroids, and (b) spherocylinders, $As = 4$, $\Theta = 0.5$. [10]

particle shapes [9]. As a result, the following approximate expression for $B(\Theta)$ can be derived for the coverage range close to the jamming state

$$B(\Theta) \cong (1-\Theta^*) e^{-(1+2\gamma_p)\frac{\Theta^*}{1-\Theta^*} - \gamma_p \left(\frac{\Theta^*}{1-\Theta^*}\right)^2} \left(\frac{\Theta'_\infty - \Theta}{\Theta'_\infty - \Theta^*}\right)^4 \tag{173}$$

valid for $\Theta > \Theta^*$, where Θ^* is the transition coverage, which can be chosen quite arbitrarily.

For $\Theta < \Theta^*$, $B(\Theta)$ is given by the expression

$$B(\Theta) \cong (1-\Theta) e^{-(1+2\gamma_p)\frac{\Theta}{1-\Theta} - \gamma_p \left(\frac{\Theta}{1-\Theta}\right)^2} \tag{174}$$

The analytical results derived from Eqs.(173,174) are compared with exact data derived from simulations in Fig. 5.33 for the side-on adsorption of spheroids of various axis ratio. As can be seen, these analytical results reflect well the $B(\Theta)$ function for the entire range of coverage. On the other hand, the low coverage expansion, Eq. (150) with the C_1, C_2 coefficients calculated from Eq. (166) remains accurate for the range of Θ, where $B(\Theta) > 0.2$.

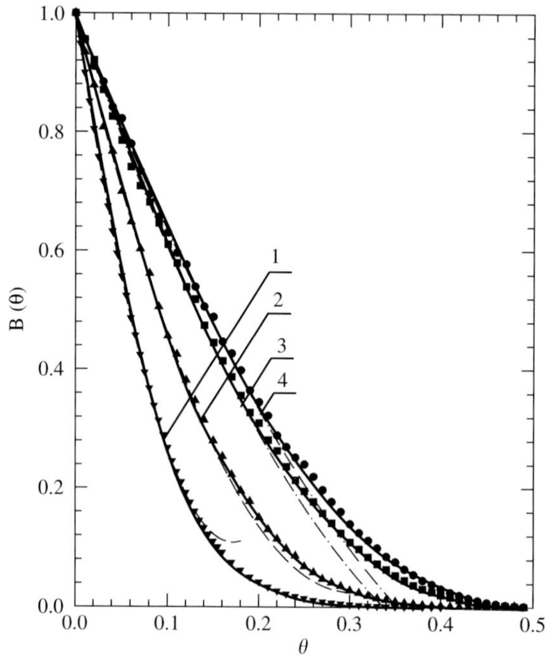

Fig. 5.33. The $B(\Theta)$ function for spheroidal particles adsorbing side-on. The points denote the numerical Monte Carlo simulations performed for 1. $As = 10$, 2. $As = 5$, 3. $As = 2$ and 4. $As = 1$ (spheres), the dashed lines denote the analytical results calculated the low coverage expansion, Eq. (150) , using two terms, and the continuous lines denote the results calculated from Eqs. (173), (174).

Except for the jamming coverage and the blocking function, one can also determine from the RSA simulations for spheroidal particles the pair correlation function and the fluctuations in particle coverage. In Figs. 5.34 and 5.35, the center-to-center and surface-to-surface pair correlation functions are shown. The latter function provides information on the surface density of particles at a given distance between their surfaces, rather than centers. As can be seen, the surface-to-surface pair correlation function for spheroids is similar to spheres at short separations, $h_m/b_s < 0.5$. Analogously as for spheres, this function diverges in the limit $h_m/b_s \rightarrow 0$.

The density fluctuations of adsorbed spheroidal particles can be assessed from the variance analysis of particle populations, generated in the RSA processes. This information is of vital significance for estimating the uniformity of particle distribution in the monolayers that are often used as masks in electronic elements production. The numerical results obtained in

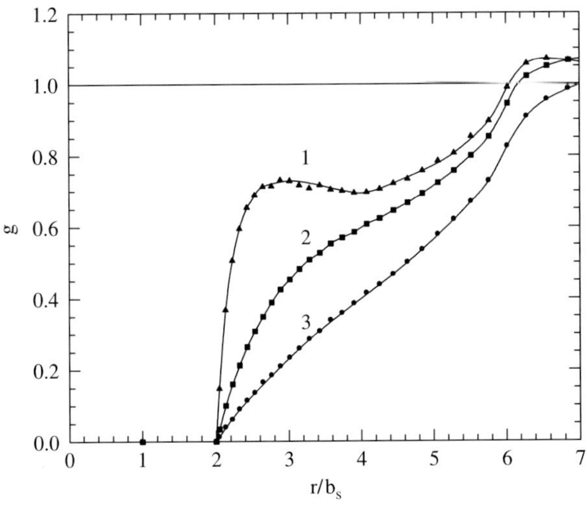

Fig. 5.34. The center-to-center pair correlation function for spheroidal particles determined from numerical simulations for $As = 5$, and various coverage: (1) $\Theta = 0.50$, triangles, (2) $\Theta = 0.35$ squares and (3) $\Theta = 0.20$ circles (the continuous lines are interpolations).

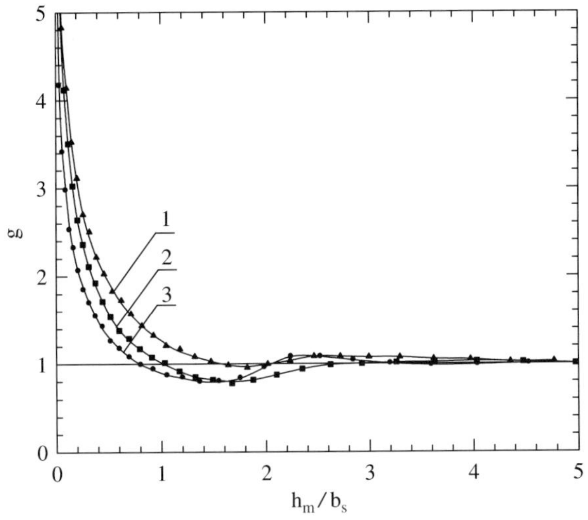

Fig. 5.35. The surface-to-surface pair correlation function for spheroidal particles determined from numerical simulations for the coverage close to jamming: (1) $\Theta = 0.53$, $As = 5$ (triangles), (2) $\Theta = 0.58$, $As = 2$ (squares), (3) $\Theta = 0.544$, $As = 1$ (circles), spheres.

RSA simulations [22] are shown in Fig. 5.36 in the form of the dependence of the reduced variance $\bar{\sigma}_v^2$ on particle coverage Θ. The variance was determined in a two-stage process where first a large population of particles was generated (exceeding 10^5), and then a statistical analysis of the number of particles adsorbed on an area of size ΔS was carried out. The number of these areas l_a of equal size was usually larger than 10^3. The averaged number of particles found over the area and the variance were calculated from the usual formula

$$\langle N_p \rangle = \frac{1}{l_a} \sum_{l_a} N_{pl} \tag{175}$$

$$\bar{\sigma}^2 = \frac{\frac{1}{l_a} \sum_{l_a} (N_{p_l} - \langle N_p \rangle)^2}{\langle N_p \rangle}$$

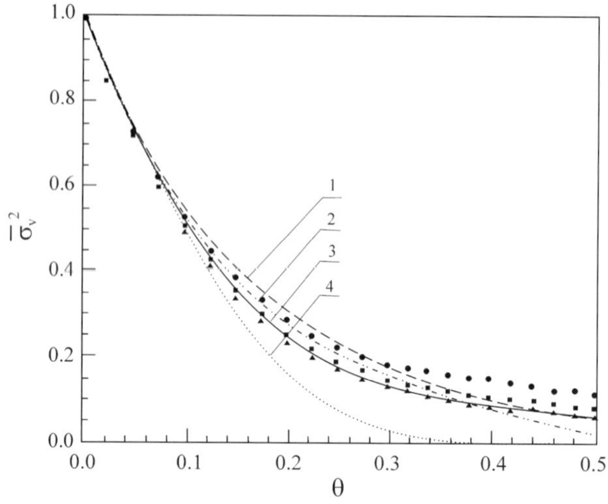

Fig. 5.36. The dependence of the reduced variance on particle coverage Θ determined by the numerical RSA simulations for spheroidal particles, $As = 5$, the circles denote the theoretical data for $\langle N_p \rangle = 30$, triangles for $\langle N_p \rangle = 180$ and the inverse triangles show the theoretical results calculated from Eq. (176). The lines are the theoretical predictions derived from: (1) SPT equilibrium theory, Eq. (38), (2) Eq. (178), (3) Fitting polynomial, Eq. (177). (4) Schaaf et al. [62] formula, $\bar{\sigma}_v^2 = B(\Theta)$. Reprinted with permission of Z. Adamczyk, P. Weroński, J. Chem. 107 (1997), 3691, Copyright 1987 by APS [22].

where N_{pl} is the number of particles found over a given area and $\langle N_p \rangle$ the average number of particles.

As one can observe in Fig. 5.36, the variance derived from computer experiments was dependent on the averaged number of particles N_p adsorbed on the area ΔS (either 30 or 180). This behavior was attributed to the fact that particle positions within the small area are not statistically equivalent [22]. This leads to the increase in the variance described by the two perturbing terms vanishing as $(S_g/\Delta S)^{1/2}$ and $S_g/\Delta S$ in the limit of $\Delta S >> S_g$ [22]. The boundary effects arising because of a finite number of particles over the area ΔS can be eliminated by calculating the variance from the known relationship [16]

$$\bar{\sigma}_v^2 = 1 + 2\Theta \int \left[g(\bar{r}) - 1 \right] \bar{r}\, d\bar{r} \tag{176}$$

where g is the center-to-center pair correlation function and $\bar{r} = r/a$.

The results obtained from Eq. (176) are shown in Fig. 5.36 as inverse triangles. For convenience they have been fitted for the entire range of coverage by the polynomial

$$\bar{\sigma}_v^2 = 1 - 2B_2\Theta + 13.98\Theta^2 + 5.418\Theta^3 - 58.86\Theta^4 + 57.35\Theta^5 \tag{177}$$

As can be noted in Fig. 5.36, the theoretical results calculated from Eq. (177) agree quite well, for $\Theta < 0.2$, with the equilibrium results obtained from Boublik's theory, Eq. (38).

The exact numerical data are also quite well reflected by the formula derived in Ref. [22] for irreversible systems using the concept of the ASF and the maximum term method

$$\bar{\sigma}_v^2 = \frac{B(\Theta)}{B(\Theta) - \Theta \dfrac{dB(\Theta)}{d\Theta}} \tag{178}$$

In contrast, the Schaaf et al. formula derived in Ref. [62], $\bar{\sigma}_v^2 = B(\Theta)$ works well for $\Theta<0.1$ only.

The results shown in Fig. 5.36 can be used as useful reference data for assessing the homogeneity of the substrates used in particle adsorption experiments.

They also have practical implication showing that the uniformity of the particle distribution in monolayers increases significantly with the coverage of particles. This suggests that the precision of direct experimental methods based on particle counting over various areas becomes much higher for increasing coverage [57]. For example, if the averaged number of particles over an area was found to be 100, the precision of the measurement is $\frac{(100\bar{\sigma}_v^2)^{1/2}}{100} \cong 10\%$, in the limit of $\Theta \to 0$. For $\Theta = 0.4$, when $\bar{\sigma}_v^2 = 0.1$, the precision becomes 3.2%.

5.4.2. The unoriented adsorption of non-spherical particles

The unoriented adsorption regime is likely to occur, as discussed above, for particles resembling oblate spheroids (flattened) or massive particles of elongated shape, resembling prolate spheroids.

Also in the case of unoriented adsorption, the low coverage expansion is given by Eq. (150). The coverage in the case of prolate spheroids was defined as $\pi\, a_s\, b_s\, N_p$, whereas for oblate spheroids $\Theta = \pi a_s^2$ in order to simplify comparison with the side-on adsorption chosen as the reference system. The C_1 and C_2 constants appearing in this expansion are given by Eqs.(151,152). Because of the appearance of multiple integrals (five-fold and nine-fold, respectively) these constants can only be calculated numerically, using the Monte Carlo integration procedure [59]. It was shown that the C_1 and C_2 constants for prolate spheroids can be well fitted by the polynomial expansions

$$C_1 = 2.07 + 0.811\frac{1}{As} + 2.37\frac{1}{As^2} - 1.25\frac{1}{As^3}$$

$$C_2 = \frac{0.670As^2 - 0.301As + 3.88}{0.283As^2 + As} \tag{179}$$

For oblate spheroids the interpolating polynomials were [59]

$$C_1 = 1.59 + 2.80\frac{1}{As} - 0.388\frac{1}{As^2}$$

$$C_2 = 0.700 + 1.83\frac{1}{As} + 0.776\frac{1}{As^2} \tag{180}$$

As can be seen in Figs. 5.37 and 5.38 these interpolating polynomials fit well the exact numerical data for the entire range of the *As* parameter. Note that both the C_1 and C_2 constants remain finite in the limit of $As \rightarrow 0$, both for the prolate and oblate spheroids. These limiting values are $C_1 = 2.07$ and $C_2 = 2.37$ for prolate spheroids, and $C_1 = 1.59$, $C_2 = 0.700$ for oblate spheroids. However, calculation of higher order expansion coefficients for anisotropic particles becomes inefficient, therefore no exact numerical data have been reported yet.

A good estimate of the $B(\Theta)$ function for the range of high coverage can be done in an analogous way as for the side-on adsorption by performing the asymptotic analysis. In the case of the unoriented adsorption the

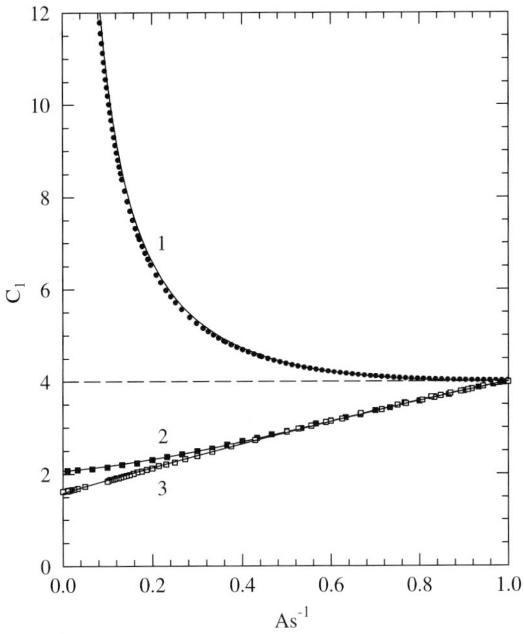

Fig. 5.37. The dependence of the C_1 constant of the expansion, Eq. (150) on the As^{-1} parameter for spheroids. The points denote the exact numerical results. (1) Side-on adsorption, with the continuous line calculated from Eq. (166); (2) Unoriented adsorption of prolate spheroids, with the continuous line calculated from Eq. (179) (3) Unoriented adsorption of oblate spheroids, with the continuous line calculated from Eq. (180).

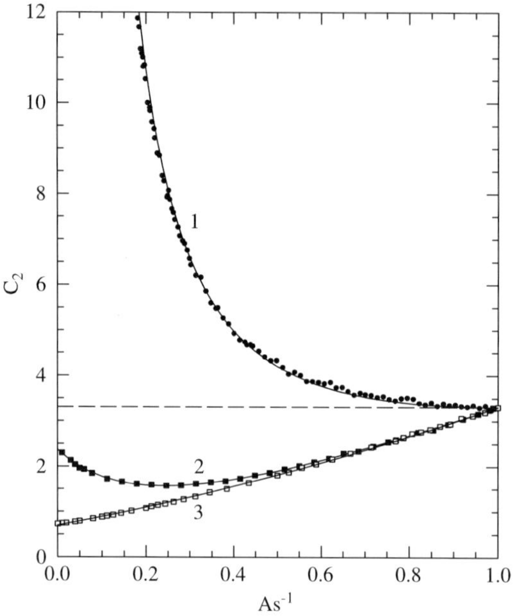

Fig. 5.38. The dependence of the C_2 constant of the expansion, Eq. (150) on the As^{-1} parameter for spheroids. The points denote the exact numerical results. (1) Side-on adsorption , with the continuous line calculated from Eq. (166); (2) Unoriented adsorption of prolate spheroids, with the continuous line calculated from Eq. (179); (3) Unoriented adsorption of oblate spheroids, with the continuous line calculated from Eq. (180).

isolated targets are 4D domains (two spacial and one two orientational coordinate), thus the adsorption probability is given by

$$p_t \sim \delta_{x_v} \delta_{y_v} \delta_{\alpha_v} \delta_{\beta_v} \cong c_p h_t^4 \tag{181}$$

where h is the size of the target and $\delta\alpha_v \sim h_t$, $\delta_{\beta v} \sim h_t$.

This results in the following kinetic equation for particle coverage evolution

$$\Theta = \Theta_\infty - C_p' \tau^{-1/4} \tag{182}$$

where C_p' is the dimensionless constant.

Thus, the $B(\Theta)$ function becomes

$$B(\Theta) = \frac{d\Theta}{d\tau} = \frac{1}{4C_p^{'4}}(\Theta_\infty - \Theta)^5 \qquad (183)$$

As can be deduced from Eq. (182) the jamming coverage Θ_∞ can be determined by extrapolating the dependence of Θ on $\tau^{-1/4}$ obtained from numerical simulations to $\tau^{-1/4} \rightarrow 0$ (infinite time)

The occurrence of this adsorption regime is clearly seen in Fig. 5.39, where the numerical data for prolate spheroids are compared with the analytical approximations derived from Eq. (182). In contrast to the side-on adsorption there is no analytical expression for the equilibrium blocking

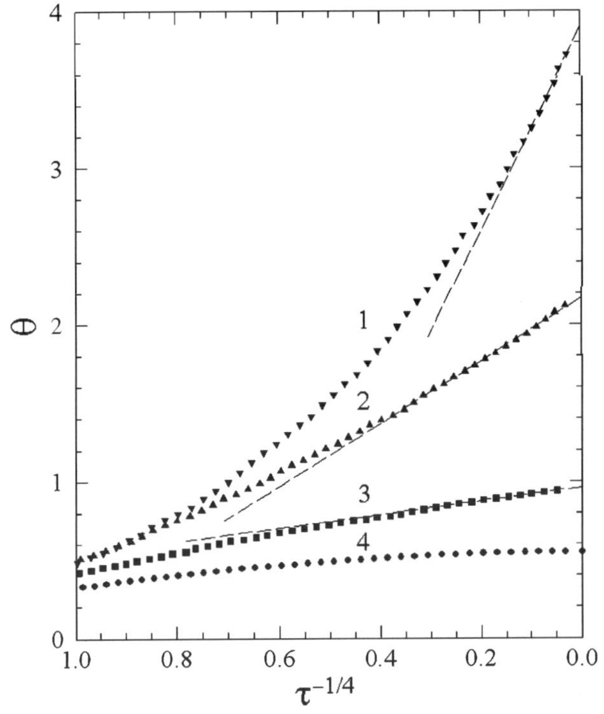

Fig. 5.39. The dependence of Θ on $\tau^{-1/4}$ for unoriented adsorption of prolate spheroids. The points denote the exact numerical results obtained for: (1) As = 10, (2) As = 5, (3) As = 2 and (4) As = 1 (spheres),The dashed lines represent the asymptotic analytical results calculated from Eq. (182).

function, which can be exploited for constructing convenient interpolation functions.

The blocking function for prolate and oblate spheroids determined from numerical simulations for the entire range of coverage is plotted in Figs. 5.40 and 5.41. As can be seen, the low coverage expansion, Eq. (150), can be used as a good approximation for quite a broad range of coverage, especially for larger values of *As* and oblate spheroids. The jamming coverage for the unoriented adsorption of spheroids was derived from numerical simulations using the extrapolation procedure described above. Selected values of the jamming limit for prolate and oblate spheroids are compiled in Table 5.3. It is interesting to note that in contrast to the side-on adsorption, the jamming coverage increases monotonically with the *As* parameter.

Alternatively, the jamming coverage can be alternatively expressed as $\Delta S\Theta'_{\infty} = \Theta'_{\infty}/As$ for prolate and oblate spheroids, i.e. relative to the

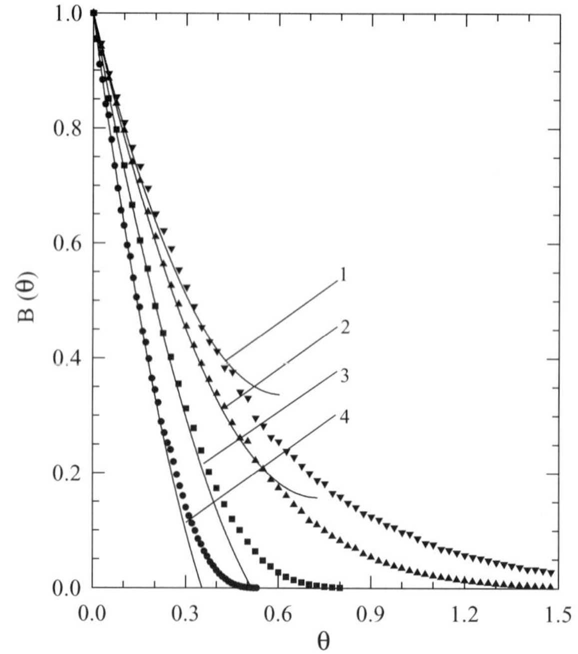

Fig. 5.40. The $B(\Theta)$ function for prolate spheroids in the case of unoriented adsorption. The points denote the numerical Monte Carlo simulations performed for (1) $As = 10$, (2) As = 5, (3) $As = 2$ and (4) $As = 1$ (spheres), the solid lines denote the analytical results calculated from the low coverage expansion, Eq. (150).

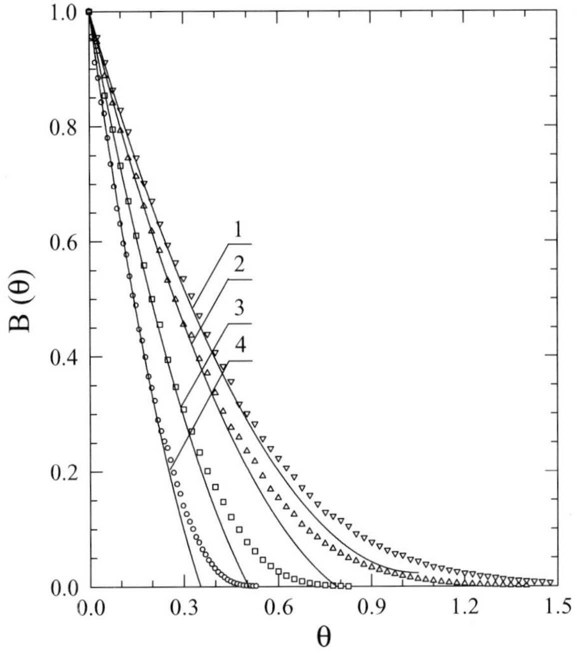

Fig. 5.41. The $B(\Theta)$ function for oblate spheroids in the case of unoriented adsorption. The points denote the numerical MC simulations performed for 1. $As = 10$, 2. $As = 5$, 3. $As = 2$ and 4. $As = 1$ (spheres), the solid lines denote the analytical results calculated for the low coverage expansion, Eq. (150).

monolayer of particles adsorbed edge-on. The jamming coverage expressed in this way decreases monotonically with *As* as can be seen in Table 5.3. The monolayers of prolate and oblate spheroids at the coverage close to the jamming state are shown in Fig. 5.42.

The dependence of the jamming coverage on the *As* parameter determined numerically for prolate spheroids is shown in Fig. 5.43. As can be seen, the exact numerical results can be well interpolated in the case of the side-on adsorption by the fitting function [63]

$$\Theta_{\infty}^{s} = 0.622\left(As + \frac{1}{As} - 1.997\right)^{0.0127} e^{-0.0274\left(As + \frac{1}{As}\right)} \tag{184}$$

Table 5.3
Jamming coverage for unoriented adsorption of hard spheroids derived from RSA simulations

Axis ratio $As = a_s/b_s$	Prolate spheroids $\Theta_\infty = \pi a_s b_s N_\infty$	Prolate spheroids $\Theta_\infty = \pi b_s^2 N_\infty$	Oblate spheroids $\Theta_\infty = \pi a_s^2 N_\infty$	Oblate spheroids $\Theta_\infty = \pi a_s b_s N_\infty$
2	0.953	0.477	1.01	0.505
3	1.38	0.461	1.38	0.461
4	1.78	0.445	1.71	0.428
5	2.17	0.434	2.02	0.404
10	3.86	0.386	3.19	0.319

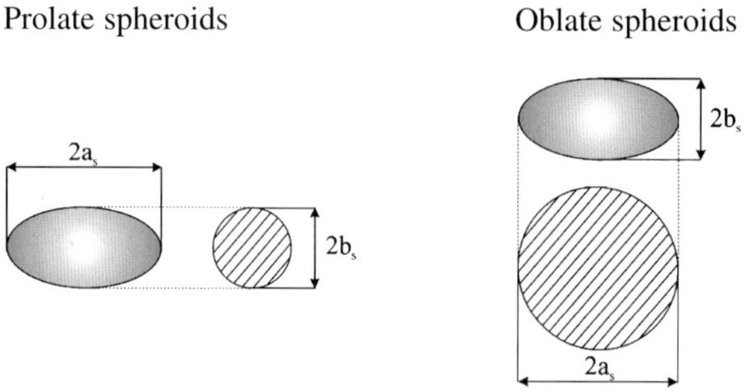

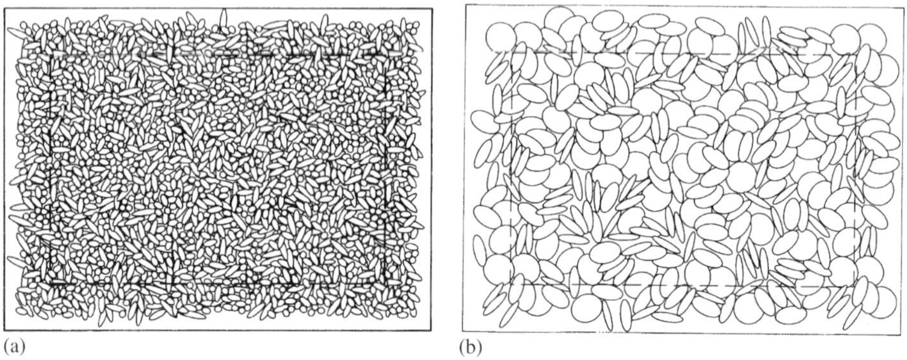

Fig. 5.42. Typical configurations of adsorbed particles generated in the RSA simulations, $A_s = 5$ of (a) prolate spheroids $\Theta = 2.14$, and (b) oblate spheroids $\Theta = 2$ [57].

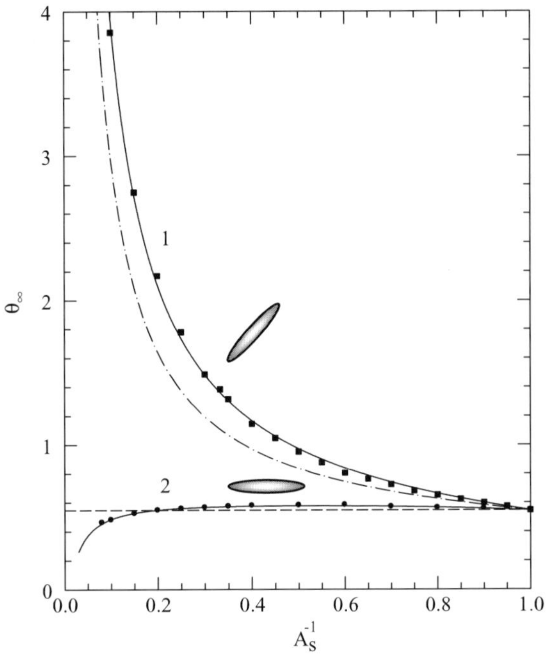

Fig. 5.43. The dependence of the jamming coverage Θ_∞ on the $b_s/a_s=As^{-1}$ parameter determined for prolate spheroids. The points denote exact numerical calculations performed for: (1) Unoriented adsorption (the solid line shows the analytical results calculated from Eq. (185)); (2) Side-on adsorption and (the solid line shows the analytical results calculated from Eq. (184)), the dashed-dotted line shows the averaged results from the side-on and edge-on orientations, Eq. (186). From Ref. [63].

For the unoriented adsorption the fitting function is [59]

$$\Theta_\infty = 0.304 + 0.365As - 0.123\frac{1}{As} \tag{185}$$

As can be seen in Fig. 5.43, the exact results are above the values of the jamming coverage calculated as the average of the side-on and edge-on orientations, given by the expression

$$\langle\Theta_\infty\rangle = \frac{1}{2}\left[0.547/As + \Theta_\infty^s(As)\right] \tag{186}$$

where Θ_∞^s is the jamming coverage for the side-on orientation. This is so because for higher coverage, the edge-on orientations of particles are preferred due to topological constraints.

Analogous results obtained for oblate spheroids are shown in Fig. 5.44. The jamming coverage for unoriented adsorption was interpolated by the function [59]

$$\Theta_\infty = 0.768 - \frac{0.473}{As} + 0.251\, As \tag{187}$$

As can be seen in Fig. 5.44, the exact results are above the values of jamming coverage calculated as the average of the side-on and edge-on orientations, given by the expression

$$\langle \Theta_\infty \rangle = \frac{1}{2}\left(0.547 + \frac{\Theta_\infty^s}{As} \right) \tag{188}$$

The data shown in Figs. 5.43 and 5.44 and approximated by Eqs.(184–188) can be exploited for determining the influence of particle shape, at a fixed volume, on the mass of the monolayer. This type of information is especially important for protein and polymer adsorption where the amount of adsorbed substance is usually expressed as mass per unit area [63]. For spherical particles the mass of adsorbed particles per unit area is given by the simple relationship

$$m_1 = \frac{4}{3}\rho_p a\, \Theta_\infty = 0.729\rho_p a \tag{189}$$

where ρ_p is the particle-specific density and a the sphere radius.

For spheroidal particles of the same volume and density one has at the jammed state

$$m_A = \frac{4}{3}\rho_p b_s \Theta_\infty (As) \tag{190}$$

Hence, one can express the m_A/m_1 ratio as

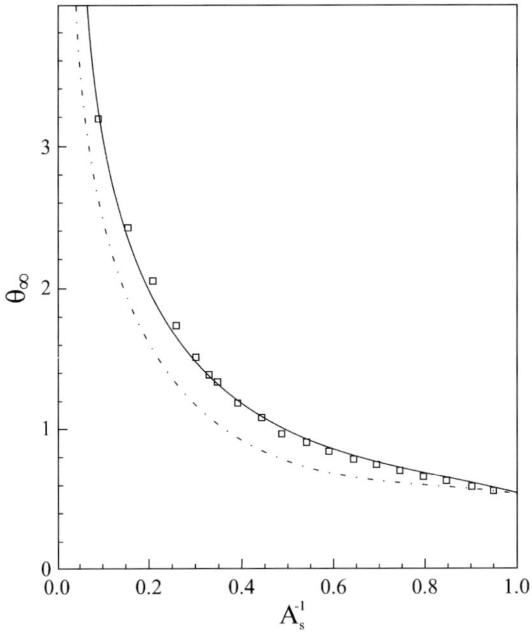

Fig. 5.44. The dependence of the jamming coverage Θ_∞ on the $b_s/a_s=A_s^{-1}$ parameter determined for oblate spheroids. The points denote exact numerical calculations performed for unoriented adsorption of oblate spheroids.The solid line shows the analytical results calculated from Eq. (187) and the dashed dotted line shows the averaged results from the side-on and edge-on orientations, Eq. (188) [63].

$$m_A/m_1 = \frac{1.292}{As^{1/3}}\left(As + \frac{1}{As} - 1.997\right)^{0.0127} e^{-0.0274\left(As+\frac{1}{As}\right)} \tag{191}$$

for the side-on adsorption of prolate spheroids and

$$m_A/m_1 = 0.667\left(\frac{a_s}{b_s}\right)^{2/3} + 0.556\left(\frac{b_s}{a_s}\right)^{1/3} - 0.225\left(\frac{b_s}{a_s}\right)^{4/3} \tag{192}$$

for unoriented adsorption. As can be deduced from these equations, in the case of unoriented adsorption, the monolayer mass in the limit of large aspect ratio diverges asymptotically as $(a_s/b_s)^{2/3}$. On the other hand, for unoriented adsorption of oblate spheroids the expression for m_A/m_1 becomes

$$m_A/m_1 = 0.459\left(\frac{a_s}{b_s}\right)^{1/3} + 1.404\left(\frac{b_s}{a_s}\right)^{2/3} - 0.865\left(\frac{b_s}{a_s}\right)^{5/3} \tag{193}$$

The results stemming from Eqs.(191–193) are plotted in Fig. 5.45. As can be seen, the possibility of an unoriented adsorption largely increases the mass of the monolayer, especially for the prolate spheroid case. For oblate spheroids, this effect becomes appreciable for $As > 10$, only, i.e., for particles in the form of thin disks.

The RSA simulations allow one to analyze structural aspects of the monolayers of spheroidal particles formed under the unoriented adsorption regime. A sequence of monolayers of prolate spheroids obtained from

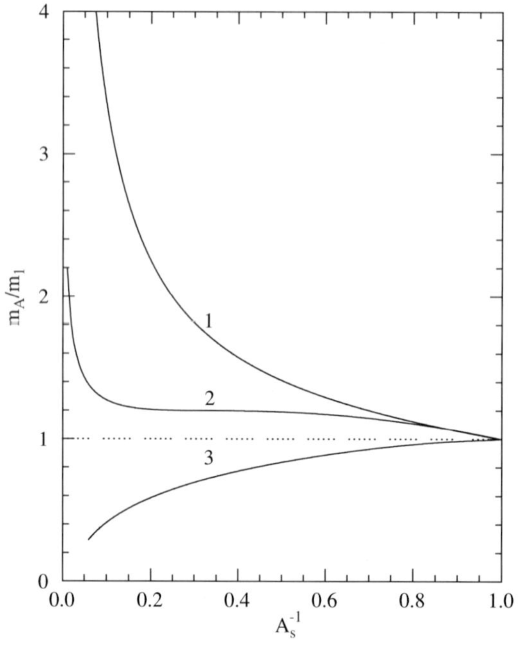

Fig. 5.45. The normalized monolayer mass m_A/m_1 calculated from Eqs.(191–193) for prolate and oblate spheroids: 1. unoriented adsorption of prolate spheroids, 2. unoriented adsorption of oblate spheroids, 3. side-on adsorption of prolate spheroids.

simulations for $As = 5$ and Θ varying between 0.543 and 2.14 (close to jamming) is shown in Fig. 5.46 with the surface-to-surface and center-to-center pair correlation functions.

As can be observed, for higher coverage, because of topological constraints, the adsorbed particles assume mostly orientations close to perpendicular. This results in the increase in the primary peak of the center-to-center pair correlation function, occurring at the distance close to $2b_s$. In consequence the surface-to-surface pair correlation function diverges for $h_m \rightarrow 0$ and Θ approaching the jamming limit. This behavior is observed in Fig. 5.47 where the surface-to-surface pair correlation function is shown for prolate spheroids of various elongation and compared with the pair correlation function derived for hard spheres. As can be seen, the pair correlation function when expressed in terms of the surface-to-surface distance exhibits a universal behavior, independent of the particle elongation parameter As.

The jamming coverage and the structural results discussed in this section have a special significance for a quantitative analysis of experimental data, e.g., in the case of protein adsorption, because they can be used as useful reference quantities, enabling their proper interpretation and detection of possible artifacts.

However, all these results have been obtained for hard (non-interacting) particles. In reality, because of the appearance of the electrostatic double-layer interactions, adsorbing particles interact over a distance, which can be comparable with their dimension for low ionic strength. Possible implications of these interactions for particle adsorption are analyzed next.

5.5. RANDOM SEQUENTIAL ADSORPTION OF INTERACTING PARTICLES

All the above results were obtained by assuming the hard body interactions among particles, becoming infinitely large upon contact and zero otherwise. This is a valid approximation in the case of higher electrolyte concentration and larger particle sizes when the Le/a parameter, characterizing the thickness of the electric double layer to particle radius is much smaller than unity. As can be deduced from the data shown in Table 5.4 this is so for colloid particles of the size larger than 200 nm and ionic strength I higher than 10^{-3} M. For smaller particle size, $a < 10$ nm, the a/Le parameter

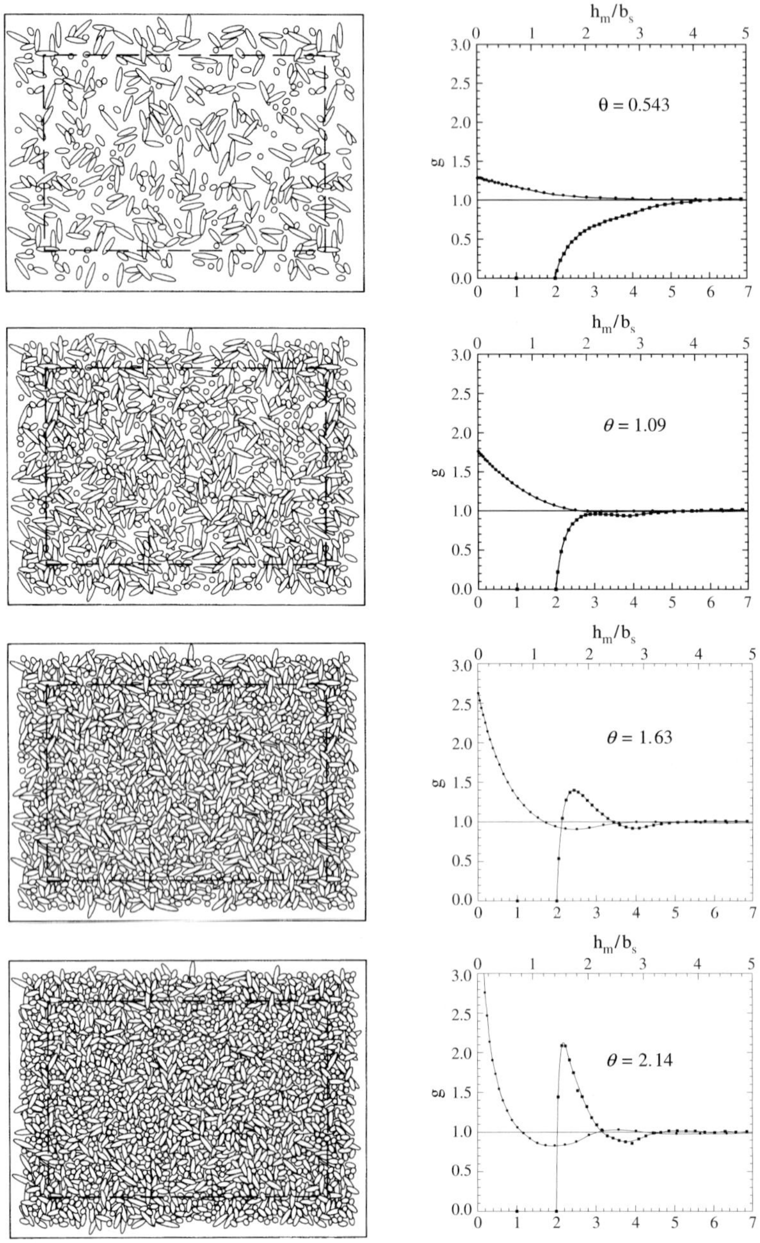

Fig. 5.46. The structure of monolayers of prolate spheroidal particles, derived from simulations with the corresponding pair correlation functions: surface to surface and center to center $As = 5$. Particle coverage (from top) $\Theta = 0.543$, 1.09, 1.63, 2.14.

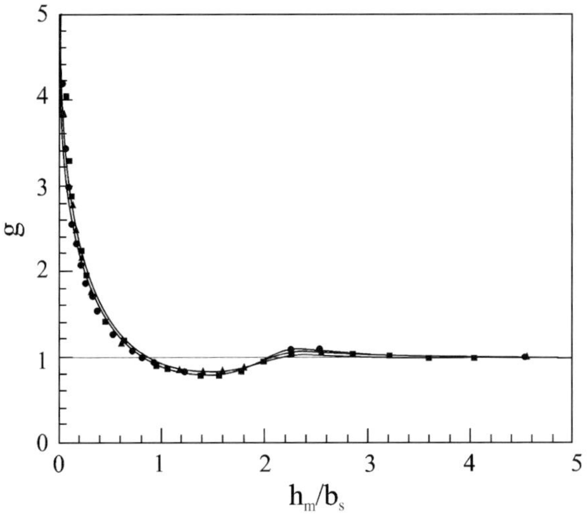

Fig. 5.47. The surface-to-surface pair correlation function for unoriented adsorption of prolate spheroids determined from numerical simulations for coverage close to jamming: triangles, $As = 1$, $\Theta = 0.544$, squares, $As = 2$, $\Theta = 0.936$, circles, $As = 5$, $\Theta = 2.14$.

becomes larger than 0.1 even at ionic strength corresponding to physiological conditions, i.e., 0.15 M. This suggests that the lateral interactions among adsorbing particles, in particular globular proteins, start to play a more significant role under these conditions.

Unfortunately, because of the complexity of interacting particle adsorption, which is a 3D problem by nature, there is no comprehensive theory available, not even for equilibrium systems. However, approximate approaches are available to deal with irreversible adsorption of interacting (soft) particles, whose usefulness was confirmed by simulations and extensive experimental data collected for model colloid suspensions [27,57,63,65]. The main assumptions of the model, referred to as the interacting RSA model, hereafter referred to as iRSA are [59,64,65]:

(i) the interactions between particles are assumed to be additive;
(ii) the interaction with the surface are neglected;
(iii) the probability of particle adsorption is calculated from the Boltzmann formula, which corresponds to the local equilibrium postulate.

In accordance with these assumptions the probability density of the virtual particle adsorption in the iRSA model is given by

Table 5.4
Values of $\kappa a = a/Le$ and $(\kappa a)^{-1} = Le/a$ (in parenthesis), for various particle size and electrolyte concentration

Electrolyte concentration (M) Particle size (nm)	1.5×10^{-1}	1×10^{-2}	1×10^{-3}	1×10^{-4}
1	1.27 (0.787)	0.328 (3.05)	0.104 (9.63)	3.28×10^{-2} (30.5)
2	2.54 (0.394)	0.65 (1.52)	0.207 (4.8)	6.56×10^{-2} (15.2)
5	6.35 (0.157)	1.64 (0.610)	0.519 (1.93)	0.164 (6.09)
10	12.7 (7.87×10^{-2})	3.28 (0.305)	1.04 (0.964)	0.328 (3.05)
20	25.4 (3.93×10^{-2})	6.56 (0.152)	2.07 (0.48)	0.656 (1.52)
50	63.5 (1.57×10^{-2})	16.4 (6.09×10^{-2})	5.19 (0.19)	1.640 (0.610)
100	127 (7.87×10^{-3})	32.8 (3.04×10^{-2})	10.4 (9.63×10^{-2})	3.28 (0.305)
200	254 (3.93×10^{-3})	65.6 (1.52×10^{-2})	20.7 (4.82×10^{-2})	6.56 (0.152)
500	635 (1.57×10^{-3})	164 (6.09×10^{-3})	51.9 (1.92×10^{-2})	16.4 (6.10×10^{-2})
1000	1270 (7.87×10^{-4})	328 (3.05×10^{-3})	104 (9.63×10^{-3})	32.8 (3.05×10^{-2})

$T = 293$ K, water, 1:1 electrolyte (KCl), $Le = \kappa^{-1} = \left(\dfrac{\varepsilon kT}{2e^2 I}\right)^{1/2}$.

$$\bar{\rho}_{p_v} = e^{-\phi_p/kT} \tag{194}$$

where the interaction energy is evaluated as

$$\phi_v = \sum_{l=1}^{\infty} \phi_{l_v} \tag{195}$$

where ϕ_{lv} is the energy of interaction of the virtual (wandering) particle with the l-th adsorbed particle.

In the 2D iRSA model, the energy is calculated by assuming that all particles contact the interface, being located at the primary minimum distance δ_m [59,64,65]. In more sophisticated models, referred to as 3D iRSA, the entire energy profile is considered, including the particle interface interactions [66,67] as discussed later on. Within the framework of the iRSA model one can perform efficient, RSA-type simulations both for spherical and anisotropic particles, mainly spheroids. Particle monolayers and their structure, blocking functions and jamming coverage have been calculated [58–60,64,65]. Before we discuss these results it is useful to present limiting analytical formulae, which allow one to formulate the robust hard particle concept.

Using the expression for the adsorption probability Eq.(194) one can calculate the low-coverage expansion coefficients of the blocking function, from Eqs.(151,152). The Mayer functions occurring in these equations are explicitly

$$f_{1v} = e^{-\phi_{1v}(\mathbf{r}_{1v},\alpha_{1v},\beta_1,\beta_v)/kT} - 1 \tag{196}$$

$$f_{2v} = e^{-\phi_{1v}(\mathbf{r}_{1v},\alpha_{1v},\beta_1,\beta_v)/kT}\, e^{-\phi(\mathbf{r}_{2v},\alpha_{2v},\beta_2,\beta_v)/kT} - 1$$

In accordance with the local equilibrium postulate, it is further assumed that the pair correlation function in the limit of low coverage is given by

$$g = e^{-\phi_{12}(\mathbf{r}_{12},\alpha_{12},\beta_1,\beta_2)/kT} \tag{197}$$

For spherical particles, the interaction energy depends only on the surface distance vectors $\mathbf{r}_{1v}$, $\mathbf{r}_{2v}$ and $\mathbf{r}_{12}$, so the expressions for the C_1 and C_2 constants derived from Eqs.(151,152) become [64,65]

$$\begin{aligned}
C_1 &= 2\int_0^{\infty}\left(1 - e^{\phi_{1v}(\bar{r})/kT}\right)\bar{r}\, d\bar{r} = 2\bar{B}_2 \\
C_2 &= 2\int_2^{\infty}\left[2C_1 - \bar{I}_2(\bar{r})\right]e^{-\phi_{12}(\bar{r})/kT}\,\bar{r}\, d\bar{r} \\
\bar{I}_2(\bar{r}) &= \frac{1}{\pi}\int_{-\infty}^{\infty}\left[1 - e^{-[\phi_{1v}(\bar{r}_{1v})+\phi_{2v}(\bar{r}_{2v})]/kT}\right]d\bar{x}\, d\bar{y}
\end{aligned} \tag{198}$$

where $\bar{r} = r_{12}/a, \bar{x} = x/a, \bar{y} = y/a, \bar{r}_{1v} = \left[\left(\bar{x}-\frac{\bar{r}}{2}\right)^2+\bar{y}^2\right]^{1/2}$,

$$\bar{r}_{2v} = \left[\left(\bar{x}+\frac{\bar{r}}{2}\right)^2+\bar{y}^2\right]^{1/2}.$$

and x, y are the co-ordinates of the virtual particle expressed in the local correlation system with the origin immobile between two absorbed particles. Note that the C_1 constant equals $2B_2$ for interacting particles at equilibrium.

In the general case of arbitrary interaction potential ϕ, the integrals appearing in Eq.(198) cannot be calculated analytically. This can only be done for the effective hard particle potential discussed in Chapter 2 (cf. Fig. 2.32), where $\phi = \infty$ if the distance between particles equals $2a+2h^*$and zero otherwise. In this case, using the previous integration procedure for hard spheres one obtains

$$C_1 = 4\left(1+\frac{h^*}{a}\right)^2 = 4(1+H^*)^2 \tag{199}$$

where $H^*=h^*/a$ can be treated as the dimensionless effective range of interactions.

Analogously, the C_2 constant is given by

$$C_2 = \frac{6(3)^{1/2}}{\pi}(1+H^*)^4 \tag{200}$$

In the case of a triangular barrier of the width $\bar{\delta} = 2H^*$ and the height ϕ_b, using Eq.(12), the C_1 constant can be expressed as

$$C_1 = 4(1+H^*)^2 - \frac{8H^*}{\bar{\phi}_b}(1-e^{-\bar{\phi}_b}) - \frac{8H^{*2}}{\bar{\phi}_b} + \frac{8H^{*2}}{\bar{\phi}_b^2}(1-e^{-\bar{\phi}_b}) \tag{201}$$

where $\bar{\phi}_b = \phi_b/\kappa T$.

For $\bar{\phi}_b >> 1$ this becomes identical to Eq.(199).

By exploiting this result a useful model of interacting particle adsorption was formulated, called the effective hard particle concept. This was done by Barker and Henderson in the perturbation theory of real fluids [68].

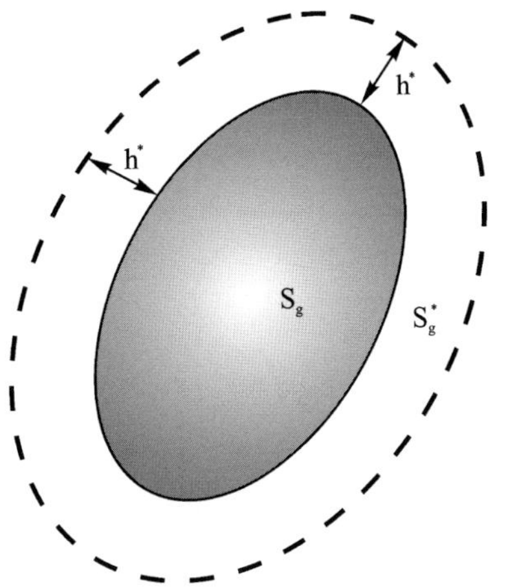

Fig. 5.48. A schematic representation of the effective hard particle concept.

The basic assumption of this model is that the behavior of particles interacting according to a short-range potential can be well reflected by hard particle behavior, whose size is uniformly increased by the increment h^* (see Fig. 5.48). The parameter h^* is often referred to as the effective interaction range.

Accordingly, the effective scaling area is increased, attaining the value $S_g^*(h^*)$. Then, by generalizing Eq.(199) it was postulated that

$$C_{1_{ex}} = C_1(h^*)\frac{S_g^*(h^*)}{S_g} \tag{202}$$

where $C_{1_{ex}}$ is the exact value of the C_1 constant known from calculations, and $C_1^*(h^*)$ the value of the constant for the effective hard particle. The major advantage of the model is that once the value of h^* is determined for any particle shape, it can be used for evaluating other quantities like the blocking function or jamming coverage. They can be calculated just by properly rescaling the hard-sphere results discussed above, which eliminates the necessity of performing tedious calculations. Despite limited accuracy, this approach can be effectively used for most practical applications. Eq.(202) can be alternatively formulated as

$$C_{1_{ex}} - \frac{S_g^*(h^*)}{S_g}\, C_1^*(h^*) = F^*(h^*) = 0 \tag{203}$$

If C_{1ex} is known, h^* can be obtained by inverting the implicit dependence, Eq.(203), which is usually in the form of a polynomial of h^*.

For example, in the case of spherical particles: $S_g^* = \pi a^2\left(1+\frac{h^*}{a}\right)^2$, $S_g = \pi a^2$, $C_1^*(h^*) = 4$, thus

$$H^* = \frac{h^*}{a} = \frac{1}{2}(C_{1_{ex}})^{1/2} - 1 \tag{204}$$

Using Eq.(198) to calculate C_{1ex} one obtains for H^* the expression

$$H^* = \left[\frac{1}{2}\int_0^{\infty}\left(1-e^{-\phi(\bar{r})/kT}\right)\bar{r}\,d\bar{r}\right]^{1/2} - 1 \tag{205}$$

For prolate spheroids adsorbing under unoriented adsorption regime

$$S_g^* = \pi a_s^* b_s^* = \pi a_s b_s (1+H^*)(1+As\,H^*) \tag{206}$$

where $H^* = h^*/a_s$
Using Eq.(206) and considering that

$$C_1^* = 2\bar{B}_2 = 2(1+\gamma_p^*) = 2\left[1+\frac{4As^*}{\pi}E^2\left(\frac{As^{*^2}-1}{As^{*^2}}\right)\right] \tag{207}$$

where

$$As^* = \frac{As(1+H^*)}{1+As\,H^*} \tag{208}$$

one can derive for H^* the implicit expression

$$2(1+H^*)(1+As\,H^*)\left[1+\frac{4As^*}{\pi}E^2\left(\frac{As^{*^2}-1}{As^{*^2}}\right)\right] = C_{1_{ex}} \tag{209}$$

Because of the appearance of the elliptic integral, Eq.(209) cannot be inverted analytically. However, by considering the approximate expression for C_1^* given by Eq.(166) one obtains

$$C_1^* = \frac{4(\pi^2-4)}{\pi^2} + \frac{8}{\pi^2}\left(\frac{1+As\,H^*}{As(1+H^*)} + \frac{As(1+H^*)}{1+As\,H^*}\right) \tag{210}$$

By substituting this into Eq.(203) one obtains after rearrangement a quadratic equation, which can be solved analytically with the result

$$H^* = \frac{1}{2}\left(\frac{1+As}{As}\right)\left\{\left[1+\frac{As}{(1+As)^2}\left(C_{1_{lx}} - C_1\right)\right]^{1/2} - 1\right\} \tag{211}$$

where $C_1 = \dfrac{4(\pi^2-4)}{\pi^2} + \dfrac{8}{\pi^2}\left(\dfrac{1}{As} + As\right)$.

For prolate spheroids adsorbing under the unoriented regime the expression for C_1^* can be derived from the fitting function, Eq.(179)

$$C_1^* = \left(2.07 + 0.811\frac{1}{As^*} + 2.37\frac{1}{As^{*2}} - 1.25\frac{1}{As^{*2}}\right)(1+H^*)(1+As\,H^*) \tag{212}$$

This produces the fourth order polynomial expression involving H^*

$$4H^{*4} + c_3 H^{*3} + c_2 H^{*2} + c_1 H^* + c_0 = 0 \tag{213}$$

where

$$\begin{aligned}
c_3 &= 5.80\frac{1}{As} + 10.2 \\
c_2 &= 0.424\frac{1}{As^2} + (16.6 - C_{1_{ex}})\frac{1}{As} + 7.01 \\
c_1 &= -2.63\frac{1}{As^3} + 8.73\frac{1}{As^2} + (7.83 - 2C_{1_{ex}})\frac{1}{As} + 2.07 \\
c_0 &= (C_1 - C_{1_{ex}})\frac{1}{As}
\end{aligned} \tag{214}$$

This equation can be solved analytically by using the improved Ferrari method.

For oblate spheroids under the unoriented regime

$$S_g = \pi\left(1+\frac{h^*}{a}\right)^2 = (1+H^*)^2 \tag{215}$$

which is identical to spheres. The C_1^* constant can be calculated from the fitting function, Eq.(180)

$$C_1^* = \left(1.59 + 2.80\frac{1}{As^*} - 0.388\,As^{*^2}\right) \tag{216}$$

By substituting this into Eq.(203) one obtains a quadratic equation whose solution is

$$H^* = \frac{2.02+5.98\,As}{8\,As}\left\{\left[1+\frac{16\left(C_{1ex}-C_1\right)As^2}{(2.02+5.98\,As)^2}\right]^{1/2} - 1\right\} \tag{217}$$

The exact values needed in Eqs.(212, 214, 218) to evaluate the effective interaction range H^* were calculated by numerical integration of Eq.(198) [58–60]. In the case of spheres, the interaction energy between particle pairs ϕ_{12} was the two parametric Yukawa-type formula derived from the LSA model (cf. Eq.(243) in chapter 2). For monodisperse spherical particles one has

$$\phi_{12} = \phi_0 \frac{a}{2a+h_m} e^{-h_m/Le} \tag{218}$$

where h_m is the surface-to-surface distance between spheres and $\phi_0 = 4\pi\varepsilon a\left(\frac{kT}{e}\right)^2 Y_1^0\,Y_2^0$ is the characteristic interaction energy.

In the case of spheroidal particles, the two previously discussed models were used, the Derjaguin approach and the equivalent sphere approach

(ESA). The former gives the expression for the interaction energy

$$\phi_{12} = \phi_0 \, G_D \, (R_1', R_1'', R_2', R_2'') e^{-h_m/Le} \tag{219}$$

where G_D is the Derjaguin factor (cf. Eq.(236) in Chapter 2) depending on the local radii of curvature of the spheroids R_1', R_1'', R_2', R_2'' evaluated near the point of minimum surface-to-surface distance h_m. For two equals spheres $G_D = 1/2$. The disadvantage of the Derjaguin model is that it becomes less accurate at large distances, comparable with the particle dimensions. A better accuracy at larger separations is obtained by using the equivalent sphere approach (ESA), which predicts the following expression for the interaction energy

$$\phi_{12} = \phi_0 \frac{\bar{R}_1 \bar{R}_2}{a_s (\bar{R}_1 + \bar{R}_2 + h_m)} e^{-h_m/Le} \tag{220}$$

where

$$\bar{R}_1 = \frac{2R_1' R_1''}{R_1' + R_1''} \qquad \bar{R}_2 = \frac{2R_2' R_2''}{R_2' + R_2''} \tag{221}$$

The minimum distance between particles was evaluated numerically by using an efficient iterative procedure.

It is interesting to note that according to the Derjaguin formula, Eq. (219) the interaction energy falls to a characteristic value ϕ_{ch} at the distance

$$h_{ch} = Le \ln\left(\frac{\phi_0 \, G_D}{\phi_{ch}} \right) \tag{222}$$

For the LSA model for spheres one can determine h_{ch} iteratively

$$h_{ch} = Le \left[\ln \frac{\phi_0}{2\phi_{ch}} - \ln\left(1 + \frac{Le}{2a} \ln \frac{\phi_0}{2\phi_{ch}} \right) \right] \tag{223}$$

For $Le/a < 1$, Eq.(223) reduces to the Derjaguin expression with $G_D = 1/2$. The h_{ch} parameter can be used as a good approximation of the effective interaction range upon a proper definition of ϕ_{ch} as discussed later on.

The validity of the effective hard particle concept was checked by performing the iRSA simulations for spheres and spheroids [58–60, 64]. The interaction energy was calculated from Eqs. (218–220). Thus, for a fixed aspect ratio As, the calculated values of the C_1^* and C_2^* constants were dependent on the Le/a and ϕ_0 parameters only. The results obtained for spherical particles are shown in Fig. 5.49 in the form of the dependence of the C_1^* and C_2^* on the κa or Le/a parameters for two values of $\phi_0 = 200$ and $\phi_0 = 2000\ kT$. As can be seen in Fig. 5.49, the C_1^* constant increases significantly with the Le/a parameter, being 2.5 times larger for $Le/a = 0.25$ ($a = 4$) than for $Le/a = 0$ (hard particles). On the other hand, the influence

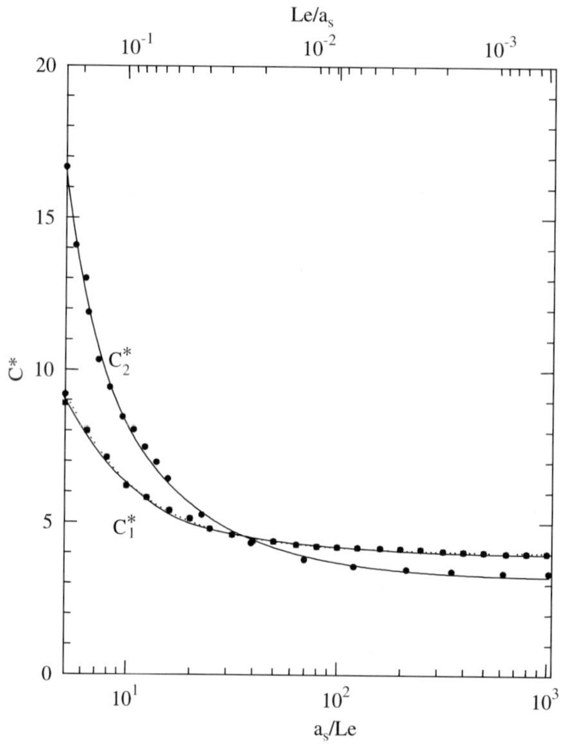

Fig. 5.49. The dependence of the C_1^* and C_2^* constant on the a_s/Le or parameters for spherical particles. The points represent exact numerical results obtained for $\phi_0 = 200\ kT$. The lines denote the analytical predictions calculated from Eqs.(224–225).

of the ϕ_0 parameter is rather moderate since the increase by an order of magnitude caused a 40% increase in C_1^* for $Le/a_s = 0.25$ [60].

The exact numerical data shown in Fig. 5.49 were reflected well by the parabolic fitting function

$$C_1^* = 4(1+H^*)^2 \tag{224}$$

where $H^* = \frac{1}{2}h_{ch}$ and h_{ch} was calculated from Eq.(222) by assuming $\phi_{ch} = 1.6\ kT$. The good agreement of the analytical data derived from Eq.(224) with the exact numerical results for both values of suggest that the effective hard particle concept works well for spherical particles. The fitting function for the C_2^* constant was calculated from the equation

$$C_2^* = \frac{6(3)^{1/2}}{\pi}(1+H^*)^4 \tag{225}$$

As can be seen in Fig. 5.49, the analytical results calculated from Eq. (225) reflect well the numerical data.

Analogous results were obtained for prolate spheroids adsorbing under the side-on and unoriented regime. The dependence of C_1^* on Le/a_s for $As = 5$ is shown in Fig. 5.50 and the dependence of C_2^* on Le/a_s in Fig. 5.51. As can be observed, the increase in the C_1 constant with Le is much more significant for the unoriented adsorption. For $Le/a_s = 0.25$ and $\phi_0 = 200$, C_1 increases more than four times for unoriented adsorption and about two times for side-on adsorption. Similarly as for spheres, the exact data were fitted well by the parabolic dependence analogous to Eqs. (224,225).

The numerical results shown in Figs. 5.50 and 5.51 can be transformed to a more universal form by calculating the effective interaction range H^* from Eqs. (204) for spheres or Eq. (211) and Eqs. (213–215) for prolate spheroids. Then, by plotting H^* against Le/a_s one can determine the range of validity of the effective hard particle concept. Numerical results transformed in this way are presented in Fig. 5.52 for spheres and prolate spheroids under the unoriented adsorption regime. As can be seen, the exact data for spheres and spheroids with $As = 2$, $As = 5$ can be well reflected for the entire range of Le/a_s parameters and both values of ϕ_0 by the linear dependence,

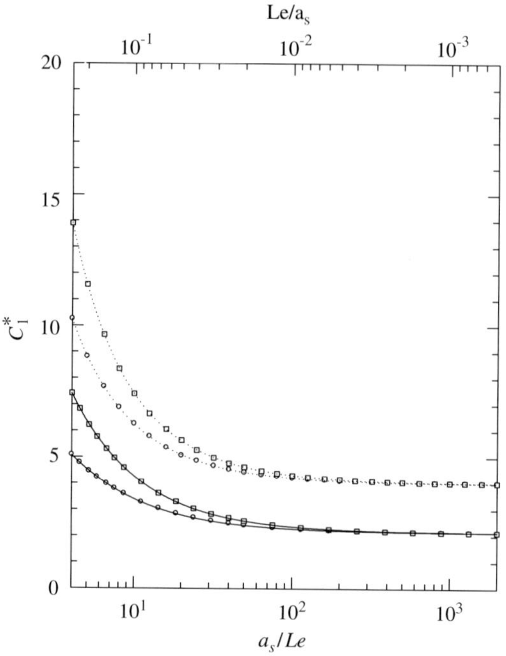

Fig. 5.50. The dependence of the C_1^* constant on the a_s/Le or Le/a_s parameters for oblate spheroids with $As = 5$. The points represent exact numerical results obtained for $\phi_0 = 2000$ (squares) and $\phi_0 = 200$ (circles). Upper curves denote the side-on adsorption and the lower curves the unoriented adsorption. The lines denote the analytical results calculated from parabolic fitting functions analogous to Eq. (224).

$$H^* = \xi_H \frac{Le}{a} \tag{226}$$

The proportionality constant ξ_H was found to be equalto 2.3 for $\phi_0 = 200$ and 3.45 for $\phi_0 = 2000$. This is consistent with the formula

$$H^* = \frac{Le}{2a_s} \ln \frac{\phi_0}{2\phi_{ch}} \tag{227}$$

where the characteristic interaction energy is $\phi_{ch} = 1.02$.

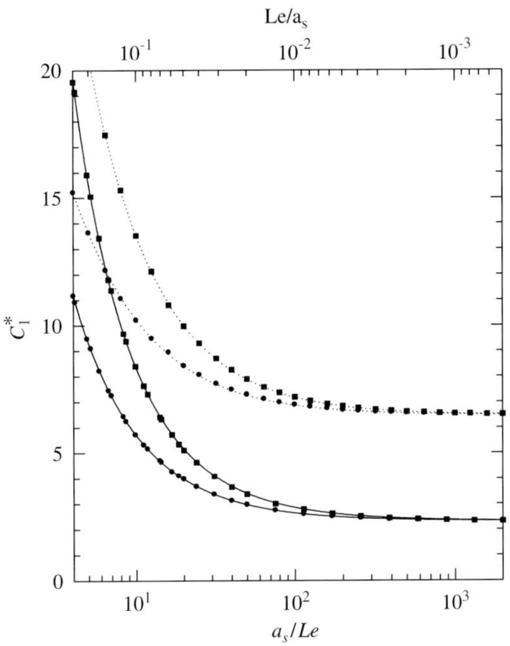

Fig. 5.51. The dependence of the C_1^* constant on the a_s/Le or Le/a_s parameters for prolate spheroids with $As = 5$. The points represent exact numerical results obtained for $\phi_0 = 2000$ (squares) and $\phi_0 = 200$ (circles). Upper curves denote the side-on adsorption and the lower curves the unoriented adsorption. The lines denote the analytical results calculated from fitting function analogous to Eq. (224).

Note that for spheroids of larger elongation, $As = 5$, the dependence of H^* on the Le/a_s parameter was better approximated by the formula

$$H^* = \frac{Le}{2a_s}\left[\ln\frac{\phi_0}{2\phi_{ch}} - \ln\left(1 + \frac{Le}{2a_s}\ln\frac{\phi_0}{2\phi_{ch}}\right)\right] \tag{228}$$

Similar results have been obtained for the side-on adsorption of prolate spheroids and unoriented adsorption of oblate spheroids. In the latter case, the results of numerical calculations are shown in Fig. 5.53. The characteristic energy was found equal to 1.6 kT for prolate spheroid adsorption and 0.862 for oblate spheroids.

The calculations shown in Figs. 5.52 and 5.53 confirm the validity of the effective hard particle concept, providing one with convenient fitting functions for analyzing adsorption of spherical and anisotropic particles.

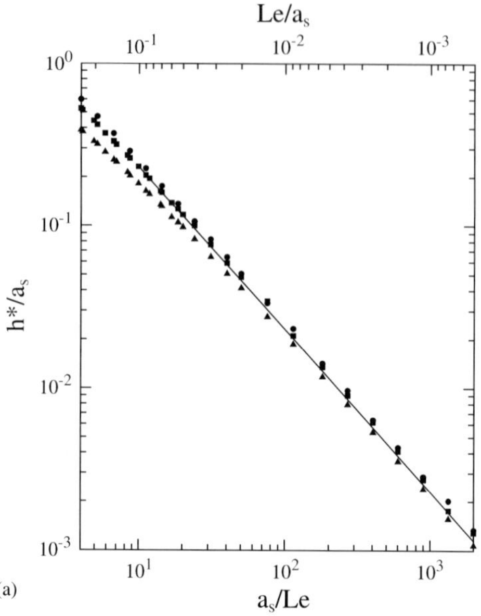

(a)

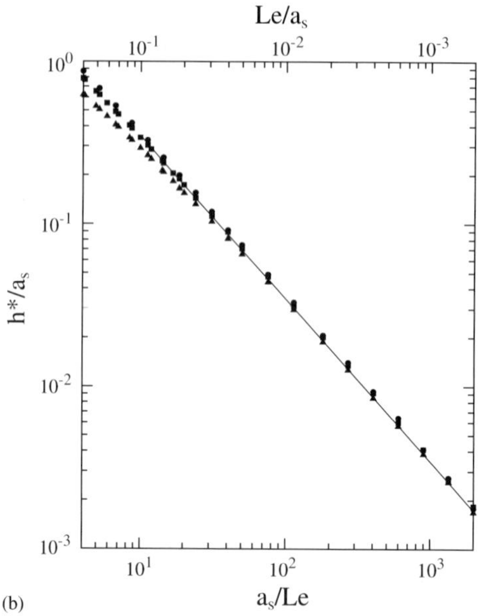

(b)

Fig. 5.52. The dependence of the effective interaction range h^*/a_s on the a_s/Le parameter; the points represent the exact results derived from numerical simulations for prolate spheroids adsorbing under the unoriented regime: circles, $As = 1$; squares, $As = 2$; triangles, $As = 5$. The line denotes the limiting analytical results calculated from Eq.(226). (a), $\phi_0 = 200\ kT$, (b), $\phi_0 = 2000\ kT$.

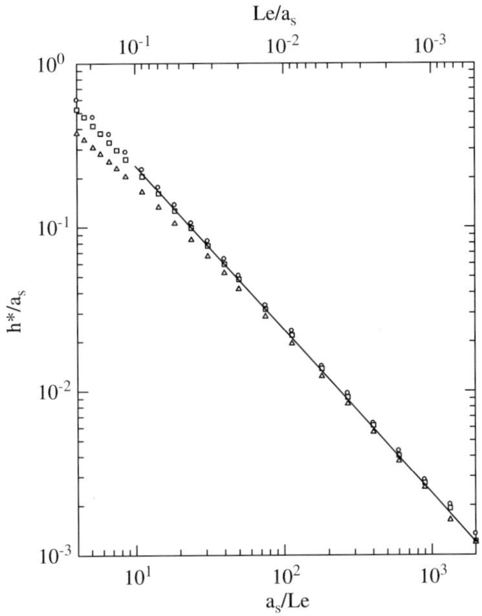

Fig. 5.53. The dependence of the effective interaction range h^*/a_s on the a_s/Le or Le/a_s parameters; the points represent the exact results derived from numerical simulations of oblate spheroids adsorbing under the unoriented regime: circles, $As = 1$; , squares; $As = 2$ $As = 5$ triangles. The line denotes the limiting analytical results calculated from Eq.(226). $\phi_0 = 200\ kT$.

These results indicate that the effective interaction range in adsorption of colloid particles exceeds the double layer thickness *Le* by a factor 2 – 3.

The equivalent hand sphere (EHS) concept can also be used for interpretation of the blocking functions of interacting particles, which governs kinetics of adsorption. In Fig. 5.54 results of numerical simulations for interacting spherical particles are shown for a series of H^* values ranging from 0 to 0.25. As can be observed, the low-coverage expansion $B(\Theta)=1-C_1\Theta+C_2\Theta^2$ with C_1 and C_2 calculated from Eqs. (199),(200) describes well the exact data for $B > 0.2$.

Similar results obtained for prolate spheroids, $As = 5$, under the unoriented adsorption regime are shown in Fig. 5.55. Interestingly, the blocking function for interacting spheroids under this regime resembles, for $H^* = 0.15$, the blocking function for hard spheres. Because this value of H^* is quite close to the globular protein adsorption conditions one can predict that the hard sphere results (colloids) can be used as useful reference states for analyzing protein adsorption kinetics.

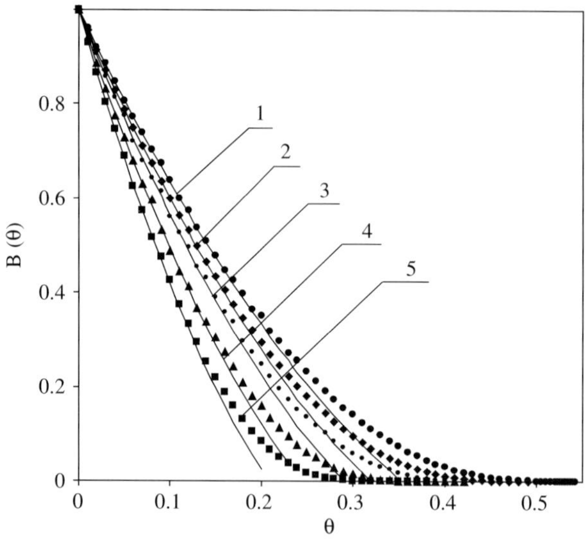

Fig. 5.54. The surface blocking function $B(\Theta)$ for soft spherical particles: the points denote the numerical simulations obtained for (1) $h^*/a = 0$ (hard particles), (2) $h^*/a = 0.05$, (3) $h^*/a = 0.1$, (4) $h^*/a = 0.15$, (5) $h^*/a = 0.25$. The solid lines denote the low-coverage analytical approximation derived from $B(\Theta)=1-C_1\Theta+C_2\Theta^2$ with C_1 and C_2 calculated from Eqs.(199),(200).

The behavior of the blocking function for high coverage can more conveniently be determined by studying the kinetics of interacting particle adsorption for the power law asymptotic regime, analogously as previously done for hard particles. In Fig. 5.56 results of simulations for interacting spheres are shown, plotted as the dependence of Θ on $\tau^{-1/2}$ for various H^*. As seen, for $0.5 < \tau^{-1/2} < 0.1$ ($\tau < 100$) these dependencies are linear with the slope decreasing gradually with H^*. As suggested in Ref. [65] these linear parts of the kinetic curves can be transformed to one universal curve by introducing the reduced coverage Θ/Θ_{mx}. The maximum coverage Θ_{mx} is calculated by postulating that the effective hard particle concept is valid for the high coverage range as well, which can be expressed as

$$N_p^* S_g^* = N_p S_g \tag{229}$$

where N_p^* is the number of interacting particles forming the monolayer.

Eq.(229) can be interpreted as the postulate that the effective area blocked at the maximum coverage by interacting particles equals the area

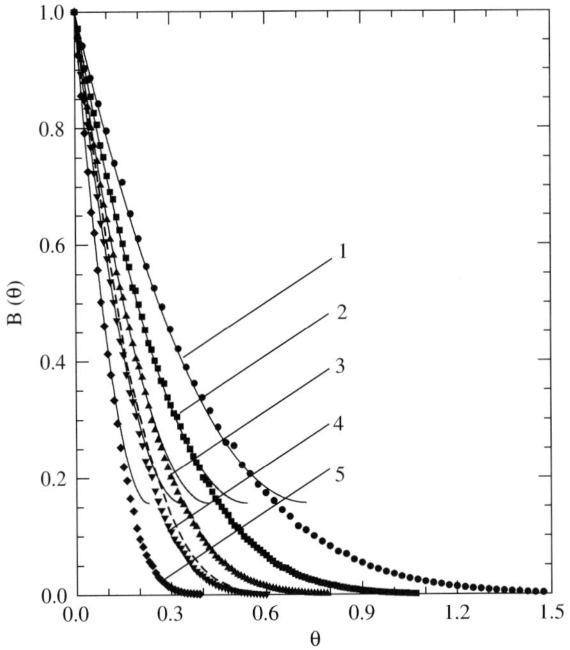

Fig. 5.55. The surface blocking function $B(\Theta)$ for soft prolate spheroids under the unoriented adsorption regime $As = 5$: the points denote the numerical simulations obtained for (1) $h_s^*/a_s = 0$ (hard particles), (2) $h_s^*/a_s = 0.05$, (3) $h_s^*/a_s = 0.1$, (4) $h_s^*/a_s = 0.15$, (5) $h_s^*/a_s = 0.25$, the solid lines denote the low-coverage analytical approximation derived from $B(\Theta)=1-C_1\Theta+C_2\Theta^2$ and the dashed line denotes the $B(\Theta)$ function for hard spherical particles.

blocked by hard particles at the jamming state. By considering that for spheres $S_g^* = S_g(1+H^*)^2$, Eq.(229) can be formulated as

$$\Theta_{mx} = \Theta_\infty \frac{S_g}{S_g^*} = \Theta_\infty \frac{1}{(1+H^*)^2} \tag{230}$$

Eq.(230) suggests the effective range of interactions h^* can be easily determined experimentally, when Θ_{mx} is measured, because

$$h^* = aH^* = a\left[\left(\frac{0.547}{\Theta_{mx}}\right)^{1/2} - 1\right] \tag{231}$$

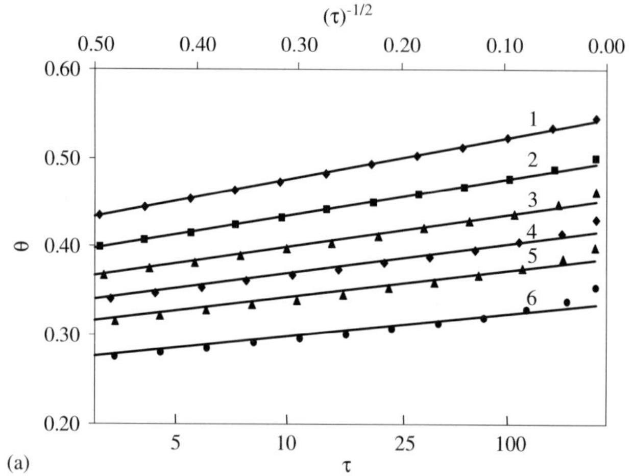

(a)

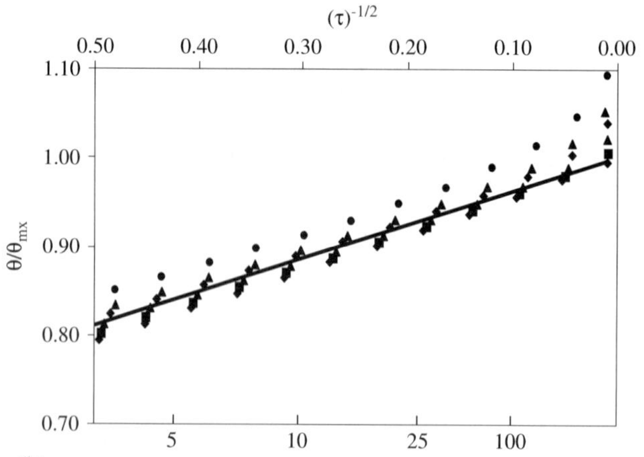

(b)

Fig. 5.56. (a), The dependence of Θ on $\tau^{-1/2}$ for soft spherical particles: the points denote the numerical simulations obtained for (1) $h^*/a = 0$ (hard particles), (2) $h^*/a = 0.05$, (3) $h^*/a = 0.1$, (4) $h^*/a = 0.15$, (5) $h^*/a = 0.2$, (6) $h^*/a = 0.3$. The solid lines denote the linear fits. (b) The dependence of Θ/Θ_{mx} on $\tau^{-1/2}$ with Θ_{mx} given by Eq.(230), the solid line denotes the analytical results calculated from Eq.(232).

As can be seen in Fig. 5.56, for $0.5 < \tau^{-1/2} < 0.1$, all the kinetic results for interacting particles expressed as Θ/Θ_{mx} vs. $\tau^{-1/2}$ dependence are close to the universal straight line describing hard particle adsorption kinetics

$$\Theta = \Theta_{mx}\left(1 - 0.37\tau^{-1/2}\right) \tag{232}$$

Eq.(232) indicates that the jamming coverage of interacting spheres, in the limit of long times, can be adequately approximated by Eq.(230).

Moreover, from Eq.(232), one can deduce that the blocking function for interacting particles is given by

$$B(\Theta) = \frac{d\Theta}{d\tau} \cong 2.31\left(1 - \frac{\Theta}{\Theta_{mx}}\right)^3 \tag{233}$$

$$for \quad \Theta \to \Theta_{mx}$$

However, Eq.(233) remains accurate for $H^* < 0.2$ only, because the Θ vs. $\tau^{-1/2}$ dependencies start to deviate from linearity for $H^* > 0.2$ and $\tau^{-1/2} < 0.2$. The coverage obtained in simulations is larger than this theoretical curve predicts, so there is no uniquely defined jamming limit. This behavior, analyzed in detail in Ref. [65] can be attributed to the finite probability of adsorption in local energy minima occurring in the vicinity of particles forming isolated targets. Because the interaction of the adsorbing particle with the interface plays a significant role in this process, it can be properly analyzed in terms of the 3D models discussed next. However, from a practical point of view this discrepancy has a marginal significance because this non-linear regime starts at $\tau > 100$, which for most situations met in practice can be treated as an infinitely long time.

Analogous dependencies are obtained for the unoriented adsorption of spheroidal particles, as can be seen in Fig. 5.57. In this the numerical results expressed in the coordinate system Θ vs. $\tau^{-1/4}$ can be well approximated by straight line dependencies

$$\Theta = \Theta_{mx} - C_s \tau^{-1/4} \tag{234}$$

where C_s is the fitting constant and the maximum coverage was calculated using the hard particle concept from the formula

$$\Theta_{mx} = \Theta_{\infty}(As^*)\frac{S_g}{S_g^*} \tag{235}$$

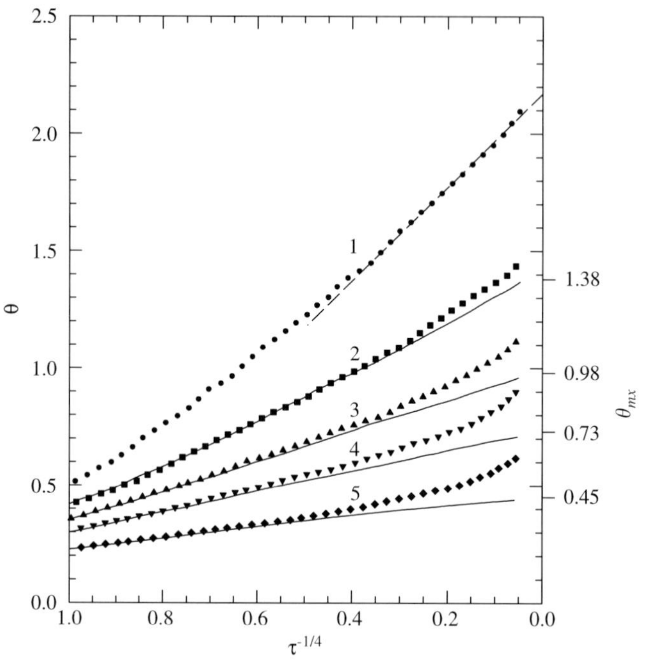

Fig. 5.57. The dependence of Θ on $\tau^{-1/4}$ for interacting prolate spheroidals under the unoriented adsorption regime: the points denote the numerical simulations obtained for 1. $h^*/a_s = 0$ (hard particles), 2. $h^*/a_s = 0.05$, 3. $h^*/a_s = 0.1$, 4. $h^*/a_s = 0.15$, 5. $h^*/a_s = 0.25$, the solid lines denote the asymptotic results calculated from Eq. (184) and Eq. (234).

where As^* is given by Eq. (208) and H^* by Eq. (227). Analogously as for the spheres, there appear deviations from the linear dependence appear as described by Eq. (234) for $\tau^{-1/4} > 0.3$ ($\tau > 100$).

The jamming coverage of spheroids, which has a practical significance for interpreting adsorption of proteins, can be estimated using exact numerical data derived from simulations. In Fig. 5.58 results obtained for spheres and prolate spheroids, characterized by $As = 2$ and $As = 5$, are shown for the side-on adsorption regime and in Fig. 5.59 for the unoriented adsorption regime. As can be noticed, Eq. (235) remains a valid approximation for the $4 < Le/a_s < \infty$ with a slight tendency to underestimate the exact data. The jamming of surfaces can be treated as a kinetically frozen limit, rather than an absolute limit as is the case for hard particles. It can also be seen in Fig. 5.58 that the decrease in the jamming coverage with decreasing the Le/a_s parameter is significantly more steep for elongated particles.

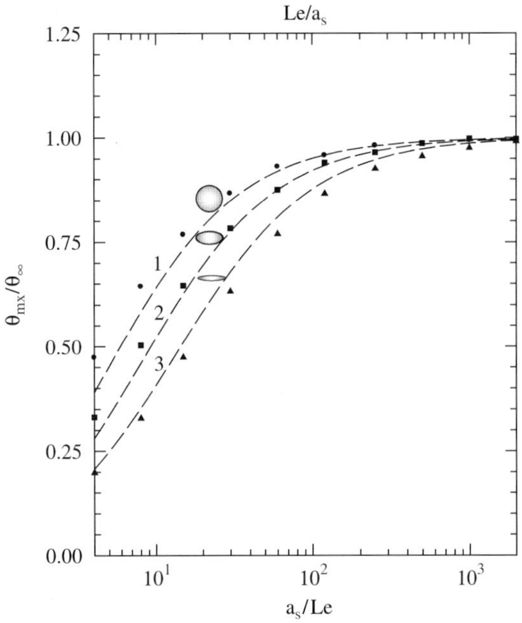

Fig. 5.58. The reduced maximum coverage of interacting particles $\Theta_{mx}/\Theta_{\infty}$ (where Θ_{∞} is the jamming coverage of hard particles) vs. the a_s/Le or Le/a_s parameters. The points denote the numerical RSA simulations performed for: (1) $As = 1$ (spheres) (2) $As = 2$ (prolate spheroids, side-on adsorption), (3) $As = 5$ (prolate spheroids, side-on adsorption). The dashed lines denote the analytical results calculated from Eq.(235). From Ref. [63].

Analogous trends are observed for the unoriented adsorption of prolate spheroids shown in Fig. 5.59. The approximate expression for the jamming coverage is given in this case by

$$\Theta_{mx} = \left(0.304 + 0.365\, As^{*} - 0.123\frac{1}{As^{*}}\right)\frac{1}{(1+H^{*})(1+As\,H^{*})} \tag{236}$$

As can be seen the analytical results are in a better agreement with the exact numerical data than for the side-on adsorption regimes.

There are no results reported in the literature for oblate spheroid adsorption under the unoriented regime. However, it can be expected that a satisfactory approximation of exact data can be derived from the analytical

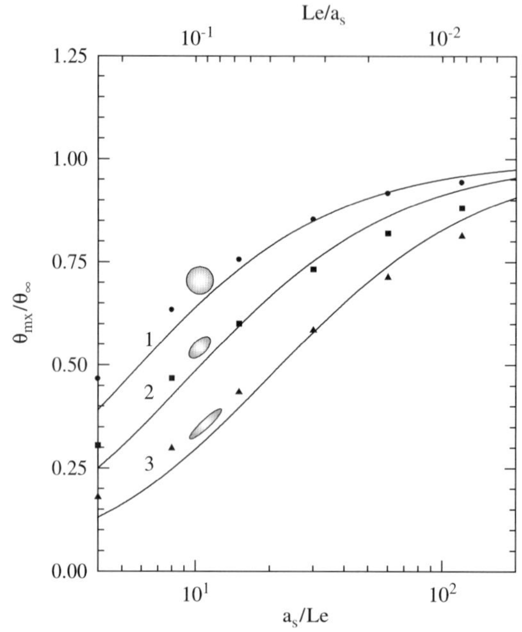

Fig. 5.59. The reduced maximum coverage of interacting particles $\Theta_{mx}/\Theta_\infty$ (where Θ_∞ is the jamming coverage of hard particles) vs. the a_s/Le or Le/a_s parameter. The points denote the numerical RSA simulations performed for: (1) $As = 1$ (spheres), (2) $As = 2$ (prolate spheroids, unoriented adsorption, (3) $As = 5$ (prolate spheroids, unoriented adsorption). The dashed lines denote the analytical results calculated from Eq. (236).

expression using the EHS concept

$$\Theta_{mx} = \left(0.768 + 0.251\, As^* - 0.473\frac{1}{As^*}\right)\frac{1}{(1+H^*)^2} \tag{237}$$

Eqs. (235–237) can be used for determining the influence of the ionic strength and the zeta potential of particles on the maximum coverage, because, according to Eq. (227)

$$H^* = \left(\frac{\varepsilon kT}{8e^2 I a_s^2}\right)^{1/2} \ln\frac{32\pi\varepsilon a_s\left(\dfrac{kT}{e}\right)^2 \tanh^2\left(\dfrac{e\zeta_p}{4kT}\right)}{\phi_{ch}} \tag{238}$$

For $Le > 0.1$ a more accurate expression for the Y_l function given in Ref. [66], leads to the formula

$$H^* = \left(\frac{\varepsilon kT}{8e^2 I a_s^2}\right)^{1/2} \ln \frac{32\pi\varepsilon a_s \left(\frac{kT}{e}\right)^2 \tanh^2\left(\frac{e\zeta_p}{4kT}\right)\left[1+\left(\frac{2\varepsilon kT}{8e^2 I a_s^2}\right)^{1/2} \tanh^2\left(\frac{e\zeta_p}{4kT}\right)\right]^2}{\phi_{ch}} \tag{239}$$

However, by deriving Eqs. (235–237) the effect of the interactions with the substrate is neglected, which reduces their accuracy for high zeta potential of the interface and for adsorption on pre-covered surfaces, for higher values of λ'. These effects can be more precisely studied using the 3D RSA model discussed in the next section.

Besides the jamming coverage one is often interested in the structure of the monolayers formed by interacting particles. The structure is quantitatively described by the pair correlation function, which is accessible experimentally, e.g., by direct imaging of adsorbed particles using optical microscopy [27,51,52,63,64]. From the analysis of the pair correlation function in the limit of low coverage one can directly estimate the range and magnitude of lateral interaction between adsorbed and adsorbing particles.

Examples of interacting particle monolayers with corresponding pair correlation functions generated in iRSA simulations are shown in Fig. 5.60. As seen, for $\Theta < 0.1$, the pair correlation function is well reflected by the Boltzmann distribution

$$g(H) = e^{-\phi_{12}/kT} \tag{240}$$

A simple inversion of this relationship gives directly the potential energy of the mean force

$$\phi_{12}(H) = -kT \ln g(H) \tag{241}$$

For increasing coverage, one can observe a formation of a primary peak on the pair correlation function, whose height increased with the coverage. The

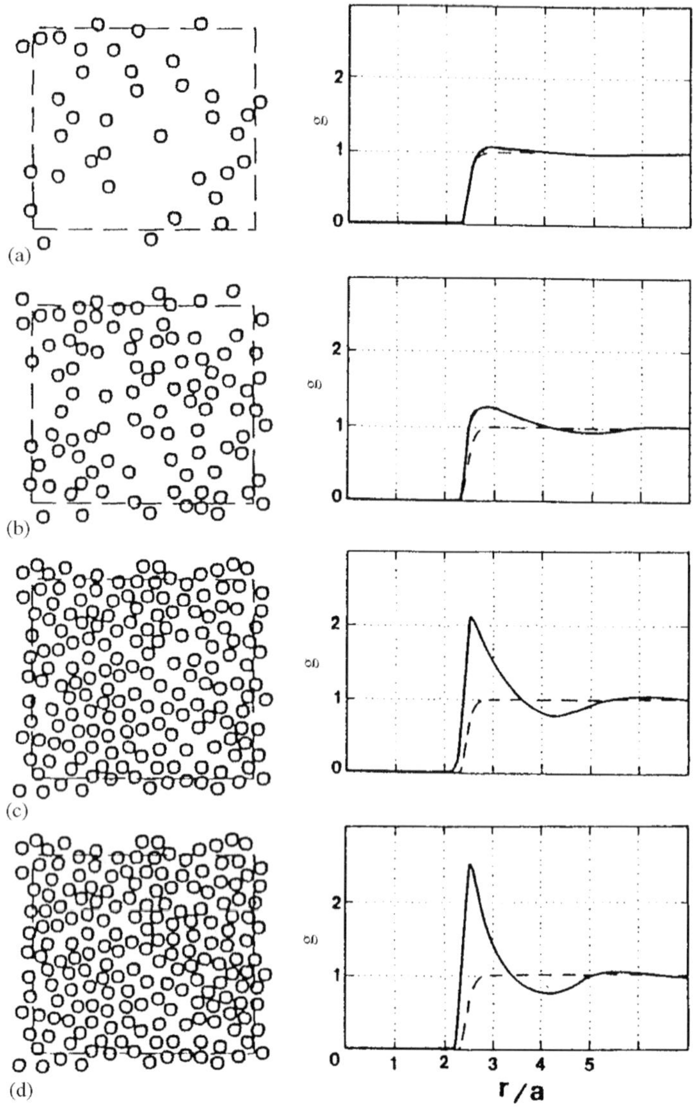

Fig. 5.60. Configurations of interacting spherical particles with corresponding pair correlation function g generated numerically for $H^* = 0.1$, (a) $\Theta = 0.082$, (b) $\Theta = 0.205$, (c) $\Theta = 0.37$, (d) $\Theta = 0.41$ (maximum coverage). The dashed lines in the pair correlation function calculated from the Boltzmann distribution, $g = e^{-\phi_{12}/kT}$.

position of the peak was found at the surface to surface distance 0.5 (for Θ = 0.41). From this estimation one can calculate the apparent interaction range $h/a = \frac{1}{2}H_p = 0.25$. This is 2.5 times larger than the effective interaction range H^* which was equal to 0.1. Hence, these results suggest that it is not proper, as often done in experiments, to identify the peak position of the pair correlation function with the real range of interactions.

A similar behavior is observed for interacting spheroidal particles adsorbing under the unoriented regime as seen in Fig. 5.61. The height of the peak on the pair correlation function increases with the increase in Θ but it becomes more diffuse than in the case of spheres. Note that because of variable orientations of particles in the monolayers, the position of the peak cannot be related directly to the interaction range. Besides there are no experimental methods which can be used to determine the positions of individual spheroidal particles adsorbing randomly.

More useful in the case of elongated particle is the analysis of the orientational ordering parameter, which is defined as

$$Q_0 = \frac{1}{N_p}\sum_{l=1}^{N_p}\cos 2\beta_l \tag{242}$$

where N_p is the number of particles in the analyzed particle monolayer and β_l are the angle between the axis of symmetry of each particle and the axis perpendicular to the interface.

For a completely disordered monolayer $Q_0 = 0$, and for ideally ordered (all particle axes parallel to each other) $Q_0 = 1$. As can be seen in Fig. 5.62 the ordering in the monolayer can be largely increased by increasing the range of lateral interaction H^*. The effect is especially pronounced for prolate spheroids as could be observed qualitatively in the monolayer shown in Fig. 5.61.

5.5.1. The 3D RSA models for interacting particles

The main advantages of the simple 2D model of interacting particles analyzed above are

(i) no adjustable parameters were necessary,
(ii) simple analytical formulae for the jamming limit and adsorption kinetics can be derived,
(iii) there is a possibility of describing anisotropic particle behavior.

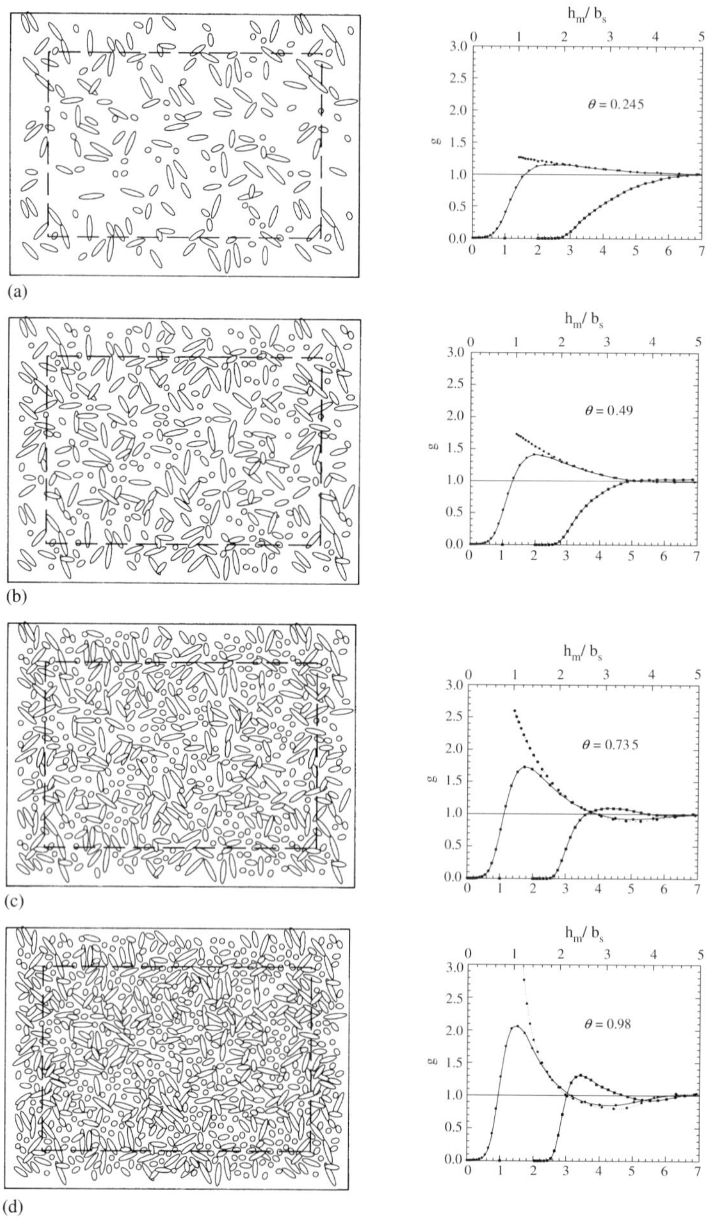

Fig. 5.61. Configurations of interacting prolate spheroids adsorbing under the unoriented regime with corresponding pair correlation function generated numerically for $As = 5$, $H^* = 0.1$, (a) $\Theta = 0.245$, (b) $\Theta = 0.49$, (c) $\Theta = 0.735$, (d) $\Theta = 0.98$ (maximum coverage). The continuous lines denote the surface-to-surface and the center-to-center pair correlation functions, and the dotted line represents the results obtained for hard spheroids.

However, a disadvantage of the model was the specific interactions of adsorbing particles with the interfaces were not considered. This may lead to inaccurate results concerning the jamming coverage if the *Le* parameter becomes comparable with unity and larger. Considering the data collected in Table 5.4 one can predict that for the ionic strength of 0.15 M (physiological conditions) $Le > 1$ for particles of the size below 1nm (at room temperature). For $I = 10^{-2}$ M the critical particle size is 2.5 nm, for $I = 10^{-3}$ M, $a < 10$ nm and for $I = 10^{-5}$ M, $a < 100$ nm. Thus, the effect of the interface is expected to play a more significant role for very low ionic strength and particle sizes below 100 nm. These estimates also show that formation of a second layer of adsorbed particles is highly unlikely, especially for proteins at physiological conditions.

In order to consider the specific effects stemming from the interface, a 3D RSA model for interacting particles was developed by Oberholzer et al. [66] and then extended to bimodal systems by Weroński [67]. Similarly to the classical 2D RSA model it is assumed that particles are moving consecutively toward the surface along rectilinear trajectories with their lateral position unchanged. The energy of interactions was calculated using the

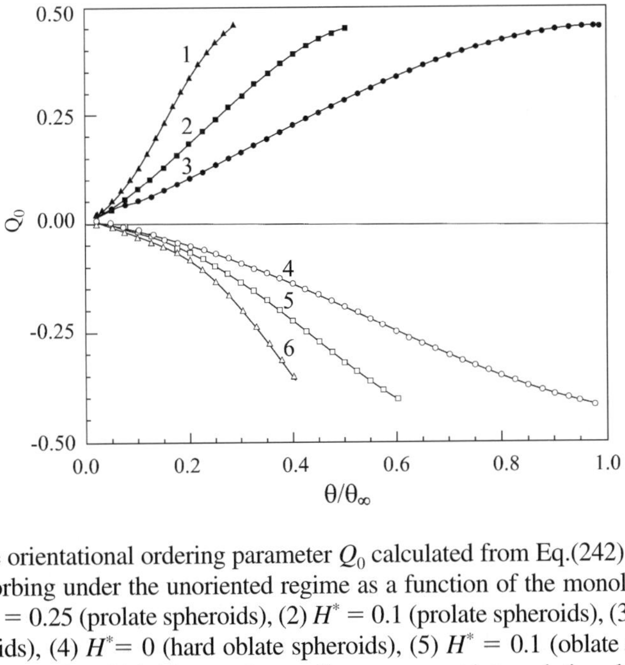

Fig. 5.62. The orientational ordering parameter Q_0 calculated from Eq.(242) for interacting spheroids adsorbing under the unoriented regime as a function of the monolayer coverage, $As = 5$, (1) $H^* = 0.25$ (prolate spheroids), (2) $H^* = 0.1$ (prolate spheroids), (3) $H^* = 0$ (hard prolate spheroids), (4) $H^* = 0$ (hard oblate spheroids), (5) $H^* = 0.1$ (oblate spheroids), (6) $H^* = 0.25$ (oblate spheroids). The continuous lines represent interpolations between points.

LSA model as the sum of the electrostatic interaction energy among particles and the interaction energy with the interface

$$\phi_v = \sum_{l=1}^{N_i} \phi_{lv}(h_l, \mathbf{r}_{vl}, \mathbf{r}_l) + \phi_{iv}(h) \tag{243}$$

where h is the distance of the virtual particle from the interface, $\mathbf{r}_{vl}$ the surface vector describing the position of the virtual particle, $\mathbf{r}_l$ are the surface vectors of the neighbors and ϕ_{iv} is the interaction energy of the virtual particle with the interface.

Because the particle/particle interaction energy is positive (by neglecting the van der Waals interactions) and the particle/interface energy ϕ_{iv} is negative, one can expect an energy maximum (barrier) to appear at a certain distance from the interface (in the case of the 2D model the total energy attained maximum at the interface).

The position of the energy maximum h_0 can be calculated from the non-linear dependence condition

$$\frac{\partial \phi_v(h_0)}{\partial h} = -F(h_0, \mathbf{r}_{v1}, \mathbf{r}_1 \cdots \mathbf{r}_{v1}) = 0 \tag{244}$$

where F is the total force on the particle.
Thus,

$$h_0 = F^{-1}(0) \tag{245}$$

The height of the barrier is given by

$$\phi_b = \phi_v(h_0) \tag{246}$$

Although the 3D model is claimed to be more realistic, it seems physically rather artificial because the repulsive forces between the adsorbing (virtual) and the pre-adsorbed particles will automatically lead to a change in the lateral position, which violates the basic assumption of the model. This deficiency was partially removed in the more sophisticated 3D model developed by Weroński [67]. The model, called curvilinear trajectory random sequential adsorption (CTRSA), is also based on the energy additivity principle with the van der Waals interactions neglected. Instead of linear, a more realistic trajectory of the adsorbing particle is calculated from the deterministic force balance equation by neglecting diffusion and hydrodynamic effects. However, by so doing the model also becomes ill-defined, because the hydrodynamic resistance tensors are highly anisotropic as demonstrated in Chapter 3, whereas in the model they are considered as isotropic quantities. Another disadvantage is the significant reduction in the speed of calculations in comparison with the classical 2D model. The comparison of the effective range of interaction derived from various models is done in Fig. 5.63. As can be noted, small differences between the 2D and the 3D models occur for $Le/a > 0.1$ only. Quite unexpectedly the CT RSA model predicts a larger value of H^* for the range of $Le/a < 0.1$, which can be attributed to the higher role of lateral interactions, causing deflection of the particle trajectory.

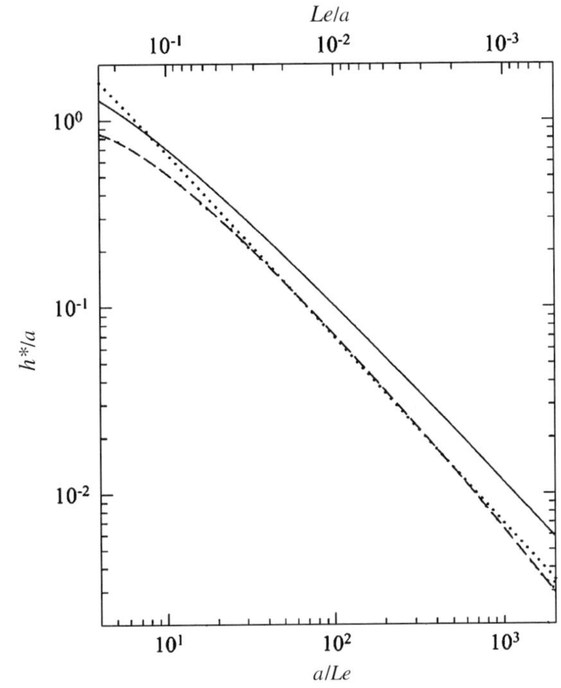

Fig. 5.63. The dependence of the effective interaction range h^*/a on the a/Le parameter determined for spherical particles from various RSA models; the dotted line denotes the 2D RSA model, the dashed line represent the 3D RSA model and the continuous line shows the results predicted by the CT RSA model. From Ref. [67].

5.6. OTHER RSA MODELS

The advantage of the classical RSA model discussed above is that it allows one to generate efficiently particle configurations and to determine the jamming or maximum coverages. As far as kinetic aspects are concerned, the classical RSA approach can be used for modeling the kinetics of an idealized process only, consisting in creation of particles at a given distance from the interface with a constant rate an in a consecutive manner. When an occupied space is found, the particle is removed with unit probability and a next particle creation attempt is undertaken, uncorrelated with previous attempts. The kinetics of this process is governed by the formula

$$\frac{1}{S_g}\frac{d\Theta}{dt} = r_c B(\Theta) \tag{247}$$

where r_c is the rate of particle creation (number of particles per unit area and unit of time).

It is not possible, within the framework of the RSA model, to find a unique relationship between the kinetics of this idealized process and the kinetics of the particle adsorption process governed by various transport mechanisms. In particular, one cannot predict how the flux of particles, governed by the bulk rate constant k_c analyzed in Chapter 4, is affected by the RSA blocking function $B(\Theta)$. It seems that any exact solution of this problem is prohibitive because of inherent multi-body physics and non-linear coupling of topological effects (monolayer structure) with the specific and hydrodynamic interactions. One has, therefore, to rely on approximate models being useful for specific transport mechanisms of particles.

One such simplified approach suitable for describing the external force driven adsorption, for example the gravitational force, is the ballistic model developed for 1D [69] and 2D [70–72] situations. The model is valid if the external force parameter $Ex >> 1$. In the case of the gravitational force one has explicitly

$$Gr = \frac{4\pi a^3 \Delta\rho|\mathbf{g}|}{3kT} = \bar{R}^4 \gg 1 \tag{248}$$

where $\bar{R}$ is the dimensional parameter often used in the literature to characterize the range of validity of the ballistic model [72].

The basic assumption of the model, in which diffusion and all hydrodynamic and specific interactions are neglected, are

(i) the particles are deposited sequentially,
(ii) the starting positions of the particles are uncorrelated,
(iii) the particles can only adsorb irreversibly on an uncovered area,
(iv) when a pre-adsorbed particle is met, the adsorbing particle follows the path of steepest descent until it reaches a stable position.

Because of the latter assumption, a cluster of at least three particles can prevent deposition of an incoming particle only. This exerts a profound effect on the blocking function $B(\Theta)$ because the two first terms in the low coverage expansion vanish. Accordingly, the blocking function for low coverage range is given by [71]

$$B(\Theta) = 1 - 9.95\Theta^3 + 0(\Theta^4) \tag{249}$$

In accordance with the formula derived previously, Eq. (178), the variance of the particle monolayer formed under the ballistic regime in the limit of low coverage is given by

$$\overline{\sigma}^2 = 1 - 29.85\Theta^3 + 0(\Theta^4) \tag{250}$$

As can be observed, in contrast to the RSA or equilibrium models, when $\overline{\sigma}^{-2} = 1 - 4\Theta$, the ballistic model predicts that the variance remains almost constant in the low coverage limit.

No simple analytical expressions for the blocking function was derived for higher coverage, especially for the asymptotic regime close to the jamming limit, which were found to be 0.610 [72]. However, the adsorption kinetics near the jamming state was described by the asymptotic dependence [72]

$$\Theta = \Theta_{\infty_B} - C_B \frac{e^{-\frac{2(3)^{1/2}}{\pi}\tau}}{\pi^2} \tag{251}$$

where $\Theta_{\infty B}$ is the jamming coverage and C_B is the dimensionless constant.

For the entire range of coverage, the deposition kinetics can be interpolated by the fitting function [72]

$$\Theta = \Theta_{\infty_B}\left[1 - \frac{e^{-\frac{2(3)^{1/2}}{\pi}\tau}(1+0.179\tau)}{1+0.714\tau+1.58\tau^2+2.47\tau^3}\right] \tag{252}$$

It is interesting to observe that because the ballistic model reflects well the basic physics of the external force driven deposition, one can assume that the rate of particle deposition is

$$k_c''(\Theta) = k_c B(\Theta) \tag{253}$$

where k_c is the constant migration rate analyzed in Chapter 4.

Consequently, particle deposition kinetics for the entire range of coverage can be described by Eq. (252) with the dimensionless time defined as

$$\tau = \pi a^2 k_c n_b t \tag{254}$$

In the case of sedimentation one has

$$\tau = \frac{2\pi a^4 \Delta\rho |\mathbf{g}|}{9\eta} n_b t \tag{255}$$

The theoretical predictions derived from the ballistic model are compared with experimental data obtained for a 8.72 μm diameter melamine particle of apparent density $1.5\text{x}10^3$ kg m^{-3} in Fig. 5.64. Note that the kinetics of particle deposition is well described for the entire range of times by Eq. (252) with the dimensionless time defined by Eq. (255).

However, the disadvantage of the ballistic model is that the possibility of particle trapping in local energy minima, which leads to multilayer formation, is neglected. The presence of these reversibly adsorbed particles can slightly enhance the transport of other particles due to the rolling over mechanism. This has been considered in the 1D case [73]. However, in experiments these particles are removed by thorough washing [73,74], which reduces the error connected with multilayer formation.

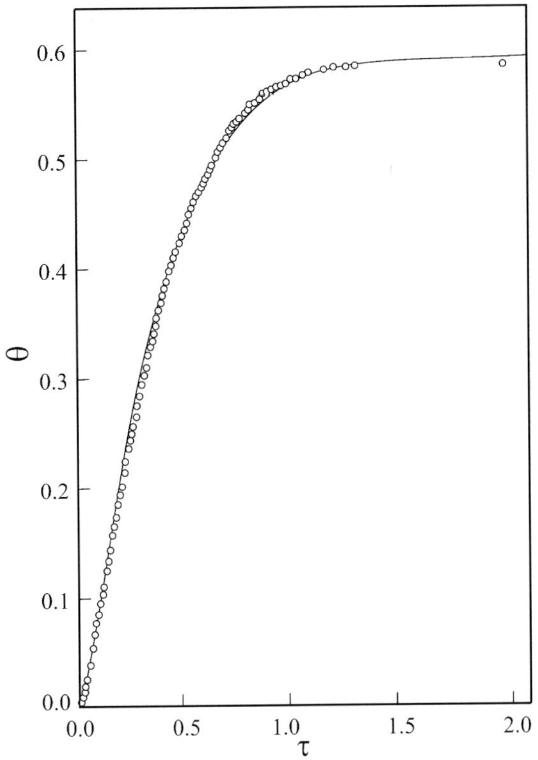

Fig. 5.64. The kinetics of particle deposition governed by sedimentation. The continuous line denotes the theoretical results calculated from Eq. (252) with the dimensionless time defined by Eq. (255). The points denote experimental results obtained for a 8.72 μm of diameter particle, having the density of 1.5×10^3 kg m^{-3}. From Ref. [72].

On the other hand, the hydrodynamic effects have been considered in the improved version of the ballistic model [75]. It was demonstrated that both the jamming limit and the blocking function were only marginally modified. The only noticeable difference occurred for the pair correlation function, which exhibited a slightly smaller peak height for the transient coverage range, about 0.15.

Another major deficiency of the RSA and the ballistic models consists in the fact that the diffusion effects are neglected, which dominates for most of the practically interesting situations. Under the diffusion-controlled adsorption regime, the particle which overlaps with any pre-adsorbed particle has another chance to adsorb in the vicinity. As a result the adsorption attempts are correlated, and should be described by more refined models. Many attempts have been undertaken in the literature to consider the diffusion effects, which is a challenging task in view of a complicated many-body hydrodynamics and ill-defined dependence on the initial and boundary conditions. In one of the first works, the lattice model was applied to mimic to some extent particle diffusion and sedimentation occurring at the same time [76]. The primary goal of these simulations was to determine the jamming coverage as a function of the *Gr* parameter. It was found that for the pure diffusion, where $Gr = 0$, the jamming coverage, within the error bounds, was the same as for the RSA model, i.e., 0.547. For $Gr > 1$ the ballistic limit of 0.6010 was attained. The transition between the diffusion and the ballistic jamming coverage was described by the fitting function [72]

$$\frac{\Theta(Gr) - \Theta_{\infty}}{\Theta_{\infty_B} - \Theta_{\infty}} = e^{-\frac{1.32}{Gr} - 3.44\frac{1}{Gr^{3/4}}} \tag{256}$$

where $\Theta(Gr)$ is the coverage for arbitrary *Gr*.

Other simulations performed for the purely diffusion transport mechanism [77] confirmed that the jamming limit was the same as for the RSA process. Small differences between the RSA and the diffusion RSA (DRSA) model were predicted in respect to the pair correlation function for the transient coverage range $\Theta < 0.2$. However, calculations done considering the hydrodynamic effects via the lubrication approximation showed that the correlation effects in the DRSA model are less pronounced. Later on, the simulations based on the Brownian dynamics method confirmed this in the case where the specific interactions between particles and interfaces were considered [78,79]. It was predicted that the adsorption probability near an adsorbed particle can be reflected with good accuracy by the Boltzmann law, which was the main assumption in the iRSA model discussed in the previous section.

However, neither the ballistic nor the DRSA model can be used for deriving links between the bulk transport processes governed by the k_c constant and the surface transport governed by the blocking function. This can be achieved in terms of the generalized RSA model discussed next.

5.7. THE GENERALIZED RSA MODEL

The main assumption of the generalized RSA approach is that the adsorbed particles located in the primary minimum δ_m form a separate phase, see Fig. 5.65. These adsorbed particles exert hydrodynamic and specific interactions on adsorbing (wandering) particles, which leads to volume exclusion effects.

Because a 3D motion of adsorbing (wandering) particles in the adsorption layer is considered, the blocking function $B(\Theta, h)$ is dependent in this case not only on the local particle coverage Θ, and the structure of the adsorbed layer but also on the distance from the interface h. By assuming a local thermodynamic equilibrium between various planes parallel to the adsorption plane one can express the chemical potential of adsorbing particles in the form [80,81]

$$\mu = \mu^0 + kT \ln\left[f(\Theta,h)n(h)\right] \tag{257}$$

where the activity coefficient f, which is a position dependent quantity, is connected with $B(\Theta, h)$, by

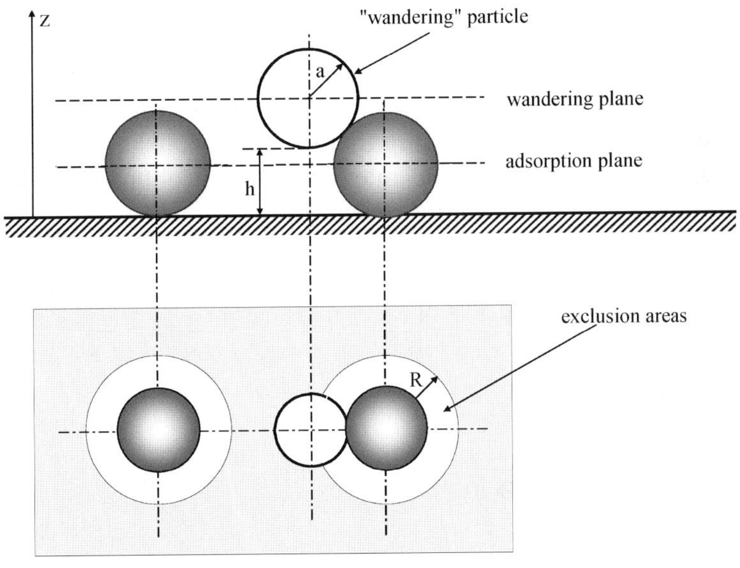

Fig. 5.65. A schematic representation of the generalized blocking function $B(\Theta, h)$. From Ref. [82].

$$f(\Theta,h)=\frac{1}{B(\Theta,h)} \tag{258}$$

An important property of the $B(\Theta, h)$ function is that it approaches for $h = \delta_m$, the value of $B(\Theta)$ derived previously for the 2D RSA model.

The expression for the chemical potential can also be written as

$$\mu=\mu^0+kT\ln n(h)+\phi_s \tag{259}$$

where

$$\phi_s=-kT\ln B(\Theta,h) \tag{260}$$

can be treated as the specific energy contribution due to adsorbed particles.

According to Eq. (259), the presence of adsorbed particles increases the interaction energy, because $B(\Theta, h) < 1$, which leads to formation of an energy barrier, of an extension comparable with particle size (see Fig. 5.66). The height of the barrier equals $-kT\ln B(\Theta)$ increasing monotonically with surface coverage of adsorbed particles.

Eq. (259) has a significance because it allows one to formulate quasi-linear transport equations of particles in the adsorption layer, analogous to Eq. (251) in Chapter 4. By assuming that the flow vanishes and neglecting the particle/particle hydrodynamic interactions, one obtains the following transport equation for the adsorption layer [82]

$$\frac{\partial n}{\partial t}=\frac{\partial}{\partial h}\left[D(h)\,e^{-\Phi/kT}\frac{\partial}{\partial h}\left(n\,e^{\Phi/kT}\right)\right] \tag{261}$$

where

$$\Phi=\phi+\phi_s=\phi-kT\ln B(\Theta,h) \tag{262}$$

can be treated as the overall interaction energy consisting of the steric contribution given by Eq. (260), the specific interactions energy of particles with the interface and the external force energy.

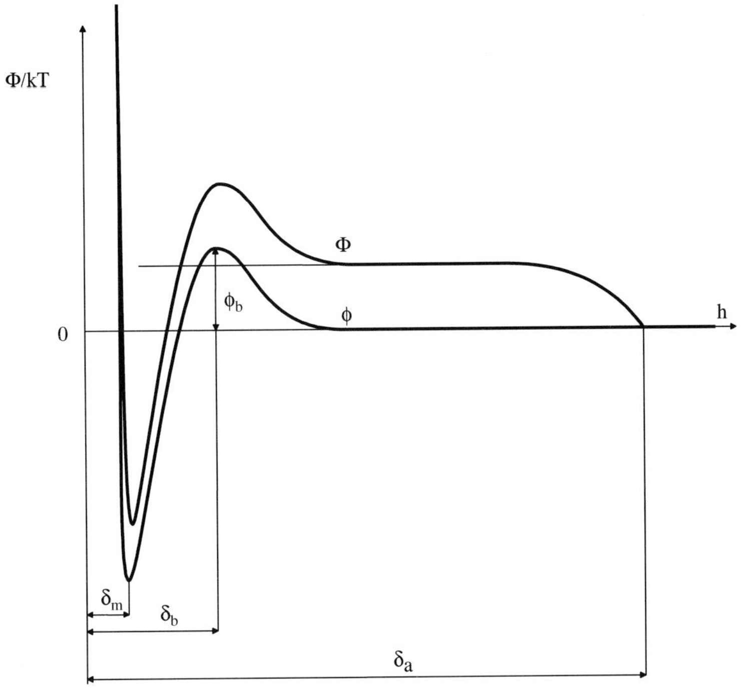

Fig. 5.66. A schematic representation of the steric barrier occurring due to adsorbed particles [82].

Because Eq. (261) has the same mathematical shape as Eq. (251) in Chapter 4, results derived previously for the barrier-controlled adsorption regime can be applied for the steric barrier case as well. In particular one can estimate that the relaxation time of establishing steady-state conditions through the steric barrier region is

$$t_{ch} \cong a^2 / D_\infty \tag{263}$$

One can estimate that for proteins, by taking $a = 10$ nm and $D = 2.15 \times 10^{-11}$ m^2 s^{-1}, the characteristic relaxation time is 4.6×10^{-6} s. For colloid particles, by taking $a = 100$ nm and $D = 2.15 \times 10^{-12}$ m^2 s^{-1}, the characteristic relaxation time is 4.6×10^{-3} s. Since these values are by a few orders of magnitude smaller than typical times of monolayer formation, (see Table 4.7), one can use a quasi-stationary version of Eq. (261), which is given by

$$D(h)e^{-\Phi/kT}\frac{\partial}{\partial h}\left(ne^{\Phi/kT}\right)=-j_a(\Theta) \tag{264}$$

Eq. (264), which is analogous to Eq. (261), in Chapter 4 gives upon integration the expression for particle concentration distribution.

$$n(\Theta,h)=\left[n(\delta_m)e^{\Phi(\delta_m)/kT}-j_a(\Theta)\int_{\delta_m}^{h}\frac{e^{\Phi/kT}}{D(h)}dh\right]e^{-\Phi/kT} \tag{265}$$

where the flux of particles j_a, depending now on the coverage Θ, is given by the expression

$$-j_a(\Theta)=\frac{1}{S_g}\frac{d\Theta}{dt}=k_a n(\delta_a)\overline{B}(\Theta)-\frac{k_d}{S_g}\frac{\overline{B}(\Theta)}{B(\Theta)}\Theta \tag{266}$$

k_a, k_d are the adsorption and desorption constants as defined previously, and are the generalized blocking functions (transport resistance of the adsorbed layer) given by

$$\overline{B}(\Theta)=\frac{1}{k_a\displaystyle\int_{\delta_m}^{\delta_a}\frac{e^{\Phi/kT}}{D(h)}dh} \tag{267}$$

Note that $\overline{B}(\Theta)$ depends not only on particle coverage but also on the particle diffusion coefficient and specific and external interactions governed by the interaction potential ϕ_s. Moreover, according to Eq. (266), the desorption flux (described by the second term on the r.h.s. of this equation) depends in a complicated manner on the blocking function $\overline{B}(\Theta)$.

Under equilibrium, when $j_a = 0$, Eq. (266) gives the isotherm equation

$$K_a S_g n(\delta_a)=\frac{\Theta_e}{B(\Theta_e)} \tag{268}$$

where Θ_e is the equilibrium coverage and $K_a = k_a/k_d$ the equilibrium adsorption constant identical as previously defined for the linear adsorption regime.

This means that K_a depends solely on the specific interaction energy distribution near the primary minimum.

Under non-equilibrium conditions, particle adsorption kinetics can be predicted by Eq. (266). This requires, however, explicit evaluation of the blocking function $\bar{B}(\Theta)$ that is rather cumbersome because $\bar{B}(\Theta)$ depends not only on particle coverage but also on particle distribution over the surface. This distribution is in turn governed by the particle transport mechanism, i.e., diffusion, or external force. At present, these effects cannot be decoupled within the framework of the generalized RSA model. However, useful approximations for $\bar{B}(\Theta)$ can be specified for limiting adsorption regimes having large practical significance. This is the case when the interaction potential exhibits a well-defined maximum described by a type II energy profile (see Fig. 2.33). Then, Eq. (267) can be approximated at $h = \delta_m$ by the expression

$$\bar{B}(\Theta) = B(\Theta, \delta_b) \cong B(\Theta) \tag{269}$$

This is so because usually the specific energy maximum appears at the distance δ_b, which is much smaller than particle dimension. One may conclude that in this case the overall blocking function can be well approximated by the classical RSA model. A physical explanation of this fact is that all moving particles that come into contact with adsorbed particles (overlap) are removed due to the presence of strong repulsive forces (energy barrier). Hence, the adsorption events are uncorrelated as postulated in the RSA model. This has practical implications because all the previous results pertinent to the classical RSA model can be used to calculate $\bar{B}(\Theta)$. Such conditions can be realized in experiments using the impinging-jet cells when the gravity force is directed outwards from the surface [57,63].

Other useful expressions for the $\bar{B}(\Theta)$ function can be derived in the form of a power series solution in the case of negligible external force and the specific energy profile of type 1b where Eq. (267) simplifies to

$$\bar{B}(\Theta) = \frac{1}{k_a \int_{\delta_m}^{\delta_a} \frac{dh}{D(h) B(\Theta, h)}} \tag{270}$$

where $B(\Theta,h)$ is given by [81],

$$B(\Theta,h) = 1 - C_1'\Theta + C_2'\Theta^2 + O(\Theta^3) \tag{271}$$

The C_1' and C_2' coefficients of this expansion were calculated analytically by considering the exclusion areas at various heights of the particle above the adsorption plane [81]

$$C_1'(\xi) = 4(1-\xi^2)$$
$$C_2'(\xi) = \frac{8}{\pi}\left\{2\xi^2(\xi^2-1)\cos^{-1}\left[\frac{1}{2(1-\xi^2)^{1/2}}\right] + \frac{1}{4}(3-4\xi^2)^{1/2}(3-2\xi^2)\right\} \tag{272}$$

where $\xi = h/2a$

Obviously, for $\xi \to \delta_m/2a \cong 0$ the C_1' and C_2' constants approach 4 and $\frac{6(3)^{1/2}}{\pi}$ as was the case for the RSA model.

Using Eq. (271) and Eq. (270) and substituting $D(h)=h/(a+h)$ one obtains for $\bar{B}(\Theta)$ the expression

$$\bar{B}(\Theta) = 1 - \langle C_1\rangle\Theta + \langle C_2\rangle\Theta^2 + O(\Theta^3) \tag{273}$$

where the $\langle C_1\rangle$ and $\langle C_2\rangle$ constants are given by

$$\langle C_1\rangle = k_a I_1 = 4\frac{\frac{3}{2}\ln\bar{\delta} - \frac{1}{4}(1-\bar{\delta})(5-7\bar{\delta}-4\bar{\delta}^2)}{\frac{3}{2}\ln\bar{\delta} - 3(1-\bar{\delta})}$$
$$\langle C_2\rangle = k_a^2 I_1^2 - k_a I_2 \tag{274}$$

where

$$k_a = \frac{1}{\frac{2a}{D}\int_{\delta_m}^{1}\left(1+\frac{1}{2\xi}\right)d\xi} = \frac{D}{2a}\frac{1}{\left(1-\bar{\delta}+\frac{1}{2}\ln\frac{1}{\bar{\delta}}\right)} \cong \frac{D}{2a}\frac{1}{\left(1+\frac{1}{2}\ln\frac{1}{\bar{\delta}}\right)} \tag{275}$$

$$I_1 = \frac{2a}{D}\int_{\bar{\delta}}^{1}\left(1+\frac{1}{2\xi}\right)C_1'(\xi)\,d\xi$$

$$I = \frac{2a}{D}\int_{\bar{\delta}}^{1}\left(1+\frac{1}{2\xi}\right)\left[C_1'^2(\xi)-C_2'(\xi)\right]d\xi$$

$$\bar{\delta} = \delta_m/2a$$

One can calculate from Eq. (264) that for $\bar{\delta}_m = 10^{-2}$, $\langle C_1\rangle = 3.3$, whereas for $\bar{\delta}_m = 10^{-3}$, $\langle C_1\rangle = 3.5$, which is only 0.125 less than the 2D RSA value. The $\langle C_2\rangle$ coefficient for $\delta_m = 10^{-2}$ calculated numerically is 1.2 and for $\delta_m = 10^{-3}$, $\langle C_2\rangle = 1.6$. The dependence of $\langle C_2\rangle$ and $\langle C_2\rangle$ on $\bar{\delta}$ calculated from Eq. (274) is shown in Fig. 5.67.

In Fig. 5.68, the 2D RSA blocking function $B(\Theta)$ and the generalized $\bar{B}(\Theta)$ function, calculated from the low coverage expansion Eq. (273), respectively, are compared. As can be seen, the differences between these two models are not too significant. For example, at $\Theta = 0.2$ one has $B(\Theta) = 0.33$ for the RSA model and $\bar{B}(\Theta) = 0.37$ for the generalized RSA model. However, because of the lack of analytical expressions for $\bar{B}(\Theta)$, a quantitative comparison of the two models for higher coverage range is not possible.

Useful estimates of the $\bar{B}(\Theta)$ function for a higher coverage range can be specified by realizing that the steric barrier in this case has the properties

$$\begin{aligned}\phi_s &\cong -kT\ln B(\Theta) \quad \text{for } h\to\delta_m\\ \phi_s &= 0 \quad \text{for } h\geq 2a\end{aligned} \tag{276}$$

Therefore, for $-\ln B(\Theta)>>1$ one can approximate the real distribution of the steric barrier $\phi_s = \ln \bar{B}(\Theta)$ by a linear distribution, given by

$$\phi_s = -\ln B(\Theta)(1-\xi) \tag{277}$$

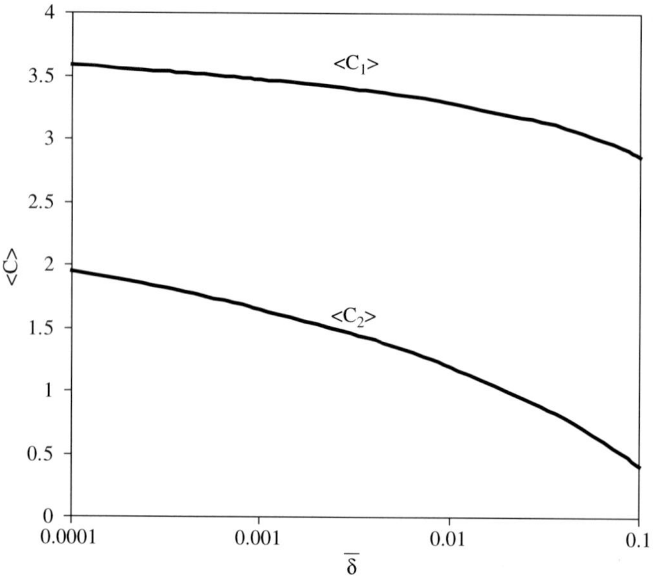

Fig. 5.67. The dependence of the $\langle C_1 \rangle$ and $\langle C_2 \rangle$ constants in the expansion for the $B(\Theta)$ function, Eq. (274).

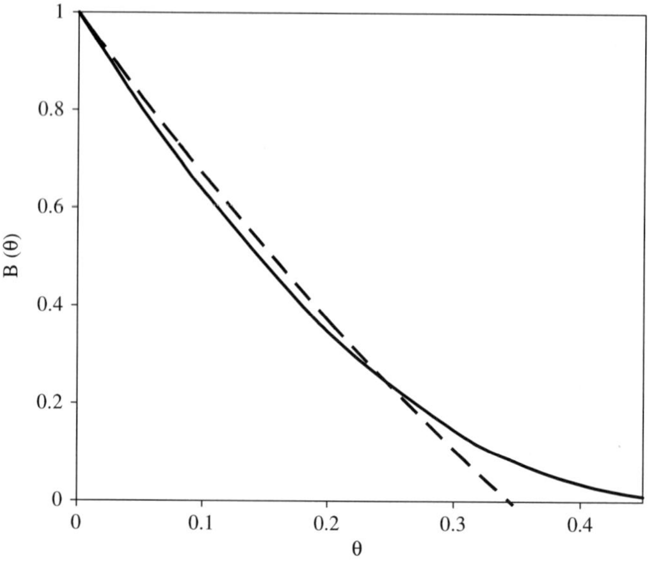

Fig. 5.68. The dependence of the $B(\Theta)$ function calculated from the classical RSA model (solid line) and the $\overline{B}(\Theta)$ function calculated from Eq. (273) for $\overline{\delta} = 10^{-3}$ (dotted line).

By substituting the steric energy distribution into Eq. (267) one obtains upon integration the dependence

$$\bar{B}(\Theta) \cong \frac{2+\ln\dfrac{1}{\bar{\delta}}}{-\ln B(\Theta)+\ln\dfrac{1}{\bar{\delta}}}\,B(\Theta) \tag{278}$$

For $\dfrac{1}{B(\Theta)} >> 1$, Eq. (278) simplifies to

$$\bar{B}(\Theta) \cong C_B^{'} B(\Theta) \tag{279}$$

where the dimensionless constant C_B varies between $0.9 \div 0.7$ for typical experimental conditions.

Using the previously derived asymptotic expressions for the blocking function $B(\Theta)$ in the limit of Θ approaching the maximum coverage Θ_{mx}, one obtains from Eq. (279) the approximation for $\bar{B}(\Theta)$

$$\bar{B}(\Theta) = C_{mx}^{'}\left(1-\frac{\Theta}{\Theta_{mx}}\right)^{l} \tag{280}$$

where C'_{mx} is the dimensionless constant and the exponent $l = 3$ for spheres, $l = 4$, for non-spherical particles adsorbing side-on and $l = 5$, for unoriented adsorption of non-spherical particles.

Because, as shown above, the $\bar{B}(\Theta)$ function does not deviate significantly from the RSA blocking function $B(\Theta)$, for the entire coverage range, it seems reasonable to use the latter in practical applications. This is the only alternative for the higher coverage range because calculating $\bar{B}(\Theta)$ in an exact way seems prohibitive.

Using the approximate form, Eq. (280), one can formulate the expression for the flux in the adsorption layer in the form

$$-j_a(\Theta) = k_a\, n(\delta_a) C_{mx}^{'}\left(1-\frac{\Theta}{\Theta_{mx}}\right)^{l} - \frac{k_d}{S_g} C_{mx}^{'}\Theta \tag{281}$$

Consequently, under equilibrium conditions, where $j_a = 0$, the isotherm equation is given by

$$K_a S_g n(\delta_a) = \frac{\Theta}{\left(1 - \frac{\Theta}{\Theta_{mx}}\right)^l} \tag{282}$$

The generalized RSA approach, whose major results are Eqs. (266),(267), can be effectively used for coupling the quasi-stationary transport through the adsorption layer with the bulk transport governed by various mechanisms like diffusion, external force or flow. This can be done, in analogy to previously discussed surface force boundary layer (SFBL) theory using Eq. (266) or its approximate counterpart, Eq. (281), as the boundary conditions for the bulk transport equations. This is a valid approach when the thickness of the adsorption layer remains much smaller than the diffusion boundary layer, which is often the case for practical situations. By using this approach one can reduce the non-linear particle adsorption problem to the linear transport equation in the bulk

$$\frac{\partial n}{\partial t} = D\nabla^2 n - \nabla\cdot(\mathbf{U}n) \tag{283}$$

and the non-linear boundary conditions at the adsorption layer

$$D(\delta_a)\left(\frac{\partial n}{\partial h}\right)_{\delta_a} = k_a n(\delta_a)\bar{B}(\Theta) - \frac{k_d}{S_g}\frac{\bar{B}(\Theta)}{B(\Theta)}\Theta \tag{284}$$

The boundary value problem expressed by Eqs. (283,284) is completed by the mass balance equation for the adsorbed phase

$$\frac{1}{S_g}\frac{d\Theta}{dt} = -j_a(\Theta) \tag{285}$$

Because the boundary condition, Eq. (284) involves the coverage Θ, the bulk transport equation is coupled with the surface mass balance equation in a non-linear way, which makes its analytical solution prohibitive.

However, a useful approximate solution can be derived by realizing that the particle flux from the bulk to the adsorption layer is given by the previously derived expression, Eq. (361) in Chapter 4

$$-j = k_c''\left[n_b - n(\delta_a)\right] \tag{286}$$

where the bulk transfer rate constant k_c'' was calculated previously for quasi-stationary transport conditions.

In the case of the diffusion-controlled transport, the bulk transfer rate is dependent on time and can be approximated by

$$k_c' = k_c = \left(\frac{D}{\pi t}\right)^{1/2} \tag{287}$$

Using Eq. (286) one can derive the expression for describing particle adsorption kinetics under an arbitrary transport mechanism [82]

$$\frac{1}{S_g}\frac{d\Theta}{dt} = \frac{k_a\bar{B}(\Theta)n_b - \dfrac{k_d}{S_g}\dfrac{\bar{B}(\Theta)}{B(\Theta)}\Theta}{k_a\bar{B}(\Theta) + k_c'(t)}k_c''(t) = -j_a(\Theta) \tag{288}$$

Under the convective diffusion transport conditions the thickness of the diffusion boundary layer remains fixed after a short transition time. Then, for the uniformly accessible surfaces, the bulk rate constant k_c becomes time-independent and one can express Eq. (288) in the form

$$\frac{1}{S_g}\frac{d\Theta}{dt} = -j_0\frac{1}{1+(K-1)\bar{B}(\Theta)}\left[K\bar{B}(\Theta) - \frac{k_d}{S_g k_c n_b}\frac{\bar{B}(\Theta)}{B(\Theta)}\Theta\right] = -j_a \tag{289}$$

where $k_c' = k_a k_c/(k_a - k_c)$, $j_0 = -k_c n_b$ is the flux for zero coverage of particles and $K = k_a/k_c$ is the dimensionless coupling constant representing the ratio of the bulk transport resistance $1/k_c$ to the adsorption layer resistance $1/k_a$.

By considering the previous expression for the k_a constant Eq. (275) and using the dependence $k_c = \dfrac{D}{a}Sh$ one obtains for a uniformly accessible surface the explicit expression for the coupling constant

$$K = \frac{k_a}{k_c} = \frac{1}{2\left(1+\frac{1}{2}\ln\bar{\delta}\right)Sh} \tag{290}$$

Because for smaller colloid particles $Sh=0.616\ Pe^{1/3}$(cf. Table 4.12) and $Pe \sim a^3/D$ one can express K as

$$K = \frac{Pe^{-1/3}}{1.232\left(1+\frac{1}{2}\ln\frac{1}{\bar{\delta}}\right)} \sim a^{-4/3} \tag{291}$$

As can be noticed, the coupling constant increases for smaller particle sizes as can be observed in Table 5.5, where values of K are collected, calculated for the impinging jet cell.

Upon integration one obtains from Eq. (289) the kinetic equation

$$\int_0^{\Theta} \frac{1+(K-1)\bar{B}(\Theta')}{K\bar{B}(\Theta) - \frac{k_d}{S_g k_b n_b}\frac{\bar{B}(\Theta')}{\bar{B}(\Theta')}\Theta'} d\Theta' = \tau \tag{292}$$

where $\tau = S_g k_c n_b t = S_g |j_0| t$ is the dimensionless adsorption time.

Eq. (292) indicates that the complex mass transfer problem under the convective diffusion transport conditions is reduced to a single quadrature.

For irreversible systems, where $k_d = 0$, Eq. (289) for the flux simplifies to

$$j_a = j_0 \frac{K\bar{B}(\Theta)}{1+(K-1)\bar{B}(\Theta)} = j_0\tilde{B}(\Theta) \tag{293}$$

where

$$\tilde{B}(\Theta) = \frac{K\bar{B}(\Theta)}{1+(K-1)\bar{B}(\Theta)} \tag{294}$$

is the overall blocking function.

Table 5.5.
Values of the dimensionless adsorption constant $\bar{k}_a$ and the coupling constant K for spherical particles

Particle radius (nm)	Sh	$\bar{k}_a$	K
2	1.27×10^{-4}	0.245	1930
5	4.34×10^{-4}	0.200	461
10	1.09×10^{-3}	0.176	161
20	2.75×10^{-3}	0.157	57.1
50	9.34×10^{-3}	0.137	14.7
100	2.35×10^{-2}	0.125	5.32
200	5.93×10^{-2}	0.115	1.94
500	0.201	0.104	0.52

$$T = 293\ \text{K},\quad K = \frac{k_a}{k_b} = \frac{1}{2\left(1+\frac{1}{2}\ln\frac{2a}{\delta_m}\right)Sh} \cong \frac{1}{8Sh};\quad Sh = 0.616\,Pe^{1/3} = 0.616\left(\frac{2\alpha_r V_\infty a^3}{R^2 D}\right)^{1/3}$$

$$\bar{k}_a = \frac{k_a L}{D} = \frac{1}{\Phi_v}\frac{2}{3\left(1+\frac{1}{2}\ln\frac{2a}{\delta_m}\right)} \cong \frac{1}{6\Phi_v}$$

$R = 10^{-3}\,\text{m},\ \alpha_r = 6\ (Re = 10)$

$V_\infty = 10^{-2}\ \text{m s}^{-1},\ \delta_m = 5\times10^{-10}\,\text{m}$

Eq. (294) has a significance because it indicates that the flux of particles from the bulk is corrected by the factor $\tilde{B}(\Theta)$ rather than by $B(\Theta)$ as commonly assumed in the literature. For $K >> 1$, which is the case for particle sizes below 100 nm, e.g., proteins, $\bar{B}(\Theta)$ is given by the expression

$$\tilde{B}(\Theta) = 1 - \frac{1}{K}\left(\frac{1}{\bar{B}} - 1\right) \tag{295}$$

By substituting $\bar{B}(\Theta) \cong 1 - \langle C_1\rangle\,\Theta$, Eq. (295) becomes

$$\tilde{B}(\Theta) \cong 1 - \frac{\langle C_1\rangle}{K}\Theta \tag{296}$$

This dependence clearly indicates that in the case of large resistance in the bulk (large diffusion boundary layer thickness) the influence of the blocking effects remains negligible, which means that a precise shape of the blocking function $\bar{B}(\Theta)$ is irrelevant. The blocking effects become only important for long times when the inequality holds

$$\frac{1}{\bar{B}(\Theta)} \gg K-1 \tag{297}$$

This is always the case when the particle coverage approaches the maximum value Θ_{mx} and the blocking function $\bar{B}(\Theta)$ is given by Eq. (280). Then, the inequality, Eq. (297) becomes

$$\Theta < \Theta_{mx}\left[1-\frac{1}{(K-1)^{1/l}}\right], \tag{298}$$

The kinetics of particle adsorption in the case of irreversible systems can be derived by integrating Eq. (293), which results in the implicit expression

$$(K-1)\Theta + \int_0^{\Theta} \frac{d\Theta'}{\bar{B}(\Theta')} = K\tau \tag{299}$$

In the special case when $K \cong 1$, which can be realized in practice for micrometer-sized particles using the impinging-jet cells, Eq. (299) simplifies to the commonly used form

$$\int_0^{\Theta} \frac{d\Theta'}{\bar{B}(\Theta')} = \tau \tag{300}$$

Assuming that $\bar{B}(\Theta)$ is in the form of the series expansion $1-\langle C_1\rangle\, \Theta+\langle C_2\rangle\, \Theta^2$ one can formulate Eq. (300) as

$$\Theta = \Theta_1 \frac{1-e^{-\bar{q}\langle C_1\rangle\tau}}{1-\dfrac{\Theta_1}{\Theta_2}e^{-\bar{q}\langle C_1\rangle\tau}} \tag{301}$$

where

$$\Theta_1 = \frac{\langle C_1 \rangle}{2\langle C_2 \rangle}(1-\bar{q}), \quad \Theta_2 = \frac{\langle C_1 \rangle}{2\langle C_2 \rangle}(1+\bar{q}), \quad \bar{q} = \left(1 - \frac{4\langle C_2 \rangle}{\langle C_1 \rangle^2}\right)^{1/2}$$

For higher coverage, substituting for $\bar{B}(\Theta)$ Eq. (280), one obtains from Eq. (300) the expression

$$\Theta = \Theta_{mx} - \frac{\Theta_{mx}^{\ l/(l-1)}}{\left[C_{mx}(l-1)\tau\right]^{1/(l-1)}} \tag{302}$$

From Eq. (302) one can deduce that the final jamming coverage is attained for long times proportionally to $\tau^{-1/2}$ for spherical particles, to $\tau^{-1/3}$ for non-spherical particles adsorbing side-on and $\tau^{-1/4}$ for the unoriented adsorption.

On the other hand, for $K \gg 1$ (low transfer rate from the bulk) which is the case for small colloid particles and proteins under forced convection transport conditions, Eq. (298) indicates that the blocking effects governed by the $\bar{B}(\Theta)$ function remain negligible if $\Theta < \Theta_{mx}[1-(C'_{mx}(K-1))^{1/l}]$, so $\Theta \cong \tau$ for this adsorption regime.

Note that the range of applicability of the linear regime increases significantly for larger values of the coupling constant K. This can be observed in Fig. 5.69 where the exact results calculated numerically from Eq. (299) are plotted for various K ranging from 1 to 50. The blocking function $\bar{B}(\Theta)$ in these calculations was approximated by the RSA blocking function given by Eq. (86). As seen in Fig. 5.69, the magnitude of the coupling constant K exerts a decisive influence on the kinetics of particle deposition. For small colloid particles and proteins, $K \gg 1$ as can be deduced from the data collected in Table 5.5. Thus, in this case, the kinetics of particle deposition is described by a curve similar to line 2 in Fig. 5.69. This means that the blocking effects are completely irrelevant until the coverage attains about 0.8 of its maximum value and the adsorption kinetics is described by straight line dependence. Then, for Θ approaching Θ_{mx}, the kinetic curves become abruptly deflected. These results show quite unequivocally that protein deposition kinetics is mainly controlled by the initial deposition rate $k_a n_b$ (fixing the initial slope of the kinetic curve) and the maximum coverage Θ_{mx}. Both these parameters can be precisely calculated by using the formulae given above.

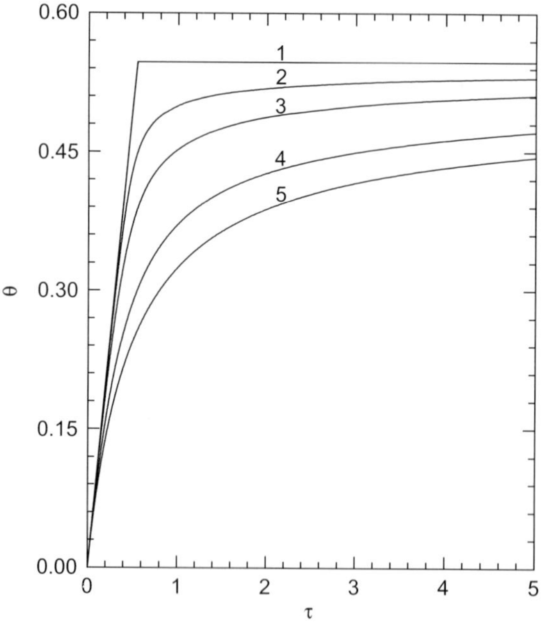

Fig. 5.69. Kinetics of irreversible adsorption of spherical particles under the forced convection transport conditions calculated numerically from Eq. (299) for: 1. $K = \infty$, 2. $K = 50$, 3. $K = 10$ 4. $K = 2$ 5. $K = 1$. From Ref. [63].

Only for larger colloid particles where $K \cong 1$, the kinetics depend specifically on the precise form of the blocking function.

In the general case of non-stationary transport, the generalized RSA approach enables one to specify the appropriate boundary condition for the bulk transport equation, which can be easily solved numerically, the finite difference method. This approach was exploited for describing the diffusion-controlled adsorption of proteins and colloid particles from an infinite volume on a planar interface [82]. The boundary value problem in this case was formulated in the dimensionless form analogous to Eq. (294) in Chapter 4.

$$\begin{aligned} \frac{\partial \bar{n}}{\partial \tau} &= \frac{\partial^2 \bar{n}}{\partial H^2} \\ \frac{d\Theta}{d\tau} &= -\bar{j}_a(\Theta, \tau) \end{aligned} \qquad (303)$$

with the boundary and initial conditions

$$
\begin{aligned}
&-\bar{j}_a(\Theta,\tau) = \left(\frac{\partial \bar{n}}{\partial H}\right)_{\bar{\delta}_a} = \bar{k}_a\,\bar{n}(\bar{\delta}_a)\bar{B}(\Theta) - k_d\,\frac{\bar{B}(\Theta)}{B(\Theta)}\Theta \\
&\bar{n} = 1 \quad \text{for} \quad \tau = 0,\ \bar{\delta}_a < H < \infty \\
&\bar{n} = 1 \quad \text{for} \quad \tau > 1,\ H \to \infty
\end{aligned}
\tag{304}
$$

where the dimensionless variables and parameters are

$$
\begin{aligned}
&\bar{n} = n/n_b, \quad \bar{n}(\bar{\delta}_a) = n(\delta_a)/n_b, \quad H = h/a, \quad \bar{\delta}_a = \delta_a/a \\
&\tau = t/t_{ch}, \quad t_{ch} = L^2/D, \quad L = 1/\pi a^2 n_b \\
&\bar{k}_a = k_a L/D, \quad k_d = k_d L^2/D, \quad \bar{j}_a = j_a L/D n_b
\end{aligned}
\tag{305}
$$

The dimensionless adsorption and desorption constants can be expressed as

$$
\begin{aligned}
&\bar{k}_a = \frac{2}{3\Phi_v}\,\frac{1}{\displaystyle\int_{\bar{\delta}_m}^{\bar{\delta}_a} \frac{e^{\phi/kT}}{F_1(H)}\,dH} \\
&\bar{k}_d = \bar{k}_a\,\frac{4}{3\Phi_v \displaystyle\int_{min} e^{-\phi/kT}\,dH}
\end{aligned}
\tag{306}
$$

In the case of no specific barrier and deep energy minimum when $|-\phi_m|/kT >> 1$, Eq. (306) reduces to

$$
\bar{k}_a = \frac{2}{3\Phi_v\left(1 + \frac{1}{2}\ln\frac{2a}{\delta_m}\right)} \tag{307}
$$

As can be observed, the constants depend solely on the volume fraction of particles $\Phi_v = \frac{4}{3}\pi a^3 n_b$, which is very low for proteins and colloid particles. It can be calculated that for a 100 ppm solution of BSA, $\Phi_v = 7.4\times10^{-5}$, and $\bar{k}_a = 4.5\times10^4$. Similar values are obtained for other proteins as well [82]. In this case, for adsorption time t, which is smaller than the characteristic relaxation time $t_{ch}=1/(S_g n_b)^2 D$, particle flux and coverage are governed by previously derived dependencies, (Eq. (291) and Eq. (292) in Chapter 4), respectively. This is illustrated in Fig. 5.70 showing the results of numerical calculations derived in Ref. [82]. The blocking function was approximated by the RSA blocking function described by Eq. (82). It is interesting to note that for $\bar{k}_a \gg 1$(which is usually the case for protein adsorption) the entire kinetic curve can be constructed as

$$\Theta = 2S_g (\pi D_\infty t)^{1/2} n^b \quad \text{for } t/t_{ch} \le \left(\frac{\Theta_{mx}^2}{4\pi}\right)$$
$$\Theta = \Theta_{mx} \quad \text{for } t/t_{ch} > \left(\frac{\Theta_{mx}^2}{4\pi}\right) \tag{308}$$

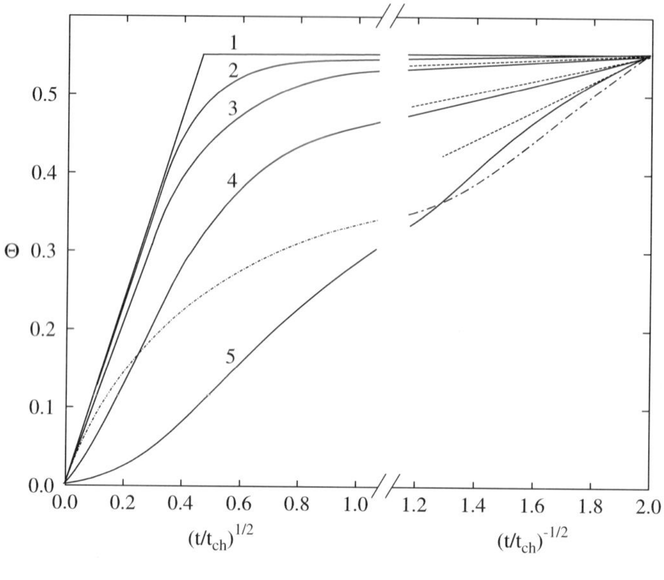

Fig. 5.70. Kinetics of irreversible adsorption under the diffusion-controlled transport at a planar surface expressed as the Θ vs. $(t/t_{ch})^{1/2}$,$(t/t_{ch})^{-1/2}$ dependencies (where $t_{ch} = \dfrac{1}{(S_g n_b)^2 D}$ is the characteristic adsorption time); the continuous lines denote the numerical solution of the diffusion equation for: 1. $\bar{k}_a = \infty$, 2. $\bar{k}_a = 1000$, 3. $\bar{k}_a = 100$, 4. $\bar{k}_a = 10$, 5. $\bar{k}_a = 1$.
The dotted lines represent the asymptotic long-time results calculated from Eq.(311) and the dashed-dotted line shows the results calculated from Eq. (314). From Ref. [82].

This indicates that the blocking effects are negligible except for $\Theta \to \Theta_{mx}$, which proves that Θ_{mx} is a parameter of primary practical significance. Hence, for larger $\bar{k}_a$, an exact knowledge of the $\bar{B}(\Theta)$ function is not required. An additional significance of Eq. (308) is that for the entire coverage range, the particle flux remains equal to the limiting diffusion flux $-j_a = (D/\pi t)^{1/2}$. This suggests that adsorption measurements can be used for determining in an accurate way the particle or protein diffusion coefficient (particle size). As can be seen in Fig. 5.70, for $\bar{k}_a < 100$ the deviations from the limiting adsorption regime, given by Eq. (308) become more pronounced. The kinetic curves in this case can only be evaluated numerically. However, it was shown in Ref. [82] that for larger coverage when the inequality $\Theta > \Theta_{mx}\left[1-\left(\frac{1}{10(l-1)\bar{k}_a}\right)^{1/l-1}\right]$ is met, particle adsorption is governed by the asymptotic law

$$\Theta = \Theta_{mx} - \frac{K_l}{(t/t_{ch})^{1/(l-1)}} \tag{309}$$

where

$$K_l = \Theta_{mx}\left(\frac{\Theta_{mx}}{(l-1)C_B'\bar{k}_a}\right)^{1/(l-1)} \tag{310}$$

Eq. (309) indicates that the maximum coverage is approached in the long time limit as $(t/t_{ch})^{-1/(l-1)}$. For spheres, where $l = 3$ one has

$$\Theta = \Theta_{mx} - \frac{K_l}{(t/t_{ch})^{1/2}} \tag{311}$$

As can be seen in Fig. 5.70, the exact numerical results are well reflected by the limiting analytical formula, Eq. (311), for longer times. This has important practical implications because a useful extrapolating formula can be derived on the basis of Eq. (311)

$$\Theta_{mx} = \Theta_l + \Theta_l^{l/(l-1)} \frac{1}{\left[(l-1)C_B' S_g k_a n^b t_l\right]^{1/(l-1)}} \tag{312}$$

where Θ_l is the coverage attained for long but finite adsorption time t_l.

Eq. (312) seems particularly useful for experimental studies because attaining the maximum (jamming) coverage would require a prohibitively long adsorption time, especially for dilute protein or colloid suspensions.

The calculations shown in Fig. 5.70 also demonstrated the inadequacy of the commonly used empirical model postulating that [83]

$$j_a = j_0 \, B(\Theta) \tag{313}$$

By integrating this dependence one obtains

$$\Theta = F_{RSA}^{-1} \tag{314}$$

where $F_{RSA} = \int_0^{\Theta} \frac{d\Theta'}{B(\Theta')}$ is the function describing the RSA kinetics. As seen in Fig. 5.70 the kinetic curve originating from Eq. (314) (dashed/dotted line) deviates from the exact results derived from numerical simulations. This is so because for the diffusion-controlled adsorption the thickness of the diffusion boundary layer grows with time. This makes the bulk transport resistance much higher than the surface layer resistance governed by the blocking function except for very long times when Θ approaches Θ_{mx}. In contrast, Eq. (311) remains a valid approximation for the long-time adsorption kinetics for arbitrary Θ as can be seen in Fig. 5.71 where the exact data derived by numerical solution of Eq. (303) are shown for Θ = 0.547, 0.4, 0.3, and 0.2, respectively. Under the experimental conditions the maximum coverage can be regulated by changing the ionic strength of the suspension as discussed above.

Because of the frequent use of the Langmuir blocking function in experimental works dealing with protein adsorption, calculations of irreversible adsorption kinetics were also performed. The blocking function was assumed to have the form

$$B(\Theta) = 1 - \frac{\Theta}{\Theta_{mx}} \tag{315}$$

The limiting, long time solution derived from Eq. (313) is given by

$$\Theta = \Theta_{mx}\left(1 - e^{-\bar{k}_a \tau / \Theta_{mx}}\right) \tag{316}$$

which is expected to be valid for $\Theta > \Theta_l = \Theta_{mx}\left(1 - \frac{0.1}{\bar{k}_a \Theta_{mx}}\right)$.

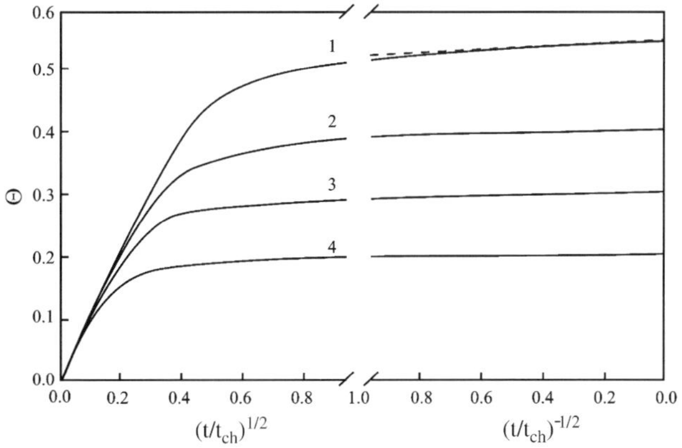

Fig. 5.71. Kinetics of irreversible adsorption under the diffusion-controlled transport at a planar surface expressed as the Θ vs. $(t/t_{ch})^{1/2}$, $(t/t_{ch})^{(1/2}$ dependencies; the continuous lines denote the numerical solution of the diffusion equation for $\bar{k}_a = 100$, and various Θ_{mx} values: 1. $\Theta_{mx} = 0.547$, 2. $\Theta_{mx} = 0.4$, 3. $\Theta_{mx} = 0.3$, 4. $\Theta_{mx} = 0.2$. The dashed line represents the asymptotic long-time results calculated from Eq. (311). From Ref. [82].

Accordingly, the extrapolating function , analogous to Eq. (312), is given by

$$\Theta_{mx} \cong \Theta_l \left(1 + e^{-\bar{k}_a \tau_l / \Theta_l}\right) \tag{317}$$

where Θ_l is the coverage attained after adsorption time τ_l.

Eq. (317) indicates that for the same $\bar{k}_a$ and τ_l, the correction is much smaller than for the RSA blocking function. On the other hand, the predictions stemming from the flux correction model expressed by $j_a = j_0\, B(\Theta)$ are described by the analytical dependence

$$\Theta = \Theta_{mx} \left(1 - e^{-\frac{2\tau}{\pi^{1/2}\Theta_{mx}}}\right) \tag{318}$$

As can be observed in Fig. 5.72, this formula, used often in the literature, poorly describes the exact data.

The theoretical results presented in Figs. 5.70–5.72 suggest that for $\bar{k}_a >> l$, which is the case for proteins and small colloid particle adsorption,

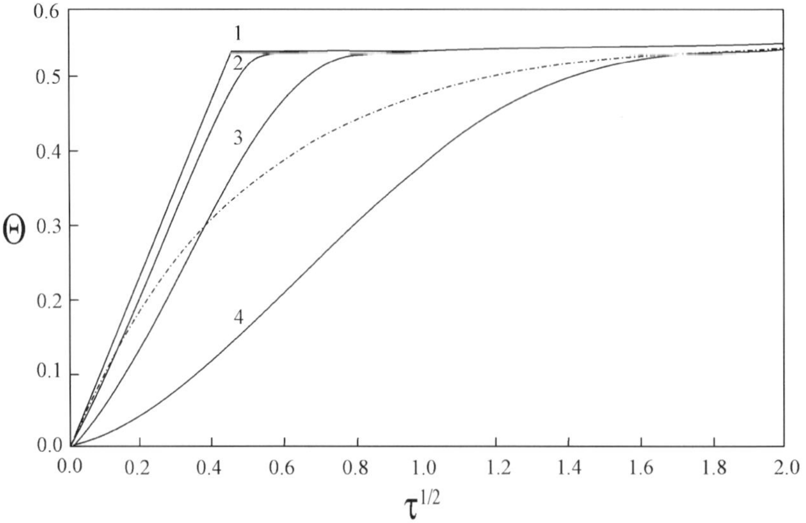

Fig. 5.72. Kinetics of irreversible adsorption under the diffusion-controlled transport at a planar surface expressed as the Θ vs. $\tau^{1/2}$ dependency; with the blocking function $B(\Theta)$ derived from the Langmuir model, Eq. (315). The continuous lines denote the numerical solution of the diffusion equation for 1. $\bar{k}_a = \infty$; 2. $\bar{k}_a = 100$; 3. $\bar{k}_a = 10$; 4. $\bar{k}_a = 1$. The dashed-dotted line shows the results calculated from Eq. (318). From Ref.[82].

and for $t < \frac{\pi}{4}\Theta_{mx}^2 \, t_{ch} = \frac{4\pi a^2}{9\phi_v^2 D}$ particle adsorption kinetics can be well described by the square root of time dependence $\Theta = 2\pi a^2 \left(\frac{Dt}{\pi}\right)^{1/2} n^b$. For $t > \frac{4\pi a^2}{9\Phi_v^2 D}$ adsorption kinetics can be well approximated by the power law dependence, Eq. (309), which is also valid for non-spherical particles and the unoriented adsorption regime.

ACKNOWLEDGMENTS

This work was supported financially by the Ministry of Education and Science. Grant No: 4T 09A 07625, and partially by the Centre of Excellence "CatColl" within the EC Grant No: G5 MA-CT-2002-04024.

LIST OF SYMBOLS

Symbol	Definition	Character	Unit	Numerical value
Ad	Dispersion interaction parameter	S	[1]	
As	Axis ratio for spheroids	S	[1]	
As^*	Axis ratio for interacting spheroids	S	[1]	
$ASF(\Theta)$, ASF_{RSA}	Available surface functions	S	[1]	
A_w	Parameter describing van der Waals model	S	[1]	
A_{132}	Hamaker constants for the interactions of two different particles in medium	S	J	
a, a_1, a_2, a_p, a_s	Radii of particles	S	m	
$\langle a \rangle$	Averaged radius of particle	S	m	
a_{max}, a_{min}	Maximum and minimum size of particles	S	m	
a_s	Longer semi-axis of spheroid	S	m	
$B(\Theta)$, $B^0(\Theta)$, B_{RSS}	Surface blocking functions	S	[1]	
$\tilde{B}(\Theta)$	Generalized blocking functions	S	[1]	
$\bar{B}(\Theta)$	Overall blocking functions	S	[1]	
B_w	Parameter describing van der Waals model	S	[1]	
$\bar{B}_2$,$\bar{B}_3$,$\bar{B}_4$ $\bar{B}_5$,$\bar{B}_6$,$\bar{B}_7$	Reduced virial coefficients	S	[1]	
b	Shorter spheroid semi-axis	S	m	
b_1, b_2, b_3	Dimensionless coefficients	S	[1]	
C', C'_0	Dimensionless constants	S	[1]	
C_b	Dimensionless constant	S	[1]	
C_c	Dimensionless constant	S	[1]	
C_h	Dimensionless parameter	S	[1]	
C'_{mx}	Kinetic constant	S	[1]	
C_p	Constant characterizing shape of target	S	m^2	
C'_p	Constant describing target distribution	S	[1]	
C_{RSA}	Kinetic constants	S	[1]	
C_1, C_2, C_3, C_4, C_1^*, C_2^*	Dimensionless constants of ASF expansion	S	[1]	

List of Symbols (continued)

Symbol	Definition	Character	Unit	Numerical value
C'_1, C'_2, C'_3, C'_4	Dimensionless constants of ASF expansion	S	[1]	
$\langle C_1 \rangle, \langle C_2 \rangle$	Averaged constants of ASF expansion	S	[1]	
c	Dimensionless constant of jamming coverage approximation	S	[1]	
c_0, c_1, c_2, c_3, c_4	Dimensionless constants describing effective interaction range	S	[1]	
$D_\perp D_\parallel$	Particle diffusion coefficients perpendicular and parallel to interfaces	S	$m^2 s^{-1}$	
$D(h)$	Translation diffusion coefficients of particle near wall	S	$m^2 s^{-1}$	
d_{RSA}	Dimensionality of RSA problem	S	[1]	
$E(x)$	Complete elliptic integral of second kind	S	[1]	
Ex	Dimensionless external force number	S	[1]	
$\hat{\mathbf{e}}_1, \hat{\mathbf{e}}_2, \hat{\mathbf{e}}_n$	Direction vectors (versors)	**V**	[1]	
e	Elementary charge	S	C	1.60218×10^{-19}
$F(\Theta)$	Function describing kinetics of particle deposition	S	[1]	
F_l	Virial coefficient function	S	[1]	
F_{RSA}	Function describing blocking effect	S	[1]	
$f(\Theta)$	Activity coefficient of adsorbed particles	S	[1]	
f_b	Bulk activity coefficient of particles	S	[1]	
f_{ij}, f_{1v}, f_{2v}	Meyer functions	S	[1]	
f_0	Dimensionless coefficient	S	[1]	
G_D	Derjaguin factor	S	m	
Gr	Dimensionless gravity number	S	[1]	

List of Symbols (continued)

Symbol	Definition	Character	Unit	Numerical value
g, $g(r)$, $g_1(r)$	Pair correlation functions (radial distribution functions)	S	[1]	
$g(2a,\Theta)$	Contact pair correlation function	S	[1]	
H	Dimensionless gap width	S	[1]	
H^*	Dimensionless effective range of interactions	S	[1]	
h	Distance from wall	S	m	
$\bar{h}$	Dimensionless distance between particles	S	[1]	
h^*	Effective range of interactions	S	m	
h_{ch}	Characteristic distance of interactions	S	m	
h_m	Distance between particle and surface	S	m	
h_t	Size of target	S	m	
I	Ionic strength	S	m^{-3}	
I_1, I_2	Surface integrals	S	[1]	
j_a	Adsorption flux of particles	S	$m^{-2}\,s^{-1}$	
$\bar{j}_a$	Reduced adsorption flux of particles	S	[1]	
j_0	Limiting flux of particles	S	$m^{-2}\,s^{-1}$	
K	Coupling constant characterizing particle adsorption	S	[1]	
K_a	Equilibrium adsorption constant	S	m	
k	Boltzmann constant	S	$J\,K^{-1}$	1.38065×10^{-23}
k_a	Adsorption rate constant	S	$m\,s^{-1}$	
$\bar{k}_a$, $\tilde{k}_a$	Dimensionless adsorption constants	S	[1]	
k_b	Bulk mass transfer rate	S	$m\,s^{-1}$	
k_c, k'_c	Mass transfer coefficients	S	$m\,s^{-1}$	
k_d	Desorption rate constants	S	s^{-1}	
$\bar{k}_d$, $\tilde{k}_d$	Dimensionless desorption constants	S	[1]	
k_d^*	Dimensionless adsorption rate parameter	S	[1]	
L	Characteristic domain dimension for RSA			

List of Symbols (continued)

Symbol	Definition	Character	Unit	Numerical value
	simulations, length of channel, cylinder, plate, particle	S	m	
Le	The diffuse double-layer thickness	S	m	
l_s	Site coordination number	S	[1]	
m_A , m_1	Particle monolayer mass per unit area	S	kg m^{-2}	
N	Surface concentration of particles	S	m^{-2}	
N_a	Number of adsorbing species	S	[1]	
N_{att}	Number of adsorption attempts in simulations	S	[1]	
N_{mx}	Characteristic monolayer concentration of particles	S	m^{-2}	
$N_0\ (h_t)$	Number of target distribution for initial time	S	[1]	
N_p	Number of adsorbed particles	S	[1]	
$\langle N_p \rangle$	Averaged number of particles	S	[1]	
N_s	Number of sites	S	[1]	
N_{suc}	Number of successful adsorption attempts in simulations	S	[1]	
N_t	Number of target distribution	S	[1]	
N_2	Surface concentration of pairs	S	m^{-2}	
n_b	Number concentration of particles in bulk	S	m^{-3}	
n_s	Averaged number of particles over surface area	S	[1]	
Pe	Peclet number	S	[1]	
P_r	Perimeter of convex particle	S	m	
$p_a, p_t, p_0(\Theta)$, δp	Probability of particle adsorption	S	[1]	
Q_0	Orientational ordering parameter	S	[1]	
q_s	Parameter describing adsorption of anisotropic particles	S	[1]	

List of Symbols (continued)

Symbol	Definition	Character	Unit	Numerical value
R	Radius of interface, cylindrical channel, cylindrical or spherical collectors, rotating disk	S	m	
$\overline{R}$	Parameter characterizing ballistic deposition regime	S	[1]	
R_c	Radius of cavity in fluid	S	m	
Re	Reynolds number	S	[1]	
R'_1, R''_2, R'_2, R''_2	Radii of curvature of bodies	S	m	
$\overline{R}_1, \overline{R}_2$	Mean radii of curvature	S	m	
r, r_1, r_2, r_v	Position vectors	**V**	m	
r	Radial distance between particles	S	m	
$\overline{r}$	Dimensionless radial distance	S	[1]	
r_c	Rate of particle creation	S	$m^{-2}\,s^{-1}$	
r_p	Radial position of particle in local flow	S	m	
$S, \Delta S$	Surface area	S	m^2	
S_g	Characteristic cross-section of particle	S	m^2	
S_g^*	Effective cross-section of particle	S	m^2	
Sh	Dimensionless mass transfer Sherwood numbers	S	[1]	
S_t	Target area	S	m^2	
S_1	Surface area blocked by one particle	S	m^2	
S_2	Common surface area of two exclusion zones	S	m^2	
T	Absolute temperature	S	K	
t	Adsorption time	S	s	
$\overline{t}$	Dimensionless time for adsorption on sites	S	[1]	
$W(R_c)$	Irreversible work of creating fluid cavity	S	J	
Greek				
α_p	Dimensionless parameter characterizing adsorption on sites	S	[1]	

List of Symbols (continued)

Symbol	Definition	Character	Unit	Numerical value
α_r	Flow parameter for radial impinging jet	S	[1]	
$\alpha_1, \alpha_2, \alpha_{1v}, \alpha_{12}$	Orientation angles of particles	S	[1]	
$\beta_1, \beta_2, \beta_v$	Orientation angles of particles	S	[1]	
$\bar{\delta}$	Reduced thickness of energy barrier	S	[1]	
δ_a	Thickness of adsorption layer	S	m	
δ_b	Thickness of energy barrier	S	m	
δ_d	Diffusion boundary layer thickness	S	m	
δ_m	Primary minimum distance	S	m	
$\bar{\delta}_m$	Dimensionless primary minimum distance	S	m	
$\Gamma(1/2)$	Gamma Euler function	S	[1]	1.77245
$\Gamma(4/3)$	Gamma Euler function	S	[1]	0.8930
γ_E	Euler's constant	S	[1]	0.577216
γ_p	Particle shape parameter	S	[1]	
ξ, ξ_H	Integration variable, dimensionless distance function	S	[1]	
ξ_H	Parameter describing effective range of interactions	S	[1]	
ζ_p	Zeta potential of particle	S	V	
Θ, Θ_p	Surface coverage of particles	S	[1]	
Θ_e	Equilibrium coverage of particles	S	[1]	
Θ_{mx}	Maximum surface coverage	S	[1]	
Θ_s	Surface coverage of sites	S	[1]	
Θ_∞, Θ_p	Jamming coverage of particles	S	[1]	
$\kappa = Le^{-1}$	Reciprocal double-layer thickness	S	m^{-1}	
λ	Disk size ratio	S	[1]	
λ'	Spherical particle size ratio	S	[1]	
$\lambda_{ij}, \lambda'_{ij}, \lambda''_{ij}$	Geometrical parameters characterizing multicomponent adsorption of particles	S	[1]	
λ_s	Dimensionless parameter	S	[1]	
π	Pi number	S	[1]	3.14159265
$\bar{\pi}$	Reduced pressure of particles (two dimensional)	S	[1]	

List of Symbols (continued)

Symbol	Definition	Character	Unit	Numerical value
π_s	Pressure of particles (two dimensional)	S	J m^{-2}	
$\sigma, \sigma, (r_s)$	Surface charge	S	C m^{-2}	
σ_p	Standard deviation of particle size distribution	S	m	
$\bar{\sigma}_p$	Reduced standard deviation of particle size distribution	S	[1]	
σ_v^2	Variance of particle coverage fluctuations	S	[1]	
$\bar{\sigma}_v^2$	Reduced variance of particle coverage fluctuations	S	[1]	
τ	Dimensionless time	S	s	
ϕ_b	Energy barrier height	S	J	
$\bar{\phi}_b$	Reduced energy barrier height	S	[1]	
ϕ_0, ϕ_{ch}	Reference energy of interaction	S	J	
ϕ_m	Depth of energy minimum	S	J	
ϕ_p	Interaction energy of wandering particle	S	J	
ϕ_p, ϕ_{12}	Interaction energy of particle pair	S	J	
ϕ_v	Interaction energy of virtual particle	S	J	
$\phi_\perp, \phi_\parallel$	Interaction energies of spheroid with surface	S	J	

Note: S = scalar; **V** = vector; **T** = tensor (matrix); [1] = dimensionless.

REFERENCES

[1] H. Reiss, H.L. Frisch and J.L. Lebowitz, J. Chem. Phys., 31 (1959) 369.
[2] H. Reiss, H.L. Frisch, E. Helfand and J.L. Lebowitz, J. Chem. Phys., 32 (1960) 119.
[3] H. Helfand, H.L. Frisch and J.L. Lebowitz, J. Chem. Phys., 34 (1961) 1037.
[4] J. L. Lebowitz, E. Helfand and E. Praestgaard, J. Chem. Phys., 43 (1965) 774.
[5] T. Boublik, Mol. Phys. 29 (1975) 421.
[6] E.L. Hinrichsen, J. Feder and T. Jossang, J. Stat. Phys., 44 (1986) 793.
[7] P. Schaaf and J. Talbot, J. Chem. Phys., 91 (1989) 4401.
[8] J. Talbot, P. Schaaf and G. Tarjus, Mol. Phys., 72 (1991) 1397.
[9] P. Viot, G. Tarjus, S.M. Ricci and J. Talbot, J. Chem. Phys., 97 (1992) 5212.
[10] S.M. Ricci, J. Talbot, G. Tarjus and P. Viot, J. Chem. Phys., 97 (1992) 5219.
[11] J. Talbot, P. Schaaf, G. Tarjus and P. Viot, Mol. Phys., 72 (1991) 1397.

[12] J.W. Evans, Rev. Mod. Phys., 65 (1993) 1281.
[13] J. Talbot, G. Tarjus, P.R. van Tassel and P.Viot, Colloids Surf. A, 165 (2000) 287.
[14] B. Widom, J.Chem.Phys., 39 (1963) 2808.
[15] B. Widom, J.Chem.Phys., 44 (1966) 3888.
[16] D. Henderson and S.G. Davison, in "Equilibrium theory of liquids and liquid mixtures" (H. Eyring, D. Henderson and W. Jost eds., Physical Chemistry, an Advanced Treatise, Vol. II, 339 pp.) Academic Press, New York,1967.
[17] F.H. Ree and W.G. Hoover, J. Chem. Phys., 40 (1964) 939.
[18] F.H. Ree and W.G. Hoover, J. Chem. Phys., 46 (1967) 4181.
[19] G. Tarjus, P. Viot, S.M. Ricci and J. Talbot, Molec. Phys., 73 (1991) 773.
[20] J.J. Talbot, X. Jin and N.-H.Wang, Langmuir 10 (1994) 1663.
[21] Y. Song and E.A. Mason, Phys. Rev. A, 41 (1990) 3121.
[22] Z. Adamczyk and P. Weroński, J. Chem. Phys., 107 (1997) 3691.
[23] A. Renyi, Publ Math. Inst. Hung.Acad. Sci., 3 (1958) 109.
[24] J.J. Gonzales, P.C. Hemmer, J.S. Hoye, J. Chem. Phys., 3 (1974) 228.
[25] Y. Pomeau, J. Phys. A: Math. Gen., 13 (1980) L193.
[26] Z. Adamczyk and P. Belouschek, J. Colloid Interface Sci., 146 (1991) 123.
[27] Z. Adamczyk, B. Siwek, M. Zembala and P. Belouschek, Adv. Colloid Interface Sci., 48 (1994) 151.
[28] Z. Adamczyk, B. Siwek, M. Zembala and P. Weroński, J. Colloid Interface Sci., 185 (1997) 236.
[29] P. Meakin and R. Jullien, Phys. Rev. A, 46 (1992) 2029.
[30] Z. Adamczyk, B. Siwek, P. Weroński and M. Zembala, Prog. Colloid Polymer. Sci., 111 (1998) 41.
[31] J. Talbot and P. Schaaf, Phys. Rev. A, 40 (1989) 422.
[32] Z. Adamczyk and P. Weroński, J. Colloid Interface Sci., 195 (1997) 261.
[33] Z. Adamczyk and P. Weroński, J. Chem. Phys., 108 (1998) 9851.
[34] M.Y. Boluk and T.G.M. van de Ven, Colloids Surf., 46 (1990) 157.
[35] Y. Lvov, K. Ariga, I. Ichinose and T. Kunitake, J. Am. Chem. Soc., 117 (1995) 6117.
[36] T. Serizawa, H. Takashita and M. Akashi, Langmuir, 14 (1998) 4088.
[37] T. Serizawa, S. Kamimura and M. Akashi, Colloids Surf., 164 (2000) 237.
[38] H.A. Chase, Chem. Eng. Sci., 39 (1984) 1099.
[39] I. Willner, M. Lion-Dagan, S. Marx-Tibbon and E. Katz, J. Am. Chem. Soc., 117 (1995) 6581.
[40] E. Katz, A.E. Buckmann and I. Willner, J. Am. Chem. Soc., 123 (2001) 10752.
[41] H.D. Inerowicz, S. Howell, F.E. Reginer and R. Reifenberger, Langmuir, 18 (2002) 5263.
[42] S.W. Howell, H.D. Inerowicz, F.F. Reginer and R. Reifenberger, Langmuir, 19 (2003) 436.
[43] Z. Adamczyk, M. Zembala, B. Siwek and J. Czarnecki, J. Colloid Interface Sci., 110 (1986) 188.
[44] M. Elimelech and C.R. O'Melia, Langmuir, 6 (1990) 1153.
[45] L. Song and M. Elimelech, J. Colloid Interface Sci., 167 (1994) 301.
[46] P.R. Johnson, N. Sun and M. Elimelech, Environ. Sci. Technol., 30 (1996) 3284.
[47] X. Jin. N.H.L.Wang, G. Tarjus and J. Talbot, J. Phys. Chem., 97 (1993) 4256.

[48] X. Jin, J. Talbot and N.H.L.Wang, AIChE J., 40 (1994) 1685.
[49] Z. Adamczyk, P. Weroński and E. Musiał, J. Chem.Phys. 116 (2002) 4665.
[50] Z. Adamczyk, P. Weroński and E. Musiał, J. Colloid Interface Sci., 248 (2002) 67.
[51] Z. Adamczyk, K. Jaszczółt, B. Siwek and P. Weroński, Langmuir, 21 (2005) 8952.
[52] Z. Adamczyk, K. Jaszczółt, A. Michna, B. Siwek, L. Szyk-Warszyńska and M. Zembala, Adv. Colloid Interface Sci., 118 (2005) 25.
[53] M. Malmsten, J. Colloid Interface Sci., 166 (1994) 333.
[54] J.Y. Yoon, H.Y. Park, J.H. Kim and W.S. Kim, J. Colloid Interface Sci., 177 (1996) 613.
[55] M. Malmsten, J. Colloid Interface Sci., 166 (1994) 333.
[56] K. Ariga and Y.M. Lvov, in "Self-assembly of functional protein multilayers, from planar films to microtemplate encapsulation" (M.Malmsten ed.), Biopolymers at Interfaces, chapter 14, Vol. 110, 367 pp. Marcel Dekker 2003.
[57] Z. Adamczyk, in "Kinetic of particle and protein adsorption" (E.Matijevic and M.Borkovec eds.), Surface and Colloid Science, Vol. 17, Chapter 5, p. 300. Kluver Academic, New York 2004.
[58] Z. Adamczyk and P. Weroński, J. Chem. Phys., 105 (1996) 5562.
[59] Z. Adamczyk and P. Weroński, J. Colloid Interface Sci., 189 (1997) 348.
[60] Z. Adamczyk and P. Weroński, Langmuir, 11 (1995) 4400.
[61] R.D. Vigil and R.M. Ziff, J. Chem. Phys., 91 (1989) 2599.
[62] P. Schaaf, P. Wojtaszczyk, E.K. Mann, B. Senger, J.C. Voegel and D. Bedeaux, J. Chem. Phys., 102 (1995) 5077.
[63] Z. Adamczyk, In "Irreversible adsorption of particles" (J. Toth ed.), Adsorption: Theory, Modeling and Analysis, 251 pp., Marcel Dekker. 2002 .
[64] Z. Adamczyk, M. Zembala, B. Siwek and P. Warszyński, J. Colloid Interface Sci., 140 (1990). 123
[65] Z. Adamczyk, B. Siwek and M. Zembala, J. Colloid Interface Sci., 151 (1992) 351.
[66] M.R. Oberholzer, J.M. Stankovich, S.L. Carnie, D.Y.C. Chan and A.M. Lenhoff, J. Colloid Interface Sci., 194 (1997) 138.
[67] P. Weroński, Adv. Colloid Interface Sci., 118 (2005) 1.
[68] J.A. Barker and D. Henderson, J. Chem. Phys., 47 (1967) 4714.
[69] J. Talbot and S.M. Ricci, Phys. Rev. Lett., 68 (1992) 958.
[70] A.P. Thompson and E.D. Glandt, Phys. Rev. A, 46 (1992) 4639.
[71] H.S. Choi, J. Talbot, G. Tarjus and P. Viot, J. Phys. Chem., 99 (1993) 9296.
[72] B. Senger, J.C. Voegel and P. Schaaf, Colloids Surf. A, 165 (2000) 255.
[73] J. Faraudo, Phys. Rev. Let., 89 (2002) 1.
[74] P. Carl, P. Schaaf, J.C. Voegel, J.F. Stolz, Z. Adamczyk and B. Senger, Langmuir, 14 (1998) 7267.
[75] I. Paganabarraga, P. Wojtaszczyk, J.M. Rubi, B. Senger and J.C. Voegel, J. Chem. Phys., 105 (1996) 7815.
[76] B. Senger, F.J. Bafaluy, P. Schaaf, A. Schmitt and J.C. Voegel, Proc. Natl. Acad. Sci. USA, 89 (1992) 9449.
[77] B. Senger, P. Schaaf, J.C. Voegel, A. Johner, A. Schmitt and J. Talbot, J. Chem. Phys., 97 (1992) 3813.
[78] F.J. Bafaluy, B. Senger, J.C. Voegel and P. Schaaf, Phys. Rev. Let., 70 (1993) 623.

[79] B. Senger, P. Schaaf, F.J. Bafaluy, F.J.G. Cuisinier, J. Talbot and J.C. Voegel, Proc. Natl. Acad. Sci. USA, 91 (1994) 3004.
[80] P. Wojtaszczyk, J. Bonet Avalos and J.M. Rubi, Europhys.Lett., 40 (1997) 299.
[81] Z. Adamczyk, B. Senger, J.C. Voegel and P. Schaaf, J. Chem. Phys., 110 (1999) 3118.
[82] Z. Adamczyk, J. Colloid Interface Sci., 229 (2000) 477.
[83] J. Ramsden, Phys. Rev. Lett., 71 (1993) 296.

Index